Contents

ACCESSIBLE PHYSICS

FOR A-LEVEL

A Guided Coursebook

Francis Azzopardi

St Ambrose College, Hale Barns, Altrincham

and

Brian Stewart

St Ambrose College, Hale Barns, Altrincham

For Edward and Ronnie

First published 1995 by
MACMILLAN PRESS LTD
Houndmills, Basingstoke, Hampshire RG21 6XS and London
Companies and representatives throughout the world

ISBN 0–333–62780–6

A catalogue record for this book is available from the British Library.

10 9 8 7 6 5 4 3 2 1
04 03 02 01 00 99 98 97 96 95

Design and page make-up by
Ascenders
92A Winchester Road, Basingstoke, Hampshire RG21 8UJ

Printed in Great Britain by
Unwin Brothers Ltd, The Gresham Press
Old Woking, Surrey
A member of the Martins Printing Group

Acknowledgements

The authors would like to express their appreciation to the following people for their constant support during the preparation for the book:

- The staff and advisers of Macmillan Press, for their perseverance, cheerful encouragement, and patience with their fledgling authors.
- Sheila Ball and Joan Binns, for their help with the typing, and Cath Lee for reprographics.
- Frank Coan, their Head of Department, for his patience and support for two rather distracted teachers.
- Eric Hester, their Headmaster, for his support and encouragement.
- All the physics students on whom the material has been tested over the past few years.
- Their wives, Doreen and Marie, for their continued forbearance.

The authors and publishers wish to thank the following sources for permission to reproduce photographs:

Addis Houseware Ltd p. 332; Allsport p. 37; Associated Press p. 108; Barnaby's Picture Library p. 43; Camera Press Ltd/NASA pp. 73, 383; Canon p. 170; J. Allan Cash Ltd pp. 22, 139, 181; Cern pp. 395, 414, 455; Colorsport pp. 15, 70, 97; Daresbury Laboratory, Rutherford-Appleton Laboratories p. 289; Esso p. 9; Ford Motor Company Ltd pp. 30, 106, 238; GR Photography p.5; GEC Alsthom p. 297; Hulton Deutsch Pictures pp. 397, 462; JET, Abingdon p. 439; Pasco Scientific/ International Instruments Ltd p. 119; Philips CED Publicity pp. 128, 292; QA Photos pp. 58, 258; Science Museum pp. 13, 66; Science Photograph Library pp. 123, 150, 161, 191, 200, 214, 240, 297, 408; John Simmons p. 26; Marie Stewart p. 11; Zefa pp. 113, 265.

The authors and publishers wish to thank the following for permission to use copyright material from past A-Level examination papers:

The Associated Examining Board; The Northern Examinations and Assessment Board (incorporating Northern Examining Association and the Joint Matriculation Board); Oxford and Cambridge Schools Examination Board; University of London Examinations and Assessment Council; and the Welsh Joint Education Committee.

Every effort has been made to trace all the copyright holders but if any have been inadvertently overlooked the publishers will be pleased to make the necessary arrangement at the first opportunity.

List of fundamental constants

Physical quantity	Symbol	Value	Unit
Speed of electromagnetic radiation in vacuum	c	2.998×10^{8}	m s^{-1}
Universal gravitational constant	G	6.672×10^{-11}	N m^{2}kg^{-2}
Planck constant	h	6.626×10^{-34}	Js
Permittivity of free space (vacuum)	ε_o	8.854×10^{-12}	F m^{-1}
Permeability of free space (vacuum)	μ_o	$4\pi \times 10^{-7}$	H m^{-1}
Avogadro constant	N_A	6.022×10^{23}	mol^{-1}
Boltzmann constant	k	1.381×10^{-23}	J K^{-1}
Universal molar gas constant	R	8.314	J K^{-1}mol^{-1}
Unified atomic mass unit	u	1.661×10^{-27}	kg
Electron: mass	m_e	9.110×10^{-31}	kg
charge	e	-1.602×10^{-19}	C
specific charge	e/m	1.759×10^{11}	C kg^{-1}
Proton mass	m_p	1.673×10^{-27}	kg
Neutron mass	m_n	1.675×10^{-27}	kg
Density at 20°C – air	P_a	1.204	kg m^{-3}
mercury	C_{Hg}	1.355×10^{4}	kg m^{-3}
water	C_w	4.190×10^{3}	kg m^{-3}
Triple point of water		273.16	K
Standard temperature – Ice point		273.15	K
Standard atmospheric pressure	P_A	1.013×10^{5}	Pa
Speed of sound in air at 20°C	v	343.6	m s^{-1}
Earth: Mass	M_E	5.98×10^{24}	kg
Mean density	P_E	5.52×10^{3}	kg m^{-3}
Equatorial radius	R_E	6.378×10^{6}	m
Mean distance from Sun	r_E	1.50×10^{11}	m
Escape velocity	v_e	1.1×10^{4}	m s^{-1}
Sun: Mass	M_s	1.99×10^{30}	kg
Radius	R_s	6.96×10^{8}	m
Power output		3.9×10^{26}	W
Moon: Mass	M_m	7.35×10^{22}	kg
Radius	R_m	1.74×10^{6}	m
Mean distance from Earth	r_m	3.84×10^{8}	m

Preface

The transition from GCSE to A-Level can present a seemingly insurmountable hurdle to even the brightest and most conscientious. *Accessible Physics* has been written to give students the help they need to bridge the gap by providing:

- concise, readable explanations of concepts and principles
- simple, clear line diagrams which can be easily understood and reproduced
- relevant examples with complete guidance
- uncluttered, logical and attractive layout
- regular self-assessment of understanding

Accessible Physics is designed as a complete course companion for students following A- or AS-Level courses in physics, leading to a terminal examination in 1996 and beyond. It has been written to incorporate the core of common topics to be encountered by students following modular courses of the latest syllabuses of all the major UK examination boards. Students in further and higher education establishments following courses with a subsidiary physics component will also find it a useful introductory text.

How to use the Book

Definitions of terms, quantities and units are highlighted in a coloured box, and there is extensive use of **bold** text to draw your attention to key facts.

The level of mathematics used throughout is equivalent to GCSE-level, while any more complex areas are highlighted in special **Mathematics Windows**. These should be particularly helpful to students not taking A-level mathematics.

Useful **derivations**, whether compulsory or otherwise are highlighted.

Guided examples are used at regular intervals. These should be attempted whenever they are encountered to reinforce the principles in the text. There are hints and suggestions as prompts, but the emphasis is on solving the problem for yourself.

Each unit ends with a **Self-assessment** section, which is divided into:

1. **Qualitative Assessment** This section contains questions of a descriptive nature, concentrating on the understanding of fundamental principles. This is the 'bread and butter' of examination preparation, and students should use this section as a regular test of their knowledge and understanding.
2. **Quantitative Assessment** This consists mainly of numerical questions. The answers to these questions are provided at the beginning of each quantitative assessment section. They are listed in ascending numerical order. This means that students can check that they have arrived at the correct answer, but will not be able to work backwards from a correct answer! The answers are generally given correct to **two significant figures** only, and are in **SI units**. The exceptions to this rule occur whenever more significant figures are justified on the basis of the accuracy of the data used in the question, and if an alternative unit is specified in the question.

Past-paper questions from a variety of examining boards are included at the end of each section to provide examination preparation.

Francis Azzopardi
Brian Stewart

About this book

Accessible Physics has been written with you, the student, in mind. We have designed the book to cover the 'end of course' syllabuses of all the major examination boards, as well as the core material of all the 'modular' syllabuses. The material is presented in a clear, precise and logical manner to support students of all levels of ability, so, whether you are a future Nobel Prize winner, or simply chose physics as the least unpalatable option on your timetable, we trust you will find inspiration and encouragement within these covers.

Talking to our own students over the years has made us very aware of the needs of those who find the need to press on through a seemingly impossible volume of information incompatible with the need to spend time on developing the understanding so necessary for success in this complex and demanding subject. Consequently, the book has been designed to be easy to read, with all the important details easy to find – in short, ***Accessible** Physics*. Important definitions and equations are highlighted by a light-coloured background, and key words emphasised in **bold** text throughout. Physical quantities, such as **mass, force, potential difference,** etc. are all defined in the text, and commonly-used symbols are quoted, together with the SI unit. Every quantity in important equations has its SI unit clearly displayed, as is the example below:

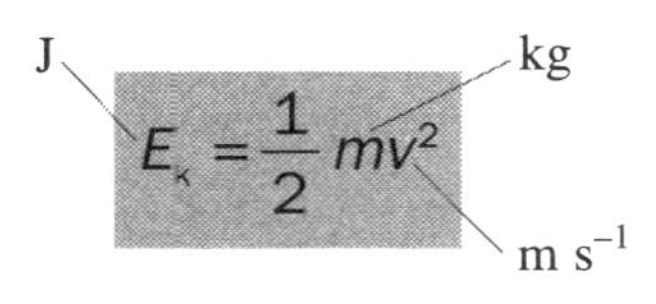

Definitions of the quantity and its unit are highlighted in coloured boxes. Throughout the book, symbols for quantities are printed in *italic* typeface. In most cases, a unique symbol is used for each quantity, but you should be aware that a symbol is only a shorthand way of writing a physical quantity. Letters and symbols can (and do) represent different quantities in different contexts. You should make a habit of defining the meaning of symbols every time you use them. In line with present trends, the mathematics has been kept to a minimum throughout the book with the more important concepts developed in the **'Maths windows'**. Lengthy derivations are no longer a major feature of examinations, but where these are still important, they are included in boxed areas of the text, labelled **'Derivation windows'**.

The guided examples which are scattered through each chapter give you an opportunity to practise what you have just learned by applying the physical principles to the solution of numerical problems. Instead of being given a model solution to follow, you are encouraged to attack the problem by following the guidance given. We feel that this method provides more motivation than merely reading someone else's solution, and you will achieve more of a sense of achievement by working through the examples as you come to them.

In the **self-assessment** sections at the end of each chapter, questions are set to provide:

- a thorough test of your knowledge and understanding
- essential practice in the use of concepts to explain physical phenomena
- essential practice in solving numerical problems and
- a solid framework for your revision.

The **self-assessment** questions are divided into two distinct types:

The **qualitative** questions are designed to lead you through the fundamental concepts covered in the chapter. They can be used at the end of a topic of study, for periodic review and for final examination revision. Unlike examination questions, these short questions can be attempted several times, as there is no 'final answer', but you should aim at a more detailed and thorough response each time. Ideally, you should attempt each one without referring to your notes or the book, then check your answer by reference to the text. This technique, if applied honestly and critically, will reveal those gaps in your knowledge which require further study, and identify areas of misunderstanding, which can be clarified in consultation with your teacher.

The **quantitative** questions are numerical questions designed to lead you progressively through the material covered in the chapter. A unique feature of this book is that all the answers to the questions are given at the start of each set of questions, but not in the correct order! This is to encourage you to work through the questions by applying the physics you have learned in the chapter, rather than working backwards from a known answer. In solutions to numerical problems, most credit is given for the techniques applied to the problem, rather than the final answer.

The answers are listed:

- in ascending numerical order
- with negative values ranked according to their modulus
- in SI units (unless otherwise stated in the question)
- correct to the appropriate number of significant figures (usually 2)
- in standard form where the figure is 1000 or more, or less than 0.01.

Past-paper questions from a selection of examination boards are printed at the end of each major section, and we would like to express our appreciation to the examination boards for their kind permission to reproduce this material.

FRANCIS AZZOPARDI
BRIAN STEWART

SECTION

Mechanics

1.1 Physical quantities

A **physical quantity** such as length or mass for example, is a measurable property whose meaning is precisely defined so that everyone can have the same understanding of the term. The meaning of a physical quantity is usually presented in two ways:

- A **word definition**. For example, 'density of a substance is the mass per unit volume of the substance'.
- A **defining equation** $\text{density} = \dfrac{\text{mass}}{\text{volume}}$

When quoting the measurement of a physical quantity, both the numerical value and the unit must be stated. This is technically good practice and also enables you to judge whether or not a quoted measurement actually makes sense. Quotations of a person's height as either 1.75 or 17.5 m are equally invalid: the first because it has no unit, and the second because experience tells you that people of this size do not exist.

Physical quantities are classified as either

- **base quantities** (fundamental quantities), or
- **derived quantities**

Base quantities and units

In the ***Systeme Internationale (SI)*** there are seven base quantities and units which are listed with their symbols in Table 1.1. You should note that these quantities and their units are chosen for convenience, and not out of necessity. The idea is that by having a small number of internationally agreed base units, meaningful and accurate exchanges of scientific information can occur on a global scale.

Table 1.1 Base quantities and units

Base quantity		Base unit	
Name	Symbol	Name	Symbol
mass	m	kilogram	kg
time	t	second	s
length	l	metre	m
electric current	I	ampere	A
temperature	T, θ	kelvin	K
amount of substance	n	mole	mol
luminous intensity		candela	cd

Except for the international prototype kilogram, the base units of the system have definitions involving measurements of physical quantities which could be reproduced in laboratories world wide. Brief descriptions of the seven base units are given in Table 1.2.

Table 1.2 SI Base units

Unit	Definition
kilogram	the mass of a particular cylinder of platinum–iridium alloy, known as the international prototype kilogram
second	the duration of 9 192 631 770 periods of the radiation emitted by a specified electron transition from an atom of caesium-133
metre	1/299 792 458 of the distance travelled in a vacuum by light in 1 second
ampere	the constant current which, if maintained in two straight, parallel conductors of infinite length and negligible circular cross-section, when placed 1 metre apart in a vacuum, would produce a force of 12×10^{-7} newton per metre length between the two conductors
kelvin	the fraction 1/273.16 of the thermodynamic temperature of the triple point of water
mole	the amount of substance of a system which contains as many elementary units as there are carbon atoms in 12×10^{-3} kilogram of carbon-12
candela*	the luminous intensity of a source of monochromatic radiation of frequency 5.4×10^{14} Hz that has a radiant intensity of 1/683 watt per steradian

NB The topic of illumination is not covered in this book. The definition of the candela is included for completeness.

Derived quantities and units

All other physical quantities (and units) which are additional to the base quantities (and units) are called **derived** quantities (and **derived** units). These additional quantities are derived from the base units by following the procedure laid down in their defining equations.

For example, the defining equation for the *volume* of a cube of side l is:

$$V = l \times l \times l$$

Since we obtain the **volume** (a derived quantity) by multiplying together three **lengths** (base quantities), the derived unit of volume is obtained by multiplying together the units of the base quantities. Thus the derived unit of volume is the **cubic metre (m^3)**.

Similarly, the defining equation for velocity is:

$$velocity = \frac{displacement}{time}$$

Displacement and time are both base quantities, so we obtain the derived unit of velocity by dividing the base units of length and time, which gives the **metre per second ($m\ s^{-1}$)**.

Some derived units have names which are just combinations of the base units. Table 1.3 shows some derived quantities whose units have combined base unit names.

Table 1.3 Derived quantities with combined name units

Derived quantity	Defining equation	Derived unit
density	$density = \frac{mass}{volume}$	$kg\ m^{-3}$
velocity	$velocity = \frac{displacement}{time}$	$m\ s^{-1}$
acceleration	$acceleration = \frac{change\ of\ velocity}{time}$	$m\ s^{-2}$
momentum	$momentum = mass \times velocity$	$kg\ m\ s^{-1}$

Some derived units may be given specific names, such as the **newton** (force) and **joule** (energy). Table 1.4 shows some of these derived quantities together with their unit names and base unit combinations.

Table 1.4 Derived quantities with special unit names

Derived quantity	Derived unit		
Name	**Name**	**Symbol**	**Base unit combination**
force	newton	N	kg m s^{-2}
pressure	pascal	Pa	kg m^{-1} s^{-2}
energy	joule	J	kg m^2 s^{-2}
power	watt	W	kg m^2 s^{-3}
electric charge	coulomb	C	A s
electric potential	volt	V	kg m^2 s^{-3} A^{-1}
electrical resistance	ohm	W	kg m^2 s^{-3} A^{-2}
electrical conductance	siemens	S	kg^{-1} m^{-2} s^3 A^2
capacitance	farad	F	kg^{-1} m^{-2} s^4 A^2
magnetic flux density	tesla	T	kg s^{-2} A^{-1}
activity	becquerel	Bq	s^{-1}

When using units in calculations you should note that

- names of units written in full begin with a lower case letter, despite being names of people (**newton**, **hertz** etc.)
- symbols for units with special names begin with an upper case letter (**N**, **Hz** etc.)
- full stops are not placed after unit symbols (except at the end of a sentence)
- the symbol for a unit should remain unaltered in the plural (8 **J** is correct; 8 **Js** is incorrect)
- it is good practice to incorporate units into calculations, for example:

$$pressure = \frac{force}{area} = \frac{500}{10}\ \frac{\text{N}}{\text{m}^2} = 50\ \text{Pa}$$

Standard prefixes for SI units

There is only one SI unit for each physical quantity, but using this could lead to cumbersome numbers when dealing with very small or large magnitudes, so we use an approved set of standard prefixes for multiples and submultiples of the base unit. These submultiples and multiples are shown in Table 1.5.

Table 1.5 Standard prefixes

Multiples			Submultiples		
Multiple	**Prefix**	**Symbol**	**Sub-multiple**	**Prefix**	**Symbol**
10^3	kilo	k	10^{-3}	milli	m
10^6	mega	M	10^{-6}	micro	μ
10^9	giga	G	10^{-9}	nano	n
10^{12}	tera	T	10^{-12}	pico	p
10^{15}	peta	P	10^{-15}	femto	f
10^{18}	exa	E	10^{-18}	atto	a

Homogeneity of physical equations

No matter how complex they may appear, all derived units can be reduced to a combination of base units, as can be seen by inspecting Tables 1.3 and 1.4. An equation is said to be **homogeneous** if the base units of every term in the equation are identical. Each term is a group of one or more quantities separated from other terms by a plus (+) sign, a minus (–) sign or an equals (=) sign. It can be useful to check for consistency of units in a freshly-derived or rearranged equation, but the method has some limitations:

- If all terms have identical units, the equation may be correct, but the method will not reveal any inaccuracies in numerical factors and dimensionless quantities.
- If the units in each term do not match, you will know there is an error in the equation, but the method will not identify where.

Checking equations for homogeneity of units

Example

The kinetic energy (E_k) of a body of mass (m) moving with velocity (v) is given by:

$$E_k = \frac{1}{2} mv^2$$

Using Tables 1.1, 1.3 and 1.4, you can check to see whether the equation is homogeneous:

Taking each term in turn:

units of E_k: kg m^2 s^{-2}

units of $\frac{1}{2}mv^2$ kg × (m s^{-1})^2 = kg m^2 s^{-2}

(the factor 1/2 is a pure number, and has no unit).

Clearly, each term has the same units, and the equation may be correct. However, if the factor 1/2 had been omitted, or was incorrect, this method would not reveal the error.

Guided example

You have derived the following equation for the energy (W) stored in a wire of cross-sectional area (A) which extends by an amount (e) when subjected to a stretching force:

$$W = \frac{EAe^2}{2l}$$

where E is called the *modulus of elasticity* and has the units N m^{-2} or Pa (pressure).

Check the equation for dimensional homogeneity.

Guidelines

Express each term as units of each quantity, and reduce these to base units where necessary:

units of W

units of $\frac{EAe^2}{2l}$

1.2 Scalar and vector quantities

Many measured quantities are not associated with direction. For example, **mass** cannot be considered to act in any particular direction, but the force of gravity on a mass (its **weight**) has a very definite line of action, towards the centre of the earth. In a similar way, **pressure** is assumed to act uniformly in all directions, but the **force** it exerts on a particular surface acts in a specific direction, at a right angle to that surface. So all the **quantities** that we can measure are classified according to whether or not a specific direction is associated with them.

Vector quantities are those which have both **magnitude** (size) and **direction**. Those which only have **magnitude** are called **scalars**.

A vector quantity is only fully defined when its magnitude (with unit) **and** direction are specified, whereas for a scalar, the direction is not needed. Although the direction of a vector may be described in many different ways, it is most important that it is stated, even when it seems quite obvious. In many cases a diagram is the simplest and clearest way to specify a direction. Frequently, there are only two possibilities – in opposite directions. In these cases, the terms **positive** and **negative** can be used. Some examples of scalars and vectors are listed in Table 1.6, but the lists are by no means exhaustive:

Table 1.6 Scalar and vector quantities

Scalars	Vectors
length, area, volume, speed, time, mass, density, pressure, energy, power, temperature, electric charge, electric potential etc.	displacement, velocity, acceleration, force, weight, momentum, torque, electric current, electric field strength, magnetic field strength etc.

In diagrams, **vector** quantities are drawn as **lines** whose length and direction represent the magnitude and direction of the quantity. These lines are referred to as vectors. In Figure 1.1(a) v is a vector which represents a velocity of magnitude 35 m s^{-1} in a direction 50° south of east; while in Figure 1.1(b) F represents a force of magnitude 5 N acting in a direction of 30° above the horizontal.

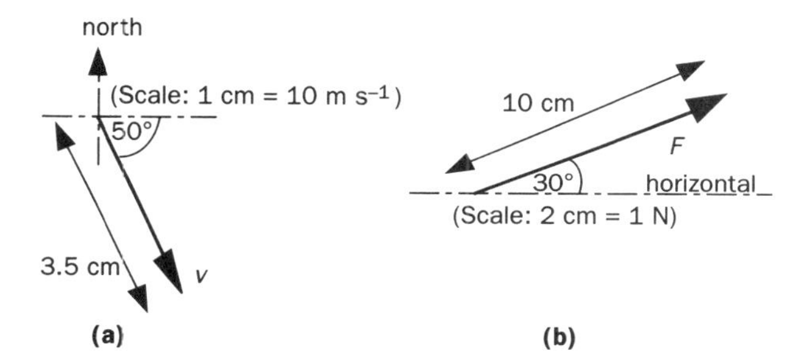

Figure 1.1 Vector representation

Vector addition

The addition of **scalar** quantities is very straightforward, and only requires that the quantities being added have the same unit. Thus the total of masses 9 kg, 5 kg and 2 kg is simply 16 kg. Scalars can be added and subtracted without difficulty. The only situation in which vectors can be added in this simple way is when they act along the same straight line.

You will understand **vector addition** more clearly by considering the following example which analyses the effect of wind on the speed of a light aircraft. To start with let us assume that the aircraft, which has a speed of 40 m s^{-1} in still air, is in level flight heading due North (0° mag.). Now we will consider the effects of a 20 m s^{-1} wind from various directions.

Vectors with the same line of action

If the directions of two or more vectors happen to be in the **same direction** along the **same straight line**, then their magnitudes can simply be added together:

Tail wind of 20 m s^{-1}

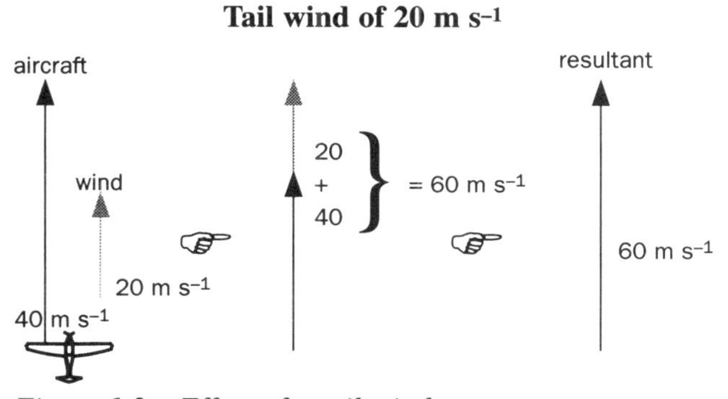

Figure 1.2 Effect of a tail wind

To calculate the resultant velocity, the two vectors are drawn in order, tip to tail, as shown in Figure 1.2. Since the directions are identical, the magnitude of their resultant is simply the algebraic sum of the two vectors, and the direction is the same as the aircraft's original path. Thus the aircraft crosses the ground below at a velocity of 60 m s^{-1}, due North. (We say its **ground speed** is 60 m s^{-1} and its **track** is 0°.)

Wind can have a marked effect on an aircraft's ground speed

If the vector directions are opposite, one vector is given a **negative** value before adding:

Head wind of 20 m s^{-1}

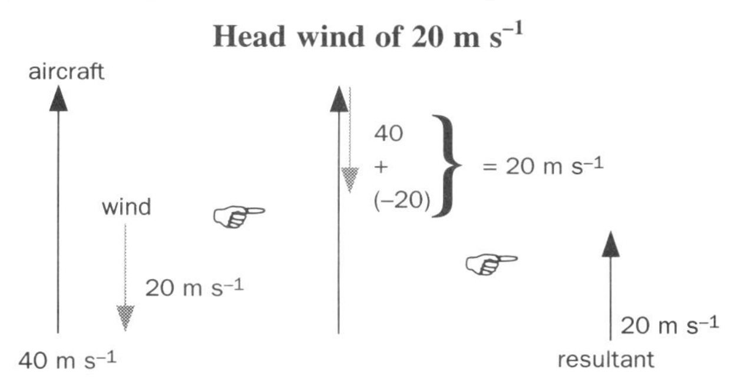

Figure 1.3 Effect of a head wind

The two vectors are again drawn in order, tip to tail, but this time the wind's velocity is in a negative direction relative to the aircraft's path, as shown in Figure 1.3. In this case, the aircraft's ground speed is reduced to only 20 m s^{-1}, but its direction is unaltered. In both cases above, only **speed** is changed. The two examples we have considered so far are special, simple cases. If the directions of the vectors do not coincide, then vector addition techniques must be used to calculate their resultant magnitude and direction.

Vectors with different lines of action

Two non-parallel vectors must be combined by use of the **parallelogram law** which states:

> If two vectors acting at a point are represented in magnitude and direction by the sides of a parallelogram drawn from that point, their resultant is represented by the diagonal of the parallelogram.

In the diagrams above, the vectors have been drawn tip to tail, rather than from the same point, which gives the same result but is more convenient. The technique can be applied to any number of vectors by adding each vector tip to tail.

Perpendicular vectors We shall examine this case by considering the effect of a cross wind at 90° to the aircraft's velocity (Figure 1.4).

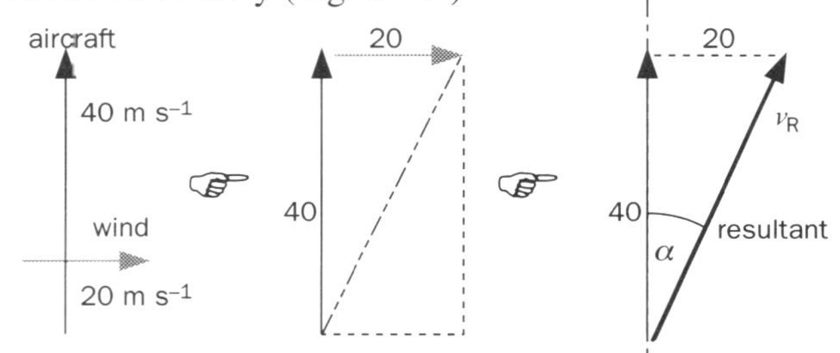

Figure 1.4 Effect of a cross wind at 90°

Since these two vectors form a right angle, the resultant is simply the **hypotenuse** of the **triangle** formed by the two vectors drawn tip to tail.

By Pythagoras, the magnitude of the resultant velocity (v_R) is given by:

$$v_R = \sqrt{20^2 + 40^2} = 44.7 \text{ m s}^{-1}$$

and the new direction of the aircraft (α) is given by:

$$\alpha = \tan^{-1}\frac{20}{40} = 26.6° \text{ East of North}$$

Notice that in each second, the aircraft will move:

- 40 m further North – we say it has a **northerly component** of velocity of 40 m s^{-1}
- 20 m further East – we say it has an **easterly component** of velocity of 20 m s^{-1}
- the combination of these two **components** produces the **resultant** motion of the aircraft, which moves a distance of 44.7 m every second, in a direction 26.6° East of North.

The motion of the aircraft can therefore be studied either by considering its **resultant** speed and direction, or its **components**.

Clearly, if the two vectors are at 90°, the situation is more complex than the in-line vectors (since **direction** is altered as well as speed), but it is still straightforward. Application of Pythagoras and simple trigonometry give the answers.

Cross wind at other angle to flight path If the wind is now blowing from a direction 210° magnetic, it makes an angle of 30° to the aircraft's original heading (due North). The vectors are again drawn in order, and the **resultant** is the vector required to close the triangle, as shown in Figure 1.5. This time, the triangle contains no convenient right angles.

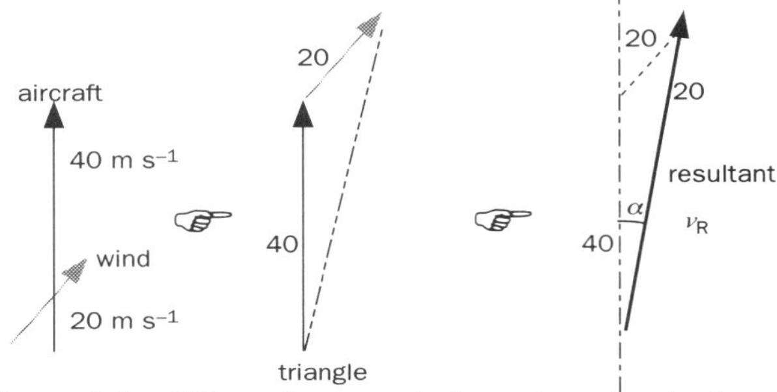

Figure 1.5 Effect of cross-wind on aircraft velocity

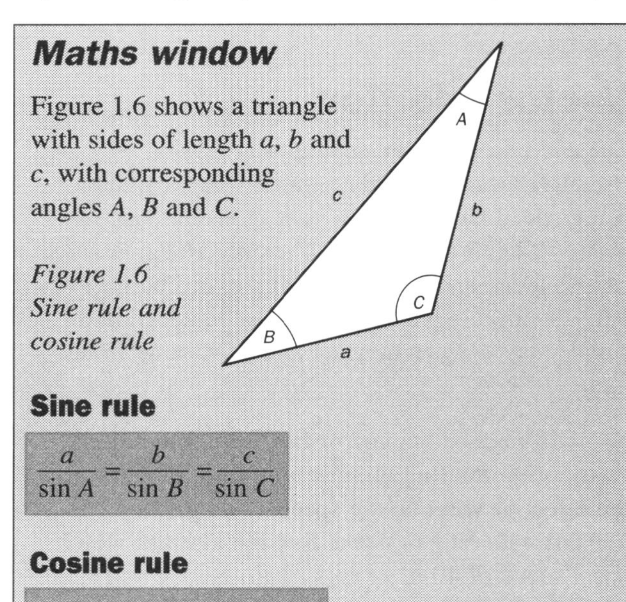

Maths window

Figure 1.6 shows a triangle with sides of length *a*, *b* and *c*, with corresponding angles *A*, *B* and *C*.

Figure 1.6 Sine rule and cosine rule

Sine rule

$$\frac{a}{\sin A} = \frac{b}{\sin B} = \frac{c}{\sin C}$$

Cosine rule

$$c^2 = a^2 + b^2 - 2ab \cos C$$

These rules can be used to calculate unknown angles and lengths of sides of triangles. They can be applied directly to **vector** triangles.

In the previous example, the length of the resultant velocity vector v_R is given by:

$$v_R^2 = 40^2 + 20^2 - 2 \times 40 \times 20 \times \cos 150°$$

$$v_R^2 = 1600 + 400 - 1600 \times (-0.866)$$

$$v_R^2 = 3386$$

$$v_R = 58.2 \text{ m s}^{-1}$$

and the aircraft's direction (α) is given by:

$$\frac{\sin \alpha}{20} = \frac{\sin 150°}{58.2}$$

$$\therefore \alpha = \sin^{-1}\left(\frac{20 \sin 150°}{58.2}\right)$$

$$\therefore \alpha = 9.9° \text{ East of North.}$$

It is possible to determine the magnitude and direction of the resultant velocity (v_R) by accurate scale drawing and measurement. Such a **graphical solution** can be somewhat tedious, but is the method used by pilots in the absence of electronic navigation aids. You can also apply geometry directly to the problem, using the **cosine** rule and the **sine** rule.

Resolution of vectors

An alternative solution involves **resolving** one of the velocities into two **components** – one **parallel** to the other vector, and one **perpendicular** to it.

A wind from the direction shown will have two effects:

- it tends to push the aircraft further north (i.e. it has a Northerly **component**)
- it tends to push it to the East (It has an Easterly **component**).

Both effects occur together, but we can examine them separately, by resolving the wind's velocity into **components** – one **parallel** to the plane's velocity, and the other **perpendicular** to it.

Maths window

It is often useful to **resolve** a vector into two **components** at 90° to each other. The effects of the two components can be calculated separately. To see how this is done, first imagine that the wind vector in the above example has been drawn to scale, as shown in Figure 1.7. The **magnitude** of the vector is represented by the **length** of the arrow (20 units in this case), and the direction is accurately represented as a bearing of 30° from the aircraft's original heading.

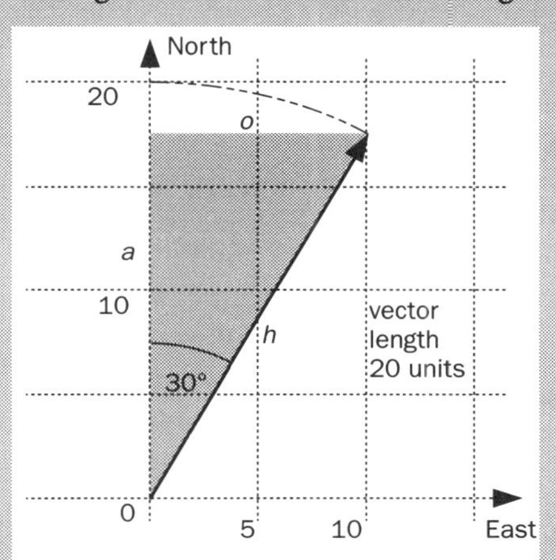

Figure 1.7 Scale drawing

To determine the wind's components **parallel** and **perpendicular** to the aircraft's heading, consider the right-angled triangle (shown shaded) where the vector is the **hypotenuse** (h), the magnitude of the northerly component is equal to the length of the **adjacent** side (a) at 30° to the vector, and the magnitude of the easterly component is equal to the length of the **opposite** side (o).

The basic trigonometric relationships are:

$$\sin 30° = \frac{o}{h} \qquad \cos 30° = \frac{a}{h} \qquad \tan 30° = \frac{o}{a}$$

Using trigonometry, the wind's velocity is resolved into components as follows:

Component **parallel** to the aircraft's velocity

$$= 20 \cos 30° = 17.3 \text{ m s}^{-1}$$

Component **perpendicular** to the aircraft's velocity

$$= 20 \sin 30° = 10 \text{ m s}^{-1}$$

This principle can be applied to any vector.

In Figure 1.8, A and B represent two vector quantities (such as velocity, force, magnetic field strength etc.) which can be added. The direction of **one** of the vectors is chosen as a reference direction, and the other vector is resolved into components: one component **parallel** to this direction, and the other component **perpendicular** to it. A and $B \cos \theta$ are now in line, and can be added; $B \sin \theta$ is at 90° to their direction, so Pythagoras can be used to calculate the magnitude and direction of the resultant.

Figure 1.8 Resolving a vector into components

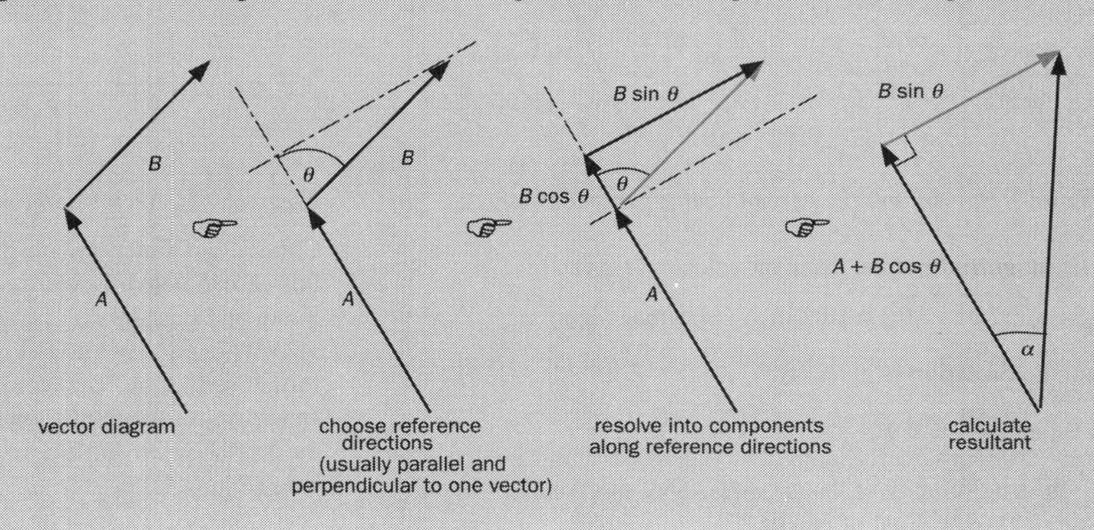

We can now apply this method to the aircraft problem. The sensible choice for the reference direction is that of the aircraft's velocity, so the wind velocity is resolved into components – one parallel and the other perpendicular to this direction, as shown in Figure 1.9.

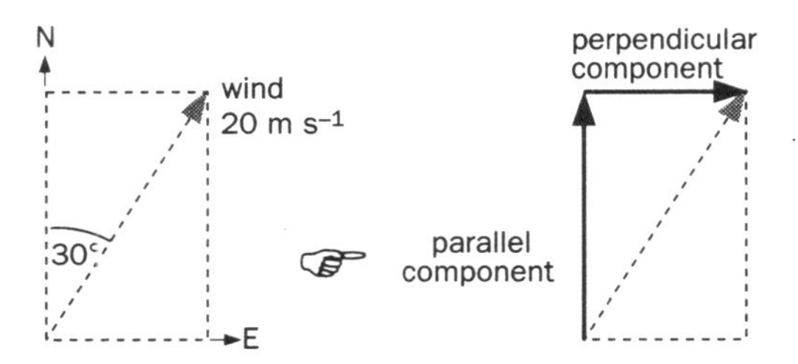

Figure 1.9 Resolution of wind velocity

We can then easily combine the **parallel** components algebraically, as shown in Figure 1.10. First, the aircraft's original velocity and the **parallel component** of the wind's velocity are added to give a net northerly component of 57.3 m s^{-1} (40.0 + 17.3).

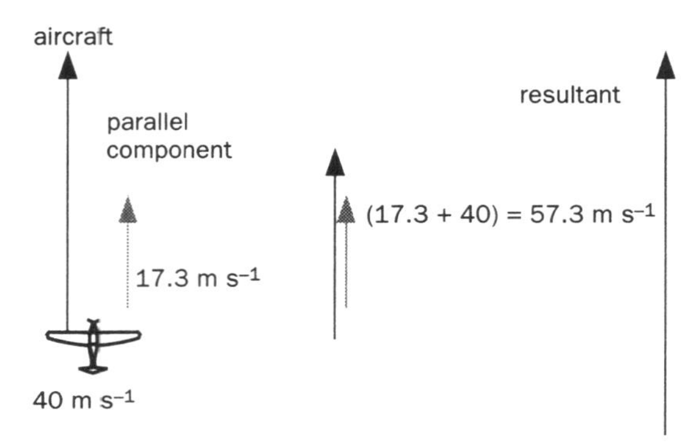

Figure 1.10 Adding the parallel components

Then this northerly component of 57.3 m s^{-1} is added vectorially to the perpendicular component of the wind's velocity of 10 m s^{-1} (Figure 1.11).

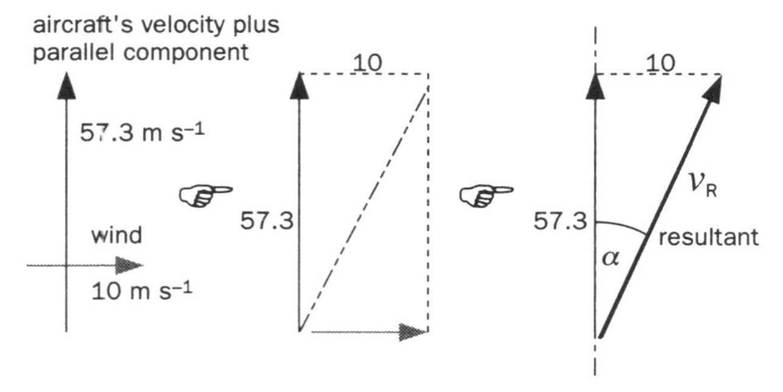

Figure 1.11 Adding the perpendicular components

The **magnitude** of the resultant velocity is given by:

$v_R = \sqrt{(57.3^2 + 10^2)} = 58.2 \text{ m s}^{-1}$ (Pythagoras)

and the **direction** is given by:

$$\alpha = \tan^{-1}\frac{10}{57.3} = 9.9°$$

So the true velocity of the aircraft is **58.2 m s^{-1}** along a direction 9.9° **East of North**.

Summary

All the techniques we have used can be applied to **ALL** vectors.

- Vector addition can be applied directly to parallel or perpendicular vectors.
- When vectors are neither parallel nor perpendicular, you can either
  (i) use the parallelogram rule,
  or
  (ii) resolve one or more of the vectors into components along two perpendicular directions.

You should practise with both methods, and choose the one with which you are most confident.

When vector quantities are stated, **both** the magnitude and direction must be identified. The easiest way to do this is with a clear diagram.

Guided examples (1)

1. An aircraft is flying at 250 m s^{-1} on a heading due East when it encounters a wind of 50 m s^{-1} blowing from a direction 60° East of North. Calculate the resultant velocity of the aircraft.

Guidelines

(i) Draw a sketch showing the two vectors v_a and v_w drawn from the same point. It is helpful to draw them approximately to scale and in the correct directions.

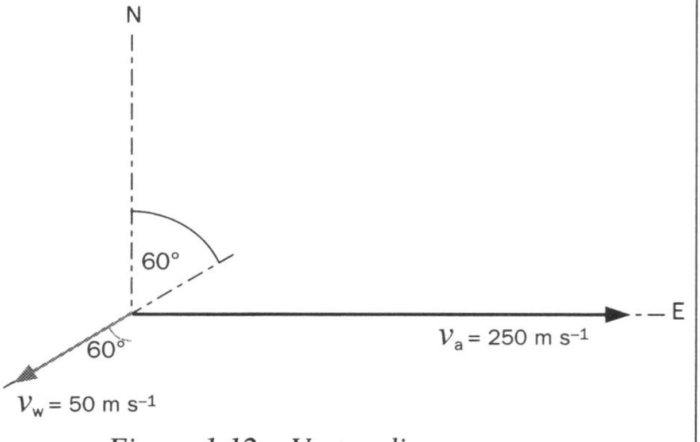

Figure 1.12 Vector diagram

(ii) Choose two directions at right angles – in this case the obvious choice is North–South and East–West. The aircraft's velocity is already East, so needs no further resolution. The wind's velocity should be resolved into components along the North–South and along the East–West axes (Figure 1.13).

Guided examples continued

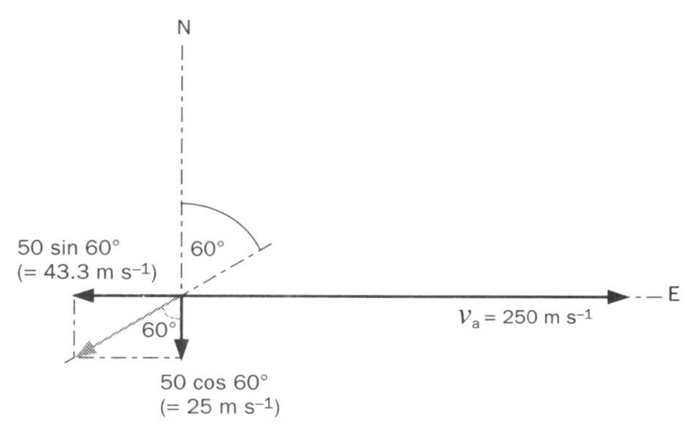

Figure 1.13 Component diagram

(iii) Combine the **East** and **West** components by algebraic addition. This leaves two vectors at 90°, which can be combined easily to find their resultant (Figure 1.14).

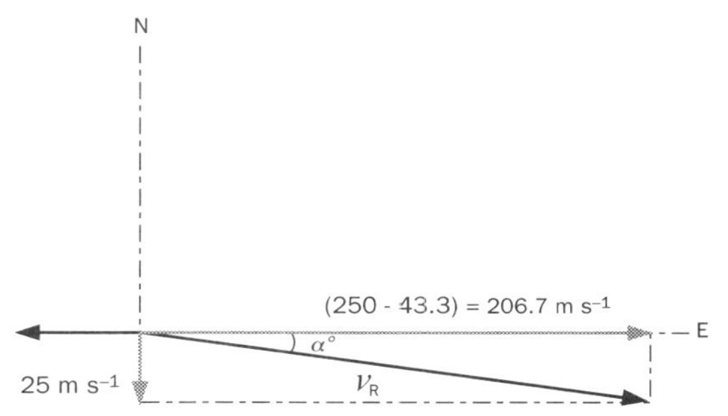

Figure 1.14 Calculating the resultant

2.

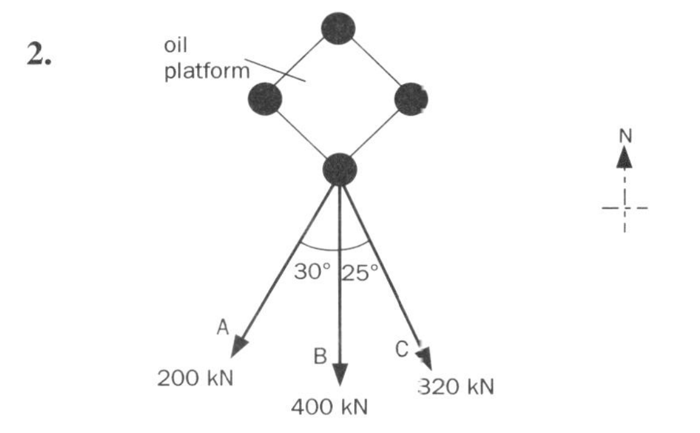

(**NB**: Length of tow line does not represent the magnitude of tension of the rope)

Figure 1.15 Vector diagram

Figure 1.15 shows three tugs A, B, and C which are being used to move a floating oil platform. Calculate the resultant force on the platform due to the three tugs.

Guidelines

(i) Choose a **reference** direction and resolve all vectors into components **parallel** to, and **perpendicular** to, this direction. Again, getting the scale approximately right is a great help (Figure 1.16):

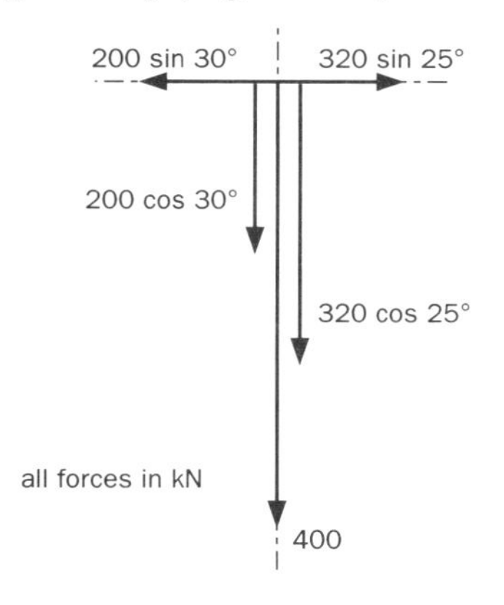

Figure 1.16 Component diagram

(ii) Add the components algebraically in each of the two directions, to produce a single component in each direction.

(iii) Finally, use Pythagoras to calculate the resultant of these two components.

The Nelson oil and gas platform was towed to its present position in the North Sea

Self-assessment

SECTION A
Qualitative assessment

1. Explain what is meant by:
   (a) a physical quantity
   (b) a defining equation.
2. Name the seven base SI quantities and units.
3. Explain what is meant by a **derived unit**.
4. Using the defining equations in each case, express the units of each of the following physical quantities in terms of the base units: force, pressure, energy, power, electric current, electric potential difference.
5. Explain what is meant by homogeneity of an equation, with respect to its units.
6. Explain how an equation shown to be homogeneous with respect to its units may still be incorrect.
7. Check each of the following equations for homogeneity:
   (a) The periodic time (T) of a simple pendulum of length (l) is given by

   $$T = 2\pi\sqrt{\frac{l}{g}}$$

   where (g) is the acceleration due to gravity.
   (b) The centripetal force (F) acting on a particle of mass (m) moving in a circular path of radius (r) at speed (v) is given by:

   $$F = mv^2r$$

   (c) The distance (s) travelled by a body in time (t) if it has initial velocity (u) and moves with constant acceleration (a) is given by:

   $$s = ut^2 + \frac{1}{2}at$$

   (d) The power dissipated (P) by an electrical component of electrical resistance (R) when the potential difference across it is (V) is given by:

   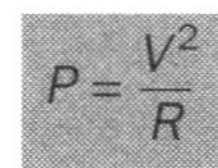

   $$P = \frac{V^2}{R}$$

8. Using the equation $F = \frac{Gm_1m_2}{r^2}$,
9. where (F) is the gravitational force acting between two bodies of masses (m_1) and (m_2) separated by a distance (r), express the unit of (G), the universal gravitational constant, in terms of its base units.

SECTION B
Quantitative assessment

(***Answers:*** 2.0; 3.6; 4.2; 4.8; 8.3; 8.6; 13; 15; 17; 25; 25; 25; 34: 41; 53; 75; 90; 400; 1.9×10^3.)

1. Hailstones fall vertically in still air with a constant velocity of 15 m s^{-1}. If a gale suddenly springs up and the wind blows horizontally at 20 m s^{-1}, calculate the resultant velocity of the hailstones (magnitude **and** direction).
2. A girl throws a javelin into the air at an angle of 38° to the horizontal. If the initial **horizontal** component of the javelin's velocity is 19.7 m s^{-1}, calculate
   (a) the initial velocity of the javelin
   (b) the initial vertical component of the javelin's velocity.
3. A paraglider is a type of high performance parachute which can glide for considerable distances. A pilot took off in perfectly still air from the top of a hill 600 m high, and landed 5.0 minutes later at sea level, at a point 2.5 km away as measured on the map. Assuming she flew at constant velocity throughout the flight, calculate her:
   (a) horizontal component of velocity (ground speed)
   (b) vertical component of velocity ('sink rate')
   (c) glide angle expressed in degrees below the horizontal and as a ratio of horizontal to vertical velocity
   (d) the true reading on her airspeed indicator (her resultant velocity)
4. It is possible to fly at a much lower 'sink rate' by flying more slowly. This particular glider has a minimum sink rate of 1.5 m s^{-1} at an airspeed of 5.0 m s^{-1}. If the pilot in question 3 repeats the flight at the new airspeed, calculate the new
   (a) groundspeed
   (b) glide angle expressed in degrees below the horizontal and as a ratio of horizontal to vertical velocity
   (c) time of flight
   (d) distance travelled (as measured on the map). Comment on the new distance and flight time.
5. A river is 50 m wide, and flows at 3.0 m s^{-1}. A man can swim at 2.0 m s^{-1} in still water. If he sets off at an angle of 90° to the bank, calculate:

A high performance paraglider being flown by Brian Stewart above the hills of Lancashire. By flying in rising columns of air, called thermals, skilled pilots can achieve flight distances in excess of 250 kilometres

(a) his true velocity (magnitude and direction)

(b) his distance downstream from the starting point when he reaches the opposite bank

(c) the actual distance he swims through the water

(d) the time taken.

6. As question 5, but the river now flows at 1.5 m s^{-1}. Calculate the direction in which the man must set off if he is to reach the opposite bank by the shortest possible route.

1.3 Forces

When two bodies interact, they do so by exerting forces on each other. Large masses such as planets exert a gravitational attraction (called weight) on neighbouring masses. (All masses attract each other due to gravitational effects, but the force can usually be neglected.) In this section, we shall not be considering the forces between nuclear particles – the so-called strong and weak forces.

When bodies are far enough apart that they do not interact, they are said to be separated. This distance may be hundreds of millions of miles for planets, or a few neutron diameters at the atomic level. Bodies in contact can exert pushes or pulls on each other.

(Strictly speaking there is no such thing as contact – what we call 'touching' simply means that the surface atoms of the bodies are close enough so that their electrostatic fields can interact. So all contact forces, including pushes, pulls, lift, drag, frictional and viscous forces, fluid upthrust, as well as electrostatic and magnetic repulsion and attraction are caused by electromagnetic fields around the charged particles of atoms.)

Forces always occur in pairs equal in magnitude and opposite in direction. Every body that exerts a force on another body experiences exactly the same force exerted on itself in the opposite direction.

In diagrams, forces are represented by arrows. The *line of action* of the force is represented by the direction of the arrow. Care should be taken in diagrams to show directions as accurately as possible, as this can indicate possible solutions to problems. Using the length of the arrow to indicate approximate magnitude can also give a valuable insight. Let us now look at some common forces.

Tensile and compressive forces

Figure 1.17 Tensile forces

Tensile forces (see Figure 1.17) tend to cause stretching along the line of action of the forces. Strings or rods can apply tensile forces to bodies. A rope in a tug-of-war is under tension.

Figure 1.18 Compressive forces

Compression is the opposite condition to tension (see Figure 1.18). Only stiff objects (like rods) can apply or resist compressive forces. A snooker cue is in compression at the moment of impact, due to the force applied by the player, and the contact force from the ball on the tip.

Weight

The distinction between **mass** and **weight** is of fundamental importance. The protons, neutrons and electrons that make up matter all have mass.

*(The most recent research suggests that mass might be something that becomes attached to nuclear particles when they collect other particles called **Higgs' Bosons,** but that is beyond the scope of this book.)*

Since all matter is made of these basic particles, all bodies possess mass.

The **mass** of a body is a measure of the amount of **matter** contained by that body.

Common symbol m, M
SI unit kilogram (kg)

The unit is based on the **standard kilogram**, which has been defined in the section on physical quantities. All measurements of mass ultimately refer back to this standard.

The **weight** of a body is defined as the **force** exerted on the body by the **gravity** of a planet.

Common symbol W
SI unit newton (N)

The **gravitational field strength** of a planet is defined as the **force** exerted by the planet's **gravity** on a mass of 1 kg placed on the planet's surface.

Common symbol g
SI unit newton per kilogram (N kg^{-1})

Defining equation:
If a body of mass m (in kg) experiences a force F (in newton) due to a gravity field, then the gravitational field strength (g) is given by:

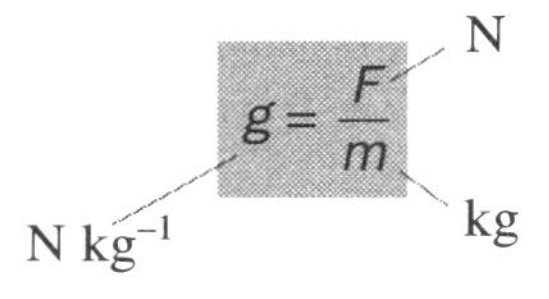

$$g = \frac{F}{m}$$

Equation 1

Measurements at the Earth's surface show that the Earth's gravitational field strength is around 9.81 N kg^{-1}, with slight variations at different places. For most purposes, this value is usually rounded up to 10 N kg^{-1} to simplify calculations.

The **weight** (W) of a body of **mass** (m) in a **gravitational field strength** (g) is given by:

$$W = mg$$

Equation 2

Mass and weight are two totally different quantities. The mass of a body is fixed, and has the same value everywhere in the universe, while the weight of the body varies greatly from place to place, according to the strength of gravity. However, in everyday life, confusion arises between the two because we usually measure the mass of a body by **weighing** it! When a person stands on the bathroom 'scales' and is horrified to discover an increase in mass of 2 kg, he or she will say 'I've put on 2 kg of weight'. This statement is meaningless, since the kilogram is not a unit of weight. The correct statement would be 'I've put on 2 kg of

An astronaut on a space walk hangs apparently weightless above the Earth. The manned maneuvering unit backpack provides the propulsion needed for movement

mass', but this is not a part of normal speech. Most weighing instruments work by measuring the strength of the force of gravity on the body, and converting this to a measure of mass, since the Earth's gravitational field strength can be assumed to be approximately constant in magnitude everywhere on the planet. Since most people are unlikely to need to measure mass on other planets, this system works well enough but, in science, we need to be much more precise in our use of these terms. The concept of gravity will be developed further in a later section.

Normal contact forces

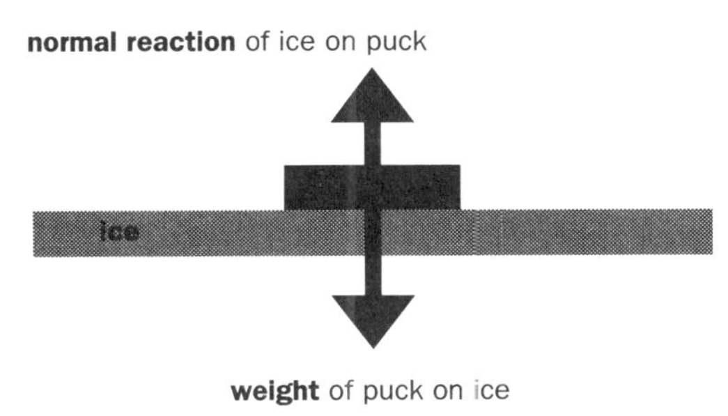

Figure 1.19 Normal (perpendicular) contact forces

These contact forces occur between bodies which are touching. In the absence of friction (friction-free surfaces are described as being **perfectly smooth**), the direction of this force is always **normal** (perpendicular) to the surface at the point of contact. An ice hockey puck sliding on ice comes close to this ideal.

Friction forces

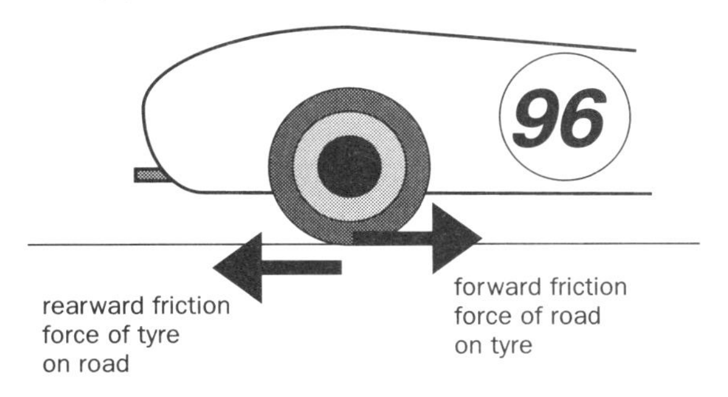

Figure 1.20 Friction forces

Friction is another contact force that may occur between bodies in contact (see Figure 1.20). It is always **parallel** to the surface at the point of contact, and acts in a direction to resist relative motion between the two bodies. Friction between the driven wheel of a car and the road provides the force to propel the car forwards. In general, the force of friction between two surfaces depends on the nature of the surface, and the magnitude of the force pressing the surfaces together. Racing cars have aerofoil surfaces to create an aerodynamic down force on the wheels. This means that a much higher friction force is available between the tyres and the road, but the mass of the car is not affected.

Combination of friction force and normal reaction

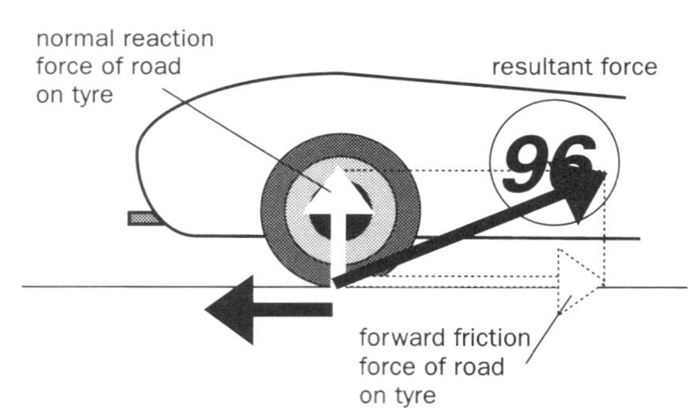

Figure 1.21 Resultant of friction and normal reaction

The true contact force between real bodies is the **resultant** of the normal contact force and the friction force, so the true force exerted by the road on the tyre is as shown in Figure 1.21. (**NB**: Only the forces exerted by the road on the tyre are shown; there would be equal and opposite forces exerted **by** the tyre **on** the road.)

Fluid upthrust

An object immersed in a fluid experiences a net upward force which tends to support it. This force is called the **upthrust** and is responsible for all **floating**, including that of ships, hot-air balloons etc., and is caused by the difference in the fluid **pressure** between the top and bottom surfaces of the object, as shown in Figure 1.22. Push a beach ball under water, and you will feel a powerful upthrust which increases as more of the ball is immersed, but remains constant once all the ball is under the surface.

The forces on the ball are due to the pressure of the water at that depth. Each force acts normally (at 90° to the surface). Pressure increases with depth, so the **magnitude** of the pressure forces increases towards the bottom of the ball. At any depth, the **horizontal** components of the pressure forces will be equal and opposite, so the resultant of all these forces can have no horizontal component. However, the vertical components do not cancel out, as the upward forces on the deeper parts of the ball are greater than the downward forces nearer the water surface. Therefore the resultant of the pressure forces is vertically upwards. This applies to an object of any shape, in any fluid.

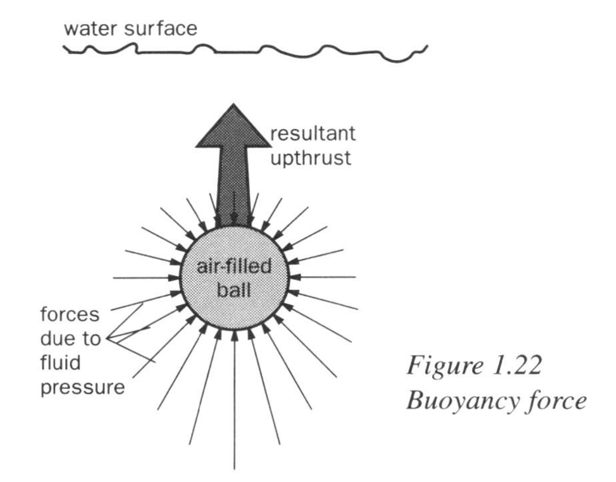

Figure 1.22 Buoyancy force

The magnitude of the upthrust on a body immersed in a fluid depends on

- the **density** of the fluid
- the **volume** of fluid **displaced** by the body

Archimedes' principle

The upthrust which acts on a body which is wholly or partly immersed in a fluid is equal to the weight of the fluid which is displaced.

The forces acting on a body immersed in a fluid (including air) are:

- the **weight** of the body. This force is caused by gravity acting on the body's mass, and always acts vertically downwards.
- the **upthrust** of the fluid on the body. This force always acts vertically upwards.

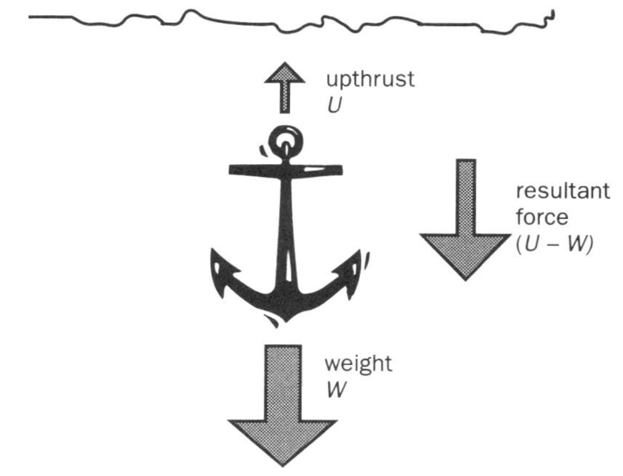

Figure 1.23 Forces on immersed bodies

If the **weight** is greater than the **upthrust**, as shown in Figure 1.23, the **resultant** of the two forces will be downwards, and the body (in this case an anchor in water) will sink.

The sky is littered with hot air balloons at the start of a race across Australia

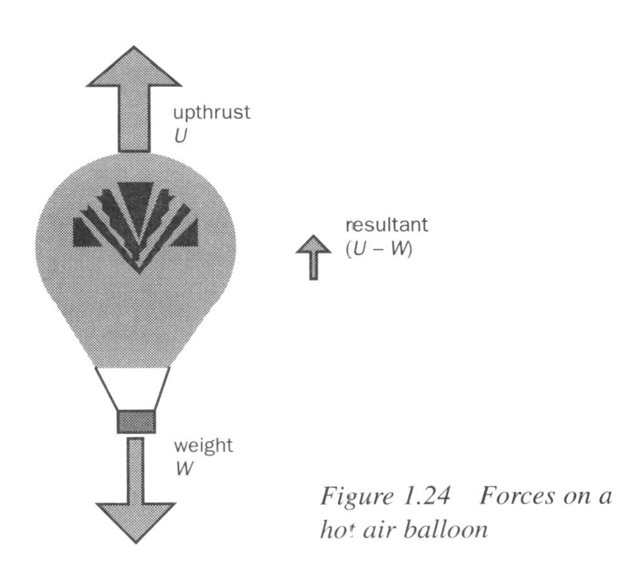

Figure 1.24 Forces on a hot air balloon

If the upthrust is greater than the weight, as shown in Figure 1.24, the resultant force will be upwards. The average density of the hot air balloon and its occupants must be less than that of the surrounding air. The weight of the displaced air (equal to the upthrust) is then greater than the weight of the body. In both of these cases, the body will accelerate in the direction of the resultant force. The balloon and anchor will reach a constant velocity due to the effect of friction opposing the motion (see page 14).

If the weight and the upthrust are exactly equal, then the body will float. Ships sink to a level where the upthrust is exactly equal to the weight of the ship. Adding more weight causes the ship to sink to a new level, until the forces balance again.

Viscous force (drag)

This is the resistive force which acts on any object as it moves through a fluid. Unlike friction between two surfaces, the magnitude of the viscous drag force depends on the velocity of the motion. The relationship between drag and velocity is complex: at low speeds, the flow of fluid over the body is smooth, and the viscous drag is proportional to the **velocity** of the body, but at higher speeds, the flow becomes turbulent, and the drag force is proportional to the **square** of the velocity. In other words, if the velocity of a body is doubled, it experiences a viscous drag force four times as great.

Figure 1.25 Forces acting on a falling body

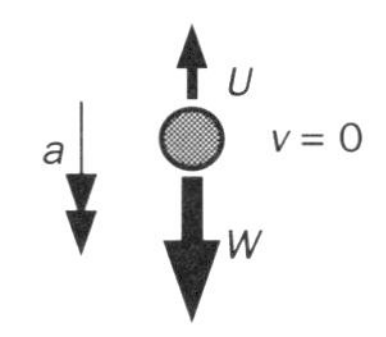

Resultant force = $W - U$

(a) Just released

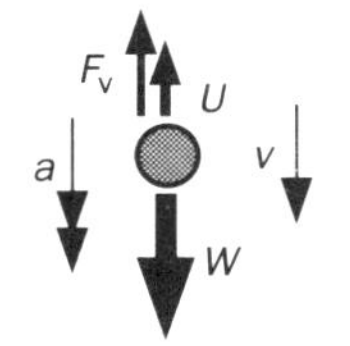

Resultant force = $W - (U + F_v)$

(b) After falling for short time

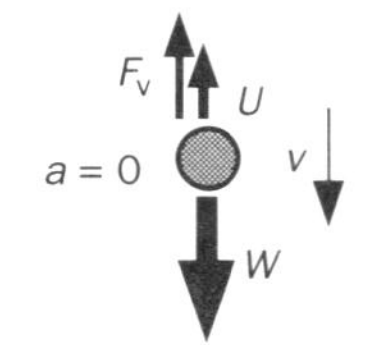

Resultant force = $W - (U + F_v) = 0$

(c) After reaching terminal velocity

Bodies accelerating through fluids under the action of a constant force, such as bodies falling under gravity, tend to reach a **terminal velocity**, i.e. they stop accelerating once they reach a particular speed.

Figure 1.25 shows the forces acting on a body falling under gravity through a fluid.

W represents the weight of the body (constant)

U represents the upthrust of the fluid on the body (constant)

F_v represents the viscous drag of the fluid on the body (varies with speed).

Part (a) shows the body just released. Viscous drag is initially zero, as the body has not yet started moving. The weight is greater than the upthrust, so the body accelerates downwards.

Part (b) shows the situation after the body has started to move. The viscous drag is no longer zero, and acts upwards. The weight is still greater than the combined upward forces (U and F_v), so the body still accelerates. However, the magnitude of the acceleration is reduced, as the resultant downward force has been reduced.

Part (c) shows the situation when the body has reached terminal velocity. The viscous drag has increased to such an extent that the combined upward forces (U and F_v) are exactly equal to the weight. The resultant force is zero, therefore the body can no longer accelerate, and continues at this velocity. Figure 1.26 shows velocity/time graphs for falling bodies.

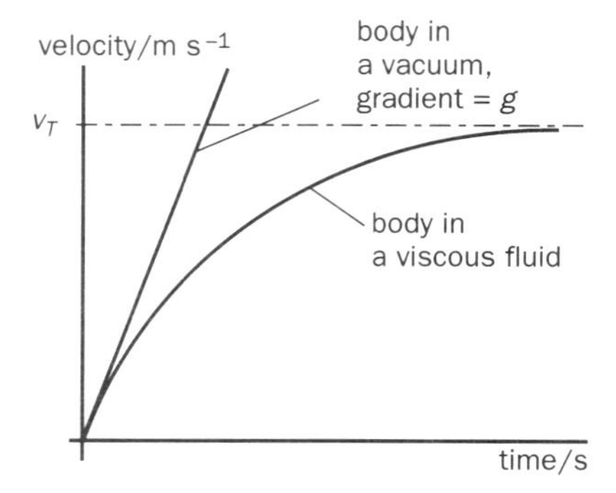

Figure 1.26 Velocity/time graphs of falling bodies

The straight line shows the motion of a body on which no viscous forces act. This can only happen in a vacuum, but can be assumed to apply to smooth, dense bodies falling in air at relatively low speeds. The curved line shows the motion of a body subject to viscous drag forces. Note that the initial acceleration is the same as for a body in a vacuum, but that the acceleration gradually decreases as the velocity increases. This is due to the effect of the increasing viscous drag opposing the weight, and therefore reducing the resultant force on the body. The body reaches its terminal velocity (^{v}T) when the resultant force is zero (shown by the graph becoming parallel to the time axis).

Self-assessment

SECTION A
Qualitative assessment

1. Distinguish between **tensile** and **compressive** forces.
2. Explain, using diagrams, why the resultant force exerted by the road on the driving wheel of a car which is accelerating acts at an angle to the horizontal other than 90°.
3. An object which is immersed in a fluid experiences an **upthrust**.
   (a) State what is meant by the term upthrust, and explain how it is produced.
   (b) State **Archimedes' principle**.
   (c) State the **principle of flotation**.
4. The three forces acting on a parachutist falling through still air are the **weight**, **upthrust** and **viscous drag**. Describe and explain how these forces change as she falls from a stationary balloon until she reaches terminal velocity.

1.4 Equilibrium

In this section, we shall examine situations in which the forces on bodies are **balanced** – we say they are **in equilibrium**. A body which is in equilibrium is not suffering any change in the way it is moving: It may be stationary or moving with **constant velocity** (i.e. it is not accelerating). Since forces tend to cause acceleration, and it is almost impossible for a body to have no forces at all acting on it, the following conditions must be satisfied if a body is to remain in equilibrium under the action of real forces.

*(In this book, we shall consider only forces in a two-dimensional plane (**coplanar** forces), but the principle applies also to three-dimensional space.)*

1. The **resultant** of all the **forces** acting on the body must be zero. So the linear acceleration of the body = 0.

The usual test for equilibrium is to pick any **two** directions (usually perpendicular to each other) and **resolve** all the forces on the body into **components** along these directions. If the **resultant** force is zero in **both** directions, the body is in equilibrium.

2. The **resultant** of all the **moments** or **torques** acting on the body must be zero. So the angular acceleration of the body = 0.

This condition is commonly called the **principle of moments** which states that:

If a body is in equilibrium under the action of coplanar forces, the algebraic sum of the moments of all the forces about any point in the plane of the forces must be zero.

Application of equilibrium

The most common application of these principles is in the calculation of unknown forces and moments. It is usually obvious whether or not a body is in equilibrium – although the body may be stationary or moving, there will be **no acceleration** (linear or rotational). If it is in equilibrium, then every force acting must be opposed by an equal and opposite force, and every moment acting must be opposed by an equal and opposite moment. This fact can be used to calculate the magnitude and direction of any unknown forces or moments acting on the body.

Parallel forces – same line of action

The simplest example is that of a body at rest, supported by a surface, as shown in Figure 1.27.

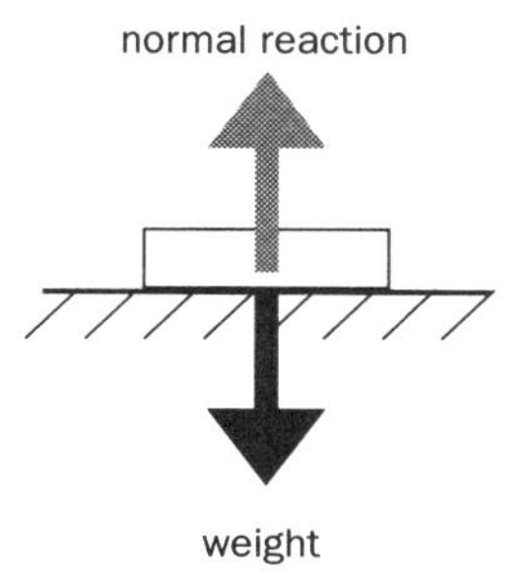

Figure 1.27 Forces on a body at rest

The forces acting on the body are:

(i) the body's **weight** acting vertically downwards and

(ii) the **normal reaction** force of the surface vertically upwards on the body.

The line of action of the weight force acts through the body's centre of gravity C_g (by definition). Applying the conditions for equilibrium:

Resultant force = 0

∴ *normal reaction – weight = 0*

∴ *normal reaction = weight*

Resultant torque = 0

Both forces must have the same line of action, and must act through the centre of mass C_m.

This result is almost trivial, and can be applied as a matter of routine whenever bodies are in equilibrium on a horizontal surface.

The free body diagram

Figure 1.27 is an example of what is called a **space diagram**, showing all the bodies (the object and the support surface) together with the forces acting between them. This is fine for simple cases, but where there is more than one point at which forces act, or there are several forces or bodies involved, the free body diagram is more useful. A single body is drawn, and all of the forces acting on it are drawn as arrows. Any other bodies are omitted, only the forces they exert are drawn. Figure 1.28 is a free body diagram of an aircraft, in level flight, at constant velocity. The earth and the air around the aircraft are responsible for the forces, but only the arrows representing the forces are shown. You should not try to combine forces at this stage, but should show each individual force separately.

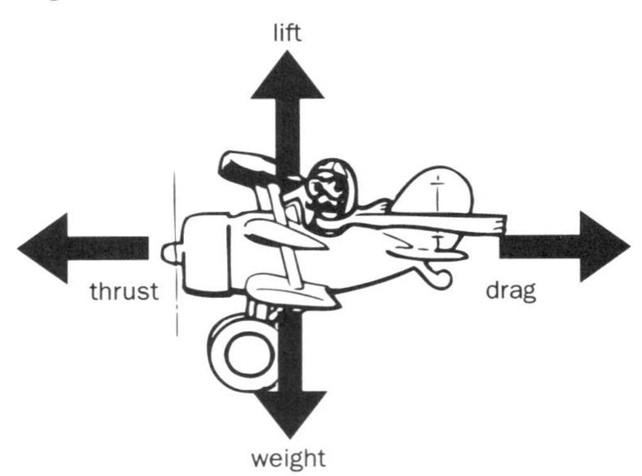

Figure 1.28 Forces on an aircraft in equilibrium

The forces on the aircraft can be considered as two separate sets acting along the horizontal and vertical directions.

The aircraft is in **vertical** equilibrium since the resultant of the **vertical** forces is zero (no vertical acceleration), so the lift force generated by the wings is equal to the weight of the aircraft. The aircraft is in **horizontal** equilibrium since the resultant of the **horizontal** forces is also zero (no horizontal acceleration) so the thrust of the engine is equal to the drag on the aircraft caused by friction.

Even if the aircraft is not in equilibrium under the action of the horizontal forces (i.e. if it were gaining or losing speed), if it continues in level flight it must be in equilibrium in the vertical plane.

Forces can have no effect in a direction perpendicular to their line of action.

This example leads us to a powerful tool for analysing more complex systems of forces.

Parallel forces – different lines of action

Consider a uniform beam of length 5 m and weight 200 N resting on supports at each end ('simply-supported'), carrying a man weighing 800 N at a distance of 1 m from one end, as shown in Figure 1.29.

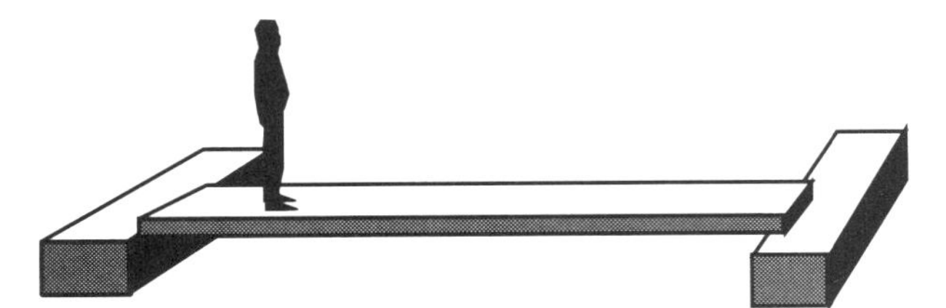

Figure 1.29 Man on a simply-supported plank

Calculate the forces exerted **by** the beam **on** the supports.

Draw a **free-body diagram** showing all the relevant dimensions and known forces acting on **one body** (the beam). Represent unknown forces and distances by symbols as shown in Figure 1.30.

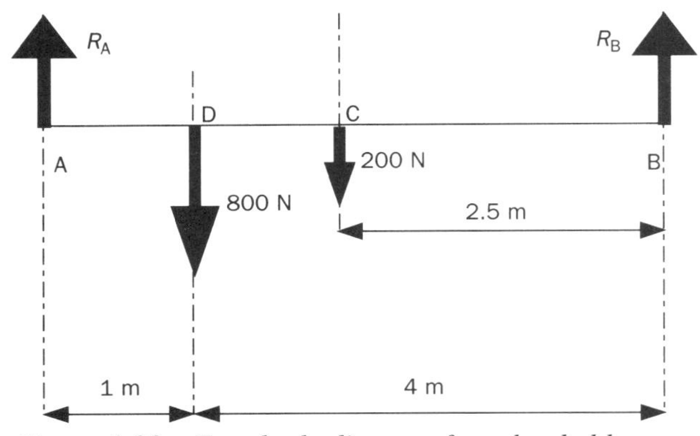

Figure 1.30 Free body diagram for a loaded beam

R_A and R_B are the **reaction forces** exerted **by** the supports **on** the beam.

The weight of the beam is shown acting through the centre of gravity of the beam, which is at the mid-point since the beam is of **uniform** construction.

Apply the conditions for equilibrium:

Resultant Force = 0

Since all the forces are vertical, we only need to consider the vertical direction,

$$R_A + R_B + (-800) + (-200) = 0$$

$$\therefore \quad R_A + R_B = 1000 \qquad \text{(a)}$$

There are two unknowns in this equation, so we need to apply the second condition for equilibrium:

Resultant moment about any point = 0

(See page 23 for more detail on moments.)

Here we pick a point in the diagram, and write expressions for the moments of all the forces about that point. Picking either point A or point B has a big advantage: the line of action of an unknown force passes through each of these points, so that force has no moment about it:

A force can have no moment about any point on its line of action.

Taking moments about point A

Both of the weight forces tend to cause clockwise rotation, which is prevented by R_B which produces an anticlockwise moment. Choosing the clockwise direction as positive, we obtain:

$$(800 \times 1) + (200 \times 2.5) - (R_B \times 5) = 0$$

$$\therefore \quad (R_B \times 5) = 1300$$

$$\therefore \quad R_B = 260 \text{ N}$$

Substituting this result in Equation (a):

$$(R_A + R_B = 1000$$

$$\therefore \quad R_A = 1000 - 260 = 740 \text{ N}$$

R_A and R_B are the forces exerted **on** the beam **by** the supports. By Newton's third law, there must be exactly equal forces exerted by the beam on the supports, and these are the answers to the set question.

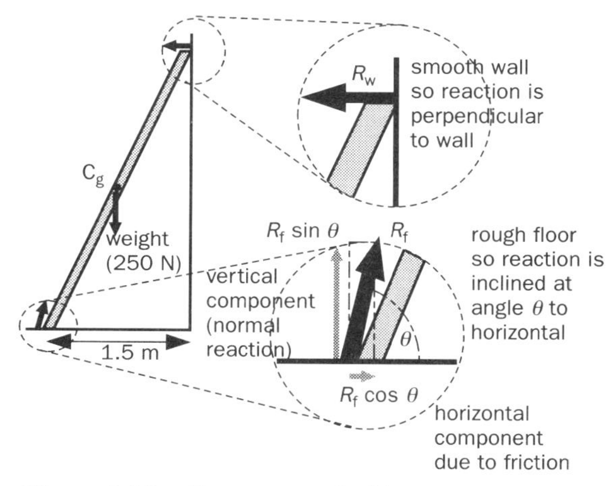

Figure 1.31 Forces on a ladder

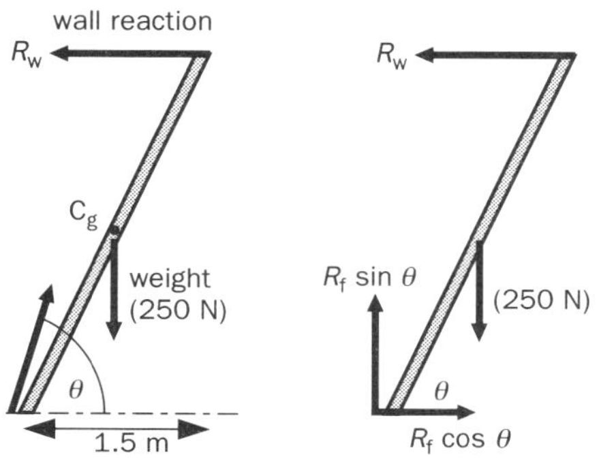

Figure 1.32 Free body diagram for a ladder

Non-parallel forces

A classic example is that of a uniform ladder of known weight on a **rough** floor, leaning against a **smooth** wall, in which you are asked to calculate the magnitude and direction of the reactions at the floor and wall.

Applying the principles of equilibrium in this situation is best illustrated by means of a worked example. Consider the problem of calculating the reactions at the wall and the floor for a ladder of weight 250 N and length 4 m when the foot of the ladder is resting on **rough** ground at a distance of 1.5 m from a **smooth** vertical wall. The words **rough** and **smooth** are very significant:

- Smoothness means no friction, so the reaction force at the wall has no component parallel to the wall which means that the reaction at this point (R_w) must be normal to the wall.
- Roughness implies friction, so at the foot of the ladder there must be a component parallel to the floor, as well as a normal reaction. This means that the resultant reaction at the floor (R_f) is inclined at some angle (θ), as shown in Figure 1.31. The value of angle θ is not known at this stage.

Since the forces are not aligned conveniently along two perpendicular directions, it is necessary to **resolve** some of the forces into **components**. Any directions could be chosen, but in this case the most obvious choice is **horizontal** and **vertical**. Draw a **free body diagram** showing the forces on the ladder (Figure 1.32(a)), then resolve these forces into vertical and horizontal components in the component diagram (Figure 1.32(b)).

For horizontal equilibrium

The only horizontal forces are R_w and $R_f \cos \theta$

$$R_w - R_f \cos \theta = 0 \qquad \text{(a)}$$

For vertical equilibrium

The only vertical forces are W and $R_f \sin \theta$

$$W - R_f \sin \theta = 0 \qquad \text{(b)}$$

There are three unknowns (R_f , R_w and θ) , which cannot be found using only two equations, so a third is needed, and is obtained by applying the **principle of moments**.

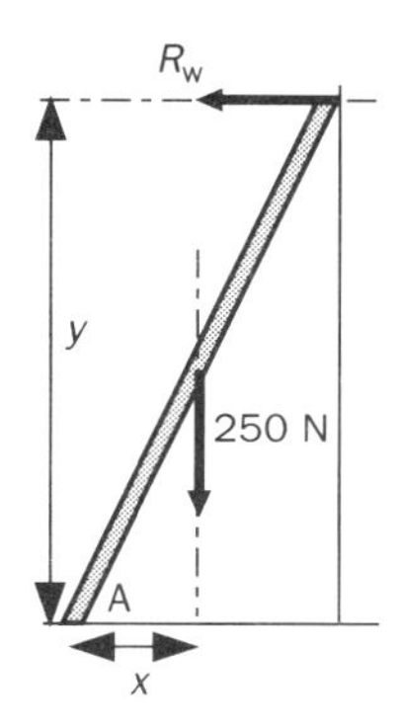

Figure 1.33 Taking moments about the foot of the ladder

By choosing the **foot** of the ladder (point A in Figure 1.33) as the point about which to take moments, we avoid having to deal with the reaction force R_f. The line of action of R_f passes through point A, and cannot therefore exert a moment about this point.

R_w produces an **anticlockwise** moment, equal to $(R_w \times y)$ (since y is the **perpendicular** distance between the line of action of R_w and point A).

The weight produces a **clockwise** moment, equal to $(250 \times x)$.

Since the ladder is in equilibrium, the sum of these moments must be zero:

$$(R_w \times y) - (250 \times x) = 0 \qquad \text{(c)}$$

The values of x and y are obtained as follows: since the position of C_g is half-way up the ladder, x must be half the distance from the ladder foot to the wall (by symmetry), i.e. $x = 0.75$ m.

y can be determined from geometry, since the ladder, the wall and the floor form a right-angled triangle:

$$y = \sqrt{5^2 - 1.5^2} = \sqrt{25 - 2.25} = 4.77 \text{ m}$$

Substituting for x and y in Equation (c) gives:

$$(R_w \times 4.77) = (250 \times 0.75)$$

$\therefore \quad R_w$ 39.3 N directed horizontally to the right

Equation (a) becomes:

$$39.3 - R_f \cos\theta = 0$$

$$\therefore \quad R_f \cos\theta = 39.3 \qquad \text{(d)}$$

Equation (b) becomes:

$$250 - R_f \sin\theta = 0$$

$$\therefore \quad R_f \sin\theta = 250 \qquad \text{(e)}$$

Maths window

Simultaneous equations involving **sin** and **cos** can often be reduced to a simple equation by using the standard relationship:

$$\frac{\sin\theta}{\cos\theta} = \tan\theta$$

Dividing Equation (e) by Equation (d) gives:

$$\frac{R_f \sin\theta}{R_f \cos\theta} = \frac{250}{39.3} = 6.36$$

$$\therefore \quad \tan = \theta = 6.36$$

$$\therefore \quad \theta = \tan^{-1}(6.36) = 81°$$

Substituting for θ in Equation (b) gives:

$$250 - R_f \sin 81° = 0$$

$$R_f = \frac{250}{\sin 81°} = \underline{253 \text{ N}}$$

Figure 1.34(a) illustrates the actual forces acting on the ladder, but Figure 1.34(b) – the component diagram – is more instructive.

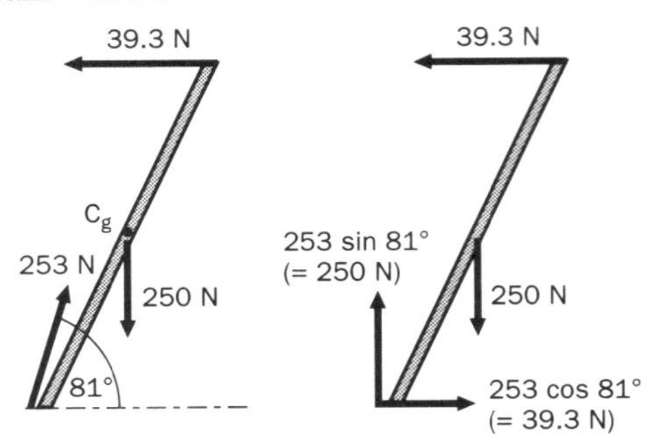

Figure 1.34 Complete free body component diagrams for the ladder problem

You should note that:

- the horizontal component of the reaction at the floor (39.3 N) is provided by **friction**. This force is necessary to prevent the foot of the ladder from slipping out away from the wall, which would happen if the floor were too slippery
- the horizontal component of the floor reaction exactly balances the wall reaction
- when using ladders, it is advisable to have a second person to hold the foot of the ladder, to provide an additional force if there is any possibility of slipping
- the vertical component of the reaction at the floor exactly balances the weight of the ladder.

Summary

To solve problems involving bodies in equilibrium:

- Draw a **space diagram** to show the dimensions of the bodies and the positions of the lines of action of the forces.
- Draw a **free body diagram** for one body. Represent **all** the forces acting by arrows, paying particular attention to the directions of the forces. (Make reasoned guesses if these are not known.)
- If necessary, choose two directions at right angles and resolve the forces into components along these directions.
- Write the equilibrium equations for forces (one equation for each of the two perpendicular directions).
- Choose a suitable point in the plane of the diagram and apply the principle of moments about that point.

The triangle of forces

If the three forces acting on the ladder in the previous example are drawn to scale (magnitude and direction), when placed in order, they form a closed triangle as shown in Figure 1.35:

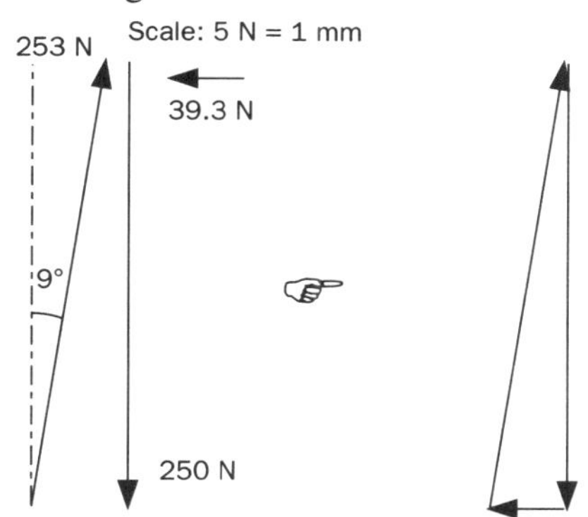

Figure 1.35 The triangle of forces

This leads us to a statement of the **triangle of forces** rule:

> If a body is in equilibrium under the action of three coplanar forces, and these forces are represented in magnitude and direction by vectors drawn to scale and drawn in order, they will form a closed triangle.

You should note that:

- the vector triangle will be closed, since forces in equilibrium have zero resultant force
- the rule can be extended to apply to any number of forces in equilibrium – the vector diagram will be a polygon with the number of sides equal to the number of forces
- the principle can be applied to *any vectors* as well as forces.

The principle can be used directly to find unknown forces by scale drawing and/or calculation. An important consequence is that if a body is in equilibrium under the action of coplanar forces, the lines of action of all the forces must pass through a single point somewhere in the plane of the forces. This can lead to very quick and easy solutions to complex problems:

Referring back to the ladder problem:

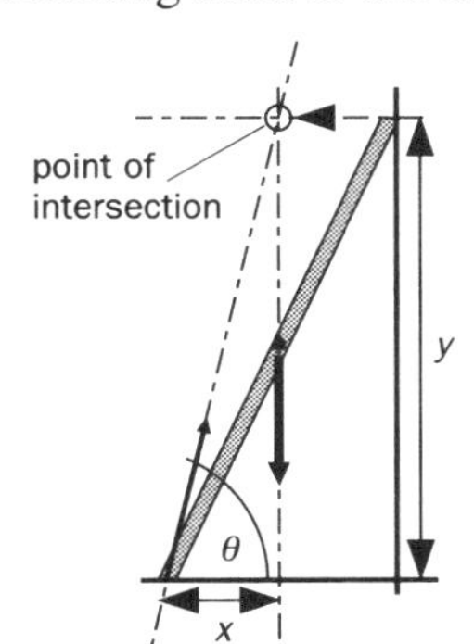

Figure 1.36 Using the triangle of forces to solve the ladder problem

The point of intersection is fixed by the directions and points of application of the weight (vertical, through C_g), and R_w (horizontal, at the ladder/wall contact) as shown in Figure 1.36.

Angle θ can now be calculated directly from the geometry of the ladder, knowing the position of the point of intersection of the forces From previous calculations:

$y = 4.77$ m (Pythagoras)

$x = 0.75$ m (symmetry)

$$\theta = \tan^{-1}\frac{y}{x} = \tan^{-1}\frac{4.77}{0.75} = 81° \text{ as before}$$

Knowing θ, R_f can be resolved into horizontal and vertical components ($R_f \cos\theta$ and $R_f \sin\theta$ respectively)

For vertical equilibrium: $R_f \sin\theta = 250$ N; hence R_f can be calculated. For horizontal equilibrium $R_f \cos\theta = R_w$ so then R_w can be calculated.

Inspection of all complex-looking statics problems will usually reveal simplifications that can be made using this powerful principle.

Where there are only three forces acting, we can say that:

> The **resultant** of any two of the forces will be the **equilibrant** of the third force (i.e. a force exactly equal in magnitude, but in the opposite direction).

Support reactions on beams – cantilevers

You saw in a previous problem how to calculate the support reactions for a simply-supported beam carrying a point load. If the load is uniformly distributed across the length of the beam, the weight of the load is considered to act through the centre of mass of the load. This will be at the centre of the beam.

The load will often be stated as a mass per unit length, e.g. 10 kg m^{-1}. If the beam is 2.5 m long, the total mass is $2.5 \times 10 = 25$ kg.

A cantilever is a beam which is only supported at one end. Therefore the support end must provide a moment to counter the turning effect of the load on the beam. A diving board is a simple example of the cantilever principle. Consider the situation shown in

This unfinished section of elevated motorway is a good example of the use of the cantilever principle in construction

Figure 1.37, where a man of mass 75 kg stands at the end of a diving board of mass 50 kg and length 2.5 m.

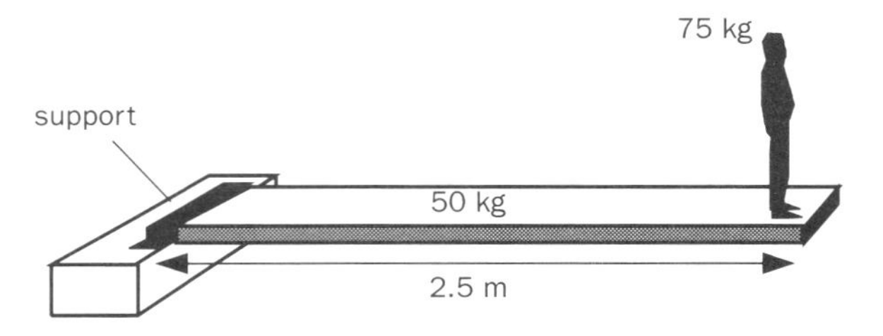

Figure 1.37 Diving board as example of a cantilever

The weight of the man and the board act vertically downwards, so the support must provide a force directed vertically upwards to balance them.

Additionally, the support must provide an anticlockwise moment to resist the clockwise turning effect of the weights of the man and the board, as shown in Figure 1.38, the free body diagram.

For equilibrium:
Resultant force = 0 (vertical plane)
$R + (-500) + (-750) = 0$
$R = 1250$ N

Resultant moment about any point = 0
$M + (-500 \times 1.25) + (-750 \times 2.5) = 0$
$M = 625 + 1875$
$M = 2500$ N m

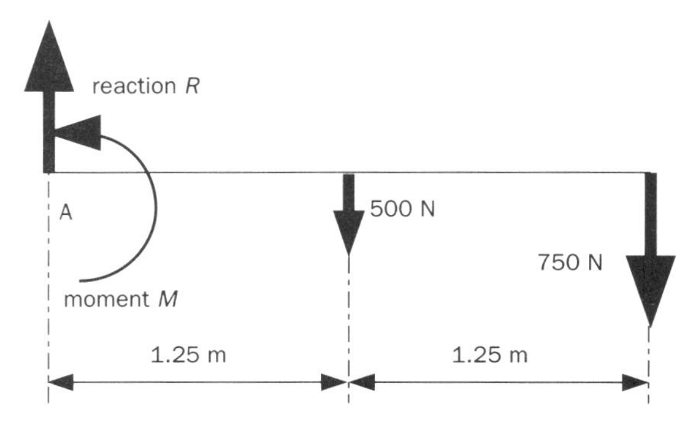

Figure 1.38 Free body diagram of the diving board

Thus the structure remains in equilibrium only if the support is capable of supplying both the vertical reaction and the anticlockwise moment.

The cantilever principle is widely used in building construction. The example in Figure 1.39 shows a cantilevered beam being used to support a wall without the need for pillars.

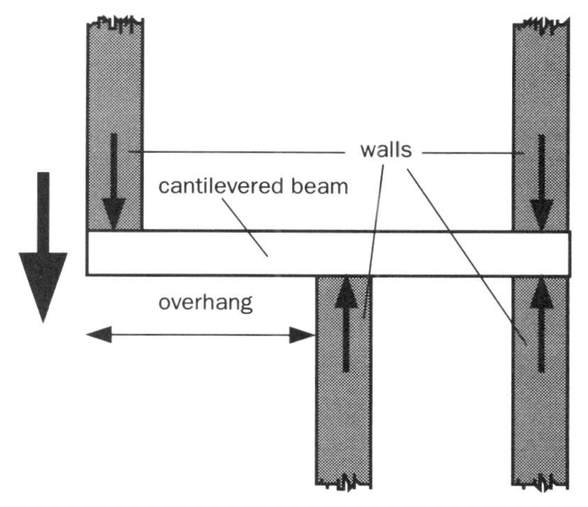

Figure 1.39 Cantilever construction in buildings

The cantilever principle is also used to extend single-span beam bridges. A cantilever bridge is built like two diving boards facing one another, as shown in Figure 1.40.

The shore ends are anchored to the bank, and in combination with the support pillars, provide the necessary **moment** to maintain the bridge in equilibrium. The advantage of this style of construction over the simple beam bridge (where there is no cantilever action) is that a wider span between the support pillars is possible. The Forth Bridge in Scotland, built in the 1890s, is a typical example of such construction. Its two cantilever spans are 520 m long in total.

Turning effect of forces

Turning effect, moment, and torque all have the same meaning, and are a measure of the ability of a force to rotate a body about a given point:

> The **moment** of a **force** about a point is calculated by multiplying the **force** by the perpendicular **distance** from the line of action of the force to the point.

Common symbol	T, M
SI unit	newton metre (N m)

Figure 1.41 shows a spanner exerting a turning force to tighten a nut.

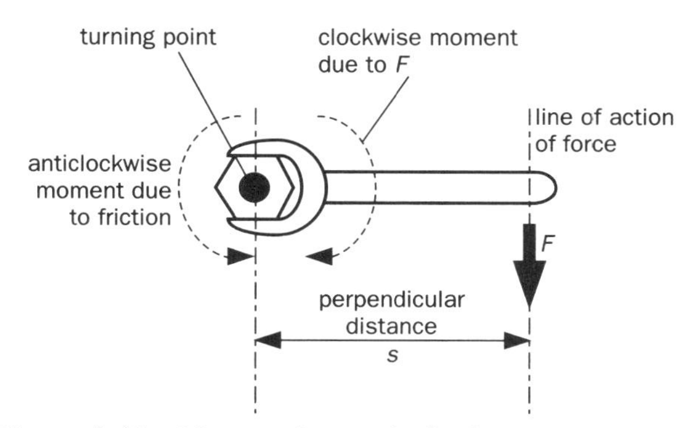

Figure 1.41 Moment (torque) of a force

The moment or torque T of the force about the centre of the nut is given by:

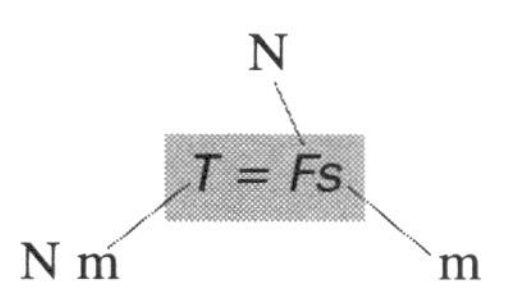

$$T = Fs$$

Equation 1

The unit is the **newton metre (N m)** – which has no special name. *(In particular, it must **not** be confused with the unit of **work** (also N m), to which it has no connection.)*

The effect of this moment is to rotate the spanner and nut in a clockwise direction, about the centre of the nut. If no rotation occurs, it is because friction exerts an equal and opposite (i.e. anticlockwise) moment.

Moments can be combined by addition – clockwise and anticlockwise moments must be given opposite signs. The distance s must be measured at a right angle to the line of action of the force (see page 24):

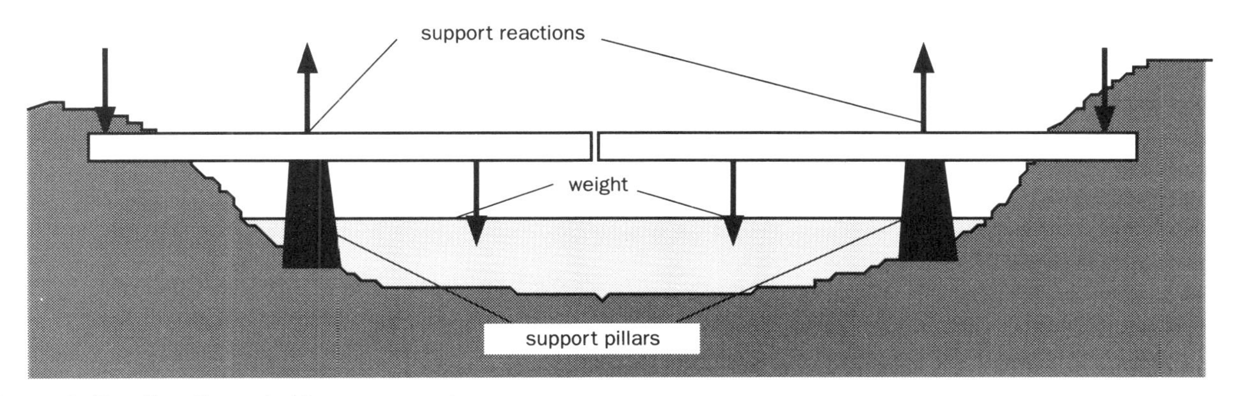

Figure 1.40 Cantilever bridge construction

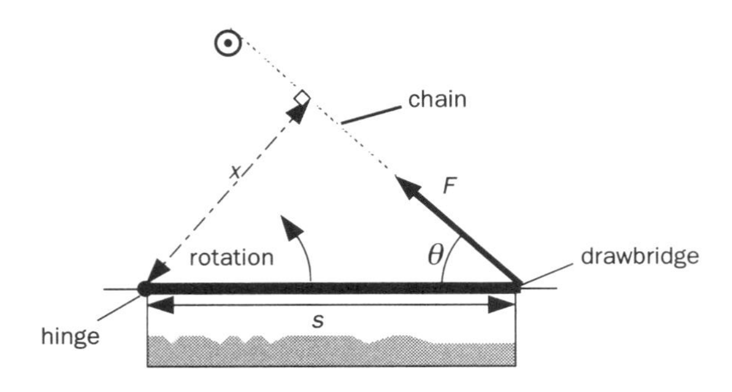

Figure 1.42 Moment of applied force on a drawbridge

In the case of a drawbridge being raised by a chain attached as shown in Figure 1.42, the perpendicular distance between the line of action of the force and the hinge is x.

As this is not an easy distance to measure, a better approach is to use components:

since $x = s \sin \theta$

then the moment or torque (T) is given by:

$$T = Fs \sin \theta$$

Equation 2

Couples

A couple:

- consists of two equal and opposite coplanar forces whose lines of action do not coincide;
- always tends to produce rotation;
- cannot produce a resultant force (since the magnitudes of the two forces are equal and opposite) and hence:
- cannot produce translational (linear) motion.

Figure 1.43 shows a car steering wheel which is free to rotate about an axis through a point O:

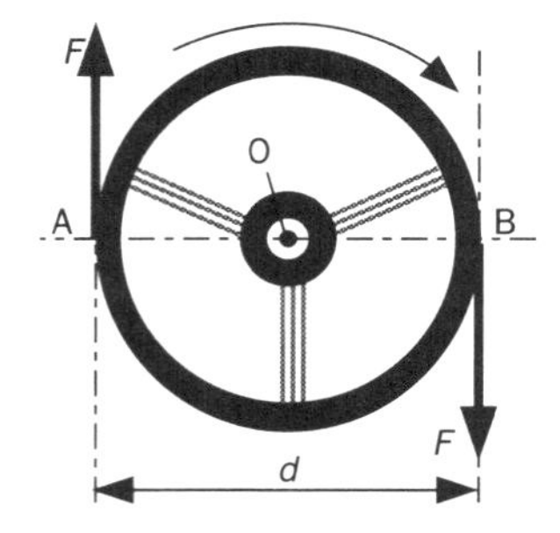

Figure 1.43 Torque due to a couple applied to a steering wheel

If two equal, opposite, parallel, coplanar forces of magnitude F act as shown, the clockwise torque (T) due to the couple is given by:

$T = (F \times \text{OA}) + (F \times \text{OB})$

$T = F \times (\text{OA} + \text{OB})$

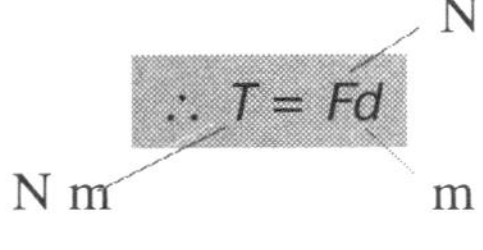

$$\therefore T = Fd$$

Equation 3

The **torque** (T) due to a couple is equal to the magnitude of **one** of the forces (F) multiplied by the **perpendicular distance** between their lines of action (d).

Centre of mass (C_m) and centre of gravity (C_g)

The centre of mass (C_m) is the point at which the total mass of a body acts, or appears to act. If the body is symmetrical and of uniform construction, C_m will be at the geometric centre of the body.

The centre of gravity (C_g) of a body is the point at which at which the weight of the body acts or appears to act.

In a uniform gravitational field, C_g and C_m will be at the same point.

At this level, we are only concerned with two-dimensional (flat) bodies or laminas, although the principle applies to all bodies. The positions of C_m for various laminar bodies are shown in Figure 1.44.

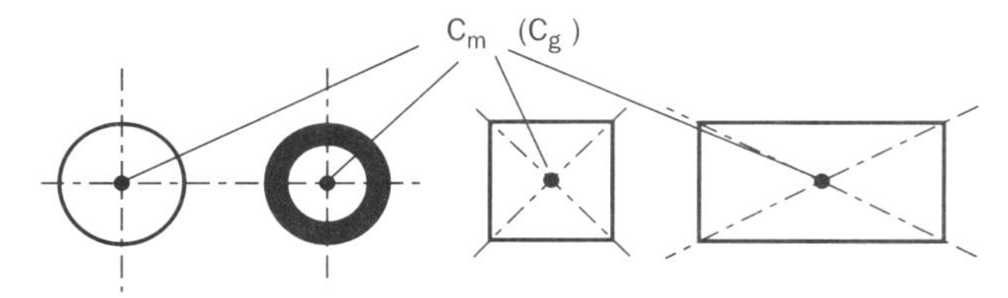

Figure 1.44 Centre of mass (and centre of gravity) for various laminar bodies

Significance of the centre of mass

When forces act on a body, they may tend to produce both linear and angular accelerations (i.e. the body may move in a straight line, it may rotate, or do both). If the line of action of the resultant of all the forces on the body acts through C_m, then angular acceleration is impossible. The body may acquire linear acceleration, but will not rotate (or change its rate of rotation).

If the line of action of the resultant does not pass through C_m then there will be a moment about C_m which may cause angular acceleration.

This is why a car may spin if one of its wheels

encounters a slippery surface when braking, as shown in Figure 1.45:

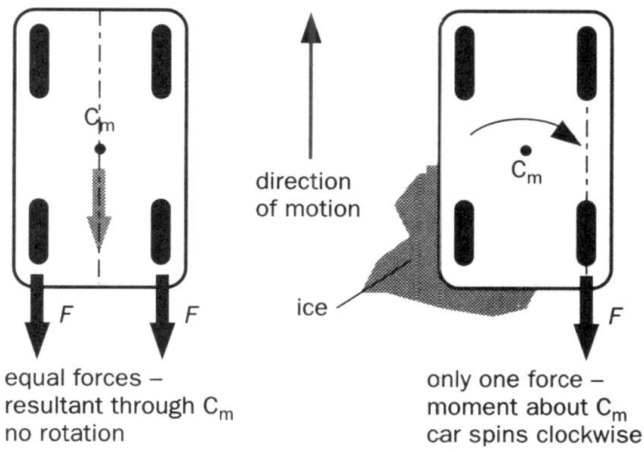

Figure 1.45 Forces and torques on a braking car

Finding the position of C_m (and C_g) for a plane lamina

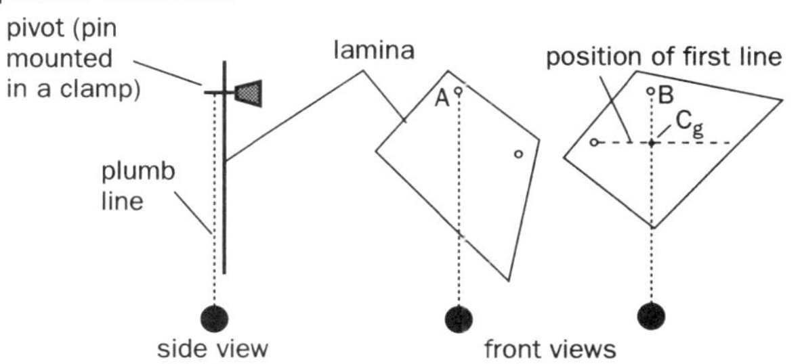

Figure 1.46 Finding the centre of gravity of a plane lamina

This method assumes that C_m and C_g are in the same position.

The lamina is suspended from a pin through a hole (A) drilled near one edge, as shown in Figure 1.46. A plumb line is hung from the pin, and its position marked on the lamina. Since the plumb line and the lamina must hang with their centres of gravity vertically below the point of suspension, the position of the centre of gravity of the lamina (C_g) must lie along the vertical line which has been marked by the plumb line. The lamina is then hung from another hole (B), and the position of the plumb line marked again. Where the two lines intersect is the position of the centre of gravity, C_g.

Guided examples (1)

1.

hinge

wind

1.5 m

15°

sign

The diagram shows an end view of a 1.5 m long uniform rectangular pub sign of mass 8.0 kg which is hinged along its upper edge. At a given instant, the sign is held in equilibrium at an angle of 15° to the vertical by the wind. Calculate the moment of the force of the wind on the sign, if a torque of 3.5 N m is needed to overcome friction at the hinge.

Guidelines

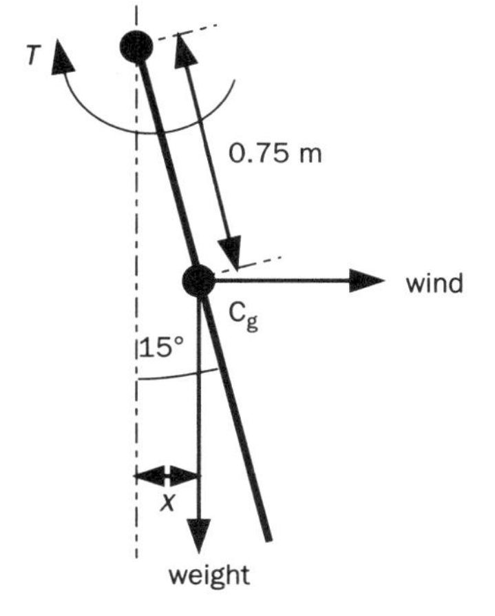

The weight of the sign acts vertically downwards through the centre of gravity (C_g), and produces a **clockwise** moment about the

Guided example continued on next page

Guided examples (1) *continued*

hinge. The friction at the hinge also produces a **clockwise** moment (T). The wind is the only force producing an **anticlockwise** moment.

Since the sign is in equilibrium, the sum of the moments must be zero.

2.

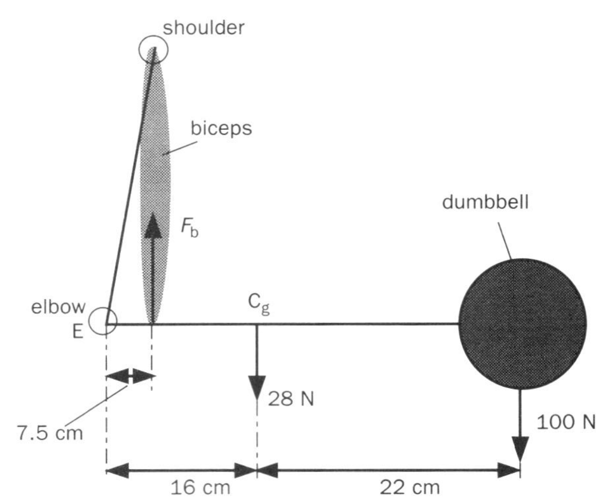

As part of an exercise routine, a body-builder holds a 10 kg dumbbell in his hand, and keeps his forearm horizontal as shown in the diagram. His forearm and hand can be assumed to weigh 28 N, and the tension force (F_b) in his biceps muscle can be assumed to act vertically. Calculate F_b.

Guidelines

Take moments about the elbow, E. Apply the principle of moments.

A young weight-trainer tries the bicep curl

3. The diagram shows a uniform solar panel 2.8 m long, weighing 1200 N. Its angle is adjusted continuously so that its face is perpendicular to the Sun's rays by a cable attached to the foot of the panel. The cable is wound over the drum of a winch, so that the foot of the panel can be moved towards or away from the vertical wall.

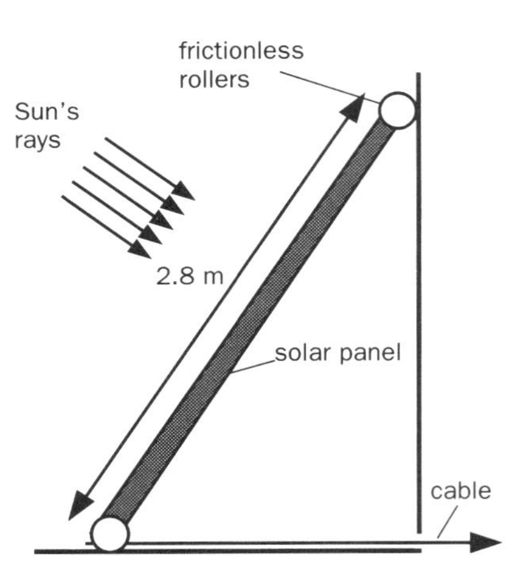

The ends of the panel are fitted with rollers so that they can slide against the wall or floor with negligible friction.

At a particular time the Sun is 40° above the horizontal. Calculate:

(a) the tension in the cable

(b) the magnitudes of the reactions at the upper and lower ends of the panel.

The magnitude and direction of the resultant force at the foot of the panel.

Guidelines

Draw a **free body diagram** showing all the forces acting on the panel. There is no friction between the ends of the panel and the wall and floor, so the reactions at the ends are perpendicular to the wall and floor.

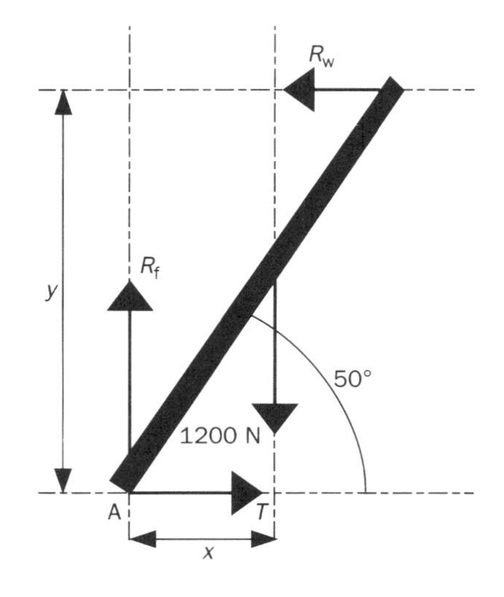

Guided examples (1) *continued*

Calculate the values of y and x from the geometry of the situation. The panel is in equilibrium so write equations for the horizontal forces, the vertical forces, and moments about point A. Solve for the unknown forces.

Self-assessment

SECTION A
Qualitative assessment

1. Distinguish between **vector** and **scalar** quantities. State as many examples of each as possible.
2. State the **parallelogram law** for the resultant of two **non-parallel** vectors acting at a point.
3.

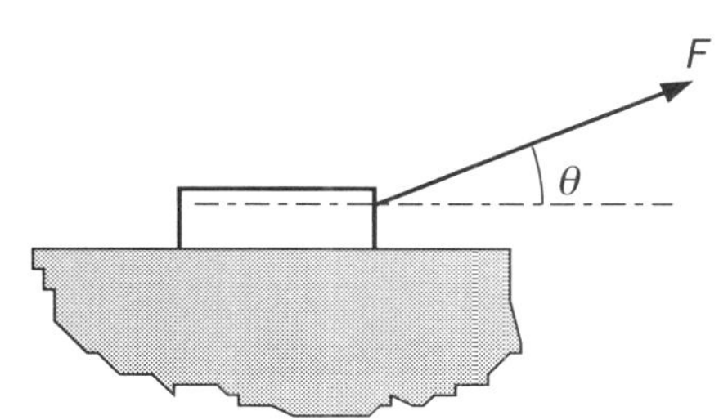

The diagram shows a force (F) acting at an angle (θ) above the horizontal. State the horizontal and vertical **components** of the force.

4. Four **coplanar** forces act at a point as shown in the diagram.
   (a) Explain what is meant by the term **coplanar.**
   (b) Explain how to determine the **resultant** of the four forces (**magnitude** and **direction**).

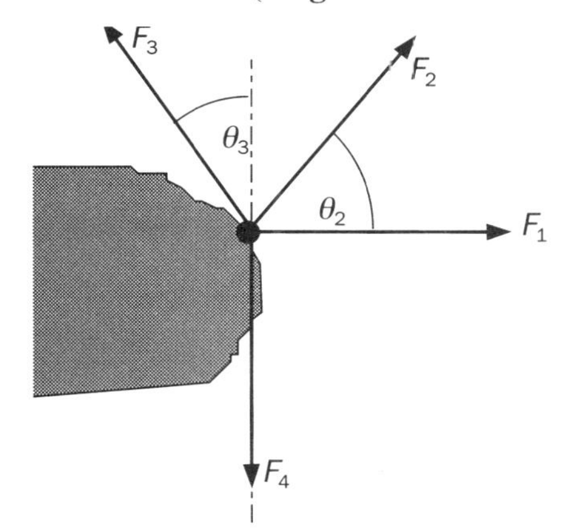

5. Explain what is meant by the **moment** or **torque** of a force. State the SI unit, and explain why this is not the same as the unit of **energy** or **work**.
6. A metal rod OA of length (s) is hinged at O, and a force (F) is applied successively at positions 1, 2, 3 and 4 as shown. State the magnitude of the **torque** F about O.

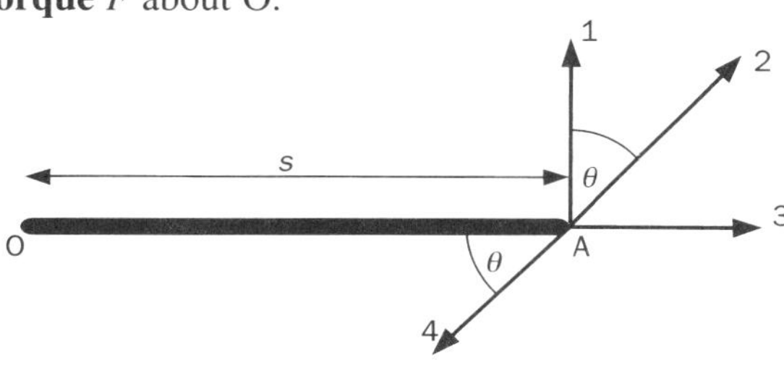

7. Define the terms **centre of mass** and **centre of gravity** of a body. Describe a simple experimental procedure for locating the position of the centre of gravity of a plane laminar body, using a plumb line.
8. What is a **couple**?
9. State the two conditions which must be satisfied for a body to remain in equilibrium under the action of coplanar forces.
10. State the **principle of moments**.
11. State the **triangle of forces** rule. If a body is in equilibrium under the action of three coplanar forces, what can be said about the lines of action of the forces?
12. Explain how the cantilever principle is used in building construction so that a balcony can project from a building without the need for support from beneath.

SECTION B
Quantitative assessment

(***Answers:*** 0.22; 0.50; 10; 29; 35; 62; 70; 320; 470; 510; 1.0×10^3; 1.0×10^3; 1.1×10^3; 1.4×10^3; 2.5×10^3; 2.5×10^3; 4.0×10^3; 5.0×10^3; 5.1×10^3; 1.7×10^4; 3.4×10^4; 4.0×10^4.)

1. A space craft is moving in a straight line as it travels away from the solar system. The rocket motor has a total thrust of 15 kN, Calculate the resultant force (magnitude and direction) on the craft while a side thruster which exerts a force of

Self-assessment *continued*

8300 N in a direction at 90° to the main engine thrust is being fired.

2. In the diagram, B represents a fallen climber, of mass 80 kg, who has pulled his partner, A, of mass 75 kg, off his ledge. Their combined weight is supported by the two belays, 1 and 2. Calculate the tensions (T_1 and T_2) in the two belay ropes.

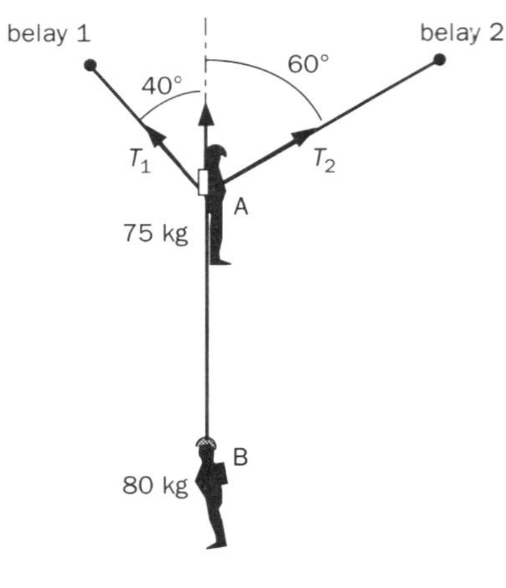

3. A hot air balloon can just raise a load of 400 kg. If this load is released, the balloon ascends until it reaches the end of its mooring wire, of length 50 m. A horizontal wind acts on the balloon, and it remains in equilibrium at a vertical height of 40 m.
   (a) Calculate the tension in the mooring wire.
   (b) If the mooring wire will break if the tension reaches 7000 N, and the wind strength increases, calculate the maximum angle between the mooring wire and the ground when the wire breaks.

4. 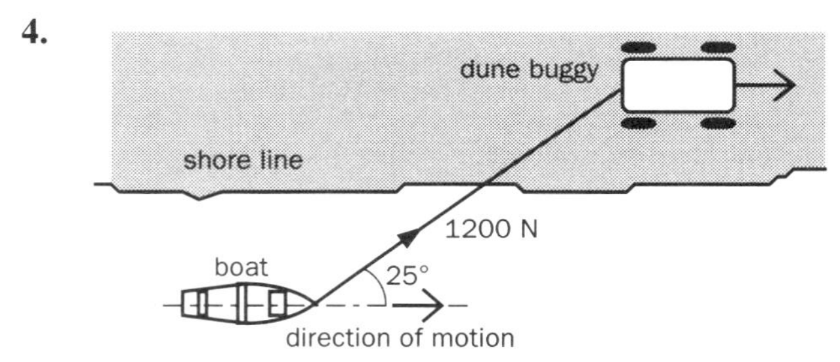

A dune buggy is used to tow a small boat at a steady speed of 2.0 m s^{-1} parallel to the shore as shown. If the tension in the rope is 1200 N, calculate:
   (a) the resistive 'drag' force of the water on the boat;
   (b) the sideways force of the water on the boat (necessary to prevent the boat being pulled toward the shore).

5. A car of mass 1200 kg is carrying a total load (passengers and luggage) of 350 kg. If it stalls on a hill where the ground is at an angle of 15° to the horizontal, calculate the minimum friction force between the wheels and the ground needed to stop the car slipping downhill.

6. 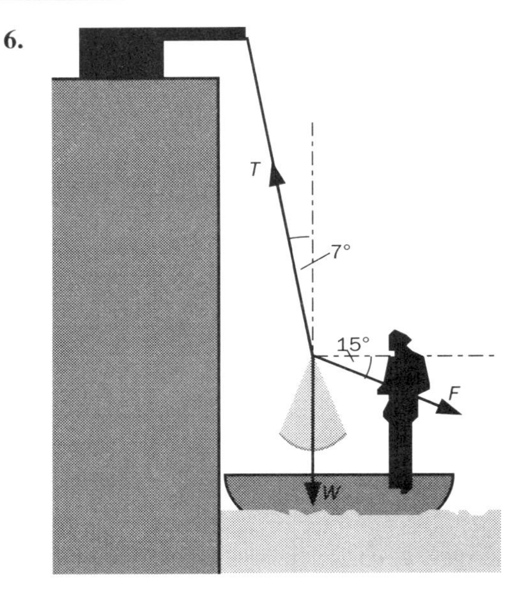

A loaded cargo net, having a total mass of 240 kg, is being lifted by a crane from a barge tied up to the harbour as shown. At the instant the net leaves the deck of the barge, the man exerts a sideways force (F) in a direction of 15° below the horizontal to stop the net swinging into the side of the harbour wall. If the net is in equilibrium at this instant, calculate
   (a) the tension (T) in the crane cable
   (b) the force (F) exerted by the man.

7. A rectangular tombstone has dimensions 1500 mm × 750 mm × 100 mm and it is made from a type of stone which has a density of 4500 kg m^{-3}. Calculate:
   (a) the weight of the tombstone
   (b) the minimum force needed to lift one end of the tombstone if it is lying flat on its largest face.

8. 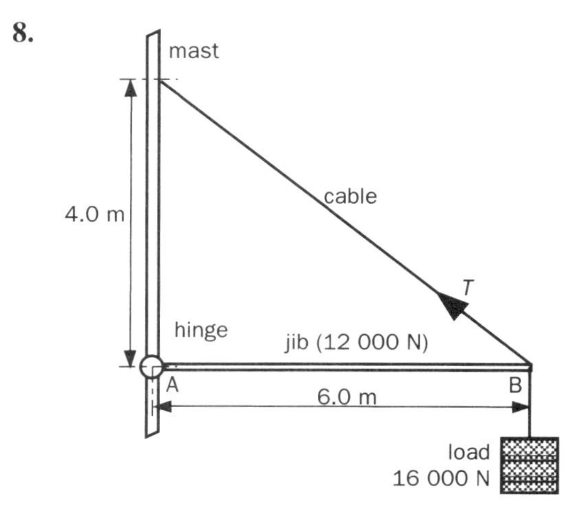

The uniform jib AB on a ship is hinged at A and a supporting cable is attached to end B from a point

Self-assessment *continued*

on the mast 4.0 m above the hinge. The jib is 6.0 m long and weighs 12000 N. If a load of 16000 N is suspended from end B, calculate:

(a) the tension T in the supporting cable

(b) the reaction force (magnitude and direction) at the hinge.

9. A uniform ladder AB of length 5.0 m and mass 20 kg is propped against a smooth, vertical wall with end B uppermost, and end A on rough, horizontal ground. End B is 4.0 m above the ground and end A is 3.0 m out from the foot of the wall. If a man of mass 70 kg stands on the ladder at a point 3 m above the ground, calculate:

(a) the reaction force at the wall

(b) the magnitude of the reaction force at the ground and the angle between this reaction force and the horizontal.

10. A uniform glass rod of length 18 cm and mass 60 g is placed inside a 12 cm deep glass tumbler as shown:

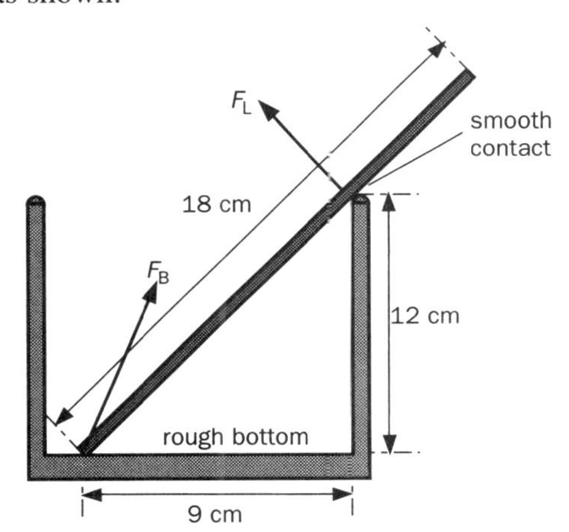

The interior lip of the tumbler is smooth, so the reaction force (F_L) at the lip acts at 90° to the surface of the rod. The bottom of the glass has been decoratively ground, and may be considered to be rough. Calculate:

(a) the magnitude of the reaction force (F_L) at the lip

(b) the magnitude of the reaction force (F_B) at the bottom of the glass, and the angle between this force and the glass bottom.

1.5 Dynamics

Dynamics is the study of bodies in motion, the forces which give rise to that motion and the associated energy changes. The fundamental principles of dynamics have evolved after years of patient observation and experimentation. Today, the sophisticated technology of particle accelerators gives ever deeper insights into the nature of matter, and sensitive probes gather data on the far reaches of space, but the basic concepts such as conservation of energy and momentum still hold true over the vastness of the universe.

We start our study of moving bodies by examining the quantities used when measuring movement and the relationships between these quantities.

Motion-related quantities

Displacement

Common symbol s
SI unit metre (m)

The displacement of a body is the **distance** moved by the body in **a specified direction**. It is a vector quantity, so its value may be either **positive** or **negative** depending on the direction from the start point. It can also be stated as the **straight line distance** between the body and some given reference point.

For example, if a runner completes one lap of a standard running track, his total **distance** travelled will be 400 m, but his total displacement will be zero. If the start line is taken as the reference point, then his **displacement** may be positive while he is in front of the line, and negative while he is behind it.

Speed

Common symbol v
SI unit metre per second (m s^{-1})

Speed is defined as the **rate of change of distance**. It is a **scalar** quantity, so it is always positive in value.

The Ford Escort Cosworth has a top speed of nearly 140 mph and can accelerate from 0 to 60 mph in just over 6 seconds

$$average\ speed = \frac{distance\ travelled}{time\ taken}$$

instantaneous speed is the speed at a particular **instant**.

During a road journey of 20 km, a car's **instantaneous speed** will change as road conditions dictate. This variation (between 0 and, say, 100 km h^{-1}) is indicated continuously by the speedometer. However, if the whole trip took 30 minutes, then the **average speed** is given by:

$$average\ speed = \frac{20}{0.5} = 40\ \text{km h}^{-1}$$

Velocity

Common symbols	u, v (it is common to use u for the **initial** value of velocity and v for the **final** value when considering **changes** of velocity)
SI unit	metre per second (m s^{-1})

The velocity of a body is defined as its **rate of change** of **displacement**. It is a vector quantity, and may be positive or negative depending on the **direction** of motion.

A body is moving with **constant** or **uniform** velocity if it suffers equal changes of **displacement** in equal time intervals. A body for which this is not true is said to be **accelerating**.

Maths window

The mathematical symbol for 'overall change of' is Δ (the upper case Greek letter 'delta'). Using the symbol (s) for displacement, change of displacement is written (Δs). Similarly the time taken is written (Δt). So the equation for average velocity can be written:

$$\text{average velocity} = \frac{\Delta s}{\Delta t}$$

The symbol $\frac{\Delta}{\Delta t}$ means 'average rate of change of'.

The measurement of **instantaneous velocity** needs some careful thought. Ideally, the displacement in an infinitely short period of time should be measured, but practically this is absurd, since no displacement can take place in zero time. In the laboratory, this means reducing the time interval for the measurement to the smallest practicable value.

Maths window *continued*

Mathematically, this is expressed using the symbol (d): instantaneous change of displacement becomes (ds), and instantaneous time interval becomes (dt), so the defining equation for instantaneous velocity becomes:

$$\text{instantaneous velocity} = \frac{ds}{dt}$$

Acceleration

Common symbol	a
SI unit	metre per second per second or metre per second^2 (m s^{-2})

The **acceleration** of a body is defined as the rate of change of **velocity** of the body. It is a **vector** quantity, and may be positive or negative depending on whether the **velocity** is increasing or decreasing, and on the **direction** of motion (positive or negative).

When deciding whether acceleration is positive or negative, you must consider both the change in magnitude of the velocity **and** the direction of motion (this will be explored in more detail later). Since displacement is a vector quantity, a change in **direction** is also regarded as acceleration. For example, a body following a circular path at constant speed is continually changing its direction of motion, and is therefore accelerating (this will also be examined in more detail in the section on **circular motion**).

If the velocity of a body changes from u to v in a time interval Δt, then the average acceleration of the body is given by:

$$average\ acceleration = \frac{velocity\ change}{time\ taken} = \frac{\Delta v}{\Delta t} = \frac{v - u}{\Delta t}$$

The **instantaneous** acceleration (a) of the body at some instant during the motion can be written:

$$a = \frac{dv}{dt}$$

A body moves with **constant** or **uniform** acceleration if its velocity changes by equal amounts in equal time intervals.

Equations of motion for uniform acceleration (kinematics equations)

Kinematics is the study of the motion of bodies without considering the **masses** of the bodies or the **forces** which produce changes in their motion. Consider a body moving with an initial velocity (u) which accelerates with a constant acceleration (a) to a final velocity (v) in time (t). During this time interval, the body's change of displacement is (s) (see Figure 1.47):

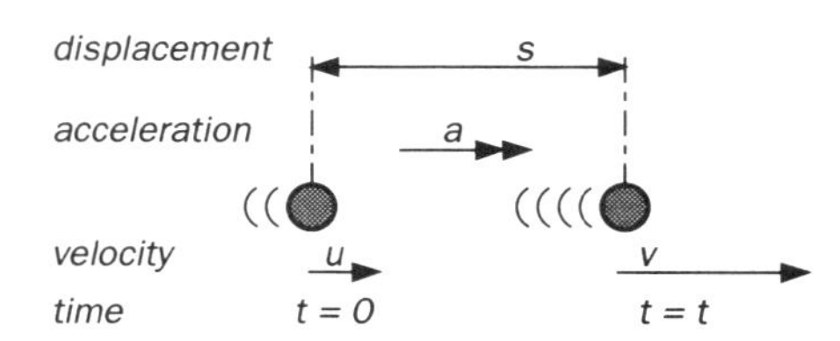

Figure 1.47 Uniformly accelerated motion

From the definition of acceleration:

$$a = \frac{(v - u)}{t}$$

$$\therefore at = v - u$$

$$\therefore \quad v = u + at$$

(units: v – m s^{-1}, u – m s^{-1}, a – m s^{-2}, t – s)

Equation 1

Since the velocity increases **uniformly**, the average velocity (V) is simply the average of the initial and final velocities, and is given by:

$$V = \frac{(u + v)}{2}$$

and since by definition *total displacement = average velocity × time*,

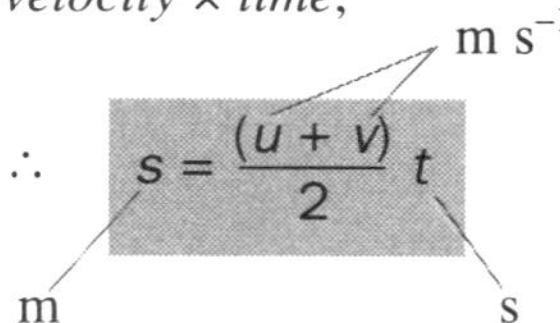

Equation 2

Since from Equation 1, $v = u + at$

then substituting for v in Equation 2 gives:

$$s = \frac{(u + u + at)}{2} t = \frac{(2ut + at^2)}{2}$$

$$\therefore \quad s = + ut\frac{at^2}{2} \text{ or } s = ut + \frac{1}{2}at^2$$

(units: s – m, u – m s^{-1}, a – m s^{-2}, t – s)

Equation 3

From the definition of acceleration,

$$t = \frac{(v - u)}{a} \text{ and } s = \frac{(v + u)}{2} t$$

$$\therefore s = \frac{(v + u)}{2} \times \frac{(v - u)}{a} = \frac{(v^2 - u^2)}{2a}$$

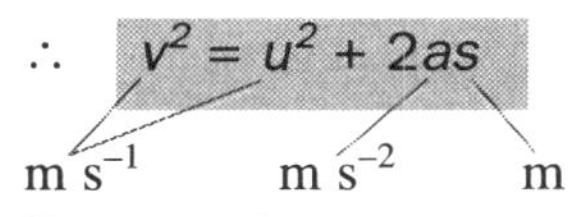

Equation 4

Summary

u = initial velocity (when time $t = 0$)
v = final velocity (when time $t = t$)
a = **constant** acceleration
s = change of displacement during the time interval
t = time interval

1. $v = u + at$

2. $s = \frac{(u + v)}{2} t$

3. $s = ut + \frac{at^2}{2}$ or $s = ut + \frac{1}{2}at^2$

4. $v^2 = u^2 + 2as$

If any **three** of the quantities u, v, a, s or t are known, the other **two** can be calculated using one or more of the equations.

To solve numerical problems:

- identify and **write down** what is known;
- identify what is missing (check for **'hidden'** values – e.g. 'at rest' means **zero** velocity);
- use unique symbols like u_1 and u_2 to separate values for **different** parts of the problem;
- finally, use the **simplest** equation to find the missing values.
- Remember that u, v, a and s are **vectors**. Choose a direction to be **positive**, and keep to this convention throughout the problem.

Guided examples (1)

1. An aircraft has a landing velocity of 50 m s^{-1} and decelerates uniformly at 10 m s^{-2} until its velocity is reduced to 10 m s^{-1}. Calculate:
   (a) the time taken to slow down to this velocity
   (b) the distance covered during the deceleration.

   Guidelines

   Write down the known values *u*, *v* and *a* (note that the acceleration is **negative**).
   (a) Use Equation 1.
   (b) Use Equation 2.

2. A car accelerates from rest at a steady rate of 1 m s^{-2}. Calculate:
   (a) the time taken to reach 15 m s^{-1}
   (b) the distance travelled during this time
   (c) the velocity of the car when it was 100 m from the start point.

   Guidelines

   Write down the known values *u*, *v* and *a*.
   (a) Use Equation 1.
   (b) Use Equation 2.
   (c) Use Equation 3.

3. A car is speeding along a straight country road of width 4 m at a speed of 25 m s^{-1} when the driver sees a farm vehicle of length 3.5 m just starting to cross the road at a point 40 m ahead. The driver's reaction time (i.e. the time interval between seeing the obstacle and actually applying the brakes) is 0.80 s, and the maximum deceleration with the brakes fully applied is 5.5 m s^{-2}. If the car can be assumed to decelerate uniformly at this rate without swerving, calculate
   (a) the distance travelled by the car during the driver's reaction time
   (b) the car's velocity when it reaches the position of the farm vehicle
   (c) the total time which has elapsed from first sighting until the car reaches the farm vehicle
   (d) the minimum constant velocity of the farm vehicle so that the car does not collide with it.

   Guidelines

   There are two stages (reaction time and braking time) and two vehicles involved. Be careful to identify these clearly when assigning symbols for use in equations.
   (a) During the reaction time, the car's velocity is constant (i.e. acceleration = 0).
   (b) Use Equation 4 to calculate the distance travelled during deceleration (**negative** acceleration).
   (c) Calculate the deceleration time using Equation 2, then add this to the reaction time.
   (d) To avoid a collision, the whole length of the farm vehicle must have completely crossed the road in the time obtained in (c).

Motion under gravity

On earth, the gravitational field between the planet and all bodies produces an attractive force on all matter. The effect of this attractive force is to impose a uniform acceleration on all bodies free to move in a vertical direction (i.e. towards the centre of the Earth). The average value of this acceleration is found by experiment to be 9.81 m s^{-2} (rounded off to 10 m s^{-2} for most work) if air resistance effects are ignored.

There are slight variations, caused by non-uniformity of the shape and density of the Earth, which are useful to geologists attempting to survey deep rock structures, and the effects of the Earth's rotation, which reduces the acceleration due to gravity near the equator.

All the kinematics equations apply, but it may be useful to use the symbols '*g*' for acceleration, and '*h*' for vertical displacement as reminders that we are dealing with acceleration due to gravity and vertical height. Remember that these are **not new equations** – the new symbols are only used as **memory aids**. With vertical motion, it is particularly important to choose a **sign convention** (i.e. which direction is regarded as **positive**) for a problem, and to **stick rigidly to it**.

The suggestion below may be found useful:

	Displacement	**Velocity**	**Acceleration**
Bodies **initially** falling	down positive	down positive	$g = +10$ m s^{-2}
Bodies projected upwards	up positive	up positive	$g = -10$ m s^{-2}

Guided examples (2)

1. Find the time taken for an object dropped from a height of 60 m to reach the ground.

   Guidelines

   The body is initially falling, so g is positive (+10 m s^{-2}). It is 'dropped' which implies that its initial velocity is zero.

2. A stone is thrown vertically upwards from the ground with an initial velocity of 30 m s^{-1}. Calculate
   (a) the maximum height reached
   (b) the time taken to return to the ground.

   Guidelines

   The stone is projected upwards, so if you adopt the **upwards positive** convention, g will be **negative**.
   (a) At the maximum height, the velocity is momentarily zero.
   (b) When the stone returns to the ground, its displacement is **zero**. Throughout the motion, the acceleration is constant. Use these values in the kinematics equations.

Graphical representation of motion

The displacement/time (S/T) graph

This is a useful way of presenting information about how the position of a body changes with time. In addition, the velocity of the body can be calculated at any instant.

In this graph, **displacement** is plotted vertically (y-axis or ordinate) against **time** horizontally (x-axis or abscissa).

The **instantaneous velocity** of the body at any instant = the **gradient** of the S/T graph.

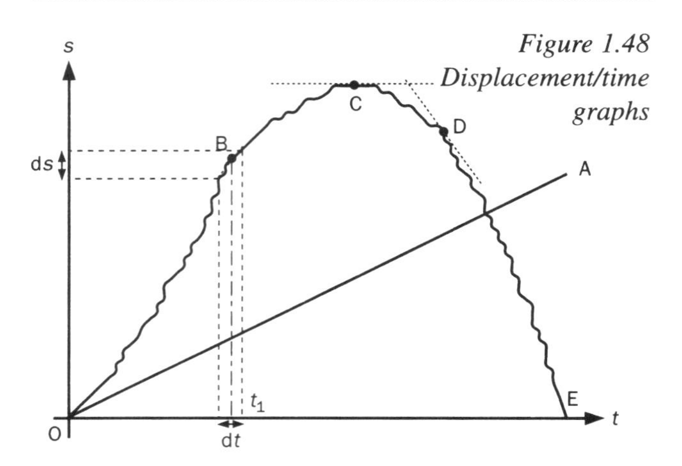

Figure 1.48 Displacement/time graphs

Two separate motions are shown in Figure 1.48:

Motion OA The **gradient** of the graph is **constant**, indicating **uniform velocity**. Since the gradient is positive, this also suggests that the displacement is increasing, i.e. the body is moving away from the start point.

Motion OBCD The **gradient** is **constantly changing**, indicating non-uniform velocity.

At time t_1 (point B on the graph), the instantaneous velocity v is given by the **gradient** of the graph at that instant;

$$\text{i.e. } v = \frac{ds}{dt} = \frac{\text{tiny change of displacement}}{\text{brief time interval}}$$

At point C, the gradient of the graph is zero (the line is parallel to the time axis), therefore the velocity of the body is zero (i.e. it has stopped).

Up to point C, the trend of the line shows that the displacement is increasing, but after point C, it is decreasing. At point D, for example, the gradient is negative, indicating that the velocity is in the opposite direction to that at B. At point E, the displacement is zero. You should note that at this point, the body has returned to its starting position, and that the total displacement for the motion illustrated by the graph is zero.

Guided example (3)

For the *s/t* graph in Figure 1.49, analyse the motion of the body by describing its behaviour during each of the sections A–B, B–C etc.

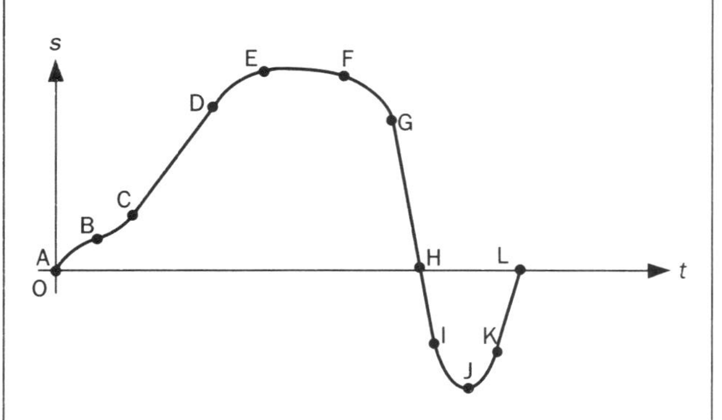

Figure 1.49 s/t graph to be analysed

Guidelines

A methodical way to approach this analysis is to draw a table of the type shown in Table 1.7, and complete the details for each part of the graph. This has been done for sections A–B and B–C.

Guided example (3) continued

Table 1.7 Analysis of s/t graph

Section	Gradient	Velocity	Direction of motion	Acceleration
A–B	positive – decreasing	positive – decreasing	away from start point	negative
B–C	positive – increasing	positive – increasing	away from start point	positive
C–D				

The velocity/time (V/T) graph

This useful graph displays information about how the **velocity** of a body is changing with time. The **acceleration** of the body at any instant, and its **displacement** can also be obtained from this graph. The body's **velocity** is plotted vertically (y - axis or ordinate) against **time** horizontally (x - axis or abscissa).

The **instantaneous acceleration** at any instant = the **gradient** of the v/t graph

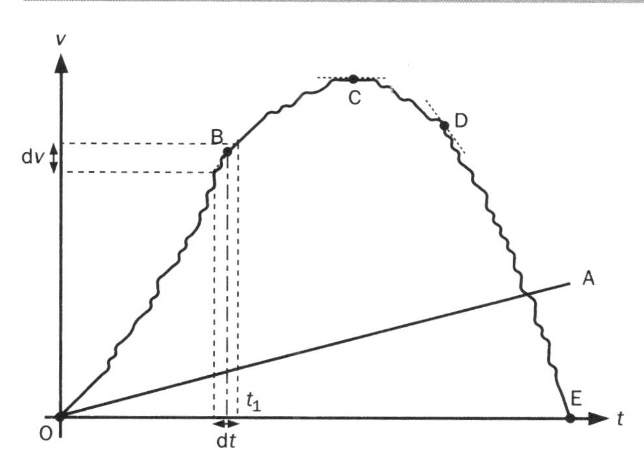

Figure 1.50 Velocity/time graphs

Figure 1.50 shows the motions of two separate bodies plotted on the same axes:

Motion OA — The **gradient is constant**, indicating uniform change of velocity with time, i.e. **uniform acceleration**.

Motion OBCDE — The **gradient** is **constantly changing**, indicating **non-uniform acceleration**. The magnitude and direction of the gradient of the v/t graph at any point gives the instantaneous value of the acceleration at that instant.

At point B on the graph, the gradient is **positive**, so the acceleration is **positive**. The magnitude of the acceleration is given by:

$$a = \frac{dv}{dt} = \frac{\textbf{tiny change} \text{ in velocity}}{\textbf{brief} \text{ time interval}}$$

At point C, the gradient is **zero**, so there is **zero** change of velocity, hence **zero acceleration**. In this case, the velocity has reached its **maximum** value.

At point D on the graph, the gradient is **negative** (since the velocity **decreases** as time **increases**), so the acceleration is **negative**.

At point E, the velocity is zero, i.e. the body has **come to rest**. Note that the position of the body has changed, there has been some **displacement**.

Area enclosed by the velocity/time graph

The shaded area between the line OBCDE and the time axis is equivalent to **average velocity** × **time**,

i.e. the area bounded by the graph and the time axis is equal to the **total displacement** of the body.

Consider a body having an initial velocity (u) which undergoes a constant acceleration (a) and achieves a final velocity (v) after a time (t). This motion is illustrated in Figure 1.51, together with the associated velocity/time graph.

displacement s
acceleration a
velocity u v
time $t = 0$ $t = t$

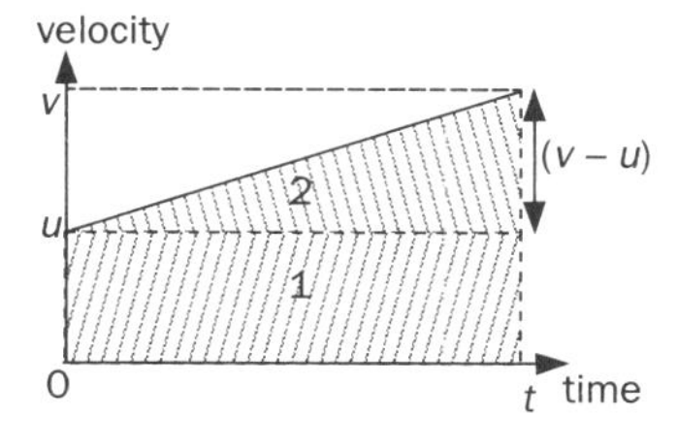

Velocity/time graph

Figure 1.51 Displacement = area enclosed by v/t graph

The equation of motion for the displacement is:

$$s = ut + \frac{1}{2}at^2$$

From the velocity/time graph:

s = area 1 + area 2

$$= ut + \frac{t(v-u)}{2}$$

$$= ut - \frac{ut}{2} + \frac{vt}{2}$$

$$s = \frac{ut}{2} + \frac{vt}{2}$$

but $v = u + at$

$$\therefore \quad s = \frac{ut}{2} + \frac{(u+at)t}{2}$$

$$\therefore \quad s = ut + \frac{t(at)}{2} = ut + \frac{1}{2}at^2$$

Guided examples (4)

1. The velocity/time graph of the motion of a body is shown in Figure 1.52. Analyse the motion by describing the behaviour of the body during each section A–B, B–C etc.

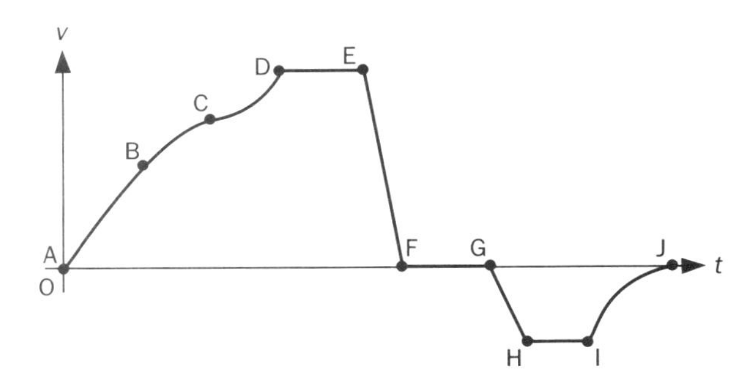

Figure 1.52 v/t graph to be analysed

Guidelines

As in the distance/time graph example, draw a table as shown in Table 1.8, and complete the information for each part of the motion. The first two rows have been completed as examples.

Table 1.8 Analysis of v/t graph

Section	Gradient	Velocity change	Direction of motion	Acceleration
A–B	constant – positive	increasing at a uniform rate	away from start	constant – positive
B–C	decreasing – positive	increasing at a non-uniform rate	away from start	decreasing – positive

2. A diesel locomotive starts from rest and accelerates for 1 minute with a uniform acceleration of 1 m s^{-2} after which the velocity is kept constant for a time, until the locomotive is finally brought to rest with a retardation of 2 m s^{-2}. If the total distance travelled is 4500 m, calculate the time taken for the whole journey.

Guidelines

Sketch a velocity/time graph for the whole journey, inserting any known numerical values.

Calculate the **velocity** reached during the acceleration period, and the **time taken** for deceleration.

Then use the fact that **displacement = area enclosed by *v/t* graph** to find the time during which the velocity is constant. The total time is the sum of the individual times for each stage.

Motion in two dimensions

The kinematics equations and graphical analysis techniques developed so far have been specifically dedicated to dealing with **rectilinear** motion. This is fine when the motion is confined to rails and tracks, or when bodies are falling vertically under gravity, but some refinement is needed when considering the motion of bodies **projected** (or thrown) at some angle other than 90° to the horizontal. Such motion is called **projectile motion**, and includes the motion of footballs, javelins, bullets etc. All such **projectiles** have the following in common:

- They are all given some **initial velocity**. This is the velocity with which they start their flight, and is produced by throwing, kicking, explosives etc.
- Throughout their flight, the only force acting is their **weight** due to **gravity**, which exerts a constant force vertically downwards. There is no force which has a component in the horizontal direction (i.e. this analysis does not apply to rockets, which have a thrust due to the rocket engine).

The basic refinement is to treat such motion **vectorially**. The velocity and displacement are resolved into horizontal and vertical components, and the kinematics equations are applied separately. Figure 1.53 shows a dramatic illustration of such motion. An arrow is aimed and fired horizontally at a target some distance away. At the instant the arrow is fired, the target is released and allowed to fall under gravity.

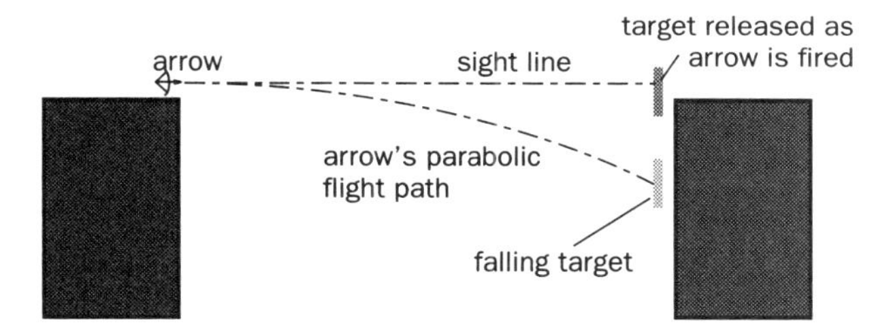

Figure 1.53 Impressive archery!

As long as air resistance is negligible, which is reasonable for a streamlined projectile such as an arrow, and the target is released at exactly the same instant as the arrow, the archer will always score a bullseye. The explanation for this remarkable performance is as follows:

- The arrow's horizontal velocity is constant, assuming no air resistance.
- Both the arrow and the target accelerate vertically at exactly the same rate. Since they both start to fall at the same instant, they will always be at the same level, falling towards the Earth with the same vertical velocity.
- The horizontal motion of the arrow has no influence on the vertical motion.

In fact, as long as the arrow is aimed accurately at the target, the initial direction can be at any angle above or below the horizontal. The arrow will still follow a parabolic path, and meet the vertically-falling target.

Denis Irwin makes use of projectile motion as he cracks a shot at goal

Summary for
Solving projectile problems

Resolve the initial velocity into horizontal and vertical components.

For the horizontal direction:

- the horizontal component of velocity can be assumed to remain constant for the whole flight
- horizontal acceleration = 0
- (final velocity) = (initial velocity)
- (horizontal displacement) = (horizontal component of velocity) × (time of flight).

For the vertical direction:

- the acceleration is constant, equal to 'g'
- kinematics equations apply
- at the highest point, the vertical velocity is momentarily zero
- when the projectile has returned to the level of the launch point, the vertical displacement is zero.

Guided example (5)

During a football match, a free kick is taken and the ball is projected with a velocity of 20 m s^{-1} at an angle of 35° to the horizontal. Assuming negligible air resistance, and ignoring any effects of spin on the ball, calculate:

(a) the initial vertical and horizontal components of velocity

(b) the time taken for the ball to reach its maximum height

(c) the maximum height reached by the ball

(d) the horizontal distance to the point where the ball hits the ground.

Guidelines

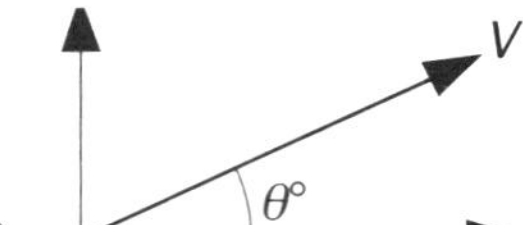

(a) horizontal component $= V \cos \theta$; vertical component $= V \sin \theta$.

(b) Consider only the vertical motion, and use the fact that the vertical velocity will be zero at the maximum height. Use $v = u + at$ to calculate t.

(c) Now you know the time taken to reach this height, calculate the height reached.

(d) The time taken to return to the ground from this height will be exactly the same as the time taken to reach it. Calculate the total time in the air.

Use *(horizontal displacement) = (horizontal velocity) × (time of flight)*

Self-assessment

SECTION A
Qualitative assessment

1. Explain the difference between
   (a) **distance moved** and **displacement**
   (b) **speed** and **velocity**
   (c) **instantaneous** velocity and **average** velocity
   (d) **uniform** and **non-uniform** acceleration

2. An aircraft which is flying level in a straight line with an initial velocity (u), moves with uniform acceleration (a) and attains a final velocity (v) in a time (t) after travelling a distance (s).
   (a) Starting with the definition of acceleration, show that $v = u + at$.
   (b) Show that $s = ut + \frac{1}{2}at^2$.
   (c) Combine the equations obtained in (a) and (b) to obtain $v^2 = u^2 + 2as$.

3. (a) Name the quantity obtained from the gradient of
   (i) a **distance**/time graph
   (ii) a **displacement**/time graph
   (iii) a **velocity**/time graph.
   (b) What is given by the **area** bounded by a **velocity**/time graph and the time axis between two given instants?

4.

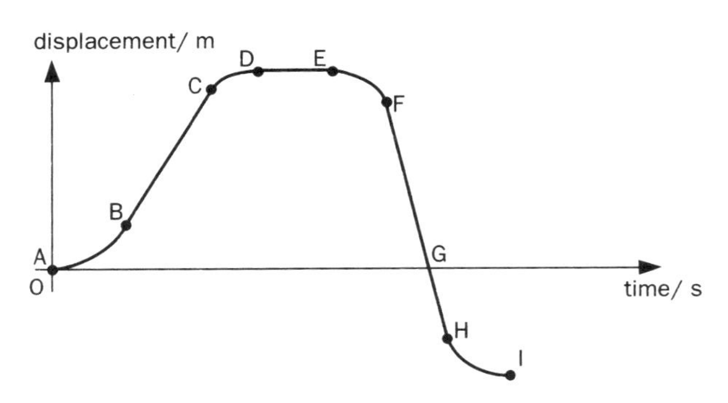

The diagram shows a **displacement/time** graph for a body. Describe the motion of the body for each of the sections AB, BC, CD etc.

5.

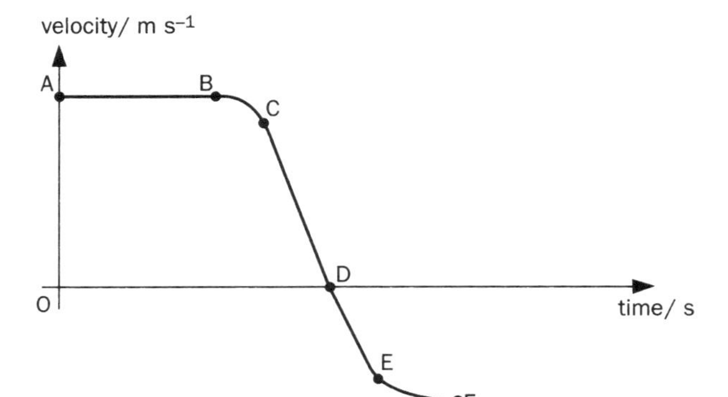

The diagram shows a **velocity/time** graph for a body. Describe the motion of the body for each of the sections AB, BC, CD etc.

Self-assessment *continued*

6. 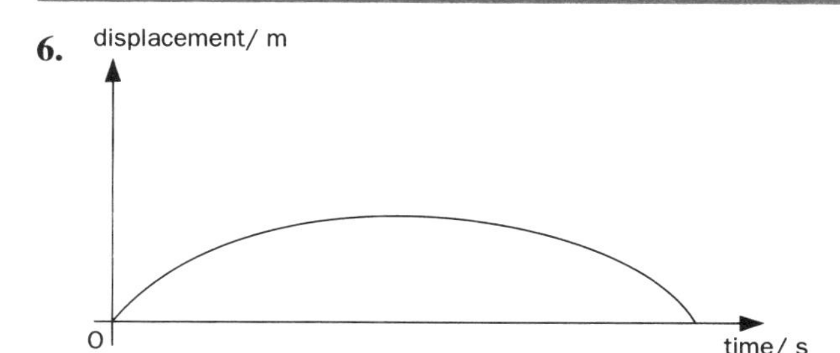

The diagram shows the displacement of a stone which is thrown vertically upwards from the ground.

(a) Explain how you would obtain a **velocity/time** graph from this graph.

(b) Copy the displacement/time graph and sketch the shape of the corresponding **velocity/time** graph beneath it.

7. A rugby ball is kicked on a level playing field and moves off with a velocity (u) at an angle (θ) above the horizontal. Neglecting air resistance:

(a) what are the **vertical** and **horizontal** components of the ball's velocity
 (i) at the moment it is kicked
 (ii) when the ball is at its greatest height
 (iii) when the ball just comes back to the ground.

(b) Derive a relationship for the time (t) taken for the ball to reach maximum height (H) in terms of (u), (θ), (H) and the acceleration due to gravity (g).

SECTION B
Quantitative assessment

(***Answers:*** 2.4×10^{-3}; −1.3; 1.4; 4.2; 4.4; 4.5; 5.8; 5.8; 6.0; 7.8; 9.7; 16; 22; 26; 30; 30; 44; 45; 50; 58; 72; 88; 90; 110; 130; 630; 1.3×10^3; 5.0×10^3; 1.4×10^5.)

(Use acceleration due to gravity = 10 m s^{-2} unless otherwise stated.)

1. Use the equations of motion to complete the following table.

u m s^{-1}	v m s^{-1}	a m s^{-2}	s m	t s
0		10		5
	300	9		30
5	120			20
56		2.1	500	
0	330		0.4	

2. An aeroplane accelerates uniformly from rest to reach its take-off speed of 126 km h^{-1} in 6 s. Sketch a velocity/time graph of the motion and calculate

(a) the acceleration

(b) the distance travelled along the runway.

3. When the brakes are applied in a car which is moving at 40 m s^{-1}, the speed is reduced to 25 m s^{-1} over a distance of 140 m. If the retardation remains constant, what further distance will the car travel before coming to rest? Sketch a velocity/time graph of the motion.

4. A sports car moves from rest with uniform acceleration to reach a velocity of 25 m s^{-1} in 6 s. It then maintains this velocity for a further 12 s after which it is uniformly retarded until it comes to rest 38 s after the start of the motion. Sketch a velocity/time graph for the whole journey and use it to calculate:

(a) the initial acceleration of the car
(b) the final retardation
(c) the total distance travelled
(d) the average velocity.

5. An arrow is fired vertically upwards, reaches a height of 100 m, then falls back to earth. Sketch a velocity/time graph of this motion and calculate:

(a) the initial velocity of the arrow
(b) the time taken to reach maximum height
(c) the velocity of the arrow when it is 55 m above the ground, moving upwards
(d) the height of the arrow 6.0 s after firing.

6. A rocket rises vertically from earth with a uniform acceleration of 30 m s^{-2}. If, after 9.0 s, the burnt-out first stage separates from the rocket, how long does it take the first stage to return to the ground?

7. A ball is kicked horizontally with a velocity of 6.0 m s^{-1} from the top of a tall building which is 96 m high. Neglecting air resistance, calculate:

(a) the time taken to reach the ground
(b) the horizontal distance travelled by the ball
(c) the **horizontal** and **vertical** components of the ball's velocity when it hits the ground.

8. A golfer makes an approach shot which travels a horizontal distance of 46 m before landing on the green, at the same level as the point from which the shot was played. If the golf ball is projected at an angle of 40° above the horizontal, calculate:

(a) the velocity of projection of the ball
(b) the time taken for the ball to reach its maximum height
(c) the maximum height reached.

(Ignore aerodynamic effects.)

1.6 Newton's laws of motion

These are the three laws published by Sir Isaac Newton in 1687, developed from careful **observation** of the real world, which describe the effects of forces on the motion of bodies. As such they form the basis of the science of **kinetics**, and are responsible for a great deal of our understanding of the way in which all the bodies of the universe interact. They are believed to apply everywhere in the universe, although as we shall see later they have limitations: e.g. they need modification to explain what happens at very high velocities (approaching the speed of light), but are quite capable of solving all the calculations needed to put men on the moon.

Newton's first law (Newton I)

All bodies will continue to be **stationary** or to move with **uniform** velocity unless they are acted on by a **resultant force**.

You should note that:

- **Stationary** means no motion relative to a stationary observer or fixed frame of reference. Obviously, a parked car appears stationary to us on Earth, but it could be booked for speeding by an intergalactic policeman who clocked it at over 18 000 mph on the 'Solar System inner ring road'.
- **Uniform** velocity implies no change of speed or direction.
- Zero **resultant** force means that either no forces at all act on the body (virtually impossible) or that the **combined effect** of all the forces acting is zero.

From Newton I:

- We define force as something which changes (or tends to change) the state of rest or uniform motion of a body, either through contact, or through 'field' action (gravity, electric and magnetic fields).
- We express the concept of inertia: all bodies possess an in-built property which resists changes in speed and/or direction. The mass of a body is a measure of its **inertia**: the larger the mass, the greater the inertia and *vice versa*.

Newton's second law (Newton II)

The rate of change of **momentum** of a body is directly proportional to the applied **resultant force**, and occurs in the **direction of the resultant force**.

Momentum

Common symbol p
SI unit kilogram metre per second (kg m s^{-1})
newton second (N s)

The **momentum** (p) of a body is defined as the product of the bodys mass (m) and its velocity (v).

Defining equation:

$$p = mv$$

(kg m s^{-1} = kg × m s^{-1})

Equation 1

Figure 1.54 shows a body of mass m which accelerates from velocity u to velocity v in time t due to a **constant** force F **acting in the direction of the acceleration**. F is the **resultant** of all forces which may be acting on the body.

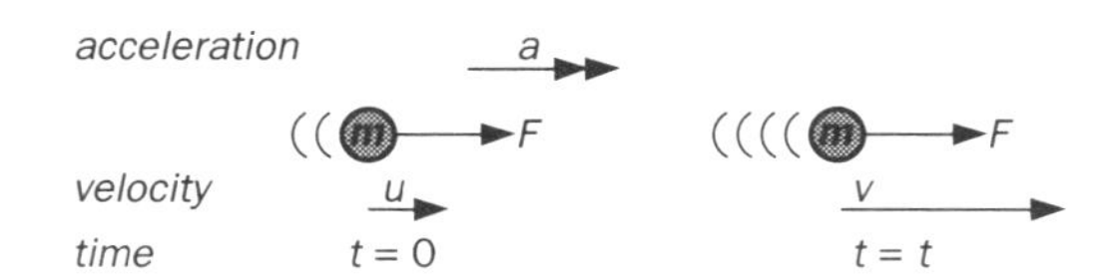

Figure 1.54 Derivation of F = ma

Initial momentum = mu Final momentum = mv

$\therefore$ momentum change = $(mv - mu)$

$$\therefore \text{ rate of change of momentum} = \frac{mv-mu}{t} = \frac{m(v-u)}{t} = ma$$

From Newton II

$F \propto ma$

$\therefore$ $F = kma$ where k is a constant of proportionality

In the SI system of units, 1 newton is **defined** as the force which gives a mass of 1 kg an acceleration of 1 m s^{-2}. Substituting these values into this equation gives k a value of 1. So provided we use only SI units, we can state that:

$$F = ma$$

(F in N; m in kg; a in m s^{-2} or N kg^{-1})

Equation 2

Equation 2 is a mathematical expression developed from Newton's second law.

Weight – the force of gravity

The weight (W) of a body is the **force** exerted on it by gravity. This is an example of an 'action-at-a-distance' force field, and is due to the interaction between the mass of the body and the mass of the planet. The acceleration given to a body by this force is called the **acceleration due to gravity** (g). Using Equation 2, we see that the weight (W) of a body of mass (m) at a place where the acceleration due to gravity is (g) is given by:

$$W = mg$$

(W in N; m in kg; g in m s^{-2} or N kg^{-1})

Equation 3

> **Maths window**
>
> Calculus approach to $F = ma$
>
> From Newton II
>
> *force* = rate of change of *momentum*
>
> This can be written:
>
> $$F = \frac{d(mv)}{dt} \quad \text{using symbols as before}$$
>
> $$\therefore \quad F = m\frac{dv}{dt} + v\frac{dm}{dt}$$
>
> For constant mass, the rate of change of mass $\frac{dm}{dt}$ is zero,
>
> $$\therefore \quad F = m\frac{dv}{dt}$$
>
> $\frac{dm}{dt}$ is rate of change of velocity, i.e. acceleration (a)
>
> $$\therefore \quad F = ma$$
>
> *Equation 4*

You should note that:

- at speeds approaching that of light, mass does change, and Newton II needs to be modified to work
- problems involving variable mass need to include the variable mass term.

Newton's third law (Newton III)

> If body A exerts a force on body B, then body B exerts an equal force in the opposite direction on body A.

This important law has implications in all walks of life. The law explains that forces always occur in equal, and opposite, pairs. The force of the Earth's gravity causes an apple to accelerate towards the Earth. But the apple also exerts an equal force on the Earth, which accelerates towards the apple! Of course, because of the difference in their masses, the acceleration of the Earth towards the apple is infinitesimally small, but it still exists.

Newton's law problem types

The application of Newton's laws is best illustrated by examining some typical problem situations and their solution.

Resultant force

Figure 1.55 shows a force of 400 N being used to pull a box of mass 40 kg against a constant frictional force of 80 N. Calculate the acceleration of the box.

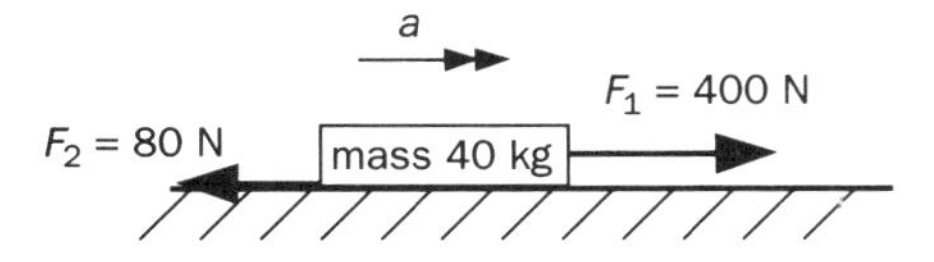

Figure 1.55 Resultant force problem

Note that the vertical forces (weight and normal reaction of the surface) are omitted for clarity, since the box is in equilibrium in the vertical direction. To apply Newton's second law, we must calculate the **resultant** force on the body:

The resultant force (F_r) is given by:

$$F_r = (F_1 - F_2) = 400 - 80 = 320 \text{ N (to the right)}$$

From Newton's second law we have:

$F_r = ma$

$\therefore \quad 320 = 40a$

$\therefore \quad a = \dfrac{320}{40} = 8 \text{ m s}^{-2}$

Lift problem (accelerating frame of reference)

Figure 1.56 shows an object of mass 2 kg attached to the hook of a spring balance which is hung from the roof of a lift. Calculate the reading on the spring balance when the lift is:

(a) accelerating upwards at 0.2 m s^{-2}

(b) accelerating downwards at 0.1 m s^{-2}

(c) going upwards with a constant velocity of 0.15 m s^{-1}.

Acceleration due to gravity (g) = 10 m s^{-2}.

There are no horizontal forces, so we only need to consider the vertical direction. There are **two** separate forces acting on the mass:

(i) the weight (W) of the body acting downwards. This force is **constant**.

(ii) the spring tension (T) acting upwards. This force varies, depending on the motion of the lift, and is equal to the reading on the spring balance. If the lift is not accelerating, the mass will be in equilibrium, and the spring balance reading will be equal to the weight of the mass (20 N).

It is the **resultant** of these two forces which causes the acceleration of the mass.

- The **magnitude** of this **resultant** force (F_r) = the **difference** between (W) and (T).
- The **direction** of this resultant force depends on the relative magnitudes of (W) and (T).

(a) **Upward acceleration of the lift**

T must be greater than W for the mass to accelerate upwards, so the resultant force is upwards:

$T - W = ma$

$\therefore \quad T - mg = ma \qquad \therefore T = mg + ma$

$\therefore T = 20 + (2 \times 0.2)$

$\therefore T = 20.4 \text{ N}$

Thus the reading on the spring balance is 20.4 N, which is greater than the weight of the mass. This result accounts for the sensation felt when a lift either **starts to ascend** or **comes to a halt while descending**. In each case, the **acceleration** is directed **upwards**, and we experience a momentary increase in the **contact force** between our feet and the lift floor.

(b) **Downward acceleration of the lift**

In this case T must be less than W as the resultant force must be downwards:

$W - T = ma$

$\therefore \quad mg - T = ma \qquad \therefore T = mg - ma$

$\therefore T = 20 - (2 \times 0.1)$

$\therefore T = 19.9 \text{ N}$

So the reading on the spring balance is 19.9 N, which is less than the weight of the mass.

Passengers in a lift which comes to a halt while ascending, or starts to descend, experience a brief decrease in the contact force between their feet and the floor, as shown in this calculation.

(c) **When the lift is moving with constant velocity,** the acceleration is zero

$\therefore \quad T - mg = 0 \qquad \therefore \quad T = mg = 20 \text{ N}$

The body is in equilibrium under the action of these two forces, and the reading on the spring balance is equal to the weight of the mass. This condition applies if the lift is stationary, or moving with constant velocity. Passengers in the lift experience only a steady contact force at the floor, equal to their own weight.

Conveyor belt (continuous flow of mass)

Figure 1.57 shows sand falling onto a horizontal conveyor belt at a constant rate of 5 kg s^{-1}. If the conveyor belt is moving at a velocity of 2.0 m s^{-1}, calculate the horizontal force needed to drive the belt.

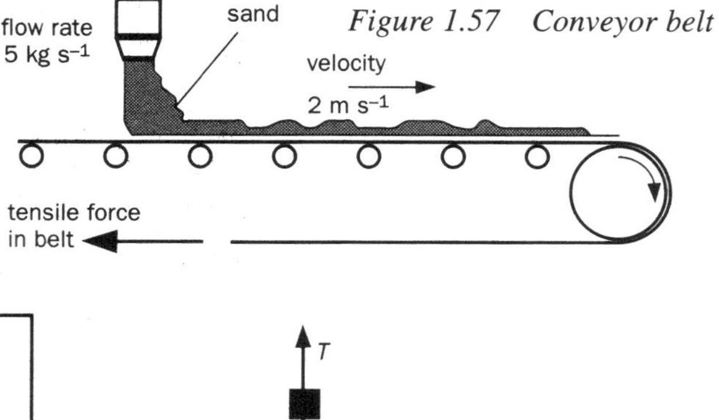

Figure 1.57 Conveyor belt

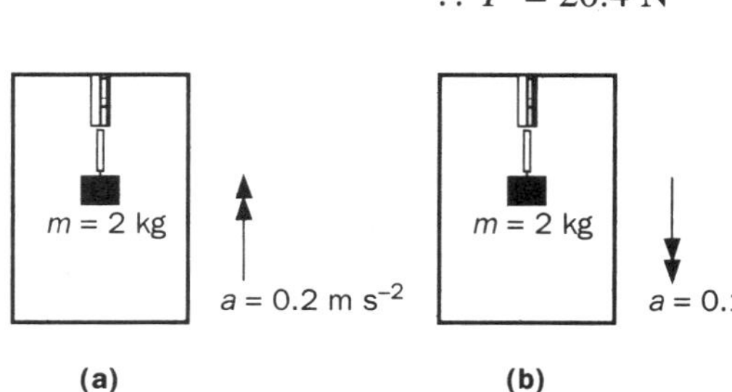

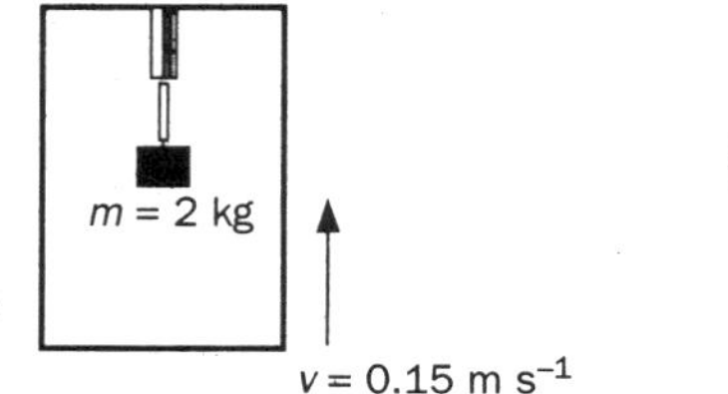

T
W (= mg)
(free body diagram)

Figure 1.56 Lift problems

In this case, we can neither calculate the individual accelerations of the sand particles, nor can we work out the sum of the individual forces on them, as we have no information on the **time taken** for the sand particles to be accelerated from rest up to the speed of the belt. However, we can use the concept of **rate of change of momentum**, and still apply Newton's second law.

Momentum is a vector quantity, and as we are only concerned with the horizontal force in the belt, we need only analyse the change of momentum in the horizontal direction. One technique is to consider what happens in unit time (1 second):

(mass deposited) = (flow rate) × (time)
$= 5\ \text{kg s}^{-1} \times 1\ \text{s} = 5\ \text{kg}$

The initial horizontal velocity of the sand falling onto the belt is zero.

(change of momentum) = (mass) × (change of velocity)
$= 5\ \text{kg} \times 2\ \text{m s}^{-1} = 10\ \text{kg m s}^{-1}$

(rate of change of momentum) = (change of momentum)/(time) $= \dfrac{10\ \text{kg m s}^{-1}}{1\text{s}} = 10\ \text{kg m s}^{-2}$

And from Newton II:

(resultant force) = (rate of change of momentum)
= 10 N

You should note that the unit for rate of change of momentum (kg m s^{-2}) is the same as the base SI unit for force (the newton).

The resultant force needed to move the gravel on this conveyor belt can be calculated using Newton's second law of motion

Hosepipe

A hosepipe ejects a jet of water horizontally at a speed of 4 m s^{-1} through an area of 100 cm^2, as shown in Figure 1.58. Assuming the water strikes the wall perpendicularly, and that the water escapes parallel to the wall after impact, calculate the force exerted on the wall. (Density of water = 1000 kg m^{-3}.)

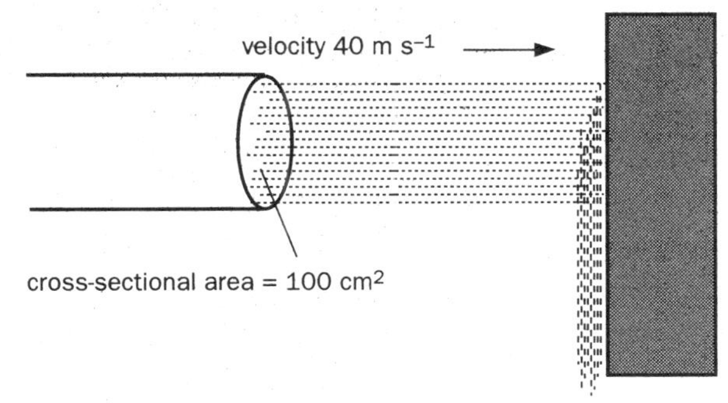

Figure 1.58 Hosepipe

This problem can be approached in a similar manner to the conveyor belt, by considering the change of momentum in 1 second. In this case, the **mass** of water must be calculated from information about the speed of flow, the area of the pipe and the density of the water.

In 1 second, a 'cylinder' of water of length 40 m is ejected from the hose. The volume of this 'cylinder' is given by:

(volume) = (length) × (area) = $40 \times (100 \times 10^{-4}) = 0.4\ \text{m}^3$
(Note the conversion of cm^2 to m^2)

The mass of this water is given by:

(mass) = (volume) × (density) = $0.4 \times 1000 = 400$ kg

The change of velocity of the water is $(40 - 0) = 40\ \text{m s}^{-1}$. Since this takes place in 1 second, the rate of change of momentum is given by:

(mass) × (velocity change) = $400 \times 40 = 16\,000\ \text{kg m s}^{-2}$

This is the rate of change of momentum of the water on impact with the wall, so (by Newton II) it is the force exerted **by** the wall **on** the water. According to Newton III, there must be an equal and opposite force exerted **by** the water **on** the wall,

∴ Force on the wall = 16 000 N

Helicopter

Rotating helicopter blades force air downwards, i.e. the air is accelerated by the rotor blades exerting a force on it. It follows from Newton III that there must be an equal and opposite (i.e. upward) force exerted by the air, on the rotor. This is the principle behind all forms of hovering flight, from the natural wonders of the bumble bee and humming bird, to the technological brilliance of the Harrier vertical takeoff aircraft.

Guided example (1)

A helicopter of mass 1000 kg is hovering (see Figure 1.59). Given that its rotor has a radius of 5 m, calculate the velocity of the air flow produced by the rotor. (Assume density of air = 1.3 kg m^{-3} and acceleration due to gravity = 10 m s^{-2}.)

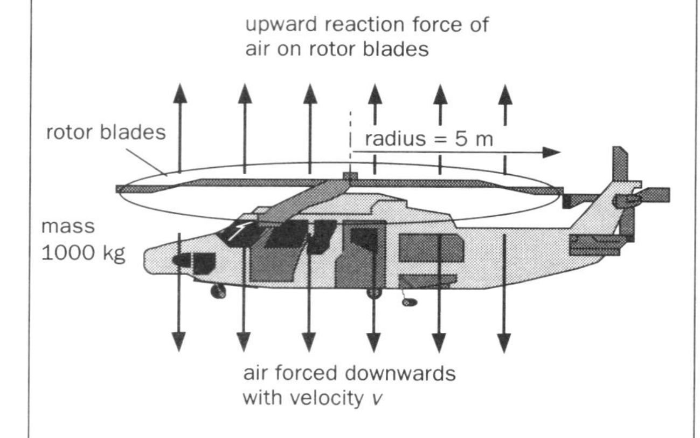

Figure 1.59 Hovering helicopter – supported by the upward reaction force of the air on the rotor blades

Guidelines

From Newton II, the rate of change of momentum of the air forced downwards by the rotor must equal the weight of the helicopter. We assume the initial downward velocity of the air above the rotor to be zero.

Consider a time of 1 second, and write expressions for the volume of air flowing in this time (using v for the unknown air velocity). Multiply by density to obtain the mass of air. Multiply by the change of air velocity (v) again to calculate rate of change of momentum.

Momentum and impulse

The **momentum** of a body has already been defined as the product of its **mass** and **velocity**.

Change of momentum produced by a constant force

Consider a body of mass m which is acted on by a **constant** force F for a time Δt, so changing its velocity from u to v (see Figure 1.60).

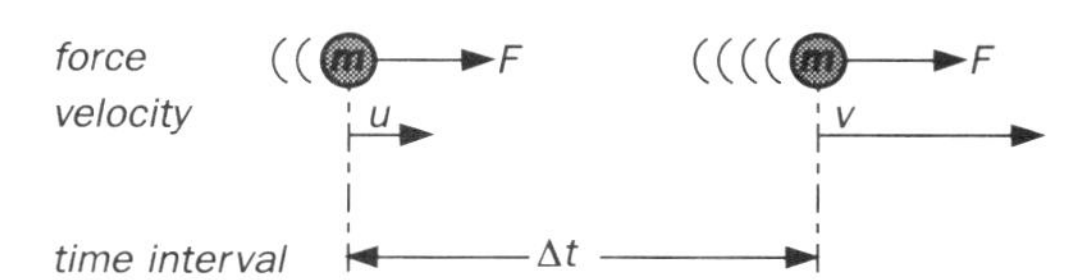

Figure 1.60 Momentum change produced by a constant force

From Newton II:

(force) = (rate of change of momentum)

$$\therefore \quad F = \frac{(mv - mu)}{\Delta t}$$

$$\therefore \quad F\Delta t = (mv - mu)$$

(units: N s = kg m s^{-1})

Equation 5

Impulse

SI unit newton second or kilogram metre per second (N s or kg m s^{-1})

The product of **force** and **time** interval ($F\Delta t$) is called the impulse of the force on the body.

It is a vector quantity. In words, Equation 5 can be written as:

(**impulse** of a force acting on a body) = (change of momentum of the body)

Change of momentum produced by a varying force

Although the above analysis was developed using the idea of a constant force, the concept also applies to varying forces, as the product $F\Delta t$ can be thought of as being equal to the **area** enclosed by a graph of **force** against **time**:

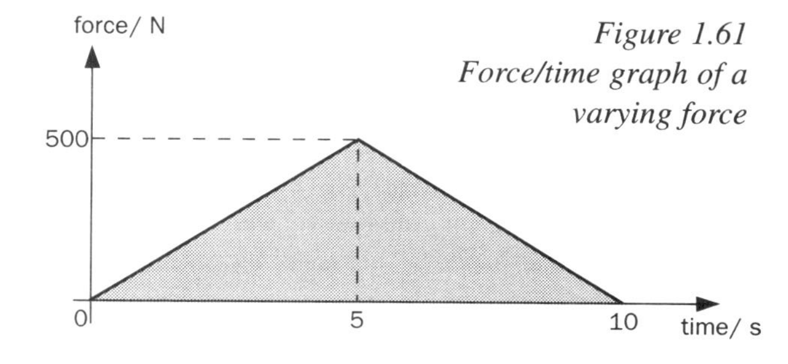

Figure 1.61 Force/time graph of a varying force

The force/time graph in Figure 1.61 shows how the force F applied to a body varies with time. The change of momentum of the body during this time is equal to the impulse of the force. The impulse is calculated by determining the area bounded by the F/t graph and the time axis:

impulse = area of triangle

$$\therefore \quad \text{impulse} = \frac{1}{2} \times 500 \times 10 = 2500 \text{ N s (or kg m s}^{-1})$$

If the force is in the direction of motion of the body, its momentum will be increased by this amount. If the force opposes the motion, this will be the decrease of momentum.

Implications of impulse and change of momentum

Obviously, when a body suffers a particular change of speed, its change of momentum is fixed, as this depends only on its mass and the change in velocity.

To decelerate a moving object and bring it to rest requires an opposing force (F) to be exerted for a time (Δt). The impulse ($F\Delta t$) can be provided either by a small force for a long time, or a large force for a short time. This is the principle behind each of the following examples:

- Cricketers try to slow down a ball **gradually** when catching it (they try to exert the smallest possible force for the longest time). Stopping it too quickly requires a large force, and it hurts!
- Cars have **crumple zones** designed to collapse progressively on impact. This increases the time taken for the car to come to rest in a crash, reducing the force exerted on the car and on the occupants.
- Airmen whose parachutes have failed to open have survived falling thousands of metres by landing in deep snow, on steep hillsides. The time taken for them to come to rest is increased, so the force on them is small enough to cause only slight injury.
- In any sport where balls are struck (tennis, golf, football, etc.), the player trying for maximum speed always tries to **follow-through** so as to increase the contact time between the ball and the racquet, club, foot, etc. This maximises the impulse given to the ball, producing a greater momentum change.

Conservation of momentum

The principle of conservation of momentum states that:

> When bodies in a system interact, the **total momentum** remains **constant** provided no **external force** acts on the system.

This principle is a deduction from Newton's second and third laws:

If two bodies collide, then by Newton's third law, each body exerts an equal and opposite force F on the other, and since this acts on each body for the same time Δt, then each body receives an equal and opposite impulse, of magnitude $F\Delta t$.

Thus each body receives equal and opposite changes of momentum, i.e. the total change of momentum of the system (the two bodies) is zero.

This has important implications in many areas. In guns, for example, the change of momentum suffered by the projectile and the gun are equal and opposite – this is the cause of **recoil**. Obviously, the gun has much more mass than the bullet, so its velocity will be much less than that of the bullet.

This principle is universally true and applies to the motion of galaxies and the interactions of the smallest sub-atomic particles, even where Newton's laws fail due to relativistic effects.

Guided examples (2)

1. Collisions

A railway truck (A) of mass 2×10^4 kg travelling at 1 m s^{-1} collides with a second truck (B) of mass 1×10^4 kg moving in the opposite direction at 0.5 m s^{-1}. If the trucks couple automatically on impact, calculate the common velocity with which they move after the collision.

Guidelines

Collision problems are usually solved by treating them as a **before/after** situation. Draw diagrams to show the bodies before and after the collision:

Assume a direction for the unknown velocity of the trucks after collision, and apply the Principle of Conservation of Momentum:

(initial momentum of truck A) + (initial momentum of truck B) = (combined momentum of A & B)

Insert values of mass and velocity (using v for the unknown velocity) remembering that velocity is a vector. Solve the equation for v.

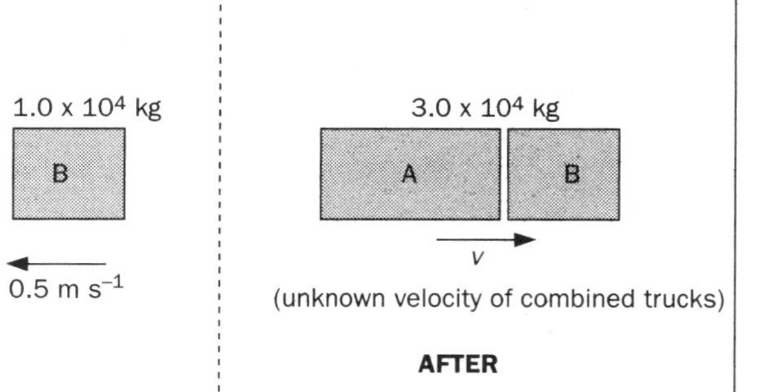

2. Explosions

A gun of mass 2 kg fires a bullet of mass 50 g. Calculate the initial recoil velocity of the gun if the bullet has a muzzle velocity of 300 m s^{-1}.

Guidelines

This is an example of an explosion, involving bodies which are initially combined, and which separate in opposite directions. Again this is analysed as a before/after situation. The special consideration here is that the total initial momentum of the bodies will be zero.

The sum of the momentum of the bullet and the momentum of the gun must be zero. The recoil velocity of the gun (V) is the only unknown quantity.

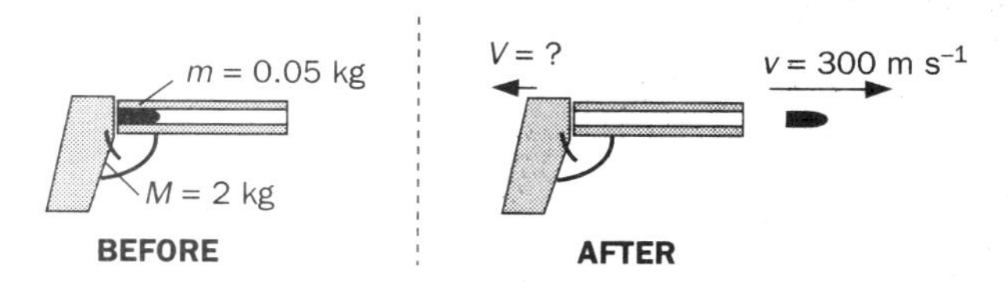

Self-assessment

SECTION A
Qualitative assessment

1. State Newton's **first** law of motion. This law expresses the concept of **inertia**. Explain what is meant by the term inertia.

2. In which of the following cases does a **resultant force** act on the body:
   (a) a bicycle travelling along a straight, level road at constant velocity
   (b) an electron orbiting a nucleus at constant velocity
   (c) a bumble bee hovering in still air
   (d) a car accelerating uniformly?

3. Define the **momentum** of a body and state its SI unit.

4. State Newton's **second law of motion in terms of momentum**.

5. Use the equation $\boxed{F = ma}$ to define the SI unit of force.

6. State Newton's **third** law of motion. Explain how it applies in the following situations:
   (a) a body falling freely towards Earth
   (b) an electron orbiting a nucleus
   (c) a sprinter starting to accelerate from rest
   (d) a rocket motor firing in outer space to accelerate a spacecraft.

7. A fireman of mass (M) slides down a vertical pole with an average acceleration (a). If the acceleration due to gravity is (g), derive an expression for the average frictional force (F) exerted on him, in terms of M, g and a.

8. State the **principle of conservation of linear momentum**, and state under what conditions it is valid.

9. Explain what is meant by the **impulse** of an applied force.

10. (a) State the unit of **impulse** and show that this unit is the same as that of linear momentum.
    (b) Explain the significance of the **area** enclosed by a **force/time** graph and the time axis.

SECTION B
Quantitative assessment

(***Answers:*** 0.020; −0.026; 0.95; −1.1; −1.5; 2.7; 5.0; 8.8; 9.6; 10; 15; 17; 18; 28; 40; 40; 50; 96; 130; 130; 1.0×10^3; 1.2×10^3; -1.3×10^5.)

1. An ice hockey puck of mass 0.75 kg is struck and moves off with an initial horizontal velocity of 24 m s^{-1} across the surface of a frozen lake. If it travels 190 m before coming to rest, calculate
   (a) the average retardation of the puck
   (b) the average friction force on the puck.

2. A submarine of mass 5.0×10^6 kg is moving with a velocity of 8.5 m s^{-1} while fully submerged. The power is suddenly shut off, and the submarine takes 5.5 minutes to come to rest. Calculate the average retardation and retarding force on the submarine.

3. An eccentric fisherman insists on weighing his fish inside the lift of a tall building. If he hooks a 4 kg sea bass onto his weighing machine, calculate the indicated weight when the lift is
   (a) stationary
   (b) accelerating upwards at 2.5 m s^{-2}
   (c) accelerating downwards at 3.0 m s^{-2}
   (d) moving upwards at a constant velocity of 4.0 m s^{-1}.

4. A hot air balloon with its basket and passengers has a total mass of 1150 kg, including the air in the balloon envelope. If, when it is stationary in still air, 100 kg of ballast is thrown out, calculate:
   (a) the resultant force on the balloon
   (b) its initial acceleration.

5. Fine salt is deposited from negligible height at a rate of 12 kg s^{-1} onto a conveyor belt moving at 8.0 m s^{-1}. Calculate the force needed to keep the belt moving at this velocity.

6. A jet of water issues horizontally from a nozzle of cross-sectional area 5.0×10^{-3} m^2 at a rate of 25 kg s^{-1}. Calculate:
   (a) the velocity of the water jet
   (b) the rate of change of momentum of the water
   (c) the force exerted by the water on the nozzle.

Self-assessment *continued*

7. The wind in a severe storm is blowing at a velocity of 30 m s^{-1} perpendicularly to the wall of a large barn, of area 80 m^2. Assuming that the air moves parallel to the wall of the barn after striking it, calculate the pressure exerted on it, assuming the air density to be 1.3 kg m^{-3}.

8. A helicopter together with its passengers has a total mass of 8500 kg and its rotor blade diameter is 30 m. Calculate the velocity of the air forced downwards by the rotating blades when the helicopter is hovering over an oil platform prior to landing. (Assume the air density to be 1.3 kg m^{-3}.)

9. A bumble bee hovers in mid-air by pushing the air downwards with its wings. If the total area swept out by the beating wings is 1.5 cm^2, and the mass of the bee is 2.0 g, calculate the downward velocity of the air.

10. A large bowling ball of mass 6.0 kg moving with velocity 3.0 m s^{-1} has a head-on collision with a single pin, of mass 0.50 kg. If the pin moves off with a velocity of 4.0 m s^{-1}, calculate the velocity of the bowling ball after the collision;

11. A fully-laden Range Rover of total mass 1800 kg travelling at 20 m s^{-1} collides with a Reliant Robin of mass 850 kg. Assuming the two vehicles separate after impact, calculate the velocity of the Range Rover after impact if:
    (a) the Reliant is initially stationary, and its velocity after impact is 10 m s^{-1}
    (b) the Reliant's initial velocity is 12 m s^{-1} in the same direction as the Rover, and its velocity after collision is 18 m s^{-1}.
    (c) If the two cars become stuck together in the impact, calculate their combined velocity after the collision if the Reliant's initial velocity is 15 m s^{-1} in the same direction as the Range Rover.

12. 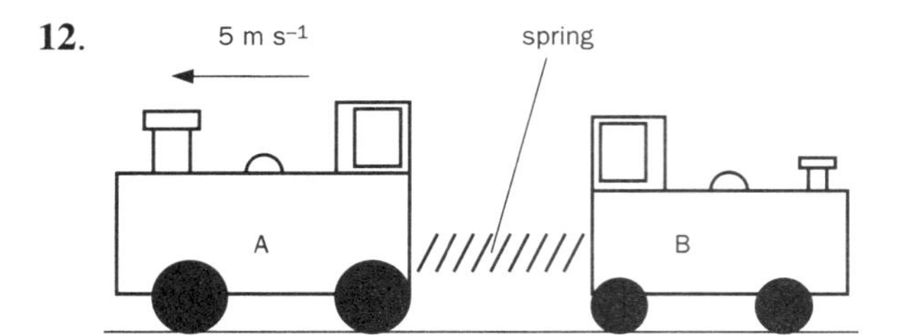

Two toy trains A and B of masses 0.70 kg and 0.40 kg respectively are held together on a level frictionless track against the force exerted by a compressed spring. When the trains are released, A moves to the left with an initial velocity of 5 m s^{-1}. Calculate the initial velocity of B.

13. 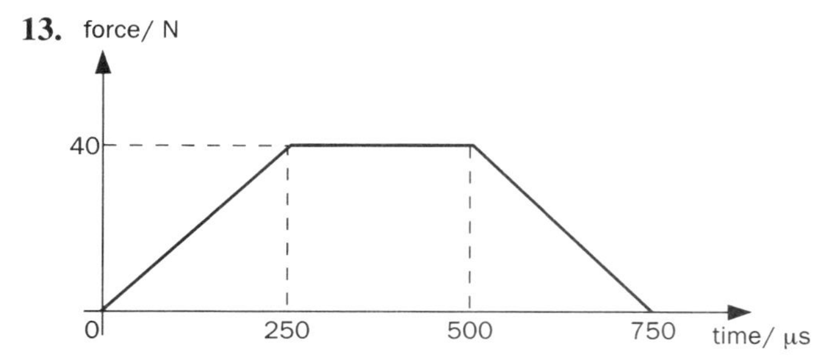

A force F acts on a body which is initially stationary. The graph shows how F varies with time t.
    (a) Sketch a velocity/time graph for the 750 μs period and **explain** its shape.
    (b) Explain what the area enclosed by the F/t graph represents.
    (c) Calculate the momentum gained by the body.

1.7 Work and energy

Work is done when a **force** moves its point of application in the direction in which the force acts.

Common symbol W

SI unit newton metre (N m); unit name: joule (J);

Defining equation

(work) = (force) × (distance moved along the line of action of the force)

1 joule is the work done when a force of **1 newton** moves its point of application a distance of **1 metre** in the direction of the line of action of the force.

Lifting a box, pushing it along the floor or carrying it up a flight of stairs are all examples of work being done. In each case a force is applied, and movement takes place in the direction of the line of action of the force. No work is done if the box is simply held stationary off the ground, or if it fails to move when pushed.

Work done by a constant force

Simple case: the force and displacement are in the same direction (Figure 1.62)

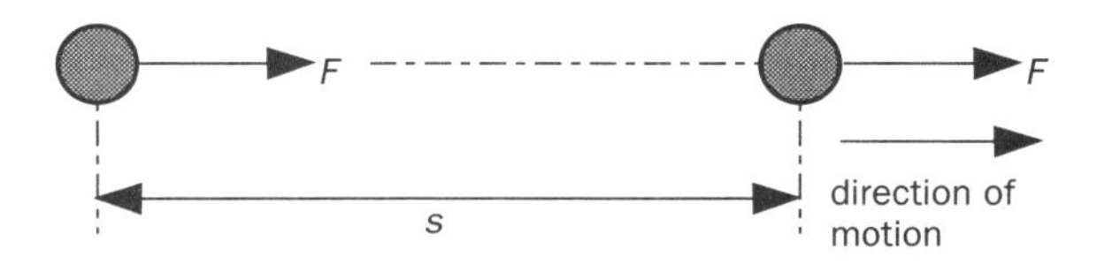

Figure 1.62 Work done by a force along the direction of motion

When a constant force (F) moves a distance (s) along its line of action, the work done (W) is given by:

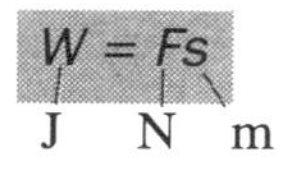

Equation 1

Force and motion are in different directions (Figure 1.63)

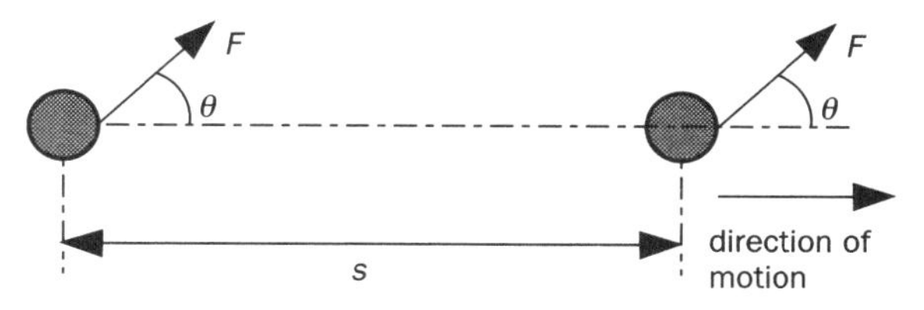

Figure 1.63 Work done by a force at an angle to the direction of motion

Here the **component** of the force along the direction of motion is $F \cos \theta$. The work done by this force is thus given by:

$$W = Fs \cos \theta$$

Equation 2

NB: If F acts in a direction perpendicular to the direction of motion, then $\theta = 90°$, i.e. $\cos \theta = 0$

$\therefore$ $Fs \cos \theta = 0$.

Thus the work done by a force is **zero** if the motion takes place at a **right angle** to its line of action.

For example, when a body moves **horizontally** along the Earth's surface, no work is done by the **gravitational force** (weight) which acts vertically, although there may well be other forces doing work.

Work done by a variable force

If the force varies in magnitude during the motion, the work done is given by:

(work) = (average force) × (distance moved along the line of action of the force)

If the varying force can be represented by a graph of force against displacement, then the **work done** by the force is equal to the **area** enclosed between the graph and the displacement axis (see Figure 1.64).

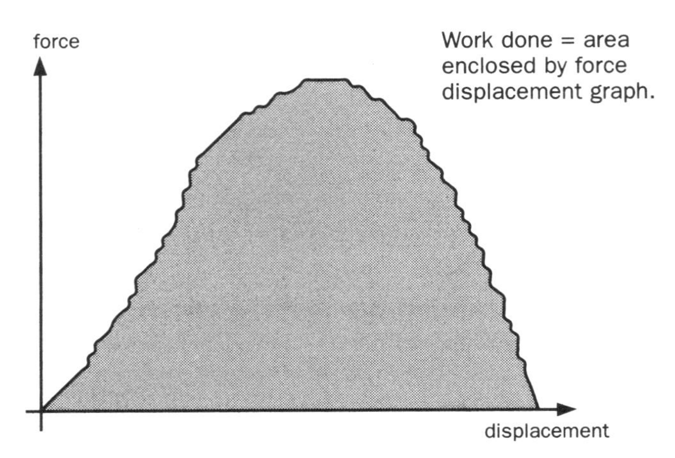

Figure 1.64 Force/displacement graph for a varying force

You should note that:

- Work is a **scalar quantity**, i.e. it is a quantity which has **magnitude**, but **not direction**. Work done against friction, for example, is not undone if the direction of motion is reversed.
- The concept of work is useful for calculations involving forces and **displacements**, where there is no knowledge of the time for which the force acts, hence accelerations may be difficult or impossible to measure or calculate using the kinematics or kinetics equations.
- If the force (or a component of that force) acts in the **same direction** as the displacement, then the work done is positive, i.e. work is done **by** the force. An example of this is the work done **on** a falling body by gravity.
- If the force (or a component of that force) acts in the **opposite direction** to the displacement, then the work done is **negative**, i.e. work is done **against** the force, e.g. the work done by friction forces is always negative, since the friction force always acts to oppose the direction of motion.
- In problem solving, you should be careful to calculate the work done by **individual** forces, not **resultant** forces. For example, a car moving at constant velocity on a level road has zero resultant force acting. However, work is being done **by** the engine pushing the car, and **against** the equal friction force in the opposite direction.

Energy

Common symbols	*W, E*
SI unit	joule (J), identical to the unit of work

Energy is defined as that which enables a body to do work.

A body which does work **on** another body suffers a **decrease** in energy; the body **on which work is done** has its energy **increased**.

We are mostly concerned in this section with mechanical energy, which takes the forms of:

Kinetic energy (E_k)	which is energy possessed by a body due to its *mass* and *velocity*;
Potential energy (E_p)	which is energy possessed by a **system** of bodies due to their relative **positions** and the **force** fields between them: gravitational, electrical and electromagnetic.

You should note that although we give names to many other forms of energy, such as *chemical*, *electrical*, *heat energy*, etc., all forms of energy may be classified as either **kinetic** or **potential** energy.

Kinetic energy

Common symbol	E_k
SI unit	joule (J)

Defining equation: $E_k = \frac{1}{2}mv^2$

for a body of mass (m) moving with velocity (v).

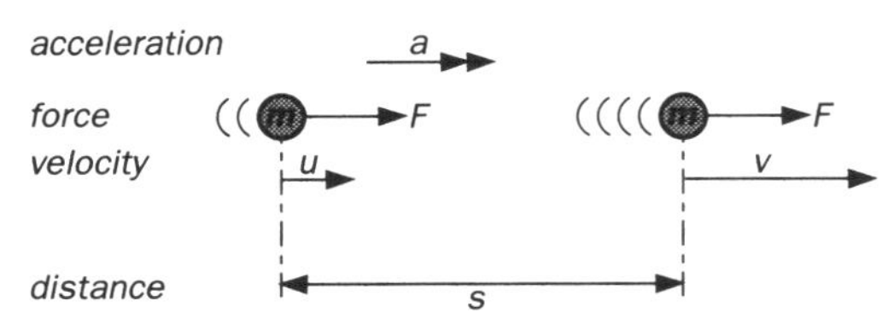

Figure 1.65 Kinetic energy

Consider a body of mass (m) acted on by a constant force (F) which gives it a uniform acceleration of (a) as shown in Figure 1.65, so that it accelerates from initial velocity (u) to final velocity (v) over a distance (s). The force does work **on** the body, which increases its velocity. We say that the body's kinetic energy has increased. There is an important principle which states that:

The work done **by** the forces acting **on** the body is equal to the **increase in energy** of the body.

Derivation window – kinetic energy

From Newton II:

$$F = ma$$

From the definition of **work**:

$$W = Fs$$

$\therefore \quad W = mas$ this is the work done by the force on the body

Since the only effect of the force is to increase the kinetic energy, the change of kinetic energy (E_k) is given by:

$$E_k = mas;$$

From kinematics:

$$v^2 = u^2 + 2as$$

$$\therefore \quad 2as = v^2 - u^2$$

$$\therefore \quad as = \frac{1}{2}v^2 - \frac{1}{2}u^2$$

Multiplying by (m):

$$\therefore \quad mas = \frac{1}{2}mv^2 - \frac{1}{2}mu^2$$

Since $E_k = mas$:

$$\therefore \quad E_k = \frac{1}{2}mv^2 - \frac{1}{2}mu^2$$

this is the change in kinetic energy of a body of mass (m) which accelerates from velocity (u) to velocity (v).

If the body were initially at rest (i.e. u = 0), then the kinetic energy of the body is given by:

$$E_k = \frac{1}{2}mv^2$$

(E_k: J; m: kg; v: m s^{-1})

Equation 3

The quantity $\frac{1}{2} \times$ (*mass*) $\times$ (*velocity*)^2 is defined as the **kinetic energy** of a body of mass (m) moving with velocity (v). It can be thought of as either:

- the amount of work done **on** the body by external forces to accelerate it to this velocity from rest; or
- the amount of work which must be done **by** the body against external forces in order to bring it to rest.

Gravitational potential energy

Common symbol E_p

SI unit joule (J)

Defining equation: $E_p = mg\Delta h$

for a body of mass (m) in a gravitational field strength (g) where Δh is the change of height.

In a gravitational field, there is a vertical force directed downwards on all bodies – called the **weight** of the body. In order to raise a body to a different height, a force equal to the weight must be applied in the upward direction. By definition, this upward force does work on the body. We say that the **gravitational potential energy** of the body has been increased, by an amount equal to the work done on it.

(Strictly speaking, it is the gravitational potential energy of the ***system****, comprising the body* ***and*** *the planet which has been increased, but we usually refer to it as the energy of the body for simplicity.)*

Derivation window – gravitational potential energy

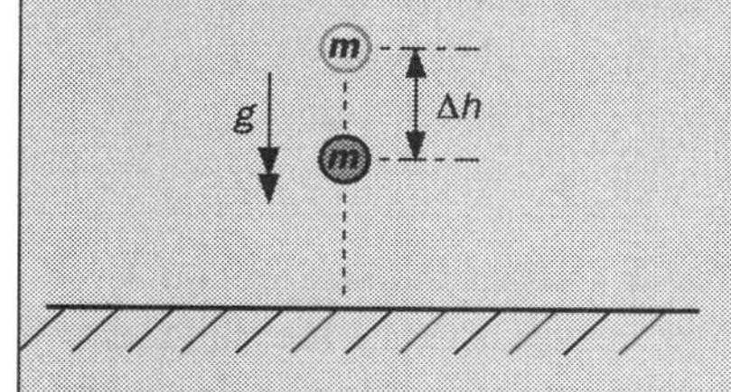

Figure 1.66 Gravitational potential energy

Figure 1.66 shows a body of mass (m) at some height above the ground. (g) is the gravitational field strength. The force required to raise the body is the **weight** of the body (mg).

By definition, the work done in raising the body through a vertical distance (Δh) to a new rest position is given by:

$W = mg\Delta h$ assuming (g) is constant

Since the body does not acquire any kinetic energy, we say that the quantity $mg\Delta h$ is equal to the gain of gravitational potential energy of the body (E_p):

$$E_p = mg\Delta h$$

(E_p: J; m: kg; g: m s^{-2} or N kg^{-1}; Δh: m)

Equation 4

If the body were released, then the **weight** would do positive **work** on the body, which would result in this energy being converted into **kinetic energy** as it accelerates towards the ground.

This equation applies only to a uniform gravitational field. On Earth, the gravitational field can be regarded as uniform, within certain limitations:

- The magnitude of the gravitational field strength decreases with distance from the centre of the planet, but within, say, 100 km or so of the surface, the variation is negligible for most purposes.
- The direction of the gravitational field varies continuously, as it is always directed radially, towards the centre of the Earth. Given the dimensions of the planet, it is reasonable to assume that this is constant over areas of, say, around 100 km^2.
- There are small changes in the magnitude of (g) at different parts of the Earth due to local variations in the density of the planet, and the fact that the Earth is not a true sphere, but these effects can be ignored for most purposes.

Elastic potential energy

Changing the shape of an elastic body causes a gain in **Elastic potential energy**, since work is done by the external forces against the **intermolecular** forces. Figure 1.67 shows a spring of elastic constant (k) extended by an amount Δl, together with a graph showing how the force exerted varies with the displacement of the spring.

Since the force exerted to extend the spring is not constant, the work done is equal to the area bounded by the graph and the displacement axis. If (F) is the maximum force exerted, when the extension is (Δl), these are related by:

$F = k\Delta l$ where (k) is the spring constant (the force needed to extend the spring by 1 metre)

The work done (W) is given by:

$$W = \frac{1}{2} F\Delta l$$

Combining these two equations gives:

$$W = \frac{1}{2} k\Delta l^2$$

J N m^{-1} m^2

Equation 5

Thus the **elastic potential energy** stored in a spring stretched within its **elastic limit** is directly proportional to the (**extension**)^2. This also applies to a spring in compression.

Conservation of energy

The **principle of conservation of energy** states that:

> Energy cannot be created or destroyed, it can only be **transformed** from one form into another.

When we speak of energy being 'lost' or 'wasted', what we really mean is that some device or process has produced a form of energy that is not wanted, such as heat energy caused by friction. This energy has not been 'lost' from the Universe, but we may regard it as being **dissipated** to the surroundings by whatever process has produced it. We can regard the universe as possessing a fixed quantity of energy, which is slowly being converted from one form into another as the Universe ages. The ultimate destiny of all this energy is that all the energy conversions will cease when every particle in the Universe is at the same temperature.

In mechanics problems, it is convenient to consider only the kinetic and gravitational potential energies; any other form of energy produced can then be regarded as being dissipated to the surroundings. The principle of conservation of energy can be modified to read:

> The total amount of mechanical energy ($E_p + E_k$) possessed by the bodies in an **isolated system** is constant.

$\therefore$ loss of E_p = gain of E_k or vice versa

When friction is involved, this represents an **external force**, and usually produces a rise in temperature due to an increase in internal energy:

i.e. loss of E_p = gain of E_k + gain of internal energy

Figure 1.67 Elastic potential energy

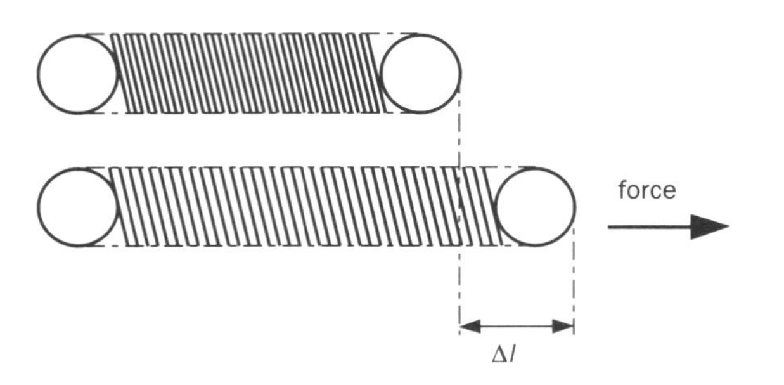

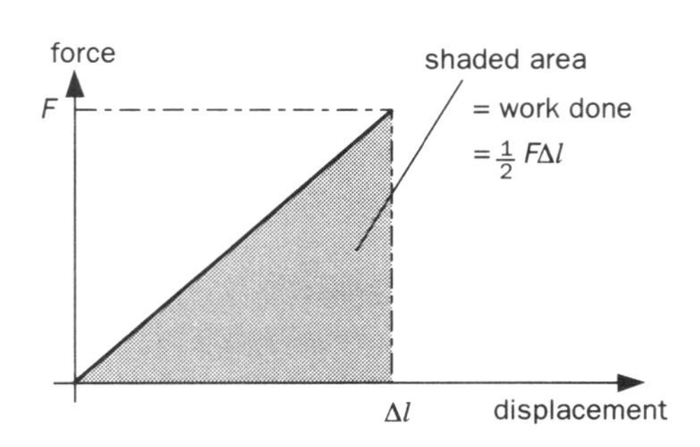

To illustrate this statement, consider the energy changes which occur when you throw a ball vertically upwards. If the ball has a mass (m) and leaves your hand with velocity (v), its initial kinetic energy is ($\frac{1}{2}mv^2$). As the ball rises its kinetic energy decreases as it does work against gravity, and its potential energy increases. Assuming no work is done against air resistance (which would be an **external** force):

(E_k lost by ball) = (work done against gravitational force) = (E_p gained by ball)

When it reaches its maximum height, the ball momentarily comes to rest ($v = 0$). At this point the kinetic energy is zero, and the increase in potential energy is exactly equal to the loss in kinetic energy. When the ball falls, this energy conversion is reversed. The gravitational force does work on the body to increase its kinetic energy, and the potential energy decreases. When the ball returns to the point of release, it will have re-acquired kinetic energy equal to its original amount. At every stage of the ball's flight, the total energy ($E_p + E_k$) is constant.

Elastic and inelastic collisions

We have already stated that when bodies collide, momentum is always conserved, as long as no external forces act. However, some of the bodies' original kinetic energy is generally dissipated to the surroundings, so mechanical energy is not conserved by the bodies. The energy is converted to heat, sound, etc.

- Perfectly elastic collisions are those in which no kinetic energy is lost. In practice, this is virtually impossible, but collisions between gas molecules approach this ideal, as do collisions between electrons and the molecules of a superconductor.
- All other collisions are inelastic.
- Perfectly inelastic collisions are a special case, and are collisions in which there is no rebound after collision (i.e. the bodies become joined).

Guided examples (1)

1. A car of mass 1×10^3 kg is travelling at 20 m s^{-1} on a horizontal road, and is brought to rest in a distance of 40 m by braking. Find

(a) the average frictional force exerted by the brakes

(b) the time taken to stop the car.

Guidelines

(a) (work done against friction) = (kinetic energy decrease of car in coming to rest).

(b) Use Newton II to calculate the acceleration of the car (negative). Then use the kinematics equations to calculate the time taken to come to rest.

2. A bullet of mass 10 g travelling horizontally at a speed of 200 m s^{-1} embeds itself in a block of wood of mass 990 g suspended by strings so that it can swing freely. Find

(a) the vertical height through which the block rises

(b) the amount of the bullet's kinetic energy that is converted to internal energy.

(Assume the acceleration due to gravity to be 10 m s^{-2})

Guidelines

(a) This is a collision problem. When the bullet strikes the block it becomes embedded in it, and friction will dissipate a great deal of the bullet's **energy**. **Momentum** is conserved, however, so the momentum of the block and bullet **after** the collision is equal to the momentum of the bullet **before** the collision (which can be calculated). The mass of the block and bullet are known, so calculate the velocity with which the block and the bullet **start** to move.

You can now calculate the initial kinetic energy of the block and bullet. This will be converted to potential energy as the block swings upwards. You can assume that energy will be conserved during this part of the motion, so calculate the height reached when the block and bullet come to rest at the top of the swing.

(b) Calculate the kinetic energy of the bullet before the collision. This will be much greater than the kinetic energy of the block and bullet just after the collision. The energy 'lost' will have been dissipated within the block.

3. A toy car of mass 250 g is released from point A on a frictionless track (Figure 1.68).

Calculate

(a) its maximum kinetic energy

(b) its maximum velocity

(c) its velocity at C

(d) Explain what happens when it reaches D.

Guided example continued on next page

Guided examples continued

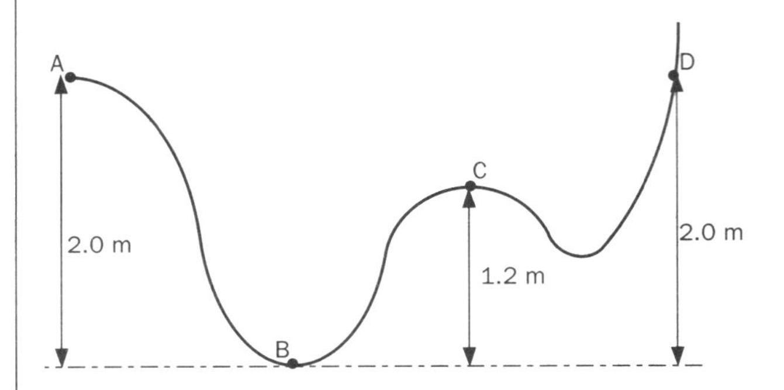

Figure 1.68
Toy car rolling down a frictionless track

Guidelines

(a) In moving from A to B, the car suffers the greatest loss of gravitational potential energy, and so the maximum kinetic energy will occur at B. Therefore the gain of kinetic energy will be equal to the loss of gravitational potential energy (as the track is frictionless, no energy is dissipated).
(b) If you know the maximum kinetic energy, you can calculate the maximum velocity.
(c) In moving from B to C, the car will gain gravitational potential energy, at the expense of its kinetic energy.
(d) Point D is at the same height as point A, therefore the potential energy of the car at point D will be the same as at point A.

Power

Common symbol P
SI unit joule per second (J s^{-1}) or watt (W)

Devices which convert energy from one form to another are called **machines**. For example, car engines convert the chemical energy stored in petrol or diesel fuel into kinetic energy and other unwanted forms such as heat and sound energy. Electric motors convert electrical energy into rotational kinetic energy, while generators do the reverse. One of the most important measures of a machine is the **rate** at which it converts energy – how many joules per second are being transformed.

The **power** of a machine is defined as the **rate** at which the machine converts **energy** into other forms or the **rate** at which **work** is done by the machine.

Defining equation

$$(\text{power}) = \frac{(\text{work done})}{(\text{time taken})} \text{ or } \frac{(\text{energy converted})}{(\text{time taken})} = \frac{\Delta W}{\Delta t}$$

If the work is being done (or energy is being converted) at a non-uniform rate, then these equations will yield the **average** power of the machine.

The **instantaneous** power of a machine could be calculated from a graph of **energy** against **time** such as that shown in Figure 1.69.

The **gradient** of the graph at any given instant gives the instantaneous power (P) at that instant.

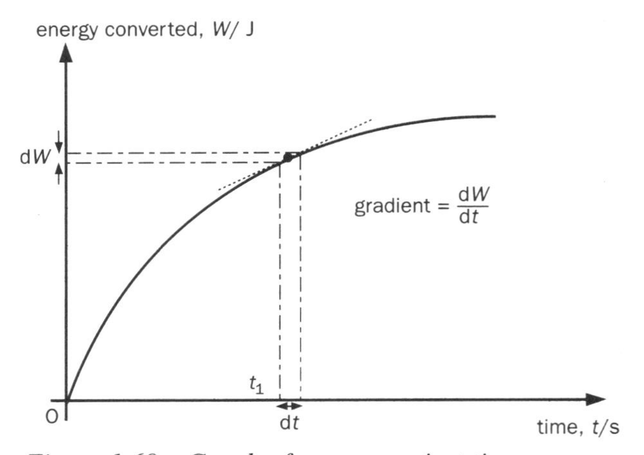

Figure 1.69 Graph of energy against time

$$P = \frac{dW}{dt}$$

If a constant force (F) is applied, and does work by moving its point of application a distance (s) in time (t), then the power is given by:

$$\text{power} = \frac{\text{work done}}{\text{time taken}} = \frac{Fs}{t}$$

but s/t is the velocity of the motion, so the power (P) is given by:

$$P = Fv$$

W N m s^{-1}

Equation 6

This is a useful expression for situations where the accelerating force is known to be constant, enabling instantaneous power to be calculated if the instantaneous velocity is known.

Efficiency

A machine which converted all of the energy **input** into **useful** energy or work would be said to be 100% efficient, as it transforms 100% of its energy without wasting any. Such a machine is impossible, mainly because friction can never be completely eliminated from any process. The **efficiency** of an energy conversion process is a measure of what proportion of the energy input is converted into desirable forms of energy.

Defining equation:

$$\% \text{ efficiency} = \frac{\text{(energy input)}}{\text{(energy output)}} \times 100\% = \frac{\text{(power input)}}{\text{(power output)}} \times 100\%$$

Equation 7

Guided examples (2)

1. A car of mass 1000 kg is moving on a horizontal road at a steady speed of 10 m s^{-1} against a constant frictional force of 400 N.

(a) Calculate the power cutput of the engine.

(b) The car now climbs a hill at an angle to the horizontal of 8°. Assuming the frictional force remains constant at 400 N, calculate the new engine power output required to maintain speed.

Guidelines

(a) You can use Equation 6, since the force on the car will be **constant**. The driving force exerted by the engine must be **equal** to the friction force, as the car is in equilibrium.

(b) As well as doing work to overcome the frictional force, the engine must increase its output to supply the increase in **gravitational potential energy** as the car climbs the hill. Equation 6 still applies, since the new force will remain constant while the car ascends the hill. Calculate the increase in this energy per second.

2. A pump which is powered by a 2.5 kW electric motor is used to raise water through a height of 8.5 m. If the system is 55% efficient, calculate the mass of water pumped per second.

Guidelines

The input power is 2500 W; use Equation 7 to calculate the output power. Since this is the useful energy converted to gravitational potential energy (of the water) per second, calculate the mass of water raised to the height given.

3. Figure 1. 70 illustrates the principle of operation of a wind turbine in which the kinetic energy of the wind is converted into electrical energy. Downwind of the turbine, the wind velocity is less than that upwind, as some of the wind's kinetic energy has been converted by doing work on the turbine blades.

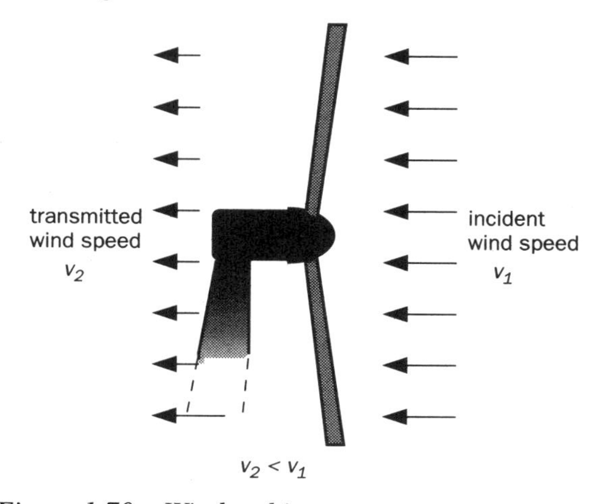

Figure 1.70 Wind turbine

The blades of the world's largest wind turbine, situated in Hawaii, have a diameter of 122 m. If at a given time the incident and transmitted wind speeds are 16 m s^{-1} and 14 m s^{-1} respectively, calculate

(a) the mass of air (of density 1.3 kg m^{-3}) flowing per second through the area swept out by the blades

(b) the power output of the turbine if the system can be assumed to have an efficiency of 70%.

Guidelines

(a) In one second, a 'cylinder' of air of length 14 m and diameter 122 m passes through the blades. Calculate its volume, and hence calculate its mass.

(b) Consider a time interval of 1 second. Calculate the energy input per second (from the wind) needed to give the quoted energy

Guided example continued on next page

Guided examples *continued*

output using Equation 7. This must be equal to the **decrease** in kinetic energy per second suffered by the wind as its velocity is reduced by passing through the blades of the turbine.

4. A fully-laden car has a mass of 1650 kg and accelerates uniformly from rest against a constant frictional force of 400 N to reach a speed of 20 m s^{-1} after 12 s. Calculate

(a) the power developed by the engine at

(i) 4 s after the start

(ii) 12 s after the start.

(b) (i) the power input from the fuel when the car is 30 m from the start position, assuming the engine has an efficiency of 35%

(ii) the flow rate (in kg s^{-1}) of petrol into the engine at this point, assuming 1 kg of petrol releases 48.2 MJ of heat energy when burned in the engine.

Guidelines

(a) Calculate the acceleration using kinematics equations, then use Newton II to calculate the resultant force on the car. Use this to work out the engine's driving force. Then use Equation 6 to calculate instantaneous power (after calculating the instantaneous velocities at the appropriate times).

(b) (i) Calculate the instantaneous power at this point, and use Equation 7 to calculate the input power.

(ii) From (i), calculate the mass of petrol required to release this energy.

Self-assessment

SECTION A
Qualitative assessment

1. (a) Define the **work done** by a force.

(b) State the SI unit of work.

(c) Explain why no work is done by the force of gravity when a body moves on a horizontal surface.

(d) Define **energy** and state its SI unit.

2. (a) Define

(i) **kinetic energy**

(ii) **gravitational potential energy.**

(b) Show that

(i) the kinetic energy (E_k) of a body of mass (m) which is initially at rest and is acted on by a constant force which accelerates it to a final velocity (v) is given by

$$E_k = \frac{1}{2}mv^2$$

(ii) the increase in gravitational potential energy (E_p) of a mass (m) which is raised through a vertical height (Δh) in a gravitational field strength (g) is given by

$$E_p = mg\Delta h$$

3. Explain what is meant by **elastic potential energy**.

4. State the **principle of conservation of energy**.

5. (a) Distinguish between **elastic** and **inelastic** collisions, and give an example of each type.

(b) What happens in a **perfectly inelastic** collision?

6. Define **power** and state its SI unit.

7. (a) Explain what is meant by the **efficiency** of a machine.

(b) Why is it impossible for any energy conversion process to be 100% efficient?

8. An aircraft is flying horizontally at velocity (v) against a constant air resistance force (F). If the engines have an efficiency of 20%, derive an expression for the rate at which energy must be supplied to them.

SECTION B
Quantitative assessment

(***Answers:*** 1.1×10^{-21}; 0.063; 1.1; 2.8; 6.0; 6.5; 15; 200; 500; 1.0×10^3; 2.3×10^3; 3.6×10^3; -5.1×10^3; 5.4×10^3; 9.3×10^3; 1.6×10^4; 8.3×10^4; 8.3×10^4; 3.0×10^5; 1.2×10^6; 1.5×10^6; 1.0×10^7; 1.1×10^7.)

(Use acceleration due to gravity g = 10 m s^{-2} unless otherwise stated.)

Self-assessment *continued*

1. A shell of mass 30 kg is fired with a velocity of 280 m s^{-1} from a gun of mass 6.5×10^3 kg. Calculate the kinetic energies of the gun and shell immediately after firing.

2. Calculate the energy converted in the following situations:
   (a) stopping a locomotive of mass 50000 kg travelling at 20 m s^{-1}
   (b) raising 20 tonne of coal up a mineshaft which rises 55 m vertically
   (c) raising the velocity of an electron of mass 9.1×10^{-28} g from rest to a velocity of 5.0×10^4 m s^{-1}.

3. A bullet of mass 45 g is travelling horizontally at 400 m s^{-1} when it strikes a wooden block of mass 16 kg which is suspended on a string so that it can swing freely. If the bullet becomes embedded in the block, calculate
   (a) the velocity with which the block starts to swing
   (b) the height to which the block rises above its initial rest position
   (c) the amount of the bullet's kinetic energy which is converted to internal energy.

4. 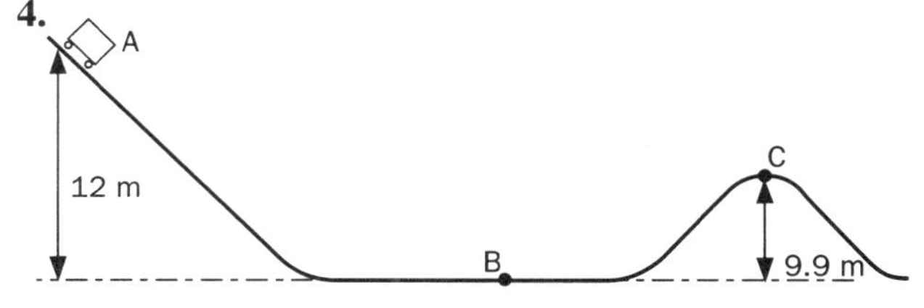

   The diagram shows a section of a roller coaster in a vertical plane. A vehicle and its occupants with a total mass of 920 kg is released from point A. Ignoring frictional and rotational effects, calculate the velocities of the vehicle at points B and C.

5. A horizontal force of 3500 N is applied to a car of mass 800 kg which is initially stationary on a horizontal surface. If the motion is resisted by a constant frictional force of 1300 N calculate:
   (a) the average acceleration of the car
   (b) the kinetic energy of the car after 10 s.

6. A coin of mass 10 g is dropped from the top of a tall building. If, after falling for 10 s, it falls into soil and penetrates to a depth of 5 cm, calculate:
   (a) the height of the building.
   (b) the average force exerted on the coin by the soil.

7. A bullet of mass 25 g has its speed reduced from 450 m s^{-1} to 150 m s^{-1} when it passes through a 14 cm thickness of timber. Calculate:
   (a) the loss of kinetic energy of the bullet
   (b) the average friction force exerted by the timber.

8. A car is involved in a collision when it is travelling at 24 m s^{-1}. The driver, of mass 85 kg, is brought to rest by the seat belt in a time of 400 ms. Calculate the average force exerted on the driver by the seat belt. Compare this force with his weight, and hence calculate the 'g-force' to which he is subjected by the crash, and comment on the likelihood of his sustaining serious injury.

9. A girl of mass 60 kg rides a mountain bike of mass 16 kg at a constant speed of 3.0 m s^{-1} up a hill which rises 1.0 m for every 15 m of its length. If air and road resistance amount to 25 N, calculate the power she is developing.

10. A ship whose engines are developing a useful power output of 500 kW is cruising at a constant velocity of 6.0 m s^{-1}. Calculate:
    (a) the push exerted by the propeller on the water
    (b) the total magnitude of the forces resisting the ship's motion
    (c) the power input if the engines have an efficiency of 33%.

11. A small dinghy has an outboard motor with a propeller diameter of 20 cm. If the dinghy is tied to the quayside and the engine is started, the propeller forces back a stream of water at a speed of 6.0 m s^{-1}. Calculate the input power to the engine if it is 40% efficient and the density of sea water is 1.1×10^3 kg m^{-3}.

1.8 Motion in a circle

An object which moves through equal distances in equal times along a circular path has constant speed, but its velocity is continuously changing (i.e. it is accelerating). This seemingly contradictory statement can be explained if we remember that velocity is a vector, and that therefore a change in direction is as much an acceleration as a change in speed.

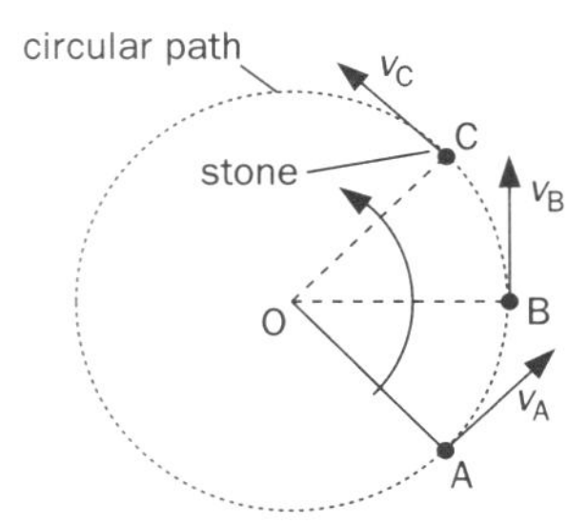

Figure 1.71 Plan view of a body moving in a circular path

Consider a small stone tied on the end of a string and whirled in a horizontal circle at constant speed, as shown in Figure 2.71.

The stone's velocity at any point on its path is always directed along the tangent to the circle, i.e. at a right angle to the string. Since the speed is constant, the magnitude of the velocity at A, B, C, etc. is constant, ($v_A = v_B = v_C$, etc.), but clearly the direction of the velocity is continuously changing. This means that the stone is **accelerating**.

For such an acceleration to happen, there must be a **resultant force**. Since the **magnitude** of the velocity is unaffected, the force cannot have any component **parallel** to the direction of the velocity. It is the tension in the string which causes the acceleration, and this tension is always directed towards the centre of the circle.

> The resultant force which must act on a body so as to make it follow a circular path is called the **centripetal force.**

It is important to remember that this centripetal force is just the name given to whatever resultant force acts on the body. It is not some new kind of force. In the above example, it is the tension in the string.

The centripetal force:

- acts at 90° to the direction of motion of the body, and is therefore always directed towards the centre of the circular path
- gives the body an acceleration towards the centre of the circular path – called the **centripetal acceleration**.

The channel tunnel was bored out by the circular motion of the many cutting edges of this gigantic drill

Maths window

Angular measure

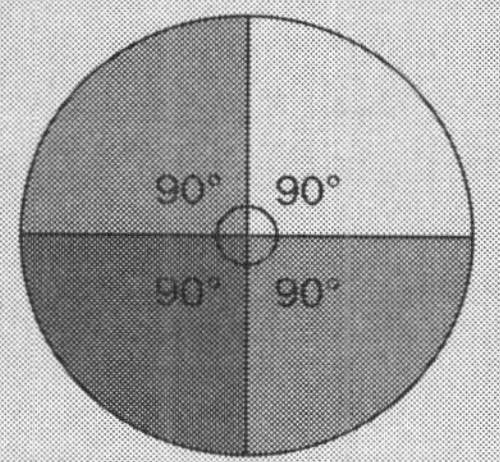

Figure 1.72 Angles in degrees

Conventionally, we divide a circle into an arbitrary number of *degrees* (360° in a circle) (see Figure 1.72). The angle between two lines can thus be expressed as a number of degrees. This is fine in most cases, but is inadequate for our purposes here.

Angular measure in radians

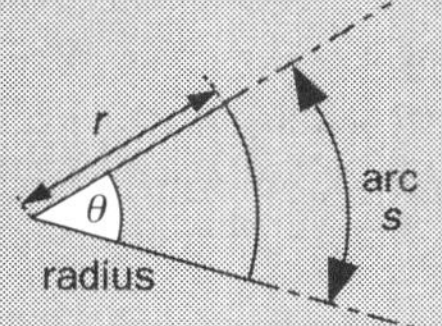

Figure 1.73 Radian measure

Where an angle is formed by two lines meeting at a point, a circular arc can be drawn with its centre on the point, and radius r (see Figure 1.73). If s is the *arc length* at radius r, then an angle in *radians* is defined by the equation:

$$\text{angle} = \frac{\text{arc length}}{\text{radius}} \qquad \theta = \frac{s}{r}$$

(s: m; r: m; θ: rad)

Equation 1

For a complete circle, the arc length (s) becomes equal to the circumference ($2\pi r$) of the circle:

$$\theta = \frac{s}{r} = \frac{2\pi r}{r} = 2\pi \text{ (rad)}$$

So in a complete circle (360°), or one revolution, there are 2π radians.

Therefore:

$$180° = \pi \text{ rad}; \quad 90° = \frac{\pi}{2}\text{rad}; \quad 45° = \frac{\pi}{4}\text{rad, etc.}$$

To convert degrees to radians: multiply by $\frac{\pi}{180}$

To convert radians to degrees: multiply by $\frac{180}{\pi}$

Thus 1 radian $= 1 \times \frac{180}{\pi} = 57.3°$

Angular velocity

Common symbol	ω
SI unit	radian per second (rad s^{-1} or simply s^{-1})

The motion of a particle rotating about a point can be can be expressed in terms of its:

- instantaneous **linear** speed or velocity
- number of **revolutions** (or orbits) per second (Hz or r.p.s.) or per minute (r.p.m.)
- **angular velocity** (sometimes called **angular frequency**)

A particle may be considered as a **point** on a rotating body, or as a **body** whose size is small compared to the radius of its orbit. For example, the earth could be considered as a particle in relation to its orbital radius around the sun.

> The **angular velocity** (ω) of a body is defined as the angle (θ) swept out per unit time (t) by the radius joining the position of the body to the centre of the circle along which it is moving.

Since 1 revolution is equal to 2π radians:

> To convert revolutions per second (r.p.s.) to rad s^{-1}: multiply by 2π
>
> To convert revolutions per minute (r.p.m.) to rad s^{-1}: multiply by $2\pi/60$

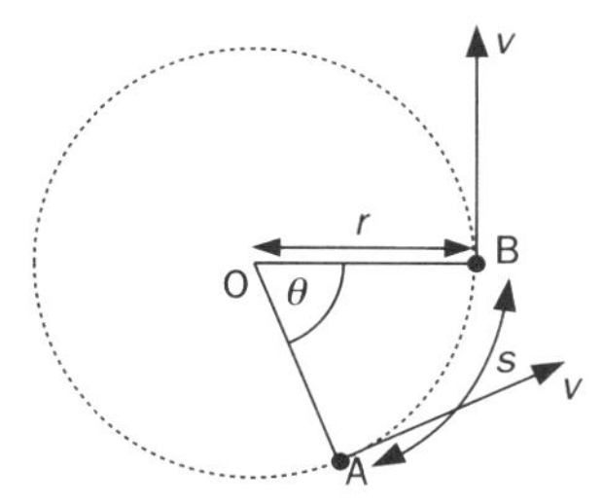

Figure 1.74 Angular velocity and linear speed

Figure 1.74 shows a particle moving uniformly a distance (s) from A to B along a circular arc, radius (r), in time (t).

The radius of the circle, OA, sweeps out an angle (θ) in this time, so the **angular velocity** (ω) is given by:

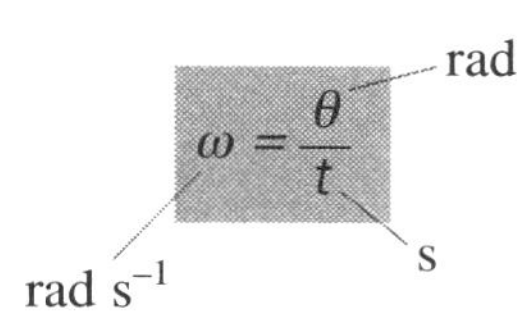

$$\omega = \frac{\theta}{t}$$

Equation 2

The **linear speed** (v) of the body is given by:

$$v = \frac{s}{t}$$

But from Equation 1:

$$s = r\theta$$

$$\therefore \quad v = \frac{r\theta}{t}$$

$$\therefore \quad v = r\omega$$

(v in m s^{-1}, r in m, ω in rad s^{-1})

Equation 3

Period and frequency

The term **period** means the time taken for 1 complete cycle of a repeated action, e.g. 1 complete wave to pass a given point.

In circular motion, the period T of the motion is **the time taken for one complete revolution** (i.e. the time taken for a radius to sweep through an angle of 360° or 2π radians). T is usually measured in seconds.

Since $\omega = \frac{\theta}{t}$, then for 1 complete revolution:

$$\omega = \frac{2\pi}{T}$$

(ω in rad s^{-1}, T in s)

Equation 4

Frequency is the number of complete cycles per unit time.

In circular motion, the frequency (f) of the motion is the number of complete revolutions per unit time. f is usually measured in revolutions per second (Hz).

An important relationship is:

$$T = \frac{1}{f}$$

(T in s, f in Hz)

Substituting in Equation 4 gives:

$$\omega = 2\pi f$$

(ω in rad s^{-1}, f in Hz)

Equation 5

Guided examples (1)

1. A racing car is travelling on a horizontal track which is part of a circle of radius 550 m at a constant speed of 198 km h^{-1}. Calculate its angular velocity.

Guidelines

Change speed to m s^{-1} and use Equation 3.

2. The drill-bit in a drilling machine has a radius of 0.5 cm and rotates at 2000 r.p.m. Calculate:
   (a) the angular velocity of the bit in rad s^{-1}
   (b) the linear velocity of a point on the periphery of the bit.

Guidelines

(a) ω (in rad s^{-1}) = $2\pi/60 \times$ (number of r.p.m.).
(b) Use Equation 3.

3. Calculate the angular velocity of the second, minute and hour hands of a clock.

Guidelines

Work out the time taken for each hand to make one complete revolution. This is the **period** of that hand.

4. A compact disc of radius 60 mm rotates at a speed of 100 r.p.m. Calculate:
   (a) its angular velocity in rad s^{-1}
   (b) the linear velocity of a point on the outer edge of the disc.

Guidelines

(Same as for guided Example 2.)

5. If the Earth's radius is 6.4×10^6 m and it takes 24 h to make one complete revolution, calculate the linear velocity of a point on the equator.

Guidelines

Remember T (period) must be in **seconds**. Use Equation 3.

6. If the radius of the Earth's orbit around the sun is 1.5×10^{11} m, and the period is 1 year, calculate
   (a) the orbital angular velocity
   (b) the orbital linear velocity.

Guidelines

(a) Use Equation 4 – remember T must be in seconds.
(b) Use Equation 3.

Centripetal acceleration and force

We saw earlier that a body moving along a circular path has an acceleration (and therefore a force) directed towards the centre of the circle. Figure 1.75 shows a particle of mass (m) moving with constant angular velocity (ω) along a circular path of radius (r). The particle has a constant linear speed (v) whose instantaneous direction is always along a tangent to the circular path.

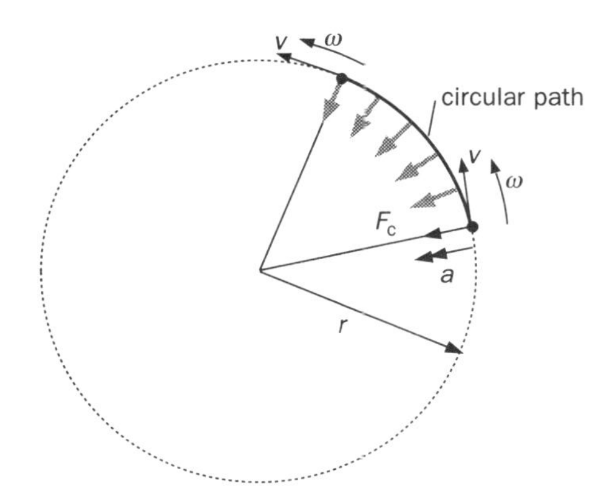

Figure 1.75 Centripetal acceleration and force

It can be shown that the **centripetal acceleration** (a) is given by:

$$a = \frac{v^2}{r}$$

(units: a in m s^{-2}, v in m s^{-1}, r in m)

Equation 6

Also, since $v = \omega r$:

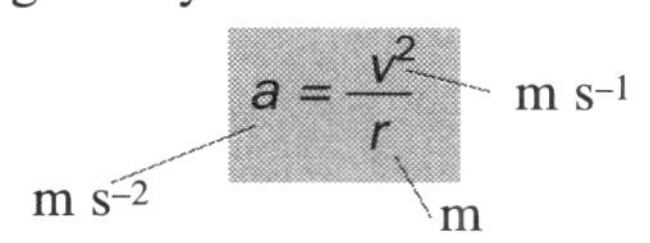

Equation 7

You should note that:

- for a body moving with uniform angular velocity, the centripetal acceleration is constant in **magnitude** but the **direction** varies continuously
- if the angular velocity is not uniform, there will be some **tangential acceleration** (i.e. an acceleration directed **along** the direction of motion) as well as the centripetal acceleration.

From Newton's second law, it follows that a resultant force must act on the particle to give it the centripetal acceleration needed to move in a circular path. If this centripetal force were to stop acting, the particle would (in the absence of other forces) continue in a straight line. This direction would be a tangent to the circle, from the point at which the force stopped acting.

Equations for centripetal force

Since $F = ma$:

Centripetal force F_c is given by:

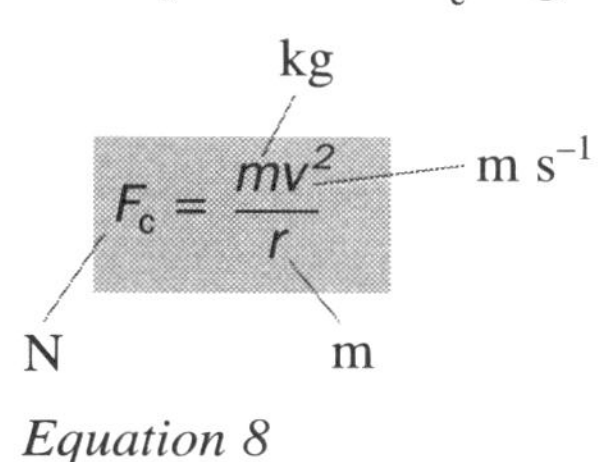

Equation 8

or

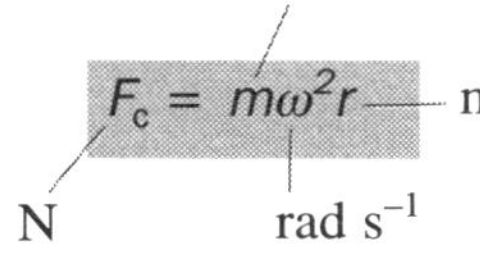

Equation 9

Table 1.9 shows some situations in which a centripetal force is necessary, and what provides that force.

Table 1.9 Centripetal force situations

Situation in which a resultant centripetal force acts	What provides the centripetal force
Earth orbiting the Sun	Gravitational force of attraction between the Earth and the Sun
An object on the Earth's surface	Force of Earth's gravity on the object (i.e. its **weight**)
Vehicle rounding a bend on a road	Friction force between the tyres and the road
Electron orbiting a nucleus	Electrostatic attraction force between the negatively-charged electron and the positively-charged nucleus

Centripetal force examples

1. Mass on a string

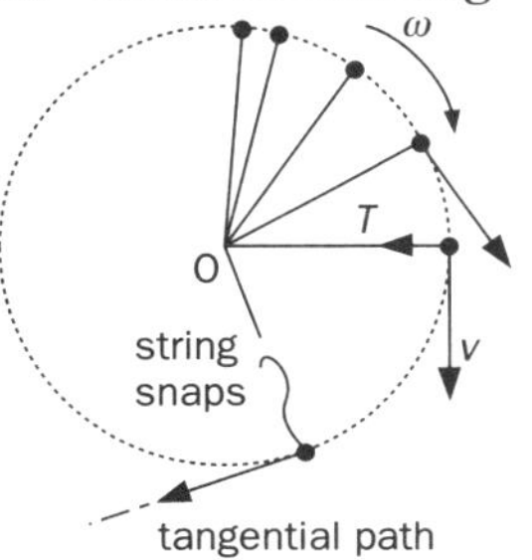

Figure 1.76 Mass on a string

If a small mass is attached to a string and whirled around in a horizontal circle as shown in Figure 1.76, the centripetal force (F_c) required to constantly re-direct its path into a circular arc is provided by the tension, (T), in the string, which must be directed towards the centre.

The magnitude of F_c (and therefore T) for a given length of string thus depends on the angular velocity of the mass (since $T = m\omega^2 r$).

If ω is gradually increased, a value is reached at which the string is not strong enough to provide this tension, and the string snaps. The mass continues to move in a straight line (in the absence of other forces), along a tangent to the circle as shown. It does not 'fly outwards from the centre' as is commonly thought!

2. Vehicle rounding a bend on a horizontal road Figure 1.77 shows a vehicle moving with constant linear speed (v) round a bend on a level road.

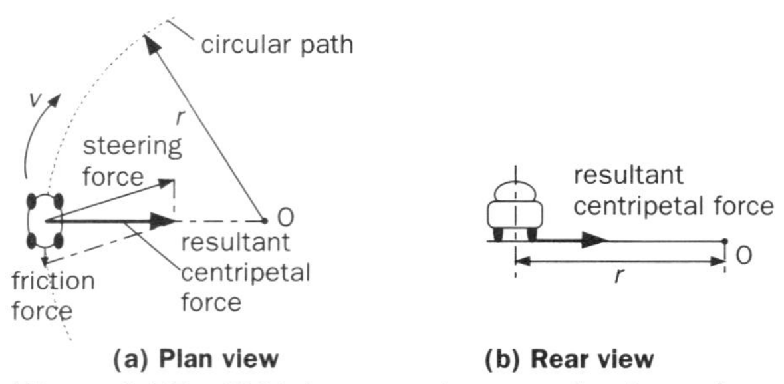

Figure 1.77 Vehicle cornering on a horizontal road

Since the vehicle must be in equilibrium in the vertical plane, we need only examine the horizontal direction. The **horizontal** forces acting on the vehicle are:

- the friction force, which is partly air resistance and partly road friction, which acts in the opposite direction to the motion
- the steering force produced by the front wheels which acts in the direction shown.

The resultant of these two forces provides the centripetal force towards the centre of the circular path of the vehicle. The magnitude of the steering force which can be exerted depends on the **friction** (grip) between the tyres and the road, so there is a maximum speed at which a vehicle can negotiate a given radius of bend safely. Another way of looking at it is to say that there is a minimum radius of curved path that a vehicle can follow at a given speed. If the bend has a smaller radius than this, the vehicle will run wide – literally 'running out of road'. On a dry surface, the maximum friction force between road and tyre is not increased by the tread on the tyre, hence the use of 'slicks' in racing. The tread is designed to channel water away from the surface of a wet road, so that the tyre is not lifted away from the road by a film of water – so called 'aquaplaning'.

Improvements in tyre design, suspension technology etc. all serve to increase this friction limit. In racing cars, aerodynamic forces press the car more firmly down on the track (without increasing weight) to increase the steering force. Wet roads reduce the friction available, hence the steering force is lower, and the maximum speed is reduced.

You should note that when solving problems involving vehicles rounding bends, an acceptable simplification can be achieved by ignoring the friction forces along the direction of motion. Therefore the steering force alone provides the centripetal force.

Banking Cornering speed can be greatly increased by the use of a 'banked' track.

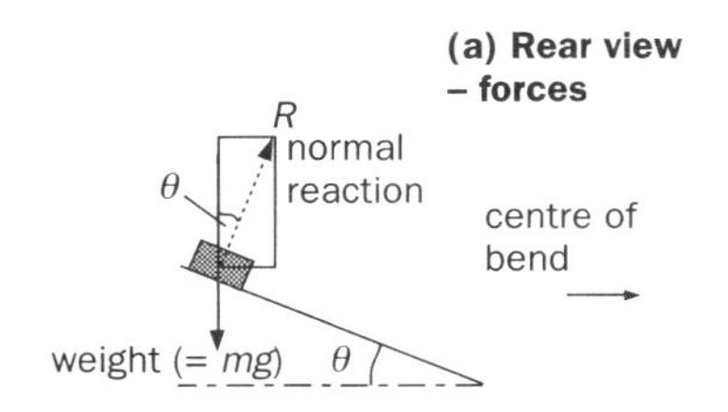

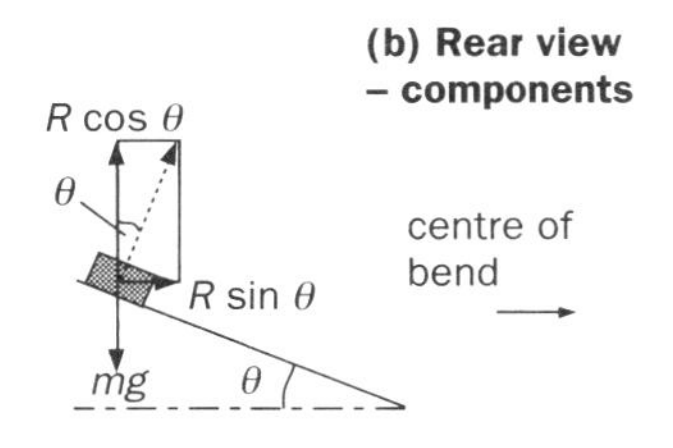

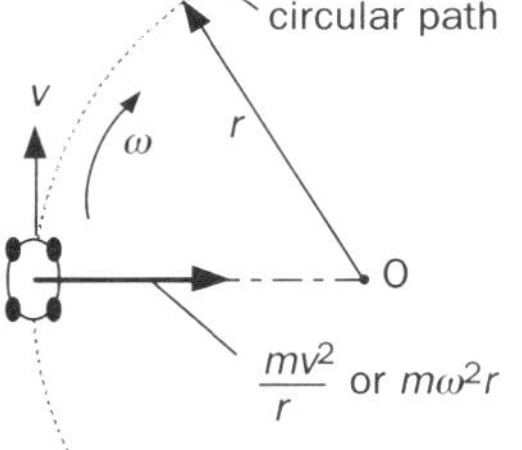

Mass acceleration diagram

Figure 1.78 Vehicle cornering on a banked track

As you can see from Figure 1.78(a) and 1.78(b), the reaction force (R) of the road on the vehicle is directed perpendicularly to the road. This can be resolved into vertical and horizontal components:

- the horizontal component $R \sin \theta$ provides the **centripetal** force necessary to make the vehicle follow the curved path
- the vertical component $R \cos \theta$ balances the weight of the vehicle.

So if a vehicle of mass (m) follows a circular path of radius (r) at a constant speed (v):

$$R \sin \theta = \frac{mv^2}{r} \quad \text{(a)}$$

$$R \cos \theta = mg \quad \text{(b)}$$

Dividing (a) by (b) gives:

$$\frac{R \sin \theta}{R \cos \theta} = \frac{mv^2/r}{mg}$$

$\therefore$

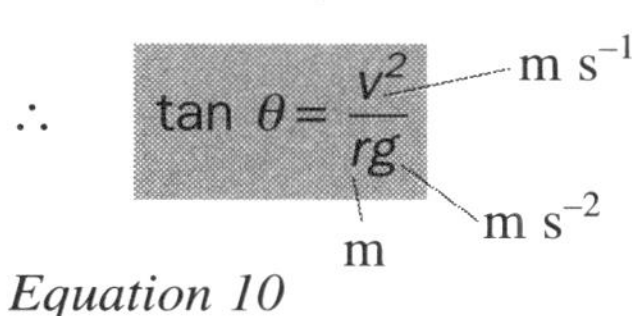

Equation 10

Equation 10 shows that for a given radius of bend (r), the banking angle (θ) is only correct for one value of velocity (v). At this speed, the centripetal force is provided entirely by the normal reaction of the road on the car, and no steering input is required. This is called the 'hands-off' speed, as the driver does not need to apply a steering force.

Car test tracks have a concave surface, so that the driver can select the required banking angle. At the outer edge of the track, the banking angle is large, and the 'hands-off' speed is high. At the inside edge, the banking angle is low, giving a low 'hands-off' speed.

Other examples of banking include:

- Railway tracks, which are tilted on bends to reduce the sideways force which the outer rail must exert on the flange of the wheel to turn the train.
- Banking of aircraft

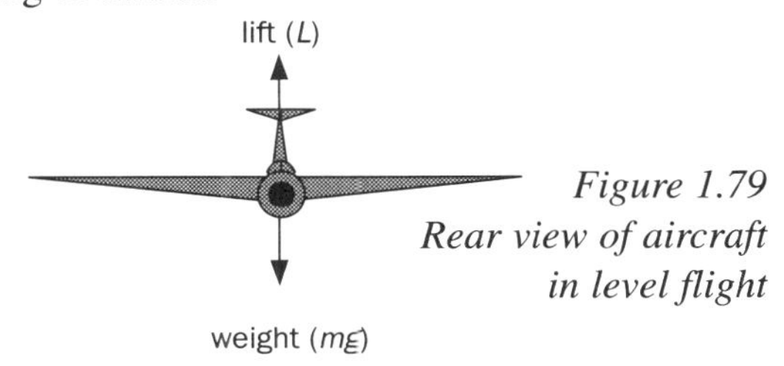

Figure 1.79 Rear view of aircraft in level flight

Figure 1.79 shows an aircraft in level flight. The **lift** force (L) developed by the wings (which acts at a right angle to the wingspan) must exactly balance the **weight** (mg).

Figure 1.80 Forces on an aircraft during a banked turn

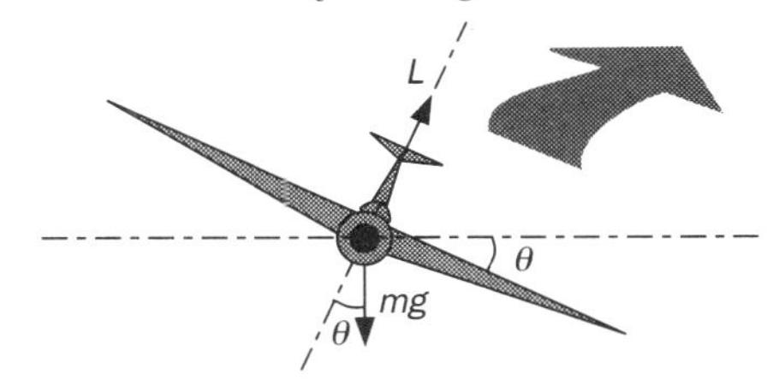

(a) Force diagram

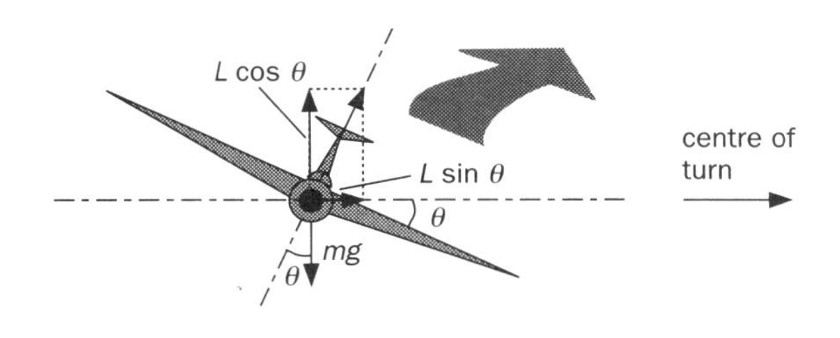

(b) Component diagram

In order to execute a properly balanced turn, the aircraft must be **banked**, as shown in Figure 1.80, so that the lift force can provide a horizontal component towards the centre of the turn.

Figure 1.80(b) shows the **lift** force resolved into vertical and horizontal components and Figure 1.81 is a plan view of the aircraft as it executes its turn.

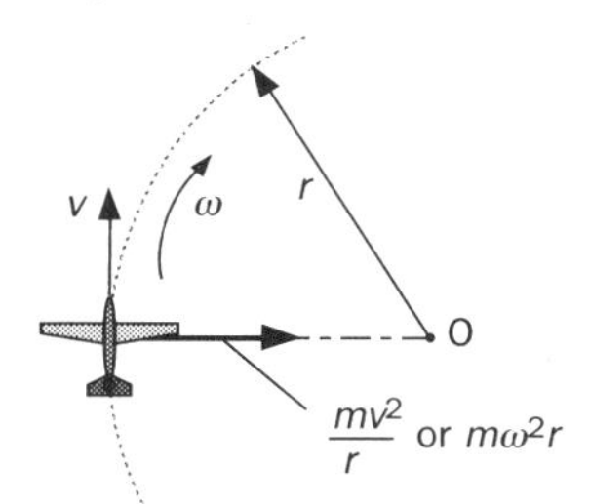

Figure 1.81 Mass–acceleration diagram for a turning aircraft

The same analysis as was used for the vehicle on a banked track yields the same result for the banking angle of a turning aircraft:

$$\tan \theta = \frac{v^2}{rg}$$

Thus the aircraft's angle of bank depends only on its speed and the radius of the turn. The mass of the aircraft has no effect.

As the aircraft banks, the weight must be balanced by the **vertical component** of the lift force produced by the wings. Clearly, as the angle of bank increases, this component will decrease. If the pilot takes no action to increase the force generated by the wings, this component will be less than the weight of the aircraft, and it will lose height. Maintaining height in a turning aircraft is a complex action involving lifting the nose of the aircraft (which increases the lift force by increasing the *angle of attack* of the wings), and increasing the engine power (to offset the increased drag produced by lifting the nose).

Motion in a vertical circle

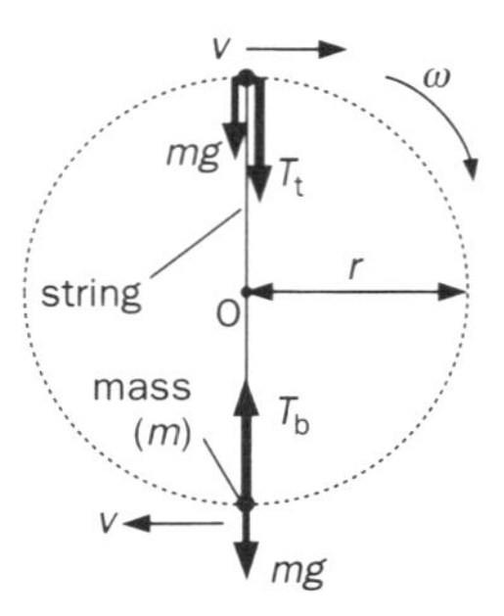

Figure 1.82 Vertical circular motion

Figure 1.82 shows a small mass *'m'* attached to the end of a light inextensible string and whirled in a vertical circle, radius *'r'*, at constant speed *'v'*. The forces acting on the mass at all times are the **weight** (mg) and the **tension** in the string. The two positions of interest are:

At the top of the circle

The tension (T_t) and the weight (mg) act in the same direction, so their resultant is ($T_t + mg$). This is the resultant force that provides the **centripetal acceleration**.

Newton II: (Resultant force = mass × acceleration)

$$\therefore \quad T_t + mg = \frac{mv^2}{r}$$

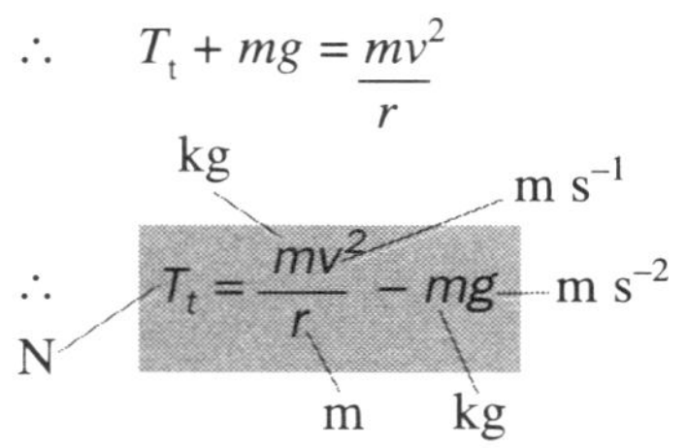

Equation 11

From this equation, it can be seen that if

$$\frac{mv^2}{r} \leq mg \quad (\text{i.e.} \frac{v^2}{r} \leq g)$$

then T_t will be zero, or become negative. This means that the string will become slack if (v^2/r) becomes less than (g).

At the bottom of the circle

T_b and mg act in opposite directions, so their resultant is given by ($T_b - mg$).

$$\therefore \quad T_b - mg = \frac{mv^2}{r}$$

$$\therefore \quad T_b = \frac{mv^2}{r} + mg$$

Equation 12

Comparing Equation 11 and Equation 12, we can see that the tension in the string is a **maximum** at the **bottom** of the circle and a **minimum** at the **top**.

The same argument applies to water in a bucket swung in a vertical circle, people in aircraft looping the loop, the 'Revolution' and 'Corkscrew' fairground rides, etc. In these cases, the string tension is replaced by the **reaction contact force** between the water and the bottom of the bucket, or between the seats and the passengers.

The paradox of circular motion

One of the difficulties with circular motion is that many of the findings **appear** to conflict with our everyday experience. In particular, when we are passengers in a vehicle rounding a bend, we appear to be **thrown outwards** (the so-called **centrifugal** force, which does not exist!).

In fact, what we are feeling is the force towards the centre exerted **on us by** the seat of the vehicle. Because we individually tend to regard ourselves as the centre of the universe, we interpret this as a force exerted **by** us **on** the seat. Apparent motion towards the outside of the car is in fact the side of the car moving towards us, as we try to continue in a straight line (Newton I).

Guided examples (2)

1. A pendulum bob of mass 250 g is attached to a light inextensible string of length 1.5 m. It is then whirled in a horizontal circle at a constant rate of 5 revolutions per second. Calculate the tension in the string.

   Guidelines

   Calculate the angular velocity in rad s^{-1}, then use Equation 9.

2. The same bob is now rotated in a vertical circle with the same constant angular velocity. Calculate the maximum and minimum tensions in the string, and state at which points on the circle they occur.

   Guidelines

   Use Equation 11 and Equation 12.

3. A space station creates its own artificial gravity in its outer ring by rotating about an axis through the centre. If the station has a radius of 120 m, calculate the angular velocity required to produce an artificial gravity equal to 75% of that on the Earth's surface, at the rim of the station. What precautions will have to be taken by an astronaut who steps out of the station at the rim on a 'spacewalk'?

Guided examples *continued*

Guidelines

The sensation of gravity is created by the rotation of the spacecraft. The acceleration due to the artificial gravity will be 75% of the Earth's gravity, i.e. 7.5 m s^{-2}. Calculate the angular velocity using Equation 7.

4. A fighter aircraft is flying vertically downwards at a speed of 210 m s^{-1} when the pilot pulls out of the dive and goes into a vertical climb without a change of speed. If the path of the aircraft is a vertical circle of radius 1200 m, draw a diagram showing the directions of all the forces acting **on the pilot** at the bottom of the circle, and calculate their magnitude. Calculate the centripetal acceleration of the pilot, and hence the *'g-force'*.

Guidelines

Use Equation 12 to calculate the centripetal force on the pilot.

'g-force' is a measure of the acceleration experienced as a multiple of the normal acceleration due to gravity. Thus an acceleration of 30 m s^{-2} would be equal to $3g$.

Self-assessment

SECTION A Qualitative assessment

1. Define the **radian**.

2. Explain why there must be a force acting on a particle which is moving with constant speed along a circular path. Name the force, and state in which direction it acts.

3. A particle moving with constant angular velocity (ω) travels a distance (s) along a circular path of radius (r) in a time (t). Show that the linear (tangential) velocity (v) of the particle is given by: $v = \omega r$.

4. (a) Define the terms
   (i) period (T)
   (ii) frequency (f).
   (b) State an expression which could be used to obtain angular velocity (ω) from:
   (i) frequency
   (ii) period.

5. State an expression for the centripetal force acting on a particle of mass (m) moving with constant speed (v) along a circular path of radius (r).

6. For a particle moving along a circular path, explain what is meant by the following statements:
   (a) the centripetal acceleration is constant in magnitude, but not direction
   (b) the centripetal acceleration should not be confused with any tangential acceleration.

7. State how the centripetal force is provided in the following cases:
   (a) a satellite in Earth orbit
   (b) an electron orbiting the nucleus of a hydrogen atom
   (c) the stone whirled around David's head in the sling when he slew Goliath
   (d) a car rounding a bend on a level road
   (e) an aircraft executing a turn in level flight
   (f) a bobsleigh in a high speed turn.

8. A small stone is tied to a string and whirled around so as to move in a circle at constant speed. For simplicity, we usually assume that the string is **horizontal**.
   (a) Draw a diagram showing the forces acting on the stone, assuming air resistance is negligible. Use the diagram to explain:
   (i) why the string cannot really be horizontal
   (ii) the direction of the **resultant** force on the stone.
   (b) If the stone is now whirled in a **vertical** circle, and the speed is gradually increased, state and explain at which points the string is
   (i) most likely to snap
   (ii) least likely to snap.

1.9 Rotational motion of rigid bodies

In our study of circular motion so far, we have considered only objects which can be assumed to be particles. This assumption is valid if the body's dimensions are small compared to the radius of their circular paths. In the case of small stones and pendulum bobs, this is easy to visualise, but even a planet travelling round the sun can be considered as a particle, due to the large radius of its orbit. However, when we come to deal with the forces and energies involved in the rotation of a body on its own axis, such as the spin of a planet, then this assumption cannot be applied. In rotational dynamics, we consider the rotation about specified axes of large bodies whose mass is distributed over a considerable volume, and whose dimensions are not small compared to the radius of rotation. Car road wheels, engine flywheels, propellers, roundabouts and spinning planets are all examples of objects which rotate about axes through their centres.

We will apply the principles used previously in our study of linear dynamics, and adapt them for rotational motion. Table 1.10 shows the link between the quantities in linear and rotational dynamics.

A working traction engine is a gleaming mass of rotating whells and axles

Table 1.10 Analogous quantities in linear and rotational motion

Linear (translational) motion			Angular (rotational) motion		
Quantity	**Symbol/ equation**	**Unit**	**Quantity**	**Symbol/ equation**	**Unit**
displacement	s	m	angular displacement	θ	rad
velocity	$v = \left\{\frac{\Delta s}{\Delta t}\right\}$	m s^{-1}	angular velocity	$\omega = \left\{\frac{\Delta \theta}{\Delta t}\right\}$	rad s^{-1}
acceleration	$a = \left\{\frac{\Delta v}{\Delta t}\right\}$	m s^{-2}	angular acceleration	$\alpha = \left\{\frac{\Delta \omega}{\Delta t}\right\}$	rad s^{-2}
mass	m	kg	moment of inertia	I	kg m^2
force	$F\ (= ma)$	N	torque	$T\ (= I\alpha)$	N m
momentum	$p\ (= mv)$	kg m s^{-1}	angular momentum	$L\ (= I\omega)$	kg m^2 s^{-1}
kinetic energy	$E_k\ (= \frac{1}{2}mv^2)$	J	rotational kinetic energy	$E_k\ (= \frac{1}{2}I\omega^2)$	J

Angular displacement, velocity and acceleration

Figure 1.83 shows a rigid body rotating with uniform angular velocity (ω) about an axis through O.

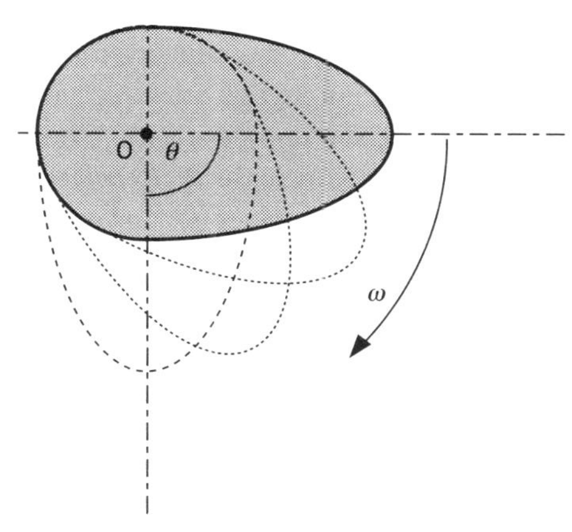

Figure 1.83 Rigid body rotating with uniform angular velocity

If the body undergoes an **angular displacement** (θ) in time (t), then its average angular velocity (ω) is given by:

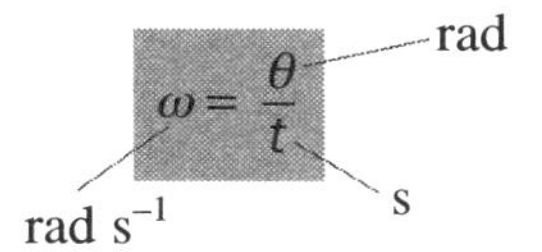

$$\omega = \frac{\theta}{t}$$

Equation 1

If the body now starts to increase its angular velocity, then it will have an angular acceleration (α) defined by:

$$\text{acceleration} = \frac{\text{change of angular velocity}}{\text{time taken}}$$

If the angular velocity increases from some initial value (ω_i) to a final value (ω_f) in time (t), then the angular acceleration (α) is given by:

$$\alpha = \frac{\omega_f - \omega_i}{t}$$

From which we obtain:

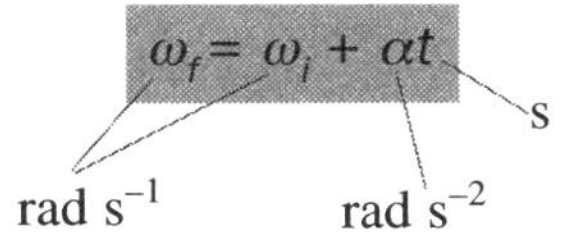

$$\omega_f = \omega_i + \alpha t$$

Equation 2

which is exactly analogous to the linear motion equation: $v = u + at$.

If the body undergoes an angular displacement (θ) during the acceleration, then the complete set of rotational dynamics equations can be obtained by analogy with the linear equations of motion for a body with acceleration (a) which increases its velocity from an initial value (u) to a final value (v) when its displacement is (s) during a time interval (t). These equations are shown in Table 1.11.

Table 1.11 Comparison of rotational and linear dynamics equations

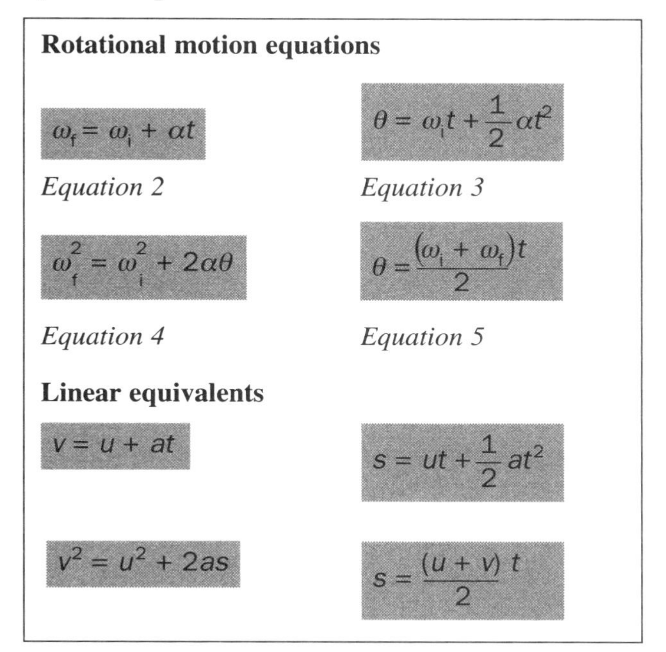

Rotational motion equations

$\omega_f = \omega_i + \alpha t$	$\theta = \omega_i t + \frac{1}{2}\alpha t^2$
Equation 2	*Equation 3*
$\omega_f^2 = \omega_i^2 + 2\alpha\theta$	$\theta = \frac{(\omega_i + \omega_f)t}{2}$
Equation 4	*Equation 5*

Linear equivalents

$v = u + at$	$s = ut + \frac{1}{2}at^2$
$v^2 = u^2 + 2as$	$s = \frac{(u + v)\,t}{2}$

Torque and angular acceleration

In our study of linear dynamics, we found that a body which is subjected to a constant **resultant force** experiences a constant **acceleration**. The rotational equivalent of this is the **angular acceleration** given to a body by application of a **resultant torque**. Lawnmower engines are commonly started by wrapping a cord several times round a pulley and pulling the cord. This exerts a torque on the engine, which starts to rotate from rest until it reaches a sufficient angular velocity for it to start and run. Figure 1.84 shows such an arrangement where a constant force (F) is applied to the pulley of radius (r), and produces an angular acceleration (α):

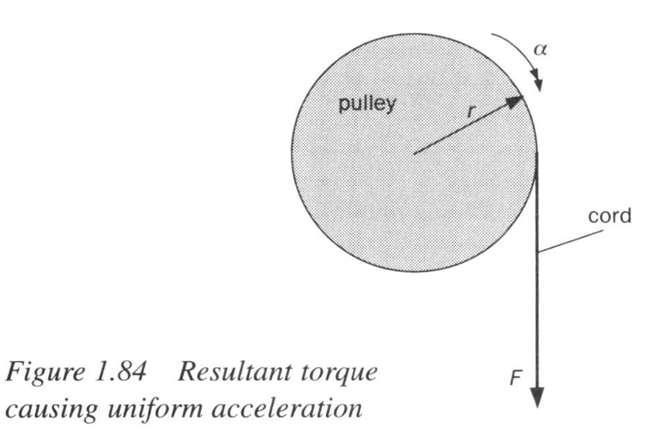

Figure 1.84 Resultant torque causing uniform acceleration

If friction effects are ignored, and F is the only force, then the resultant torque (T) is given by:

$$T = Fr$$

In linear dynamics, it is found by experiment that the linear acceleration (a) of a body is directly proportional to the resultant force (F) on the body. The constant of proportionality in the equation is the **mass** (m) of the body, which leads to the equation:

$F = ma$ (which comes from Newton's second law)

If the angular acceleration of a rotating body is measured when various constant torques (T) are applied, then the **angular acceleration** (α) is found to be proportional to the **resultant torque.**

However, in this case, the constant of proportionality is not just the mass of the body. Experiments with bodies of various masses and shapes show that the angular acceleration of a body subjected to a constant torque depends not only on the magnitude of the torque and the mass of the body, but also on its **shape**, and the position of the **axis of rotation**.

For a particular body, with a particular axis of rotation, Newton's second law equation becomes:

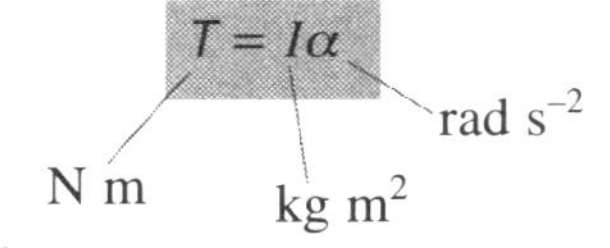

Equation 6

where I is a constant for a particular body for a specific axis of rotation, and is known as the **moment of inertia** of the body.

Comparing: $T = I\alpha$ (rotational dynamics)
with $F = ma$ (linear dynamics)

leads to the conclusion that **moment of inertia** (I) in the rotational equation is analogous to the mass (m) in the linear version. You are already familiar with the concept of mass as a measure of **inertia**, which can be expressed as the opposition of a body to acceleration, so in rotational dynamics:

> The **moment of inertia** (I) of a rigid rotating body can be expressed as a measure of its opposition to angular acceleration.

Units of I

From Equation 7, we obtain:

$$I = \frac{T}{\alpha} \text{ so the units of } I \text{ are } \frac{\text{N m}}{\text{rad s}^{-2}}$$

The newton can be expressed in its base units as kg m s^{-2};the radian has no base unit, as it is simply a ratio, so the moment of inertia can be expressed in its base units as kg m^2.

The moment of inertia depends not only on the mass of the body, but also on the way that mass is distributed about the axis of rotation. In general, shapes where the mass is concentrated close to the axis of rotation have low moments of inertia. Rally cars need to be able to change direction very quickly. Changing the direction of a car means exerting a torque about a vertical axis through the centre of the car (by turning the front wheels). To make the car responsive to steering inputs, the moment of inertia about this axis must be as low as possible. One of the most effective ways to achieve this is to mount the heavy items like the engine and gearbox near the centre of the car. The mid-engine layout is one of the classic sports and racing car designs.

Rotational kinetic energy and momentum

The energy possessed by a rotating body due to its rotation is known as the **rotational kinetic energy** to distinguish it from the kinetic energy which the body may also possess as a result of its translational motion. Many of the rotating objects in our everyday experience are both rotating and translating, such as bicycle wheels, bowling balls, etc., and so possess both rotational and translational kinetic energy. Figure 1.85 shows a bowling ball of mass (m) moving with linear velocity (v) and angular velocity (ω).

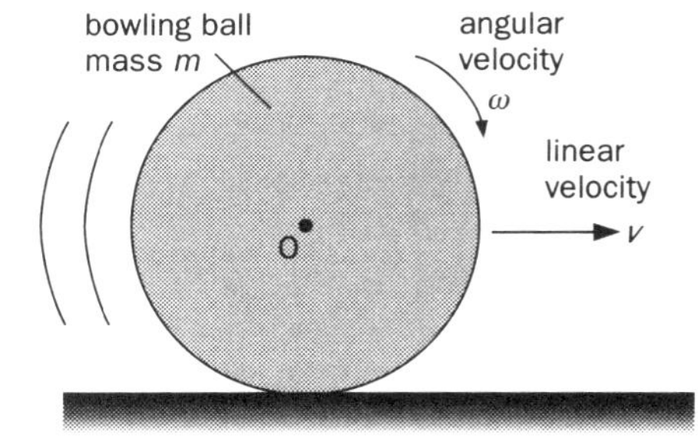

Figure 1.85 Bowling ball

If the moment of inertia of the bowling ball about an axis through its centre is (I), then by analogy with the equation for the translational kinetic energy, the rotational kinetic energy is given by:

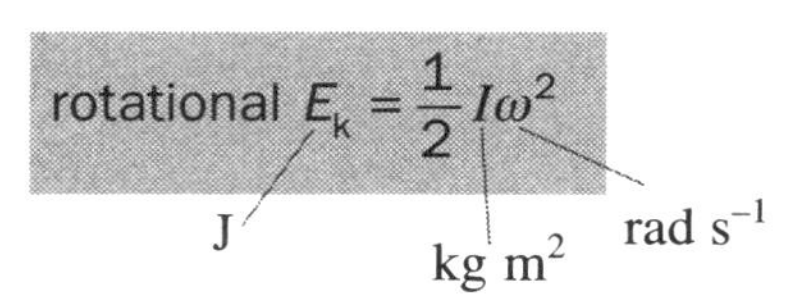

Equation 7

The ball will also have **angular momentum** due to its angular velocity in addition to its linear momentum.

The angular momentum (L) is given by:

$$L = I\omega$$

(L in kg m^2 s^{-1}; I in kg m^2; ω in rad s^{-1})

Equation 9

A spinning bowling ball causes more disruption to the pins than one which arrives at the same linear velocity, but still sliding on the floor rather than rolling. In cricket, the bowler can give the ball angular velocity in various directions. The angular momentum possessed by the ball can be used to change its direction of flight when it bounces.

Conservation of angular momentum

We have previously stated the principle of conservation of momentum, and applied it to linear motion. The principle can be extended to rotational motion:

> The principle of conservation of angular momentum states that the total angular momentum of a system of bodies is constant, provided no external torques act.

Examples of angular momentum conservation

A spinning ice skater can speed up or slow down by merely moving her arms closer to or further from her body. By drawing her arms in, she is reducing her moment of inertia (I), and by holding them out, she increases it. Since there is no external torque acting, the angular momentum remains constant, so reducing I produces a proportional increase in angular velocity (ω) and vice versa. This principle is illustrated in Figure 1.86.

You should also note that the skater has to do work to bring her arms closer in to her body. This work increases her rotational kinetic energy.

High divers, gymnasts and trampolinists use this principle when performing somersaults about a horizontal axis through their centre of gravity. Their angular velocity can be increased by adopting the 'tuck' position, which reduces the body's moment of inertia about this axis to the minimum, so they complete a greater number of rotations during their flight. For landing, the athlete extends his body again, increasing the moment of inertia, and reducing the angular velocity.

Figure 1.86 Spinning ice skater

A spinning ice skater uses the principle of conservation of angular momentum to increase or decrease the spin speed

Guided examples (1)

1. The wheels on an accelerating car increase their angular velocity from 150 r.p.m. to 750 r.p.m. in a 10 s interval. Calculate:
   (a) the angular acceleration of the wheels
   (b) their angular displacement (in radian) during the 10 s interval
   (c) the number of rotations of the wheels in this time.

Guidelines

(a) Convert angular velocities to rad s^{-1} then use Equation 2.
(b) Use Equation 5.
(c) Convert the angular displacement obtained in (b) to rotations (revolutions).

2. A flywheel has a moment of inertia 12 kg m^2 about a horizontal axis through its centre. There is a constant frictional torque of 15 N m which opposes the motion.
   (a) If an external torque is applied to increase the flywheel's angular velocity from 0 to 75 rad s^{-1} in 10 s, calculate
      (i) the magnitude of the applied torque
      (ii) the angular momentum after 10 s
      (iii) the rotational kinetic energy after this time.
   (b) If the external torque is now removed, calculate:
      (i) the time taken for the flywheel to come to rest
      (ii) the angular velocity after 20 rotations of the flywheel.

Guidelines

(a) Use Equation 2 to calculate the angular acceleration, then Equation 6 to calculate the resultant torque. Use (resultant torque = external torque – frictional torque) to calculate the external torque, then use Equation 8 to calculate the angular momentum and Equation 7 to calculate the rotational kinetic energy.
(b) When the external torque is removed, the friction torque will cause a steady deceleration.
   (i) Use Equation 6 to calculate this deceleration, and hence calculate the time taken to come to rest.
   (ii) Use Equation 2 to calculate the angular velocity.

3. A playground roundabout of radius 3.0 m has a moment of inertia of 1.8×10^3 kg m^2, and rotates with an angular velocity of 2.4 rad s^{-1} when there are four children standing close to the rim. The children all move as close to the centre as possible, and the moment of inertia decreases to 8.0×10^2 kg m^2. Assuming that friction can be ignored, calculate:
   (a) the new angular velocity of the roundabout
   (b) the initial and final kinetic energies of the system.

Guidelines

(a) Use the principle of conservation of angular momentum.
(b) Use Equation 7.

Self-assessment

SECTION A
Qualitative assessment

1. Define
(a) angular velocity
(b) angular acceleration
State the unit for each quantity.

2. State the equivalent rotational motion equation for each of the following linear motion equations (symbols as defined in the text).
(a) $v = u + at$
(b) $s = ut + \frac{1}{2}at^2$
(c) $v^2 = u^2 + 2as$

3. Define what is meant by the **moment of inertia** of a rigid body, and explain why most bodies do not have a unique moment of inertia.

4. A constant tangential force (F) is applied to the rim of a wheel of radius (r) and moment of inertia (I) giving it a uniform angular acceleration (α).
(a) State an expression for the torque on the wheel (T) in terms of the force (F) and the radius (r).
(b) State the relationship between T and α.

5. A ball which is spinning with constant **angular** velocity (ω) about an axis through its centre has a moment of inertia (I). State the expressions for
(a) the rotational kinetic energy of the ball
(b) the angular momentum of the ball.

6. State the principle of conservation of **angular** momentum and use it to explain:
(a) how an ice skater can change her spin speed by a simple arm movement
(b) why it is very difficult to stay upright on a stationary bicycle, and very easy once the wheels are spinning.

SECTION B
Quantitative assessment

(***Answers:*** 2.4×10^{-9}; 1.0×10^{-7}; 1.1×10^{-7}; 8.3×10^{-7}; 1.5×10^{-5}; -0.16; 0.52; 0.56; 2.6; 3.5; 6.3; 11; 15; 32; 55; 94; 160; 160; 240; 370; 570; -1.6×10^{3}; 2.5×10^{3}; 4.0×10^{3}; 4.0×10^{3}; 5.4×10^{3}; 6.8×10^{3}; 7.6×10^{3}; 4.8×10^{4}; 5.6×10^{4}; 8.2×10^{4}; 3.3×10^{5}; 5.7×10^{7}; 1.1×10^{15}; 1.4×10^{21}; 9.6×10^{21}; 1.6×10^{25}.)

1. A long playing record of diameter 300 mm is played at $33\frac{1}{3}$ rev. min^{-1}. Calculate:
(a) the angular velocity in rad s^{-1}
(b) the linear velocity of a point on the rim.

2. The minute hand on a watch is 8.5 mm long. Calculate
(a) the time taken to move through 1 radian
(b) the linear speed of the tip of the hand.

3. A proton enters a uniform magnetic field at right angles to its path, and as a result moves with a constant speed of 2×10^7m s^{-1} in a circular path of radius 350 mm. Calculate for the proton:
(a) the time taken for 1 orbit
(b) the angular velocity
(c) the centripetal acceleration.

4. The rotation rate of the blades in a food mixer is 3500 rev. min^{-1}. If the blade length is 6 cm calculate:
(a) the angular velocity of the blades in rad s^{-1}
(b) the linear velocity of a point 3 cm from the centre on one of the blades and
(c) the centripetal acceleration of a point on the tip of a blade.

5. See table below. Assuming that the orbit of each planet around the sun is circular, use the astronomical data below to calculate for each of the plantes:
(a) the angular velocity in rad s^{-1}
(b) the linear velocity in m s^{-1}
(c) the centripetal force.

Planetary body	Distance from sun (m)	Mass (kg)	Period of revolution around sun (s)
Mercury	0.58×10^{11}	0.24×10^{24}	7.57×10^{6}
Jupiter	7.79×10^{11}	1.90×10^{27}	5.99×10^{7}
Uranus	2.87×10^{12}	8.97×10^{25}	2.65×10^{9}

Self-assessment *continued*

6. An astronaut of mass 86 kg is strapped into a seat at the end of a 6.5 m long horizontal arm. In tests, the arm is rotated with gradually increasing angular velocity. Calculate the maximum rotation rate in rev. s^{-1} which the astronaut can withstand if he is found to black out when the force exerted on him by the seat is equal to 6900 N.

7. (a) A fully-laden lorry of total mass 12 000 kg enters a horizontal circular bend of radius 350 m at a constant speed of 45 km h^{-1}. Calculate the magnitude and direction of the resultant horizontal force acting on the lorry as it travels round the bend.
   (b) Calculate the angle to which the road in (a) must be banked in order to provide the same resultant force without the need to rely on friction.

8. A passenger aircraft of total mass 29 000 kg is cruising at a constant speed of 210 m s^{-1} when the pilot executes a correctly banked horizontal turn at an angle of 30°. Calculate
   (a) the lift-force on the wings
   (b) the radius of the circular path in which the aircraft flies.

9. A small aircraft of mass 1200 kg which is flying horizontally at 150 m s^{-1} is taken into a vertical circular path of radius 500 m and it is kept in this path until the aircraft makes a complete loop. If the pilot has a mass of 72 kg, calculate the force exerted on him by the seat when the aircraft is at the top and bottom of the loop.

10. The wheel of a large dumper truck has a moment of inertia of 10 000 kg m^2 and is being tested by rotating it at 60 r.p.m. If it is brought to rest in 40 s by a constant frictional torque, calculate:
    (a) the initial angular velocity in rad s^{-1}
    (b) the angular acceleration
    (c) the angular displacement during the first 20 s^{-1}
    (d) the magnitude of the applied torque.

11. At a particular instant during a race, the wheels on a racing car are rotating at 1500 r.p.m. Each wheel and tyre has a mass of 25 kg, radius 0.35 m, and moment of inertia 1.5 kg m^2. Calculate:
    (a) the angular velocity of the wheels in rad s
    (b) the translational velocity of the car
    (c) the total kinetic energy of a wheel
    (d) the angular momentum of each wheel.

12. A skater rotates with an angular velocity of 3.0 rad s^{-1} about a vertical axis with both arms and one leg outstretched. When she is in this position, her moment of inertia about the vertical axis of rotation is 7.0 kg m^2. If she draws in her arms and leg and as a result her moment of inertia decreases to 1.4 kg m^2, calculate
    (a) the new angular velocity
    (b) the rotational kinetic energy
        (i) with arms and leg outstretched
        (ii) with arms and leg drawn in
    (c) Account for the difference in rotational kinetic energy between the two positions.

1.10 Gravitation

We are born, and live our lives, under the influence of the Earth's gravitational pull. Every upward movement we make involves energy dissipation as a result of the work done against gravity's constant opposition. Even asleep our hearts continue to do work by pumping blood against the force of gravity. To the pilot, climber or bungee-jumper, the pull of gravity is a necessity which sometimes, through miscalculation, can become a fatal attraction.

Surprisingly, little is known about the fundamental origins of the force field we call gravity. However, we may not fully understand what causes it, or how it is transmitted, but, thanks to Isaac Newton, the mechanics of gravitational attraction are well enough known to safely land astronauts on the moon, place communications satellites in precise locations and send unmanned probes to the far corners of the solar system.

Newton's theories quantify diverse gravitational events, from falling apples to the complexity of planetary motion. His great insight was to realise that these different phenomena were linked by a single cause – gravity. His realisation that every mass attracts every other mass with a predictable force gave his law of gravity, published in 1666 at the ripe old age of 24, a truly universal status.

Newton's law of gravitational attraction

> Every particle in the universe attracts every other particle with a force which is directly proportional to the product of their masses, and inversely proportional to the square of the distance between them.

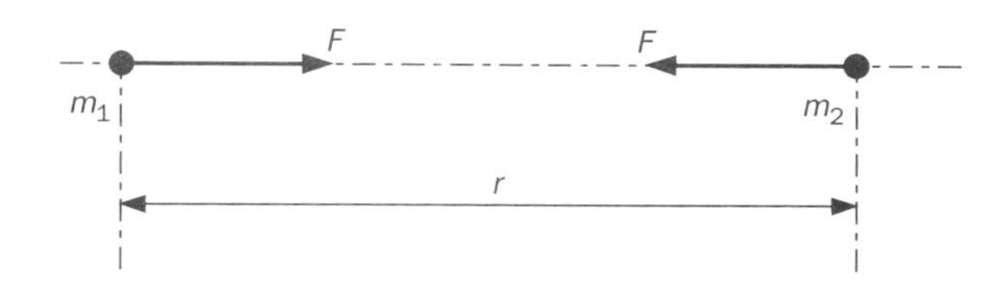

Figure 1.87 Newton's law of universal gravitation

One of the earliest weather satellites which was launched in the late 1960s and paved the way for the more reliable long-range forecasting of today

Figure 1.87 shows two point masses (m_1 and m_2) whose centres are a distance (r) apart. The gravitational attraction force (F) between them is given by (from Newton's law above):

$$F \propto m_1 m_2 \quad \text{and} \quad F \propto \frac{1}{r^2}$$

$$\therefore \quad F \propto \frac{m_1 m_2}{r^2}$$

Inserting a constant of proportionality turns this into an equation which expresses Newton's law:

$$F = -\frac{Gm_1m_2}{r^2}$$

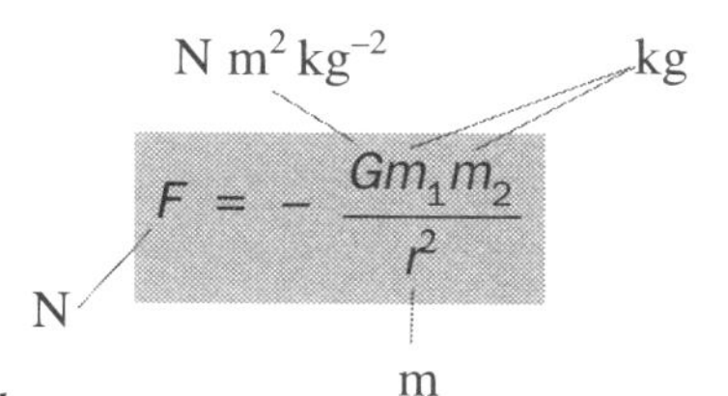

Equation 1

You should note that:

- G is called the **universal gravitational constant**, and has the value $G = 6.67 \times 10^{-11}$ N m^2 kg^{-2}.
- The **minus sign** in the equation is there since it is conventional in field theory to regard forces exerted by **attractive** fields as **negative**, and gravity is attractive everywhere in the universe.
- The force exerted on a mass by Earth's gravity is called its **weight**, and acts in a direction towards the centre of the Earth. The direction of this force defines what we call **vertical**.

- The gravitational field acts at a distance, and without the need for any medium (the planets and stars attract each other across millions of kilometres of empty space). Gravity is unaffected by the presence of an intervening medium – there is no 'shield' against it.
- Newton's law is expressed in terms of particles. For real bodies, the law can be applied by assuming all the mass of a body to be concentrated at its centre of mass. The distance between bodies should be measured between their centres of mass as shown in Figure 1.88.

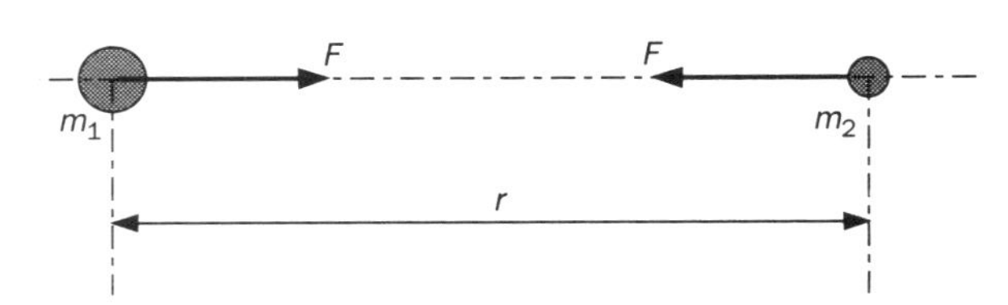

Figure 1.88 Newton's Law of gravity for spherical bodies

- The gravitational attraction force only becomes significant if at least one of the bodies has a very large mass.

Guided example (1)

Calculate the gravitational attraction force between the following pairs of bodies. Assume that the universal gravitational constant (G) has the value 6.7×10^{-11} N m^2 kg^{-2}.

(a) A man of mass 75 kg standing on the surface of the Earth (mass 6.0×10^{24} kg), assuming the radius of the Earth is 6.4×10^6 m.

Guided example (1) continued

(b) The same man at a point 12.8×10^6 m from the centre of the Earth.

(c) The Moon, of mass 7.4×10^{22} kg and the Earth, if the distance between their centres is 3.8×10^8 m.

(d) Two spacecraft, of masses 2500 kg and 3000 kg, when their centres of mass are 10 m apart.

(e) Two protons, each of mass 1.7×10^{-27} kg, whose centres are 1.0×10^{-15} m apart.

Guidelines

Simply apply Equation 1, remembering that r is squared, and taking great care with powers of 10. Note that only where planets are involved does

Inverse square law

Equation 1 (and a bit of common sense), tells us that the force of attraction between two objects **decreases** as the distance between them **increases**. This is an example of an **inverse** relationship. In this case, the force due to gravitational attraction is inversely proportional to the **square** of the distance between the bodies. In words, this means that if the distance is increased by a factor of **two**, the force becomes **four** times smaller; increasing the distance **three** times, produces a force **nine** times smaller, and so on. The forces due to electric fields, and the intensity of light, sound, gamma radiation etc. all obey the same kind of relationship, which we refer to as an **inverse square law**. The following guided example should help to clarify this.

Guided example (2)

The diagram shows a spacecraft of mass 3000 kg at various distances from the Earth, corresponding to R, $2R$, $4R$ and $8R$, where R is the radius of the Earth (6.4×10^6 m). Calculate the gravitational force on the spacecraft on the Earth's surface, assuming the mass of the Earth to be 6.0×10^{24} kg, and G to have a value of 6.7×10^{-11} N m^2 kg^{-2}.

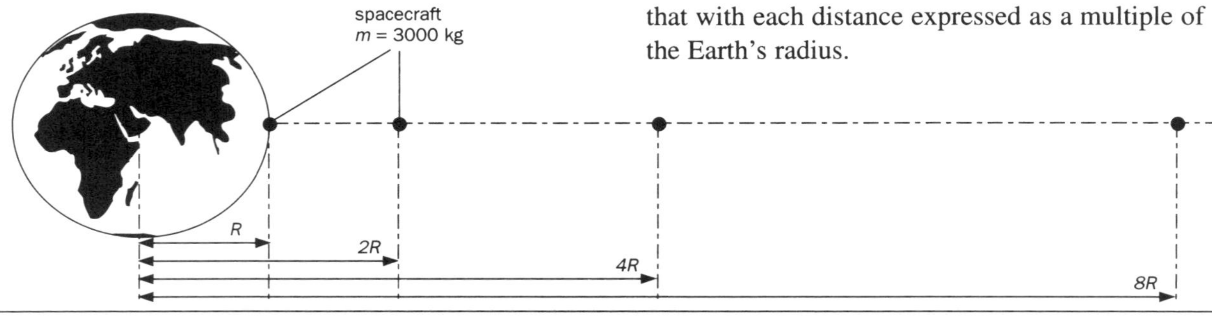

Calculate the force on the spacecraft at each position shown, and express these forces as **fractions** of the force at the Earth's surface. Do your answers support the inverse square law of gravitation?

Guidelines

Use Equation 1 for each calculation. Work out each force as a fraction of that at the surface. Compare that with each distance expressed as a multiple of the Earth's radius.

Satellite orbits

Any body orbiting a planet is a **satellite** of that planet. Our Moon, for example, is a natural satellite of the planet Earth, while Earth itself is a satellite of the Sun. Some planets have many natural satellites, and even the particles which make up the rings of planets like Saturn can be thought of as satellites of their planet. Artificial satellites in Earth orbits are becoming increasingly numerous, providing communication links between ground stations, observational data on the weather, land use, military activity, etc. and navigation information. These satellites maintain their orbits due to the gravitational attraction between themselves and the Earth, at sufficient heights to escape atmospheric friction that would dissipate their energy and send them crashing back to the Earth's surface.

The orbits of many of these artificial satellites are as circular as possible, giving them a reasonably constant height above the Earth's surface, but most natural orbits are elliptical to a greater or lesser degree. (A circle is simply a special case of an ellipse.) Halley's comet is a good example of a body orbiting the Sun with a highly elliptical orbit. Its closest distance of approach to the Sun is only 0.59 AU*, while at its furthest point it is 35 AU away, which is 59 times greater. In this section we will examine the factors which control satellite orbits, making the simplifying assumption that orbits are circular.

Figure 1.89 shows the Earth in orbit around the Sun, but could also apply to any satellite orbiting a star or planet.

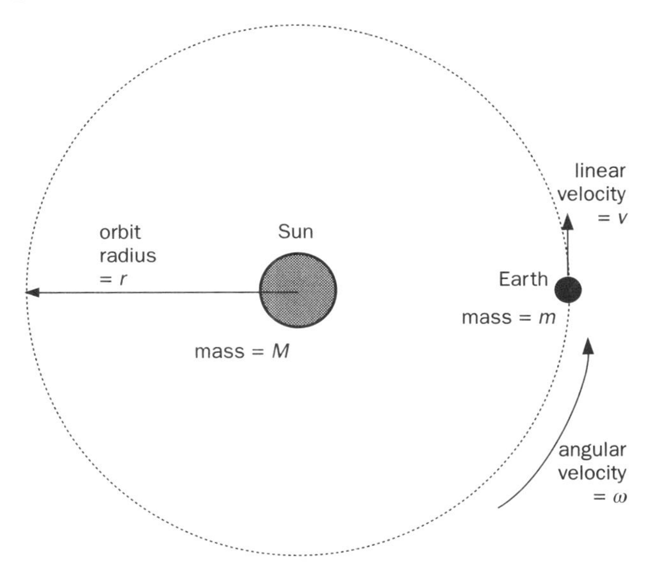

Figure 1.89 The Earth as a satellite of the Sun

* AU is the abbreviation for the **astonomical unit**
1 AU = 1.496×10^{11} m.

Newton's second law tells us that the **resultant** force on the Earth is equal to its **mass** multiplied by its **acceleration**. Since the only force acting is the **gravitational attraction** to the Sun, and its only acceleration is the **centripetal acceleration**, we have:

$$\frac{GMm}{r^2} = \frac{mv^2}{r}$$

Re-arranging gives:

$$M = \frac{rv^2}{G} \quad \text{(i)}$$

so the velocity (v) of a satellite in a circular orbit of radius (r) about a planet of mass (M) is given by:

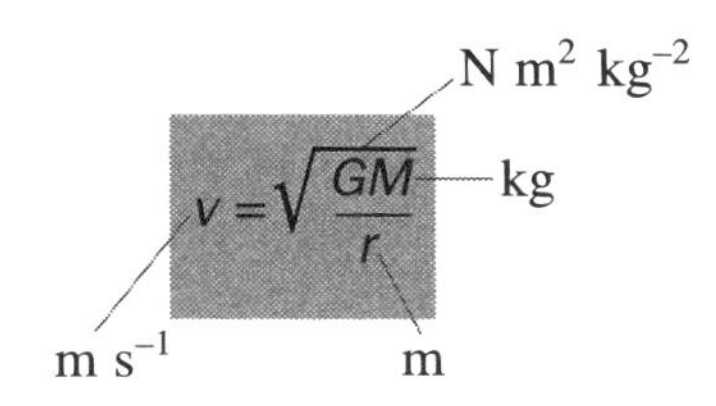

Equation 2

The distance travelled during one orbit is $2\pi r$, so the **period** (T) is given by:

$$T = \frac{2\pi r}{v} \quad \left\{\text{since time} = \frac{\text{distance}}{\text{speed}}\right\}$$

$$\therefore \quad v = \frac{2\pi r}{T} \quad \text{(ii)}$$

Substituting for v in (i) gives:

$$M = \frac{rv^2}{G} = \frac{r}{G} \times \frac{4\pi^2 r^2}{T^2}$$

$$\therefore \quad \frac{T^2}{r^3} = \frac{4\pi^2}{GM}$$

Equation 3

Equation 3 shows that for a given planet or star, the ratio T^2/r^3 should be constant for all of its satellites, regardless of their mass. Kepler in 1596 made observcations of the orbit radius and time period of the planets in our Solar System which confirmed this ratio, and he published his findings as his third law of planetary motion.

Re-arranging Equation 3 gives the following forms:

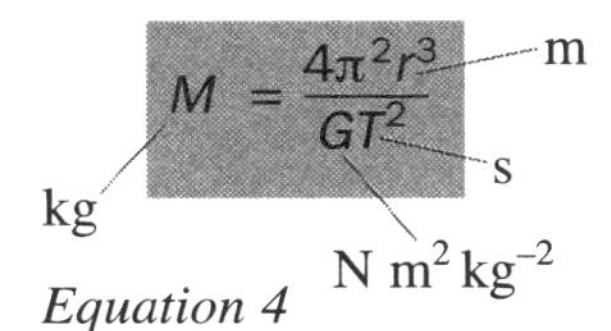

Equation 4

Equation 4 allows the mass (M) of the central planet or star to be calculated from measurements of the period (T) and orbit radius (r) of one or more of its satellites.

$$T = 2\pi\sqrt{\frac{r^3}{GM}}$$

Equation 5

Equation 5 allows the period (T) of a satellite to be calculated for any orbit radius (r).

Guided examples (3)

1. Calculate the mass of the Sun from the following data:

   Earth's orbit radius = 1.5×10^{11} m

   Earth's period = 3.0×10^{7} s (1 year)

   Gravitational constant (G) = 6.7×10^{-11} N m^2 kg^{-2}

   Guidelines

   Use Equation 4, taking care to identify each quantity.

2. Calculate the mass of the Earth given that the Moon's orbital radius is 4.0×10^{8} m and its rotational period is 28 days.

   Guidelines

   Use Equation 4 again, remembering to convert the period to seconds, and identifying clearly all the necessary quantities.

Close earth orbit

The majority of satellites orbiting the Earth are at heights of only 100 km or so above the surface. This means that their orbit radii are approximately the same as the radius of the planet.

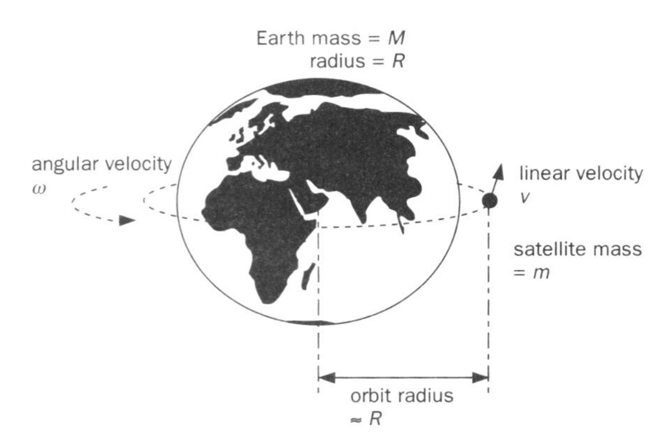

Figure 1.90 Artificial satellite orbit

Figure 1.90 shows an artificial satellite of mass (m) moving with constant speed (v) in a circular orbit around the Earth which has mass (M) and radius (R). If the satellite is orbiting close to the Earth, its orbit radius is approximately the radius of the Earth (R).

The equations for the linear speed (v) and period (T) of the satellite become:

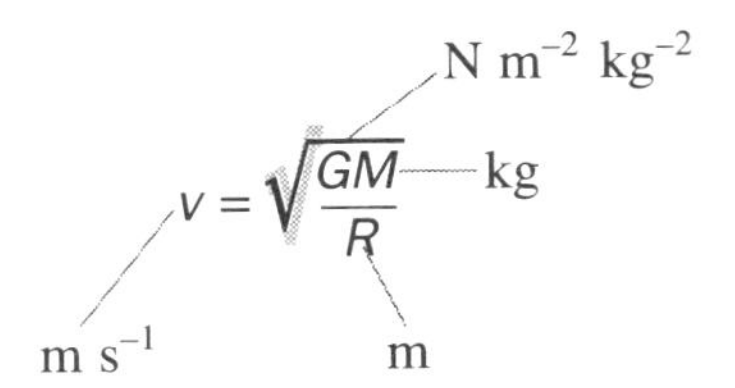

Equation 6

and

$$T = 2\pi\sqrt{\frac{R^3}{GM}}$$

Equation 7

Geosynchronous (parking) orbit

In the early days of satellite communication, the satellites were in fairly close Earth orbits, and so were 'visible' over the horizon for only short periods of time. Broadcasts could only be made when the satellite was in range of both the transmitter and receiver, which was of limited value. The writer Arthur C. Clarke in 1945 predicted the value of satellites which orbited with the same angular velocity and direction as the Earth. These would appear to be stationary over a point on the Earth's surface, and therefore always available for receiving or transmitting radio waves to anywhere on the side of the planet facing the satellite.

A **geostationary** satellite has an orbital period the same as that of the Earth's rotation about its own axis (24 hours), and travels over the equator in the same direction as the Earth. Therefore it always appears to be above the same point on the Earth's surface. Such an orbit is called **geosynchronous.**

Guided example (4)

Given that the mass of the Earth is 6.0×10^{24} kg, and the time period for one revolution is 24 hours, calculate the orbital radius of a satellite in geosynchronous orbit.

Guidelines

Re-arrange Equation 3 to calculate the radius of the satellite orbit.

Gravitational field strength

Common symbol g
SI unit newton per kilogram (N kg^{-1}) equivalent to metre per second^2 (m s^{-2})

The **field strength** (g) at a point in a gravitational field is defined as the gravitational **force** (F) per unit **mass** (m) experienced by a small test mass placed at that point.

Note that the field strength is defined with reference to a small test mass. 'Small' in this sense means not big enough to cause a significant change in the field being measured. A unit mass (1 kg) will hardly affect the Earth's gravitational field, but a body like the Moon causes a significant effect, being partly responsible for tidal movements in our oceans.

Field strength may be expressed mathematically as:

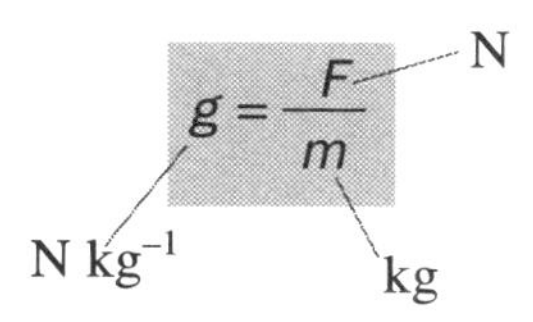

Equation 8

Here on Earth, a mass of 1 kg experiences a force due to gravity of around 9.8 N. The actual value varies slightly from place to place due to:

- non-uniformities in the shape and composition of the planet
- the effect of rotation of the Earth on its axis, which reduces the measured value of g by an amount varying from zero at the poles to a maximum at the equator.

The average value of the Earth's gravitational field strength is measured at 9.81 N kg^{-1}.

For most quantitative work at A-level, a value of 10 N kg^{-1} is sufficiently accurate.

You should note that:

- the field strength at any point in the gravitational field is equal to the acceleration due to gravity experienced by a body at that point
- field strength is a **vector** quantity, whose direction is defined by the direction of the force produced by the field.

Laboratory determination of gravitational field strength

The following method involves timing the oscillation of a simple pendulum to enable a calculation of acceleration due to gravity, which is numerically equal to gravitational field strength.

The simple pendulum consists of a dense, spherical metal bob on the end of a light, inextensible string (see the section on Vibrations and waves for a

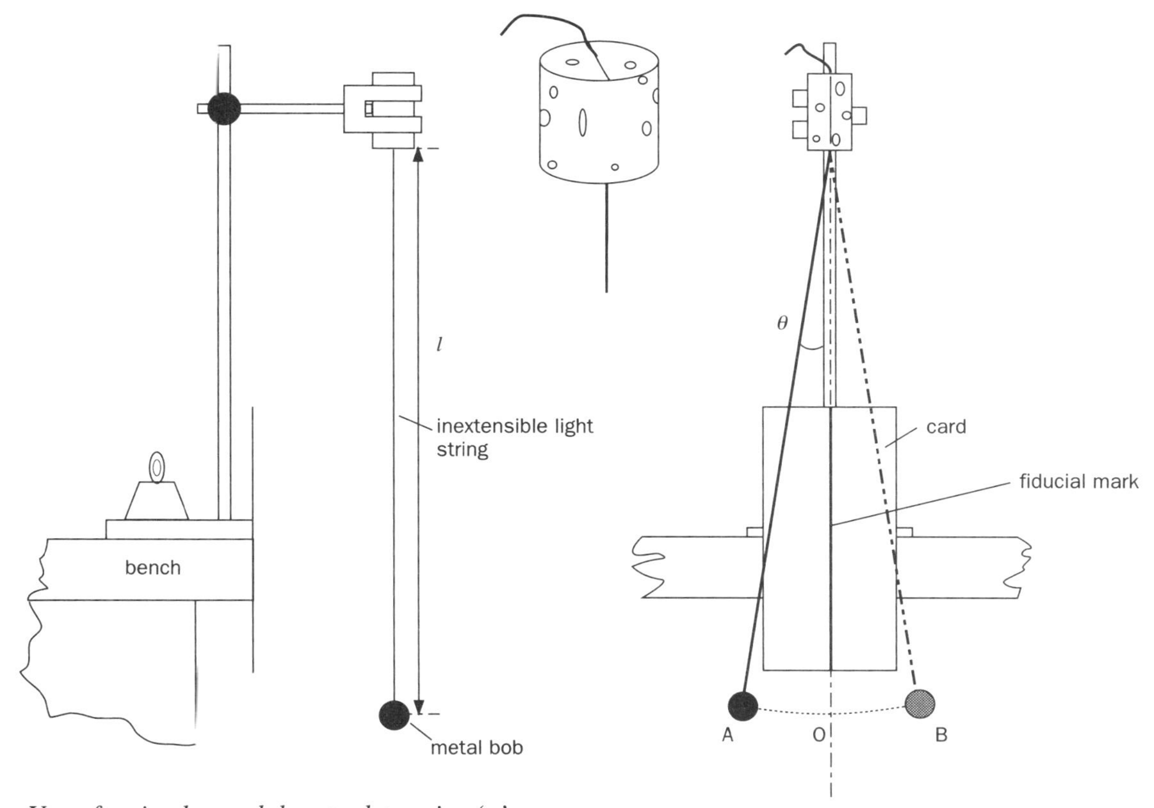

Figure 1.91 Use of a simple pendulum to determine 'g'

description of the theory). A suitable laboratory version is shown in Figure 1.91. Starting with a length (l) of around 1.5 m, the bob is given a slight sideways displacement and released, so that it performs oscillations of small amplitude. The **period** (T) of oscillation is determined by measuring the time taken for a large number of complete oscillations (say 40) and calculating the average time for one oscillation. This timing should be repeated for a wide range of values of l. The **measurement errors** in this procedure include:

Error		Minimised by:
Pendulum length	String cannot be massless, nor completely inextensible. 'Length' of pendulum is from pivot to centre of mass of string and bob	Use best compromise of lightness and stiffness available. Measure from bottom of cork to centre of mass of metal bob
Period	Human reaction time operating stopwatch	Time a sufficiently large number of complete oscillations to make percentage error reasonable. Use a 'fiducial mark' as timing reference mark
Angular displacement	Pendulum theory is only true for small angles of displacement	Keep $\theta < 10°$
Air friction	Weight and string tension are not the only forces acting	Keep string as thin as possible; make bob small and dense to minimise drag

A graph of (T^2) against (l) is plotted, and should yield a straight line through the origin, as shown in Figure 1.92.

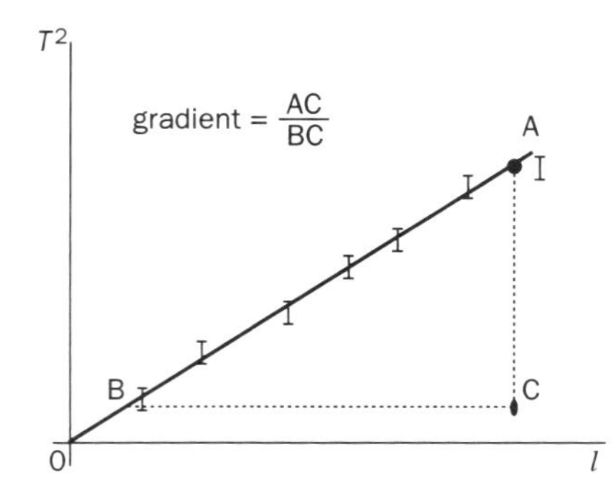

Figure 1.92 Graph of T^2 against l for a simple pendulum

Maths window

The relationship between the periodic time (T) and the length (l) of a simple pendulum is:

$$T = 2\pi\sqrt{\frac{l}{g}} \qquad \therefore\ T^2 = \frac{4\pi^2}{g}\,l$$

where (g) is the gravitational field strength.

A graph of T^2 against l gives a straight line through the origin, of gradient $4\pi^2/g$.

The gradient of the graph is determined by direct measurement, and the average value of g is obtained from:

$$\frac{4\pi^2}{g} = gradient$$

$$\therefore \quad g = \frac{4\pi^2}{gradient}$$

Self assessment

SECTION A
Qualitative Assessment

1. (a) Write down an equation which expresses Newton's law of gravitation, defining each term, and explaining the significance of the negative sign in the equation.
   (b) Derive the unit for the gravitational constant (G).

2. On Earth, the weight of particular object is (W) at the surface. Assuming the Earth's radius to be (R) and its mass to be (M), what is the weight of this object at a point:
   (a) $3R$ from the centre of the planet?
   (b) on the surface of another planet, radius $4R$ and mass $2M$?

3. Use gravitational and circular motion theory to derive an expression for the mass (M) of a planet which has a moon in a circular orbit of radius (r). Assume that the moon moves with constant angular velocity (ω) and has a period (T).

4. (a) Explain why the speed of an artificial satellite which is travelling in a circular orbit around the Earth would decrease if it were moved to an orbit of larger radius.
   (b) Communications satellites are placed in **geostationary** orbits around the Earth.
      (i) Explain the meaning of the term geostationary.
      (ii) State the period of such a satellite.
      (iii) Why must the satellite orbit above the equator?

SECTION B
Quantitative assessment

(***Answers:*** 1.0×10^{-47}; 1.5×10^{-42}; 3.5×10^{-10}; 5.7×10^{-6}; 0.21; 6.1; 870; 1.1×10^{3}; 3.8×10^{3}; 4.9×10^{3}; 7.5×10^{3}; 1.2×10^{5}; 2.1×10^{5}; 3.6×10^{7}; 2.0×10^{20}; 2.0×10^{30}.)

Gravitational constant, $G = 6.67 \times 10^{-11}$ Nm^2 kg^{-2}

1. Calculate the gravitational attraction force between:
   (a) The Earth and the Moon, given that the mass of the earth (m_E) is 6×10^{24} kg, the mass of the moon is equal to $0.012 \times m_E$ and the orbital radius of the moon = 3.8×10^5 km.
   (b) Two molecules of mass 3×10^{-25} kg at a distance apart of 2×10^{-9} m.
   (c) The proton and electron in a hydrogen atom, given that the rest mass of one electron (m_e) = 9×10^{-31} kg, the mass of one proton (m_p) = $1837m_e$ and the electron's orbital radius = 10^{-10} m.

2. Three particles A, B and C having masses of 0.50 kg, 0.10 kg and 0.75 kg respectively lie along a straight line separated from each other by distances as shown in the diagram below:

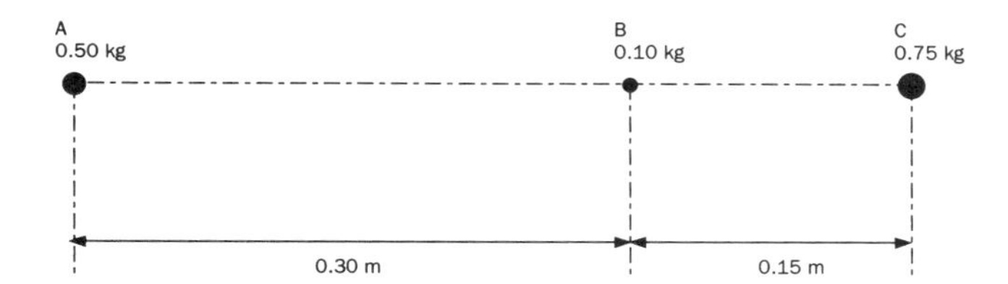

   Calculate
   (a) the resultant gravitational force acting on C
   (b) the distance from A along the line AB to which C must be moved so that there is no resultant gravitational force acting on it.

3. Given the data below, estimate
   (a) the mass
   (b) the mean density of the Sun.
   Mean radius of Earth orbit around the Sun, $R_E = 1.5 \times 10^{11}$ m.
   Period of Earth orbit around the Sun, $T_E = 365$ days.
   $R_E/R = 200$, where R = radius of the Sun.

4. A TV relay satellite has a mass of 150 kg and revolves around the Earth in an orbit of mean radius 1.3 × Earth's radius. If the Earth's radius is taken as 6.4×10^6 m, calculate
   (a) the centripetal force acting on the satellite,
   (b) the orbital period of the satellite. (Earth mass, $m_E = 6 \times 10^{24}$ kg).

5. An Olympic high jumper can clear 2.3 m on Earth. Estimate the height he could jump on a planet whose radius is three-quarters that of the Earth and whose mean density is half that of the Earth.

6. (a) Calculate the orbital height above the Earth at which a communications satellite would have to be placed in order to have a period of 24 hours (Earth mass = 6×10^{24} kg; Earth radius = 6.4×10^6 m).

Self-assessment *continued*

(b) What is the name given to the satellite orbit described in (a)?

7. Uranus has a mass of 9.6×10^{25} kg and the orbital radii of its five moons are given below. Calculate:

(a) the orbital period for
  (i) Miranda
  (ii) Ariel

(b) the orbital linear velocity for
  (i) Umbriel
  (ii) Titania

(c) the orbital angular velocity for Oberon.

	Miranda	Ariel	Umbriel	Titania	Oberon
Orbital radius (m) $\times 10^{85}$	**1.3**	**1.9**	**2.7**	**4.4**	**5.8**

1.11 Gravitational fields

The idea of a **field** helps us to understand the interactions of bodies which are not in contact. All masses can be thought of as having a gravitational field around them which is a region of space in which an **attractive** force is exerted on any other **mass** which is in that space. The Sun, the planets, moons and asteroids of the Solar System all have their own gravitational fields, and the forces which they exert on each other keep the whole system in a state of dynamic equilibrium. This system, in turn, is only a minute part of the whole system of bodies which we call the Universe.

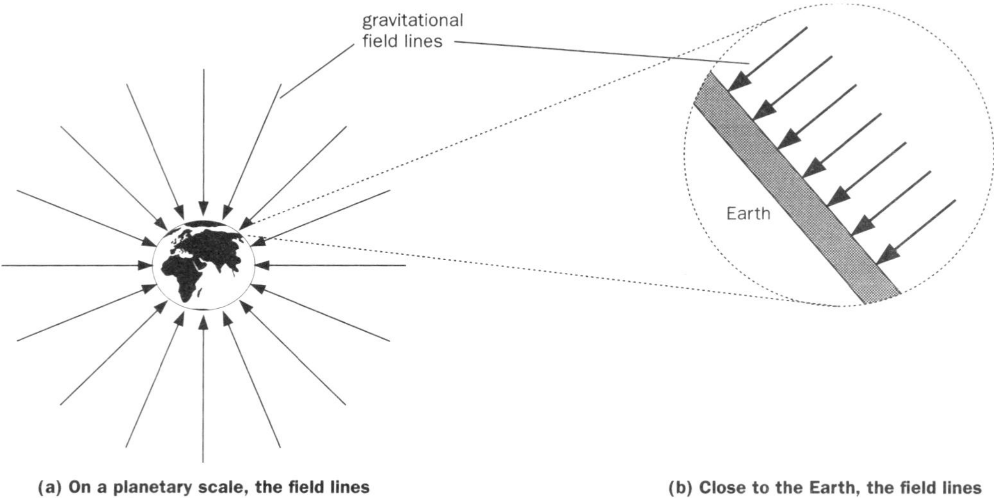

(a) On a planetary scale, the field lines are radial, and the field strength decreases with distance from Earth

(b) Close to the Earth, the field lines can be assumed parallel, and the field strength uniform

Figure 1.93 The Earth's gravitational field

The concept of **field strength** gives us a measure of the force involved in any particular interaction, and **field lines** enable us to picture the shape of the field and the direction of the forces around the body. Figure 1.93 uses field lines to show the gravitational field around our own planet. Both diagrams are approximations, since they assume the Earth to be a perfect sphere of constant density.

You should note that:

- Figure 1.93(a) is an overall picture of the Earth's gravitational field whereas Figure 1.93(b) shows a close-up view over a limited area of the Earth's surface.
- The direction of the field lines indicates the direction of the gravitational force on a mass in the field (i.e. its weight). This is the direction in which a freely-falling mass will accelerate, and defines the vertical direction.
- The field lines are directed towards the centre of the planet, which indicates that the gravitational field is **attractive.**
- The **strength** of the field can be inferred from the separation of the lines of force. In a **radial** field, the separation of the lines increases with distance from the centre, indicating that the field strength is decreasing as distance increases, in agreement with Newton's law of gravitation.

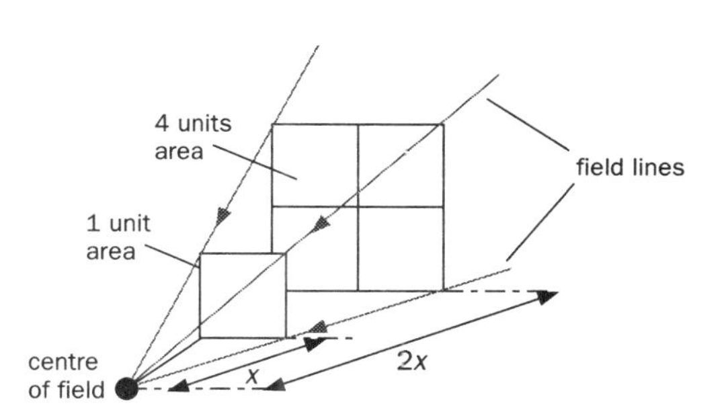

Figure 1.94 Inverse square law

Figure 1.94 shows how the gravitational field lines which cover one unit of area at a given distance from the centre of the field, cover an area four times as great at twice the distance. This is an illustration of the **inverse square law** – the field strength is inversely proportional to the square of the distance from the centre of the field.

- Close to the surface, and over an area small in comparison with the overall area of the planet, the field can be assumed to be uniform (in both strength and direction).

Field strength in a radial field (see Figure 1.95)

The gravitational field pattern of a real body such as a planet can be assumed to be identical to the pattern which would be produced if all the mass of the planet were concentrated at a point at its centre (i.e. the planet behaves as a **point mass**, whose mass is equal to that of the planet, situated at the centre of the planet).

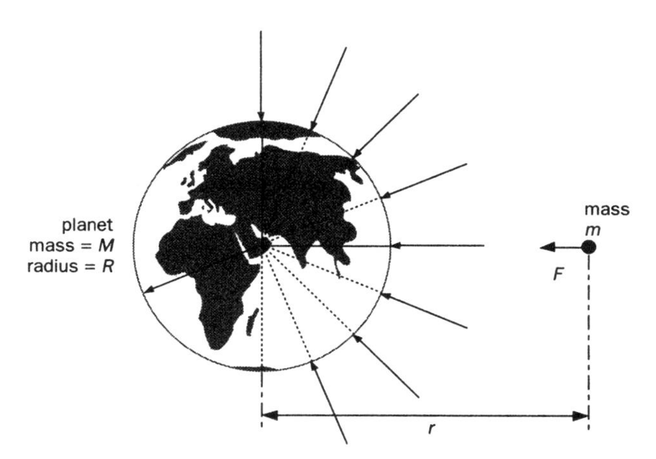

Figure 1.95 Radial field due to a planet

Relationship between *g* and *G*

By definition of gravitational field strength (g), the force (F) acting on the mass (m) in Figure 1.95 is given by:

$$F = mg$$

also, from Newton's law of gravitation, the force on this mass at distance r from a planet of mass (M) is given by:

$$F = -\frac{GMm}{r^2}$$

$$\therefore \quad g = -\frac{GMm/r^2}{m}$$

$$\therefore \quad g = -\frac{GM}{r^2}$$

(g: m s^{-2}; G: N m^2 kg^{-2}; M: kg; r: m)

Equation 1

Equation 1 shows that away from the surface of a planet, the field strength decreases with the **square** of the distance from the centre of the planet, i.e. it follows an inverse square relationship. The **minus** sign indicates that g acts in a direction opposite to that in which r is measured. For a **point source**, the field strength would increase to infinity as the source was approached. A planet, however, only behaves as a point source at distances greater than the radius of the planet (i.e. above its surface). Below the surface, the field strength decreases. This is an example of where the behaviour of a real body differs from that of the theoretical model.

Figure 1.96 below shows the nature of the relationship between gravitational field strength and distance for the planet Earth.

You should note that:

- below the surface, field strength is directly proportional to distance from the centre
- at the centre, the field strength is zero
- at any distance r from the centre (where r is less than r_e) the field strength is equal to the field strength that would be produced by a planet of radius r, and the same density as the Earth
- all the above can be applied to any planet.

Guided examples (1)

1. Calculate the mass of the moon given that its radius is 1.74×10^6 m and the gravitational field strength at its surface, measured by the Apollo astronauts, is 1.70 N kg^{-1}. (Assume G, the universal gravitational constant, to be 6.67×10^{-11} N m^2 kg^{-2}.)

Guidelines

Use Equation 1. Take care to adapt it by using appropriate values for the Moon, rather than Earth.

Guided examples (1) *continued*

2. Calculate the gravitational field strength on the surface of a planet whose mean density is $0.5 \times$ that of the Earth, and whose radius is $15 \times$ the Earth's radius. Assume that g on the Earth's surface is 10 N kg^{-1}.

Guidelines

Use Equation 1.

mass = density × volume

$$M = \rho \times \frac{4}{3}\pi r^3$$

where M, ρ and r are the planet's mass, density and radius respectively.

Substitute for M_e in Equation 1, which shows that g is proportional to the product of ρ and r.

Energy changes in a gravitational field

Builders and rocket scientists are equally concerned about the energy needed to increase the potential energy of a load, whether the load is a hod full of bricks to be carried up a ladder, or a new telescope being placed in orbit. A man carrying bricks up a ladder is using the chemical energy in food to do

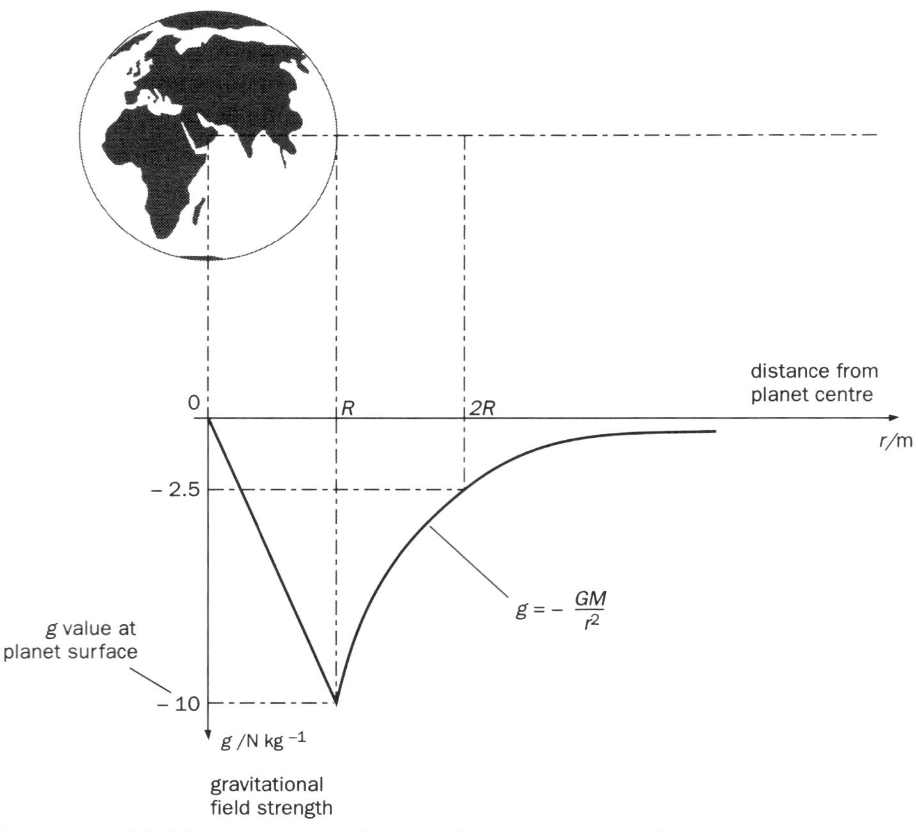

Figure 1.96 Variation of field strength with distance from the Centre of the Earth

work on the bricks to increase their potential energy, while putting the telescope into orbit will require vast amounts of liquid hydrogen and oxygen to be burned in a rocket's engines. Close to the Earth (generally speaking within our atmosphere), the gravitational field can be assumed uniform. This means that the force necessary to move the load against the force of gravity is constant, and the work done is simply equal to the constant force multiplied by the height increase. The rocket scientist, however, has to deal with much greater distances, where the changes in the field due to distance from the Earth are significant.

Energy changes in a uniform field – close to a planet's surface (see Figure 1.97)

In Section 1.7 (Work and energy), the concept of gravitational potential energy was introduced, with the following simplifying assumptions:

- the gravitational field is **uniform** over the change of height involved (i.e. the **weight** of the body is **constant**)
- the gravitational potential energy is **zero** at **ground level.**

Figure 1.97(a) shows a mass (m) being raised through a height Δh. The force needed to raise the mass against the force of gravity (i.e. the weight (mg) of the mass) does work on the mass, which increases its gravitational potential energy by an amount $mg\Delta h$ (where g is the gravitational field strength).

Figure 1.97(b) shows the same mass released and allowed to fall back to the surface. In falling, the gravitational field does work on the body, so its kinetic energy increases and the gravitational potential energy decreases. At every point during its fall, the gain of kinetic energy of the body is equal to the decrease of gravitational potential energy of the system. At ground level, just before impact, the kinetic energy of the body (neglecting friction) is equal to $mg\Delta h$. This energy would be dissipated on impact with the ground.

True zero of gravitational potential energy

So far in our study of energy, we have made the assumption that when a body is on the Earth's surface, the system comprising the body and the Earth has zero gravitational potential energy. This is a very convenient assumption to make, and is to be recommended whenever bodies close to the Earth's surface are being considered. However, the system clearly does not have zero gravitational potential energy, as there is a large force of attraction between the Earth and the body, and this has the **potential** to do work. The system can only have zero potential energy when there is no force of attraction between the planet and the body, and this can only be when they are infinitely far apart. This means that the true position in which a mass has zero gravitational potential energy is when it is removed to infinity.

If a body is on the Earth's surface, work has to be done **on** that body in order to increase the distance between the body and the planet. This increases the gravitational potential energy stored. If enough work is done, the distance between body and planet can be increased to infinity, and the energy stored will have increased to zero! This apparent paradox can be resolved if we treat the potential energy stored in the attractive gravity field as **negative** at all distances less than infinity.

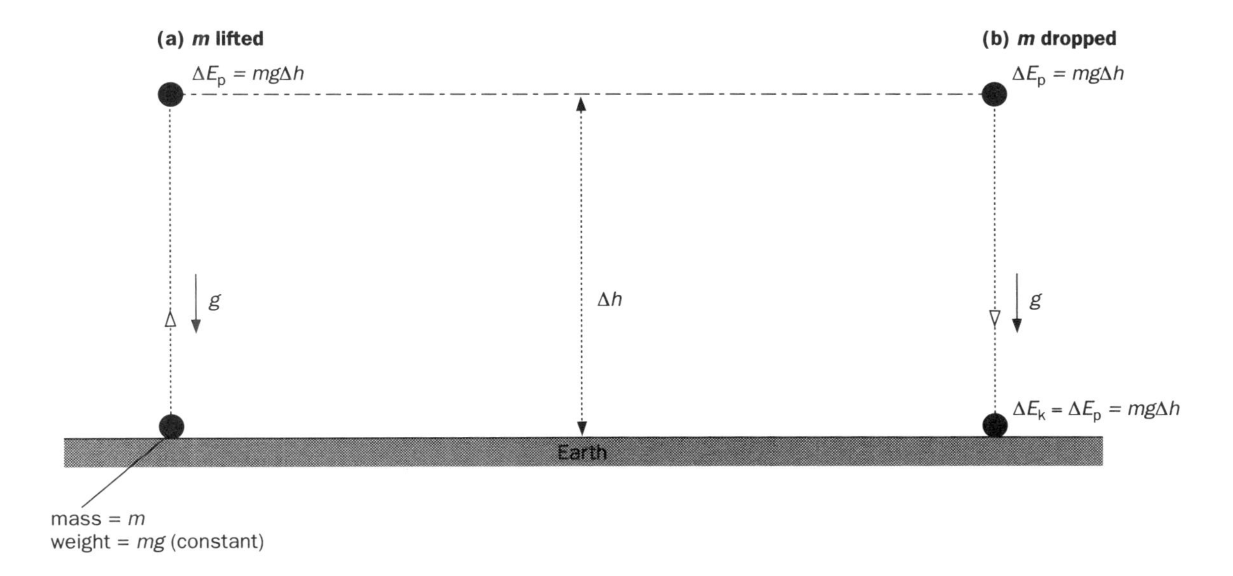

Figure 1.97 Gravitational potential energy in a uniform field

Energy changes in a non-uniform field

The gravitational field around a planet is **radial**, and so the field strength varies with distance from the centre of the field. Thus when we are considering the energy changes of any mass when moved over large distances in a planetary or stellar gravitational field, the following points must be noted:

- the gravitational field strength is **inversely** proportional to the **square** of the distance from the centre of the planet
- the gravitational potential energy is **zero** when the distance between the body and the planet is **infinity**
- the gravitational potential energy is **negative** at all distances less than infinity.

Calculating the gravitational potential energy

Figure 1.98 shows how the gravitational force (mg) on a body varies with distance from the centre of the planet. At the planet's surface, (radius R), the gravitational field strength is g_s and the force on the body is mg_s. If the body is raised through a small distance Δr, so that g_s can be assumed constant, the work done on the body is given by $mg_s\Delta r$. This is equal to the area of the shaded strip in Figure 1.98.

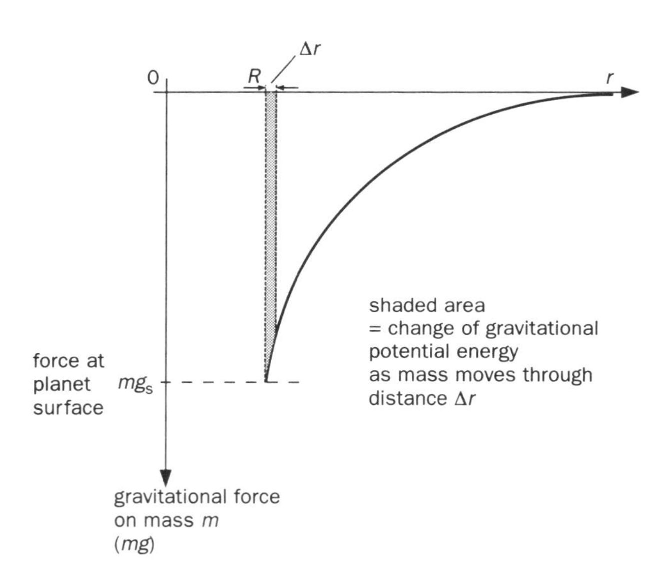

Figure 1.98 Variation of gravitational force with distance

If the body is then moved away from the planet to infinity in a series of small steps, the work done during each small step is equal to the area of each strip under the graph, as shown in Figure 1.99.

So the total work done in taking mass m from the surface of the planet to infinity is equal to the total shaded area. This is equal to the increase in gravitational potential energy of the mass.

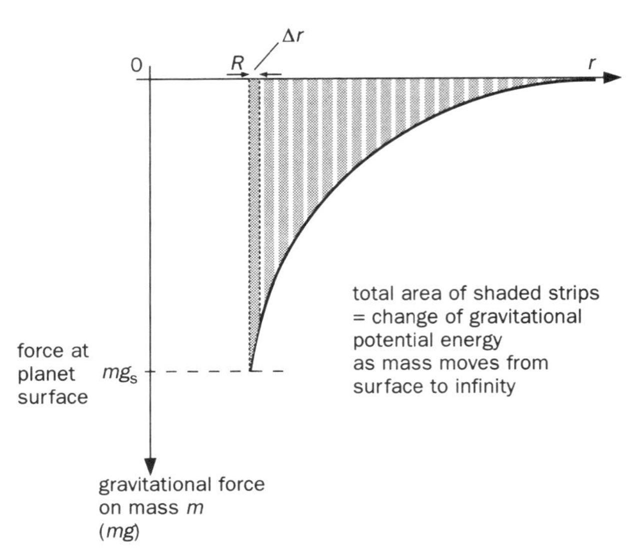

Figure 1.99 Gravitational potential energy of a mass at the surface of a planet = work done in removing the mass to infinity

Maths window

Finding the **area** under a curve can be done in two ways:

- physically counting the squares on the graph paper
- integration

The equation of this curve is:

$$mg = \frac{GMm}{r^2}$$

The increase in potential energy (ΔE_p) is given by the area under the curve and the r-axis between the limits $r = R$ and $r = \infty$.

$$\therefore \quad \Delta E_p = \int_R^\infty -\frac{GMm}{r^2}\,dr = -GMm\int_R^\infty \frac{1}{r^2}\,dr$$

$$= GMm\left[\frac{1}{\infty} - \frac{1}{R}\right]$$

$$\therefore \quad \Delta E_p = -\frac{GMm}{R}$$

Equation 2

In general, if a body of mass (m) moves from a point, distance r_1 from the centre of the planet of mass (M), to a point at distance r_2, the change in gravitational potential energy (ΔE_p) is given by:

$$\Delta E_p = GMm\left(\frac{1}{r_2} - \frac{1}{r_1}\right)$$

Equation 3

If r_2 is infinity (where the potential energy is zero), Equation 3 calculates the energy required to take a mass (m) to infinity. This can be thought of as the **absolute potential energy** (E_p) of the mass at distance (r_1) from the centre of the planet of mass (M). In general, for any distance (r):

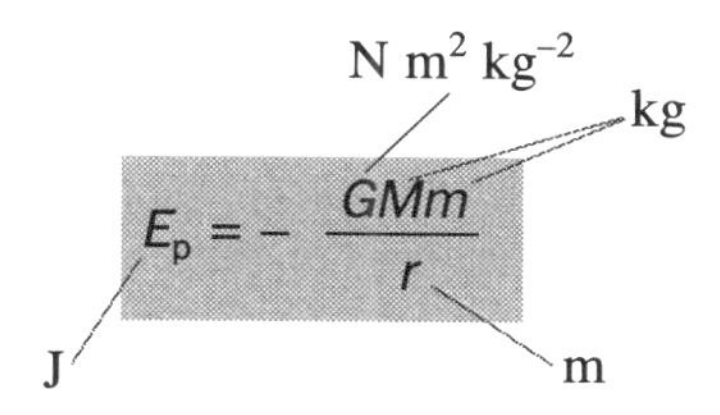

$$E_p = -\frac{GMm}{r}$$

Equation 4

These equations have been developed by considering a mass (m) at various distances (r) from the centre of a gravitational field due to a planet of mass (M). A very useful form of these equations can be obtained by considering a **unit mass** (1 kg in SI units), which leads us to the concept of **potential**

Gravitational potential

The **gravitational potential** (V) at any point in a field is the potential energy (E_p) **per unit mass** at that point.

Common symbol	V
SI unit	joule per kilogram (J kg^{-1})

Defining equation:

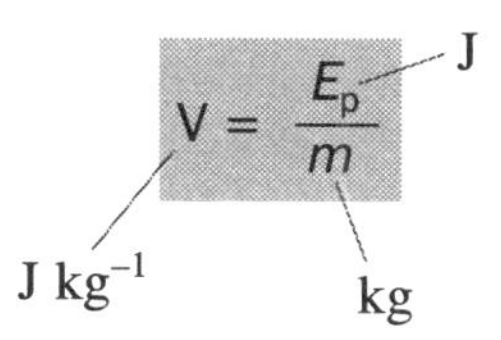

$$V = \frac{E_p}{m}$$

Equation 5

Rearranging Equation 5:

$$E_p = mV$$

If a unit mass is taken from a point 1 to point 2 where the gravitational potentials are V_1 and V_2 respectively, then the **change** of potential energy (ΔE_p) is given by:

$$\Delta E_p = mV_2 - mV_1$$

$$\therefore \quad \Delta E_p = m(V_2 - V_1)$$

In words:

(change of gravitational potential energy) = (mass) × (potential difference)

This is a very useful equation when dealing with space travel.

Since the potential at an infinite distance from the planet is zero, it follows that:

The **potential** (V) at a point in a gravitational field can be defined as the **work done** (W) in bringing a **unit mass** (m) from infinity to that point.

Gravitational potential is a **scalar** quantity.

Working through the following guided example will help you to understand how potential is used to calculate energy changes in a gravitational field, and illustrates the range of distances over which g can be assumed constant.

Guided example (2)

Imagine you are a space scientist calculating the energy needed to carry an extra 2 kg payload to various heights above the Earth's surface. The diagram shows the gravitational potential at various heights:

potential / MJ kg^{-1}	height / km
0	∞
– 2.45	10000
– 54.3	1000
– 61.8	100
– 62.7	10
– 62.8	0

Calculate, using gravitational potential, the energy required to send this mass to each of the heights shown.

Then calculate the energy required using the $mg\Delta h$ formula (assuming g is constant at 10 N kg^{-1}) and compare your answers.

Guidelines

(increase of potential energy)
= (mass) × (potential difference)

So to rise to a height of 10 km, the increase of potential energy is given by:
(increase of potential energy)
= (mass) × (potential difference)
= 2 × (–62.7 – (–62.8))kg MJ kg^{-1}
= **0.20 MJ**

Guided example continued on next page

Guided example (2) *continued*

Using the formula:

(increase of potential energy)
$= mg\,\Delta h$

(increase of potential energy)
$= 2.0 \times 10 \times 10\,000$ kg N kg^{-1} m
$= 2.0 \times 10^5$ N m
$=$ **0.20 MJ**

These answers are identical. Work out the rest of the problem for yourself. Your solutions should show an increasing disagreement between the two methods, and according to the $mg\,\Delta h$ formula, an infinite amount of energy is needed to escape from the planet's gravity field (i.e. to reach infinity). Clearly, this formula is perfectly adequate for distances within the Earth's atmosphere, but is inadequate for space travel.

Gravitational potential and field strength

The idea of potential is a useful one, as it allows us to calculate the energy needed to get from point A to point B in a gravitational field. In space travel, the gravitational field strength at any point will be the resultant of the effects due to all the neighbouring bodies which have an influence on that point. Gravitational field strength is a vector quantity, so both magnitude and direction need to be considered. However, as potential is a scalar quantity, knowing the potential at two points enables a simple calculation to be made of the energy needed to move between the two points.

The gravitational potential energy (E_p) of a mass (m) at a point, distance (r) from the centre of the gravitational field due to a planet of mass (M) is given by:

$$E_p = -\frac{GMm}{r}$$

By definition of potential, the gravitational potential (V) at this point is given by:

$$V = \frac{E_p}{m}$$

$$\therefore \quad V = -\frac{GM}{r}$$

(units: V in J kg^{-1}; G in N m^2 kg^{-2}; M in kg; r in m)

Equation 6

Maths window

In the radial gravitational field surrounding a planet of mass (M), the potential (V) at any distance (r) from the centre of planet is given by:

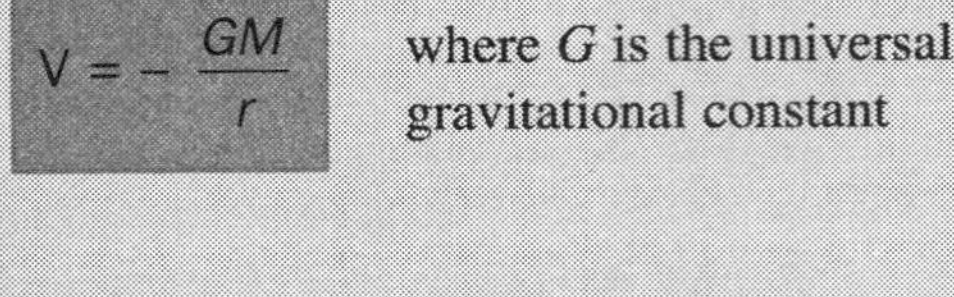

$V = -\frac{GM}{r}$ where G is the universal gravitational constant

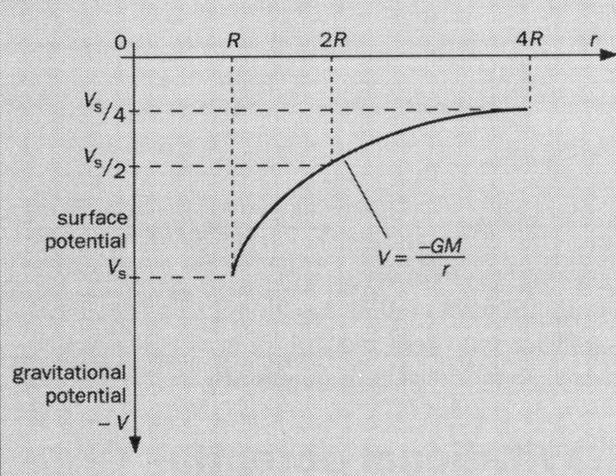

Figure 1.100 Variation of potential with distance in a radial attractive field

Figure 1.100 shows how the gravitational potential due to a spherical planet varies with distance from its centre. At the surface of the planet (radius R) the potential is (V_s).

V is inversely proportional to r. Thus when r is equal to $2R$, V is equal to $V_s/2$; and when r is equal to $4R$, V is equal to $V_s/4$ etc.

Link between field strength (*G*) and potential (*V*)

Figure 1.101 shows the graphical relationship between gravitational potential (V), gravitational field strength (g) and distance (r) from the centre of a spherical mass such as a planet.

- The value of the gravitational field strength (g) at a given distance from the centre of the planet is equal to the gradient of the potential/distance (V/r) graph at that distance.

i.e. 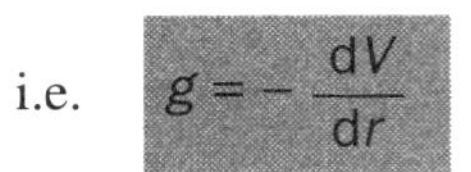

$$g = -\frac{dV}{dr}$$

Equation 7

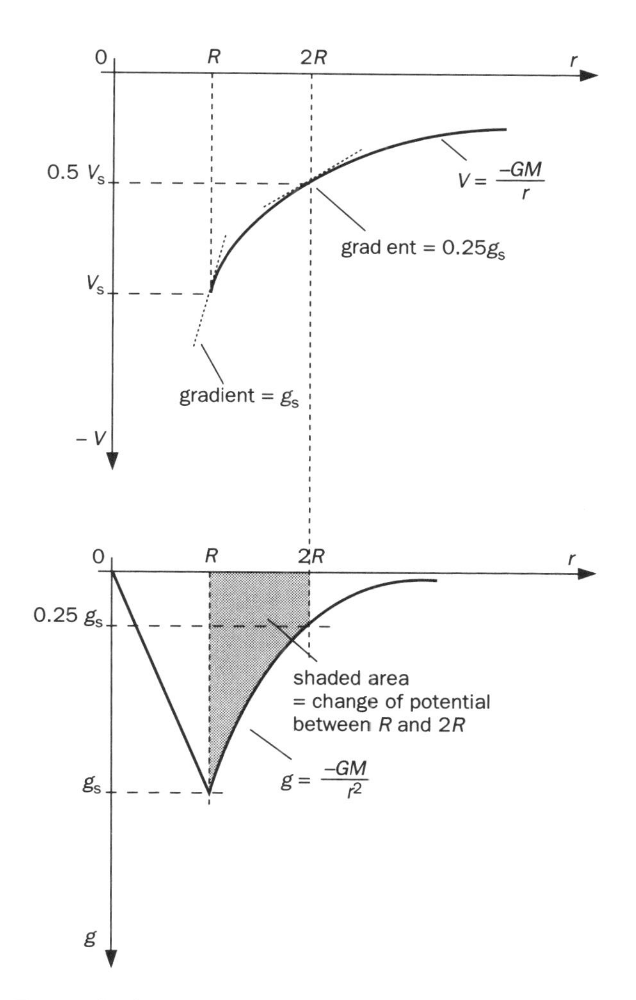

Figure 1.101 Graphs of V against r and g against r for a spherical planet

Maths window

We can show the relationship between g and V by differentiating Equation 6 with respect to r:

$$V = -\frac{GM}{r}$$

Then differentiating with respect to r yields:

$$\frac{dV}{dr} = \frac{GM}{r^2}$$

But from Equation 1, $g = -\frac{GM}{r^2}$

$$\therefore \quad g = -\frac{dV}{dr}$$

- The **area** bounded by the g/r curve and the r-axis between two values of r is equal to the **change of potential** between those two positions. In the above example, the shaded area of the g/r graph (Figure 1.101) is equal to the change of potential between a point on the surface of the planet, radius (R), and a point at a distance equal to $2R$.

Maths window

The change of potential (V) between any two points distances r_1 and r_2 from the centre of a gravitational field is obtained mathematically by the process of integration, which is equivalent to measuring the area between the graph and the r-axis between the ordinates at $r = r_1$ and $r = r_2$.

i.e. $V = \int_{r_1}^{r_2} -g\,dr$

Escape velocity

We have seen that a spacecraft needs a certain amount of energy in order to escape from the gravitational field of a planet. This energy could be carried in the form of fuel to be burned continuously on the journey into space, but this is not economical as the fuel tanks have to be taken along, which means extra expense. The usual method is to accelerate the craft to very high speeds in a short time, giving it a great deal of kinetic energy, then jettison the fuel tanks and engines to be recovered later. If the spacecraft has sufficient kinetic energy, it will be able to leave the gravitational field of the planet.

The **escape velocity** of a planet is the velocity which a projectile must be given at the surface of the planet so that it will escape completely from the planet's gravitational field due to its kinetic energy alone.

To calculate the escape velocity of a planet, consider a body near the surface of the planet, and calculate its gravitational potential energy. Using the previous Guided Example of a body of mass 2 kg on the Earth (p. 85), the potential at the surface is –62.8 MJ kg^{-1} so the body's gravitational potential energy at the surface is –125.6 MJ. Therefore it needs exactly 125.6 MJ of kinetic energy to reach infinity, where its gravitational potential energy will have increased to zero. The velocity (v) required is given by:

$$\frac{1}{2}mv^2 = 125.6 \times 10^6$$

$$\therefore \quad v = \sqrt{\frac{2 \times 125.6 \times 10^6}{2}} = 11200 \text{ m s}^{-1}$$

Thus to escape completely from the Earth's gravitational field, a spacecraft needs to reach a velocity of 11200 m s^{-1}. (This calculation ignores any extra energy needed to overcome friction in the atmosphere if it is launched from the surface.)

In general, for a body of mass (m) distance (r) from the centre of a planet of mass (M), the escape velocity (v_e) is calculated by:

$$\frac{1}{2} m v_e^2 = \frac{GMm}{r}$$

$$\therefore \quad v_e = \sqrt{\frac{2GM}{r}}$$

(v_e in m s^{-1}; G in N m^2 kg^{-2}; M in kg; r in m)

Equation 8

Since at distance r from the centre of the planet, the magnitude of the gravitational field strength (g) is given by:

$$g = \frac{GM}{r^2}$$

$$gr = \frac{GM}{r}$$

$$2gr = \frac{2GM}{r}$$

$$\therefore \quad v_e = \sqrt{2gr}$$

(v_e in m s^{-1}; g in m s^{-2}; r in m)

Equation 9

The escape velocity of the Earth was previously calculated to be around 11000 m s^{-1}. This accounts for the fact that the Earth retains an atmosphere. Typical speeds of oxygen and nitrogen molecules in the atmosphere are of the order of hundreds of metres per second, so there will be very few molecules which reach the required escape velocity.

Guided examples (3)

1. Use the data given below to calculate the gravitational potential on the surface of the planet Mercury due to
   (a) Mercury itself
   (b) the Sun
   (c) the Earth

   Sun's mass M_s = 2.0×10^{30} kg
   Mercury's mass M_m = 3.3×10^{23} kg
   Earth's mass M_e = 6.0×10^{24} kg
   Mean radius of Mercury R_m = 2.4×10^6 m
   Mean Mercury–Sun distance r_m = 5.8×10^{10}m
   Distance from Earth to Mercury = 9.2×10^{10} m

Guidelines

Use Equation 6 in each case to calculate the potential due to each of the three gravity fields.

2. Given that the mass and mean radius of the Earth are 6.0×10^{24} kg and 6.4×10^6 m respectively, calculate the gravitational potential difference between a point at sea level and the summit of Mt. Everest which is 8.9 km above sea level.

Guidelines

Gravitational potential **increases** with distance from the planet. Use Equation 6 and be careful with the signs.

3. A rocket of mass 5.0×10^5 kg is standing on the surface of a planet whose mass and mean radius are 6.6×10^{23} kg and 3.4×10^6 m respectively. Calculate:
   (a) the minimum velocity with which the rocket must be projected in order to escape from the planet's gravitational field
   (b) the initial kinetic energy of the rocket
   (c) the gravitational field strength on the surface of the planet.

Guidelines

(a) The velocity required is the **escape velocity** of the planet.
(b) The kinetic energy required is equal to $\frac{1}{2}mv^2$, where v is the escape velocity.
(c) Use $g = GM/r^2$.

4. Calculate the escape velocity of the Moon if the gravitational field strength at its surface is 1.7 N kg^{-1} and its radius is 1.6×10^6 m. Comment on the feasibility of creating an artificial atmosphere by extracting oxygen and nitrogen from minerals on the Moon.

Guidelines

Use Equation 9.

Satellite orbits and energies

Here we examine the factors which control the **shape** of a satellite orbit. The general shape of an orbit is an **ellipse**. (A circle is simply a special case of an ellipse.) Controllers of space missions devote a great deal of effort towards making the orbits of communications and observation satellites as circular as possible. Most of the planets of the solar system have nearly circular orbits, but the shape of the orbit of any body depends on the total energy of the body.

Close, circular orbit

A satellite of mass (m) orbiting close to a planet of mass (M), radius (R), has both **gravitational potential energy** (E_p) and **kinetic energy** (E_k):

$$E_p = -\frac{GMm}{R} \quad \text{(from Equation 4)}$$

In section 1.10 (Gravitation) we saw that the velocity (v) of such a satellite in close orbit was given by:

$$v = \sqrt{\frac{GM}{R}}$$

and since kinetic energy $= \frac{1}{2}mv^2$, the kinetic energy (E_k) of the satellite is given by:

$$E_k = -\frac{GMm}{2R}$$

The total energy (E_t) of the satellite is therefore the sum of these two:

$$E_t = E_p + E_k$$

$$\therefore \quad E_t = -\frac{GMm}{R} + \frac{GMm}{2R} = -\frac{GMm}{2R}$$

The proportions of kinetic and potential energies will be constant throughout the orbit, since its radius is constant. This is illustrated in Figure 1.102.

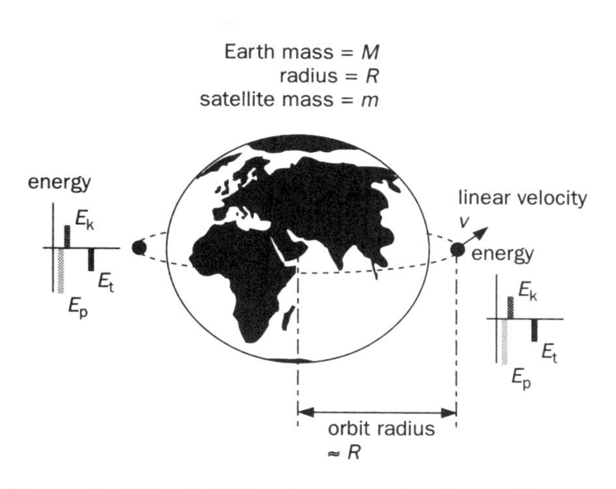

Figure 1.102 Satellite in close, circular orbit

Now think about what happens if the satellite's thrusters are switched on for a **short time**, then switched off, increasing the velocity and, therefore, kinetic energy of the satellite. If the total energy is still negative, the satellite will follow an elliptical path.

In Figure 1.103, the effect of increasing the kinetic energy of a satellite can be seen. As the satellite moves away from the planet, its potential energy increases, so its kinetic energy decreases. At the furthest distance from the planet (the **apogee**) the potential energy is a maximum, and the kinetic energy is a minimum. The satellite will then continue on the elliptical path shown, moving closer to the planet. As a result, its potential energy decreases and its kinetic energy increases as it accelerates. At the point where it is closest to the planet (the **perigee**) it has maximum kinetic energy and minimum potential energy. The total energy (potential plus kinetic) remains constant throughout the entire orbit.

When the satellite is at the apogee, the orbit could be circularised at this new radius by using the thrusters

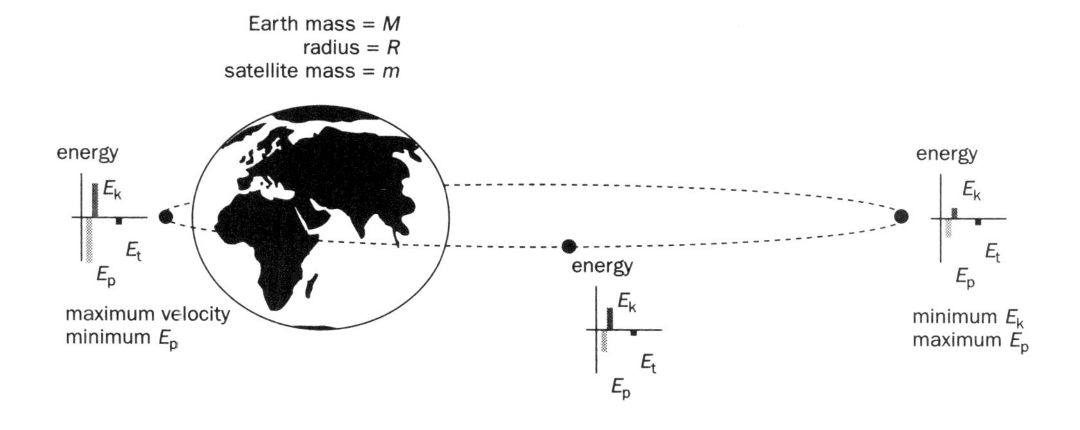

Figure 1.103 Elliptical satellite orbit

again. This would increase the kinetic energy (and total energy) of the satellite. If the increase of energy is exactly right, the satellite will follow a circular orbit of the correct radius. This is a delicate operation, demanding considerable skill on the part of the mission controllers.

Now consider what happens if, during a circular orbit, the force of the thrusters increases the velocity to such an extent that the total energy becomes zero (i.e the kinetic energy of the satellite is equal in magnitude to the potential energy). In this case, as the satellite moves away from the planet, its kinetic energy will decrease and its potential energy will increase as before, but this time, the satellite will continue to move away until it reaches infinity, as shown in Figure 1.104. This is the **escape velocity** condition, discussed above (p. 87).

The path followed by the satellite is a parabola. At infinity, the kinetic energy will be zero as the satellite will come to rest. The potential energy will also be zero, as it is beyond the range of the planet's gravitational field.

If the thrusters gave the satellite even greater kinetic energy, then the total energy would be greater than zero, so at infinity the satellite would still have some positive energy. This means that it would still have some kinetic energy, i.e. it would still be moving away from the planet. The path followed is called a **hyperbola**.

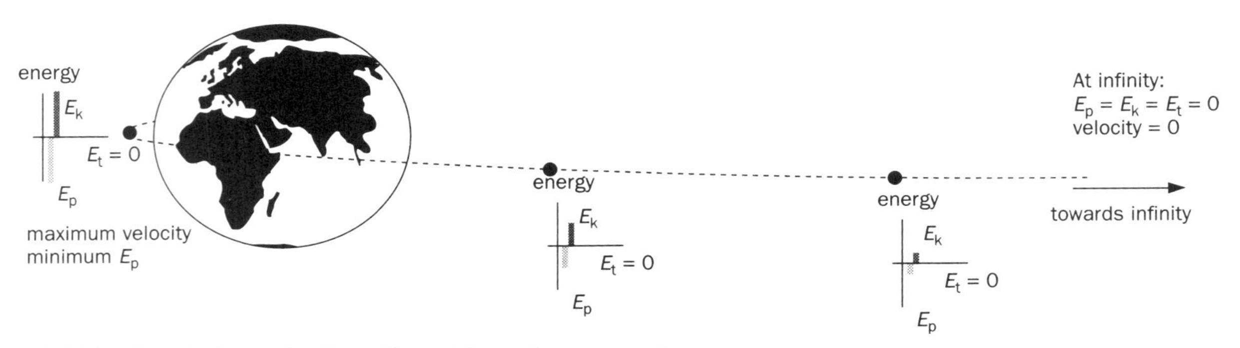

Figure 1.104 Parabolic path of satellite with total energy = 0

Guided examples (4)

1. A satellite of mass 100 kg is in a circular orbit around the Earth (mass 6×10^{24} kg, radius 6.4×10^6 m) at a height of 300 km. Assuming G to be 6.7×10^{-11} N m^2 kg^{-2} calculate its

(a) gravitational potential energy

(b) velocity and kinetic energy

(c) total energy.

Guidelines

(a) Use Equation 4. Take care when adding the height above the surface to the planet radius.

(b) Since the satellite is in a circular orbit, use Equation 2 from Section 1.10 on Gravitation.

(c) Add the two energies together (take care with signs!).

2. If the satellite's thrusters are briefly fired, giving the satellite an extra 1 GJ of kinetic energy, calculate

(a) the new velocity of the satellite immediately after firing the thrusters

(b) the velocity of the satellite when at a distance of 2000 km above the Earth's surface.

Guidelines

(a) Add the extra energy to the kinetic energy, and use this to calculate the new initial velocity of the satellite.

(b) As the satellite moves away from the Earth, its potential energy will increase. Calculate the potential energy at the new position. Since the total energy is constant, use this to calculate the new kinetic energy, and hence calculate its velocity.

Self-assessment

SECTION A
Qualitative assessment

1. Explain what is meant by a **gravitational field** and define **gravitational field strength** (g).

2. Explain what is meant by the term **uniform** field and discuss to what extent the gravitational field of the Earth can be considered uniform.

3. (a) Show that in the **radial** gravitational field around a planet of mass (M), the field strength (g) at any distance (r) from the centre of the planet is given by $\boxed{g = -GM/r^2}$.
   (b) Draw a graph to show how g varies with distance from the centre of a planet. Explain the shape of your graph.

4. Describe a laboratory method for determining the gravitational field strength (g).

5. Define **gravitational potential** (V) and state how it is related to **gravitational potential energy** (E_p).

6. Show that the difference in gravitational potential energy (ΔE_p) between a mass (m) at a point on the Earth's surface and at a point at some height (Δh) above it is given by $\boxed{\Delta E_p = mg\Delta h,}$ where (g) is the gravitational field strength at the Earth's surface. State clearly any assumptions you make.

7. (a) Give an expression for the **gravitational potential** (V) at some distance (r) from the centre of planet of mass (M).
   (b) Draw a graph to show how V varies with distance from the centre of the planet.
   (c) State the relationship between **gravitational field strength** (g) and **gravitational potential** (V).

8. (a) What is meant by the **escape velocity** of a planet?
   (b) Derive an expression for the escape velocity of a planet of mass (M) and radius (R).

SECTION B
Quantitative assessment

(***Answers:*** 0.61; 2.2; 2.4×10^3; 7.5×10^7; 2.3×10^{11}; -1.2×10^{12}; -2.8×10^{15}; 6.9×10^{16}).

(Universal gravitational constant, $G = 6.67 \times 10^{-11}$ N m^2 kg^{-2})

1. The gravitational field strength on the Earth's surface is 9.8 N kg^{-1}. If the Earth has a mass (M) and a mean radius (R), calculate the field strength
   (a) at a point which is at a distance equal to $4R$ from the centre of the Earth
   (b) on the surface of a planet having a mass $2M$ and radius $3R$.

2. A neutron star has a mass of 5.0×10^{29} kg and a radius of 12 km. Calculate
   (a) the mean density of the neutron star
   (b) the magnitutde of the gravitational field strength and potential at its surface
   (c) the impact velocity of a meteorite on its surface (assuming the meteorite was initially at rest at the extreme limit of the star's gravitational field).

3. The mass and radius of the Moon are 7.4×10^{22} kg and 1.7×10^6 m respectively. Calculate
   (a) the minimum energy needed to project a space vehicle of mass 4.0×10^5 kg from the surface of the Moon so that it escapes completely from the Moon's gravitational field
   (b) the escape velocity of the Moon.

Past-paper questions

1 A muscle exerciser consists of two steel ropes attached to the ends of a strong spring contained in a telescopic tube. When the ropes are pulled sideways in opposite directions, as shown in the simplified diagram, the spring is compressed. The spring has an uncompressed length of 0.80 m. The force F (in N) required to compress the spring to a length x (in m) is calculated from the equation $F = 500(0.80 - x)$.

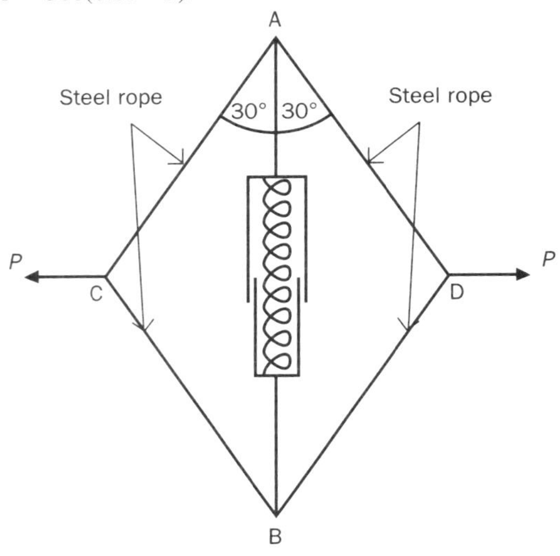

The ropes are pulled with equal and opposite forces P so that the spring is compressed to a length of 0.60 m and the ropes make an angle of 30° with the length of the spring.

(a) Calculate
(i) the force F
(ii) the work done in compressing the spring.

(b) By considering the forces at A or B, calculate the tension in each rope.

(c) By considering the forces at C or D, calculate the force, P.

(NEAB, June 1991)

2 (a) State the difference between a vector and a scalar quantity.

(b) A device for removing tightly fitting screw tops from jars and bottles is show in Figure 1.

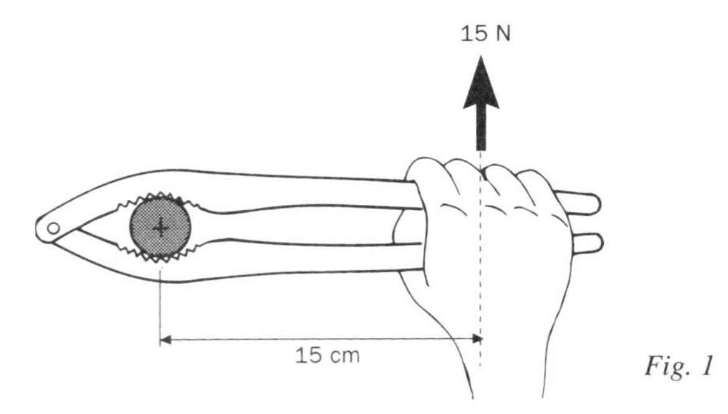

Fig. 1

In one case a constant force of magnitude 15 N has to be applied, as shown in Figure 1. The force is applied for one complete turn to remove the top.

(i) Calculate the torque which has to be applied to remove the top.

(ii) Calculate the work done when opening the bottle.

(c) Tight fitting corks can be removed by pumping air into the bottle using the device shown in Figure 2.

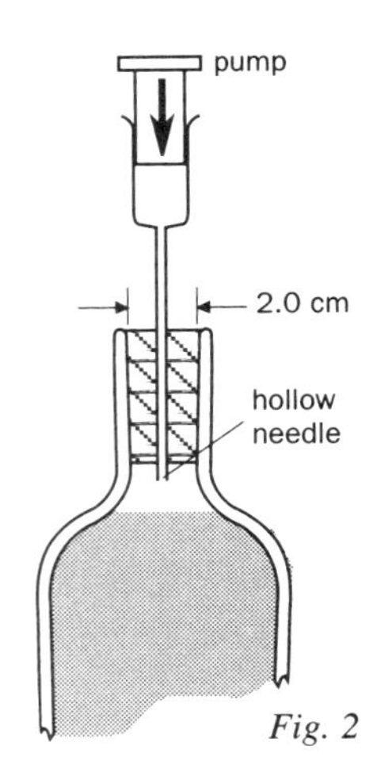

Fig. 2

Assuming that the force which has to be overcome to remove the cork is 30 N, calculate the pressure inside the bottle when the cork begins to move.

Atmospheric pressure $= 1.0 \times 10^5$ Pa.

(AEB, June 1993)

3 A bus travelling steadily at 30 m s^{-1} along a straight road passes a stationary car which, 5 s later, begins to move with a uniform acceleration of 2 m s^{-2} in the same direction as the bus.

(a) How long does it take the car to acquire the same speed as the bus?

(b) How far has the car travelled when it is level with the bus?

(WJEC, June 1992)

4 The sketch shows a water skier approaching a ramp at 10 m s^{-1}. As the skier reaches the foot of the ramp, he releases the tow line.

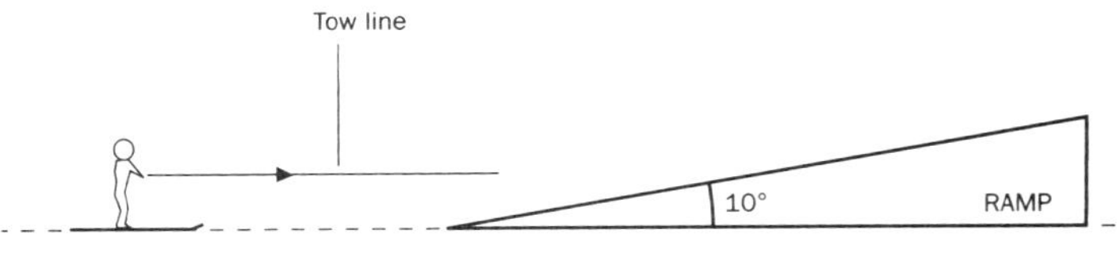

Find

(a) the deceleration of the skier as he moves up the ramp

(b) the length L of the inclined face of the ramp if the skier is to come to rest at the top.

Ignore friction.

(WJEC, June 1993)

5 An aircraft is travelling horizontally at 250 m s^{-1} at a height of 4500 m when a part of the fuselage becomes detached. Find

(a) the time taken for the detached part to reach the ground

(b) the horizontal distance travelled from the point of detachment to striking the ground

(c) the velocity on striking the ground.

(Neglect air resistance.)

(WJEC, June 1994)

6 A light string carrying a small bob of mass 5.0×10^{-2} kg hangs from the roof of a moving vehicle.

(a) What can be said about the motion of the vehicle if the string hangs vertically?

(b) The vehicle moves in a horizontal straight line from left to right, with a constant acceleration of 2.0 m s^{-2}.
   (i) Show in a sketch the forces acting on the bob.
   (ii) By resolving horizontally and vertically or by scale drawing, determine the angle which the string makes with the vertical.

(c) The vehicle moves down an incline making an angle of 30° with the horizontal with a constant acceleration of 3.0 m s^{-2}. Determine the angle which the string makes with the vertical.

(NEAB, June 1992)

7 An astronaut is outside her space capsule in a region where the effect of gravity can be neglected. She uses a gas gun to move herself relative to the capsule. The gas gun fires gas from a muzzle of area 160 mm^2 at a speed of 150 m s^{-1}. The density of the gas is 0.800 kg m^{-3} and the mass of the astronaut, including her space suit, is 130 kg.

Calculate

(a) the acceleration of gas leaving the gun per second

(b) the acceleration of the astronaut due to the gun, assuming that the change in mass is negligible.

(NEAB, June 1992)

8 A student whose mass is 40 kg is an expert with a skateboard. He stands with his left foot on the board and pushes himself along a hard level surface with his right foot. The following graph shows approximately how the horizontal force, F, exerted by his foot on the ground varies during the first few seconds.

Calculate the student's speed after 2 seconds, 4 seconds and 5.8 seconds. Use these values to draw a speed–time graph for the skater's motion from $t = 0$ to $t = 6$ s.

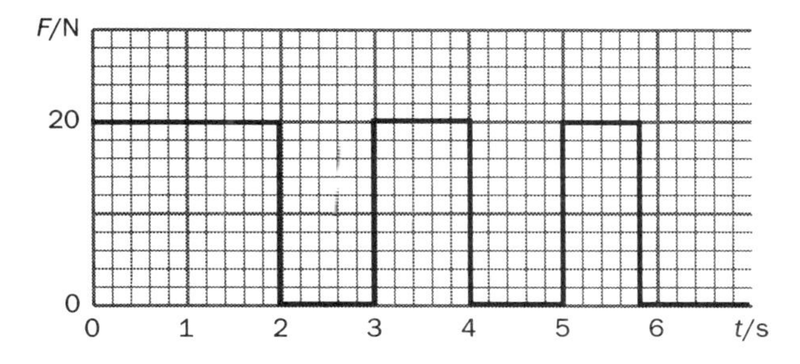

Why does the time for which the student's right foot is in contact with the ground get shorter and shorter?

(Lond., January 1992)

9 State Newton's second law of motion.

A wooden block on four wheels can be pulled along a laboratory bench over a distance of about 2 m by a constant accelerating force. How would you find by measurement the acceleration of the block?

A car of mass 950 kg slows down uniformly from a speed of 20 m s^{-1} to rest in 4.5 s.

Calculate

(i) the initial momentum of the car

(ii) the force applied to the car, and

(iii) the distance it travels while coming to rest.

(Lond., January 1994, part question)

10 On a linear air-track the gliders float on a cushion of air and move with negligible friction. One such glider of mass 0.5 kg is at rest on a level track. A student fires an air rifle pellet of mass 1.5×10^{-3} kg at the glider along the line of the track. The pellet imbeds itself in the glider which recoils with a velocity of 0.33 m s^{-1}.

(a) State the principle you will use to calculate the velocity at which the pellet struck the glider.

Calculate the velocity at which the pellet struck.

(b) Another student repeats the experiment with the air-track inclined at an angle of 2° to the horizontal. Initially, the glider is at the bottom of the track. After the impact the glider recoils with the same initial velocity but slows down and stops momentarily further along the track.

Explain *in words* how to calculate how far along the track the glider moves before stopping instantaneously.

Calculate how far the glider moves along the air-track before stopping momentarily.

(O & C, June 1992)

11 A bullet of mass 0.020 kg is fired horizontally at 150 m s^{-1} at a wooden block of mass 2.0 kg resting on a smooth horizontal plane. The bullet passes through the block and emerges undeviated with a velocity of 90 m s^{-1}.

Calculate

(a) the velocity acquired by the block

(b) the total kinetic energy before and after penetration and account for their difference.

(WJEC, June 1991)

12 Sand is poured at a steady rate of 5.0 g s^{-1} onto the pan of a direct reading balance calibrated in grams. If the sand falls from a height of 0.2 m onto the pan and it does not bounce off the pan then, neglecting any motion of the pan, calculate the reading of the balance 10 s after the sand first hits the pan.

(WJEC, June 1992)

13 The engine of a car of mass 1200 kg works at a constant rate of 18 kW. The top speed of the car is 30 m s^{-1}.

(a) Find the resistance to its motion at the top speed.

(b) Assuming that the resistance to the motion is proportional to the speed of the car, calculate the acceleration of the car at the instant when its speed is 10 m s^{-1}.

(WJEC, June 1994)

14 (a) (i) Define linear momentum.

(ii) State the principle of conservation of linear momentum making clear the condition under which it can be applied.

(b) A spacecraft of mass 20 000 kg is travelling at 1500 m s^{-1}. Its rockets eject hot gases at a speed of 1200 m s^{-1} relative to the spacecraft. During one burn, the rockets are fired for a 5.0 s period. In this time the speed of the spacecraft increases by 3.0 m s^{-1}.

(i) What is the acceleration of the spacecraft?

(ii) Assuming that the mass of fuel ejected is negligible compared with the mass of the spacecraft determine the distance travelled during the burn. Give your answer to four significant figures.

(iii) What is the thrust produced by the rocket?

(iv) Determine the mass of gas ejected by the rocket during the burn.

(AEB, June 1992)

15 (a) The law of conservation of momentum suggests momentum is conserved in any collision. A tennis ball dropped onto a hard floor rebounds to about 60% of its initial height. State how momentum is conserved in this event.

(b) (i) A top class tennis player can serve the ball, of mass 57 g, at an initial horizontal speed of 50 m s^{-1}. The ball remains in contact with the racket for 0.050 s. Calculate the average force exerted on the ball during the serve.

(ii) **Sketch** a graph showing how the horizontal acceleration of the ball might possibly vary with time during the serve, giving the axes suitable scales.

(iii) Explain how this graph would be used to show that the speed of the ball on leaving the racket is 50 m s^{-1}.

(AEB, June 1990)

16 A tennis player drops a ball of mass 0.10 kg. It falls a distance of 0.60 m before hitting the ground and rebounds vertically to a height of 0.24 m.

Assuming that air resistance can be neglected, calculate

(a) the kinetic energy lost in the bounce

(b) (i) the change in momentum of the ball in the impact with the ground

(ii) the average force on the ball producing this change in momentum if the ball is in contact with the ground for 50 ms.

(NEAB, June 1993)

17 State the law of conservation of energy.

Describe an experiment to test the law of conservation of energy in a situation involving a collision between two bodies. List the measurements that you would make to and show how you would use these measurements to check the law.

A shell of mass 7.0 kg travelling at 550 m s^{-1} explodes into two equal fragments travelling at twice this speed. Calculate the energy released by the explosion.

(Lond., January 1994, part question)

31 A pile-driver consists of a heavy block of weight 2000 N. In use, the block falls freely under gravity through a height of 3.5 m. It strikes the upper end of a vertical metal girder, the lower end of which is positioned in soft ground. After impact, the block girder remain in contact and both move down a further 0.5 m as the girder is driven into the ground.

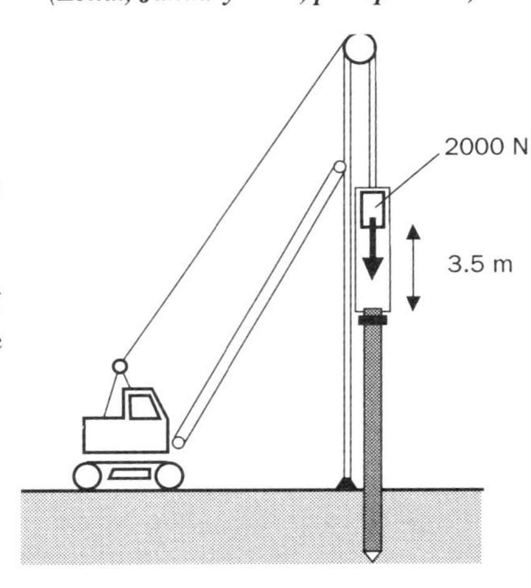

In this question neglect the mass of the girder in comparison to the mass of the driver.

(a) Find the average e resistive force of the ground to the penetration of the girder. Assume that half of the potential energy lost by the pile-driver is used to drive in the girder.

(b) At each blow of the driver the girder sinks a depth d. It is suggested that d decreases exponentially, that is, $d = d_0e^{-kn}$ where d_0 and k are constants and n is the number of blows.

(i) Use the data below to test this hypothesis. Assume that the pile-driver is dropped from 3.5 m above the top of the girder for each blow.

d/m	0.50	0.35	0.24	0.17	0.12
n	1	2	3	4	5

(ii) Suggest a possible reason for the decreasing value of d.

(O & C, June 1994)

34 The diagram illustrates part of a roller-coaster 'loop-the-loop' ride as at some theme parks. The car starts from rest at A. It moves down the slope and then travels in a vertical circle on the inside of the track and exits onto the incline at the right hand side of the diagram.

Treat the car as a particle of mass m sliding on a frictionless track. The radius of the loop is R.

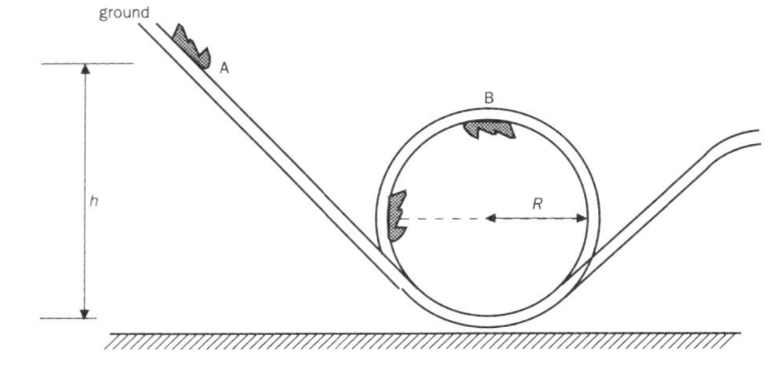

(a) (i) The speed of the car is v when it reaches B, the highest point of the loop. Show that the *minimum* value of v for the car to remain in contact with the tracks is $\sqrt{gR}$.

(ii) Obtain an expression in terms of m, g and R for the minimum kinetic energy of the car at B.

(b) (i) State the law of conservation of energy. What can you deduce about the sum of the kinetic energy and the potential energy of the car at any point in its motion?

(ii) The point A, from which the car is released, is a vertical height h above the lowest point of the track and $R = 12.0$ m. Use the law of conservation of energy to find the minimum value of h which allows the car to loop-the-loop in contact with the track.

(c) The car has $m = 150$ kg. It is now released from rest at $h = 3R$.

(i) Calculate the magnitude of the horizontal force acting on the car when it passes point C, which lies on a horizontal diameter of the loop.

(ii) Calculate the magnitude of the total force on the car as it passes C.

(O & C, June 1993)

20 The design of a space station looks like a giant rotating bicycle wheel with the laboratory area for the crew house within the 'bicycle tyre'. See Figure 1. The radius of the wheel, that is, from the hub to the floor of the laboratories, is 200 m. The crew's living quarters are situated halfway to the hub of the station and are connected to the laboratories by a radial tunnel.

(a) Why does the rotation of the space station about the hub produce artificial gravity for the crew?

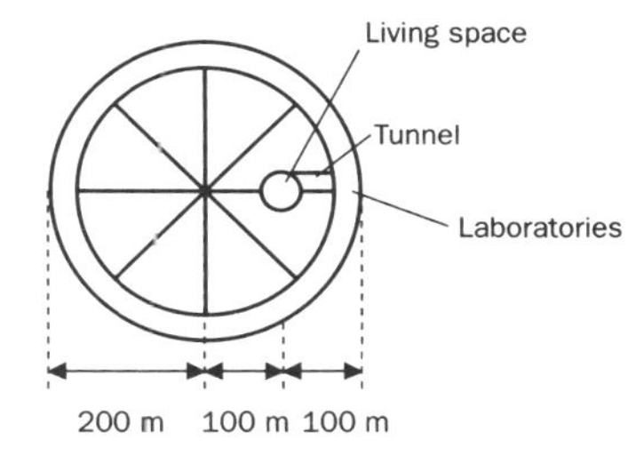

(b) Find the rate of rotation for the crew to experience artificial gravity equal to g (= 9.8 m s^{-2}) in the laboratories.

(c) What is the value of artificial gravity on the floor of the crew's quarters 100 m from the hub of the station?

(d) Sketch a graph of the strength of the artificial gravitational field felt by an astronaut in the radial tunnel between the laboratory and the living quarters. Label your axes appropriately.

(e) Find the minimum energy required for a man of mass 70 kg to move from the laboratory to the living quarters.

(O & C, June 1994)

21 (a) (i) Explain what is meant by the moment of inertia of a body.

(ii) Why is there no unique value for the moment of inertia of a given body?

(iii) A rigid body rotates about an axis with an angular velocity ω. If the relevant moment of inertia of the body is I, show that its rotational kinetic energy is $\frac{1}{2}I\omega^2$.

(b) (i) A motor car is designed to run off the rotational kinetic energy stored in a flywheel in the car. The flywheel is to be accelerated up to some maximum rotational speed by electric motors placed at various stations along the route. If the flywheel has a moment of inertia of 300 kg m^2 and is accelerated to 4200 revolutions per minute at a station, calculate the kinetic energy stored in the flywheel.
Assuming that at an average speed of 54 km h^{-1} the power required by the car is 15 kW, what is the maximum possible distance between stations on the car's route?

(ii) What assumption did you make in the last calculation? Comment on the feasibility of the design.

(WJEC, June 1990)

22 (a) In problems involving linear motion the following equations are often used:

(i) Force = mass × acceleration
(ii) Kinetic energy = $\frac{1}{2}$ × mass × (velocity)^2
(iii) Work = force × distance

Using words, write down the corresponding equations for rotational motion.

(b) A couple of torque 5 N m is applied to a flywheel initially at rest. Calculate its kinetic energy after it has completed 5 revolutions. Ignore friction.

(WJEC, June 1992)

23 State Newton's law of gravitation as applied to two particles of masses M_1 and M_2 at a distance d apart.

(a) Use the law to obtain an expression for the total energy of a satellite of mass m in a circular orbit of radius r about a planet of mass M. Account for the sign you give to this quantity.

(b) (i) What is a geostationary satellite?

(ii) What are the characteristics of a geostationary satellite?

(iii) Calculate the radius of orbit of a geostationary satellite orbiting the Earth.
(mean radius of the Earth is 6.37×10^6 m)

(O & C, June 1994)

24 (a) Distinguish between speed and velocity.

(b) Explain why a body moving in a circle with constant speed must have an acceleration. State the magnitude and direction of this acceleration.

(c) A satellite is to be placed in geostationary orbit, i.e. it always appears to be in the same place when viewed from a given point on Earth. Assuming a value for the periodic time of orbit of the satellite, calculate the height of the orbit above the equator. Why must such an orbit be in the equatorial plane of the Earth?

(d) Explain carefully why an astronaut in an orbiting space station feels 'weightless'.
(Radius of the Earth = 6400 km)

(WJEC June 1991)

25 The rings of the planet Saturn consist of a vast number of small particles, each in a circular orbit about the planet. Two of the rings are shown in the diagram.

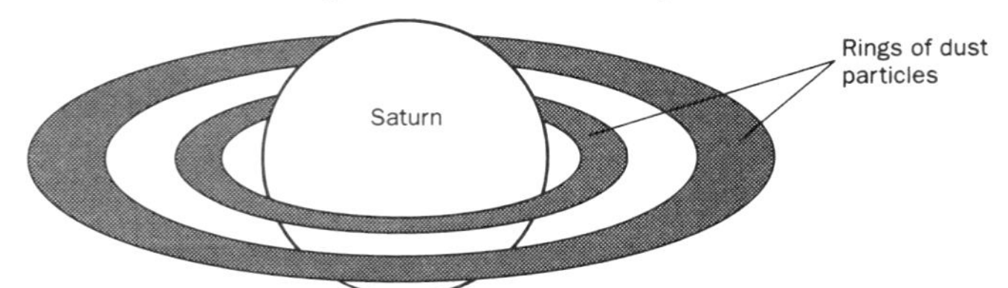

The inner edge of the inner ring is 70 000 km from the centre of the planet and the outermost edge of the outer ring is 140 000 km from the centre. The speed of the outermost particles is 17 km s^{-1}.

(a) Show that the speed, v, of a particle in an orbit of radius r around a planet of mass M is given by

$$v = \sqrt{\frac{GM}{r}}$$

where G is the universal gravitational constant, 6.7×10^{-11} N m^2 kg^{-2}.

(b) Determine the mass of Saturn.

(c) How long does it take for the outermost particles to complete an orbit?

(d) Calculate the orbital speed of the particles nearest to Saturn.

(AEB, June 1994)

26 (a) Define gravitational potential.

(b) Find the gravitational potential 1000 km above the Earth's surface.

(c) An object is released from rest at this height. Find its velocity as it reaches the Earth's atmosphere at 10 km above the Earth's surface.
(The radius of the Earth is 6400 km. The gravitational potential is given by

$$v_g = \sqrt{\frac{GM}{r}}$$

where the symbols have their usual meanings and the product GM has the value 4.0×10^{14} m^3 s^{-2}.)

(WJEC, June 1993)

46 (a) State what is meant by gravitational potential at a point and the gravitational potential energy of a body in a gravitational field due to a point mass.

(b) A deuteron consists of a proton and a neutron about 1.5×10^{-15} m apart.

(i) Calculate the gravitational force of attraction between the two particles.

(ii) Calculate the energy required to separate the two particles completely, assuming that the only force involved is the gravitational force.

(iii) The actual energy involved is about 2.2 MeV. Explain why this suggests that there are forces other than gravitational forces involved.

The universal gravitational constant
$G = 6.7 \times 10^{-11}$ N m^2 kg^{-2}

The mass of a proton = mass of a neutron = 1.7×10^{-27} kg
The charge on an electron
$e = -1.6 \times 10^{-19}$ C

(AEB, June 1991)

28 (a) The circle shown in Figure 1 represents a planet of radius 4.8×10^6 m with a gravitational field strength of 16 N kg^{-1}.

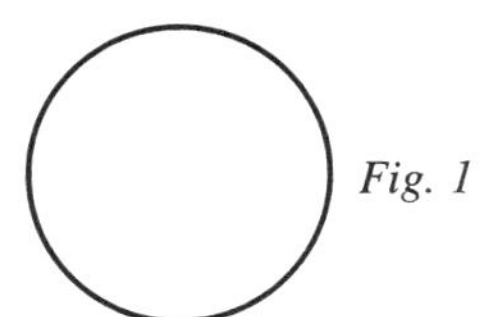
Fig. 1

(i) Draw on the diagram a set of solid lines to represent the gravitational field of the planet. Add a set of dotted lines to represent gravitational equipotentials.

(ii) Calculate the mass of the planet.

(b) Two points, P and Q, lie above the surface of the planet. The gravitational potential at P is -5.2×10^7 J kg^{-1}. and that Q is -6.9×10^7 J kg^{-1}.

(i) By comparison of these gravitational potentials, deduce which point, P or Q, is nearer to the surface of the planet. Explain your reasoning.

(ii) Calculate the difference in heights of points P and Q above the surface of the planet.

(O & C, June 1992)

SECTION

2 Oscillations, Vibrations and Waves

In this section we shall examine repetitive motion. Sewing machine needles, cyclists' legs, pistons in car engines, children on swings, etc. all **oscillate**. That is, they all exhibit a repeating cycle of motion. Some of these motions repeat in a regular fashion while others are much more complex. The study of wave motion is important to many areas of physics and is developed from the study of these oscillatory motions. For example, radio communication depends on our understanding of the properties of electromagnetic radiation. The destructive action of an earthquake is due to the distribution of energy from the epicentre by seismic waves – study of the interaction of these waves with buildings enables architects and civil engineers to design safer houses and structures. We shall also study light and sound as examples of wave motion.

Chris Boardman's legs execute continuous, repetitive motions as he hurtles round the track on his famous Lotus bike

2.1 Simple harmonic motion

In a previous chapter, we studied linear and circular motion. Another kind, called 'oscillatory motion', involves **periodic oscillation** or vibration about a fixed position. Periodic oscillation describes the **repeated** retracing of a path in a fixed **time interval**. The motion of such systems is apparently complex – the direction changes abruptly, velocity varies widely and acceleration is far from constant. A great many systems in nature exhibit a wide variety of types of periodic motion. We are concerned here with the simplest of these motions:

Simple harmonic motion (SHM) is defined as motion in which the **acceleration** of the body is directly proportional to its **displacement** from a fixed point and is always directed **towards** that point.

In other words, the body is oscillating equal distances either side of some fixed point. The further the body is from this point, the greater its acceleration back towards that point. This tells us (Newton's second law) that the **resultant force** exerted on the body must increase with distance from this point, and always be acting in a direction towards that point, so as to **restore** the body to its undisturbed position. In this undisturbed position, the resultant force on the body will be zero – it will be in **equilibrium**.

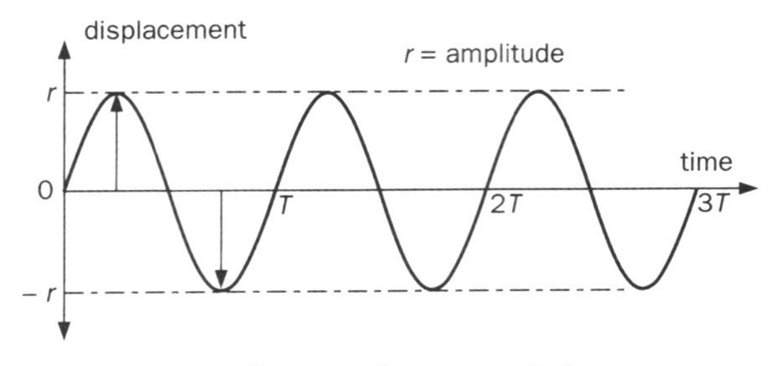

Figure 2.1 Oscillation of a tuning fork prong

We shall examine various ways of analysing this motion, which will show that for a particle undergoing SHM, the variation of displacement with time has the same shape as a **sine wave**. The prongs of a vibrating tuning fork (over a short time interval) oscillate in a pure 'sinusoidal' manner as can be seen in Figure 2.1.

The equation for a **sinusoidal** oscillation is of the form:

$$x = r \sin \omega t$$

Equation 1

where x = displacement at any time (t).
r = maximum displacement (amplitude)
ω = angular frequency (in rad s^{-1})

Maths window

The shape of a **sine** curve (Figure 2.2) should be familiar to you:

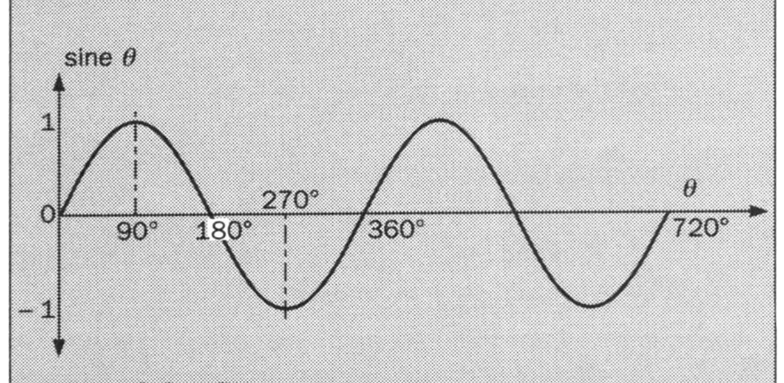

Figure 2.2 Sine curve

The maximum and minimum values of sin θ are +1 and –1 respectively. The cycle repeats after 360°.

An angle of 360° is equal to 2π radian.

At this point you should revise Section 1.8 (Motion in a circle), as the relationships between *frequency*, *period* and *angular frequency* are very important here.

Period

Common symbol T
SI unit second (s)

The **period** of an oscillatory motion is defined as the time taken to complete one oscillation.

Frequency

Common symbol f
SI unit hertz (Hz; 1 Hz = 1 cycle s^{-1})

The **frequency** of an oscillatory motion is defined as the number of complete cycles occurring in **unit time** (usually 1 s).

Angular frequency

Common symbol ω

SI unit radian per second (rad s^{-1})

The **angular frequency** of an oscillatory motion is **frequency** expressed in **radians per second** (rad s^{-1}).

Since one complete cycle of an oscillation can be represented by 360° or 2π rad:

$$\omega = 2\pi f$$

Equation 2

Analytical treatment

Here we shall look at an object moving in a circular path and examine the connection between this and simple harmonic motion.

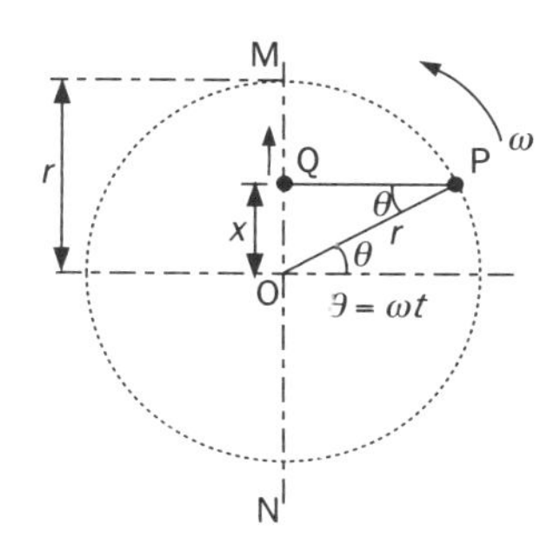

Figure 2.3 Circular motion and SHM

Point P is a particle moving along a circular path of radius (r) with constant angular velocity (ω) (see Figure 2.3). (Think of it as a piece of chewing gum on a bicycle tyre.) If we consider only its **height** above or below the axle, then this is what we call its vertical **displacement**. This is represented by the point Q in Figure 2.3, which is at the foot of the perpendicular from P to MN.

As P moves through one revolution, Q moves O → M → O → N → O. We will show that this oscillation of Q about point O is **simple harmonic motion**.

The motion of Q is **linked** to that of P, which is relatively easy to analyse since it is circular motion at constant speed.

Displacement

The displacement (x) of Q is given by:

$$x = r \sin \theta$$

Equation 3

and since $\theta = \omega t$
(angular displacement = angular velocity × time)

∴

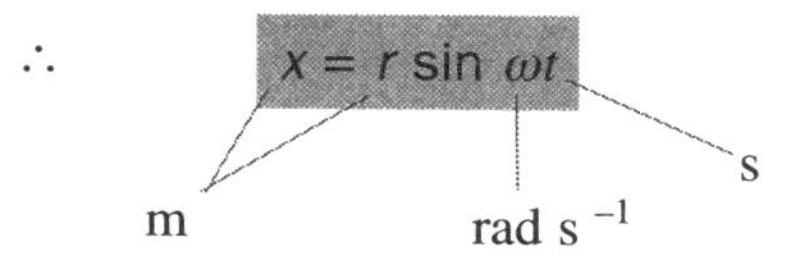

Equation 4 (sine wave equation)

And since $\omega = 2\pi f$

Equation 4 can also be expressed as:

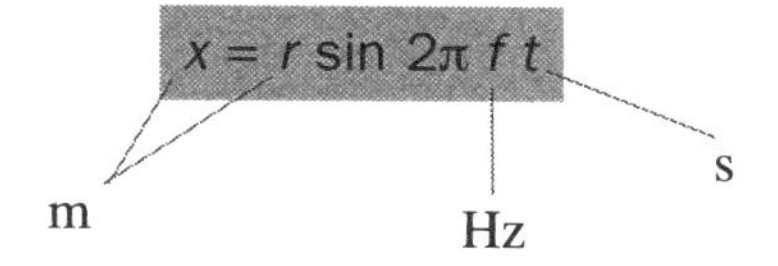

Equation 5

This can be a useful alternative if you are working with frequency (f) in Hz rather than angular frequency (ω) in rad s^{-1}.

Velocity

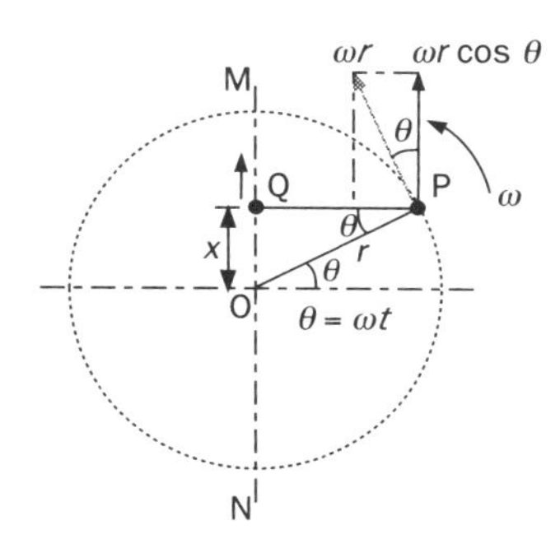

Figure 2.4 Velocity of a particle moving with SHM

Here we look at the same particle P moving in a circle of radius (r) with constant angular velocity (ω) as shown in Figure 2.4. This time we will examine how the **velocity** of point Q is connected to the motion of P.

The linear speed of P is equal to (ωr), and is directed along a **tangent** to the circle at P.

The velocity (v) of Q is the **component** of P's velocity parallel to MN (i.e. the vertical component).

$$v = \omega r \cos \theta$$

and since $\theta = \omega t$,

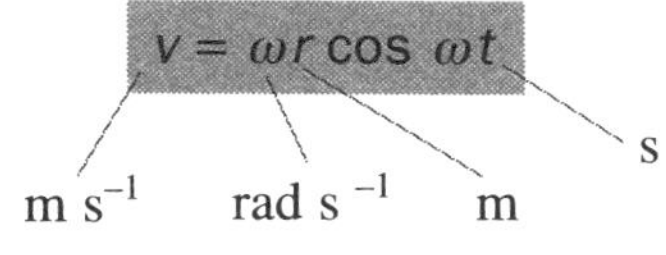

Equation 6

> **Maths window**
>
> Velocity (v) is defined as *'rate of change of displacement'*. The mathematical shorthand for *'rate of change of'* is $\frac{dx}{dt}$
>
> Equation 6 may also be written as:
>
> $$\frac{dx}{dt} = \omega r \cos \omega t$$
>
> This is a commonly-used notation, and is introduced here for completeness.

Acceleration

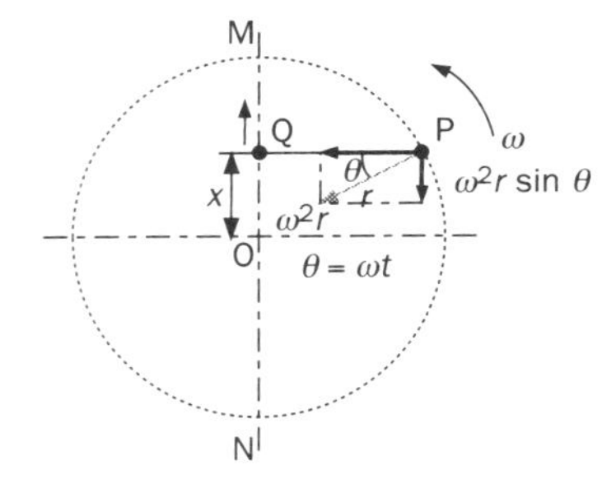

Figure 2.5 Acceleration of a particle moving with SHM

Since P is moving in a circle, its acceleration is equal to $\omega^2 r$ and is directed towards the centre, along PO (see Figure 2.5).

Acceleration of Q towards O (a) is given by the **component** of P's acceleration parallel to the direction of motion of Q (i.e. the **vertical** component of $\omega^2 r$):

$$a = -\omega^2 r \sin \theta$$

and since $\theta = \omega t$:

$$a = -\omega^2 r \sin \omega t$$

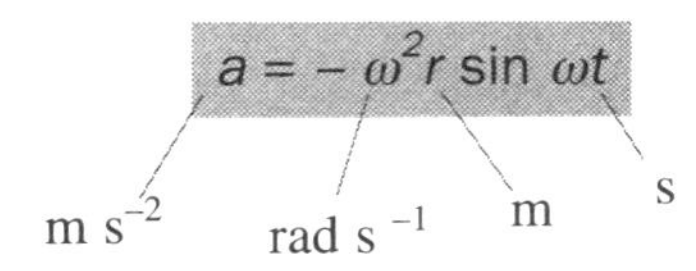

Equation 7

NB: The **negative** sign indicates that the direction of the *acceleration* is **opposite** to the direction of the *displacement.*

Since $x = r \sin \theta$

then

$$a = -\omega^2 x$$

m s^{-2} rad s^{-1} m

Equation 8

> **Maths window**
>
> Acceleration (a) is defined as *'rate of change of velocity'* and velocity is defined as *'rate of change of displacement'*. Using mathematical shorthand, this may also be written:
>
> $$a = \frac{dv}{dt} \quad \text{or} \quad a = \frac{d^2x}{dt^2}$$
>
> $$\therefore \quad \frac{d^2x}{dt^2} = -\omega^2 x$$

Equation 8 is the general equation for a body performing SHM. It shows that in this type of motion, the acceleration of the body is:

- directly proportional to the displacement (x) from a fixed point (the equilibrium or undisturbed position)
- always directed towards the fixed point (shown by the negative sign).

Significance of the constant (ω^2)

Since it is a number **squared**, it always has a **positive** value, even if the number is negative. Thus whatever the value of ω, acceleration and displacement will always have opposite signs. This means that whatever the value of x (the *displacement* from the fixed point), the acceleration is **proportional** to displacement from a fixed point, and always directed towards it.

Full analysis of one cycle

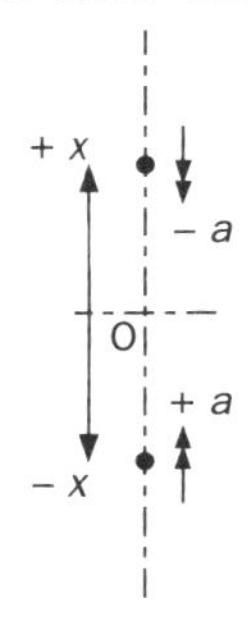

Figure 2.6 Particle moving with SHM of amplitude x

To reinforce our understanding of SHM, a full analysis is needed of the motion of point Q as P undergoes one revolution. Consider Figure 2.6, which shows only the motion of point Q.

Passing O and slowing down towards M

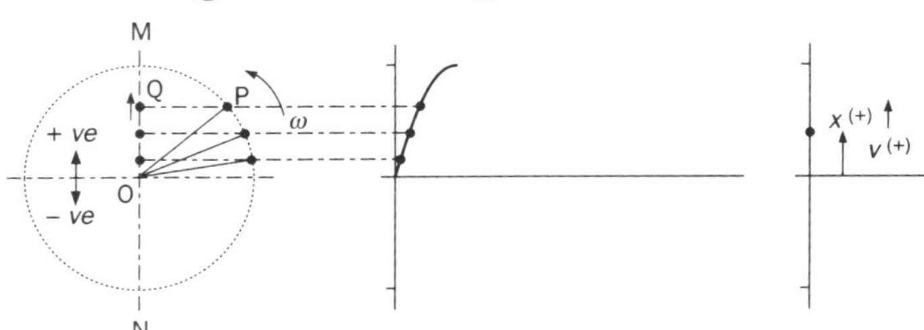

Figure 2.7 First quarter-cycle

We start at time $t = 0$ with Q moving upwards through O, and consider displacements above O to be positive as shown in Figure 2.7.

During the first quarter-cycle Q is moving upwards past O, and slowing down to come to rest (momentarily) at M, so its displacement is **positive**. Its velocity is **positive**, because it is moving in the **positive** direction. However, when we determine whether the acceleration is positive or negative, we must look at how the velocity is **changing**. Here, the speed is decreasing (a **negative** change in magnitude), but moving in the **positive** direction. (The combination of **negative** change of magnitude and **positive** direction results in a **negative** acceleration.)

Speeding up towards O

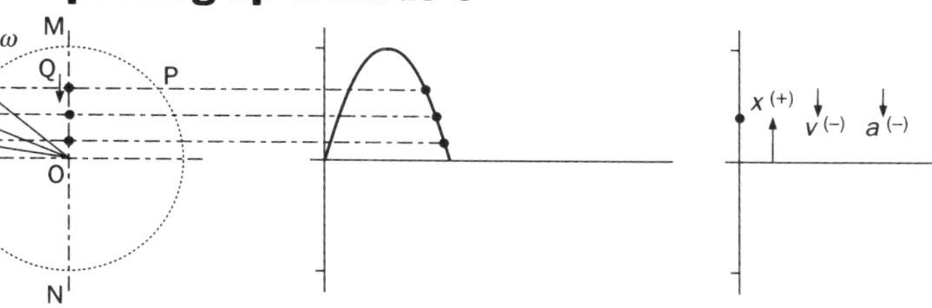

Figure 2.8 Second quarter-cycle

As Q moves down towards O, the displacement (x) is still positive but the velocity is now **negative** since it is moving in the **negative** direction.

The acceleration is still **negative** as although Q is increasing its velocity (positive **magnitude**) its direction is now **negative** as it moves towards O, as shown in Figure 2.8.

Passing O, slowing down towards N

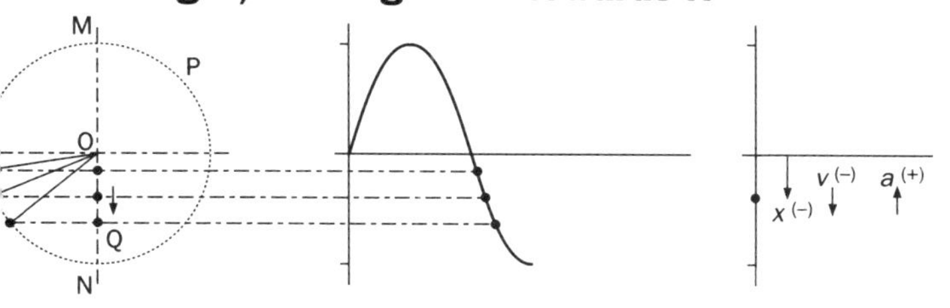

Figure 2.9 Third quarter-cycle

As Q passes through O it has **maximum** velocity. Between O and N both the displacement and velocity are **negative**, but since it is slowing down, the combination of **negative** magnitude and **negative** direction produces a **positive** acceleration, see Figure 2.9.

Speeding up towards O

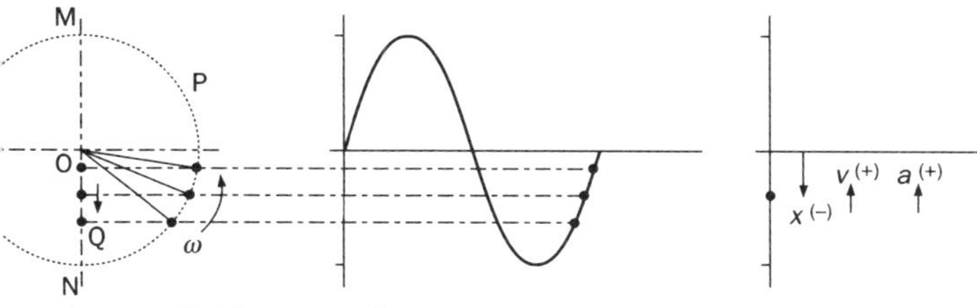

Figure 2.10 Fourth quarter-cycle

At N the velocity of Q is momentarily zero, before it starts speeding up towards O again. The displacement is still **negative**, but its velocity is **positive** in direction (since its displacement as measured from O is becoming less negative), and its magnitude is increasing. The combination of increasing magnitude of velocity and positive direction of motion results in **positive** acceleration, as shown in Figure 2.10. At this point, the cycle is complete, and will repeat.

It is vital to study these diagrams carefully, and to consider fully the implications of the terms **positive** and **negative** as applied to displacement, velocity and acceleration. Confusion here is responsible for a great deal of misunderstanding of this whole topic.

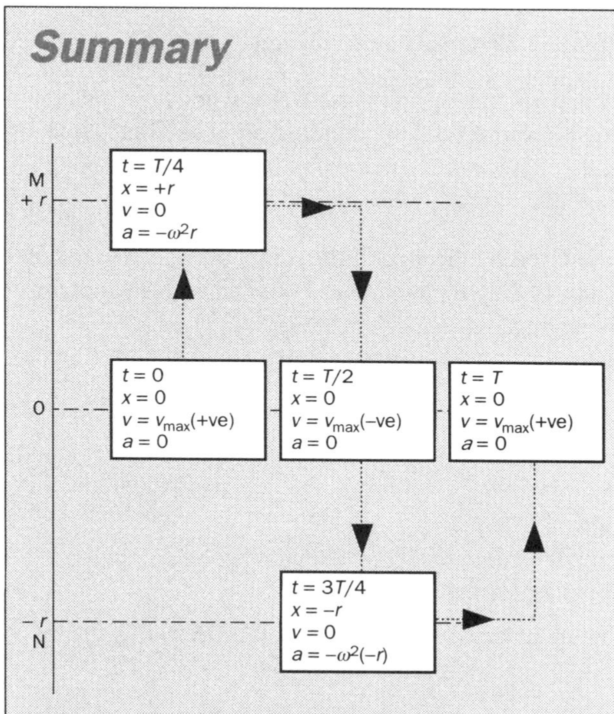

Figure 2.11 Summary of SHM analysis over one complete cycle

The motion is O → M → O → N → O as shown in Figure 2.11.

x = displacement at any instant.

r = amplitude of motion.

T = period of motion.

Graphical treatment

Consider a long spring with a mass (m) attached to one end which is fixed vertically, as shown in Figure 2.12.

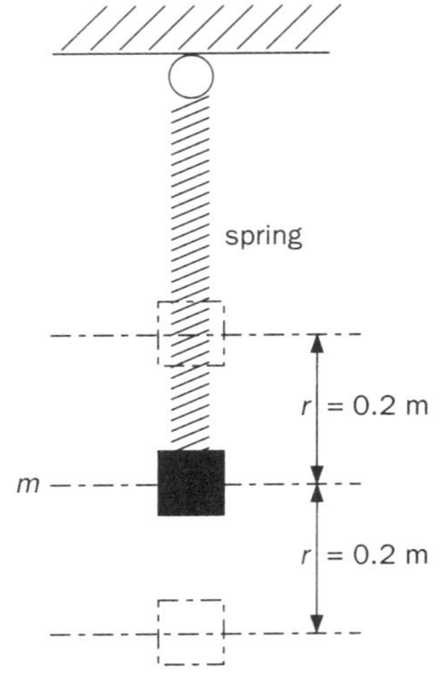

Figure 2.12 Mass on a spring

This particular spring is such that when m is pulled down a further 0.2 m and released, it oscillates in a vertical plane with period $T = 0.4$ s and amplitude $r = 0.2$ m.

(r is fixed by the size of the initial displacement; T is controlled by the magnitude of m and the stiffness of the spring.)

The frequency of the oscillation is given by:

$$f = \frac{1}{T} = \frac{1}{0.4} = 2.5 \text{ Hz}$$

and $\omega = 2\pi f =$ **15.7 rad s^{-1}**

Displacement

The displacement (x) of m at any time (t) is given by:

$$x = r \sin \omega t$$

Equation 9

$\therefore \quad x = 0.2 \sin(15.7t)$

By substituting t-values of 0.1, 0.2, 0.3 s, etc. into this equation, we obtain the corresponding x-values and a displacement–time graph can be plotted as shown in Figure 2.13.

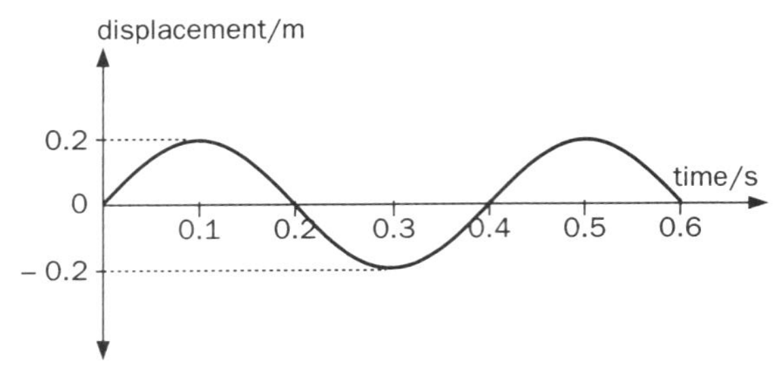

Figure 2.13 Displacement–time graph for SHM

Velocity

The velocity of m at any instant = the **gradient** of the displacement–time graph at that instant.

Looking at Figure 2.13 we can see that the gradient is:

- **zero** when the displacement is a **maximum** or a **minimum**, and
- a **maximum** when the displacement is **zero**.

Since $\quad x = r \sin \omega t$

$$v = \frac{dx}{dt} = \omega r \cos \omega t$$

Equation 10

Since the maximum value of $\cos \omega t$ is 1 (or –1), then the maximum value of the velocity is given by:

$v_m = \omega r \cos \omega t = 15.7 \times 0.2 \times (\pm 1) =$ **± 3.14 m s^{-1}**

By substituting t-values as before into Equation 10 a velocity–time graph can be plotted (Figure 2.14).

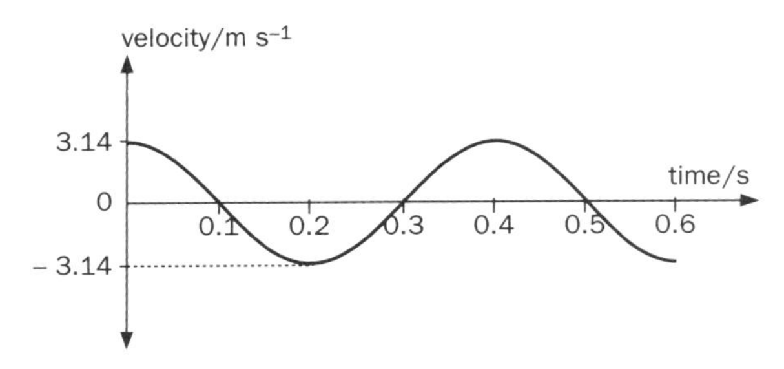

Figure 2.14 Velocity–time graph for SHM

Acceleration

The acceleration of m at any instant = the **gradient** of the velocity–time graph at that instant. From Figure 2.13 and Figure 2.14 we see that the gradient of the velocity–time graph is:

- **zero** when the displacement is zero (velocity is **maximum**)
- **maximum** when the displacement is a **maximum** or **minimum** (velocity is zero).

$$a = \frac{d^2x}{dt^2} = -\omega^2 r \sin \omega t$$

Equation 11

Since the maximum value of $\sin \omega t$ is 1 (or –1), then the maximum value of the acceleration of m (a_{max}) is given by:

$$a_{max} = \omega^2 r \sin \omega t = 15.7^2 \times 0.2 \times (\pm 1)$$

$\therefore \quad a_{max} =$ **± 49.3 m s^{-2}**

Substituting t-values into Equation 11 enables an acceleration–time graph to be plotted (Figure 2.15).

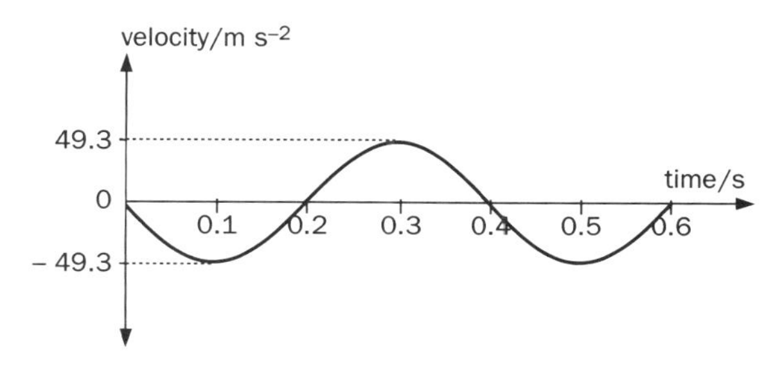

Figure 2.15 Acceleration–time graph for SHM

Comparison of the displacement–time and acceleration–time graphs shows that these graphs have the same shape but opposite sign. This may be stated mathematically as:

$$a \propto -x$$

Hence $a = -\omega^2 x$

(where ω^2 is the constant of proportionality)

This is the general equation for SHM.

Summary (time relationships)

So far we have examined how *displacement*, *velocity* and *acceleration* of an object performing SHM vary with *time*:

$x = r \sin \omega t$ Equation 9

$v = \omega r \cos \omega t$ Equation 10

$a = -\omega^2 r \sin \omega t$ Equation 11

(Symbols as defined in the text.)

Variation of velocity and acceleration with displacement

It is also important to examine how *velocity* and *acceleration* vary with the *displacement* of the body from the equilibrium position.

Velocity

As we have already seen, the velocity (v) of Q at any instant is given by $v = \omega r \cos \omega t$ (Figure 2.16).

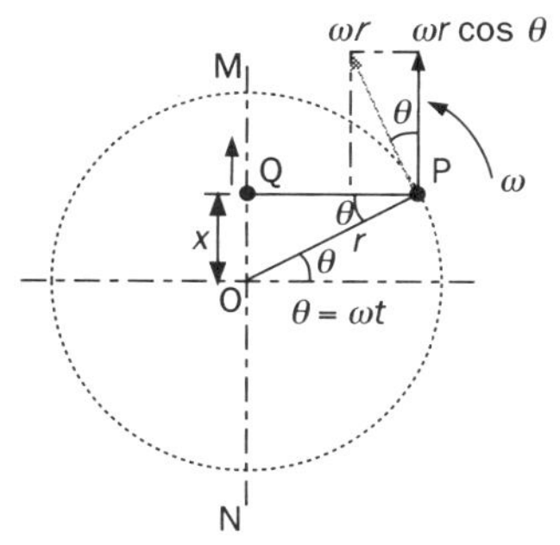

Figure 2.16 Velocity of a particle moving with SHM.

Maths window

Now, $\cos^2\theta + \sin^2\theta = 1$ . . . . (basic trigonometry)

$$\cos\theta = \sqrt{1-\sin^2\theta} \quad \text{but } \sin\theta = \frac{x}{r}$$

$$\cos\theta = \sqrt{1-\frac{x^2}{r^2}}$$

$$\cos\theta = \sqrt{\frac{r^2-x^2}{r^2}}$$

$$\cos\theta = \frac{1}{r}\sqrt{r^2-x^2}$$

Substituting for $\cos\theta$ in Equation 10

$$v = \omega r \times \frac{1}{r}\sqrt{r^2-x^2}$$

$$\therefore \quad v = \pm\omega\sqrt{r^2-x^2}$$

m s^{-1} rad s^{-1} m

Equation 12

You should note that

- When $x = 0$, $v = \pm\,\omega r$. (This means that the velocity of a body performing SHM is a **maximum** as the body passes the equilibrium position.)
- When $x = \pm\, r$, $v = 0$. (This means that the velocity of a body performing SHM is **zero** at the extremities of the motion.)

Acceleration

We have already seen that $a = -\,\omega^2 x$ (Equation 11)

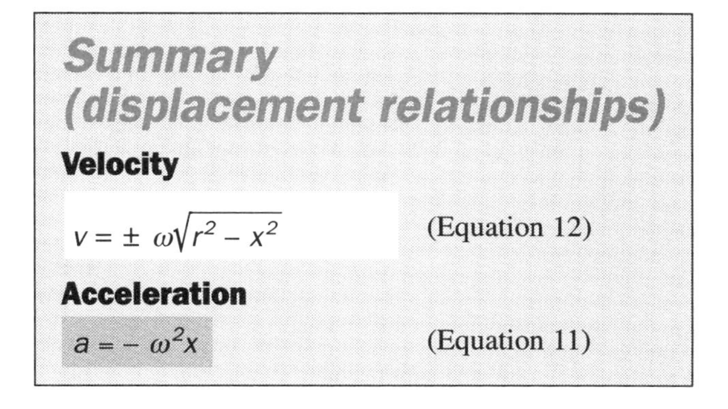

Summary (displacement relationships)

Velocity

$v = \pm\,\omega\sqrt{r^2 - x^2}$ (Equation 12)

Acceleration

$a = -\,\omega^2 x$ (Equation 11)

Maths window

Calculus Approach to SHM Equations

The relation between the acceleration of a body performing SHM and its displacement from the point about which the motion occurs can be deduced by mathematical differentiation.

From Figure 2.15 we can see that:

(i) Displacement, $x = r \sin \theta = r \sin \omega t$.

(ii) Velocity, $\frac{dx}{dt} = \omega r \cos \omega t$.

(iii) Acceleration, $\frac{d^2x}{dt^2} = \frac{dv}{dt} = -\omega \times \omega r \sin \omega t$

$$\frac{d^2x}{dt^2} = -\omega^2 r \sin \omega t$$

$a = -\omega^2 x$. General SHM Equation

Simple pendulum

The simple pendulum consists of a small bob of mass (m) suspended by a light, inextensible string of length (l) from a fixed point O. When the bob is slightly displaced, it oscillates along an arc with centre O and radius l. Consider the instant shown in Figure 2.17 below when m is at point A, distance x from the equilibrium position B:

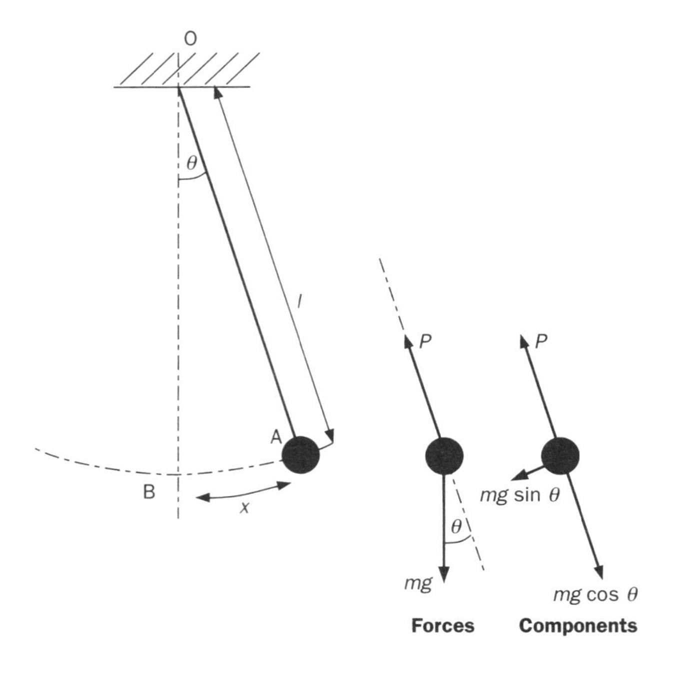

Figure 2.17 Forces on a simple pendulum

Derivation window

The forces acting on the bob are:

- *weight* (mg) vertically downwards
- string *tension* (T) along the string towards O

Resolving radially (i.e. along the string)

For equilibrium:

$P - mg \cos \theta = 0$

Resolving tangentially

Applying Newton II:

$-mg \sin \theta = ma$

Equation 13

You should note that:

- this tangential force ($-mg \sin \theta$) is always directed **towards** the equilibrium position: it is always trying to **restore** the system to its stable state – hence it is called a **restoring force.**
- the negative sign indicates that the restoring force acts towards B while x is measured towards A.

When θ is small ($\approx 5°$), $\sin \theta \approx \theta$ in radian measure.

By definition: $\sin \theta \approx \theta \approx \frac{x}{l}$

$\therefore$ Equation 13 becomes: $-mg\frac{x}{l} = ma$

$\therefore \quad a = -\frac{g}{l}x$

The motion of the bob is thus SHM, since g/l is constant (so long as the oscillations are of small amplitude so that $\sin \theta \approx \theta$).

Comparing $a = -\frac{g}{l}x$ with $a = -\omega^2 x$

we see that: $\omega^2 = \frac{g}{l} \quad \therefore \quad \omega = \sqrt{\frac{g}{l}}$

$\therefore$ Period (T) of the oscillation is given by:

$$T = \frac{2\pi}{\omega} = \frac{2\pi}{\sqrt{g/l}}$$

$$\therefore T = 2\pi\sqrt{\frac{l}{g}}$$

(T: s; l: m; g: m s^{-2})

Equation 14

What does Equation 14 tell us about the connection between the **period**, and (i) the **amplitude** of the oscillations (if they remain small) and (ii) the **mass** of the bob?

Mass–spring system

For a spiral spring which obeys Hooke's law, the extension is directly proportional to the applied tension.

i.e. *tension* = k × *extension*

(k is the **spring constant** which is equal to the tension per unit extension (unit – N m^{-1}).)

Simple harmonic motion describes a particularly simple form of repetitive oscillation. Two important characteristics of this simplicity are:

- the *time period* is constant
- the *amplitude* is constant.

In most real oscillating systems, these are not achieved. In particular, all real systems will dissipate energy to the surroundings, usually by doing work against viscous and friction forces which produce an increase in the thermal energy of the environment. This results in a decrease in amplitude. In the next section we shall examine qualitatively some of these points.

Derivation window

Consider a suspended spring with mass (m) attached to its lower end. Figure 2.18(b) shows the mass and spring in equilibrium.

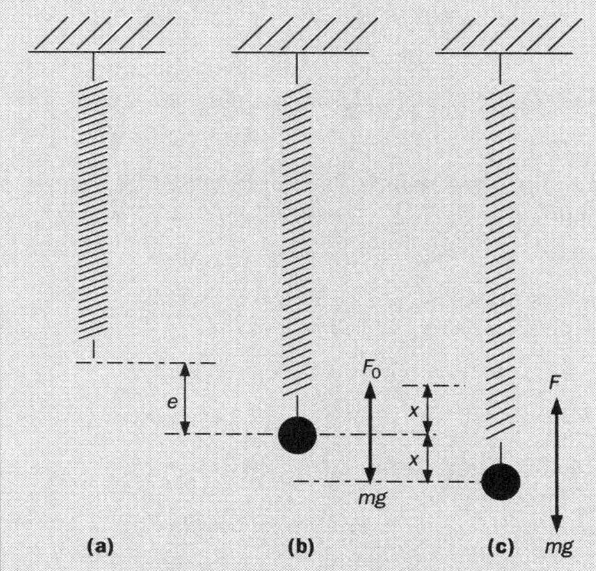

Figure 2.18 Forces in mass–spring system

Then, $F_0 = mg$ where F_0 is the spring tension when the system is in equilibrium.

But, $F_0 = ke$

$\therefore \quad mg = ke$

If (m) is now pulled down a further distance (x) below the equilibrium position, the tension in the spring will increase to a value (F) so that the net upward force on the mass will be given by:

$$F - mg$$

and $\quad F = k(e + x)$

since ($e + x$) is the total extension of the spring.

Applying Newton II:

$$-[k(e + x) - mg] = ma$$

The minus sign must be present because the net force and the displacement are always in opposite directions.

But $\quad ke = mg$

$$\therefore \quad ma = -(mg + kx - mg)$$

$$\therefore \quad ma = -kx$$

$$\therefore \quad a = -\frac{k}{m}x$$

Equation 15

The motion of the oscillating mass is thus SHM, since k/m is constant.

Comparing $\quad a = -\frac{k}{m}x \quad$ with $\quad a = -\omega^2 x$

we see that: $\omega^2 = \frac{k}{m} \quad \therefore \quad \omega = \sqrt{\frac{k}{m}}$

$\therefore$ Period (T) of the motion is given by:

$$T = \frac{2\pi}{\omega} = \frac{2\pi}{\sqrt{\frac{k}{m}}}$$

$$\therefore \quad T = 2\pi\sqrt{\frac{m}{k}}$$

(s; m – kg; k – N m^{-1})

Equation 16

Also, since $\quad mg = ke$

$$\frac{m}{k} = \frac{e}{g}$$

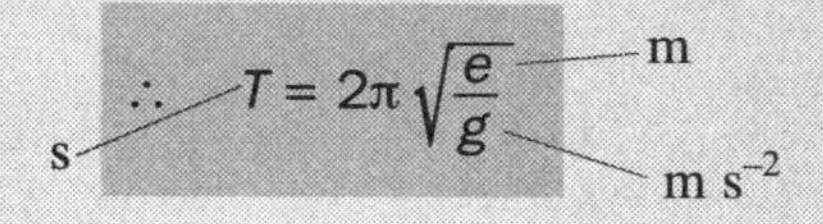

Equation 17

Damped oscillations

Free oscillation describes a system in which the total energy of the system does not decrease with time. Thus the amplitude of vibration remains constant (see Figure 2.19).

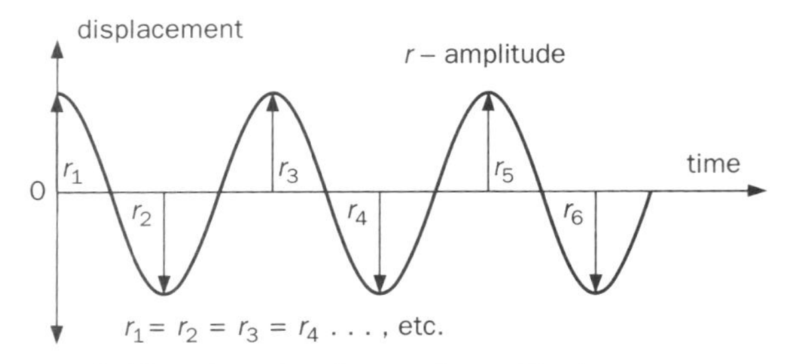

Figure 2.19 Free (undamped) oscillations

In real cases, such as a simple pendulum, a mass on a spring or a floating body bobbing up and down, the amplitude of the oscillation will decay to zero as the system dissipates its energy to the surroundings, mostly through the action of friction and viscous forces. This action is described as **damping**.

In many cases, this damping is desirable, and is maximised, as in the case of **dampers** (usually called, incorrectly, *shock absorbers*) on vehicles. In vehicle suspension systems, springs are used to absorb shocks from bumps in the road, and to allow the wheels to follow an uneven surface while the vehicle follows a more-or-less horizontal path. Without damping, the vehicle would oscillate in a vertical plane long after the first bump. Dampers dissipate the energy of oscillation, preventing the wild pitching that would otherwise occur.

When a car goes over bumps in the road the oscillations are controlled by the springs and dampers which constitute the suspension system

The paper cone of a loudspeaker is heavily damped so that it dissipates almost all of its energy (as sound energy) to the air.

In a lightly-damped system the oscillations decrease in amplitude very gradually, and over a short time interval the displacement/time graph approximates to the ideal case shown in Figure 2.19.

Heavier damping causes a more obvious decrease of amplitude with time, and the resulting graph of displacement against time is as shown in Figure 2.20.

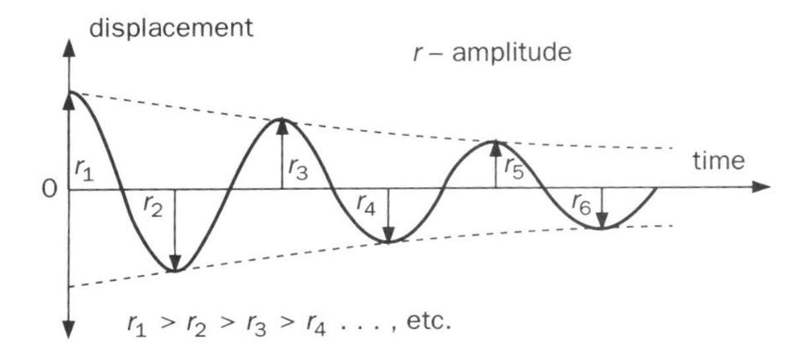

Figure 2.20 Effect of light damping

Increasing the amount of damping causes the amplitude of the oscillations to decay to zero very rapidly:

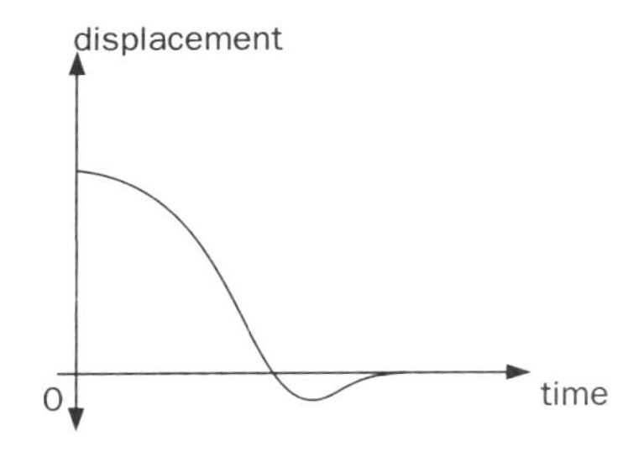

Figure 2.21 Effect of heavy damping

Figure 2.21 shows the effect of heavy, but not excessive damping. When released, the system barely overshoots its equilibrium position, before coming to rest in equilibrium. Notice that the time taken to first return to equilibrium has increased.

It is possible for the system to just return to its equilibrium position with no overshoot.

Critical damping

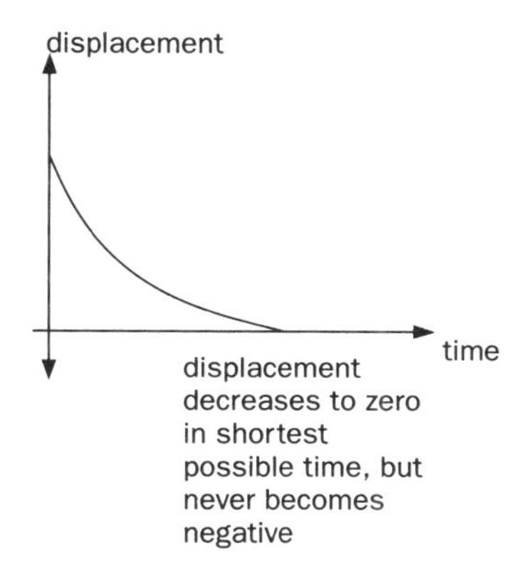

Figure 2.22 Effect of critical damping

A system which returns to its equilibrium position with no overshoot and in the shortest possible time is said to be **critically damped**, as shown in Figure 2.22.

Any further increases in damping will result in the system becoming **'overdamped'** and the system will take longer to come to rest, as shown in Figure 2.23.

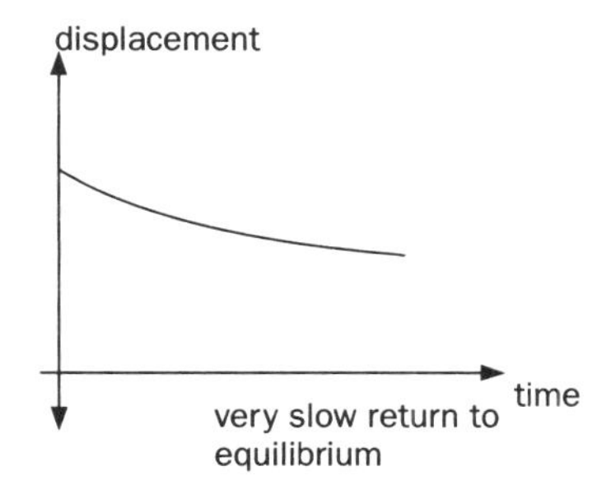

Figure 2.23 An overdamped system

Many systems rely on critical damping – car suspension systems must return the car to its desired ride height as quickly as possible without oscillation; meters and pointer instruments are usually critically damped to allow the needle to reach its correct position on the scale without oscillating. Meters which need to show rapid fluctuations (such as 'level meters' on tape recorders) may be lightly damped while meters which are required to ignore transient changes (such as car fuel gauges) are overdamped.

Forced oscillations and resonance

Systems which are free to oscillate will have one or more **natural frequencies** of vibration. Systems like masses on springs and simple pendulums, where the mass is concentrated in lumps, have a single natural frequency, whereas systems where the mass is distributed, such as stretched strings or the air in organ pipes, have several natural frequencies.

If a system is stimulated by a periodic force it will be **forced** to vibrate at the **forcing frequency**. If the forcing frequency is equal to a **natural frequency** of the system, then the amplitude of vibration will be large. (If there were no damping to dissipate energy then the amplitude of vibration would increase continuously as the system absorbed energy from whatever was stimulating it.) If the amplitude of oscillation is plotted against forcing frequency, the resulting graph is called the **frequency response** of the system, and typically looks as shown in Figure 2.24.

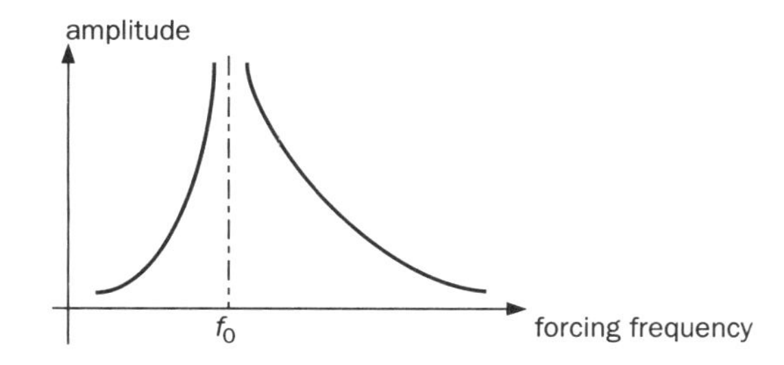

Figure 2.24 Frequency response curve

Resonant frequency (f_0)

The frequency at which maximum response occurs, f_0, is known as the **resonant frequency** of that system. If the system is undamped, it will be equal to the natural frequency of the system. With no damping, the amplitude (and energy) of the system will increase continuously. With damping, the amplitude and energy will increase until energy is being dissipated at the same rate as it is being supplied.

Effect of damping on natural frequency

Adding damping to the system has two effects:

- the amplitude of the peak oscillation decreases
- the frequency at which maximum response occurs also decreases.

These decreases in amplitude and frequency are more marked with heavier damping. With light damping, the effect on resonant frequency is negligible, as shown in Figure 2.25.

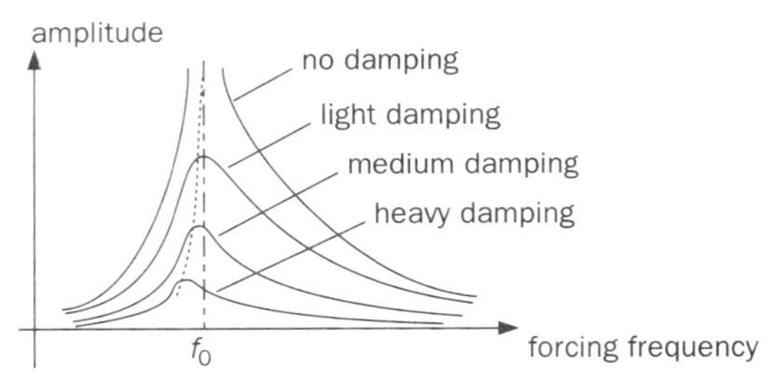

Figure 2.25 Effect of damping on frequency response

Examples of resonance

Any system with mass and elasticity can be set into oscillation.

- When we push a child on a swing, we time our pushes so that our **forcing frequency** is equal to the **natural frequency** of the swing (a pendulum).
- A diver times his bounces on a springboard so as to build up a large amplitude oscillation.

Resonance can also be a problem:

- At certain speeds, car body panels can resonate with the engine or vibrations from the road wheels.
- Soldiers marching in step with the same frequency as a natural frequency of a bridge in France in 1850 fell to their deaths when the large amplitude oscillations they caused collapsed the bridge.
- In the USA, the Tacoma Narrows suspension bridge was spectacularly destroyed when vortex eddies (caused by the wind) excited the bridge at a natural frequency.

The Tacoma Narrows bridge collapsed into Puget Sound in 1940 when wind conditions caused severe resonant oscillations to set in

Guided examples (1)

1. The pistons of a racing car engine each have a mass of 550 g and can be assumed to move with SHM of amplitude 105 mm. If the engine is designed to have a maximum piston speed of 125 m s^{-1}, calculate

(a) the theoretical maximum number of revolutions per minute of the engine

(b) the maximum acceleration of a piston

(c) the force exerted on the piston to produce this acceleration

(d) the piston speed when it is 40 mm from the middle of its stroke.

Guidelines

(a) Max. speed = $\pm\,\omega r$

speed in r.p.m. = $\dfrac{\omega}{2\pi} \times 60$

(b) Max. acceleration = $-\,\omega^2 r$

(c) Max. force = $m \times (a_{max})$.

2. The motion of the tide at a particular point can be assumed to be approximately sinusoidal, with a period of 12.5 hours. If the difference between high water and low water is 6 m, calculate

(a) the amount of time in one cycle during which the water level is more than 4.5 m above the low water mark

(b) the rate at which the tide is rising when the water level is 4.0 m above the low water mark

(c) the tidal acceleration at this instant.

Guidelines

(a) Amplitude (r) of the motion is 3 m. Period (T) is 12.5 hours (= 12.5 × 60 × 60 s).

angular frequency $\omega = \dfrac{2\pi}{T}$

Note that 360° = 2π rad.

Assume the cycle starts when the water level is half-way between low and high water, and rising.

Find the time when the displacement is 1.5 m (i.e. 4.5 m above low water).

This is t_1.

Find the time when this next occurs.

This is t_2.

The answer is $t_2 - t_1$.

(b) The rate of tide rise is the velocity, given by: $v = \pm\omega\sqrt{r^2 - x^2}$.

(c) The tidal acceleration is $-\omega^2 x$.

Self-assessment

SECTION A
Qualitative assessment

1. (a) Define **simple harmonic motion** (SHM).
   (b) For a mechanical system performing SHM, which quantity needs to be measured in order to determine its angular frequency (ω)?

2. (a) State the general equation for a body performing SHM and define any symbols used.
   (b) Explain how this equation is a mathematical representation of SHM.

3. For a particle executing SHM, the displacement (x), at any time (t) is given by:

   $x = r \sin \omega t$

   (a) Using axes aligned with each other, sketch graphs of (i) displacement; (ii) velocity and (iii) acceleration against time for this motion.
   (b) Explain how comparison of the displacement/time and acceleration/time graphs can lead to the general SHM equation.

4. A particle performs SHM along a straight line XOY, where O is the centre of the motion and X and Y are the two extremities, equidistant from O.
   (a) What can be said about the direction of the acceleration as the particle moves through one cycle?
   (b) When is the **acceleration** (i) maximum (ii) zero?
   (c) When is the **velocity** (i) maximum (ii) zero?

5. Show, from first principles, that the time period (T) of a simple pendulum of length (l), performing small amplitude oscillations is given by: $T = 2\pi\sqrt{\frac{l}{g}}$ where (g) is the acceleration due to gravity.

6. Show, from first principles, that the time period (T) of small amplitude oscillations of a mass-spring system is given by:

   $T = 2\pi\sqrt{\frac{m}{k}}$ where m is the mass and k is the spring elastic constant.

7. A simple pendulum of length (l) has a time period (T) on Earth. Give an expression in terms of T for the period:
   (a) using a string of length $3l$
   (b) on a planet with gravitational field strength equal to one sixth that of Earth.

8. A mass (m) on a spring of stiffness (k) has a time period (T) on Earth. Give an expression in terms of T for the period
   (a) using a body of mass $8m$ on the same spring
   (b) on a planet with gravitational field strength equal to 3 times that of Earth
   (c) using two springs of stiffness (k) connected in series (end-to-end)
   (d) using two springs of stiffness (k) connected in parallel (side-by-side).

9. Explain what is meant by **free** vibrations, **forced** vibrations, **resonance, damping**, and **critical** damping.

10. (a) Describe the effect of damping on the resonant frequency of a system.
    (b) Give two examples of systems in which damping is (i) desirable and (ii) undesirable.
    (c) Sketch the form of displacement/time graphs for systems which are (i) lightly damped (ii) critically damped and (iii) heavily damped.

SECTION B
Quantitative assessment

(***Answers:*** 0.032; 0.045; 0.25; 0.40; 0.47; 0.80; 0.99; 1.4; 2.0; 2.7; 4.9; 15; 24; 39; 91; 2.3×10^3; 3.5×10^3; 3.7×10^3.)

1. The displacement (x), in centimetres, of a body executing SHM is represented by the equation $x = 4.5 \cos 2.5\,\pi t$, where (t) is the time in seconds after the body passes through the mid-point of the motion. Use the equation to find
   (a) the amplitude and period of the motion
   (b) the displacement, velocity and acceleration of the body when (t) = 0.1 s.

2. The engine on a fishing boat has a piston of mass 950 g which can be assumed to perform SHM of amplitude 15 cm. At full speed the engine works at 1500 revolutions per minute. Calculate
   (a) the maximum piston speed and acceleration
   (b) the maximum force exerted on the piston by this motion (ignoring internal pressure forces)
   (c) the force on the piston when it is 10 cm from the mid-point of the motion.

Self-assessment *continued*

3. An electic sewing machine needle performs oscillations of amplitude 15 mm and frequency 5 Hz. The material being stitched is positioned 7 mm above the lowest position reached by the needle point. Calculate
(a) the maximum velocity and acceleration of the needle point
(b) the velocity of the needle point as it punctures the cloth.

4. A 'baby-bouncer' consists of a light cloth seat suspended from the top of a door frame by four identical elasticated cords of negligible mass. When a baby of mass 6.5 kg is placed gently into the seat, the cords extend by 25 cm. If the baby is pulled down a further 10 cm and released, she can be assumed to perform SHM. If her feet do not touch the floor, calculate
(a) the period of the motion
(b) the maximum and minimum forces exerted by the baby. What is the maximum **'g-force'** on her? Is this reasonable?

5. A ship's captain requires a minimum depth of 16 m for the safe passage of his ship over a submerged reef lying just outside his harbour. At high tide, the highest point of this reef lies at a depth of 18 m, and at low tide, this point is just exposed. If low tide occurs at midnight, and assuming the tidal motion is simple harmonic, with period 12.5 hours, calculate
(a) the earliest time he can leave the harbour
(b) how much time he has before the water becomes too shallow again.

2.2 Description of waves

Waves can be described according to the direction of motion of the vibrating particles as either **transverse** or **longitudinal**

Transverse waves

These are defined as waves in which the **displacements** of the particles of the medium (or, in the case of electromagnetic radiation, the **directions** of the electric and magnetic fields) are **perpendicular** (i.e. at 90°) to the direction of wave travel.

Examples of transverse waves include water waves and all electromagnetic waves – radio, microwaves, infra-red, visible light, ultra-violet, X-rays and gamma rays.

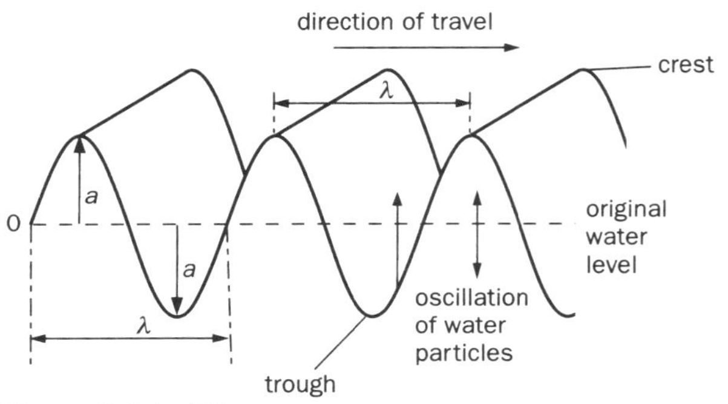

Figure 2.26 Water waves

In the example of a water wave shown in Figure 2.26 the water particles simply oscillate in a **vertical** direction while the wave travels **horizontally** along the water surface.

Description of terms associated with waves

Displacement

Common symbol	*s, x*
SI unit	metre (m)

The **displacement** of a particle is its **instantaneous** distance from the **equilibrium** (undisturbed) level.

Displacement is a vector, so may be positive or negative. We see that the transverse wave travels as a sequence of **crests** and **troughs**. Floating objects merely bob up and down as the wave passes; they do not travel horizontally with the wave. However, they can extract energy from the wave. Floats can be used to turn generators to produce electrical energy from sea waves and tidal bores. Downstream of such a generator, the wave energy would be reduced, so the wave amplitude would be lower than upstream.

Amplitude

Common symbol a, r
SI unit metre (m)

The **amplitude** of a wave is defined as the **maximum** displacement (positive or negative) of a particle from its **equilibrium** (undisturbed) position.

Intensity

Common symbol I
SI unit watt per metre^2 (W m^{-2})

The **intensity** of a wave motion at a point is defined as the **power** per unit **area** at that point.

The energy associated with a wave is directly proportional to the **square** of the amplitude of the wave,

i.e. $I \propto a^2$

so $I = ka^2$

where k is a constant of proportionality.

The highest instrumentally measured sea wave was recorded in the North Atlantic in 1972 by the vessel 'Weather Reporter' – its amplitude was 13 m! This is impressive enough, but the energy associated with such a wave is truly awesome – its intensity is 13^2 (i.e. 169) times greater than that of a wave of amplitude 1 m!

Wavelength

Common symbol λ
SI unit metre (m)

The **wavelength** of a wave is defined as the distance between any two points on **adjacent** cycles which are vibrating **in phase**.

A convenient measure of wavelength is the distance between adjacent crests or adjacent troughs, as these are readily identifiable points which are in phase.

Phase

The particles of a medium through which a wave is travelling are in a state of continuous motion. As the wave disturbance passes a point, the motion which that point has just undergone is repeated in the next point, and so on. At any given instant, different particles in the medium may have different displacements, velocities and directions.

The **phase** of an oscillating particle is a measure of what fraction of a cycle the particle has completed, as measured from some chosen starting point.

As we already know, one complete cycle of the wave is called 'one wavelength'. Therefore one way of specifying phase is to use fractions of a wavelength. For example the **phase difference** between a crest and an adjacent trough is equal to one half-wavelength ($\lambda/2$).

Since one complete revolution of a rotating body measures 360°, or 2π radians, either of these is equivalent to one complete cycle of a wave. This gives us an alternative method of specifying phase. For example the phase difference between a crest and an adjacent trough is equal to 180° or π radians. The idea of phase can be further clarified by considering Figure 2.27 which shows an instantaneous picture of part of a wave, together with the phase of some points marked on it. The 'starting point' of a cycle is taken to be point A in this case (see also Table 2.1).

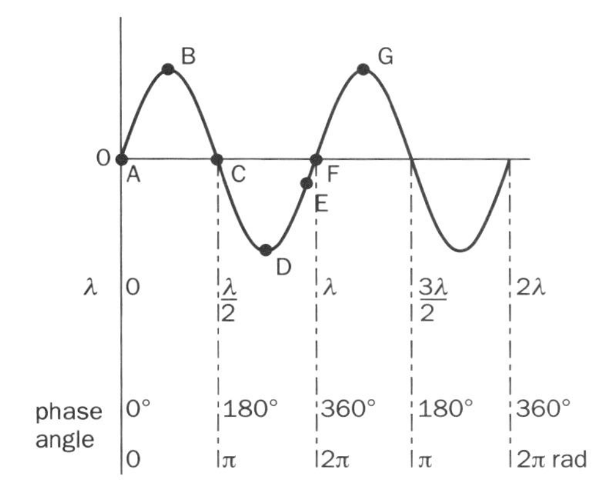

Figure 2.27 Phase angle

Table 2.1 Phase angle

Point	λ	Degrees	Radians
A	0	0	0
B	$\lambda/4$	90°	$\pi/2$
C	$\lambda/2$	180°	π
D	$3\lambda/4$	270°	$3\pi/2$
E	$0.9\,\lambda$	324°	$1.8\,\pi$

You should note that:

- From point F in Figure 2.27, the wave disturbance repeats, so point F has the same phase as point A; point G has the same phase as point B. Points which are separated by a whole number of wavelengths (in other words an **even** number of half-wavelengths) are said to be **in phase**. Particles which are in phase are always moving in the same direction throughout a cycle and they reach their maximum and minimum displacements simultaneously.
- Particles which are separated by an **odd** number of half-wavelengths are always moving in opposite directions throughout a cycle, and are said to be in **antiphase**.

Phase change on reflection

If the free end of a rope tied to a fixed object is flicked, a transverse wave pulse travels along the rope. When the pulse reaches the fixed end, it will be reflected back towards the free end. If you observe this process carefully, you will see that the reflected pulse is in **antiphase** with the incident pulse: a **crest** will be reflected as a **trough** and vice versa. A phase change of π radians has occurred on reflection at the fixed boundary as shown in Figure 2.28.

More surprisingly, even when the rope is not fixed at one end, reflection can also occur at the free end. This time there is no phase change: a crest is reflected as a crest, as shown in Figure 2.29.

In the case of sound waves, reflection at a dense medium, such as the end of a closed pipe or the fixed end of a guitar string, causes a phase change of π rad, so a **compression** is reflected as a **rarefaction**. Reflection at a less dense boundary, such as the open end of an organ pipe, produces no phase change.

Light suffers a phase change of π radian when reflected at a more optically dense medium, such as a mirror, but there is no phase change if the reflecting medium is less optically dense than the incident medium, as in the case of **total internal reflection**.

Longitudinal waves

Longitudinal waves are defined as waves in which the oscillation of the particles of the medium are **parallel** to the direction of wave travel.

Examples include sound waves and some forms of seismic waves. Longitudinal waves require a material medium – they cannot travel in a vacuum.

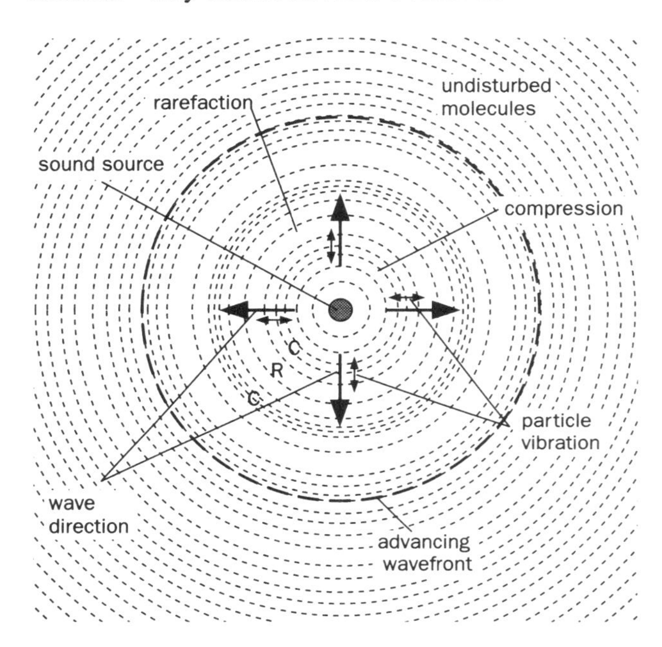

Figure 2.30 Longitudinal wave

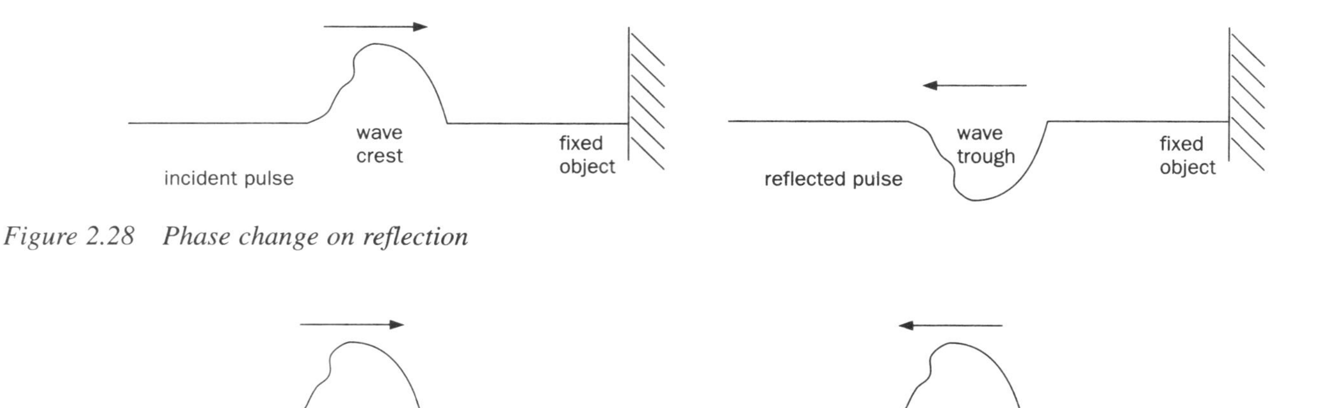

Figure 2.28 Phase change on reflection

Figure 2.29 No phase change on reflection at the free end

Figure 2.30 shows a source emitting sound waves in all directions into a region of undisturbed air molecules. The molecules of the air are represented by the dotted lines. Of course, real air molecules do not have such order and symmetry, but the model is adequate for our purposes. Along any direction from the source, the molecules are alternately pushed forwards and backwards (that is, in a direction parallel to the direction of travel of the wave). The initial disturbance advances equal distances in unit time in all directions, so the sound travels as a circular **wavefront** (shown by the heavier line). The wave energy spreads into the surroundings as a series of **compressions** (where molecules are pushed together, causing an increase in pressure), and **rarefactions** (where the molecules are forced further apart than normal, causing a reduction in pressure). Thus a sound wave travels as cycles of increasing and decreasing pressure. In Figure 2.30, the compressions and rarefactions appear as two-dimensional concentric circles. In reality, these would be three-dimensional spheres.

Further waves classification

Waves can also classified according to the nature of the vibrations.

Mechanical waves

These are produced by a disturbance in a **medium**, and are transmitted by the oscillations of the **particles** of the medium. They can be **longitudinal** or **transverse** waves (see above). Examples of mechanical waves include sound, water waves and seismic waves.

Electromagnetic waves

In the case of electromagnetic waves, the disturbance is in the form of oscillating electrical and magnetic **fields**. They are regarded as **transverse**. They do not require a medium and always travel at the same speed in a vacuum (approx. 3.0×10^8 m s^{-1}). In any material medium, the speed is lower, which accounts for the phenomenon of **refraction**. All of the radiations in the electromagnetic spectrum fall into this category.

*Water waves are **progressive**, **mechanical** waves produced by a disturbance in the water and transmitted by **transverse** oscillations of the water molecules*

Graphical representation of waves

Displacement–distance graphs

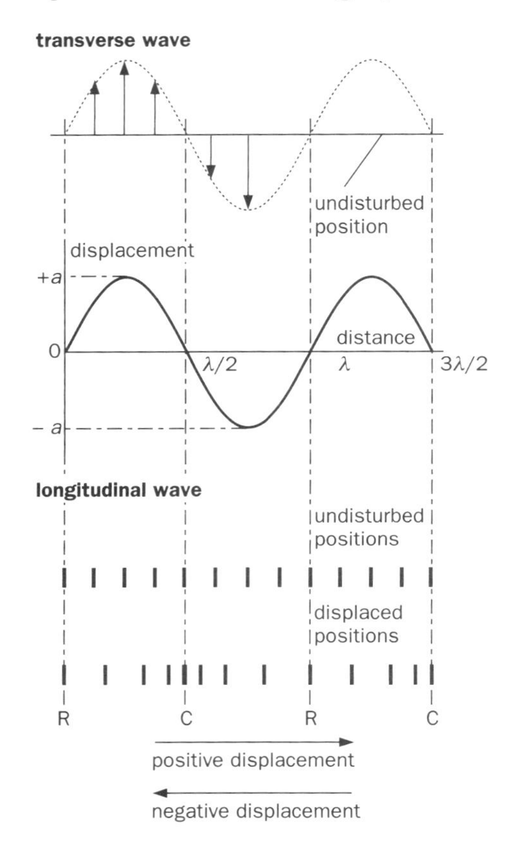

Figure 2.31 Displacement–distance graph

Figure 2.31 shows how both **transverse** and **longitudinal** waves can be represented by the same form of *displacement–distance* graph. The upper diagram shows the **instantaneous position** of particles of a medium through which a transverse wave is passing.

The central diagram shows the *displacement* (y-axis) of particles of the medium (measured from their undisturbed position) plotted against the *distance* (x-axis) of the particles from the source **at a certain instant in time**. Clearly, this diagram is easy to relate to the physical situation, as it resembles the shape of the actual wave.

The lower diagram shows a **longitudinal** wave. The particles of the medium are shown evenly-spaced in their undisturbed positions. The passage of the wave displaces the particles to their new positions. This pattern of displacements can also be represented by the same graph as we used for the transverse wave. In this case, the *displacement* of the particles of the medium is **parallel** to the direction of travel of the wave. However, the displacements are still plotted on the y-axis, to give the conventional graph.

Points 'C' represent **compressions**; points 'R' represent **rarefactions**.

This form of graph gives a useful representation of the wave, in that it is a 'snapshot' of the wave at a particular instant. Wavelength and amplitude can be represented, but the graph gives no information about how the displacement changes with time.

Displacement–time graphs

If the displacement of a **single particle** is plotted against time, a similar **looking** (but very different) graph is produced as shown in Figure 2.32.

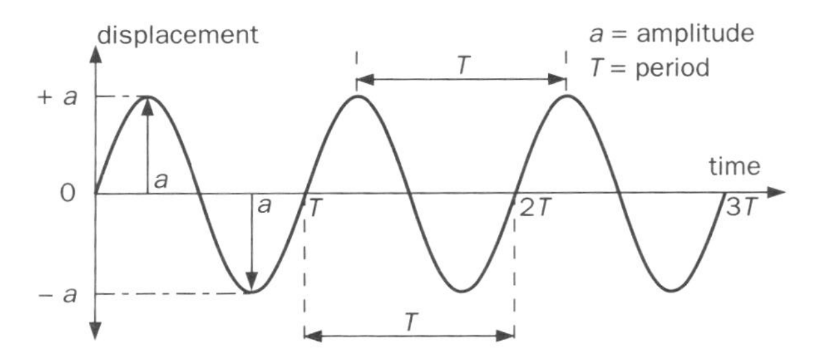

Figure 2.32 Displacement–time graph

This graph should not be confused with the displacement–distance graph: it gives no information about the wavelength of the wave. Its usefulness, however, is that it allows the **period** and **frequency** of the wave to be represented. If the strength of the electric or magnetic field in an electromagnetic wave is plotted against time, a similar graph is produced. This graph can be used to define the following quantities.

Frequency

Common symbol	f
SI unit	cycles per second (s^{-1})
	or hertz (Hz)

The frequency of the oscillator producing the wave, and the frequency of the wave itself, have the same value.

> The **frequency** of a wave is defined as the **number** of complete waves passing a given point per **second**.

Period

Common symbol	T
SI unit	second (s)

> The **period** of a wave is defined as the **time** taken for a particle of the medium through which the wave travels to make **one** complete **oscillation**.

Wave velocity

Common symbols c, v
SI unit metre per second (m s^{-1})

Derivation window

From the definitions of period and frequency:

$$f = \frac{1}{T}$$

In the time taken by the source to complete one cycle, the wave advances by a distance equal to one wavelength (λ).

Therefore, if f cycles are produced in 1 second, the wave moves forward a distance equal to $f\lambda$ each second. This, by definition, is the **velocity** v of the wave:

$$\therefore \quad v = f\lambda$$

(v in m s^{-1}; f in Hz (or s^{-1}); λ in m)

Equation 1

This relationship is true for all wave motions.

Guided examples (1)

1. Use the data below to estimate:
(a) the frequency range of light which is detectable by the human eye
(b) the range of wavelengths of sound which is audible to the human ear.

speed of light in air = 3.0×10^8 m s^{-1}
speed of sound in air = 340 m s^{-1}
approx. wavelengths of visible light = 400 nm (violet) to 700 nm (red)
audible frequency range = 20 Hz to 20 kHz.

Guidelines

Use $c = f\lambda$ (Take care to convert to SI units)

2.

displacement/mm
80
0
– 80
40
80
120
distance/cm

The diagram shows a graph of displacement against distance for a wave moving across the surface of a pond at a speed of 0.4 m s^{-1}. Calculate :

Guided examples (1) *continued*

(a) the amplitude, wavelength, frequency and period of the wave;
(b) the phase difference between two fishing floats separated by a distance of 60 cm in the direction of travel of the wave.

Guidelines

(a) Amplitude and wavelength can be obtained directly from the graph. Frequency and period require calculation (care with units!).
(b) Relate the phase difference to the wavelength, then express it in terms of a fraction of the wavelength, and give phase angle in degrees (1 cycle = 180°) and phase angle in rad (1 cycle = 2π rad).

3. Calculate the velocity of a progressive wave of frequency 350 Hz if the least distance between two points having a phase difference of $\pi/6$ rad is 0.08 m.

Guidelines

The phase difference is stated in terms of phase angle. Convert this to a fraction of a wavelength, and so calculate the wavelength and velocity.

Polarisation

Transverse and longitudinal waves can be distinguished in that transverse waves can be **polarised**, whereas longitudinal waves cannot.

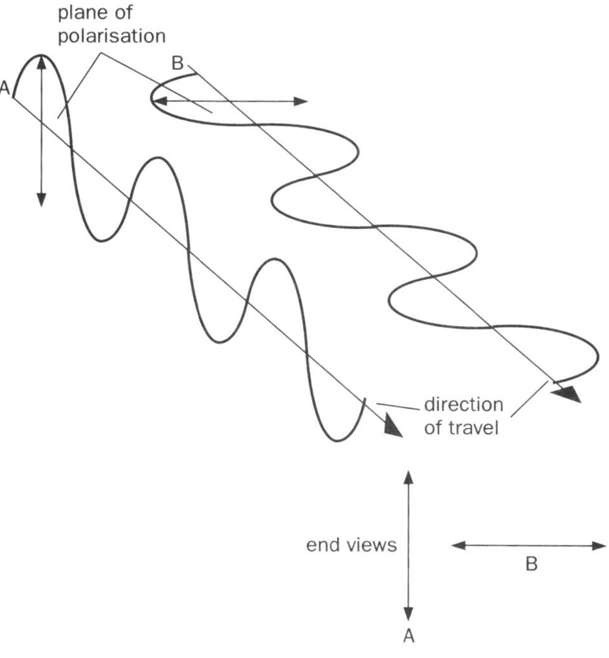

Figure 2.33 Horizontally and vertically polarised waves

Figure 2.33 shows two waves travelling in the same direction. Wave A is **vertically** polarised, as all motion is confined to the vertical plane, while wave B is **horizontally** polarised. However, both waves satisfy the condition for transverse waves, as the **displacements** of the particles are always **perpendicular** to the **direction of travel** of the wave.

A transverse wave is said to be **plane polarised** if all the vibrations in the wave are in a **single plane** which contains the direction of propagation of the wave.

In fact, these two waves are only two examples of an infinite number of possibilities. The plane of polarisation could be inclined at any angle to the direction of travel. Waves may contain many planes of polarisation simultaneously. 'Polaroid' is the trade name of a product which acts as a filter, only transmitting light polarised in a particular plane. All other planes are absorbed. Two pieces of Polaroid with their transmission planes at 90° will effectively block all transmission, as shown in Figure 2.34.

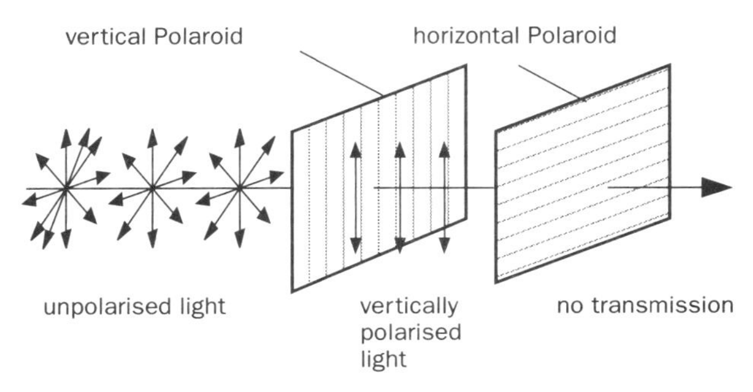

Figure 2.34 Effect of crossed Polaroids

Light reflected off a surface (water, snow, sand, etc.) may be predominantly polarised in one direction. Polaroid sunglasses effectively reduce reflected glare by filtering out the light polarised in this plane. Examples of plane polarised waves include many radio waves, the 3 cm microwaves used in laboratory experiments, and light waves after reflection.

Liquid crystal displays (LCDs)

Liquid crystal displays (LCDs) make use of the property of 'liquid' crystals to rotate the plane of polarisation through 90° when a p.d. is applied to them, as shown in Figure 2.35. When the crystal is 'turned off', it no longer rotates the plane of polarisation, so the second filter transmits the light, making the display segment no longer dark, as shown in Figure 2.36.

Engineering stress analysis

This rotation of the plane of polarisation can also be produced when polarised light passes through glass or transparent plastics material which is under strain (i.e. distorted). The amount of rotation depends on the amount of strain, so this technique can be used by engineers to visualise the stresses being suffered by components.

Figure 2.37 shows such a stress viewer. Light is first plane polarised before being passed through the model. If the model is unstressed, it has no effect on the light. The second polaroid is aligned at 90° to the first, so no light will be transmitted.

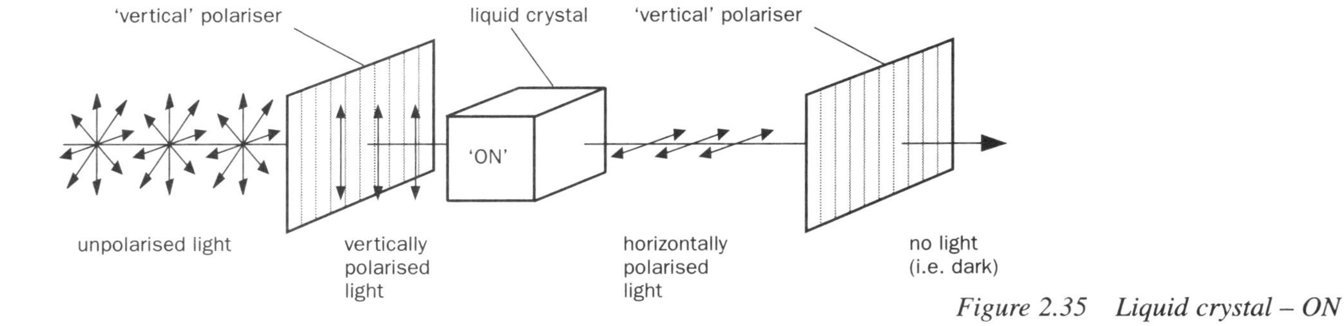

Figure 2.35 Liquid crystal – ON

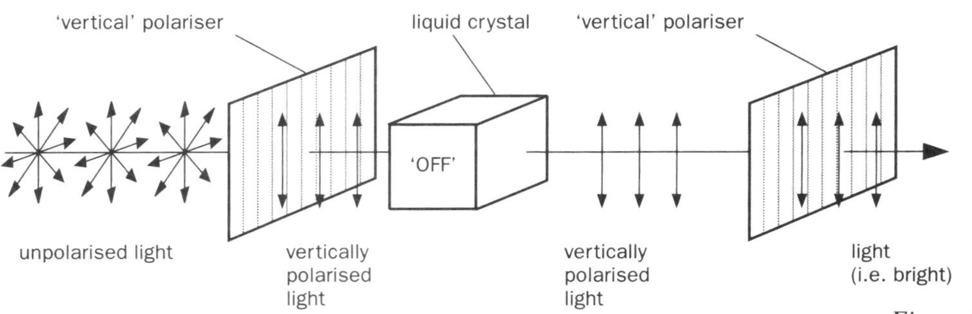

Figure 2.36 Liquid crystal – OFF

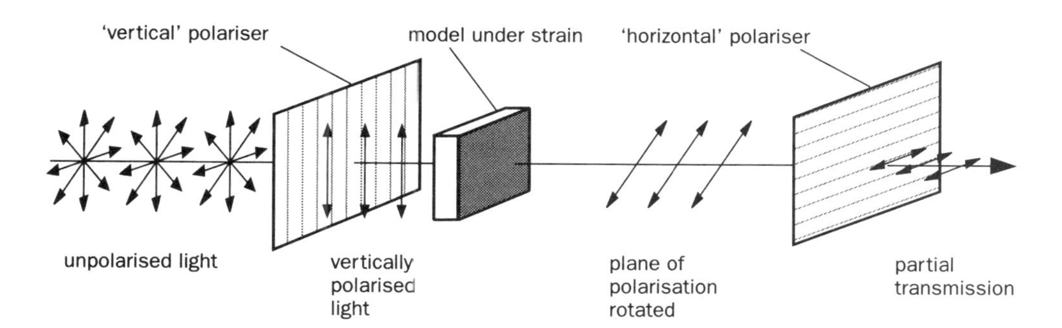

Figure 2.37 Stress analyser

When the model is stressed, different parts will rotate the plane of polarisation through an angle which depends on the amount of strain suffered. This will produce partial transmission at the second Polaroid, and the 'stress pattern' will appear as light and dark regions on the model. Coloured light can produce dramatic coloured fringes on the model.

You should note that:

- In the case of an electromagnetic wave, it is the electric or magnetic fields, not the displacements of particles, which are represented on the 'displacement' axis. The planes of the electric *(E)* and magnetic (*B*) fields are perpendicular to each other, and are in phase, as shown in Figure 2.38.
- If an electromagnetic wave is polarised, the plane of polarisation is taken to be that of the **electric field** since, when electromagnetic waves interact with matter, it is the electric field effects which dominate.

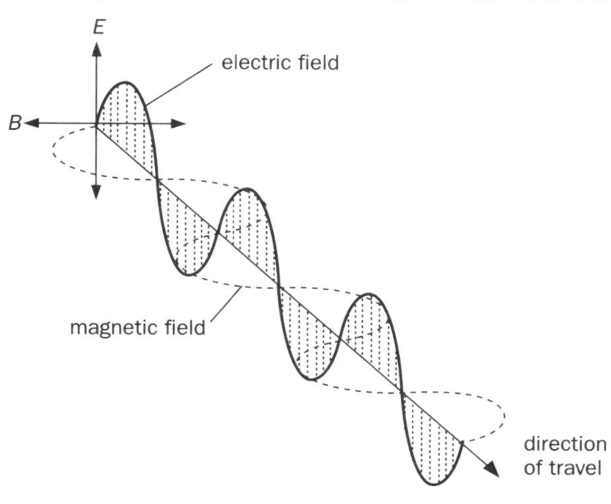

Figure 2.38 Electromagnetic radiation

Self-assessment

SECTION A Qualitative assessment

1. Explain what is meant by a
   (a) progressive wave
   (b) stationary wave.
2. Distinguish between **transverse** and **longitudinal** waves and state *two* examples of each.
3. Distinguish between **mechanical** and **electromagnetic** waves and state **two** examples of each.
4. (a) Explain the meaning of each of the following terms. State the SI unit for each:
   (i) amplitude
   (ii) wavelength
   (iii) period
   (iv) frequency.
   (b) Explain the relationship between (i) intensity and amplitude (ii) period and frequency.
5. Sketch a **displacement–distance** graph for (i) a **transverse** wave and (ii) a **longitudinal** wave.
   For part (ii), relate the positions of **compressions** and **rarefactions** to the graph.
6. Sketch a graph of displacement against time for a particle disturbed by a progressive wave. On the same axes, indicate the motion of a particle separated from the first by a distance equal to one quarter of a wavelength.
7. Use the definitions of the terms **wavelength, frequency** and **period** of a wave motion to derive the relationship between **velocity, frequency** and **wavelength** of a wave.
8. Using diagrams, explain:
   (a) what is meant by (i) a **plane-polarised** wave and (ii) an **unpolarised** wave
   (b) the effect of two crossed 'polaroids' on unpolarised light
   (c) how a stress analyser can be used to visualise areas of high stress in engineering structures.

Self-assessment *continued*

SECTION B
Quantitative assessment

(***Answers:*** 6.0×10^{-7}; 0.12; 0.25; 0.50; 1.6; 2.0; 90; 1.5×10^{3}; 1.2×10^{7}; 4.3×10^{14}.)

(Unless otherwise stated, assume the velocity of electromagnetic waves in a vacuum to be 3.0×10^{8} m s^{-1})

1. A progressive wave travels a distance of 18 cm in a time of 1.5 s. If the distance between successive crests is 60 mm, calculate (a) the velocity, (b) the frequency, and (c) the period of the wave.
2. Calculate:
   (a) the wavelength of yellow light of frequency 5.0×10^{14} Hz
   (b) the frequency of red light of wavelength 700 nm
   (c) the wavelength of radio waves of frequency 200 kHz.
3. A laser emits a single burst of light lasting 0.020 μs. If the wavelength of the light is 500 nm, calculate the number of complete waves emitted.
4. A wave of frequency 2.5 Hz travels along a stretched cable at 20 m s^{-1}. Calculate the phase difference between two points on the cable which are 2 m apart. Express your answers in terms of (a) fractions of a wavelength, (b) degrees and (c) radians.

2.3 Interference and diffraction

Principle of superposition

If two streams of particles, such as two jets of water, pass through the same region of space, any interaction or collision between the particles will produce changes in their direction and/or speed. The energy of individual particles is permanently changed. Unlike particles, waves can share a region of space and suffer no lasting changes. In the region where they overlap, a new wave is temporarily formed. After interaction, the two waves continue with exactly the same properties as before. The waves carry on as if nothing had happened. Where the waves meet, they are **superposed**. The **principle of superposition** states that:

> At a point where two or more waves meet, the instantaneous displacement is the **vector sum** of the individual displacements due to each wave at that point.

This principle only applies to waves, and the waves must be of the same type. A light wave and a sound wave cannot be combined in this way, but light and a beam of X-rays (which are both electromagnetic waves) can. Figure 2.39 shows how the form of the new wave can be obtained from two different waves travelling in the same region and crossing through each other.

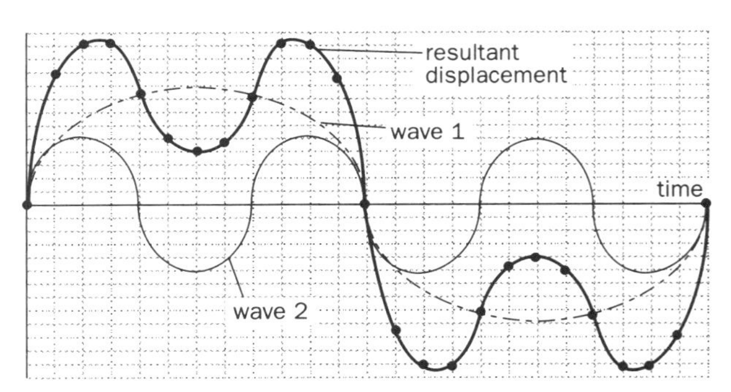

Figure 2.39 Superposition of waves

The resultant displacement is obtained by plotting the **vector sum** of the two waveforms, and the envelope of the new wave is shown by the **bold** line in Figure 2.39. This principle can be used to explain many wave phenomena including the formation of stationary waves, beats in sound, interference, diffraction etc.

Interference

The principle of superposition tells us that at the point where waves meet, the amplitude of the resultant wave will depend not only on the amplitudes of the individual waves, but also on their phase relationship. If two waves of equal amplitude are superposed, when crests coincide (i.e. the waves are in phase), the resulting amplitude will be equal to twice the individual amplitude as shown in Figure 2.40(a).

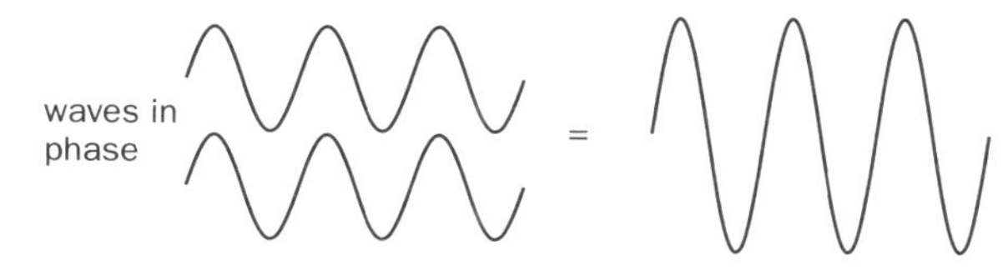

(a) Constructive interference

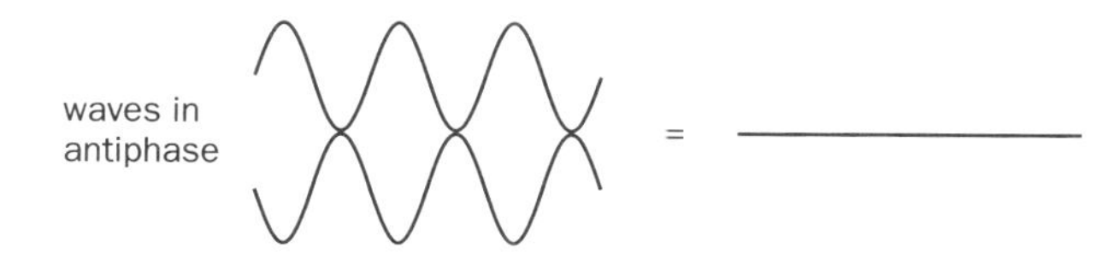

(b) Destructive interference

Figure 2.40 Interference

Figure 2.40(b) shows the effect of destructive interference, where a crest coincides with a trough (antiphase) producing zero amplitude. Other phase differences should produce amplitudes in between these two extremes. The production of variations in intensity in this way is known as **interference**. **Constructive** interference, also called **reinforcement**, describes the production of a resultant wave of increased intensity, while **destructive** interference (or **cancellation**) results in reduced intensity.

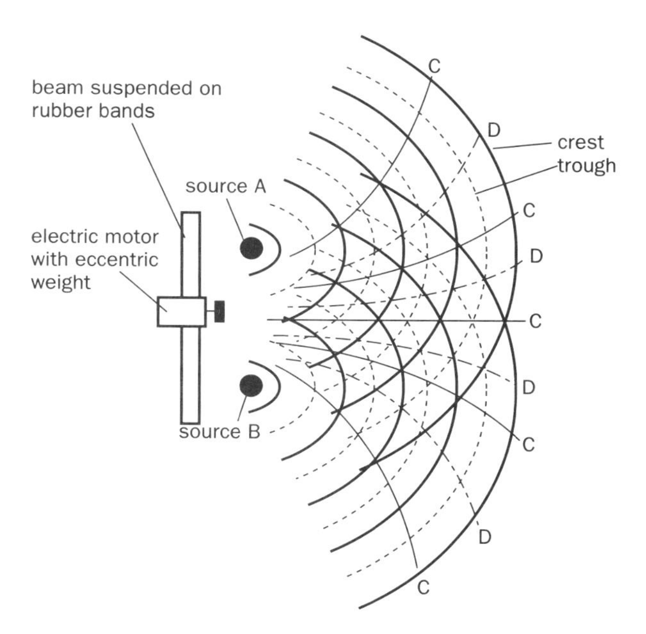

Figure 2.41 Interference of water waves

Interference of water waves – ripple tank investigation

Interference of water waves can be readily observed using a ripple tank, as shown in Figure 2.41. Two dippers hang from a beam which carries a motor with an eccentrically mounted weight on its shaft. The whole beam vibrates, causing the dippers to act as two separate sources with identical frequency and amplitude. You should note in particular that the two dippers are **phase linked**, and that in this case they are vibrating in phase. The resultant pattern of waves shows interference effects. The interface pattern can

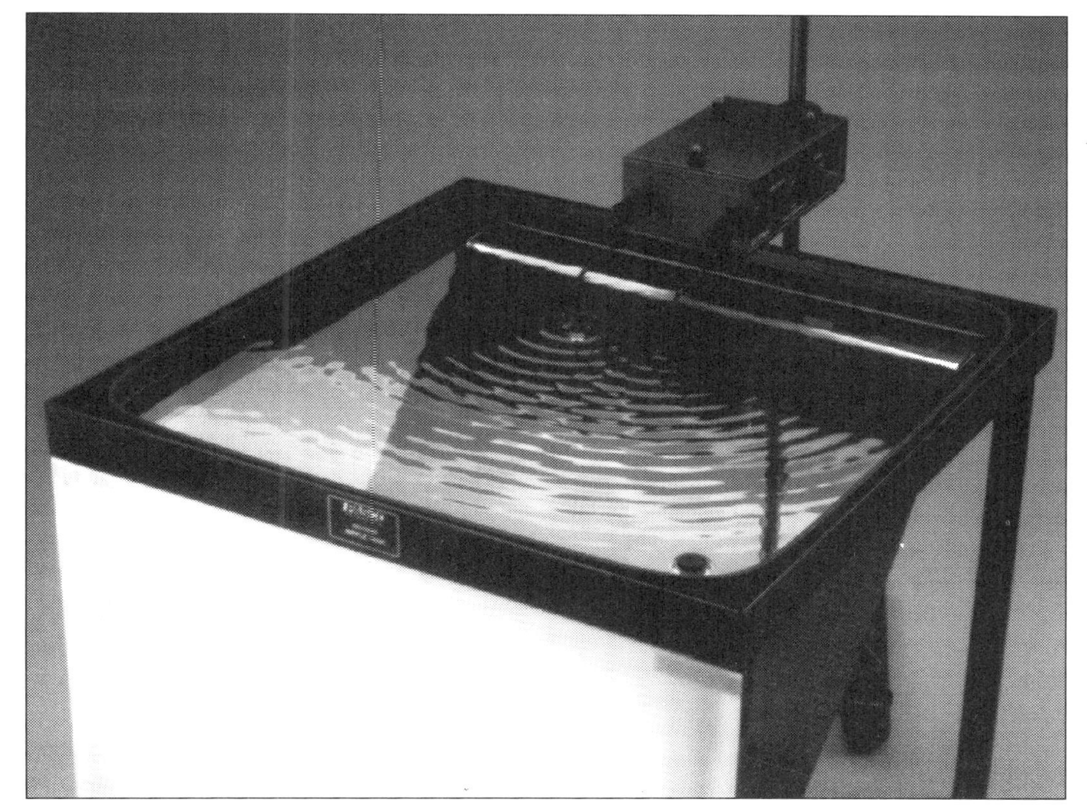

The ripple tank is an excellent tool for demonstrating wave phenomena

be interpreted as follows:

- There are certain regions on the water surface where the two sets of waves are always **in phase**, and at these regions **constructive** interference occurs, producing a large amplitude disturbance. These regions occur along lines such as those labelled C in Figure 2.41, symmetrical about the centre line between the dippers. The centre line itself is a region of constructive interference.
- Equally, there are regions where the two sets of waves are **in antiphase**, and since their amplitudes will be very similar, interference causes almost complete cancellation, and the water is calm in these places. These regions lie between the regions of reinforcement, shown by the lines labelled D in Figure 2.41.

You should note that:

- At any point on a region of constructive interference, the water surface is still rising and falling, at the same frequency as the individual waves, but with greater amplitude.
- The position of these regions (the interference pattern) does not change with time (i.e. a steady pattern is obtained).
- The total energy of the system remains constant. The energy 'lost' from the calm regions is re-distributed to the areas of reinforcement.

Interference with microwaves

With microwaves (waves of wavelength ≈ 3 cm) interference effects are also easy to demonstrate. Metal plates act as barriers to the microwaves, so slits in the plates act as narrow sources (just like dippers A and B above).

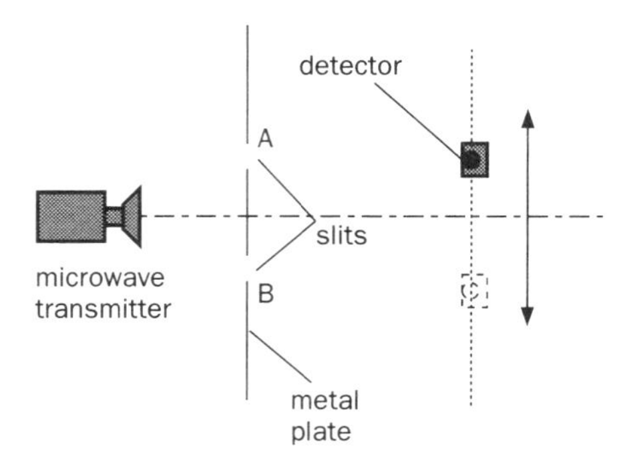

Figure 2.42 Interference of microwaves

Figure 2.42 shows how the two sources A and B are created from a single transmitter, and as long as the two slits are equal distances from the transmitter, the two sources will be in phase. The detector is moved along a line perpendicular to the centre line between the two slits as shown. Maxima and minima of intensity are revealed as corresponding high and low meter readings on the detector.

Interference with sound waves

The effect is identical with sound waves. Two loud-speakers driven by a single signal generator replace sources A and B. A microphone (or the ear) detects regions of **loudness** (reinforcement) and **quietness** (cancellation) in the space where the waves interfere.

Interference effects only become noticeable when the distance between the sources is of a similar order of magnitude to the wavelength of the waves. In the case of sound, wavelengths of the order of 1 m are common, so interference is easily observed when the speakers are closer than this. Water waves in ripple tanks and 3 cm microwaves have similar wavelengths, so the sources need to be only a few centimetres apart.

Effect of separation of the sources

If the sources are far apart compared to the wavelength of the waves, then the resulting pattern consists of a large number of regions of constructive and destructive interference (labelled C and D respectively) which are closely-spaced, as shown in Figure 2.43.

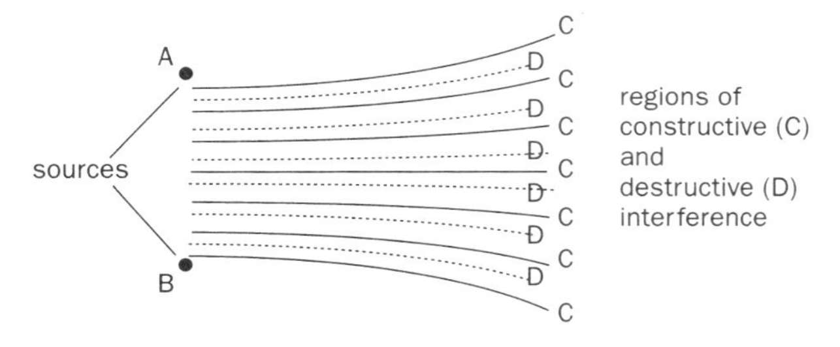

Figure 2.43 Interference effect when the separation is large compared to wavelength

Figure 2.44 shows the effect produced when the separation of the sources is small compared to the wavelength of the waves. The interference pattern shows much wider spacing and fewer regions of reinforcement and cancellation.

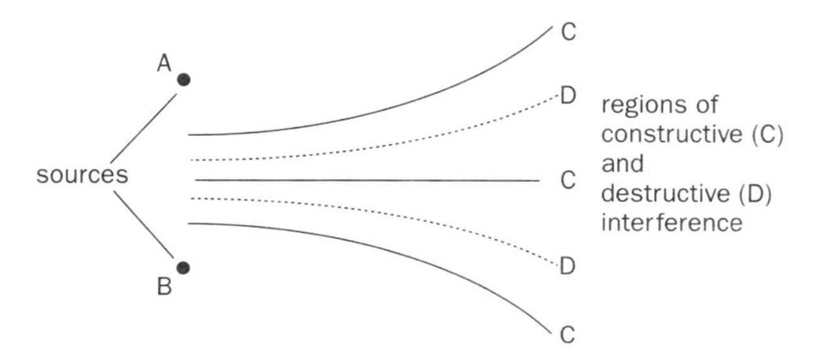

Figure 2.44 Interference effect when the separation is small compared to wavelength

Explanation of two-source interference

Phase difference and path difference

Referring to Figure 2.41, showing the interference of water waves, we see that in this simple situation, the wave produced at any point is the resultant of only two separate waves. If we look first at the regions of **constructive** interference, we note that one of these regions lies along the centre line between the two sources. Any point on this line is an equal distance from each dipper. We say that the **path difference**, which is the difference between the distances travelled by the two sets of waves, is zero anywhere along this line. Since the two sets of waves were produced in phase, and they travel with equal velocity along the water surface, they must still be in phase when they arrive anywhere along this line, as shown in Figure 2.45.

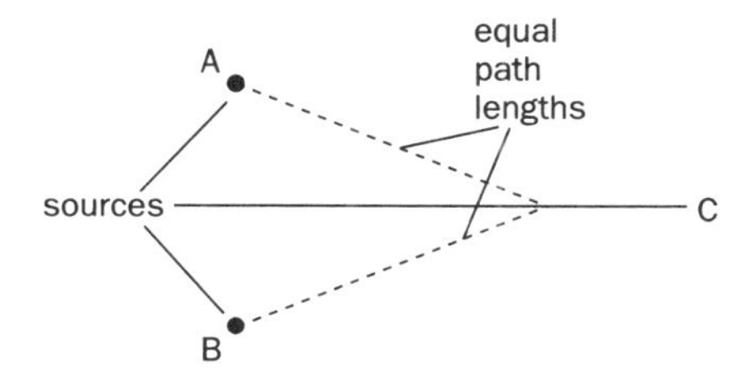

Figure 2.45 Zero path difference

Adjacent to the centre line, and symmetrically on either side of it, we find further regions of constructive interference. We know this means the waves from each source are in phase, but how can this be so? The reason is that at any point on these regions, the waves from one source have travelled one complete wavelength further than waves from the other, producing a **path difference** of one wavelength. It is convenient to call this two half-wavelengths, and we say the path difference is two half-wavelengths as shown in Figure 2.46.

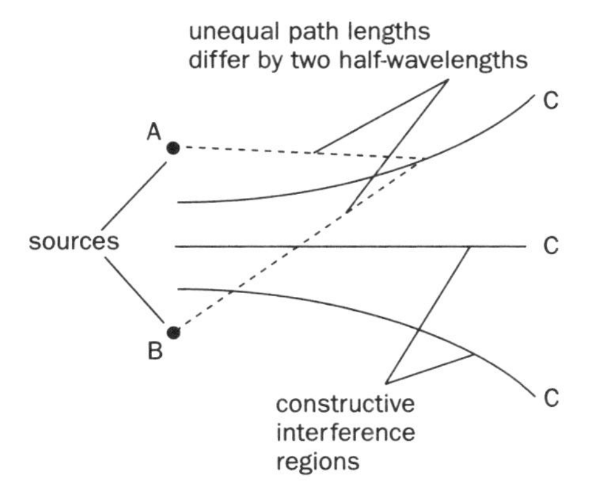

Figure 2.46 Path difference = two half-wavelengths

If we consider each region of constructive interference in turn, we find that the path difference at the next region is four half-wavelengths, then six, eight and so on.

In general, for **constructive interference** from two sources **in phase**, the path difference must be equal to an **even number** of half-wavelengths (or zero).

In between these regions are the areas of calm water produced by destructive interference, where the waves from the two sources are in antiphase. If we look at the first regions on either side of the centre line, the path difference must be just equal to one half-wavelength. Along each region of destructive interference in turn, the path difference must be three half-wavelengths, then five, seven, etc.

In general, for **destructive interference** from two sources **in phase**, the path difference must be equal to an **odd number** of half-wavelengths.

Interference of water waves is not just a laboratory phenomenon. Maritime insurers speak of the 'hundred-year wave' – a freak wave large enough to swamp a ship completely. Such a wave can be produced where several waves coincide, in phase, with the position of the vessel. The amplitudes of each wave are added, to create a single wave of enormous height. It is equally possible that the resultant wave could have zero amplitude, producing a brief calm area, but this would hardly be of concern to the insurers (or the sailors). In light, the corresponding effect would be sudden, freak flashes of brightness, while loud bursts of noise would be the equivalent in sound, yet in everyday life we are not aware of such effects.

The main reason sound and light do not appear to display these variations in intensity is due to the way in which the waves are produced. At sea, for example, water waves are produced by interactions of the wind, tides and currents. If there is only one source of waves, especially far from land, the wave pattern is simple, and can be observed to have nearly constant amplitude, frequency and wavelength, and a boat in these waters will rise and fall with monotonous regularity. However, there are usually several factors producing waves of different frequency and amplitude, and this will result in complex patterns of individual waves. At any point, the resultant amplitude will be the vector sum of several waves, and so at some times there will be constructive interference, giving a large amplitude wave, and at other times destructive interference will create relative calm. Each effect is transient and short-lived, producing what a sailor would describe as 'choppy' water. However, because the frequency of water waves is low, the human eye and brain can follow this changing intensity of the waves fairly easily, and we can observe the moments of calm and rough water.

The light illuminating this piece of paper is probably coming from several sources, including reflected light from the surroundings, producing a complex pattern of waves of varying frequencies, just as with water waves at sea. However, we do not observe variations in brightness. This is because light waves have a high frequency (of the order of 10^{14} Hz), and, although interference effects are produced at random all over the page, the changes in brightness change position too rapidly for the eye and brain to follow, so we see only the average intensity. This gives the illusion of steady illumination. Similarly, sound waves reach our ears from a variety of sources simultaneously, but any interference effects produced will normally be so fleeting as to go unnoticed.

Creating silence with sound!

The advent of high speed computers has led to the development of devices which use sound to create silence! For example, the sound reaching a car driver's ears is picked up by a suitably-placed microphone. This is amplified and played back through speakers at the same amplitude as the original sound. The difference is that this sound has been delayed by the computer, so that it arrives at the driver's ears in antiphase with the original sound, so the driver hears nothing! At the time of going to press, such systems have been shown to work, but are not considered ready for commercial production.

Source coherence

If interference effects are to be easily observable, the pattern must not change with time. This is only possible if the sources of the waves are linked in some way so that the phase difference between them is constant. This was the case with the ripple tank experiment (p.119), since the two dippers were driven by the same source. Equally with the microwaves (p.120), the two slits emit waves derived from the same source.

Two wave sources are said to be **coherent** if the waves emitted from them are '**phase-linked**' (i.e. have a zero or constant phase difference). This implies that **coherent sources must have the same frequency**.

Conditions for a steady interference pattern

If an interference pattern is to be steady and observable the following conditions must be met:

- The sources must be **coherent**. If the sources were not coherent, the phase relationship between the two sets of waves at any point would continuously alter, producing a constantly changing interference pattern. Interference is not commonly observed with light and sound as most light and sound sources are incoherent. In practice, coherent sources must be derived from a **single source** in order to guarantee a constant phase relationship.
- For good contrast between maxima and minima of intensity, the wave **amplitudes** must be approximately equal. If this is not the case, then complete cancellation cannot occur, since a small amplitude crest cannot completely cancel a large amplitude trough; the resulting interference pattern would lack contrast.
- The distance between the sources must be comparable to the wavelength of the waves to ensure adequate separation of the maxima and minima.

Interference of light

Producing two coherent sources of light is difficult because light from conventional sources is emitted in short bursts from individual groups of atoms, lasting only about 1 ns (1×10^{-9} s). These bursts are emitted at random, so abrupt phase changes in the light emitted from even a small region of a lamp occur constantly. Therefore there can be no constant phase linkage between different points, even on the same light source, and the light from such sources is said to be incoherent. An analogy can be drawn with a crowd emerging from a stadium. Being composed of individuals, there will be no connection between their strides, so any phase relationships between individuals will be short-lived.

To produce a steady interference pattern using conventional light, it is thus necessary to create two coherent sources by splitting the bursts of light emitted from a single point on the source. Since both sources receive light from the same point, they will have a constant phase relationship.

In addition, the nature of conventional light emission puts a severe limitation on the possible range of path differences. If a burst of light lasts for 1 ns, the light wave will have travelled about 0.3 m (distance = speed × time) and contain something like half a million cycles. In theory, all the waves in this wave train have a constant phase relationship. In practice, atomic collisions during the emission of the wave train mean that abrupt phase shifts occur every millimetre or so. We say that the 'coherence length' of light from conventional sources is only around 1 mm. Consequently, if the path difference from the two sources is less than this, interference effects may be observed. At greater path differences, the phase shifts will destroy any phase linkage between light from the two sources, and a steady interference pattern will not be possible.

The **laser** produces an intense beam of monochromatic light (single frequency) in which all the waves across the beam are in phase. If a laser beam is split into two, then the two beams will be coherent since they will be phase linked. To compare laser light with that from a conventional source, it is useful to make an analogy with a squad of soldiers marching in step to represent the laser, and a crowd leaving a stadium representing the conventional source. The soldiers are all in step, marching in the same direction, whereas the members of the crowd have many different step frequencies, no phase linkage and different directions.

Diffraction

We are taught from an early age that light travels in straight lines. Evidence for this includes the production of sharp-edged shadows when light is interrupted by an obstacle. However, observation of water waves passing through a narrow aperture or past a small obstacle reveals that waves tend to spread into the 'shadow' region, as shown in Figure 2.47. This effect is called **diffraction**.

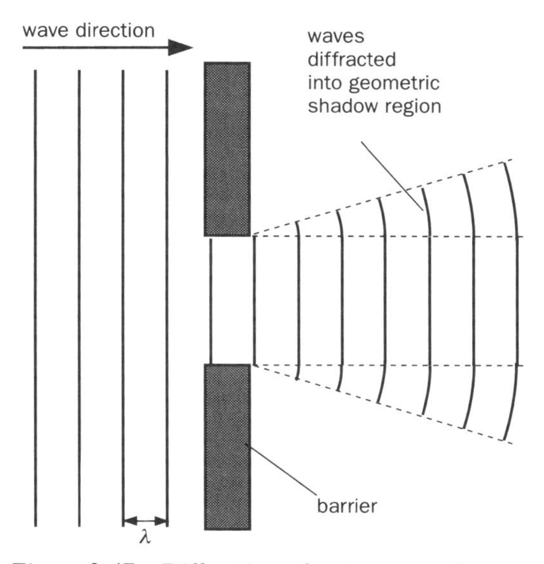

Figure 2.47 Diffraction of water waves by an aperture

A scientist using an argon ion laser to determine the flow of coal ash particles in an exhaust gas model

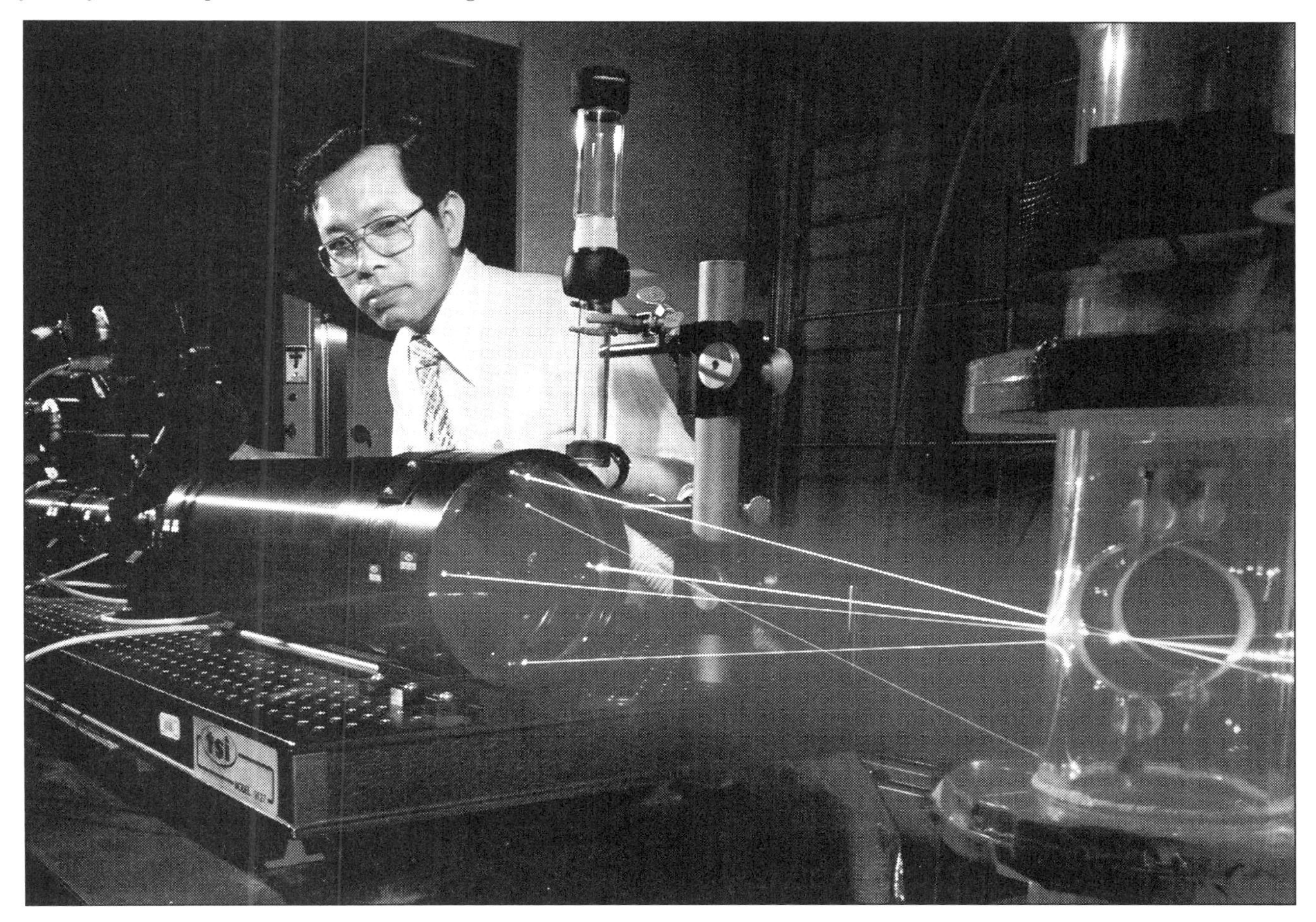

The amount of diffraction depends on the size of the opening. If this is comparable in size to the wavelength, then the diffraction can be so great as to almost fill the shadow region, as shown in Figure 2.48.

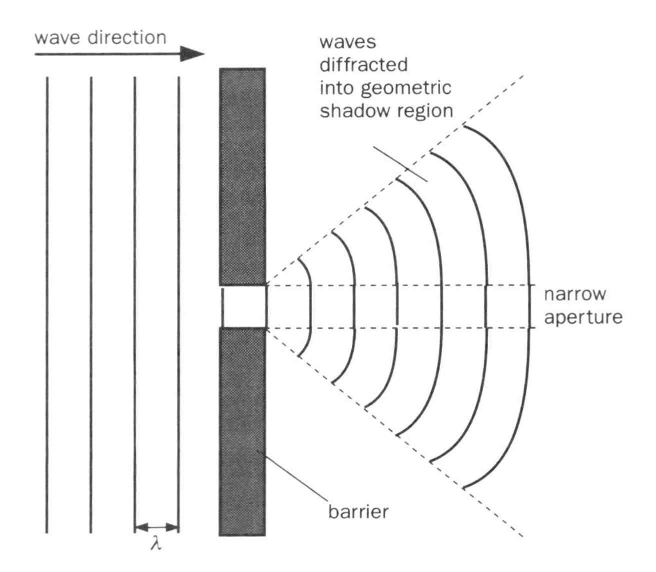

Figure 2.48 Diffraction – effect of aperture width

Historical significance of diffraction

In the early 17th century, there was much debate about the nature of light – was it a wave or a particle? Water waves could be seen, and their ability to curve around obstacles was taken to be a property of waves. Sound was also known to show diffraction, and was assumed to be a wave, but light appeared to cast sharp shadows, which suggested that it was not a wave. Careful observation by an Italian, Grimaldi, in a paper published in 1665, revealed slight diffraction of light passing through narrow apertures and led to the hypothesis that light waves must be very small. However, his work went largely unnoticed until the beginning of the 19th century when Thomas Young reproduced Grimaldi's observations. It was not until then that the wave nature of light gained favour over the then widely-accepted *corpuscular* theory, which held that light behaved as a stream of corpuscles (particles).

We now know that diffraction also occurs with all forms of wave including sound and light. With sound waves, typical wavelengths are of the order of a metre or so, and we experience diffraction of sound as a matter of course. Doorways etc. are narrow compared to the wavelength of sound, so sound waves passing through a doorway are diffracted to fill the space beyond the door. We rarely experience sound 'shadows'. Diffraction of light only becomes significant when the aperture or obstacle is of the same order of magnitude as the wavelength of light. Since this is very rare, light diffraction does not form part of our everyday experience.

Diffraction of light by a single narrow slit

If the diffraction pattern is produced by a slit of adjustable width, the following observations are made as the slit width is varied:

- If the slit is more than a few millimetres wide, only a sharp rectangular shadow will be seen, the result of near-linear propagation of the light through the slit (see Figure 2.49(a)).
- As the slit is narrowed, a pattern starts to emerge of a wide central bright band having alternating dark and bright narrow bands (fringes) on either side. The central bright band is observed to be twice the width of the outer bands (see Figure 2.49(b)).
- As the slit approaches its narrowest, the central bright band widens considerably, showing illumination well into the geometric shadow of the slit. This indicates the large diffraction effect as the slit width approaches the wavelength of light. The pattern also becomes fainter, due to the reduced amount of light energy being transmitted through the narrow gap.

The effect of changing the colour of the light can also be observed:

- If monochromatic light is used, the bright bands are the same colour as the light used.

Figure 2.49 Single-slit diffraction

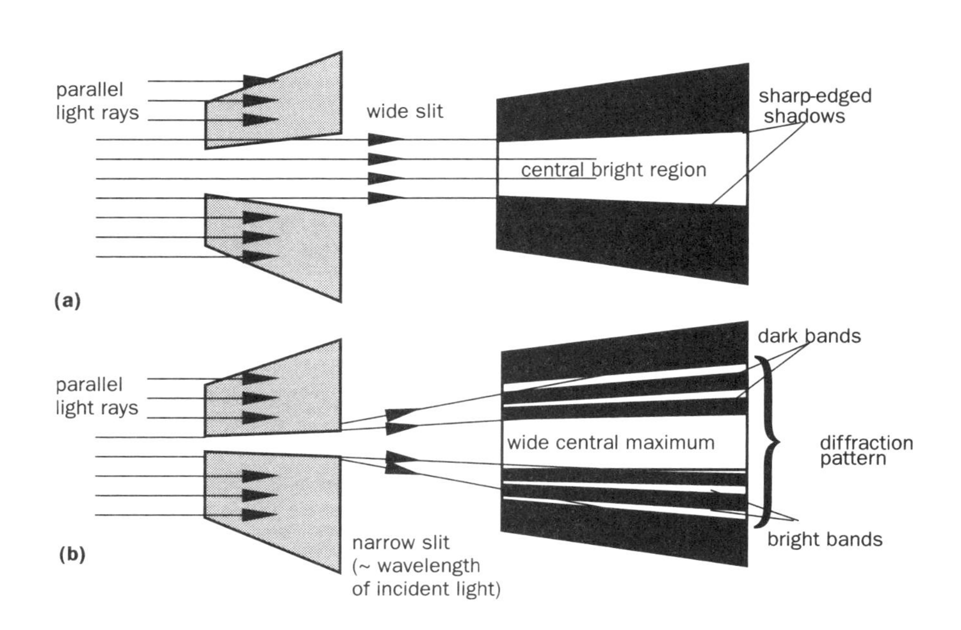

- Blue light gives bands closer together than red light. Blue light has a shorter wavelength than red, suggesting that the amount of diffraction increases with wavelength.
- With white light, the central band is white, but the fringes are tinged with overlapping colours.

Intensity distribution of single-slit diffraction pattern

Figure 2.50 shows the typical diffraction pattern obtained when monochromatic light passes through a single narrow slit. Also shown is a graph of light intensity plotted against distance along the diffraction pattern. The graph shows a wide central peak of intensity corresponding to the position of the central maximum, with narrower subsidiary maxima, of decreasing intensity, on either side.

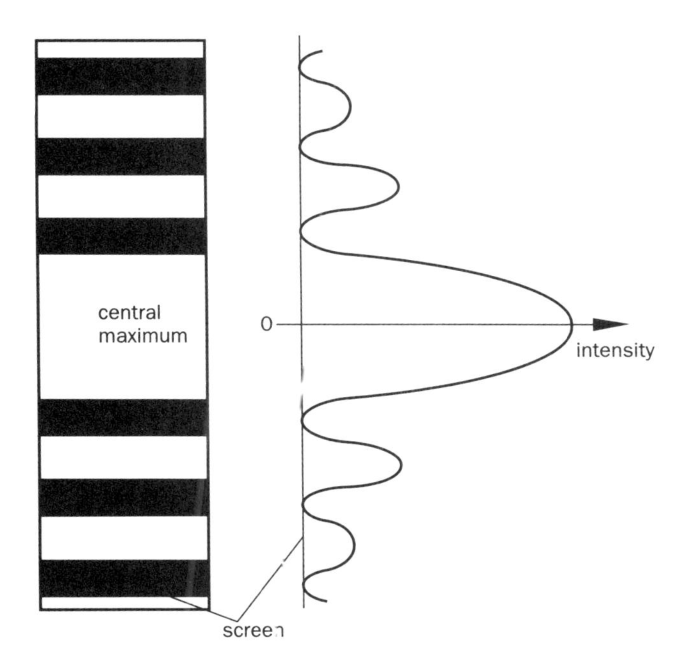

Figure 2.50 Intensity distribution of single-slit diffraction pattern

Secondary wavelets

The way in which light spreads from a source can be explained using the idea of **secondary wavelets**. Each point on a wavefront is considered to act as a new source of secondary wavelets, which spread out as circular waves from that point. The new wavefront is the **envelope** of all the secondary wavelets. If they all travel at the same speed, then the new wavefront will have the same shape as the previous one, and be travelling in the same direction. Diffraction is the name given to the effect where the wavefront is restricted (i.e. partially obstructed), and each point on the transmitted wavefront acts as a source of secondary wavelets, as shown in Figure 2.51. Secondary wavelets emitted from the extreme edges of the aperture give rise to waves spreading into the 'shadow' region. If the aperture is wide compared to the wavelength, interference between the wavelets from the edge of the aperture and the rest of the wavelets from the unrestricted wavefront produces almost complete cancellation in the shadow region, and the diffraction effect is minimal. With a narrow aperture, the interference between the wavelets is reduced and wide angle diffraction occurs. In addition, interference causes the alternating bright and dark fringes on either side of the central maximum.

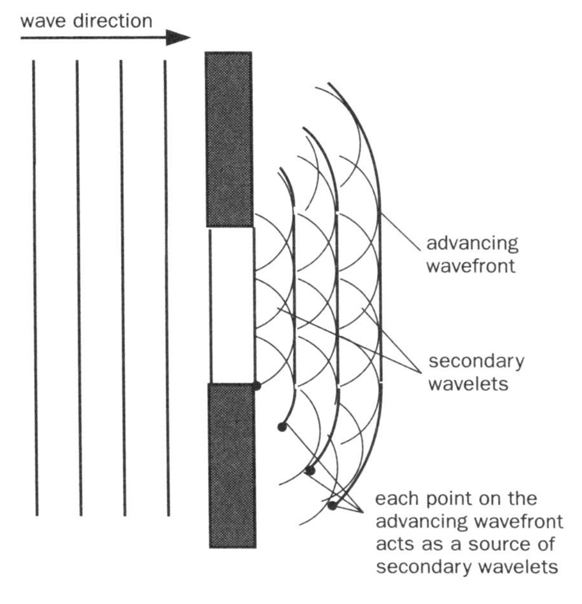

Figure 2.51 Generation of the diffracted wavefront

Double-slit interference

Thomas Young first demonstrated interference of light in 1801, providing the first widely-accepted evidence in favour of the wave theory of light.

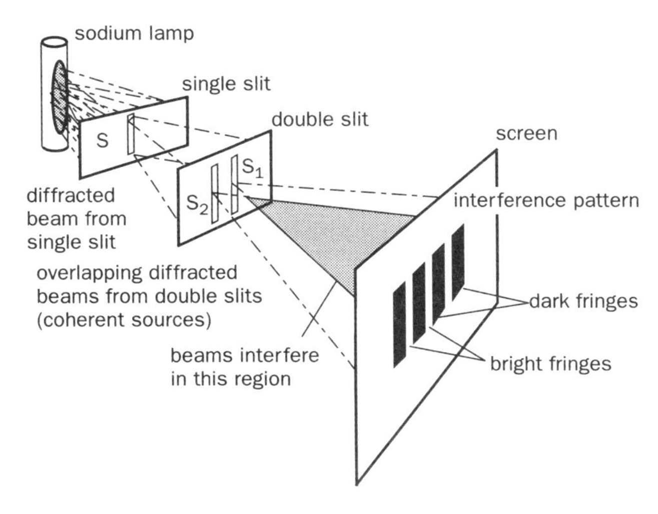

Figure 2.52 Double-slit interference – conventional light source

Figure 2.52 shows how interference effects can be produced using a conventional light source. Monochromatic light (i.e. light of a single colour) illuminates the single narrow slit S. Diffraction at this slit causes a diverging beam of light to fall on the two slits S_1 and S_2 which are very close together and parallel to S. Since S_1 and S_2 are derived from the single source S, they act as coherent sources of light. As the slits are narrow, diffraction causes diverging beams of light to spread into the region beyond the slits. Superposition of the waves in the two beams occurs where they overlap, and interference effects can be observed in this region. The experiment must be performed in darkness, and even then, the interference pattern will be almost undetectable. This is because the slits must be extremely narrow, reducing the amount of light energy transmitted. A travelling microscope eyepiece may replace the screen. This makes the pattern easier to see, and enables measurements to be made of the distance between the fringes.

You should note that the single slit must be narrow to:

- ensure that light reaching the two slits has come from a single point – the two slits will act as **coherent** sources
- **diffract** the light to produce a diverging beam to illuminate the double slit.

The double slits are also narrow to ensure adequate diffraction. This causes a large area of overlap, where the interference takes place. Narrower slits produce wider diffraction which increases the area over which interference occurs, but the pattern is fainter due to the reduced transmission of light energy.

Using the conventional light source, the interference pattern observed is one of alternating bright and dark bands (often referred to as fringes), parallel to S_1 and S_2. The fringes are equally spaced and the bright fringes are the same colour as the light source. A sodium light source produces mostly yellow light, so the bright fringes are yellow.

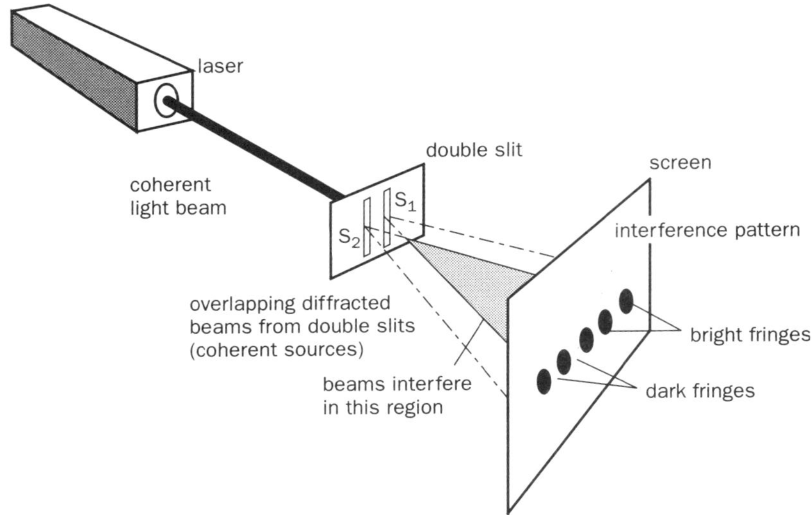

Figure 2.53 Double-slit interference – laser light

Figure 2.53 shows a simpler arrangement which produces an interference pattern bright enough to be visible in daylight. The laser replaces the conventional source and single slit. All points across a laser beam are in phase, so if this beam is split by the double slit, all light emitted from S_1 and S_2 will be phase-linked. Thus S_1 and S_2 become two coherent sources. The laser beam is narrow, and only illuminates a small region of the slits, so the resulting fringes are smaller, almost spots of light. However, the intensity of the beam is high, so a bright pattern is projected on a screen.

Explanation of double-slit interference

In Figure 2.54, point O on the screen is on the centre line of the beams from S_1 and S_2. It is equi-distant from the two slits, so light from each slit has travelled exactly the same distance. This means that the two beams arrive at O **in phase**, constructive interference occurs, and a **bright fringe** is produced.

Figure 2.54 Explanation of double-slit interference

equi-spaced light and dark fringes of equal width
diffracted beam from S_1
coherent laser beam
S_1
S_2
Q
P
O
beam overlap region
(interference effects)
double slit
diffracted beam from S_2
screen
appearance of pattern on screen

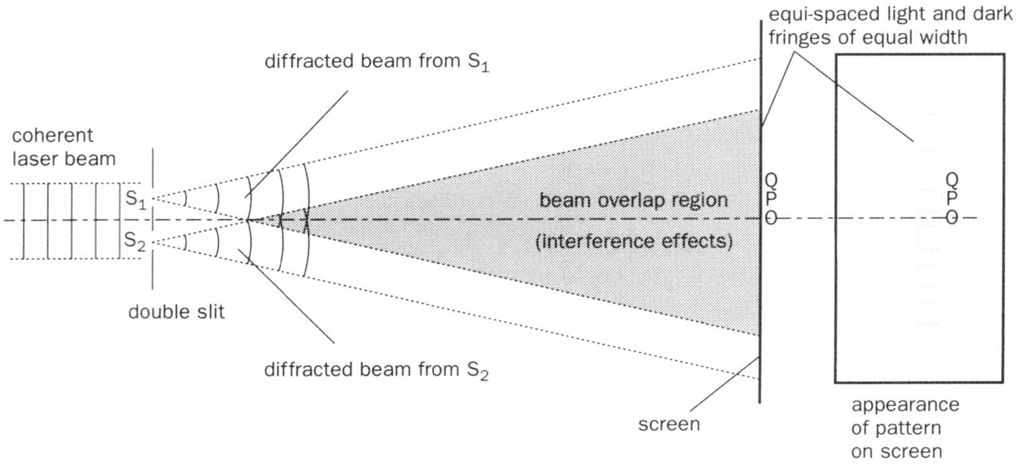

At point P, there is a dark fringe, so there can be no light energy reaching the screen at this point. The dark fringes are caused by destructive interference, i.e. the light waves from S_1 and S_2 arrive at the screen in antiphase and so cancel each other, producing zero (or minimum) intensity. Light from S_2 has had to travel further than that from S_1. If this path difference is exactly equal to 1, 3, 5 (or any odd number) of half-wavelengths, then the waves will be in **antiphase**, producing a **dark fringe**.

At point Q there is another bright fringe, produced by constructive interference. In this case the path difference must be equal to 2, 4, 6 (or any even number) of half-wavelengths of the light. The two sets of waves arrive in phase and reinforce each other.

Derivation window – Young's double-slit formula (see Figure 2.55)

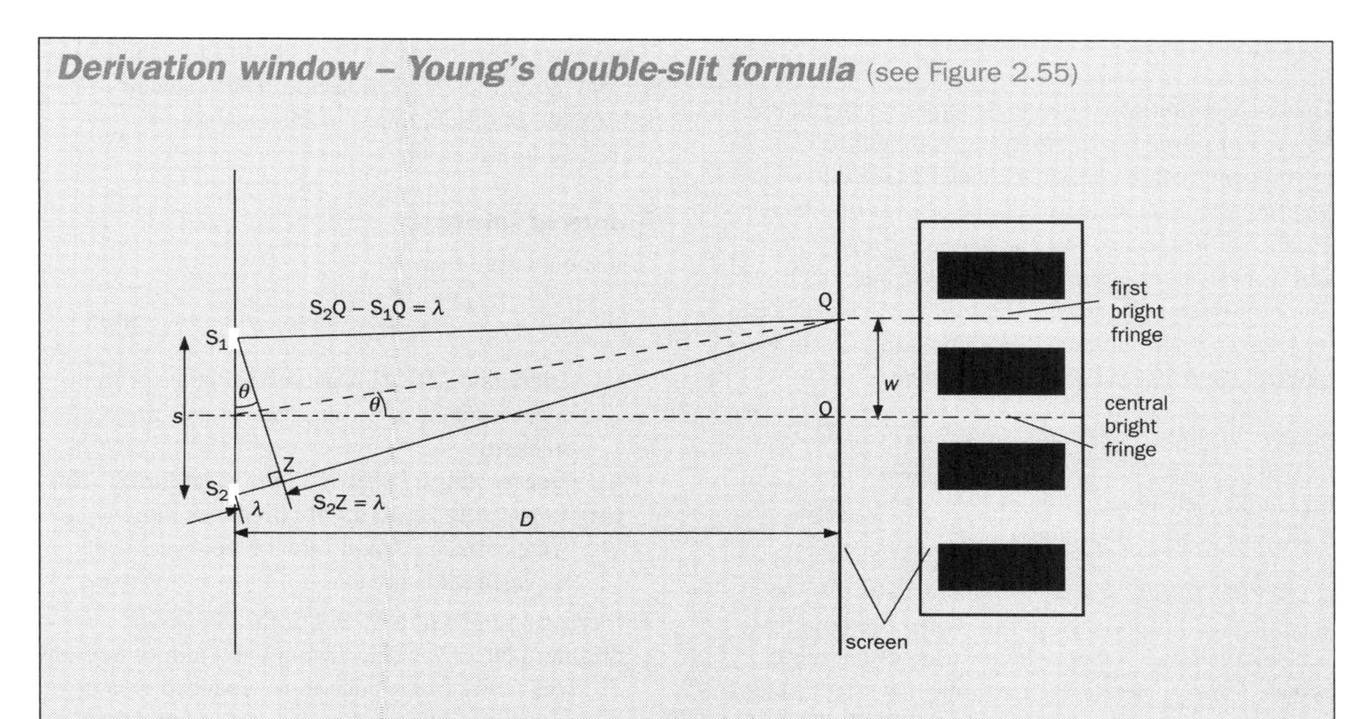

Figure 2.55 Derivation of double-slit formula

For the **central** bright fringe, the optical path difference = 0 ($S_2Q - S_1Q = 0$)

For the **first** bright fringe, the optical path difference = λ ($S_2Q - S_1Q = \lambda$)

$$\sin\theta = \frac{\lambda}{s} \quad \text{and} \quad \tan\theta = \frac{w}{D}$$

w = distance between adjacent fringes (fringe width)

s = distance between slits (slit width)

D = distance from slits to screen

for small angles $\sin\theta = \tan\theta = \dfrac{\lambda}{s} = \dfrac{w}{D}$

$$\therefore \quad \lambda = \frac{ws}{D}$$

Equation 1

This formula has been developed using w as the distance between the central fringe and the first bright fringe in the pattern. It can be extended to apply to any pair of fringes, bright or dark, in the pattern.

To produce easily viewable fringes (i.e. large w), D must be large (metres) and s very small (< 1 mm).

Experimental details

Measuring *w* Using a laser as the light source, the interference pattern is bright enough to be easily visible on a screen a few metres from the double slit. This has the advantage of making the distance between fringes large enough to be measured with a ruler, to a sufficient degree of accuracy. Again, the distance between the centres of a large number of fringes is measured, and the average calculated.

Using a conventional source, the fringe pattern must be viewed through a microscope. A travelling microscope with a vernier scale can be used to measure the distance between the centres of a large number of fringes, so that the average fringe width can be calculated. Figure 2.56 shows a simulated view of the interference pattern. The two vernier scales can be read, and one subtracted from the other to enable the average fringe width to be calculated.

Measuring *D* *Laser light source* – The distance from the slits to the screen is large, and can be measured with a metre rule or tape measure to a sufficient degree of accuracy.

Conventional light source – Without disturbing the apparatus, an illuminated pin is placed between the slits and the microscope. The position of the pin is adjusted until a clearly focused image is seen through the microscope. The pin is now in the focal plane of the microscope, and thus corresponds to the position of the image of the slits. The distance from the slits to the pin is therefore equal to *D*.

Measuring *s* A travelling microscope may be used to measure *s* directly, but the method illustrated in Figure 2.57 is easier.

The slits are illuminated with a bright light source, and a converging lens is used to focus a magnified, real image on a distant screen.

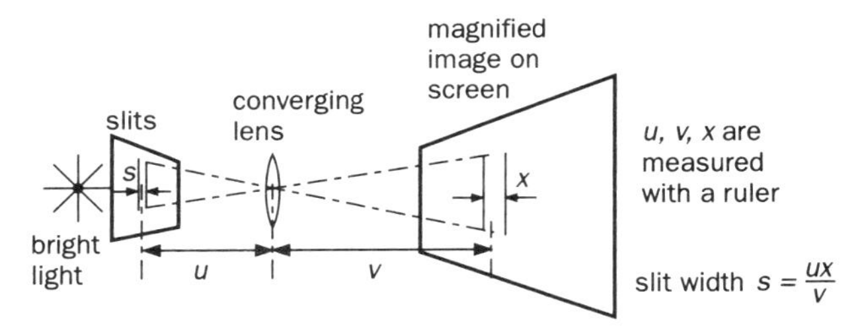

Figure 2.57 Measuring slit width

The values of *w*, *D*, and *s* have thus been obtained by simple measurements with a rule. These can be substituted into Equation 1 to calculate the wavelength of the light.

Points of interest

You should note that:

- Since $w = \frac{\lambda D}{s}$, fringe width ***w*** can be **increased** by:
  - (i) **Increasing *D*.** This decreases the error in the measurement of *w*, but reduces the fringe intensity.
  - (ii) **Decreasing *s*.** There is a practical limit to this.
  - (iii) **Increasing λ.** Red light produces more widely-spaced fringes than blue, because its wavelength is greater.
- Increasing the width of any of the slits gives a brighter pattern, but increasing the width of S_1 and S_2 gives fewer fringes due to the reduced area of overlap of the diffracted beams. In practice the slits have to be many wavelengths wide in order to transmit sufficient light energy to produce a visible interference pattern.
- White light produces fewer fringes, and only the central bright fringe is white. The rest are tinged with colour as each colour in the spectrum produces differently-spaced fringes. This leads to overlapping, confused fringes.
- There is no loss of energy from the dark fringes; it is merely re-distributed to the bright areas.
- Fringes can be seen anywhere in the region where the two diffracted beams from the two slits overlap.

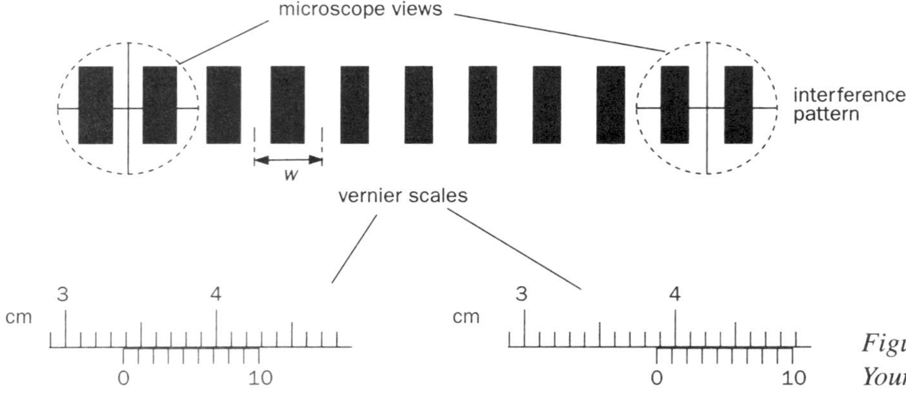

Figure 2.56 Young's double-slit experiment – measurement of fringe width

Guided example (1)

In a Young's double-slit experiment, using a conventional light source, an interference pattern is viewed through a travelling microscope as shown in Figure 2.56.

(a) Using the data from this diagram, calculate the average fringe width.

The apparatus is left undisturbed, and a brightly-lit object is positioned so that it is in the focal plane of the travelling microscope. The pin is found to be 60.0 cm from the double slit. The apparatus is then dismantled, and the set-up shown in Figure 2.57 is used to determine the slit width. The converging lens was placed 20 cm from the slits, and a sharp image of the slits was produced on a screen 3.8 m from the lens. The distance between the images of each slit was measured carefully with a ruler, and found to be 10.5 mm.

(b) Calculate the slit width.

(c) Using this data, calculate the wavelength of the source used.

Guidelines

(a) Read the two vernier scales, count the number of fringe **widths** and calculate the average fringe width.

(b) Slit width is obtained directly from the formula given in Figure 2.57.

(c) You now have enough data to calculate the wavelength of the source used.

The diffraction grating

Commercially, splitting light (and other electromagnetic radiations, particularly ultra-violet) into spectra has great importance. Spectroscopic analysis involves studying the characteristic wavelengths given off or absorbed by each element. This requires the ability to produce accurately defined monochromatic radiation, and to measure precisely radiations produced. Prisms can be used, but the diffraction grating produces a far purer spectrum for accurate work.

A diffraction grating is usually a piece of glass or plastic with closely-spaced lines on it. A transmission grating has clear spaces between the lines where the light can pass through, while a reflection grating has shiny surfaces between the lines which reflect the light.

Compact discs behave as reflection gratings. The tracks on the disc are very close together, and light is reflected from the gaps between the tracks, leading to the colourful spectra visible when light is reflected from the disc surface.

Transmission gratings are used in spectrometers and spectrographs to measure the wavelength of light to a very high degree of accuracy, in a great deal of

A compact disc behaves as a reflection grating

analytical laboratory work. Substances can be identified by their unique patterns of absorption, emission and transmission of specific wavelengths. Chemical flame tests are a crude form of spectral analysis. Our eyes make a rough estimate of the colour of the light, whereas spectrometers measure the wavelengths and intensities of each individual colour in the light. The proportion of individual elements in a substance can be calculated from such observations.

Typically, a transmission grating is a piece of glass or plastic with the lines etched on it, in numbers ranging from tens to thousands per mm. The gaps between the lines are thus very narrow, similar to light wavelengths, and so act as diffracting elements (slits). Light passing through each slit is very widely diffracted, so a great many overlapping beams are transmitted. Interference takes place everywhere in the transmitted region.

Complete cancellation (destructive interference) takes place in almost every direction. Along certain clearly defined directions, constructive interference occurs, leading to bright images. The exact directions depend on the wavelength of the light, so if the direction of a particular colour can be measured accurately, then its wavelength can be calculated. The usual procedure is to use a single narrow slit, illuminated by the light being studied. Light rays from the slit are made parallel, and passed through the grating. A telescope is used to observe and measure the angular deflections of the slit image.

With white light, each image will be a spectrum of all the colours, since each wavelength (and therefore each colour) will be diffracted by a slightly different angle. With monochromatic light, each image will be a line (an image of the slit). However, it is common to refer to these lines also as spectra, even though they are lines of a single colour. In practice, it is rare for a source to be truly monochromatic. Laser light approaches the ideal, while gas-discharge tubes usually emit one dominant wavelength, but several fainter lines of various colours will also be seen.

While it would be possible to use a double-slit arrangement to measure the wavelength of a beam of light, diffraction gratings have the following advantages:

- The maxima are much more sharply defined.
- The beam will pass through maybe hundreds of slits, instead of just two, so the intensity of the maxima is much higher.
- The large angles between the maxima can be measured with much greater precision than the distances between double-slit interference fringes.

Derivation window – diffraction grating formula (see Figure 2.58)

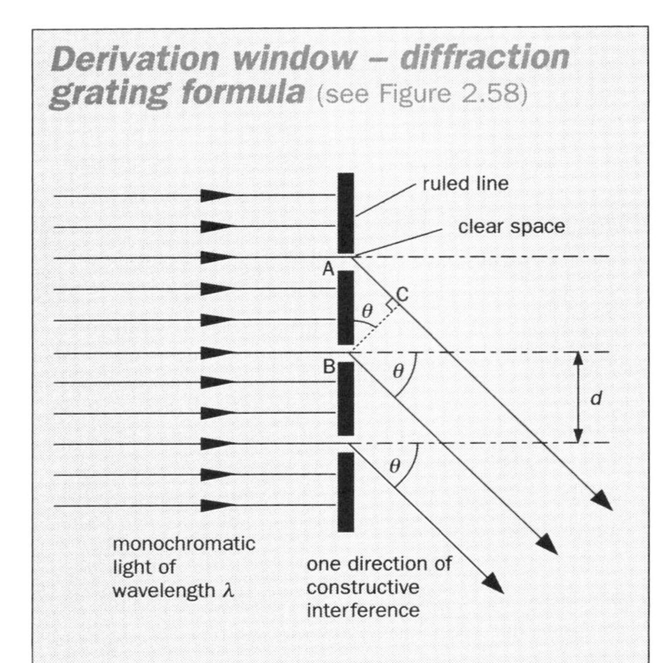

Figure 2.58 Diffraction grating

Parallel rays of monochromatic light of wavelength λ are incident on a diffraction grating in which the slit separation (distance between slits, or grating spacing) = d. If the grating has N lines per metre, then the grating spacing is given by:

$$d = \frac{1}{N}$$

Constructive interference only occurs along a few precise directions, one of which is shown above. For constructive interference, light from A must be in phase with light from B (and from every other slit). This can only happen if the path difference AC is equal to a whole number of complete wavelengths.

i.e. $AC = n\lambda$ where $n = 0, 1, 2, 3 \ldots$ etc.

but $AC = d \sin\theta$

where θ = angle of diffraction

d = grating spacing

$$\therefore \quad d \sin\theta = n\lambda$$

m m

Equation 2

where n is called the spectrum order (e.g. $n = 1$ is the *first* order diffraction image).

NB: Since $\sin\theta$ cannot be greater than 1 (i.e. θ cannot be greater than 90°) there is a limit to the number of spectra that can be obtained.

Measurement of the wavelength of light using a diffraction grating and a spectrometer

A spectrometer contains three main components:

- the turntable for mounting and rotating the grating, prism, etc.
- the collimator for producing a narrow, parallel light beam
- the telescope for viewing the beam and measuring its angular position (see Figure 2.59).

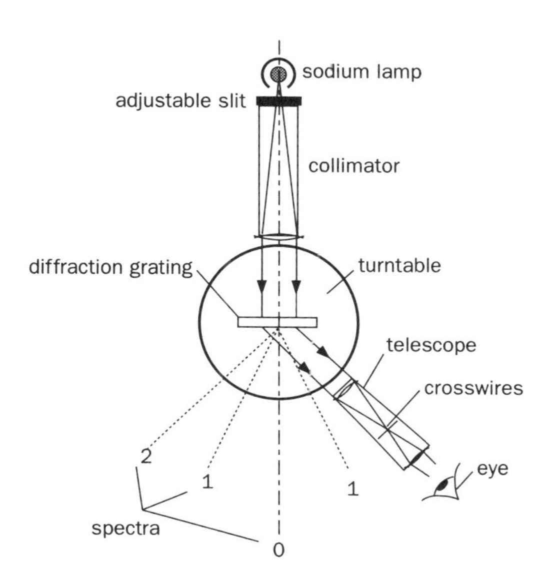

Figure 2.59 Spectrometer adjustments

Setting up the spectrometer

The eyepiece of the telescope is first focused on the crosswires, then the telescope itself is focused on infinity. This ensures that the telescope is set up to receive parallel rays, and that the intermediate image of the slit will fall on the crosswires.

The telescope is then lined up with the collimator, and the position of the slit adjusted until there is no parallax between its image, seen through the telescope, and the crosswires. The slit is now at the principal focus of the collimator lens, and a parallel beam is produced.

Mounting the grating

The grating must be positioned perpendicular to the beam of light. One way of achieving this is to set the telescope at 90° to the axis of the collimator, as shown in Figure 2.60. The turntable is rotated until an image of the slit is seen **by reflection** from the face of the grating. This sets the grating at 45° to the light beam, and it is then a simple matter to rotate the turntable 45° to set the grating in the correct position.

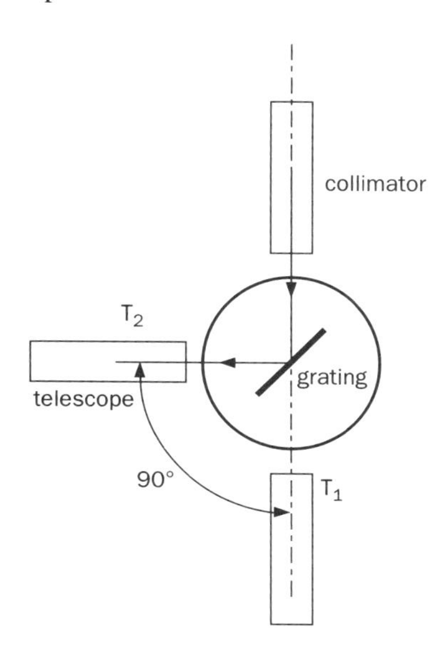

Figure 2.60 Positioning the grating

Viewing the spectra

(see Figure 2.61)

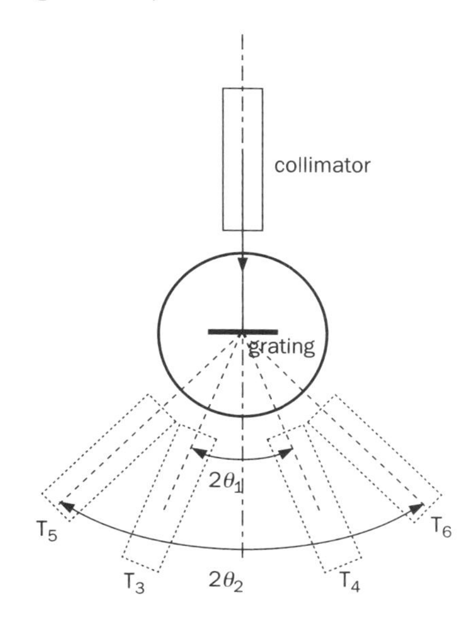

Figure 2.61 Viewing the spectra

With the grating locked in position, the telescope is rotated to the left of the straight-through position until the first-order spectrum is lined up with the crosswires (position T_3 in Figure 2.61). The angular position is measured and recorded before rotating the telescope to the right, past the straight-through position to position T_4, where the first-order spectrum is again seen. The angular displacement of the telescope between these two positions is measured. This is equal to $2\theta_1$. The process is repeated for the second order, third order, etc. spectra.

Guided example (2)

Grating: 500 lines mm^{-1}

Spectrum order	Angular position	
	Left	Right
1	198° 57'	158° 14'
2	220° 12'	136° 12'

Calculate the diffraction angle for each spectrum order, and use this to calculate a value for the wavelength of the light in each case. They should be equal (or very nearly), so calculate the average value.

Guidelines

Use the angular displacements to calculate $2\theta_1$ and $2\theta_2$ and hence θ_1 and θ_2. Substitute in Equation 2 to calculate wavelength.

Self-assessment

SECTION A
Qualitative assessment

1. Describe with the aid of diagrams how two waves can produce **constructive** and **destructive** interference.
2. Explain the meaning of **coherent** as applied to light sources, and explain why coherent sources are necessary to obtain a **steady** interference pattern.
3. Explain why interference effects are not seen in the light from a pair of car headlamps.
4. Describe and explain what happens to an interference pattern if the **distance** between the two sources producing the pattern is **increased**.
5. A beam of **monochromatic** light passes through a narrow slit and onto a screen. Describe how the appearance of the screen changes as the slit is made **narrower**. Sketch a graph of the energy distribution across the diffraction pattern on the screen.
6. Describe Young's double-slit experiment, stating clearly how each quantity is measured, and show how the results can be used to calculate the wavelength of light.
7. An interference pattern is produced on a screen in Young's double-slit experiment using **red** light. Describe and explain the effect of the following changes:
   (i) **increasing** the **distance** between the double slits
   (ii) covering **one** of the slits
   (iii) making each slit **wider**
   (iv) using **blue** light
   (v) using **white** light
   (vi) moving the screen **further away** from the double slits.
8. Derive Young's double slit formula from first principles, stating the meaning of any **terms** used and stating clearly any **assumptions** or **simplifications** made.
9. Derive the diffraction grating formula from first principles, stating the meaning of any **terms** used and stating clearly any **assumptions** or **simplifications** made.
10. Describe how a spectrometer and diffraction grating can be used to determine the wavelength of a monchromatic light source.

SECTION B
Quantitative assessment

(***Answers:*** 5.2×10^{-7}; 3.2×10^{-3}; 0.30; 6; 9.4; 5.0×10^{5}.)

1. Assuming the light from a lamp is emitted in bursts lasting 1 ns (1×10^{-9} s) calculate the distance travelled by the light emitted in a single burst and the number of cycles contained in a burst of wavelength 600 nm. Assume the speed of light to be 3.0×10^{8} m s^{-1}.
2. In a Young's double-slit experiment using a laser of wavelength 638 nm, the screen is placed 2.5 m from the double slit. If the slit separation is 0.50 mm, calculate the distance between fringes. If the red laser is replaced, and the fringes are now observed to be separated by a distance of 2.6 mm, calculate the wavelength of the light emitted by the new laser.
3. A diffraction grating has 300 lines per mm. When it is illuminated normally by light of wavelength 530 nm, calculate the angle between the first and second order maxima. What is the highest order maximum that can be obtained?

2.4 Stationary waves

All the discussion so far has concerned waves which are moving through space – i.e. progressive waves. However, there are circumstances when waves can appear to be standing still: canoeists are rightly worried about 'stoppers', where a large amplitude crest and trough can form downstream of a submerged boulder. These stay in one place, and can trap an unwary canoeist. Strings on musical instruments can be seen to be vibrating, but the positions of the points of maximum amplitude do not change with time, so the wave is stationary.

Conditions for formation of a stationary wave

Stationary waves are produced when two progressive waves of **equal frequency** and **speed** and nearly **equal amplitude** travelling in opposite directions are **superposed.**

When a guitar string is plucked, **transverse** waves travel to the ends of the string, where they are reflected. Since the reflected waves will travel backwards and forwards along the string, they will **interfere** with each other. The resultant wave is thus the sum of the two waves travelling in **opposite** directions along the string.

The frequencies of the incident and reflected waves will be **identical** since frequency is controlled by factors such as string length, tension, mass, etc. which are constant for each wave. The **amplitudes** will be nearly equal since little energy will be lost in reflection. Thus the conditions for a stationary wave are satisfied. In the simplest mode of vibration, the amplitude of vibration at the centre of the string is a maximum, and decreases to zero at the fixed ends.

Stationary waves are an example of **resonance**.

Any mechanical system which possesses mass and elasticity can vibrate. Such systems have **natural** frequencies of vibration, governed by the size of the system and the speed of travel of the waves. There is a constant interchange of energy between the **kinetic** energy of the moving particles and the **strain** energy stored in the system. The total energy is gradually dissipated to the surroundings (e.g. as sound energy in the case of musical instruments), and the amplitude of the stationary wave decays to zero.

You should note that stationary waves have:

- one or more points at which the amplitude of vibration is **zero** (called **nodes**)
- one or more points at which the amplitude of vibration is a **maximum** (called **antinodes**).

Melde's apparatus

The apparatus shown in Figure 2.62 (Melde's apparatus) can be used to study the behaviour of vibrating strings. The string is tensioned by the hanging weights, while the length of the string is controlled by the position of the movable bridge.

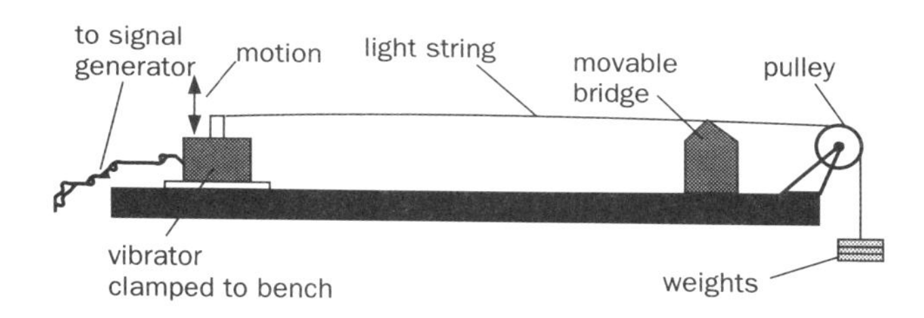

Figure 2.62 Melde's apparatus

The disturbance is provided by the vibrator at one end of the string. (Strictly speaking, the end should be a node, but this is impossible, since the vibrator will not have zero amplitude. However, provided the amplitude of the vibrator is small compared to the amplitude of the antinodes, this error can be ignored.) If the frequency of the vibrator is increased steadily from a low value, little is observed until the frequency reaches a certain value (which depends on the tension, length, etc. of the string). At this frequency, the string will start to vibrate with a large amplitude stationary wave, with one antinode at the centre, and a node at each end. This is the lowest frequency at which a stationary wave is formed. It consists of only one vibrating loop, as shown in Figure 2.63, and is known as the **fundamental** mode of vibration. The frequency at which this occurs is called the **fundamental frequency**.

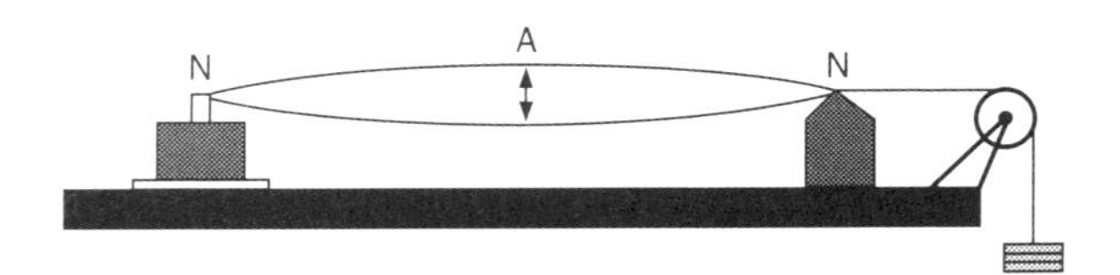

Figure 2.63 Fundamental mode of vibration of a stretched string

Explanation: If, in the time taken for the wave to reach the end and return, the vibrator is just ready to send a second wave, this will **reinforce** the first wave. Thus the amplitude of vibration builds up as each new input from the vibrator is **in phase** with the wave in the medium, and the energy is added to it.

If the frequency is further increased, the stationary wave will disappear, until a frequency equal to 2 × the fundamental frequency is reached. This time the stationary wave will have **two** vibrating loops, with a point in the middle which is not vibrating (a **node**), as shown in Figure 2.64.

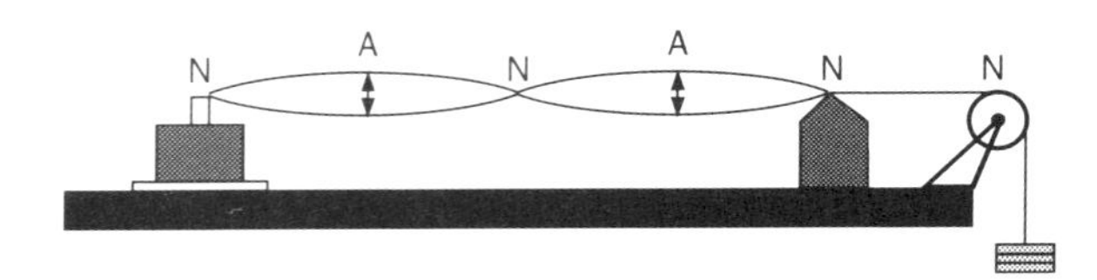

Figure 2.64 2× fundamental frequency

Further stationary waves will be observed at 3×, 4×, 5× the fundamental frequency, with 3, 4, 5 ..., etc. vibrating loops. The frequencies at which these stationary waves occur are the **resonant frequencies** of this particular string under these conditions.

Viewing the stationary wave

The frequency of vibration will be invisible to the eye if it exceeds 20 Hz or so. In order to view the motion of the wave, the apparatus can be illuminated with a **stroboscope**, a flashing light whose frequency can be controlled. If the stroboscope frequency exactly matches the frequency of the stationary wave, then the string will be in exactly the same position each time the light flashes, and so appears stationary. If the stroboscope frequency is changed slightly, the wave will appear to vibrate in slow motion. Careful adjustment of the flashing rate will reveal the nature of the string's motion.

Figure 2.65 shows the appearance of the string during **half** a cycle. Each line shows the appearance of the string after **one-eighth** of a cycle.

In this case, the frequency (f) is equal to twice the fundamental frequency.

You should note that:

- There are points where the displacement of the string is always zero, called **nodes** (N in Figure 2.65). There is always a node at each end (where the string is fixed).
- Adjacent nodes are separated by a distance equal to half a wavelength ($\lambda/2$).
- There are points where the displacement of the string is always a **maximum**, called **antinodes** (A in Figure 2.65).
- Adjacent antinodes are also separated by a distance equal to half a wavelength ($\lambda/2$).
- In the region between successive nodes, all particles are moving **in phase** with differing **amplitudes**. All points in a single vibrating loop thus reach their maximum displacements simultaneously.
- The oscillations in one loop are in **antiphase** (i.e. 180° or π radians or $\lambda/2$ out of phase) with adjacent loops.
- The frequency of vibration of the particles in the stationary wave and in the progressive waves from which it is produced is the same;
  the wavelength of both the stationary and progressive waves is also the same. Thus it is convenient to measure the wavelength of a progressive wave by producing a stationary wave from it, and measuring the distance between successive nodes.

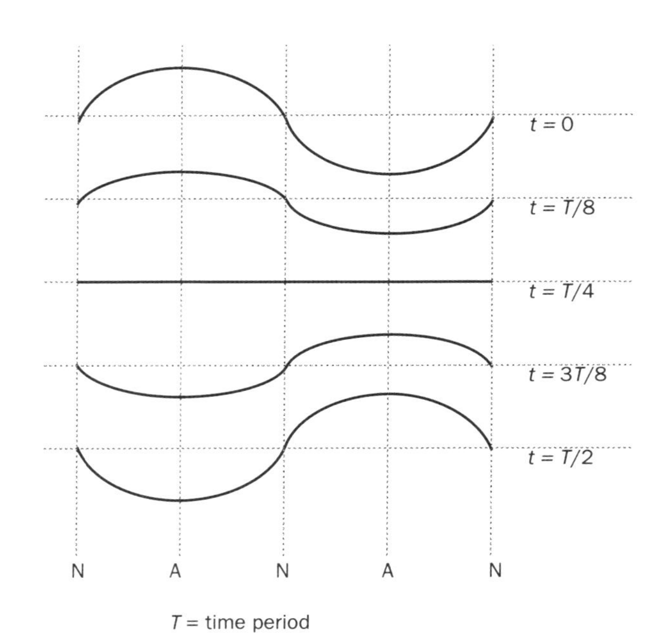

Figure 2.65 Appearance of string over 1 half-cycle

Stationary longitudinal waves

These can be readily demonstrated using a 'slinky' spring suspended vertically, excited at the lower end by a vibrator. This time, the direction of oscillation of the spring coils is longitudinal, but antinodes and nodes are easily seen.

Again, if the frequency of the vibrator is increased from a low value, stationary waves can be observed at certain resonant frequencies. The lowest frequency produces a single antinode in the centre of the slinky. This is the fundamental frequency. At twice this frequency, two antinodes are seen, as shown in Figure 2.66.

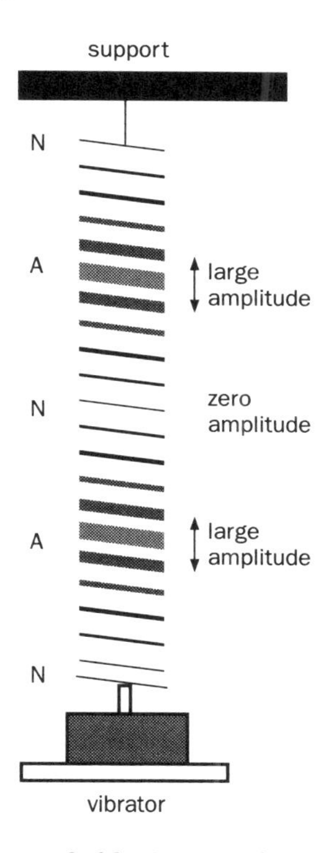

Figure 2.66 Longitudinal stationary wave

The frequency of oscillation is very low (2–4 Hz) and enables the phase relationship between adjacent vibrating loops to be seen. In the case shown, the top half of the spring will be vibrating in antiphase with the lower half. (Each coil in the upper half will be moving in the opposite direction to the coils in the lower half at every instant.) Longitudinal stationary waves also occur in organ pipes and all other wind instruments, where the molecules of air in the pipe vibrate in the same manner as the coils of the slinky.

Vibrating strings

Stretched strings may be set vibrating by being plucked, struck, or bowed. **Transverse** waves travel along the string and are reflected at the fixed ends. Since the original wave and the reflected wave have the same **speed, frequency** and **amplitude**, the conditions for the formation of a stationary wave are met, and a **stationary transverse** wave pattern is formed in the string. This generates a **progressive, longitudinal** sound wave in the air in contact with the string, which is transmitted into the surroundings, with the same frequency as the stationary wave. If the string is stretched across a 'sounding board' (as in a guitar), then this also vibrates and a louder sound is produced due to the larger mass of air being set vibrating. (The sounding board may vibrate in many complex modes simultaneously, giving rise to the variety and richness of tone associated with musical instruments.)

The energy initially given to the string is gradually dissipated to the atmosphere. The design of the instrument is a compromise between the loudness of the sound produced and the 'sustain' of the note (the time taken for the sound to die away). A loud sound will quickly dissipate the energy, reducing the sustain time, and vice versa.

If the string is assumed to be perfectly elastic, then the velocity (v) of the transverse wave along it depends on (i) the **tension**, and (ii) the **mass per unit length** of the string. The velocity v of the transverse wave along a stretched string is given by:

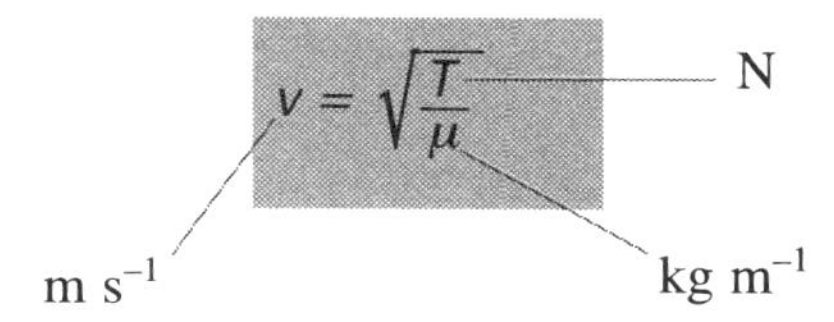

where T = tension in the string
and μ = mass per unit length of the string ($\mu = m/l$)

Increasing the tension of the string causes the wave to travel along it at a higher speed. Conversely, the wave speed is reduced in a string with higher mass per unit length. A higher wave speed produces a higher **frequency** of vibration in a given length of string, so **tighter** strings produce notes of **higher** pitch, while **heavier** strings produce **lower** pitched notes. This can be seen in the design of a guitar, where six equal-length strings all have approximately equal tensions, for comfort when playing. They produce a wide range of notes due to the fact that the mass per unit length of each string is very different. The lowest-pitched strings are wound with dense brass wire to produce a high mass per unit length (such a 'wound' string is much more flexible than a solid string of equivalent mass per unit length). In a piano or harp, wide variations in string length are also used to produce a wide range of notes.

Modes of vibration of strings

When a stretched string such as a guitar or piano string is set into vibration, a progressive wave travels to the fixed ends where it is reflected. The reflected waves will have equal amplitude and frequency, so a **stationary wave** will be established on the string by **superposition** of the reflected waves. Depending on factors such as the way the string was made to vibrate, the material of the instrument, etc. the string may vibrate in many different **modes**.

First harmonic (or fundamental frequency)
This is the simplest mode of vibration of a string, and is shown in Figure 2.67. If a guitar string is plucked at its centre, it will produce its lowest possible note, and the vibration consists of an **antinode** (A) mid-way between the two **nodes** (N) at the fixed ends of the string. This is the **lowest** possible frequency of vibration for a string, of this length, tension and mass per unit length, and is known as the **first harmonic** (or fundamental frequency) (f_1).

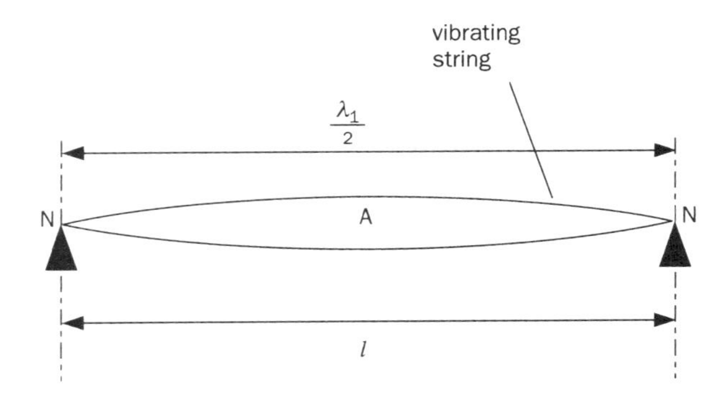

Figure 2.67 Fundamental mode of vibration

Since the distance between adjacent nodes (or adjacent antinodes) is equal to half the wavelength, then, if l = string *length*

$$\therefore \quad l = \frac{\lambda_1}{2} \qquad \therefore \quad \lambda_1 = 2l$$

where λ_1 is the *wavelength* of the **first harmonic** (or fundamental mode of vibration).

Second harmonic (or first overtone) This is the next possible mode of vibration. If a guitar string is plucked in the normal way, then **lightly** touched at its exact mid-point, a note of exactly **twice** the fundamental frequency will be heard. Careful examination of the string will show that the mid-point of the string is not vibrating, so in this case, as well as the two nodes at the fixed ends, there is a central node and two antinodes, as shown in Figure 2.68.

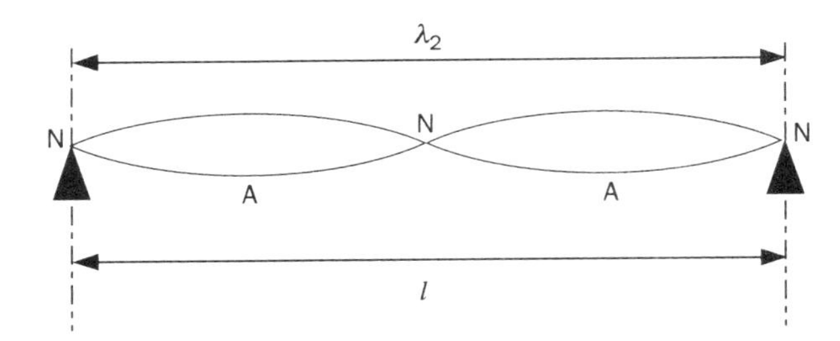

Figure 2.68 2nd harmonic (first overtone)

In this case: $l = \lambda_2$ as there is a node in the centre of the string, where λ_2 is the *wavelength* of the second harmonic (first overtone) mode of vibration.

Since this is **half** the wavelength of the first harmonic, it follows that the note produced will have **twice** the frequency of that of the first harmonic.

Third harmonic (or second overtone) Plucking the string, and lightly damping at exactly one-third of its length will produce the mode of vibration shown in Figure 2.69. This time we see three antinodes.

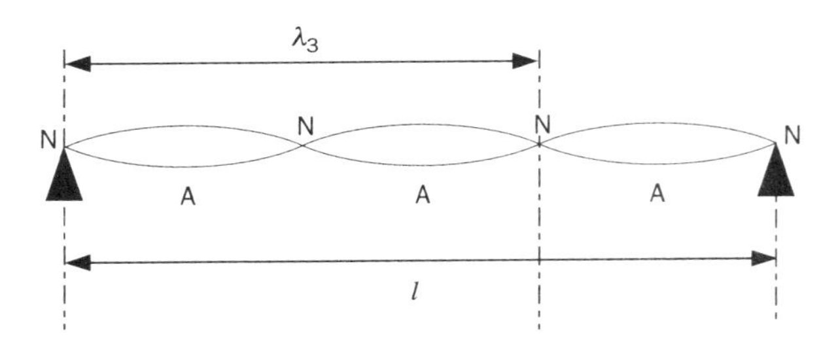

Figure 2.69 3rd harmonic (second overtone)

Following a similar analysis to that of the second harmonic, we find that the frequency of vibration of the third harmonic is equal to three times that of the first harmonic.

From the above examples, you should see that a pattern of possible overtones for a stretched string exists. The possible frequencies of vibration are f_1, $2f_1$, $3f_1$, $4f_1$, . . . , etc., i.e. each higher frequency is an **integer** multiple of the first harmonic frequency, f_1.

You should also note that a string can vibrate in several modes simultaneously. Musical instruments produce their own unique sound 'quality' due to the number and relative amplitudes of harmonics present. This is affected by the materials from which the instrument is constructed, and the way these feed back the vibrations to the string. Other factors affecting the final sound produced include the shape and construction of the instrument and, not least, the skill of the musician.

Vibrations in air columns (pipes)

When the air at one end of a tube or pipe is caused to vibrate, a **progressive, longitudinal** wave travels down the tube, and is reflected at the opposite end (which may be open to the outside air or closed). The superposed incident and reflected waves thus **interfere**, and, as they have the same **frequency**, **amplitude** and **speed**, they produce a stationary, longitudinal wave. This is the principle behind wind instruments. Sounds are produced in wind instruments as a result of the **stationary** longitudinal waves which are set up in the air column inside the instrument. These in turn transfer energy to the air outside, and produce **progressive** longitudinal sound waves.

The air in the tube is excited into vibration in many ways: by reeds, trumpet mouthpieces, or simply a sharp edge over which air is blown. These produce vibrations of almost every conceivable audible frequency, so the resonant frequency of the air column must be included. All gas filled pipes have **resonant** frequencies, which depend on the length of the pipe (and the temperature and density of the gas). If the molecules of the gas can be made to vibrate at this frequency, a note of that frequency will be heard. Higher **harmonics** will also be produced at frequencies which are integer multiples of the fundamental frequency.

The modes of vibration of the air inside 'open' and 'closed' pipes are very different, and must be studied separately.

Modes of vibration in 'closed' pipes

'Closed' refers to the type of pipe in which only one end is open to the atmosphere. Some organ pipes are made in this way.

Demonstration of stationary waves in pipes

A glass tube closed at one end, such as a measuring cylinder, is mounted horizontally as shown in Figure 2.70.

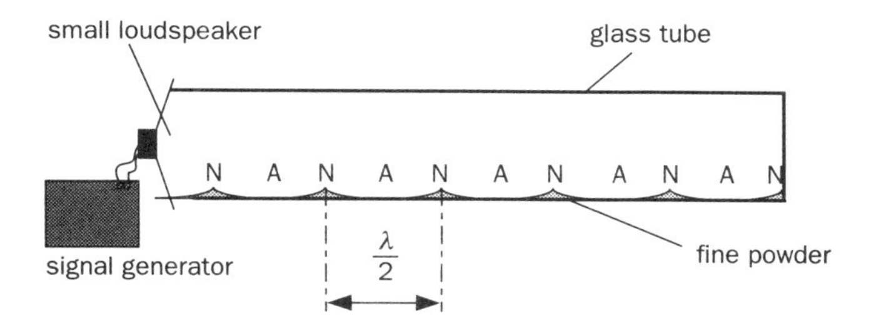

Figure 2.70 Demonstration of stationary waves in pipes

A small amount of a fine powder such as lycopodium powder is placed along the length of the tube, and a small loudspeaker is clamped at the open end. When the loudspeaker is driven by a signal generator, and the frequency is steadily increased from a low value, the sound produced is heard to increase to a maximum intensity at certain frequencies. In between these frequencies, the volume of sound heard is equivalent to that produced by the loudspeaker alone, without the tube.

The peaks of sound intensity are caused when the air in the tube **resonates** with the frequency of the loudspeaker. At these resonant frequencies, the light powder in the tube will be seen to form into equally-spaced peaks. The presence of these peaks can be explained by considering what is happening to the molecules of the air in the tube. The molecules vibrate **along** the axis of the tube, and the amplitude of vibration varies from a **maximum** at the **antinodes** (A) to **zero** at the **nodes** (N). At the positions of the antinodes, the large-amplitude vibration of the molecules disturbs the light powder, shifting it away from these regions and causing it to accumulate near the positions of the nodes, where the amplitude of vibration of the molecules is zero.

For each mode of vibration:

- The wave in the pipe is **longitudinal**. That is, the air particles vibrate **parallel** to (or along) the longitudinal axis of the pipe.
- The amplitude of vibration of the molecules is always a **maximum** at the **open** end of the pipe. There is a displacement **antinode** at the open end.
- The amplitude of vibration of the molecules must be **zero** at the **closed** end of the pipe, which means that the closed end is the position of a displacement **node**.
- All molecules in the region between two adjacent nodes vibrate **in phase**.
- All molecules on either side of a node vibrate **in antiphase**.
- Adjacent nodes (and adjacent antinodes) are separated by a distance equal to **half a wavelength**.

You should be clear that the vibration of the molecules of air in the tube is **longitudinal** and not transverse, as might be thought from looking at the shape of the powder heaps.

First harmonic (fundamental frequency) The simplest mode of vibration for a closed pipe has a single **antinode** at the open end and a single **node** at the closed end, as shown in Figure 2.71. The first diagram shows the direction and relative amplitude of vibration (highly exaggerated) of the molecules.

A node and an adjacent antinode are one quarter-wave apart, so the tube is one quarter of a wavelength long. The middle diagram demonstrates how a displacement–distance graph can be drawn to represent this wave. The vibrations of each molecule are plotted on the **vertical** axis, and the dotted line is the **envelope** of all these individual motions. The third diagram is a graph of **amplitude** against distance along the tube. Remember that these graphs are not pictures of the wave, they are only methods of describing what is happening to the molecules in the tube.

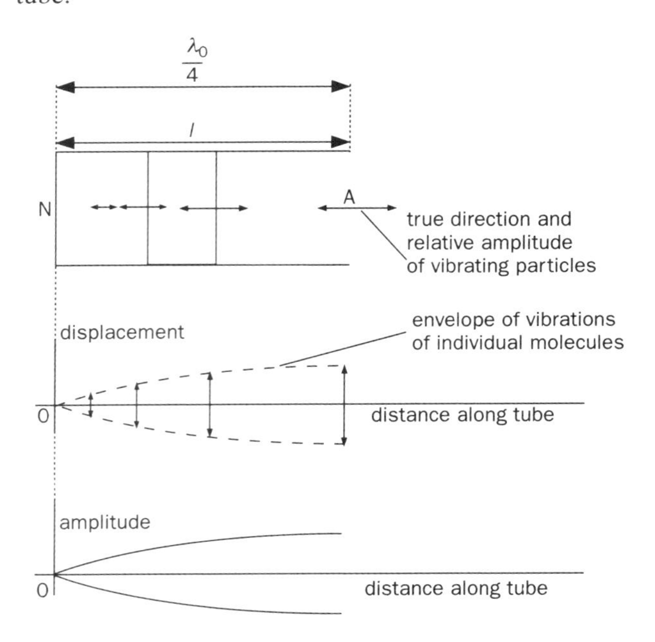

Figure 2.71 First harmonic (fundamental) mode of vibration for a closed pipe

Since a node and adjacent antinode are only a quarter wave apart, for a closed pipe of length l, vibrating at its *fundamental frequency*:

$$l = \frac{\lambda_1}{4} \qquad \therefore \quad \lambda_1 = 4l$$

where λ_1 is the wavelength of the stationary wave in the pipe.

This is the longest wavelength possible for this length of closed pipe, so the note produced has the lowest possible frequency.

Third harmonic (first overtone) If the air in the tube is excited by frequencies higher than the fundamental frequency (wind instrument players do this by changing the way in which they blow into the mouthpiece), the air in the pipe can be made to resonate in more complex modes. As there must be an antinode at the open end and a node at the closed end, the next resonant frequency occurs at **three** times the fundamental frequency, so is referred to as the **third** harmonic, as shown in Figure 2.72. Note that as this is the **first** resonant frequency higher than the fundamental, it is also called the **first overtone**.

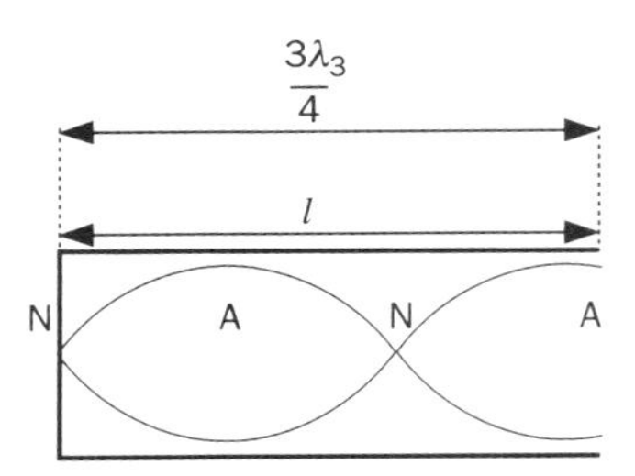

Figure 2.72 Third harmonic (first overtone) for a closed pipe

Since the length of the pipe is equal to three-quarters of a wavelength when vibrating in this mode, the wavelength (λ_3) is given by:

$$l = \frac{3\lambda_3}{4} \qquad \therefore \quad \lambda_3 = \frac{4l}{3}$$

This wavelength is one-third of the wavelength of the first harmonic (fundamental mode), so the note produced has a frequency **three** times greater than the fundamental frequency, hence it is called the **third** harmonic. The second harmonic does not exist for a closed pipe.

Fifth harmonic (second overtone) Forcing the air in the tube to vibrate at even higher frequencies, the second resonance occurs at a frequency equal to **five** times that of the fundamental, so it is called the fifth harmonic (see Figure 2.73).

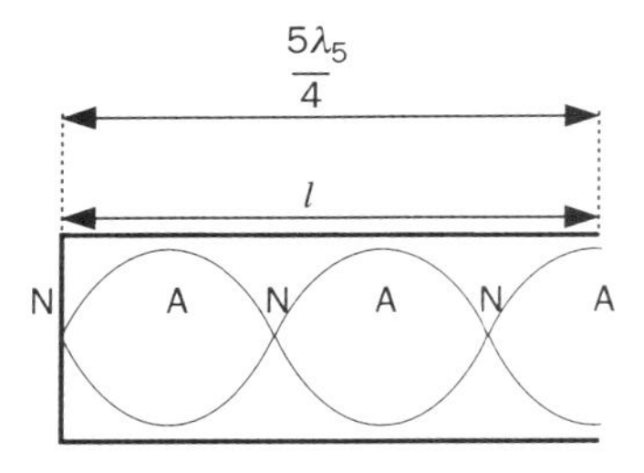

Figure 2.73 Fifth harmonic (second overtone) for a closed pipe

A similar analysis shows that the frequency of this mode of vibration is **five** times that of the fundamental mode.

The seventh, ninth etc. overtones are formed with increasing numbers of antinodes (and increasing frequency and decreasing wavelength).

Modes of vibration in open pipes

By an 'open' pipe, we mean one that is open to the atmosphere at both ends. Most wind musical instruments are of this pattern, one of the simplest being the recorder, which consists of a mouthpiece and a tube with a series of holes along its length.

The holes can be covered by the fingers. Air is excited at one end where air is blown over an edge. A progressive sound wave travels down the tube of the instrument, and is reflected, either at the open end or at the first open hole. As with the closed pipe, a stationary wave is produced, which creates a progressive sound wave in the air. The player uses her fingers on the holes to control the length of the vibrating air column, and therefore the frequency of the note produced.

The actual frequency produced depends on the **mode** of vibration, which in turn can be influenced by how hard the player blows, among other factors. We note that for each vibration mode:

- As in the closed pipe, the air molecules vibrate **along** the pipe axis (i.e. the stationary wave is **longitudinal**).
- The air column vibrates with maximum amplitude at **both** ends (i.e. there is a displacement **antinode** at each end).

First harmonic (fundamental frequency) Since each end must be an antinode, the simplest mode of vibration has a single **node** at the centre, as shown in Figure 2.74.

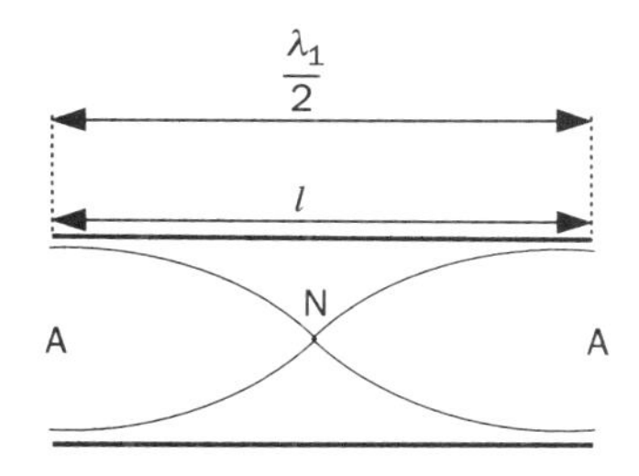

Figure 2.74 First harmonic (fundamental mode) – for an open pipe

Since adjacent antinodes are separated by a distance of half a wavelength ($\lambda_1/2$):

$$l = \frac{\lambda_1}{2} \qquad \therefore \quad \lambda_1 = 2l$$

Since this wavelength is **half** of that produced by a closed pipe sounding in its fundamental mode, we see that the fundamental frequency of an open pipe is **twice** that of a closed pipe of the same length.

Second harmonic (first overtone) The next higher frequency with which the air in the tube can vibrate has an antinode in the centre of the pipe, as shown in Figure 2.75.

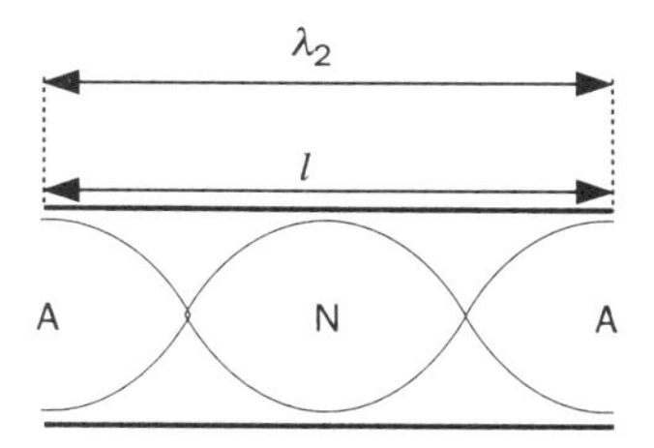

Figure 2.75 Second harmonic (first overtone) for an open pipe

Since in this case there is an antinode in the centre of the pipe, the length of the pipe is equal to the wavelength of this mode (λ_2):

$$l = \lambda_2$$

Since this is **half** the wavelength of the fundamental mode, the frequency of this note is **twice** the fundamental frequency. Note that, unlike the closed pipe, a frequency of two times the fundamental frequency is possible.

In an organ the notes are produced by forcing the air in pipes of different length to resonate

Third harmonic (second overtone)

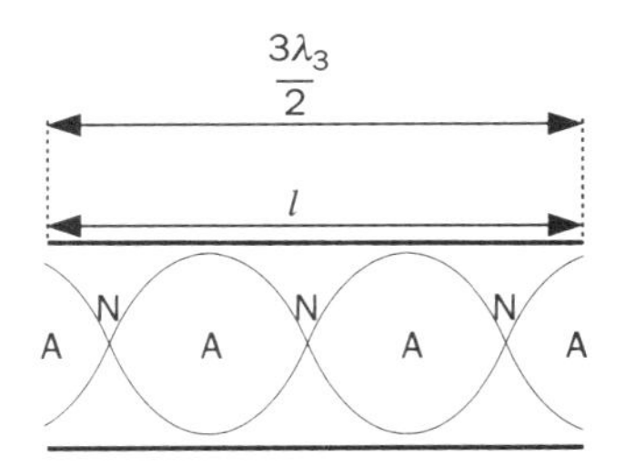

Figure 2.76 Third harmonic (second overtone) for an open pipe

In this case, the wavelength (λ_3) is given by:

$$l = \frac{3\lambda_3}{2} \qquad \therefore \quad \lambda_3 = \frac{2l}{3}$$

As this is equal to **one-third** of the fundamental wavelength, the frequency of this rate is **three times** the fundamental frequency. Thus an open pipe can produce all harmonics: f_1, $2f_1$, $3f_1$, $4f_1$ . . . , etc. (both **even** and **odd** harmonics are possible).

You should note that:

- The fundamental frequency for an open pipe is twice the fundamental frequency for a closed pipe. In musical terms, this means that the lowest note obtainable from an open pipe is one octave higher than the lowest note that can be produced from a closed pipe of the same length.
- Closed pipes can produce only odd harmonics, whereas open pipes can produce all possible harmonics. This means that open pipes are preferable for musical instruments due to the greater richness of the sound produced.
- Since the velocity of sound is dependent on factors such as the temperature of the air in the pipe, it follows that the frequencies of vibration of the air will also depend on the temperature. This is one reason why wind instruments need **tuning** to adapt to different environments.

Variation of pressure along the pipe

The vibration of the molecules gives rise to a fluctuating pressure which varies at the same frequency as the displacement. However, at a displacement **node**, the molecules on either side of the node are always in **antiphase**, which means that they are either moving towards or away from each other. When moving towards each other, the pressure rises above the average pressure in the tube, and when they separate, it decreases. This means that the greatest variations in pressure take place at the displacement nodes, while at the displacement antinodes, the variation in pressure is zero. Thus **displacement nodes** are **pressure antinodes**, and vice versa.

Phase change on reflection

As we saw on page 112, waves may suffer a **phase change** on reflection at a boundary. In the case of sound waves, a phase change of half a wavelength means that a compression is reflected as a rarefaction. Such a phase change only occurs with sound if the wave is reflected at a dense boundary, so at the **closed** end of a pipe, there is a phase change of half a wavelength (π radian). In the case of reflection at an **open** end, the boundary is not denser than the medium in which the wave travels, so there is no phase change.

Summary of wavelengths and positions of nodes (N) and antinodes (A) for strings and pipes

Harmonic	Overtone	String	Closed pipe	Open pipe
First Harmonic	Fundamental	$\lambda_1 = 2l$	$\lambda_1 = 4l$	$\lambda_1 = 2l$
Second Harmonic	1st Overtone	$\lambda_2 = l$	Impossible	$\lambda_2 = l$
Third Harmonic	2nd Overtone	$\lambda_3 = 2l/3$	$\lambda_3 = 4l/3$	$\lambda_3 = 2l/3$
Fourth Harmonic	3rd Overtone	$\lambda_4 = l/2$	Impossible	$\lambda_4 = l/2$

Guided example (1)

An organ pipe of length 1.4 m is open at one end and closed at the other. Given that the speed of sound is 340 m s^{-1}, calculate the wavelengths and frequencies of:

(a) the **fundamental** mode of vibration
(b) the **fifth** harmonic
(c) the **second** overtone.

Sketch the stationary wave pattern formed in each case.

Guidelines

(a) The wavelength (λ_1) of the fundamental note is given by $\lambda_1 = 4l$ where l = pipe length. Then use $v = f\lambda$ to calculate the frequency (f_1) of this mode (where v = wave speed).
(b) The fifth harmonic frequency is five times the fundamental frequency.
(c) The second overtone corresponds to the third harmonic.

Measuring the speed of sound

A loudspeaker is connected to a signal generator, and emits a tone of known frequency, as shown in Figure 2.77. It is also connected to channel Y_1 of a double-beam oscilloscope. This is the **trigger channel** of the instrument, which means that the screen trace always starts from the same point on the cycle of the wave from the signal generator. Channel Y_1 therefore shows the wave as it is emitted from the speaker.

A microphone placed some distance in front of the speaker is connected to channel Y_2. The Y-gains of the oscilloscope are adjusted so that the screen amplitudes are similar, and the traces are positioned one below the other on the screen. Sound travels through the air to the microphone, and channel Y_2 shows the waveform received at the microphone. As the sound takes a measurable time to travel the distance between the speaker and the microphone, the two signals will be **out of phase** (unless the distance between speaker and microphone happens to be equal to an exact number of wavelengths of the sound).

If the speaker is moved towards or away from the microphone, the phase difference between the traces on the screen will change, moving in and out of phase. The speaker is first positioned so that the two traces are exactly in phase. The speaker is then moved a distance along the bench so that the traces move in and out of phase an exact number of times (say 10 times). The distance moved is measured using a metre rule. The distance between one position where the traces are in phase, and the next position at which this occurs will be equal to one wavelength of the sound. If 10 complete phase shifts are observed, then the distance moved will be equal to 10 wavelengths. Thus the average wavelength (λ) can be calculated:

$$\lambda = \frac{\text{distance moved by speaker}}{\text{number of complete phase shifts}}$$

Since the frequency (f) of the sound is known (or measured using a digital frequency meter), the speed of sound (c) can be calculated using the standard equation:

$$c = f\lambda \qquad \textit{Equation 1}$$

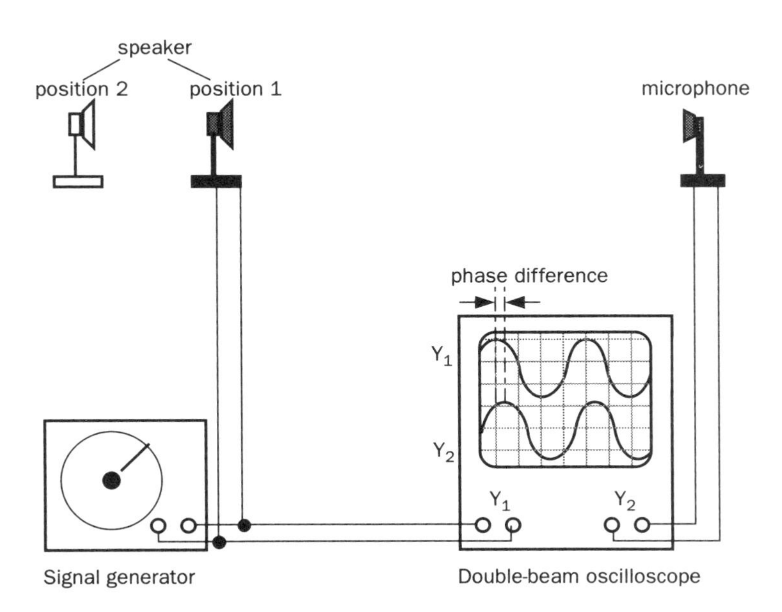

Figure 2.77 Measuring the speed of sound – progressive wave method

Guided example (2)

Using apparatus similar to that shown in Figure 2.77, with the loudspeaker giving a note of frequency 256 Hz, the microphone was positioned so that the traces were exactly in phase. When the microphone was moved a distance of 65 cm along the bench, the traces had become out of phase by half a cycle (antiphase). Calculate the velocity of sound.

Guidelines

If the traces have only moved from **in phase** to **antiphase**, then the microphone has only moved a distance equal to half a wavelength. Use this fact to calculate the wavelength, and hence the velocity of sound using the standard wave velocity formula.

Self-assessment

SECTION A
Qualitative assessment

1. How are **stationary** waves and **progressive** waves different? Refer to the terms energy, amplitude, phase and wave profile in your answer.

2. State the conditions necessary for the formation of a **stationary** wave. Explain how these conditions are satisfied in the case of
   (a) a stretched string
   (b) a pipe closed at one end
   (c) a pipe open at both ends.

3. Explain the meaning of the terms
   (a) displacement **node** and **antinode**
   (b) **fundamental**
   (c) **harmonic**
   (d) **overtone**
   as used when describing vibrating strings.

4. Sketch diagrams to show the positions of the **nodes** and **antinodes** for a stretched string vibrating in
   (a) **fundamental** mode
   (b) **third harmonic** mode.

5. Describe with the aid of clear diagrams how the **amplitude** and **phase** of vibration of air molecules vary with **distance** along the longitudinal axis of a pipe which is **closed at one end** vibrating with
   (a) its **fundamental** frequency
   (b) its **first overtone**.

6. Repeat Question 5 for a pipe **open at both ends**.

7. An **open** pipe produces 'richer' sounds than a **closed** pipe. Why?

8. Explain what is meant by a **pressure antinode** and a **pressure node**. Where would you expect to find a pressure node in a **closed** pipe vibrating in its **fundamental** mode?

9. Describe in detail an experiment to measure the velocity of sound in the laboratory. State clearly how the velocity of sound would be determined from your measurements.

SECTION B
Quantitative assessment

(***Answers:*** 3.0×10^{-2}; 0.11; 0.56; 0.64; 0.67; 1.1; 1.7; 3.2; 4.0; 110; 320; 430; 530; 1.0×10^{10}.)

1. Calculate the wavelength of the stationary wave in a guitar string of length 0.84 m when it is caused to vibrate:
   (a) in its fundamental mode
   (b) in its second overtone mode.
   In each case, draw a diagram to show the appearance of the string.
   (c) For the fundamental mode, calculate the velocity of the wave along the wire if the note emitted is middle C (256 Hz).

2. A circus high wire is fixed at both ends and vibrates transversely with a fundamental frequency of 2.0 Hz. What would be the new fundamental frequency if the length of the same wire under the same tension were:
   (a) halved
   (b) trebled.

3. A closed organ pipe has a length of 0.80 m. Draw diagrams to show the variation of displacement of the air molecules with distance along the pipe for the three lowest frequency modes of vibration. Calculate the wavelength and frequency of the notes emitted in each case. On each diagram, indicate the positions of any **pressure** nodes and antinodes. (Speed of sound in air = 340 m s^{-1})

Self-assessment *continued*

4. A microwave transmitter is positioned facing a plane metal reflector set at right angles to the direction of travel of the waves. A microwave detector connected to a microammeter is positioned at a point between the transmitter and the reflector and moving the detector along the line joining the transmitter to the receiver causes the reading on the microammeter to vary continuously between zero and a maximum. Explain why this happens. If the distance between five successive maxima is 60 mm, calculate the wavelength and frequency of the microwaves, given that the speed of electromagnetic radiation in air is 3.0×10^8 m s^{-1}.

5. A small loudspeaker which is continuously emitting a note of frequency 1600 Hz is placed 4.0 m in front of a flat metal sheet. When a microphone connected to a cathode ray oscilloscope is moved from the loudspeaker directly towards the sheet, several points of minimum amplitude are noted. If the speed of sound in air is 336 m s^{-1}, calculate the distance between two adjacent minima.

2.5 The Doppler effect

Whenever there is relative motion between a source of waves and an observer, the **frequency** (and **wavelength**) of the waves measured by the observer is found to differ from the actual **frequency** (and **wavelength**) of the emitted waves. The frequency is said to be **shifted** by so many hertz, and this phenomenon is called the **Doppler effect**. It is most readily noticeable in the case of sound waves, and accounts for the sudden change of pitch (i.e. frequency) observed by a stationary person whenever a sound source such as an ambulance sounding its siren speeds past. The same effect is experienced if a moving observer speeds past a stationary sound source, but this is a less common experience. We will examine these two situations separately, starting with a source moving towards a stationary observer.

Moving source – stationary observer

Figure 2.78 shows a wave source (sound, light, radar, etc.) moving from A to E with velocity (v) as it emits waves of frequency (f) which travel through the medium with speed (c). O_x is the position of a stationary observer ahead of the source, and O_y is the position of a stationary observer behind the source.

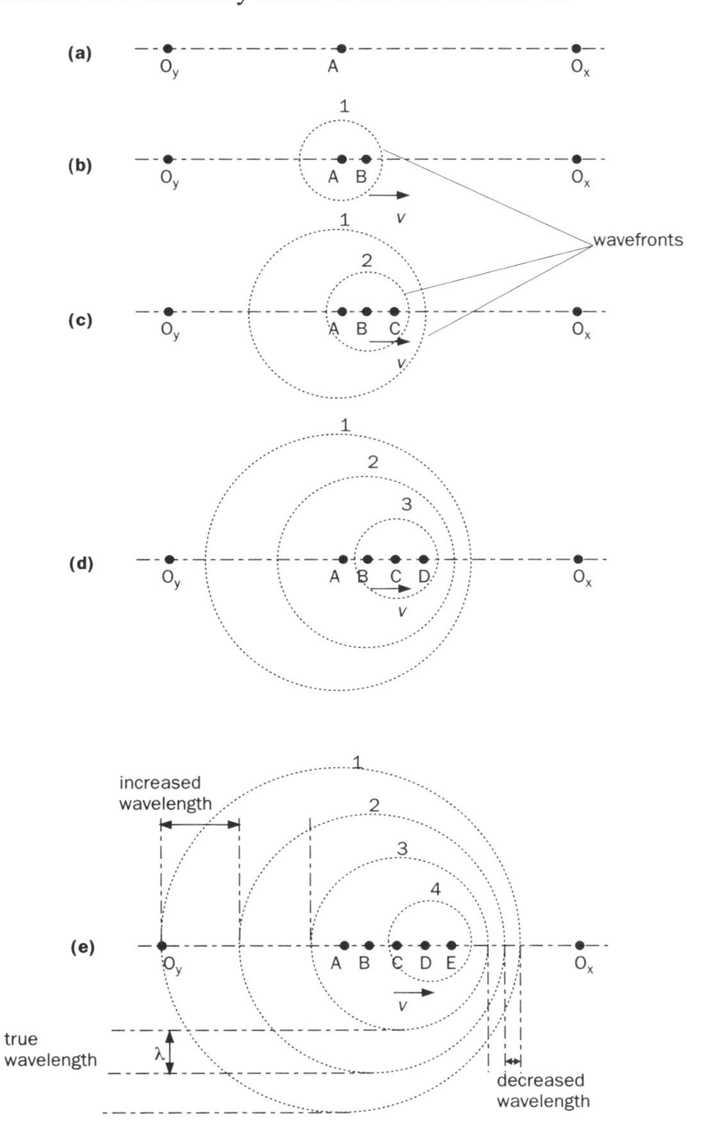

Figure 2.78 Effect of motion of source on observed wavelengths ahead of and behind the source

In (a), the source is at position A, where it is just about to emit a new wavefront.

In (b), the source has moved to position B, where it is just about to emit another new wavefront. The dotted line 1 shows the position of the first wavefront emitted when the source was at position A. The wavefront is spherical (shown in Figure 2.78 as circular), since the waves travel with the same speed in all directions. Note that this wavefront is centred on point A, but the source has moved on to point B. In (c), the source has moved on to point C, and the dotted line 2 shows the position of the wavefront just emitted when the source was at B. Wavefront 1 has expanded to the new size shown.

In (d) and (e), the source has moved on to positions D and E respectively, emitting new waves at each point. Diagram (e) clearly shows that, ahead of the source, the wavefronts become crowded together, while behind it they are spread out over a greater distance.

Because of this effect of wavefronts crowding together, the observer at O_x ahead of the source perceives a **shorter wavelength** than the true value. In one second, the source emits f waves, and these will occupy a distance $(c - v)$, (where (c) is the velocity of the waves through the medium). Thus the observed wavelength (λ_o) ahead of the source is given by:

$$\lambda_o = \frac{(c - v)}{f}$$

Equation 1

The **observed frequency** (f_o) is therefore given by:

$$f_o = \frac{\text{speed of waves relative to observer}}{\text{observed wavelength}}$$

$$\therefore \quad f_o = \frac{c}{(c - v)/f}$$

$$\therefore \quad f_o = \frac{c}{(c - v)} f$$

where f is the **true frequency** of the source.

Equation 2

Equation 2 shows that for a source moving **towards** an observer, the observed frequency (f_o) is **higher** than the true value (f).

To an observer O_y **behind** the source, the wavefronts appear **further apart** than they should, and in this case, the observed wavelength (λ_o) is given by:

$$\lambda_o = \frac{(c + v)}{f}$$

Equation 3

The **observed frequency** (f_o) is therefore given by:

$$f_o = \frac{\text{speed of waves relative to observer}}{\text{observed wavelength}}$$

$$\therefore \quad f_o = \frac{c}{(c + v)/f}$$

$$\therefore \quad f_o = \frac{c}{(c + v)} f$$

Equation 4

Equation 4 shows that for a source moving **away from** an observer, there is a **decrease** in frequency. This phenomenon accounts for the observation that the pitch of a train's horn appears to decrease as it rushes past an observer standing at the side of the track. As it approaches the observer, the pitch is higher than the true value, and as it departs, the pitch is lower.

Moving observer

If the source is stationary, then its wavefronts will be circular, and centred on the stationary source. The Doppler effect is caused by the fact that an observer approaching the source will encounter more wavefronts per second than if he were stationary. The result is a **change of frequency** as experienced by the observer.

If the observer is moving **towards** the stationary source (of frequency f) with speed (v), the relative speed of the waves to the observer is $(c + v)$ and the observed frequency (f_o) is given by:

$$f_o = \frac{(c + v)}{\lambda}$$

where λ = the true wavelength of the waves = $\frac{c}{f}$

$$\therefore \quad f_o = \frac{(c + v)}{c} f$$

Equation 5

Equation 5 shows that for an observer **approaching** a source, the frequency is **increased**.

For an observer moving **away from** the source at velocity (v), the observed frequency is given by:

$$f_o = \frac{(c - v)}{c} f$$

Equation 6

Equation 6 shows that for an observer moving **away from** a source, the observed frequency is **less than** the true value.

Approximate relationship

In most cases, particularly with electromagnetic waves, the source or observer velocity (v) will be very much less than the wave velocity (c) in the medium. If the source velocity is low, then the change of frequency (Δf) will be small. It can be shown that provided Δf is small:

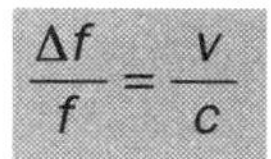

$$\frac{\Delta f}{f} = \frac{v}{c}$$

Equation 7

Equation 7 can be applied whether the source is moving towards or away from the observer. Remember that relative **approach** produces an **increase in frequency**, while **separation** causes a **decrease.**

A similar relationship can be shown for the change of wavelength ($\Delta\lambda$):

$$\frac{\Delta\lambda}{\lambda} = \frac{v}{c}$$

Equation 8

Equation 8 also applies to both relative approach and relative separation, but remember that **approach** produces a **decrease** in wavelength, while separation causes an **increase**.

The **change** of frequency or wavelength is called the **Doppler shift**.

Applications of Doppler shift

Astronomy

When hot elements emit light, the emitted frequencies are characteristic of that element, and can be used to identify it. Spectroscopic examination involves splitting the light into a **line spectrum** using a diffraction grating. If this is done with the light from a star, the spectral lines are found to be Doppler-shifted, occurring at slightly different wavelengths to those produced from the same element in the laboratory. If the wavelength shift ($\Delta\lambda$) is measured, the velocity of the star relative to the Earth can be calculated.

The expanding universe

Stars moving away from the Earth produce spectral lines at a lower frequency (longer wavelength). When compared to a reference spectrum, the lines are closer to the **red** end of the spectrum, and the light is said to be **red-shifted**. Most of the light from galaxies in all parts of the universe shows a red shift, indicating that the universe is expanding in all directions. Careful measurement of the red shift, and calculation of the velocities, shows that the recession velocity of a galaxy is directly proportional to its distance from the Earth. This is **Hubble's law**, and is part of the evidence for the Big Bang Theory of the creation of the universe.

Rotational speed of the Sun

The Doppler effect can be used to determine the rotational speed of the Sun, as shown in Figure 2.79. At one side of the Sun's disc, it is effectively moving towards the Earth, causing the light to be **blue-shifted,** while at the other side, it is moving away, causing a **red shift**. Thus the linear velocity (v) of the edge of the Sun can be calculated from measurements of the Doppler shift, and the rotational velocity (ω) can be calculated using the relationship $\omega = v/r$, where r is the radius of the Sun. This method has also been used to determine the rotational velocity of Saturn's rings.

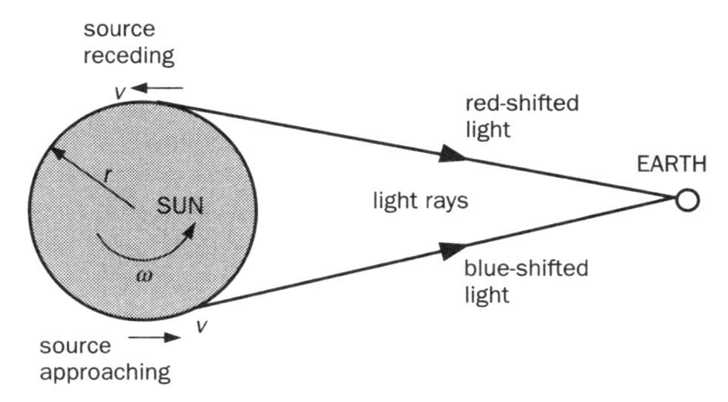

Figure 2.79 Rotational velocity of the sun using the Doppler effect

If the linear equatorial velocity (v) is calculated, and the Sun's radius is r, the the rotational angular velocity (ω) is given by:

$$\omega = \frac{v}{r}$$

Radar

RADAR is an acronym made up from the words '**RA**dio **D**etection **A**nd **R**anging', and employs the principle that electromagnetic radiation can be reflected from objects. If the time elapsed between transmission of a pulse of radio waves and reception of the reflected beam is known, the distance to the reflecting body can be calculated. If the body is moving there will be a **Doppler shift,** which means that the reflected waves will have a different frequency and wavelength to the incident beam. An example of this is the radar 'Speed Trap' used by the police to measure the speed of road vehicles, as shown in Figure 2.80. The device can be held in the hand, and consists of a microwave transmitter and receiver mounted close together. The transmitter emits brief pulses of 3 cm microwaves at the target vehicle.

Some of the transmitted radiation is reflected from the moving car, and is detected by the receiver, where it is compared with the transmitted radiation. (A small reflector reflects some of the transmitted energy directly into the receiver.)

If the emitted microwaves have a frequency (f – typically 10 GHz) and the vehicle is moving towards the observer with speed (v), the Doppler shift (Δf) of the reflected waves will be given by:

$$\Delta f = \frac{2v}{c} f$$

The factor 2 in this equation is due to the fact that the waves appear to originate from the 'image' of the transmitter, which is twice as far from the receiver as the vehicle. The image appears to be moving towards the receiver at twice the vehicle speed.

In practice, the reflected signal from the vehicle is compared with that reflected from the small fixed reflector on the radar gun. The two signals are superposed, producing **beats**, whose frequency is equal to the frequency shift (Δf).

Guided examples (1)

1. The wavelength of a particular line in the emission spectrum of an element in a distant star is 524 nm when it is observed from Earth. When the emission spectrum is produced in the laboratory, the corresponding line is found to have a wavelength of 522.8 nm. Calculate the velocity of the star, and state whether it is approaching the Earth or receding from it.

Guidelines

Use Equation 8 to find the star's velocity. Consider whether the spectral line is 'red-shifted' or 'blue-shifted' to decide the direction of motion.

Guided examples (1) *continued*

2. A terrestrial observer measures a total Doppler shift of 8.24×10^{-3} nm in light of true wavelength 600 nm when emitted from opposite sides of the Sun at its equator. If the Sun has a radius of 6.96×10^{5} km, calculate:

(a) the linear velocity of a point on the equator
(b) the Sun's angular velocity
(c) its period of rotation.

Guidelines

One side of the solar disc is moving **towards** the observer with velocity (v), the other side is moving **away** with this velocity, so the **difference** in linear velocities from one side to the other is equal to $2v$. Thus the total Doppler shift is given by:

$$\Delta\lambda = 2\left(\frac{v}{c}\lambda\right)$$

(don't forget to convert to SI units!)

Use the standard angular velocity formula: $\omega = \frac{v}{r}$

The period is given by: $T = \frac{2\pi}{\omega}$

3. The speed limit on a particular stretch of road is 40 mph (17.7 m s^{-1}). The police are using radar speed detection equipment which emits microwaves of wavelength 2.8 cm. Calculate the maximum Doppler frequency shift for these microwaves reflected from a moving car approaching the detector if it is to stay within the speed limit. (Velocity of electromagnetic radiation = 3.0×10^{8} m s^{-1}.)

Guidelines

Calculate the emitted frequency of the microwaves and then use $\Delta f = \frac{2v}{c} f$ to calculate the Doppler shift.

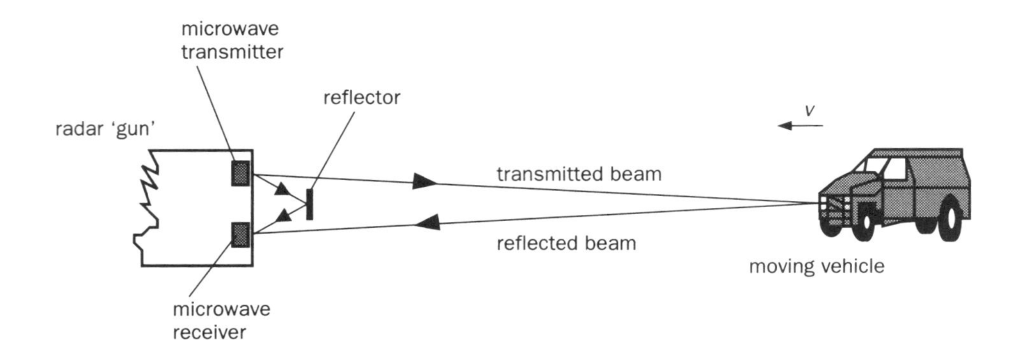

Figure 2.80 Radar speed detection

Self-assessment

SECTION A
Qualitative assessment

1. Explain what is meant by the **Doppler effect**, and describe a situation in which the effect is observed with sound waves.
2. You are a **stationary** observer in line with a **moving** source of waves. Explain the effect on the **observed** wavelength if you are
   (a) **ahead of** the source
   (b) **behind** the source.
3. What is meant by the term **Doppler shift**?
4. Explain the terms **red shift** and **blue shift** as applied to the light from distant stars.
5. Explain how the Doppler effect can be used to determine the rotational speed of one of Saturn's rings.
6. Explain the operation of a radar speed detector.
7. Examination of the spectra of light emitted from stars and galaxies shows spectral lines characteristic of elements on Earth, but with wavelengths which are either slightly longer or slightly shorter than those observed from the same elements on Earth. How does the Doppler effect explain these observations?

SECTION B
Quantitative assessment

(***Answers:*** 5.01×10^{-7}; 5.39×10^{-7}; 5.41×10^{-7}; 15 (6.6); 5.45×10^{7}.)

(Velocity of electromagnetic waves in a vacuum = 3.0×10^{8} m s^{-1})

1. A star is moving away from Earth with a relative speed of 4.58×10^{5} m s^{-1}. Calculate the observed wavelength of light of frequency 6.00×10^{14} Hz emitted by the star, as measured by an astronomer on earth.
2. Your defence against a charge of driving through traffic lights on red is that your speed of approach caused a Doppler shift, making the light appear green. Calculate the speed of approach necessary for this to occur. (Assume the wavelengths to be: red light = 650 nm, green light = 550 nm.)
3. A planetary system in another galaxy has a sun with a rotational period of 9.5 earth days and an equatorial radius of 8.4×10^{10} m. Calculate the observed wavelengths of light of true wavelength 540 nm emitted from either side of the sun's equator, as viewed from one of the planets in the system.
4. Radar speed detection equipment is used to enforce a speed limit in the grounds of a hospital. The sets emit microwaves of frequency 10.7 GHz, and will trigger a camera to photograph the offending vehicle if the reflected waves from it are Doppler shifted by more than 474 Hz. What is the speed limit being enforced? (Assume 1 m s^{-1} = 2.25 mph)

2.6 The electromagnetic spectrum

In 1666, Isaac Newton succeeded in identifying that white light was not a pure colour when he investigated the *visible spectrum* formed when a beam of sunlight was refracted by a triangular prism. When Thomas Young first demonstrated interference of light waves in 1801, Herschel had just identified a region, beyond the red end of the Sun's visible spectrum, that caused heating effects. At first this was explained by assuming that *heat rays* were emitted by the Sun, and were different to light rays. This new region of the Sun's spectrum was called *infra-red*. Around the same time, a German physicist, Ritter, noticed that the darkening of silver nitrate was faster with violet light than red light, and was even faster in the invisible region of the Sun's spectrum beyond the violet end. This was the first observation of *ultra-violet* radiation.

In 1865, the mathematician Maxwell, in his work on unifying the theories of electricity and magnetism, devised equations for the speed of radiation emitted by an oscillating charge. This turned out to be the same as the measured speed of light, leading Maxwell to claim that light was a form of electromagnetic radiation. He also predicted that the range of frequencies and wavelengths of such radiations would be enormous. Towards the end of the 19th century, Heinrich Hertz succeeded in showing that oscillating electric charges emitted a radiation that could be detected some distance away, was reflected by flat metal plates, could be focused by curved plates, and passed through other materials. At first, these waves were called *Hertzian* waves, but the term *radio* waves came into general use. Listening to the *Hertz* doesn't sound quite the same as listening to the *radio!*

Table 2.2 The electromagnetic spectrum – wavelengths, sources and detectors

Wave type	Wavelength range (m)	Sources	Detectors
Radio	> 0.001	Oscillating charges	Antennae and tuned circuits
Microwave	$0.1 - 10^{-5}$	Magnetrons and klystrons	Antennae and tuned circuits
Infra-red	$10^{-4} - 7 \times 10^{-7}$	Thermal vibrations of atoms in hot bodies	Blackened thermometer Thermopile, bolometer Photographic film Semiconductors
Visible	$7 \times 10^{-7} - 4 \times 10^{-7}$	Energy level changes of electrons in atoms	Eye Photographic film Semiconductors
Ultra-violet	$4 \times 10^{-7} - 10^{-9}$	Energy level changes of electrons in atoms	Fluorescent chemicals Photographic film Photoelectric effect
X-rays	$< 5 \times 10^{-8}$	X-ray tubes (Energy level changes in innermost shells)	Photographic film Ionisation detectors
Gamma	$< 10^{-10}$	Radioactive nuclei	Photographic film Ionisation detectors

Today, we recognise that all of these radiations, together with more recently discovered microwaves, X-rays, gamma rays and cosmic rays, form the *electromagnetic spectrum* spanning wavelengths from as long as many kilometres to as short as 10^{-15} metre and beyond. In this section we examine some of the main broad regions of this important group of waves, looking in particular at how these waves are produced, detected and used. Table 2.2 lists these regions, together with an indication of their range of wavelengths, origin and means of detection.

The wavelength ranges in Figure 2.81 and Table 2.2 are only approximate, and show considerable overlap as there are no hard and fast boundaries between the regions. They are included to give a feeling for the range of frequencies and wavelengths covered by each description. Visible light is an exception to this, as the visible wavelengths are precisely defined as those which can be seen by the human eye, but even this is subject to some small variation in individuals.

The **photon energy** is calculated by multiplying the *frequency* (in Hz) by *Planck's constant, h* (see page 402). Planck's constant, $h = 6.63 \times 10^{-34}$ J s. Thus high frequency waves have high energy photons. The higher the photon energy, the greater the penetrating power, ionising ability, etc.

Properties of electromagnetic waves

You should note that:

- All electromagnetic radiation travels at the same **speed** (c) in a vacuum. This speed is one of the universal constants in science, and for our purposes can be assumed to have a value of 3.0×10^8 m s^{-1}.
- The electromagnetic spectrum is **continuous**, i.e. there are no gaps in it. The different kinds of radiation merge into one another, with no abrupt change of properties.
- Radiations with the same wavelength and frequency have the same properties, but they may be named differently according to their origins.
- All electromagnetic waves:
  - transfer **energy** from one place to another
  - can be **emitted**, **absorbed** or **transmitted** by matter
  - can travel through a **vacuum**
  - can be **polarised** and are therefore assumed to be transverse, and
  - can be superposed to produce **interference** and **diffraction** effects
  - obey the laws of **reflection** and **refraction**
  - carry no electric charge
  - are unaffected by electric, magnetic or gravitational fields. (In fact, intense gravitational fields can cause deflection of electromagnetic radiation, and this phenomenon has led to the discovery of many of the mysterious, massive objects in the universe such as black holes, which are themselves invisible but can be detected by their effect on the light coming from more distant stars.)

Radio waves

These are produced by oscillating charges. The electrons in a cable through which an alternating current is flowing are oscillating, and are thus giving out radio waves. Most of the bodies in the universe emit radiation in the radio frequency range. The study of these emissions is the basis of radio astronomy, and enables a great deal of information about the nature of these bodies to be obtained. Radio waves are of obvious use in communications. Radio, television, satellite communications, cellular telephones, CB radio equipment, etc. all use their own particular

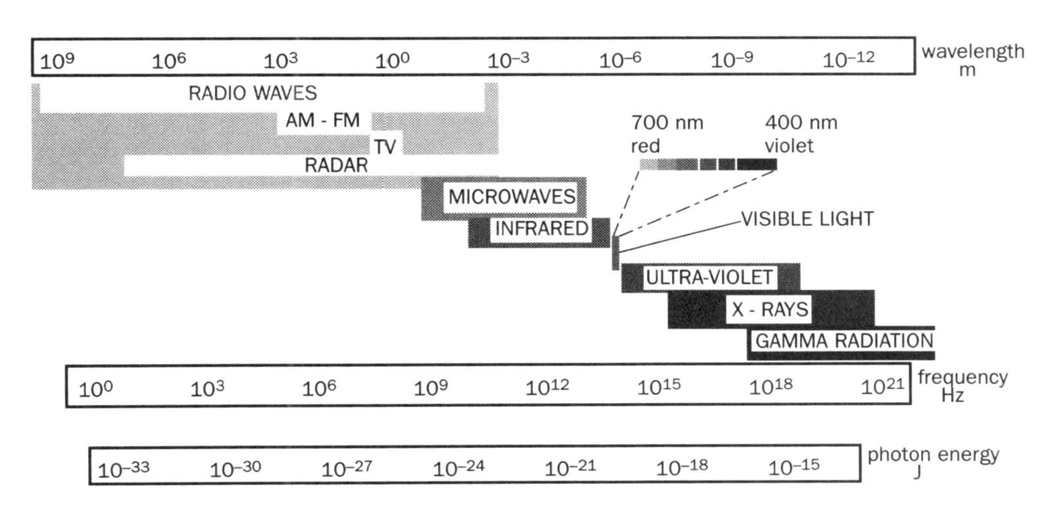

Figure 2.81 The electromagnetic spectrum

regions (or wavebands) of the spectrum, as do radar and navigation systems.

When a radio wave interacts with a conductor, the alternating electric and magnetic fields in the radio wave exert forces on the electrons in the conductor, causing them to oscillate. This oscillation of the charges constitutes an alternating electric current, of the same frequency as the electromagnetic wave. Using tuned circuits, particular oscillating frequencies may be selectively amplified.

Microwaves

There is considerable overlap between the wavelengths of radio waves, microwaves and infra-red radiation, and this causes considerable confusion. Generally speaking, radio waves are produced by tuned circuits, microwaves by magnetrons and infra-red by hot bodies. Microwave radiation finds its greatest application in communications and radar. Lying at the highest frequency end of the radio wave region, microwave transmissions can carry significantly more information than lower frequencies. Radar equipment emits high-energy beams of microwaves. The key to producing compact radar systems is to design a high-power magnetron operating at a very short wavelength. This enables small antennae to produce a highly directional beam of radar waves. Information about objects in the path of the beam is obtained by analysing the reflected radiation – position and range information can be readily deduced.

Radar screens in the control room of the coastguard and marine rescue centre at Treasure Island, California

In the home, we have become familiar with microwave cooking. A magnetron, typically operating at 2.45 GHz, delivers microwave energy to a sealed cavity containing the food to be cooked. Molecules in the food, particularly liquid molecules, absorb this energy, and become heated due to the *dielectric heating effect* whereby electrical insulators absorb energy from rapidly alternating electric fields, causing increased vibrational energy and a consequent rise in temperature. The manufacturers always issue dire warnings about the dangers of placing metals in the cooker. High voltages can be induced in metals by the microwave energy, resulting in sparking and damage.

Infra-red radiation

All objects at temperatures greater than absolute zero emit radiation in this part of the electromagnetic spectrum. The greater the temperature, the higher the intensity and maximum frequency emitted. Hot bodies can also emit in the visible spectrum (e.g. lamp filaments) and the emissions from hot bodies are often lumped together as thermal radiation. Since black surfaces are the best absorbers of infra-red radiation (and the best emitters), it can be easily detected using thermometers with blackened bulbs, thermocouples mounted behind blackened surfaces (thermopiles) or blackened thermistors (bolometers). The infra-red radiation is absorbed, raising the temperature of the material. This rise in temperature is detected by the thermometer, produces a p.d. in the thermocouple or changes the resistance of the thermistor. Photographic film can have a dye inserted which is sensitive to infra-red wavelengths, and semiconducting materials can release electrons when they absorb electromagnetic radiation of these frequencies.

The infra-red radiation emitted from warm bodies (especially human bodies) is distinguishable from the background infra-red emissions from cooler bodies, due to its higher wavelength. Burglar alarms, military night-vision equipment, and thermal imaging cameras used to detect people buried in collapsed buildings all rely on the detection of this radiation, which is focused on a CCD (charge-coupled device) detector. In the burglar alarm, the elements of the CCD are scanned, and the alarm sounds if a change is detected in the infra-red radiation received. Thermal imaging cameras operate rather like TV cameras, the only difference being that infra-red radiation is focused, not visible light.

Ultra-violet radiation

Ultra-violet (UV) radiation is a major component of the radiation received from the Sun. Most of it is filtered out before reaching ground level, particularly by ozone high in the atmosphere, but enough gets through to be responsible for chemical changes in the skin, leading not only to suntans but also to skin cancers, including the dangerous *melanoma* so much in the news. Glass absorbs a great deal of ultra-violet radiation, so it is very difficult to get a suntan behind a glass window, even on the sunniest of days. Ultra-violet radiation is also produced by electrical discharges: fluorescent lamps produce ultra-violet radiation by passing an electric current through a gas at low pressure containing mercury vapour. The mercury atoms become excited and absorb energy, re-emitting it as photons of ultra-violet radiation. This is then absorbed by chemicals on the inside of the tube, which re-emit the radiation as photons of lower frequency, in the visible part of the spectrum. Sun-beds use tubes designed to emit a great deal of their energy as ultra-violet radiation. Arc welders must protect all of their skin when working, not just against sparks and hot metal, but against the intense ultra-violet radiation produced by the electric arc used in the process. Serious damage can be done to skin and eyes if correct precautions are not taken.

Fluorescent chemicals absorb ultra-violet radiation and re-emit it as visible light. This effect is used in dyes to enhance the brightness of clothing in sunlight (which contains ultra-violet radiation). Photographic film can also be enhanced with dyes to make it sensitive to UV. The photoelectric effect is the name given to the phenomenon in which electrons are ejected from metal surfaces exposed to ultra-violet radiation. These electrons can be detected, leading to instruments for detecting and measuring ultra-violet radiation (see Photoelectric effect on pages 456–8).

X-rays

X-rays are highly penetrating electromagnetic waves produced by the interaction of high-speed electrons with matter. Photographic film can be designed to be sensitive to X-rays, leading to uses in medicine and industry. At the high frequency end of the range, they can be intensely ionising, and can be used to destroy cancer cells in the body.

Gamma rays

Of similar wavelengths to X-rays, gamma rays originate in the nuclei of radioactive atoms. Highly penetrating, they can be detected by photographic film, and by their ionising effect in Geiger Muller tubes and other ionisation detectors (see Section 5.2 on Radioactivity).

Self-assessment

SECTION A
Qualitative assessment

1. Name, in order of **increasing frequency**, the main groups of radiations which form the electromagnetic spectrum. For each group, state
   (a) a source
   (b) a detector
   (c) a **typical** wavelength.
2. Explain why radiations having the **same** wave-lengths and properties are given **different** names (e.g. UV and X-rays of wavelength 10^{17} Hz).
3. Give two similarities and two differences between visible light waves and radio waves.

Self-assessment *continued*

4. State the evidence for the assumption that all electromagnetic waves are **transverse**.
5. Briefly describe the physical processes occurring when a radio broadcast is received by an antenna connected to a tuned circuit.
6. Identify the type of radiation from each of the following descriptions:
   (a) produces fluorescence in some chemical dyes used in washing powders
   (b) produced by the interaction of high-speed electrons with matter
   (c) emitted by most astronomical bodies, used in cellular telephones
   (d) have high penetrating power and originate in the nuclei of atoms
   (e) detectable by the human skin and used in night-time surveillance
   (f) can produce suntan and skin cancer.

2.7 Reflection and refraction

In this section, we consider light to be a form of energy whose direction of travel is represented by straight lines called **light rays**. A beam of light is a collection of rays. Figure 2.82 shows the propagation of light energy from a point source of light. The dotted lines represent the position of light wavefronts at uniform time intervals. Since the light travels at uniform velocity in all directions, the wavefronts are concentric circles. The direction of travel of the light at any point is perpendicular to the wavefront, as shown by the light rays.

When considering light in this form, the straight-line travel assumes that shadows formed will be sharp. In Section 2.4 we saw how the wave theory of light predicts less sharp shadows, and that bending of rays takes place. For the time being, it is sufficiently accurate to assume straight-line travel as long as the objects causing the shadows are not too small. This is just one example of the use of different models in physics, which get more complex as you explore the topic to a deeper level. It is not that the simple models are wrong, just that for simple events there is no point in using an over-complex theory to explain it. You don't need to be a master of Keynesian economics to work out if you've been short-changed at the bar!

When light strikes a surface, it can be absorbed, transmitted or reflected, or some combination of these. Light can only be transmitted by a transparent medium. If the surface is opaque, the light can only be reflected or absorbed. When light is absorbed by a medium, the energy is dissipated as increased internal energy of the molecules of the medium. Their increased vibrational kinetic energy is detectable as a rise in temperature of the material.

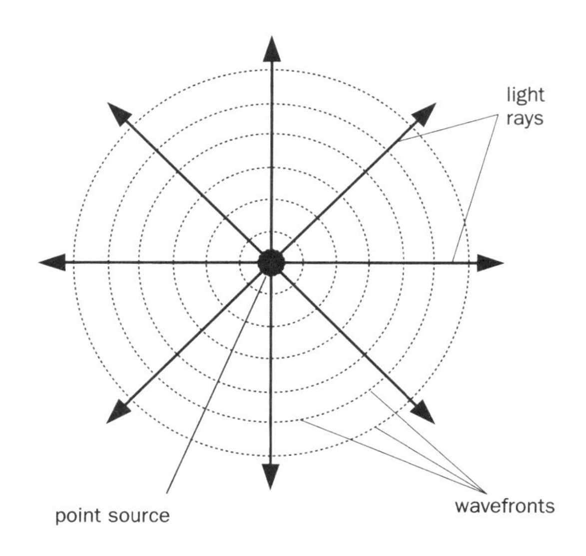

Figure 2.82 Wavefronts and rays

Regular (or specular) reflection

The nature of reflection at a surface depends on the smoothness of the surface. All surfaces have imperfections when examined at the molecular level, but a surface is considered smooth if its surface irregularities are small compared to the wavelength of light. A smooth surface produces regular (or specular) reflection. This kind of reflection is summarised by the following, experimentally observed, laws:

- **The angle of incidence equals the angle of reflection (i.e. $\theta_1 = \theta_2$).**
- **The reflected ray is in the same plane as both the incident ray and the normal to the mirror at the point of incidence.**

Figure 2.83 shows the regular reflection of a single light ray at a smooth reflecting surface.

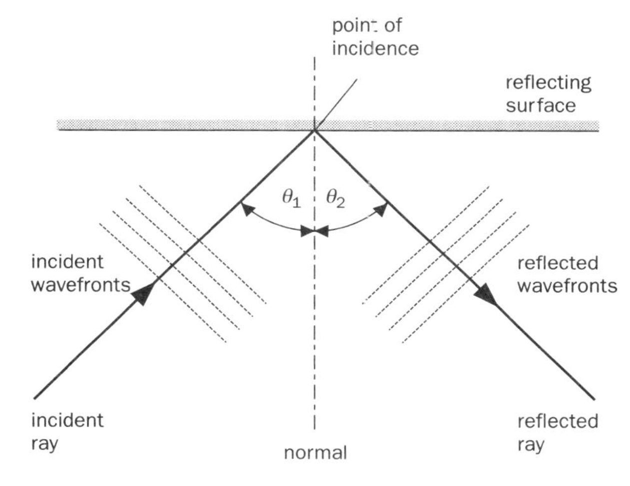

Figure 2.83 Regular reflection

You should note that:

- The normal is a line drawn perpendicular to the reflecting surface.
- The angles of incidence are measured from the normal, not the reflecting surface. This is a common convention in optical work.
- A snooker ball rebounding from the cushion is a good model for regular reflection. Both laws are illustrated.
- Mirrors are frequently made by depositing a layer of metal onto a glass surface – 'silvering'. Glass is chosen for its smoothness, rigidity and the ease with which an acceptably flat surface can be obtained. High-quality mirrors are front-silvered to eliminate the problem of multiple image formation.
- The laws of reflection apply whenever reflection takes place, regardless of the shape of the reflecting surface or the nature of the wave being reflected.

Distinction between regular and diffuse reflection

Figure 2.84 shows a parallel beam of light incident on a smooth, plane reflecting surface (a plane mirror). In this case, a parallel beam will be reflected, as all rays will have the same angle of reflection. This is regular (or specular) reflection.

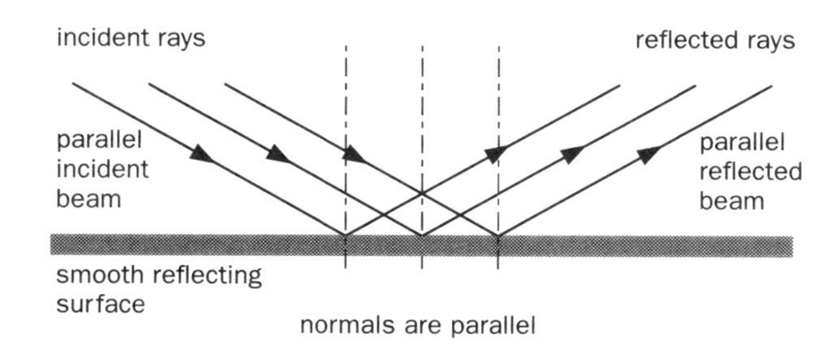

Figure 2.84 Regular (specular) reflection

Most surfaces, however, are not smooth, and the surface irregularities create different angles of incidence for the rays in the incident beam. Thus each ray may have a different angle of reflection from the others in the reflected beam. This results in the reflected beam being diffused. For each ray, the laws of reflection are observed, but the angle of incidence will vary from point to point on the surface because of its roughness. This is **diffuse reflection**, as shown in Figure 2.85.

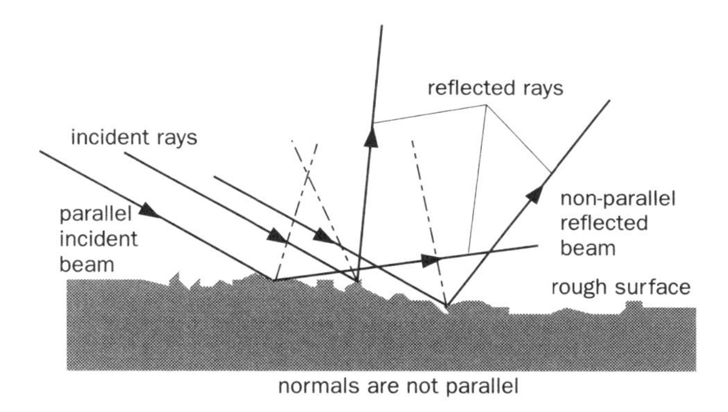

Figure 2.85 Diffuse reflection

Image formation by a plane mirror

Figure 2.86 shows how an image of a point is formed in a plane mirror. Light rays from the point O (a spot on the man's chin) diverge in all possible directions. Two rays are shown striking the mirror surface, where they undergo specular reflection. Both reflected rays continue to diverge, with the same angle between them as before reflection, and enter the observer's eye. To him, these rays appear to come from the point I, behind the mirror. The point I is said to be the image of point O. This process can be applied to every point on the man's face, so a complete image is produced.

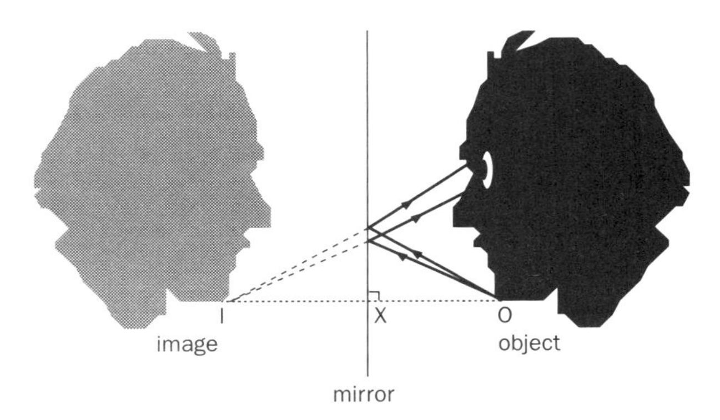

Figure 2.86 Image formation in a plane mirror

Position and properties of the image

Any image is described by considering its position, size, orientation and nature.

In the case of the plane mirror, the image is:

1. **Behind the mirror** (same distance behind as object is in front, see Figure 2.87).

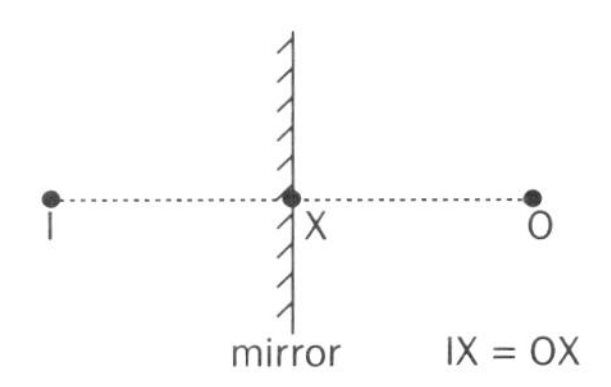

Figure 2.87 Relation between object and image distance for a plane mirror

2. The **same size** as the object, and the same way up (erect). Everyday observation and experiment bears this out.
3. **Laterally inverted**, as illustrated in Figure 2.88 which shows the appearance of print as seen in a plane mirror.

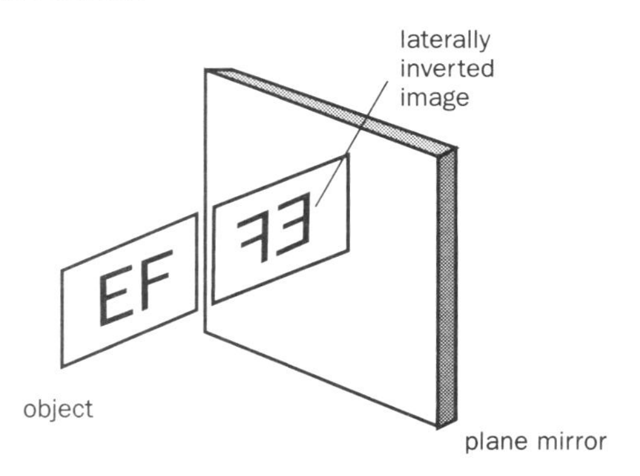

Figure 2.88 Lateral inversion in a plane mirror

4. **Virtual**. Referring to Figure 2.86, there are no actual light rays passing through the image. A screen placed behind the mirror would show nothing. The light beam from the point O is divergent when it enters the eye. A virtual image is formed at the point where these rays appear to have come from. It is the function of the mirror, and any optical instrument, to make the rays appear to originate from some point other than the true object.

Refraction of water waves

Simple experiments using plane waves in a ripple tank (see Figure 2.89) show that:

- When the depth of the water changes, the speed and wavelength of the waves is observed to change. A decrease in depth causes a decrease in both speed and wavelength. This effect is called **refraction**.
- The frequency of the waves remains unaltered. Frequency is a property of the source.

It is important to note that if the waves were to pass from a shallow region to a deeper one, the velocity would increase.

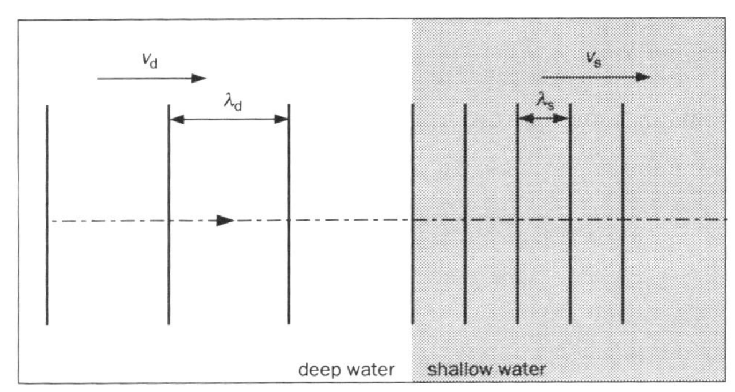

Figure 2.89 Change of speed and wavelength of water waves at a deep/shallow water boundary

When the waves are incident on the deep/shallow boundary at an angle, the wave direction changes, as shown in Figure 2.90.

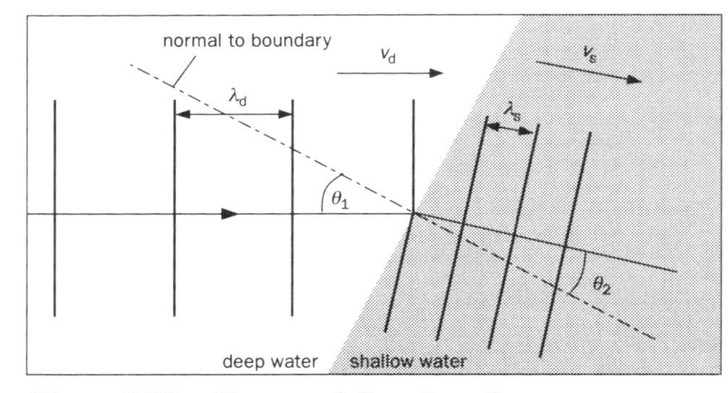

Figure 2.90 Change of direction of water waves incident at an angle to a deep/shallow boundary

If $v_d > v_s$ and f is constant

$\lambda_d > \lambda_s$

$\therefore \quad \theta_1 > \theta_2$

The change of direction is a direct result of the change of speed which occurs as the wave crosses the boundary. Looking in the direction of travel of the incident waves, the right-hand end of a wave enters the shallow region before the rest of the wavefront, and is slowed down to the new speed. Since the portion of the wave in the deeper water continues at the original speed, there must be a change of direction as shown.

Refraction of light

Refraction is a property of all waves, and the refraction of light is of particular interest as it enables the design of optical instruments like telescopes, microscopes, etc. to be studied. The speed of electromagnetic radiation depends on the medium through which it is passing. In the vacuum of free space, electromagnetic radiation travels at about 3.00×10^8 m s^{-1} (this is one of the fundamental constants of the universe), but this speed is reduced whenever the radiation travels through a material medium. For most practical purposes, the speed in air can also be taken to be 3.00×10^8 m s^{-1}.

This change of velocity causes a change in direction whenever light (or any other wave) is incident at an angle at a boundary between two media in which the velocity is different. If the boundary is distinct (e.g. between air and a glass block as shown in Figure 2.91) then the light will suffer an abrupt change of direction at the boundary.

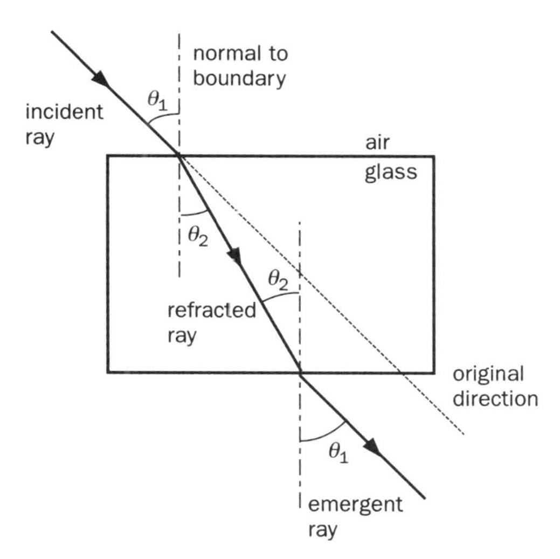

Figure 2.91 Refraction of light at an air/glass boundary

The light is refracted towards the normal as it passes from the less optically dense medium (air) into the optically denser medium (glass). On leaving the parallel-sided block, the ray is refracted away from the normal as the speed of the light returns to its higher value in the air.

If the boundary is indistinct, for example between layers of air at different densities in the atmosphere, then the refraction is gradual and the change of direction of the ray of light will occur over a considerable distance.

You should also note that the different wavelengths (and hence colours of light) are refracted by different amounts. The further towards the violet end of the visible spectrum you go, the greater the refraction (**violet** is more **violently** refracted).

Laws of refraction

The ancient Greeks observed the connection between the angle of incidence and the angle of refraction, but thought there was a fixed connection between the angles. Snell in 1621 showed that the relationship concerned the **sines** of the angles.

The refraction of light is governed by the following laws:

- **At the boundary between any two optical media, the ratio of the sine of the angle of incidence to the sine of the angle of refraction is constant for light of a specific wavelength**. This is sometimes referred to as Snell's law, and may be mathematically expressed as:

$$\frac{\sin \theta_1}{\sin \theta_2} = \text{a constant}$$ See Figure 2.92

Equation 1

- **The refracted ray, the incident ray and the normal to the boundary between the media at the point of incidence are all in the same plane.**

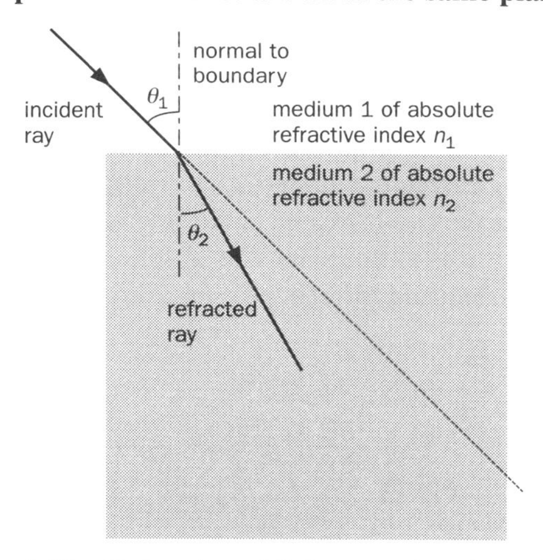

Figure 2.92 Refraction of light at a plane boundary

Refractive index

Snell's law of refraction for light passing from medium 1 to medium 2 states that:

$$\frac{\sin \theta_1}{\sin \theta_2} = \text{a constant}$$

The constant in this case is known as the **refractive index** ($_1n_2$) for light passing from medium 1 to medium 2.

$$\therefore \quad {}_1n_2 = \frac{\sin \theta_1}{\sin \theta_2}$$

Equation 2

If the speeds of light in medium 1 and medium 2 are c_1 and c_2 respectively:

$${}_1n_2 = \frac{c_1}{c_2}$$

Equation 3

Dispersion Since the speed of light in a medium is slightly different for different *frequencies* (colours) of the light, it follows that the refractive index, and hence the angle of refraction, will be slightly different for different colours. A refractive index is usually quoted for a specific colour for accurate work, otherwise the figure for yellow light is stated. Violet light travels more slowly in matter than red light, so its refractive index is slightly greater. This is the reason why white light can be split into its constituent colours when it is refracted – each colour follows a slightly different path. The effect is particularly strong when the angle of deviation is large, as in a triangular prism, shown in Figure 2.93.

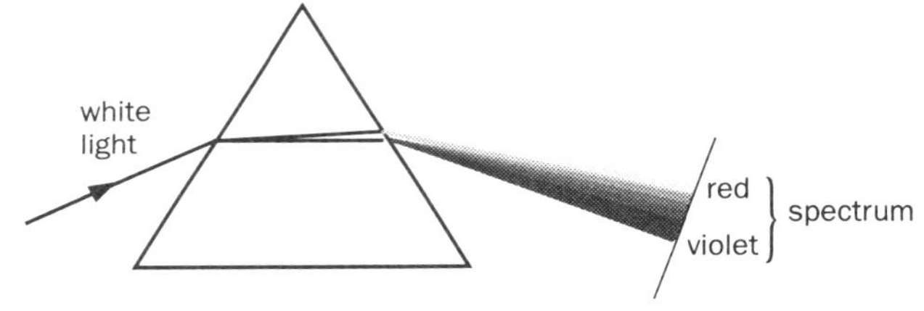

Figure 2.93 Dispersion by a triangular prism

Absolute refractive index As the refractive index relates to the two media on either side of the boundary, it is useful to have an absolute standard. This is calculated for light passing from a vacuum into the medium, and is then quoted as the *absolute refractive index* for that medium. Some typical values are shown in Table 2.3.

Table 2.3 Absolute refractive index values

Medium	Symbol	Absolute refractive index
Air	n_a	1.003
Water	n_w	1.33
Ice	n_i	1.30
Glass	n_g	1.50
Diamond	n_d	2.42

Note the very high value of refractive index for diamond, which produces a very high degree of dispersion of white light. It is this property that gives diamond its unique appearance.

Refractive index relationships

Consider a ray of light AO passing from medium 1, of absolute refractive index n_1, to medium 2, of absolute refractive index n_2, and being refracted along OB (Figure 2.94(a)).

Since light rays are reversible, a light ray passing along BO in medium 2 would be refracted along OA in medium 1 (Figure 2.94(b)).

For the light ray travelling from medium 1 to medium 2: $\quad {}_1n_2 = \dfrac{\sin \theta_1}{\sin \theta_2} \quad$ (i)

For the light ray travelling from medium 2 to medium 1: $\quad {}_2n_1 = \dfrac{\sin \theta_2}{\sin \theta_1} \quad$ (ii)

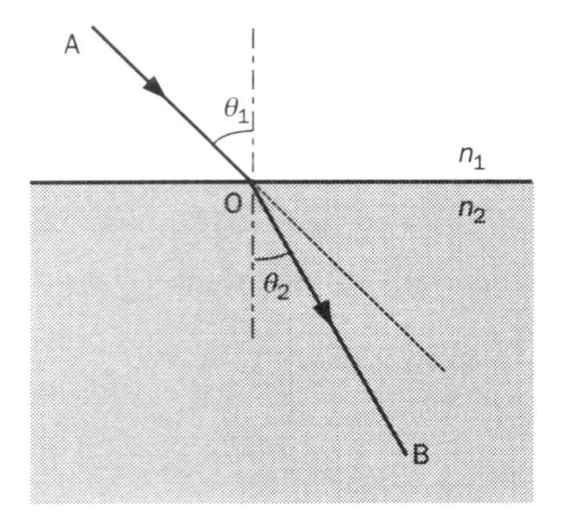

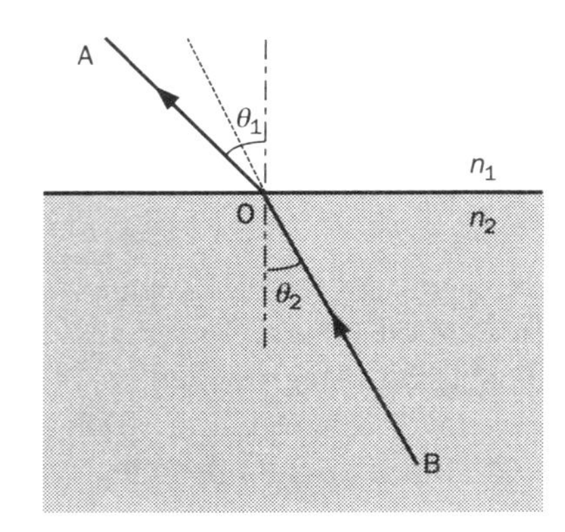

Figure 2.94 Relationships involving refractive index

From (i) and (ii)

$$_1n_2 = \frac{1}{_2n_1}$$

Equation 4

For example, if the refractive index from air to water ($_an_w$) is 1.33 then the refractive index from water to air ($_wn_a$)is given by:

$$_wn_a = \frac{1}{_an_w} = \frac{1}{1.33} = 0.75$$

Let c, c_1 and c_2 be the velocities of light in a vacuum, medium 1 and medium 2 respectively.

Then $\quad n_1 = \frac{c}{c_1}$

and $\quad n_2 = \frac{c}{c_2}$

$$\therefore \quad \frac{n_2}{n_1} = \frac{c/c_2}{c/c_1} = \frac{c_1}{c_2}$$

but from Equation 2 and Equation 3:

$$\frac{c_1}{c_2} = \frac{\sin\theta_1}{\sin\theta_2}$$

$$\therefore \quad \frac{n_2}{n_1} = \frac{\sin\theta_1}{\sin\theta_2}$$

$$\therefore \quad n_1 \sin\theta_1 = n_2 \sin\theta_2$$

Equation 5

This useful result tells us that:

- For a ray passing from one medium to another of different optical density, the product of the absolute refractive index and the sine of the angle between the ray and the normal is constant either side of the boundary.
- If $n_2 > n_1$, then $\theta_2 < \theta_1$, which agrees with the observation that light bends towards the normal when entering an optically denser medium.

Measurements of refractive index can be made to a high degree of accuracy, and are used in analytical chemistry both to identify substances and to ascertain their purity.

Total internal reflection

When light travels from one medium to another of lower refractive index (for example, from glass or water into air, or from glass to water) it is possible that no light is transmitted across the boundary. Instead, all of the light energy is reflected back into the first medium. This is known as **total internal**

Guided examples (1)

1. Calculate the angle of refraction and the angle through which the ray is deviated when a light ray travels

(a) from air to water with an angle of incidence of 48°

(b) from water to glass at an angle of incidence of 24°.

(The absolute refractive indices of water and glass are 1.33 and 1.60 respectively.)

Guidelines

(a) Use $_an_w = \frac{\sin\theta_1}{\sin\theta_2}$ and assume that the refractive index from air to water is equal to the absolute refractive index of water.

(b) Use the relation $n_1 \sin\theta_1 = n_2 \sin\theta_2$ where medium 1 is water, and medium 2 is glass.

2.

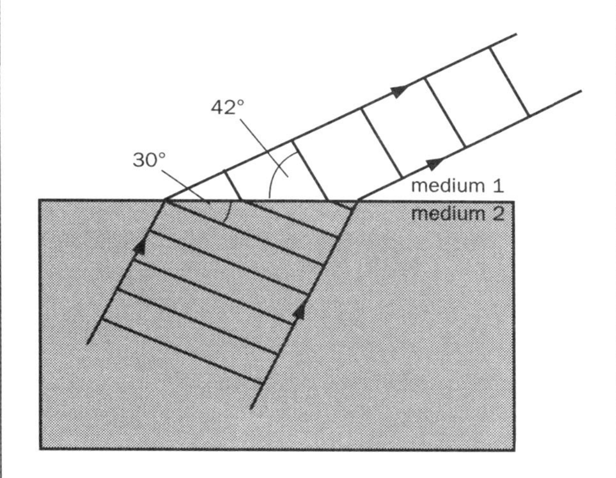

The diagram shows a beam of waves in medium 1 suffering refraction at the boundary with medium 2. If the speed of the waves in medium 1 is 2.25×10^8 m s^{-1}, calculate their speed in medium 2.

Guidelines

Use $_1n_2 = \frac{\sin\theta_1}{\sin\theta_2} = \frac{c_1}{c_2}$

(take care to identify θ_1 and θ_2 correctly)

reflection (TIR). Total internal reflection occurs whenever the angle of incidence in the medium of higher refractive index exceeds a certain value, known as the **critical angle**. The effect of varying the angle of incidence is shown in Figure 2.95.

(a) For small angles of incidence, a refracted ray emerges, bent away from the normal, and a weak reflected ray is also seen.

(b) If the angle of incidence is increased, a position will be reached where the refracted ray emerges along the plane boundary, so that the angle of refraction (θ_2) is equal to 90°. The angle of incidence is now equal to the critical angle (θ_c).

(c) If the angle of incidence in the denser medium exceeds the critical angle, there is a strong, internally reflected ray, but no refracted ray. This is total internal reflection.

Total internal reflection occurs when light passing from one medium to another of lower refractive index is incident on the boundary at an angle greater than the critical angle for the two media. All of the light is reflected from the boundary, and there is no transmission.

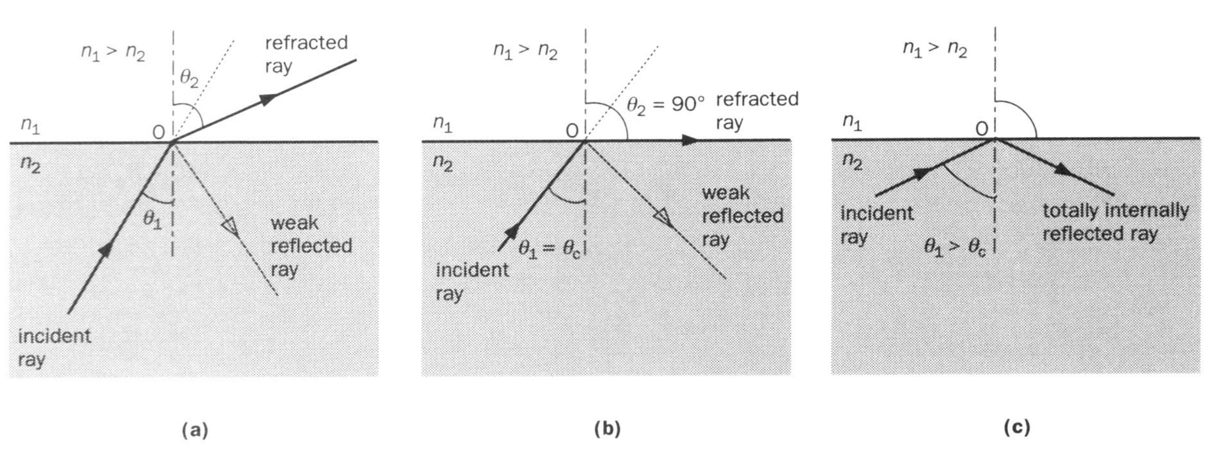

Figure 2.95 Total internal reflection

Derivation window

Figure 2.96 shows a ray of light in medium 1 striking the boundary with medium 2 of lower refractive index. The angle of refraction is equal to 90° so the angle of refraction must be the critical angle (θ_c) for the combination of the two media.

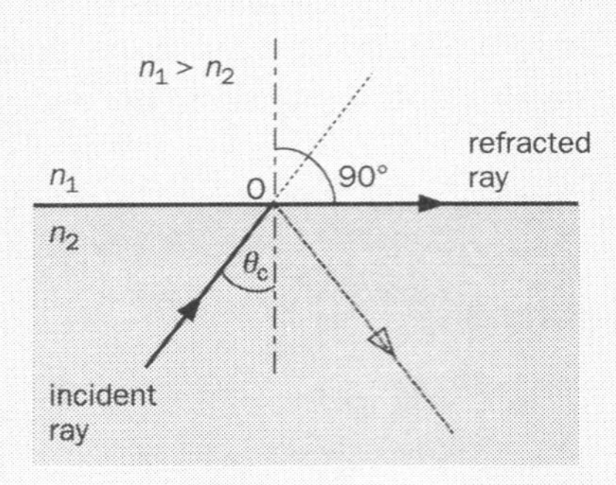

Figure 2.96 Light ray incident at the boundary between a denser and a less dense optical medium at the critical angle

$$n_1 \sin \theta_1 = n_2 \sin \theta_2$$

$$\therefore \quad n_1 \sin \theta_c = n_2 \sin 90°$$

Since sin 90° = 1

$$n_1 \sin \theta_c = n_2$$

$$\therefore \quad \sin \theta_c = \frac{n_2}{n_1}$$

$$\therefore \quad \sin \theta_c = \frac{1}{\frac{n_1}{n_2}}$$

$$\therefore \quad \sin \theta_c = \frac{1}{{}_1n_2}$$

Equation 6

Guided examples (2)

1. Given that the absolute refractive indices for glass and water are 1.5 and 1.33 respectively, calculate the critical angle for light travelling from
   (a) glass to air
   (b) water to air
   (c) glass to water.

Guidelines

(a) and (b) Use $\sin \theta_c = \frac{1}{n}$

(c) Calculate the refractive index for light passing from water to glass by using

$_1n_2 = \frac{n_1}{n_2}$ then use $\sin \theta_c = \frac{1}{n}$ as before

2. Rays from a point source of light at the bottom of a 1.8 m deep swimming pool strike the water surface, and only emerge through a circle of radius r. If the refractive index for water and air is 1.33, calculate the value of r.

Guidelines

This is a relatively simple problem, the difficulty lies in visualising what is happening in order to solve it.

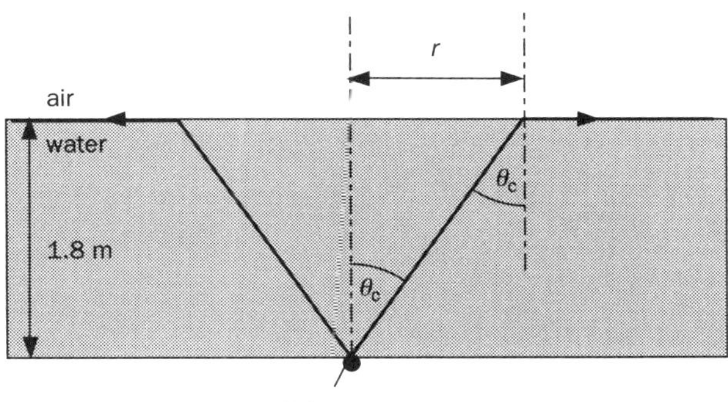

The diagram shows light rays striking the pool surface at the critical angle. Rays striking at smaller angles will escape, those with greater angles of incidence will suffer total internal reflection. Calculate the critical angle for water, then use basic trigonometry to calculate r.

3.

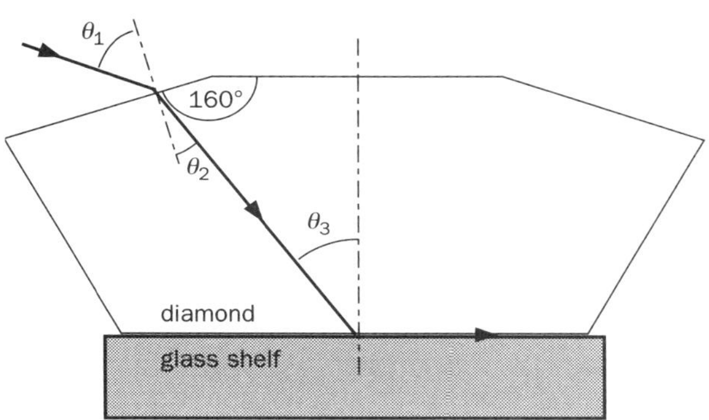

The diagram shows a light ray incident in air on the angled face of a diamond resting on a glass shelf. The top and bottom faces of the diamond are parallel. If the speeds of light in air, glass and diamond are 3.00×10^8, 2.00×10^8 and 1.24×10^8 m s^{-1} respectively, calculate:

(a) the refractive indices for light travelling
   (i) from air into diamond
   (ii) from diamond into glass

(b) the values of θ_1 and θ_3 when total internal reflection just occurs at the shelf as shown.

Guidelines

(a) Use Equation 3.

(b) When total internal reflection occurs, θ_3 is equal to the critical angle (θ_c) for light passing from diamond to glass.

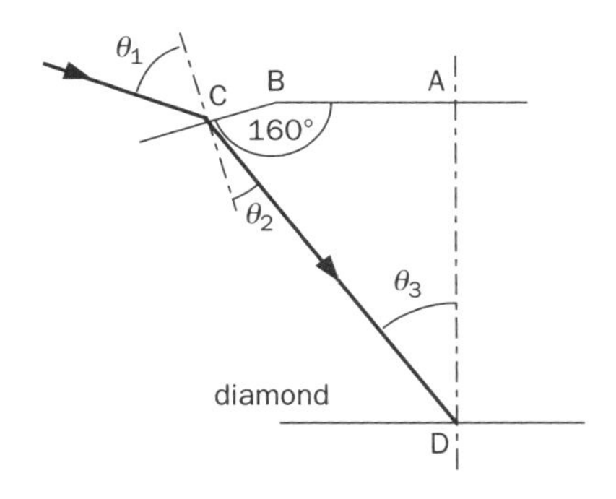

Angle BCD = 360° – (160° + 90° + θ_3)

Hence find θ_2, then use the standard refractive index formula to calculate θ_1.

Other examples of wave refraction

Sound waves Sound waves can be refracted as they travel at different speeds through different media. The range of speeds of sound through materials is very wide, so large changes of direction can occur. Submarines use sonar – pulses of sound energy are emitted, and the reflected waves give information about the presence of underwater obstacles and other submarines. Changes in density of the water, due to temperature gradients or differences in salt content, create boundaries which can refract the sound waves, or even produce total internal reflection. Consequently, sonar echoes can be misleading, and submarines in layers of water of different density may be 'invisible' to each other.

Seismic waves Shock waves from earthquakes or explosions travel through the Earth at speeds which depend on the density of the material. As a result of discontinuities, fractures and general density variations, these waves suffer refraction. World-wide study of seismic waves reveals vital information on the structure of the Earth, and observations on total internal reflection at the Earth's core revealed its high density.

Fibre optics

Optical fibres are extremely fine strands of highly transparent glass along which light can be transmitted even when the fibre is bent or twisted. Due to the construction of the fibre, light striking the walls of the fibre cannot escape, but suffers total internal reflection. One type of fibre, known as a **step-index** fibre, consists of a very small diameter (around 50 μm) glass core surrounded by a cladding made from glass of lower refractive index. The boundary between the inner core and cladding provides a sudden change or step in the refractive index, to ensure total internal reflection. The overall fibre diameter, including a protective plastic coating, is around 120 μm. Figure 2.97 shows the passage of a single ray through such a fibre.

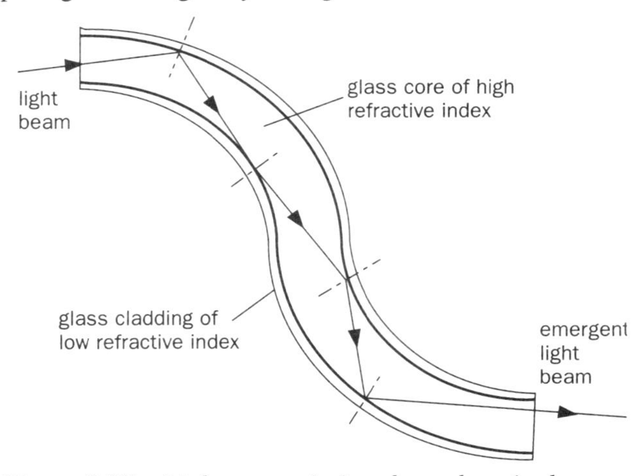

Figure 2.97 Light transmission through a single optical fibre

Most of the light incident on one end of the fibre strikes the core/cladding interface at an angle considerably greater than the critical angle, and so undergoes repeated internal reflections until it emerges at the other end. The loss of energy is very low, achieved by the use of high quality glass of great transparency.

Light pipes

Optical light pipes are made from thousands of fibres bundled together. They are used to illuminate inaccessible places from a single remote source of light. Such light pipes are referred to as **incoherent**. The spatial relationship between the optical fibres bundled in the cable is not maintained, so although they can carry light energy, they cannot be used for carrying images. Other significant applications occur in telecommunications.

The endoscope

This is a device (variously called a gastroscope, bronchoscope, cystoscope, etc.), becoming very popular, used for viewing internal organs of the body. Due to its small size it can be inserted through the mouth and other orifices, so eliminating the need for invasive surgery. The endoscope consists of two bundles of fibres mounted coaxially. One bundle is used as a light pipe to illuminate the area under investigation. The reflected light passes back up the other bundle of fibres, which must be a **coherent** bundle in which the individual fibres maintain their spatial relationship to each other through the length of the instrument. In this way, an image can be transmitted along the bundle of fibres. The image can be viewed directly by the surgeon, or a video camera mounted on the end of the bundle can be used.

Telecommunications

Optical fibres are finding increasing use in telecommunications systems. The signals are transmitted digitally by converting them to light pulses in an analogue to digital converter. Speech, data, video signals, etc. can all be digitised in this way. The digitised information is then converted into a binary code. This binary code is used to pulse a laser rapidly on and off, in a sequence corresponding to the digitised signal. These light pulses are then transmitted through the optical fibre. At the receiver, a digital to analogue converter reconstructs the original signal. Figure 2.98 shows a schematic diagram of a system for transmitting both sound and computer data via an optical fibre link.

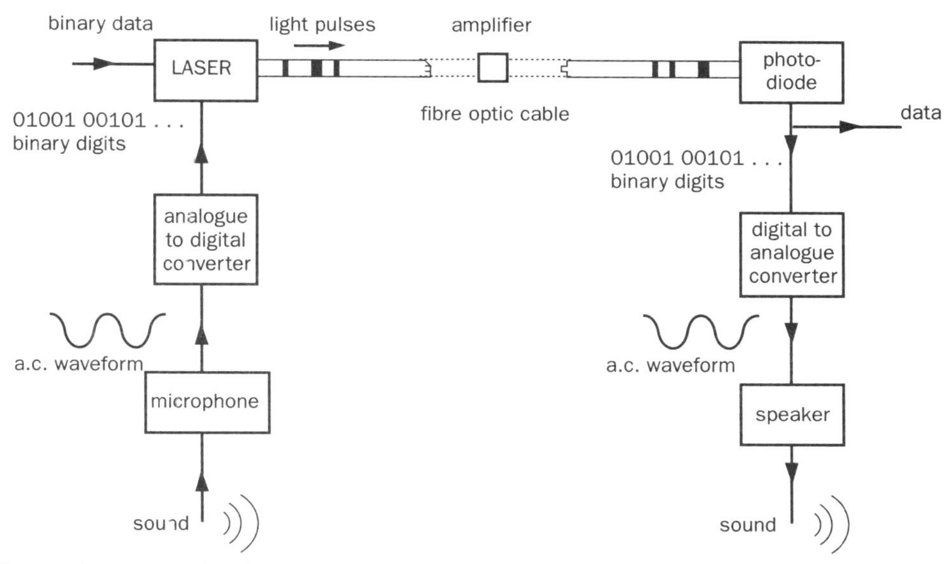

Figure 2.98 Fibre optic communication

The microphone converts the sound energy into an electrical signal of varying voltage – the analogue signal. The analogue to digital converter converts this analogue signal into a stream of digital information. This digital stream is in binary form, as a stream of 1s and 0s. This is used to turn on and off the laser beam, so that pulses of light are transmitted down the fibre optic cable. The semiconductor lasers used can switch at frequencies as high as 140 MHz. This means that the digital pulses from up to 1000 conversations can be combined and sent simultaneously down the same fibre.

The light suffers some loss of energy as it travels down the fibre, so every 50 km or so the signal is amplified. As the signal is digital, there is no loss of data quality at each stage. The choice of light wavelength can minimise the loss of energy and increase the distance between amplifiers. The wavelength chosen is usually in the infra-red range.

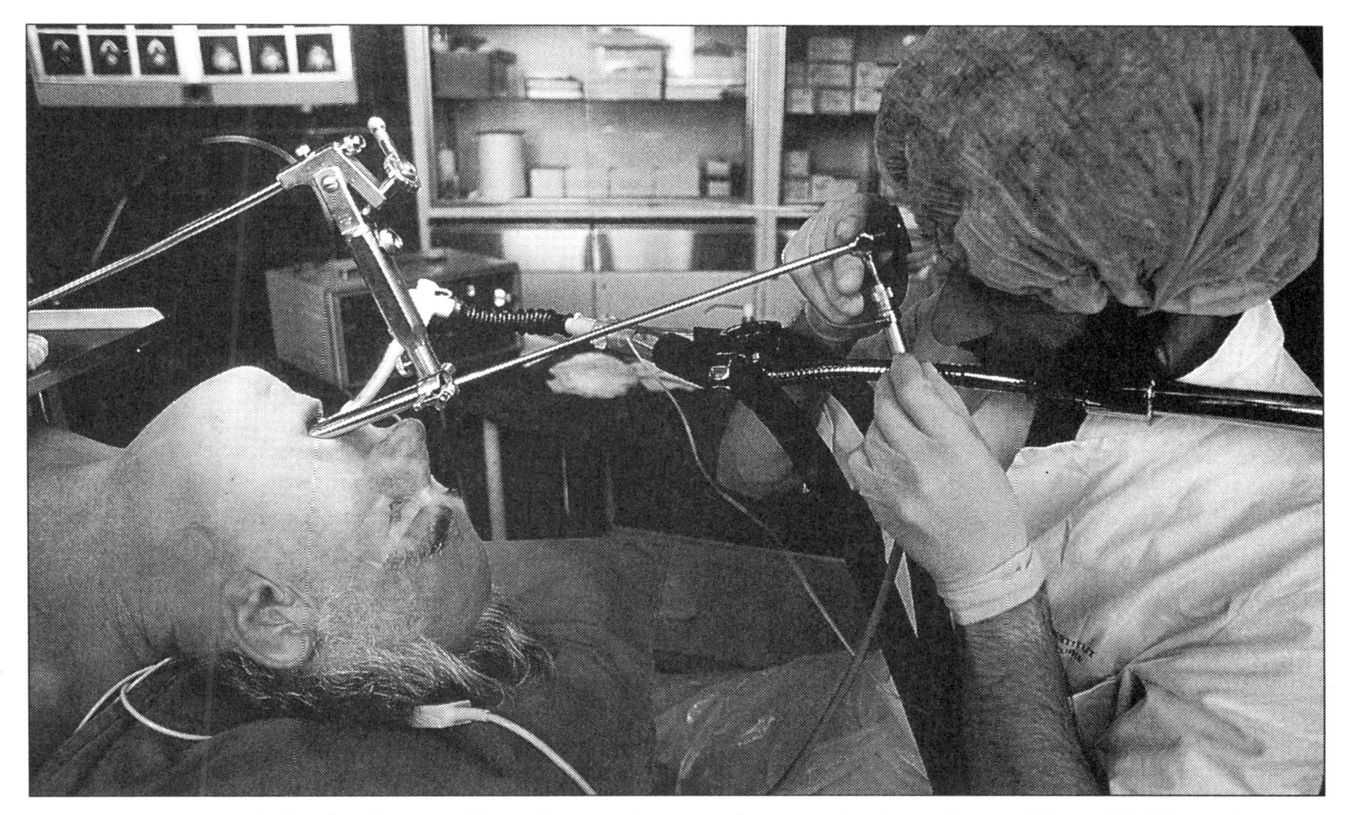

Doctor using a specialised endoscope (bronchoscope) to examine a patient's trachea and bronchi. The endoscope provides a very clear view down these airways and tissue samples etc. can be removed using long forceps which are inserted down the metal tube

At the receiving end, a photodiode converts the light pulses into electrical pulses. A digital to analogue converter recreates the original analogue signal from the digital data. When fed to a loudspeaker, the original sound is reproduced. Computer data in binary form can be fed directly to the laser to be transmitted and received in the same way.

Types of optical fibre

We are concerned here with **step-index** fibres, where there is an abrupt change of refractive index from the core to the cladding material.

Single mode fibres

Single mode (or monomode) fibres have an extremely narrow core, typically only a few μm (say 4×10^{-6} m), with a thick cladding layer of around ten times this diameter. Such a narrow core minimises the problem of **pulse broadening**, which sets a limit on the frequency of transmission of digital pulses. Pulse broadening can be explained by considering Figure 2.99 which shows two rays of light, A and B, travelling a short distance through an optical fibre. Clearly ray B has had to travel much further in reaching point C than ray A, which is parallel to the axis. A pulse of light therefore will contain rays which travel different distances and so take different times for the journey.

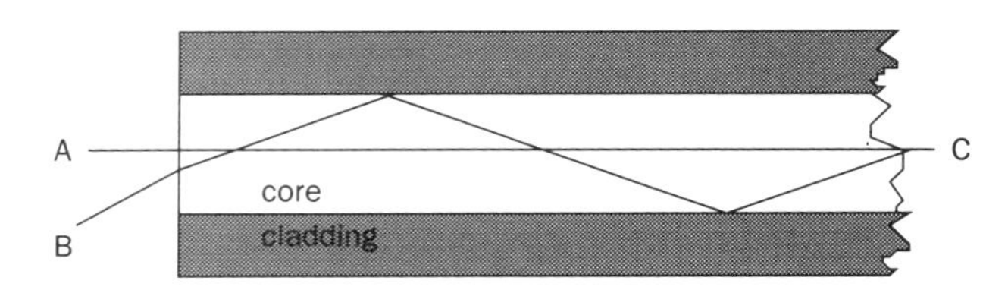

Figure 2.99 Pulse broadening in optical fibre

This **broadening** effect makes the duration of the pulse increase with distance travelled along the fibre. This means that there must be a suitable delay between pulses, so that there is always a gap between the end of one pulse and the start of the next. In effect, this limits the frequency with which pulses can be transmitted, and hence the amount of information carried per second.

The core of a single mode fibre is made as narrow as possible in order to reduce the range of path lengths. Narrow cores mean that all the rays are as parallel to the axis as possible. The disadvantages of such fibres include:

- Expense: any impurities, discontinuities, ovality of the cable will profoundly affect the transmission of light, so manufacturing cost is high.
- Narrow fibres carry less energy than wider fibres, so the signal is fainter.

Multimode fibres

The core of a multimode fibre is much wider than that of the monomode, typically 100 μm, with a cladding thickness of only 10 μm or so. There are many more paths for the light to follow, so pulse broadening is more apparent, but these fibres are useful for short distance, due to their much greater transmission of energy per fibre.

Graded index fibres

A detailed treatment is beyond the scope of this book, but briefly, the refractive index of the core of a graded index fibre decreases from the centre outwards, and is a minimum at the edge. Axial rays of light travel the shortest distance along the fibre, but at the slowest speed, as they travel through the region of highest refractive index. Off-axis rays spend more time in the regions of lower refractive index, between internal reflections, but travel faster. With careful design, the pulse broadening of the step-index fibre can be greatly reduced.

Advantages of fibre optics over copper cables

- For the same signal-carrying capacity, a fibre-optic cable will be many times narrower and lighter than the equivalent copper cable. This means a great saving in weight, and the lighter, thinner cable will be easier to handle.
- The glass is considerably cheaper than the copper.
- Fewer amplifiers are needed.
- Electrical interference is eliminated since the fibre cannot absorb electromagnetic radiation from outside.
- Since no electricity is carried along the fibres, a severed cable presents no fire risk. This is particularly important in sensitive environments such as on board aircraft. Increasing use is being made of optical fibres to transmit the pilots' commands and data around aircraft.
- There is almost no possibility of 'crosstalk' (signals from one cable being transferred to another).
- Transmission security is much enhanced, since optical circuits are almost impossible to 'tap' without breaking the cable.

Self-assessment

SECTION A
Qualitative assessment

1. State the laws of reflection of light.
2. Draw a labelled ray diagram showing how an image is formed in a plane mirror and state four properties of the image.
3. Explain the difference between **regular** and **diffuse** reflection of light.
4. Distinguish between **real** and **virtual** images.
5. Draw a labelled diagram showing the passage of a ray of light through a parallel-sided glass block when the ray strikes the block at an angle to the normal.
6. State the laws of refraction of light.
7. Define refractive index.
8. Explain, using diagrams, the terms **total internal reflection**, and **critical angle**. State the conditions necessary for total internal reflection to occur and state the relationship between refractive index and critical angle.
9. Draw a labelled diagram to show the passage of a ray of light through an optical fibre.
10. Describe how optical fibres are used in **endoscopes**.
11. Briefly explain how a telephone conversation can be transmitted using optical fibres, and state the main advantages of this system over conventional copper cables.
12. Outline the distinction between **monomode** and **multimode** optical fibres, and state the main advantages and disadvantages of each, with particular reference to **pulse broadening**.

SECTION B
Quantitative assessment

(***Answers:*** 6.7×10^{-8}; 1.36; 2.4; 2.6; 14; 14; 22; 44; 1.014×10^{3}; 1.7×10^{6}.)

1. At a height of 2.72 m, Robert Wadlow is the tallest person recorded in medical history. Assuming that his eyes were 16 cm below the top of his head, calculate the minimum length of a vertically mounted plane mirror in which he could view his entire body. What is the height of the top of the mirror above the ground?
2. A ray of light is incident on the flat surface of an ice block. If the angle between the ray and the surface is 24° calculate:
   (a) the angle of refraction of the ray
   (b) the angle through which the ray is deviated on entering the block.
   (Refractive index of ice = 1.31.)
3. If the speed of sound is 330 m s^{-1} in air and 1400 m s^{-1} in water, calculate the critical angle for sound waves passing from air to water. Hence explain why sonar transmitters (sources of ultrasound used for submarine detection) are often suspended beneath the surface of the water rather than above it.
4. A narrow beam of white light strikes one face of an equilateral glass prism as shown in the diagram.

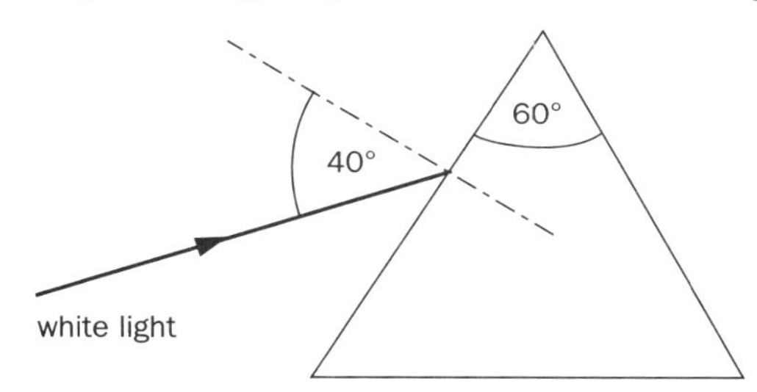

   If the refractive index of red light is 1.53, and that of violet light is 1.55, calculate the angular separation of the red and violet rays in the spectrum produced by refraction through the prism.
5. A ray of light is incident on the end of an optical fibre as shown in the diagram.

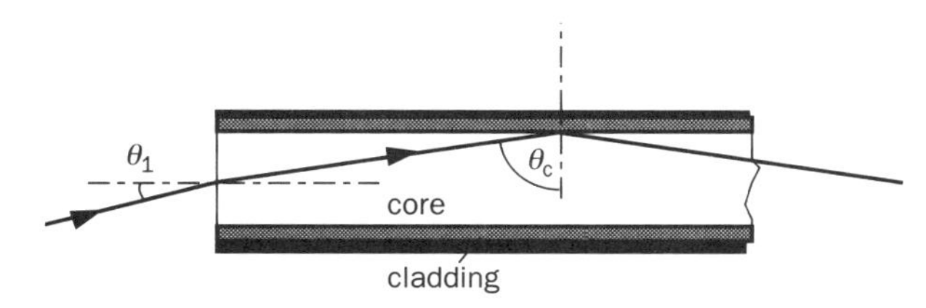

   If the refractive index of the core material is 1.472 and that of the cladding is 1.452, calculate the maximum possible value of θ_1 for which total internal reflection can take place at the core/cladding boundary. In an optical fibre of diameter 100×10^{-6} m and length 1 km, how many reflections would take place for a ray striking the boundary at this angle? (Assume the fibre is perfectly straight.) Hence calculate the path length of such a ray, and the time difference between the arrival of this ray and the arrival of an axial ray, which suffers no reflections. (This figure sets the limit on the frequency of pulses that can be transmitted.) (Speed of light in the core material = 2.04×10^{8} m s^{-1})

2.8 Lenses

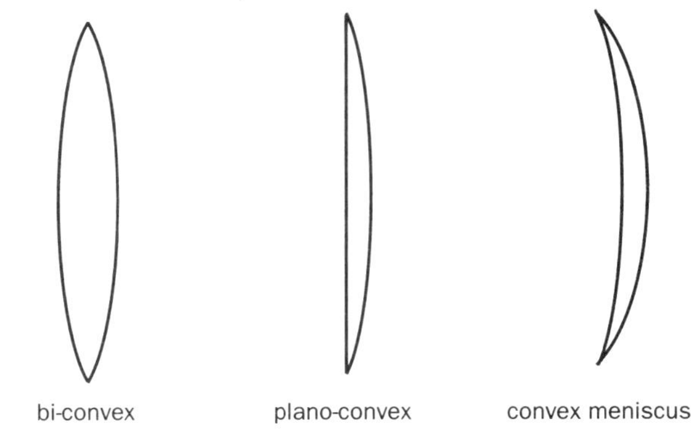

Figure 2.101 Converging lens shapes

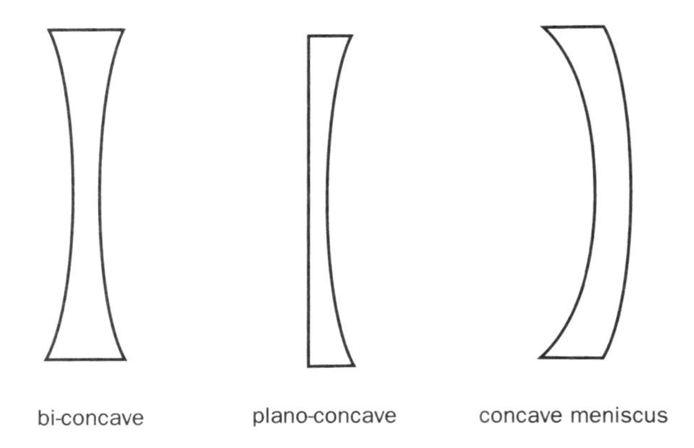

Figure 2.102 Diverging lens shapes

Lenses form the vital components of many optical systems: telescopes, cameras, film projectors, etc. All make use of the ability of curved pieces of glass to re-direct light rays and produce images. Even as you read this, the flexible lens in each of your eyes will be making small adjustments so as to maintain a sharp image on the retina.

Lenses are made from transparent materials (glass, perspex, etc.), have one or more curved faces, and are classified as either **converging** or **diverging**, according to the effect the lens has on **parallel** rays of light. In diagrams, the lens shape is represented by imagining the lens cut in half, so that its cross-section is visible, as shown in Figure 2.100.

- **Converging** lenses are thicker in the middle than at the edges, as shown in Figure 2.101.
- **Diverging** lenses are thinner in the middle than at the edges, as shown in Figure 2.102.

Lens terminology

In this book, we shall be dealing only with **thin** lenses. This implies that, for regions near the centre of the lens, the faces are nearly parallel, and close together. This is a simplification that nonetheless provides us with the ability to predict very accurately the behaviour of actual lenses. Before going any further, you should learn the meaning of the following terms:

- **Centre of curvature**. Each surface of the lens is part of a **sphere**, and so has a centre of curvature (which is at infinity for a plane surface).
- The **principal axis (PA)** of the lens is an imaginary line joining the two centres of curvature.
- **Paraxial rays** are rays close to the principal axis, and very nearly parallel to it. The analysis of lenses is based on considering these rays only,

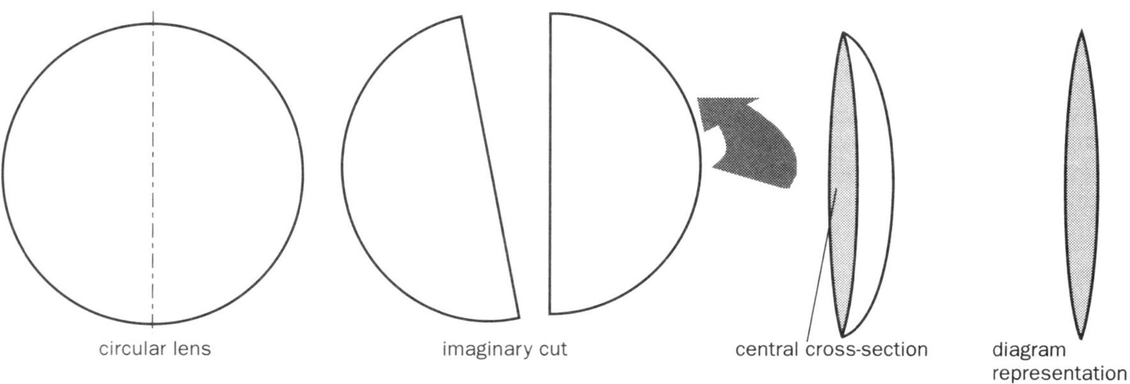

Figure 2.100 Diagrammatic representation of lens shape

which in effect means only considering lenses of very narrow **aperture** (small diameter). Diagrams show large lenses and rays at large angles to the principal axis for clarity.

- The **principal focus (F)** of a thin **converging** lens is the **point** on the principal axis to which all **paraxial** rays **converge** after refraction through the lens (see Figure 2.103(a)).
- The **principal focus (F)** of a thin **diverging** lens is the **point** on the principal axis from which all **paraxial** rays appear to **diverge** after refraction through the lens (see Figure 2.103(b)).

Since light can fall on either surface, a lens has two principal foci, one on each side, and these are at equal distances from the optical centre of the lens, C.

- **Parallel rays.** Light rays from any point on an object must always be **diverging** as soon as they leave the object. However, if the object is far away from a lens, then the rays collected by the lens will have such a small angle of divergence that we can consider them to be parallel. So we always consider rays from a point on a **distant** object to be **parallel** (see Figure 2.104).
- The **focal length** (f) of the lens is the distance from the **optical centre** (C) to the **principal focus** (F) of the lens (i.e. the distance CF in Figure 2.103). This is a **real** distance for a converging lens, but is **virtual** for a diverging lens.
- The **power** of a lens is defined by the formula

$$\text{power} = \frac{1}{\text{focal length in metres}}$$

The SI unit is the metre^{-1} (m^{-1}), sometimes called the **dioptre.**

The power of a converging lens is **positive**, since f is a **real** distance; that of a diverging lens is **negative** since f is a **virtual** distance.

Thus the power of a converging lens of focal length 2 m is given by:

$$power = \frac{1}{+2} = +0.5 \text{ m}^{-1}$$

and the power of a diverging lens of focal length 25 cm is given by:

$$power = \frac{1}{-0.25} = -4.0 \text{ m}^{-1}$$

- The **focal plane** is a plane passing through the principal focus and perpendicular to the principal axis. Parallel rays at a **small angle** to the principal axis (paraxial rays) will converge to a point on the focal plane (or appear to diverge from a point on the focal plane in the case of a diverging lens) as shown in Figure 2.105.

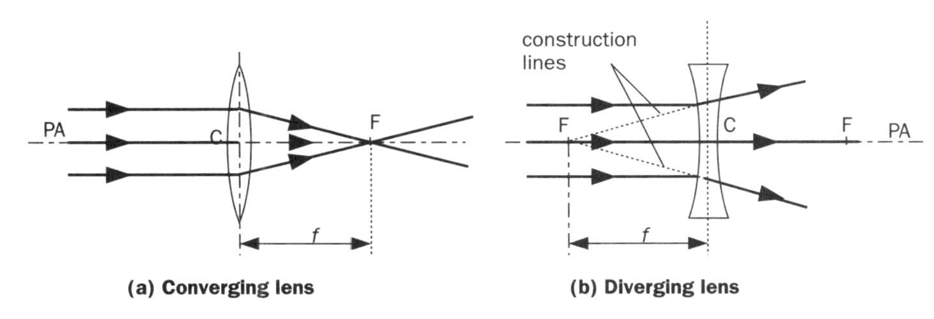

Figure 2.103 Principal focus and focal length

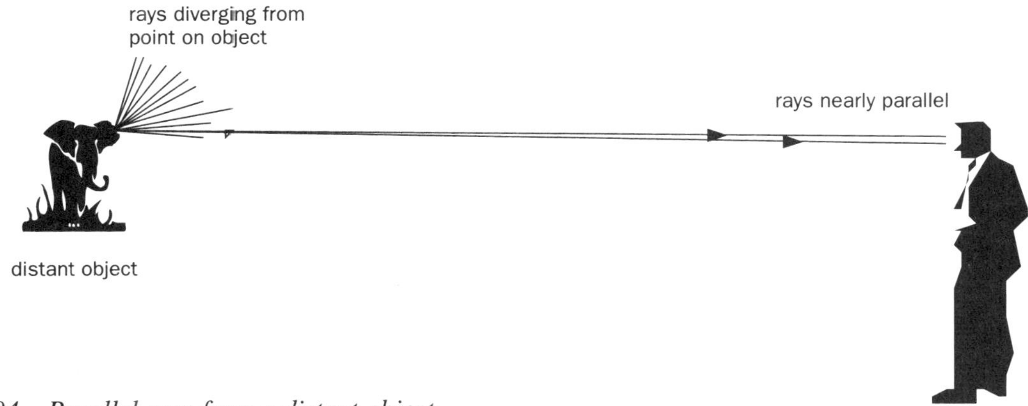

Figure 2.104 Parallel rays from a distant object

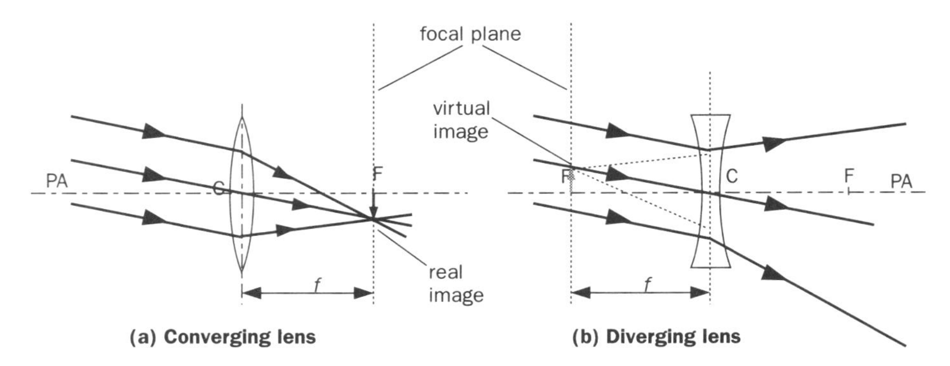

Figure 2.105 Focal plane

Formation of images by lenses

The **size**, **orientation**, **nature** and **position** of the image formed by a lens depend on:

- the size and position of the object and
- the focal length of the lens.

Ray diagrams

Rays from all points on an object spread out in all directions. To produce an image it is necessary to make these rays converge to, or appear to diverge from, a different point. To locate the image, it is only necessary to draw the paths of any **two** of these rays, and note the point at which their paths cross. Some rays are easier to locate than others, and three special rays are particularly useful:

- A ray **parallel to the principal axis** will pass through the principal focus (or, in the case of a diverging lens, appear to diverge from it) after refraction by the lens.
- A ray **through the optical centre (C)** of the lens will continue undeviated. (Strictly speaking, this only applies to very thin lenses. The central portion of such a lens will have virtually parallel sides, so any ray passing through the centre will not suffer any change in direction, and the displacement of its path will be negligible.)
- A ray **through a principal focus** (or heading towards it) will be refracted to travel parallel to the principal axis after passing through the lens.

Figure 2.106 illustrates the use of these three rays to locate the image formed by converging and diverging lenses. Note that it is only necessary to locate the image of the **tip** of the arrow. Since the base of the arrow is on the principal axis, the base of the image will also be on the principal axis. If the drawing is done to scale, the size (and so **magnification**), orientation (**erect** or **inverted**), nature (**real** or **virtual**) and position (**distance** from the lens) of the image can all be measured.

A **real** image is one through which real rays of light actually pass, and it can therefore be formed on a screen or affect photographic film. A **virtual** image is one from which rays of light only **appear** to have come, and it cannot be formed on a screen.

Figures 2.107(a) to 2.107(f) show how the image properties change as an object is brought closer to the lens. Remember that if the object is at infinity, the image will be formed in the focal plane, as shown in Figure 2.107(a).

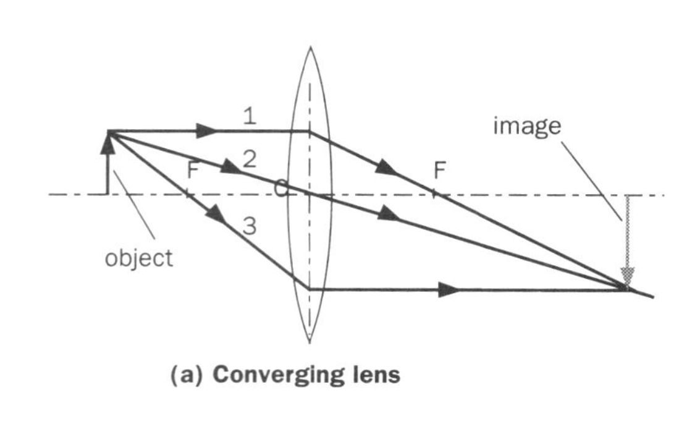

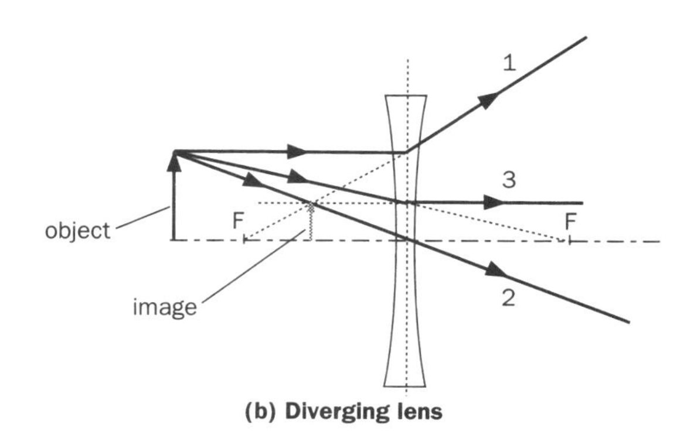

Figure 2.106 Ray construction

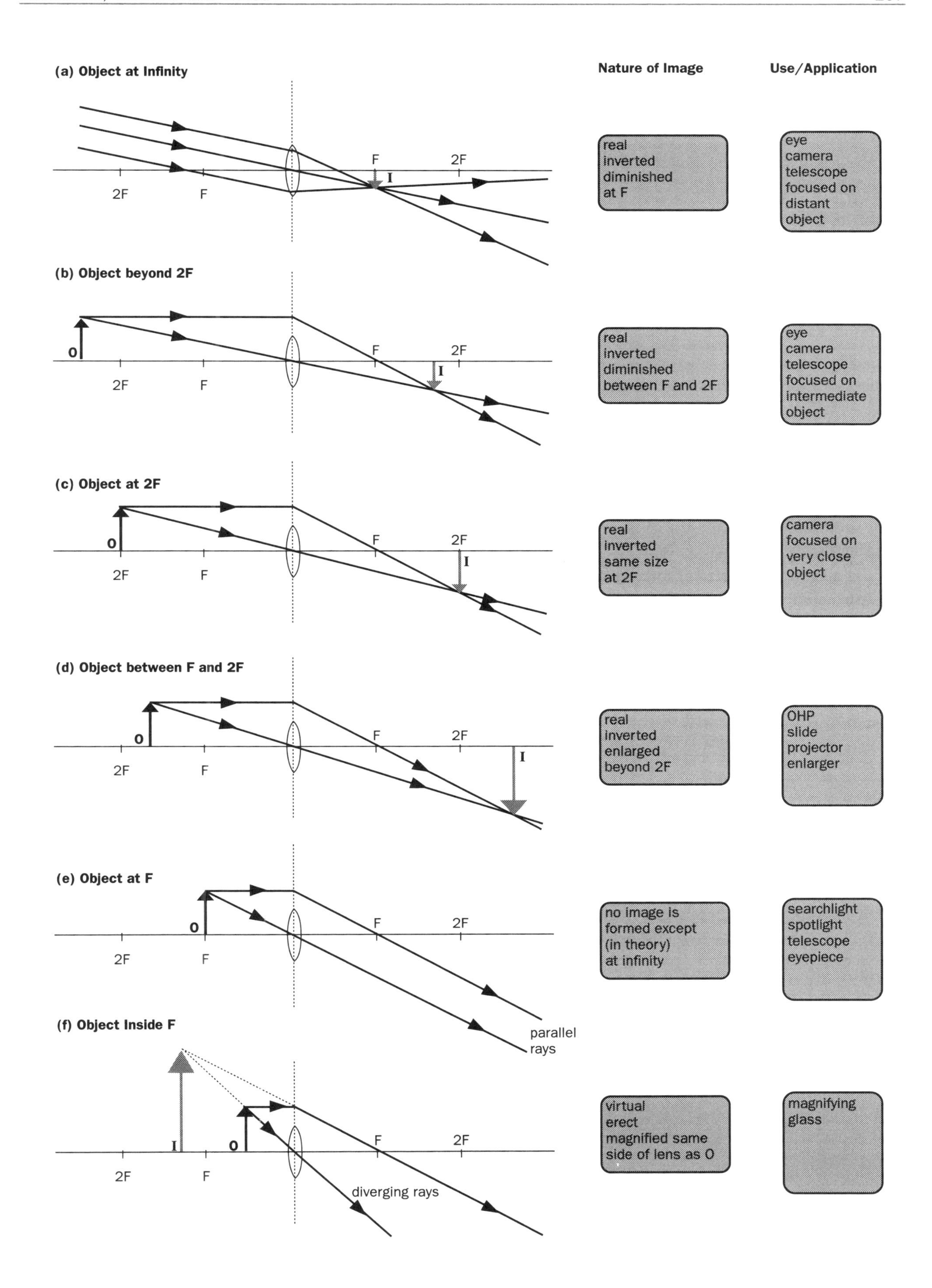

Figure 2.107 Ray diagrams – showing image formation by a convex (or converging) lens

Figure 2.108 shows how an image is formed by a diverging lens. In this case the image is always **virtual, upright, diminished** and formed somewhere between F and the centre of the lens. Possible applications are in spectacles, the security door-viewer and as elements in compound lenses.

The lens formula

For a thin lens, it can be shown that:

$$\frac{1}{u} + \frac{1}{v} = \frac{1}{f}$$

Equation 1

where: u = *object distance* (distance from optical centre of lens to **object**)
v = *image distance* (distance from optical centre of lens to **image**)
and f = *focal length* of **lens**.

Sign convention

When using this formula, you must be consistent in applying the correct **sign** to each term in the equation. One convention is called the **real is positive** convention:

- Distances from **real** objects and images must be entered as **positive** values
- Distances from **virtual** objects and images must be entered as **negative** values
- The focal length of a **converging lens** is entered as a **positive** value, whereas the focal length of a **diverging lens** is entered as a **negative** value.

Linear magnification

Figure 2.109 shows the formation by a converging lens of a real image at I (of height h_i) from an object at O (of height h_o). The **linear magnification** (m) is defined as

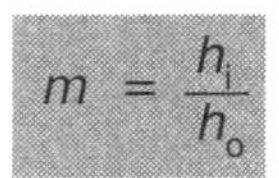

$$m = \frac{h_i}{h_o}$$

Equation 2

Note that m is simply a **ratio** of the height of the image to the **height** of the object, and so is **dimensionless** (has no **units**).

Also, since triangles OCX and ICY are similar (shown shaded in Figure 2.109):

$$\frac{h_i}{h_o} = \frac{v}{u}$$

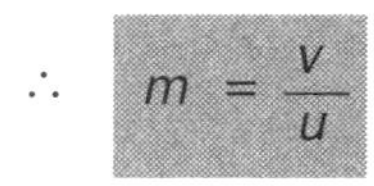

$$\therefore \quad m = \frac{v}{u}$$

Equation 3

Linear magnification equations apply to **all** lenses.

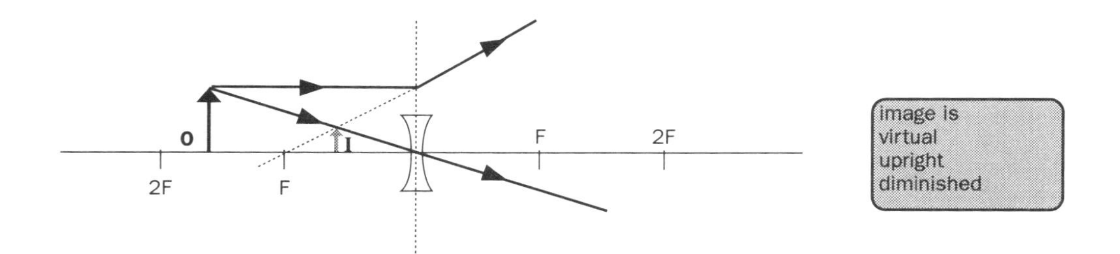

Figure 2.108 Ray diagrams – showing image formation by a concave (or diverging) lens

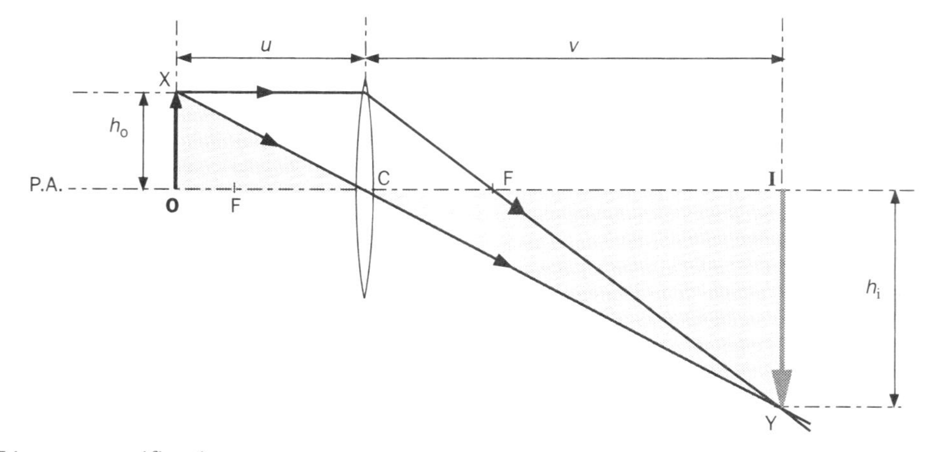

Figure 2.109 Linear magnification

Guided examples (1)

1. An object is placed 20 cm from (a) a converging lens and (b) a diverging lens, each of focal length 15 cm. Calculate the image position and magnification in each case.

Guidelines

A simple diagram, roughly to scale, is always a great help in lens problems.

In each case, write down the numerical values of u and f, which are given. Remember to apply the **sign convention**! Then re-arrange Equation 1 and use it to find v, and calculate m from Equation 3.

2. The filament of a lamp is 80 cm from a screen, and a converging lens forms an image of it on the screen, magnified 3 times. Find:
   (a) the distance between the lens and the filament
   (b) the focal length of the lens.

Guidelines

This time, use Equation 3 to find v, then substitute in Equation 1 to find f.

3. An erect image 2 cm high is formed 12 cm from a lens, the object being 0.5 cm high. Calculate the focal length of the lens.

Guidelines

The key word in the question is **erect**, and from the sizes of the image and object, you should spot that the image is **magnified**. There is only one situation with a single lens that gives rise to an erect, magnified image: the magnifying glass. Use Equation 3 to calculate u, then use Equation 1 to find f. (Remember the sign convention – **virtual** distances are **negative**.)

4. A camera fitted with a portrait lens of focal length 35 mm is used to photograph a girl whose face is 235 mm long. If she is seated 500 mm from the lens when the picture is taken, calculate:
   (a) the **power** of the lens
   (b) the distance from the lens to the film (image distance)
   (c) the length of the image of the girl's face on the developed negative.

5. A photographic enlarger is a device for projecting an image (usually magnified) of a transparent slide onto light sensitive paper in order to produce a photographic print. A particular enlarger has a maximum distance between slide and print of 90 cm. If it is desired to be able to produce a print 315 mm long from a slide of length 35 mm, calculate the focal length of the lens to be used.

Guidelines

The enlarger's maximum magnification will occur when the slide is furthest from the print (90 cm in this case). 90 cm is the **sum** of u and v.

Use Equation 2 to calculate m, then write two simultaneous equations:

$$u + v = 90$$

$$m = \frac{v}{u}$$

and solve these two equations for u and v.

Measurement of focal length

Approximate method (distant object method)

For most purposes, it is sufficient to produce an image on a screen of an object at infinity, as shown in Figure 2.110. Since the rays from any point on the object are parallel, the image will be formed in the focal plane. Simply measure the distance from the lens to the image to obtain the focal length.

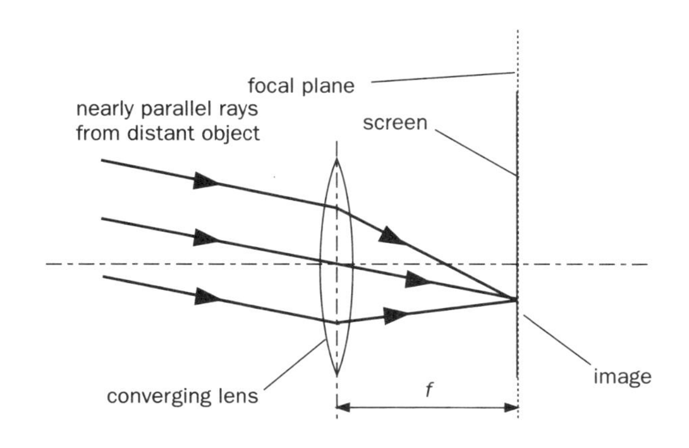

Figure 2.110 Measurement of focal length (approximate method)

Lens formula method

This method involves making a series of measurements of corresponding values of object distance (u) and image distance (v). First, an approximate measurement of focal length (f) is made using the previous method. Real images cannot be obtained if the object distance is set at less than f, so it is helpful to be aware of this limit when carrying out the experiment. A brightly-lit object, the converging lens and a plane screen are arranged as shown schematically in Figure 2.111. For highly accurate work, the components should be mounted on an **optical bench**, which facilitates measurement of the distances between object, image and lens.

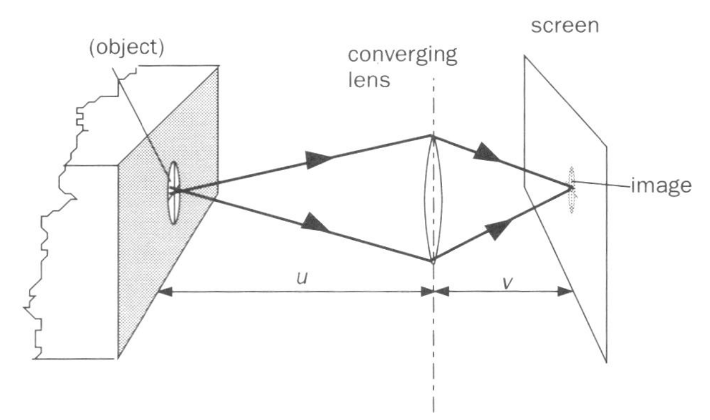

Figure 2.111 Measurement of focal length (lens formula method)

Starting with an object distance just greater than f, the position of the screen is adjusted until a sharp image of the object is obtained on it. There will usually be a considerable range of positions over which the image may be judged sharp, and some experimental skill is needed here to minimise measurement error. One method is to move the screen as far as possible away from the lens, while keeping the image acceptably focused, and measure the image distance. Then bring the screen as close as possible to the lens, again keeping the image acceptably focused, and measure this, slightly smaller, image distance. The **average** of these two measurements is a reliable estimate of the true image distance.

The procedure should be repeated, using different object distances, to obtain a suitable number of values of u and v. The **range** of values of u to be used should be as wide as possible, from near the focal length, up to the greatest distance available within the limitations of the apparatus used. From the table of values of u and v, a graph of $1/u$ against $1/v$ is plotted, as shown in Figure 2.112.

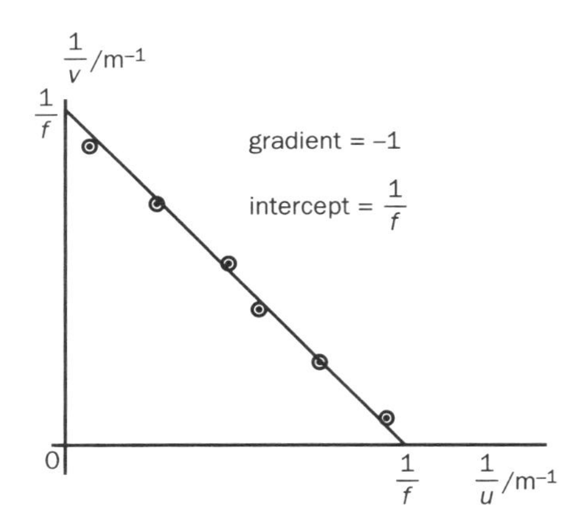

Figure 2.112 Graph of 1/v against 1/u

A lens of variable focal length fitted to a modern 35 mm camera

Maths window

The *lens formula* $\frac{1}{u}+\frac{1}{v}=\frac{1}{f}$ can be re-written:

$$\frac{1}{v}=-\frac{1}{u}+\frac{1}{f}$$

Comparing this to the standard *straight line* equation: $y = mx + c$ we see that plotting $1/v$ on the y-axis against $1/u$ on the x-axis should give a straight line graph of **gradient –1** and **y-intercept $1/f$**. (Since the gradient is –1, the x-intercept is also equal to $1/f$.)

The eye as an optical instrument

Figure 2.113 shows a cross-section of a person's right eye, as seen from above. The function of each of the main parts of the eye is as follows:

Sclerotic coat This is the tough, outer layer of the eye. It is completely opaque, except for a transparent portion at the front, through which light enters the eye.

Cornea This is the curved, transparent window in the sclerotic. Together with the **aqueous humour**, this forms a converging lens of **fixed focal length**. This is where most of the refraction of light occurs.

Lens The lens is made of flexible gelatinous tissue, and its thickness can be varied by controlling the tension of the **suspensory ligaments** using the **ciliary muscles** within the eye. This gives the lens a variable focal length, allowing the eye to **accommodate** (i.e. bring into clear focus) objects over a wide range of distances.

Distant objects are viewed with the lens at its thinnest, i.e. with the ciliary muscles relaxed. When the object being viewed is close to the eye, the ciliary muscles contract, allowing the lens to take up a **thicker** shape, giving it a **shorter** focal length (i.e. more refracting ability), as shown in Figure 2.114.

Iris/pupil The iris controls the size of the **pupil**, the aperture through which light enters the eye. In **bright** light, the iris **contracts** the size of the pupil, **reducing** the amount of light energy

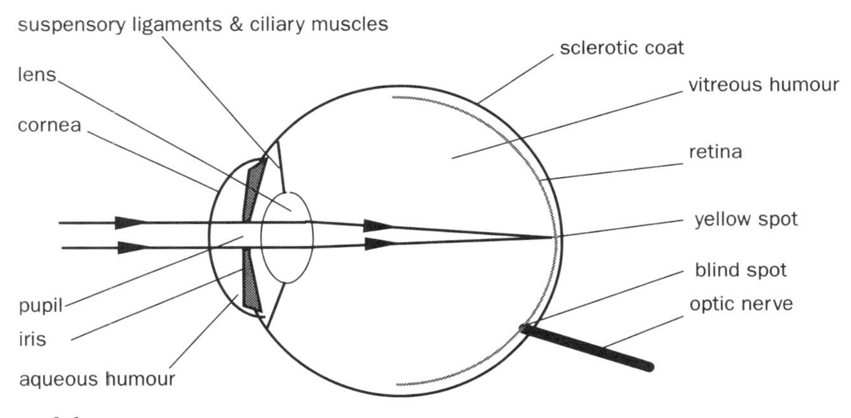

Figure 2.113 Structure of the eye

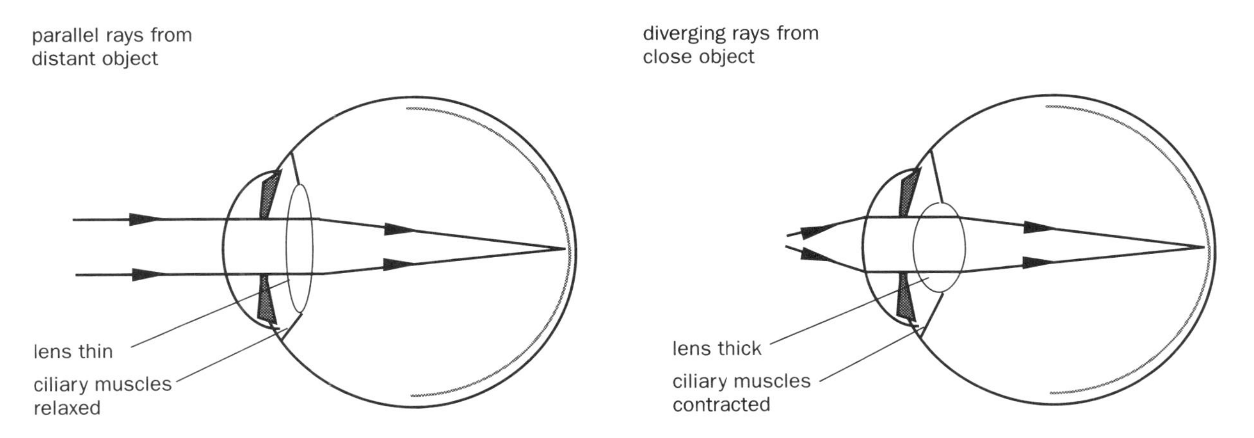

Figure 2.114 Accommodation of the eye

entering the eye. In **dim** light, the pupil is **enlarged** to **increase** the amount of light energy entering (see Figure 2.115).

Retina This is the layer of light-sensitive cells on which the **real, inverted, diminished** image is formed. When looking straight ahead, light rays fall on a particularly sensitive area of the retina where there is a greater concentration of cells. This area is called the **yellow spot**, and is used for detailed vision. When you are concentrating on some task, such as threading a needle, you see most detail if the image of the eye of the needle is focused here. Incidentally, this area is deficient in those cells which work better in low light, which is why you see objects better at night if you don't look directly at them, but slightly to one side.

Optic nerve The light falling on the retina creates electrical impulses in the cells, which are transmitted to the visual cortex of the brain by the optic nerve. The area of the retina where the optic nerve exits contains no light-sensitive cells, and is called the **blind spot**, since nothing is seen when light falls on it. The effect of the blind spot is not normally noticeable, since our brains 'fill-in' the missing detail.

Some points of interest

- The eye cannot focus objects clearly if they are too close.

The **near point** is defined as the **closest** distance of distinct vision. For the **normal** eye, it is usually assumed to be **25 cm**.

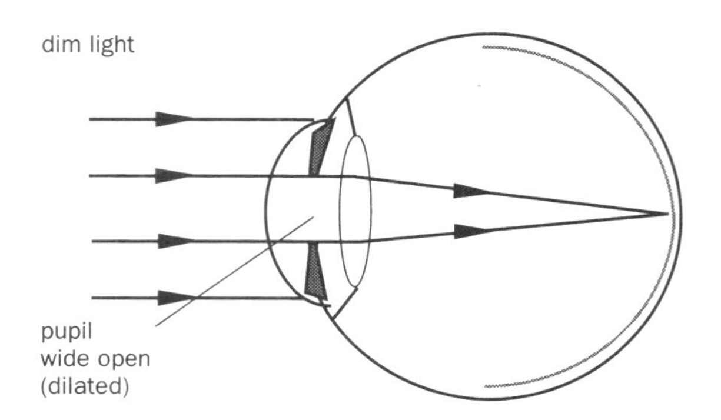

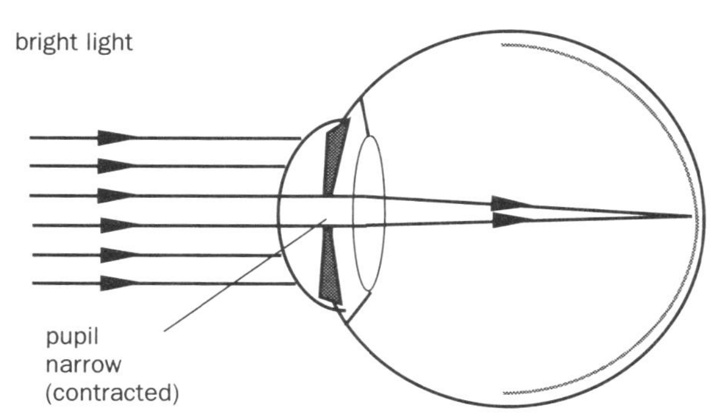

Figure 2.115 Adjustment for differing light conditions

- The normal eye can accommodate objects at infinity.

The **far point** is defined as the **furthest** distance of distinct vision. For the **normal** eye, it is usually assumed to be **infinity**.

- The retina is black and absorbs (almost) all light falling on it. Some red light is reflected, hence the 'red-eye' effect in flash photographs, where red light is reflected from the retina, straight back towards the flash gun, and enters the camera lens. Skilled photographers do not mount their flash guns near the camera lens, which eliminates this problem.
- The range of light sensitivity of the eye is of the order 10^9. (Full sunlight is about $10^6 \times$ brighter than full moonlight.)
- The flexibility of the lens decreases with age, and gradually the near point recedes.
- Cataract is a disease where the cornea and/or lens becomes clouded and opaque, leading to blindness. Cornea transplants are becoming commonplace, and synthetic lenses are available to treat this condition.

Defects of vision and their correction

There are many people whose eyes cannot accommodate the normal range of object distances (25 cm to infinity), and need corrective lenses. The two main types of defect are short sight and long sight.

Short sight (myopia)

A **short**-sighted person can see clearly at short distances, but distant objects go out of focus. (The far point is closer to the eye than infinity.) Either the eyeball is too long, or the lens is too powerful (focal length is too short and causes too much refraction). Whatever the cause, the image of a distant object is formed in front of the retina as shown in Figure 2.116(a).

The defect is corrected by placing a **diverging** lens in front of the eye, as shown in Figure 2.116(b). The effect of this lens is to cause rays entering the eye to diverge more, so the image is formed further from the eye lens (i.e. on the retina). In effect, the diverging lens forms a **virtual image** of the distant object. This virtual image (of an object at infinity) is formed at the far point of the defective eye. Closer objects will produce virtual images closer than the far point, so all object distances can be accommodated.

Long sight (hypermetropia)

A **long**-sighted person can see objects clearly which are a **long** distance away, but close objects are not focused. The near point is more than 25 cm from the eye. The eyeball may be too short, or the lens lacking in power (i.e. the focal length of the eye lens is too long). In either case, rays from a close object would be brought to focus **behind** the retina, so only a blurred image is seen (see Figure 2.117(a)).

By placing a **converging** lens in front of the eyeball, rays from close objects are made less diverging before entering the eye, as shown in Figure 2.117(b). Thus the eye lens is able to bring them to a focus on the retina. In effect, the converging lens produces a **virtual image** of an object placed 25 cm from the eye (the **near point** of a **normal** eye). This image is formed at the **near point** of the defective eye. Any object placed more than 25 cm from the eye will thus fall within the range of accommodation of the **corrected** eye.

Guided examples (2)

1. A short-sighted person has a far point of 2.0 m. What power of spectacle lens is needed to correct this defect, so that very distant objects can be seen clearly?

Guidelines

You should remember that short sight is corrected by a **diverging** lens, which makes distant objects appear closer by creating a **virtual image** of the object. To correct this defect, an object at infinity must **appear** to the person to be only 2.0 m away. The diverging lens must produce a virtual image 2.0 m from the lens.

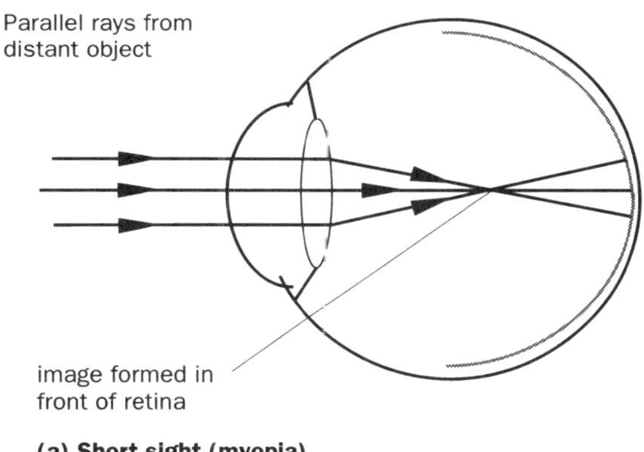

(a) Short sight (myopia)

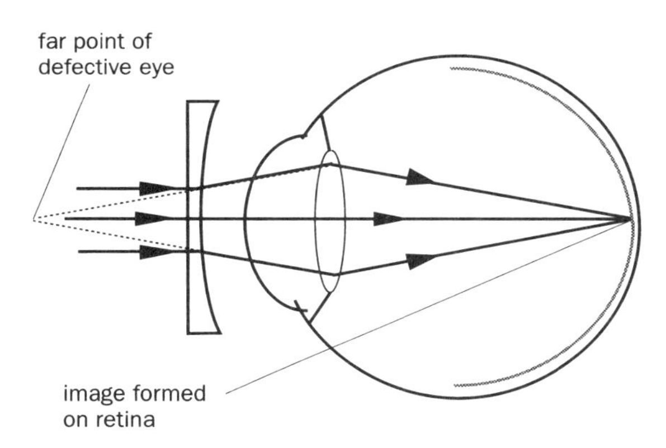

(b) Correction of myopia with a diverging lens

Figure 2.116 Short sight and its correction

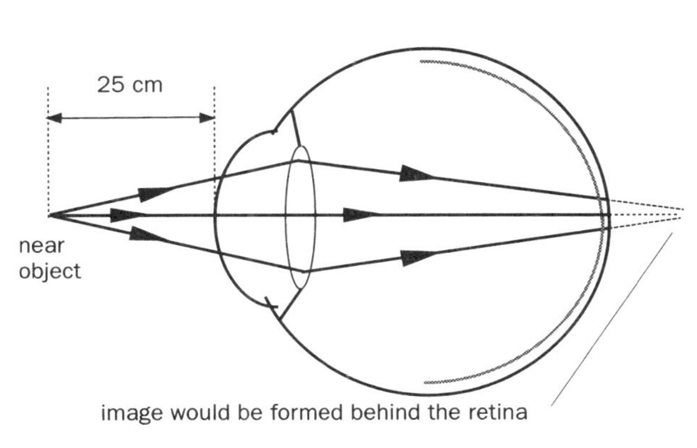

(a) Long sight (hypermetropia)

(b) Correction of hypermetropia with a converging lens

Figure 2.117 Long sight and its correction

Guided examples (2) *continued*

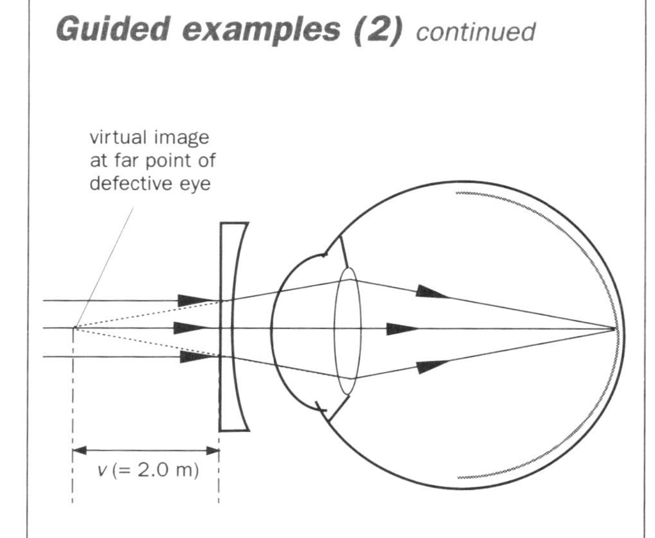

Figure 2.118 Short-sight correction for a person whose far point is at 2.0 m

The object distance is infinity, the image distance is shown in Figure 2.118. Substitute these values in the lens formula (remember the sign convention for virtual images!) to calculate the **focal length**, and hence the **power** of the corrective lens. (This should have a **negative** focal length, since it is a **diverging** lens.)

2. An old lady has a near point of 150 cm. Calculate the focal length of the spectacle lens she must wear to correct this defect, and enable her to see objects clearly which are only 25 cm from her eye.

Guidelines

You should recognise from the description of her condition that the lady is **long sighted**, and therefore needs a **converging** lens. The effect of the converging lens is to produce a **virtual** image further away from the lens than the object. To correct this lady's problem, objects at 25 cm need to look as if they are at 150 cm (her true near point), so the virtual image must be at a distance of 150 cm from the lens.

As in the previous problem, substitute values for u and v in the lens formula to calculate f. Again be careful to apply the sign convention.

Variable-focus single-lens camera

Figure 2.119 shows paraxial rays from a point on a distant object forming a **real, inverted, diminished** image on the photographic film in a simple, single-lens camera.

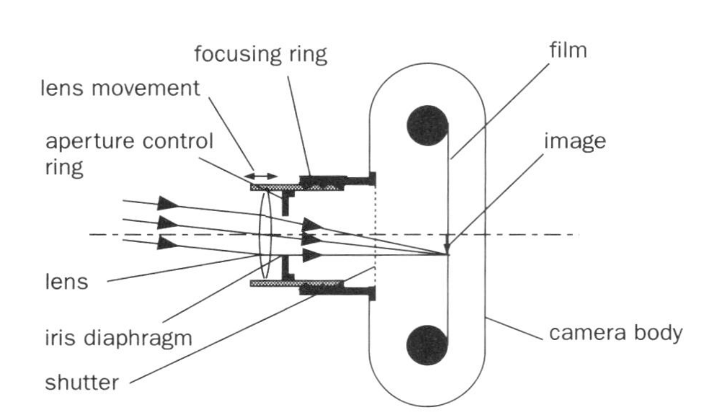

Figure 2.119 Variable-focus single-lens camera

The function of the various components of the camera is as follows:

Lens In the simplest cameras, this will be a single converging lens. Its distance from the film can be adjusted using the screw-threaded **focusing ring**. This allows objects at distances ranging from infinity to a few tens of centimetres to be focused on the film.

Aperture ring/iris The aperture control ring (or stop) controls the iris diaphragm, which consists of thin inter-leaving metal sheets which can be moved so as to vary the size of the circular aperture through which light enters the camera.

Shutter Normally blocks light from reaching the film. When taking a photograph, the shutter briefly opens to allow the image to appear on the film. The **exposure time** is the length of time during which the shutter opens to expose the film. It may vary between several minutes for night-time photography to less than one thousandth of a second for high-speed work.

The combination of aperture size and shutter speed is chosen to suit the lighting conditions, the speed of motion of the subject, and the photographer's desire for certain effects. A short exposure time demands a wide aperture, in order that sufficient light energy falls on the film, and vice versa. Aperture size (denoted by the f-number on cameras) and shutter speed are usually chosen from a range of pre-set values. These values are chosen so that an increase (or decrease) of one unit doubles (or halves) the amount of light energy reaching the film.

Comparing the eye and the camera

Eye	Camera
Convex lens system produces a **real inverted, diminished** image on the **retina**	Convex lens system produces a **real inverted, diminished** image on the **film**
Light incident on the **retina** stimulates electrical impulses to the **brain**	Light incident on the **film** stimulates chemical changes in the **photographic emulsion**
Iris controls the amount of light energy entering the eye, **pupil** is open continuously	**Aperture diaphragm** controls the amount of light energy reaching the film, in conjunction with the **shutter**, which is only opened for a pre-determined time
Can focus objects between the **near point** and **infinity**	Can focus objects between a **few centimetres** from the lens and **infinity**
Focusing is achieved by changing the **thickness** of the eye lens	Focusing is achieved by changing the **distance** between the lens and the film

Guided example (3)

A camera has a single converging lens of focal length 75 mm. Calculate the necessary range of movement of this lens if the camera is to be able to produce clear pictures of objects between infinity and 50 cm from the lens.

Guidelines

When the camera is being used to photograph an object at infinity, u_1 (the *object distance*) is equal to infinity, so v_1 (the *image distance*) is equal to f (the focal length). You can confirm this by substituting these values into Equation 1 (the lens formula).

Use the equation to calculate v_2 (the *image distance*) when u_2 (the *object distance*) is 50 cm. The lens movement range is the difference between v_1 and v_2.

Visual angle

It is common experience that objects appear smaller when they are further away. The apparent size of an object depends on the size of the image formed on the retina, which in turn depends on the visual angle – the angle subtended at the eye by the object (see Figure 2.120).

Since a is constant, the size of the image produced on the retina depends only on the **visual angle,** α. Thus the greater the visual angle, the greater the **apparent size** of the object.

Magnifying power (or angular magnification) (*M*)

Optical magnifying instruments work by increasing the visual angle. Thus if the angle subtended at the eye by the **final image** is β, then the magnifying power is given by:

$$M = \frac{\beta}{\alpha} = \frac{\text{angle subtended at the eye by the final image}}{\text{angle subtended at the unaided eye by the object}}$$

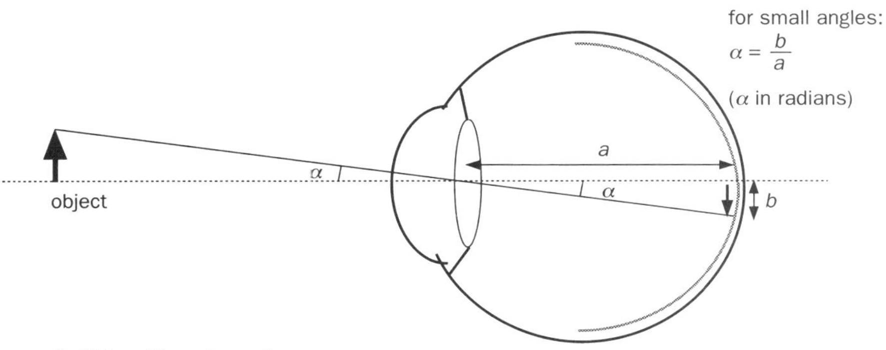

Figure 2.120 Visual angle

Self-assessment

SECTION A
Qualitative assessment

1. Explain what is meant by each of the following terms as used to describe the action of lenses.
   (a) Principal axis.
   (b) Paraxial rays.
   (c) Centre of curvature.

2. Draw diagrams to show what happens when **paraxial** rays pass through a
   (a) converging lens
   (b) diverging lens.

3. Define the **power** of a lens. State the SI unit.

4. Diagram 1 below shows a converging lens, with the principal foci marked F. The points marked 2F represent points on the principal axis at distances of 2 × the focal length of the lens. Draw ray diagrams to locate the image for each of the object positions labelled 1 to 5. In each case, state the nature, size, and orientation of the image.

5. Diagram 2 below shows a diverging lens with principal foci marked F. Draw ray diagrams to locate the image for each of the object positions 1 and 2. In each case state the nature, size and orientation of the image.

6. Explain what is meant by the **real is positive** sign convention.

7. Define the term **linear magnification** as applied to thin lenses.

8. Describe in detail an accurate method for the determination of the focal length of a thin converging lens. Include a diagram of the apparatus, a complete statement of the measurements to be made and the instruments used to obtain them, a discussion of the errors involved and their minimisation, and a description of how the results will be analysed to determine the focal length.

9. Draw a fully labelled diagram of the human eye, and explain the function of its major components.

10. Define the terms **near point** and **far point**.

11. Describe, using diagrams to illustrate your answers, the defects of **short sight** and **long sight** together with their correction using lenses.

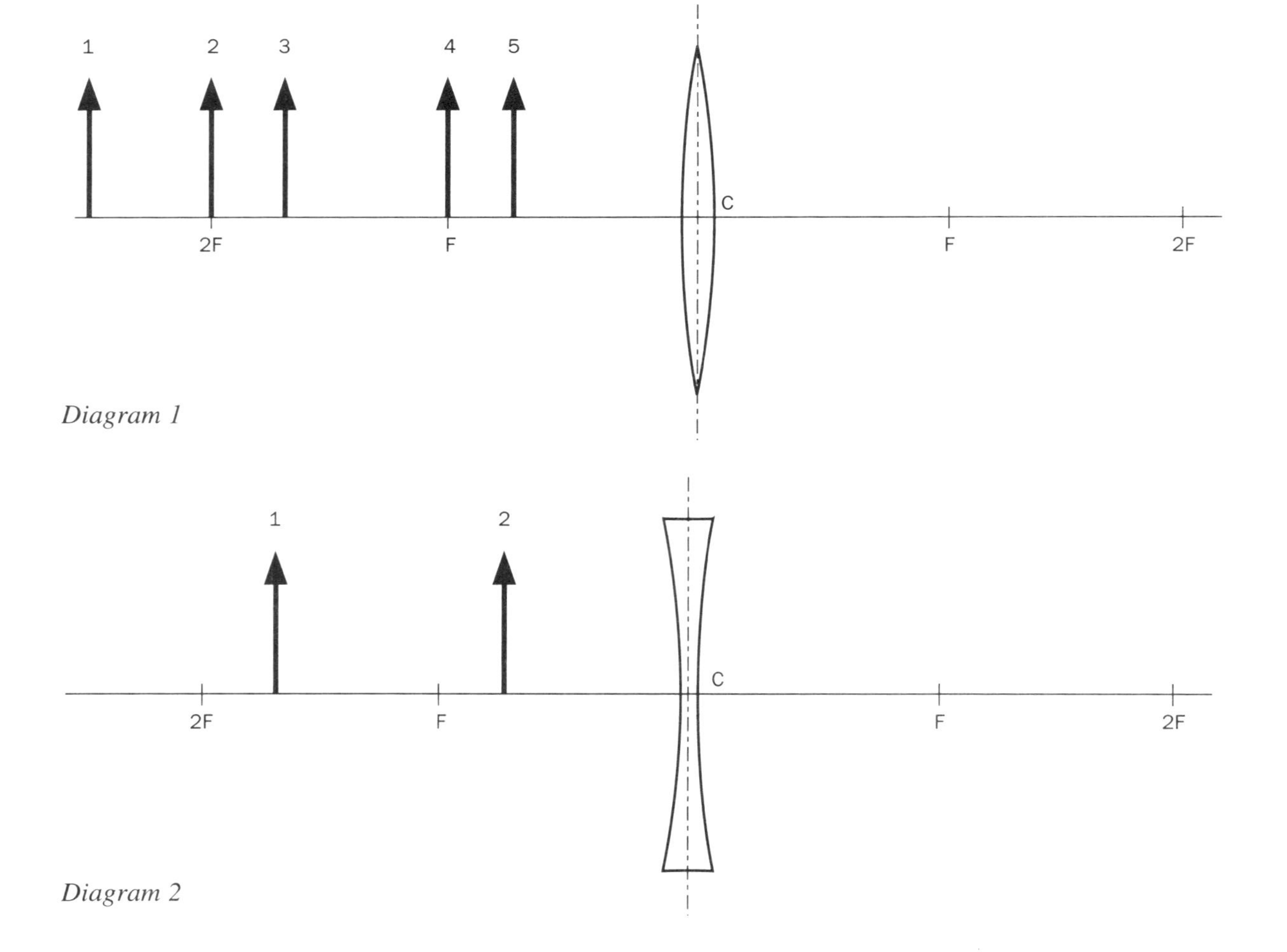

Diagram 1

Diagram 2

Self-assessment *continued*

12. Draw a labelled diagram of a simple variable-focus single-lens camera. Explain how
(a) a clear, focused image is obtained on the film
(b) the amount of light reaching the film is controlled.

13. State the similarities and differences between the human eye and the camera.

SECTION B
Quantitative assessment

(***Answers:*** 3.2×10^{-3}; 4.0×10^{-3}; 6.1×10^{-3}; 8.0×10^{-3}; 0.042 to 0.050; 0.060; 0.10; 0.36; 0.38 to infinity; –0.50; 0.60; –1.5; –8.8; 10.)

1. A projector is used to produce a magnified image of a photographic slide of height 2.5 cm. An image 100 cm high is obtained on a screen placed 410 cm from the slide. Calculate the power of the projector lens.

2. A luminous object of length 15 cm is placed **along** the principal axis of a converging lens of focal length 10 cm. If the closest end of the object is 20 cm from the lens, calculate the length of the image formed.

3. A photographic enlarger has a converging lens of focal length 5.4 cm and it is used to produce a magnified image of a 4 cm × 4 cm negative which is placed 6.0 cm from the lens. Calculate:
(a) the distance between the negative and its image
(b) the size of the image formed.

4. A small diverging lens is to be fitted in a door to be used as a 'spy-hole' to enable the occupant to see who is on the other side. If the 24 cm long face of a man standing 80 cm from the door appears to be 3 cm long when seen through the spy-hole, calculate the power of the lens used.

5. Following an eye-test, a man is prescribed spectacles to correct his **near point** to the normal 250 mm. If the lenses have a power of $+2.0 \text{ m}^{-1}$, what is his **near point distance** without the glasses?

6. A girl has a **near point** of 30 cm and a **far point** of 150 cm.
(a) What is the type and focal length of the lens she needs to enable her to see clearly objects at infinity?
(b) What is her range of vision when wearing these glasses?

7. A jeweller with normal vision (near point at 250 mm and far point at infinity) uses a thin converging lens of focal length 50 mm as a magnifying glass. If he wears the lens close to his eye when examining a diamond, calculate the range of distances of the diamond from the lens for which he sees a clear image.

8. A camera is fitted with a converging lens of focal length 100 mm.
(a) What is the lens to film distance when the camera is used to photograph a distant landscape?
(b) The camera is then used to photograph a flower which is 1.75 m from the lens. Calculate how far, and in which direction, the lens is moved in order to obtain a clear image on the film.

9. The lens on a camera has a focal length of 40 mm. Calculate:
(a) the size of the image on the film if a distant oil platform subtends an angle of 0.080 radian at the lens
(b) the range of movement of the lens if the camera is to be able to accommodate objects between infinity and 440 mm
(c) the size of the image of a medallion of diameter 80 mm placed 440 mm from the lens.

2.9 Telescopes

The astronomical refracting telescope

This consists of two separated converging lenses. It is used for looking at distant objects (i.e. at infinity), and for comfort it should be adjusted so that the final image is also at infinity – this is **normal adjustment**. If the final image is produced closer than this, slightly higher magnification is achieved, at the expense of eye-strain and discomfort for the observer.

The **objective** lens has a long focal length. It produces a real image of a distant object at its principal focus. This intermediate image acts as an object for the **eyepiece** lens. The eyepiece is a short focal length lens which acts as a magnifying glass. The intermediate image is arranged to fall on the principal focus of the eyepiece lens, so the final virtual, magnified image is at infinity, as shown in Figure 2.121.

It can be shown that when the telescope is in **normal adjustment**, the angular magnification (M) is given by:

$$M = \frac{f_o}{f_e}$$

Equation 1

You should note that:

- Separation distance between the lenses = $f_o + f_e$
- For large M, f_o must be **large** and f_e must be **small**.
- The final image is virtual and inverted.
- The angular magnification M is also given by

$$M = \frac{\beta}{\alpha}$$

- Since the final image is formed at infinity, the eye of the person using the telescope will be relaxed, as the ciliary muscles in the normal human eye will be relaxed. This means that the long periods of viewing needed in astronomical observation can be carried out with minimum eyestrain or discomfort. Such a telescope is also useful in a spectrometer. A set of cross-wires is placed internally, at the point where the principal foci of the two lenses coincide. Parallel rays entering the objective are thus focused on the cross-wires, which are themselves sharply focused by the eyepiece.

The eye ring

The above analysis describes the behaviour of rays of light through the two lenses of the telescope. However, such a telescope is always used in conjunction with another optical instrument, the human eye (or sometimes a camera). To form an image on the retina, rays of light passing through the telescope have to pass through the pupil of the eye. It is pointless having rays refracted through the telescope which fail to enter the eye because they are too widely separated to pass through the pupil. Ideally, all of the rays leaving the eyepiece should be confined to a circular area just equal in diameter to the diameter of the average pupil. If this circular area is narrower than the pupil diameter, an opportunity to maximise the operation of the telescope has been lost.

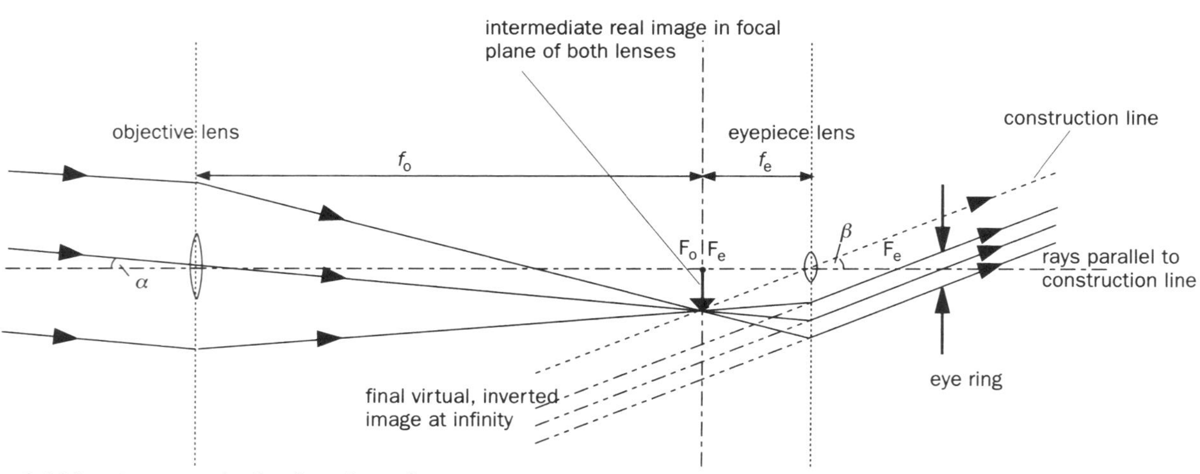

Figure 2.121 Astronomical refracting telescope

Figure 2.122 below shows the paths of **two** sets of parallel rays from points on opposite sides of a distant object. After refraction through the telescope, each set of rays emerges from the eyepiece lens as parallel rays. The circular area with diameter YZ is the smallest area containing all the rays from the object. This diameter should be the same size as the observer's pupil, and the observer's pupil should coincide with the position of YZ. The **eye ring** is the circular aperture placed at this position. It can be shown that, for a refracting astronomical telescope in normal adjustment:

$$\frac{\text{objective lens diameter}}{\text{eye ring diameter}} = \frac{f_o}{f_e} = M$$

Equation 2

You should also note that the position and diameter of the eye ring are exactly the position and diameter of the image of the objective lens formed by the eyepiece lens.

Aberrations

Refraction through lenses producing high magnification can lead to distortions or **aberrations** in the image.

Chromatic aberration

The amount of refraction produced by a lens depends on the colour of the light. The shorter the wavelength, the greater the refraction, so light from the violet end of the visible spectrum is refracted more than that from the red end. This means that when white light passes through a simple lens, some dispersion will occur, and the different colours will be focused at different distances from the lens, as shown in Figure 2.123. It follows that the focal length of a lens will be different for different colours of light.

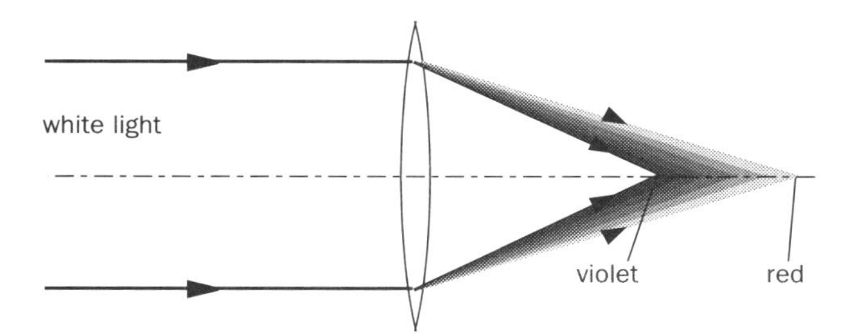

Figure 2.123 Chromatic aberration

This lens defect is called **chromatic aberration**. Optical instruments correct this by using **compound lenses** which contain several separate lenses made of different types of glass whose refractive indices are chosen so as to eliminate the separation of the colours. This makes the lens heavy and expensive, especially if the lens aperture is wide, as is necessary for a telescope.

Spherical aberration

This is a distortion caused by the fact that lenses produce more refraction near their edges than at the centre. This is particularly marked at high magnifications and large apertures, which again causes problems for telescopes.

In view of these and other defects inherent in the design of refracting telescopes, large, high-power astronomical telescopes are usually of the reflecting type. Reflection of light does not induce any dispersion, so chromatic aberration can be eliminated, and using a parabolic reflector eliminates spherical aberration. In addition, grinding a large mirror to the necessary degree of accuracy is considerably easier and cheaper than grinding a lens large and powerful enough to do the same job.

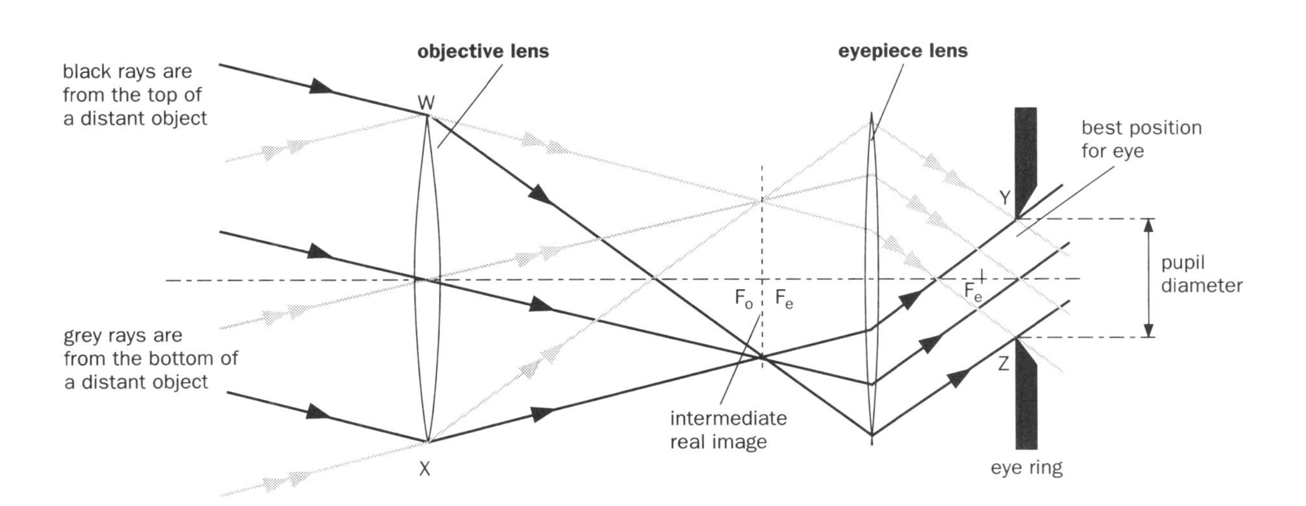

Figure 2.122 Position and diameter of the eye ring

Reflecting telescopes

The Newtonian reflector

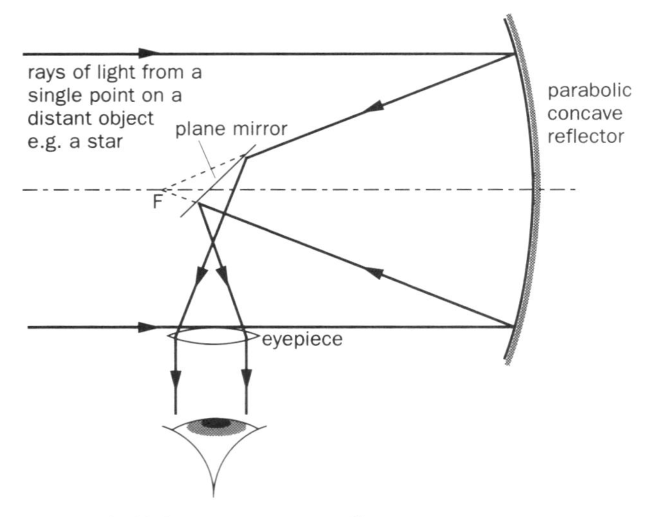

Figure 2.124 Newtonian reflector

The Newtonian reflecting telescope, illustrated in Figure 2.124, is the type most commonly employed for the small telescopes used by amateur astronomers. Light rays from a point on a distant star are reflected by the parabolic concave reflector towards its **principal focus**, F. These rays are deflected by the small plane mirror to an eyepiece, mounted at a right angle to the telescope axis. The image is then viewed through the eyepiece. The presence of the plane mirror has no appreciable effect on the quality of the image.

Cassegrain reflector

In the Cassegrain reflecting telescope shown in Figure 2.125, the light rays from the main concave reflector are reflected back along the axis of the telescope by a small convex mirror. These rays pass through a hole in the centre of the concave mirror, where they enter the eyepiece. The effect of the convex mirror is to increase the overall focal length of the telescope by a factor of about two.

Reflecting telescopes have been constructed with mirror diameters up to 6 m, and larger ones are under construction. Such large apertures mean that enormous amounts of light energy can be captured, enabling very faint, distant objects to be studied. Astronomical telescopes are normally mounted on top of mountains, away from industrialised countries, to reduce the distorting effect of the Earth's atmosphere. The ultimate platform for an astronomical telescope is in space, and the Hubble telescope, launched in 1990 by NASA, is expected to greatly enhance our ability to view the most distant and faint objects in the universe.

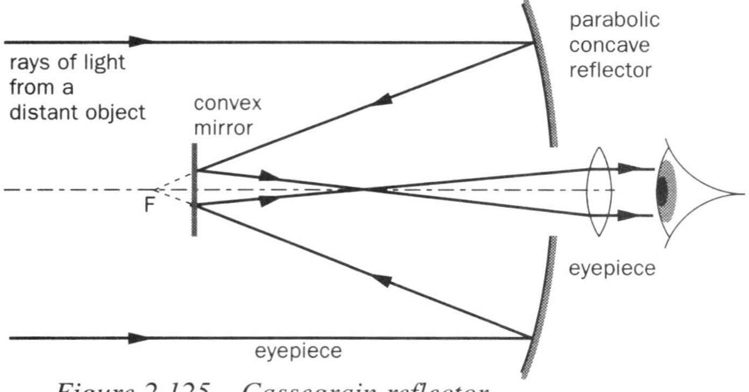

Figure 2.125 Cassegrain reflector

Image detection

In most simple optical telescopes the eye is the detector, giving the advantage of direct observation, but with the disadvantages of being insensitive at low intensities, and not providing a permanent record of the image. Photographic film provides a permanent record, and can work at much lower light intensities than the eye. Film effectively integrates the light falling on it, so that the longer the exposure time, the brighter the image, provided the telescope can track the source accurately. Film, however, is still not an efficient detector, as only a few percent of photons are recorded. In electronic detectors, a photosensitive material which emits electrons replaces the film. The electrons released can be accelerated through a *photomultiplier*, releasing many more electrons which are then focused on a phosphor screen to produce an image up to a million times brighter than the original. Charge coupled devices are increasingly popular for image production. These can be produced on a single chip of silicon, divided into a grid of hundreds of thousands of elements or pixels. Electrons liberated by the arrival of a photon on an element are stored by it, and build up a charge proportional to the number of photons detected (up to 60 or 70% of those arriving). After a chosen time, the elements are scanned by computer, and the charge recorded. The data can then be used to create an image, which can be further manipulated and enhanced electronically.

Radio telescopes

Light is not the only form of electromagnetic radiation. Indeed, it is not even the most important, from the point of view of the astronomer, since a great deal of information about the stars and planets would be missed if only the visible spectrum were studied. The Earth is bathed in radiation from across the whole of the electromagnetic spectrum, and radio astronomy is concerned with that part of the spectrum with wavelengths longer than that of light, mainly the wavelengths between about 1 mm and 30 m. The radio telescope at Jodrell Bank in Cheshire is a good example of an instrument designed to collect data in the radio frequency range.

Electromagnetic radiation in these wavebands is caused in two ways:

- Thermal radiation is the emission of electromagnetic radiation when fast-moving free electrons are influenced by the electrostatic field of other ions, as happens in the hot gas clouds around stars
- Synchrotron radiation is produced when high-speed electrons are accelerated by a magnetic field. The nature of the forces producing this acceleration will be discussed in Section 3.5 (Electromagnetic induction).

The classic radio telescope is a concave dish, with a detector mounted at its principal focus. The detector is in the form of a dipole antenna, a metal rod in which the electrons are made to vibrate by the incoming electromagnetic radiation. A TV satellite dish works on this principle. The intensity of the energy received is very low, so large diameter dishes and high-gain amplifiers to boost the weak signal are necessary. Large diameters also increase the resolving power of the telescope (see below). In order to be able to 'point' at different parts of the sky, and compensate for the rotation of the earth, most radio telescopes are steerable. However, this requirement puts a limit on the diameter, as dishes of more than 100 m in diameter would tend to sag under their own weight.

Larger diameters can be achieved by using a fixed mirror, but such a dish can only 'sweep' across the sky as the earth rotates. Some novel solutions have been found to overcome this limitation of fixed dishes:

- At the Arecibo Observatory, Puerto Rico, a fixed 305 m dish is built into a natural depression in the ground. Some 'pointability' is achieved by the ability to move the position of the receiver, which is suspended on cables over the dish.
- The RATAN 600 is a novel design using a ring of movable metal plates about 10 m high. The diameter of the ring is 600 m, and the position of each plate is computer-controlled to reflect radio waves from selected parts of the sky onto the receiving antenna.

Dish construction

The surface of the dish needs to be an efficient reflector of the wavelength being received. One requirement is that the surface of the dish appear 'smooth' to the radiation. If the wavelength of the radiation were, say, 1 m, then the dish would appear smooth if the smallest imperfections were around 5 cm or so. Consequently, most radio telescopes are

One of the many radio telescope dishes seen at Jodrell Bank in Cheshire. The incident radio waves from outer space are brought to a sharp focus after reflection from the concave dish

constructed as an open mesh rather than having a closed, shiny surface, to save weight. Microwave astronomy involves the study of wavelengths less than 1 mm, which requires very smooth reflecting surfaces. Such dishes can work efficiently with much smaller diameters since the resolution is enhanced at shorter wavelengths.

Amplification

The energy collected by the dish of a radio telescope is focused on the receiver where it produces tiny electric currents. These currents must be greatly amplified, which introduces many problems. The amplifier must be extremely selective in its ability to amplify only selected frequencies and reject the background 'noise'. Noise is defined as transmissions at unwanted frequencies, but what is unwanted noise to one astronomer could be the subject of a Ph.D. thesis to another. In addition, high-gain amplifiers can amplify the noise created by the random motion of electrons present in the conducting wires of the apparatus. This can be minimised by cooling the amplifiers with liquefied gases, but the plant and equipment required is expensive.

The 'image' in a radio telescope

A radio telescope does not produce an image in the conventional sense. Instead, the telescope 'scans' back and forth across a small region of the sky, and the intensity distribution of a particular wavelength is recorded from each point over the selected area. From this data, an 'image' of this region of the sky can be constructed by plotting the intensity on a diagram. The resulting intensity distribution pattern can be interpreted as a kind of 'contour map' of the strength of the source, so that centres of emission can be identified.

Resolving power

The resolving power of a telescope is a measure of its ability to distinguish between separate objects which are close together. Both optical and radio frequency telescopes need to have as wide a diameter as possible to maximise the amount of energy received from the distant objects. In addition, a wide diameter decreases the reduction in image sharpness in optical telescopes caused by diffraction. Light entering an aperture, such as the objective of a telescope, suffers diffraction. A circular hole produces a diffraction pattern consisting of a central bright spot, surrounded by alternating bright and dark fringes as shown in Figure 2.126. The effect of diffraction is to cause the image to be 'smeared'. Although a radio telescope does not produce an image in the same way, the diffraction of the radiation can cause the emissions from two sources close together to overlap and become indistinguishable.

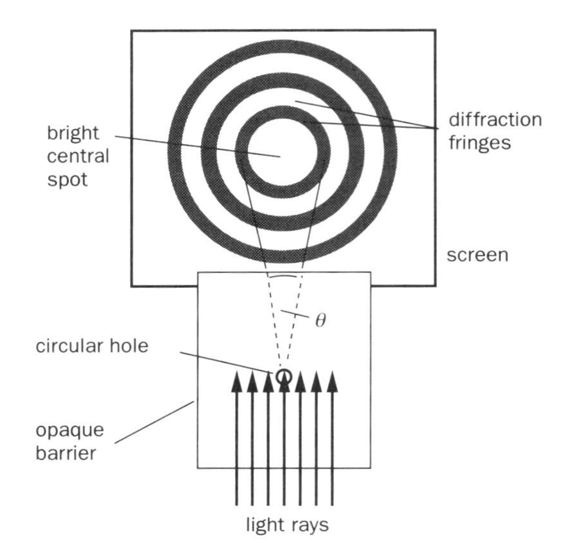

Figure 2.126 Diffraction by a circular aperture

It can be shown that the angle (θ) of the first dark ring produced by a hole of diameter (w) with light of wavelength (λ) is given by:

$$\sin\theta = \frac{1.22\lambda}{w}$$

Equation 3

For high resolution, θ should be as small as possible to reduce the smearing of the image. The hole diameter (w) should therefore be as large as possible, which is consistent with the need for large apertures to obtain bright images.

Limit of resolution

If two points on a distant object are close together, it follows that the images of these points seen through a telescope will be close together. If diffraction causes one image to be obscured by the smearing of the other, then we say that these objects are not **resolved** by the telescope, that is, they cannot be distinguished. *Rayleigh's* criterion for resolution of two points states that if their angular separation (θ) is equal to that given by $\sin\theta = \frac{1.22\lambda}{w}$ then they will just be resolved. If θ is less than this, they cannot be distinguished. Clearly, wider objectives reduce the size of this minimum angle, and improve the resolution of the instrument. Using a blue filter can also produce slightly sharper images in optical instruments, since $\sin\theta$ is proportional to λ. Blue light has a shorter wavelength than other colours, so if all other colours are filtered out, the image formed from only the blue light will be sharper as it will suffer less smearing than other colours.

Guided examples (1)

1. The refracting telescope used by an amateur astronomer has an objective lens of diameter 100 mm and focal length 900 mm. When it is used in normal adjustment, the telescope has a magnifying power of 15. Calculate
   (a) the focal length of the eyepiece, and the separation of the lenses
   (b) the position and diameter of the eye ring
   (c) the smallest crater size that the astronomer would be able to distinguish on the Moon given that the average wavelength of visible light is 550 nm, and the distance from the Moon to Earth is 3.8×10^8 m.
   (d) In order to see a slightly clearer image, the astronomer uses a coloured filter. State which colour filter would give the greatest improvement and explain your answer.

Guidelines

(a) Since the telescope is in normal adjustment, use the standard magnification formula (Equation 1) to calculate the magnification from the focal lengths of the objective and eyepiece lenses. Separation of the lenses is equal to the sum of the focal lengths.

(b) The eye ring is situated at the position of the **image** of the **objective lens** formed by the **eyepiece lens**. Use the standard lens formula to calculate the distance of this image from the eyepiece lens.
Then use the **eye ring** formula (Equation 2) to calculate the diameter of the eye ring.

(c) Use the **resolving power** formula (Equation 3) to calculate the minimum angle of resolution at this wavelength and objective lens diameter.

(d) Consider the resolving power formula (Equation 3). Would a shorter or longer wavelength produce greater resolution?

2. The Effelsberg radio telescope in Germany has a 100 m diameter dish.
   (a) Calculate the theoretical resolving power of this telescope when it is receiving radio waves of 0.2 m wavelength.
   (b) If the telescope is being used to study a galaxy 2.2×10^6 light years away, what is the minimum distance between two adjacent stars in the galaxy if they are to be resolved at this wavelength (i.e. their radio emissions at this wavelength will be observed as originating from two separate stars). 1 light year is the distance travelled by light in 1 earth year, 9.46×10^{15} m.

Guidelines

(a) Use the resolving power formula (Equation 3).

(b) Use $x = r\theta$
where x is the distance between the stars
r is the distance to the galaxy
θ is the resolving power of the telescope

Think carefully about the units!

Self-assessment

SECTION A
Qualitative assessment

1. An astronomical telescope consisting of two thin converging lenses is set up in **normal adjustment** to view a distant planet.
   (a) Draw a labelled ray diagram showing the paths through the telescope of three rays from a non-axial point on the planet.
   (b) Describe the position, magnification, orientation and nature of the **intermediate** image and the **final** image formed by the telescope.
   (c) What is the **eye ring** and where should it be situated?
   (d) Why is it important to match the diameter of the eye ring to that of the pupil in the observer's eye?
   (e) For an astronomical telescope in normal adjustment, what is the relationship between the diameter of the objective lens and the diameter of the eye ring?
   (f) Why is it advantageous to use the telescope in **normal adjustment**?
   (g) State the formula for calculating the **magnifying power** and the **lens separation**, based on the focal lengths of the lenses.

2. Explain, with the aid of a diagram, what is meant by **chromatic aberration**. Describe how the

Self-assessment *continued*

effect of this defect is minimised in practical optical instruments.

3. Describe **spherical aberration**.
4. Describe, with the aid of a diagram, how an image of a distant star is formed by a reflecting optical telescope of
   (a) the **Newtonian** type
   (b) the **Cassegrain** type.
5. State three advantages which a reflecting telescope might have over a refracting telescope of similar magnifying power.
6. Explain what is meant by the **resolving power** of a telescope.
7. Explain how the resolving power of a telescope depends on
   (a) the diameter of the objective
   (b) the wavelength of the radiation received.
8. Explain why the dish of a radio telescope does not have to be of the same highly-polished quality as that of the reflector of an optical telescope.

SECTION B
Quantitative assessment

(***Answers:*** 2.8×10^{-7}; 6.6×10^{-6}; 1.0×10^{-3}; 6.0×10^{-3}; 0.072; 0.091; 1.3; 7.5; 12; 2.3×10^{3}; 1.2×10^{5}; 5.4×10^{15}.)

1. An astronomical telescope consists of two separated converging lenses of focal lengths 120 cm and 10 cm. If the telescope is used in normal adjustment, calculate:
   (a) the separation of the lenses
   (b) the magnifying power of the instrument
   (c) the minimum objective lens diameter required so that the maximum amount of light enters the eye of an observer whose pupil is 6 mm in diameter, when placed at the eye ring position.
2. An astronomical telescope has an objective lens of focal length 600 mm and diameter 45 mm, and an eyepiece lens of focal length 80 mm. When it is being used in normal adjustment, calculate:
   (a) the magnifying power
   (b) the eye ring position
   (c) the diameter of the observer's pupil if the maximum possible amount of light enters his eye.
3. Calculate the smallest distance between two points on the surface of the Moon which are reflecting light of wavelength 600 nm if they are to be distinguishable through an optical telescope with an objective lens diameter of 120 mm. (Earth–Moon distance = 3.8×10^{8} m.)
4. Calculate the theoretical resolving power of:
   (a) the Arecibo Observatory radio telescope (diameter 305 m) when receiving radio waves of wavelength 25 cm
   (b) the Yerke's Observatory refracting telescope (objective lens diameter 101.6 cm) when receiving light of wavelength 550 nm, and allowing for a 10 × reduction in resolving power due to atmospheric distortion
   (c) the Hubble space telescope (reflector diameter 2.4 m) when receiving light of wavelength 550 nm.
5. The Mauna Kea optical reflecting telescope in Hawaii has an aperture of 224 cm. Assuming that the resolving power is reduced by a factor of ten due to atmospheric distortion, and that visible light wavelengths lie in the range 400 to 700 nm, calculate:
   (a) the size of the smallest detail that can be resolved on Mars when it is at a distance of 56 million km
   (b) the minimum distance between two adjacent stars in the Draco galaxy (2.6×10^{5} light years away) if they are to be resolved as two separate stars (1 light year = 9.46×10^{15} m).

Past-paper questions

1

displacement

time

Graph A

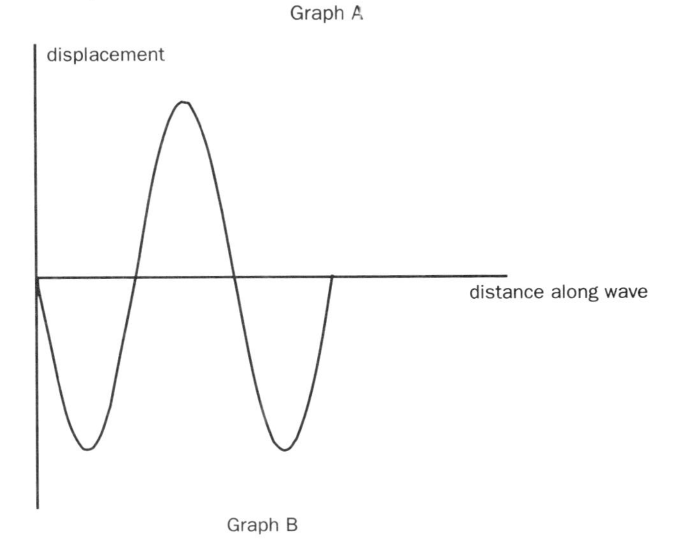

Graph B

Graph A shows the variation of displacement with **time** at a point on the path of a progressive wave of constant amplitude.
Graph B shows the variation of displacement with **distance** along the same wave at a particular instant of time.

(a) Sketch (with labelled axes) both graphs in your answer book and mark on an appropriate graph:
  (i) the wave amplitude a
  (ii) the period T of the vibrations producing the wave,
  (iii) the wavelength of the wave, λ,
  (iv) two points, P and Q, which are always π out of phase.

(b) The intensity of a wave is defined as the energy passing perpendicularly through unit area per second. Use this idea to show that, for a point source emitting waves in a uniform medium, the intensity of the wave at any point is inversely proportional to the square of the distance from the source to the point. You may assume that the medium does not absorb any of the wave energy.

(NEAB, 1994)

2 A ray of light travelling through air strikes the middle of one face of an equilateral glass prism. State what happens to the frequency, wavelength and speed of the light as it goes from the air into the glass.

The angle of incidence is 30° and the refractive index of the glass is 1.5. Draw a diagram to show the path of the ray through the prism. Calculate the angle between the ray leaving the prism and the normal to the glass surface at the point at which the ray leaves.

(Lond., 1994)

3 Refraction occurs when waves pass from one medium into another of different refractive index. The table below gives data for sound waves and light waves.

Type of wave	Speed in air $m\ s^{-1}$	Speed in water $m\ s^{-1}$
Sound	340	1400
Light	3.00×10^8	2.25×10^8

(a) Plane waves travelling in air meet a plane water surface.
  (i) Use the data in the table above to calculate the angle of refraction in the water for light waves and for sound waves if the angle of incidence of the plane waves is 10°.
  (ii) Explain what happens in each case when the angle of incidence is 15°.

(b) Submarine detection helicopters are fitted with transmitters which send out sound waves.
The reflected signals identify the location of the submarine. Explain why it is common for this transmitter to be suspended from the helicopter by a wire so that it is under water.

(NEAB, 1992)

4 Figure 1 below shows a cross-section of a pool, 2.4 m deep, with a lamp at the bottom. Light from the lamp is reflected at the surface of the pool and illuminates the bottom of the pool. Figure 2 shows how the intensity of the light at the *bottom* of the pool at night time varies with distance from the lamp.

Draw a diagram of the pool and use this to show how the lamp illuminates the bottom.

Explain why
(a) the light intensity is a maximum around the circumference of a circle of diameter AB and centre L
(b) the light intensity gets smaller as we move from this circle towards the lamp.

(Lond., 1992)

Fig 2

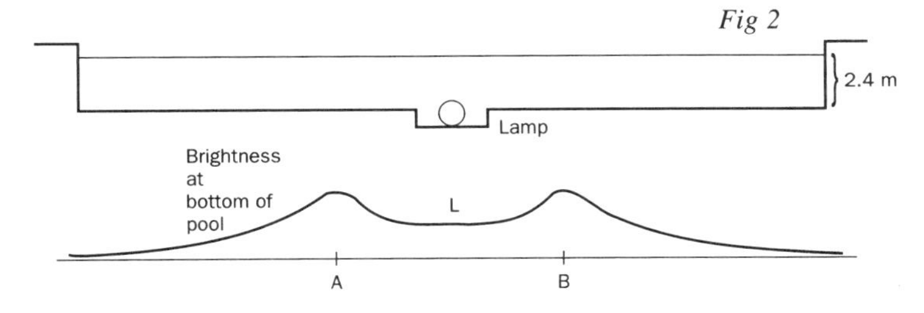

5 L is a loudspeaker emitting sound waves of a single frequency and P is a metal plate. A small microphone, M, is positioned between L and P and connected to the Y-input of a CRO with the time base turned off. As M is moved towards L, the trace height passes through a series of minimum and maximum readings. The distance moved by M between adjacent minima is 0.100 m.

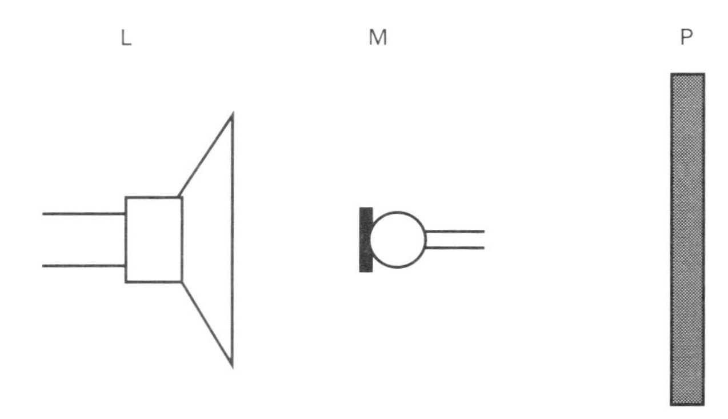

(a) (i) Explain the variation in trace height and calculate the frequency of the sound waves.

(ii) The frequency of the signal is doubled and the experiment repeated. What will be the distance between adjacent minima? Explain your answer.

Speed of sound waves in air = 340 m s^{-1}.

(b) P is replaced by a different material which reflects a smaller proportion of the incident waves. Describe how the variation in the trace would be different from that described in (a) (i) when M is moved towards L.

(NEAB, 1992)

6 (a) Using a labelled diagram in each case

(i) explain what is meant by *short sight*

(ii) show how it is corrected.

(b) A man can see clearly only objects which lie between 0.50 m and 0.18 m from his eye.

(i) What is the power of the lens which when placed close to the eye would enable him to see distant objects clearly?

(ii) Calculate his least distance of distinct vision when using this lens.

(NEAB, 1991)

7 (a) The velocity v of a transverse wave on a stretched string is given by $v = \sqrt{\frac{T}{\mu}}$ where T is the tension and μ is the mass per unit length. Show that the equation is dimensionally correct.

(b) (i) Derive an expression for the fundamental frequency of a vibrating wire.

(ii) Describe how you would use a sonometer to determine experimentally how the fundamental frequency of a vibrating wire depends on **one** of the variables in the expression (b) (i) above.

(c) (i) Write down an expression for the frequency of the nth overtone of a vibrating wire of fundamental frequency f.

(ii) Why does the note sounded by a vibrating sonometer wire sound different when it is plucked in the middle from when it is plucked at the end?

(d) The mass of the vibrating length of a sonometer wire is 1.20 g and it is found that a note of frequency 512 Hz is produced when the wire is sounding its second overtone. If the tension in the wire is 100 N, calculate the vibrating length of the wire.

(WJEC, 1990)

8 A church organ contains many pipes, which are not all made the same way.

(a) The basic design of one type of pipe, used in organs, is shown in Figure 1 below.

Air is blown in through the end P and passes a sharp lip L. The effect is to create a disturbance at Q, near the end of the trapped column of air in the pipe. The other end of the pipe is closed by an adjustable plunger at R.

(i) Indicate why stationary waves are set up in the pipe. Assume that the pipe behaves as if the end Q is open.

(ii) Draw two labelled sketch-graphs for the stationary wave at the fundamental frequency of the pipe, the first showing the variation in amplitude of the motion of air particles between Q and R, and the second showing the variation of air pressure between Q and R.

(b) In another type of pipe used in an organ, air enters in a similar way but there is no plunger closing the pipe at R.

(i) Describe and explain the difference in the fundamental notes produced by two pipes of equal length, one being the closed pipe of the type described in part (a) and the other the open pipe in part (b).

(ii) Frequencies other than the fundamental can be produced when a pipe is blown. State what series of frequencies are possible, using both closed and open pipes, and explain briefly how they arise in each case.

(c) The limited space available for a particular organ

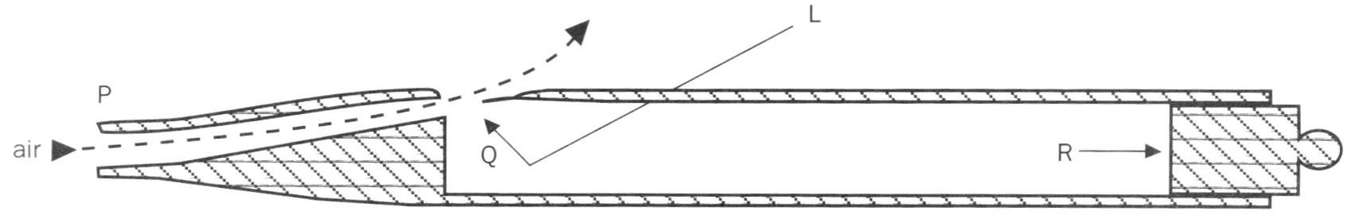

Fig. 1

means that the longest pipe which can be fitted has an effective length of 3.0 m. Calculate the lowest frequency which can be played on this organ.

(Velocity of sound in air = 330 m s^{-1})

(O & C, 1993)

9 A piano manufacturer's handbook states: 'Each string in the piano is normally stretched to a *pressure* of about 90 *kilograms*'.

Re-write this sentence and correct the two words that are in *italics*. Give reasons for the changes that you make.

The top note on the piano has a frequency of 3520 Hz. The string which produces this note is made of steel wire 6 cm long and 0.9 mm in diameter. Calculate the speed of waves in the wire and the mass per unit length of the wire. Use these answers to calculate the value for a numerical quantity that can be compared with that given in the manufacturer's statement.

(Density of steel = 7700 kg m^{-3}.)

(Lond., 1992)

10 Two identical progressive waves, travelling in opposite directions, will form a stationary wave where they superpose. Explain with the aid of diagrams how two such progressive waves combine when they meet (i) in phase, (ii) in antiphase.

Explain the terms *node* and *antinode*. How do you account for their formation?

Identify one situation that makes use of stationary waves. State whether these waves are longitudinal or transverse.

(Lond., 1993, part question)

11 (a) In a camera, the lens has a focal length of 50 mm and focused images can be obtained for objects at distances from infinity down to 1.0 m. Find the minimum and maximum separation between the lens and film needed to provide this facility.

(b) A life-size image of an insect is to be recorded on film.
(i) Find the lens–film separation required using a lens of focal length 50 mm.
(ii) State whether the camera in (a) would be suitable for this purpose. Give a reason for your answer.

(NEAB, 1994)

12 (a) A spectrometer has been set up to produce and receive parallel light. A diffraction grating has been mounted on the table and adjusted for normal incidence. The value of the grating spacing is **not** available to you. You are provided with a sodium light source of known wavelength and the usual apparatus of a physics laboratory. Describe how you would use this apparatus to determine the wavelength of light from a second monochromatic source.

(b) A spectrometer and diffraction grating were set up as described in (a). The emission spectrum of an element was viewed in the **second** order and five visible lines were observed. The angular positions of these lines measured against the scale on the spectrometer are shown in the table below. The angular position of the zero order = 126.40°; number of rulings per unit length = 4.5 × 10 m^{-15}.
(i) State and explain one advantage and one disadvantage of using the second order spectrum instead of the first order spectrum.
(ii) Calculate the wavelength of the orange line from the angular position given.

(NEAB, 1994, part question)

spectral line	1	2	3	4	5
colour	violet	green	green	orange	red
angular position	147.94°	154.54°	154.83°	162.65°	168.37°

13 Two small loudspeakers, S_1 and S_2 are positioned 1.5 m apart in a large room and are connected to the same signal generator. The loudspeakers emit a note of frequency 3400 Hz.

A microphone connected to an oscilloscope is placed at P which is equidistant from the two speakers as shown in the diagram. As the microphone is moved along the line AB, the trace height passes through a series of maxima and minima, with a maximum at P.

Speed of sound in air at room temperature = 340 m s^{-1}

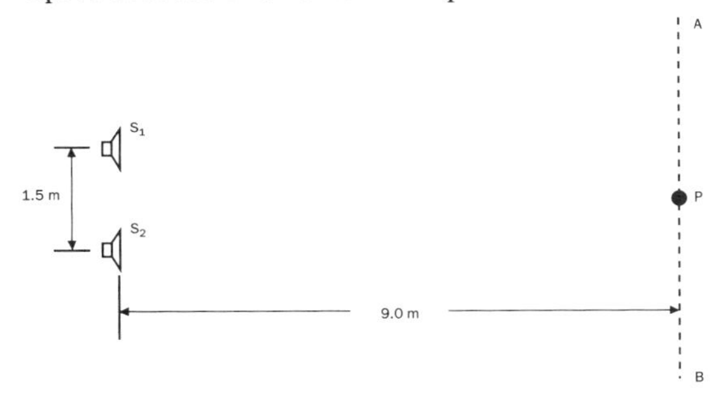

(a) (i) Explain the variation in trace height (intensity).
(ii) Calculate the approximate distance from P to the first minimum.

(b) Explain how you would expect the distance you have calculated in (a) to change if the temperature of the air in the room falls, assuming that the speed of sound waves in air decreases with decrease in temperature.

(NEAB, 1994)

14 A diffraction grating shows five orders of spectral lines.

(a) What is the spacing between consecutive lines of the grating? (Take the red wavelength to be 700 nm.)

(b) A green line has a wavelength of 552 nm and a red line one of 690 nm. At what angle do the lines overlap?

(WJEC, 1989)

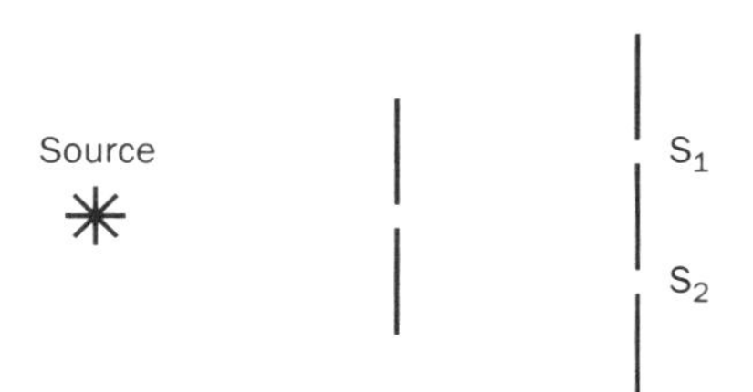

Screen

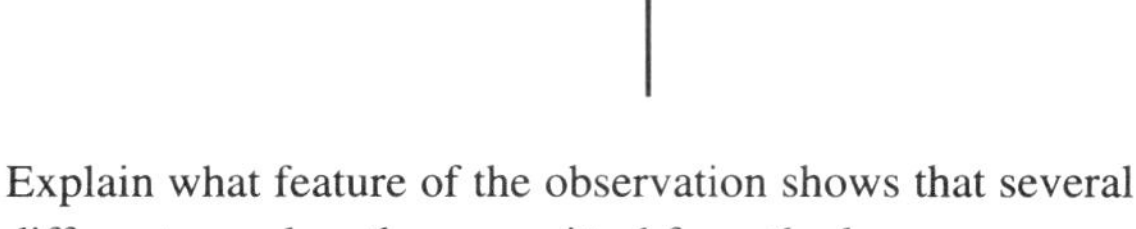

15 (a) A plane diffraction grating of width 2 cm is set up on a spectrometer in the usual way. It produces the first-order spectrum from a monochromatic source of wavelength 5.9×10^{-7} m at diffraction angle $\theta = 8.48°$. Find the number of lines on the grating.

(b) White light is now substituted and the telescope set at zero diffraction angle. Without further calculation, describe what you would expect to see through the telescope as it is slowly rotated.

(WJEC, 1993)

16 The diagram above shows the arrangement known as Young's slits. Describe what is seen on the screen when a source of monochromatic light is used.

Describe and explain what is seen on the screen when the slits, S_1 and S_2 are each covered by a narrow strip of Polaroid and

(a) the strips have their preferred (i.e. polarisation) directions parallel to each other

(b) the strips have their preferred directions at right angles to each other.

(WJEC, 1992)

17 (a) A loudspeaker, X, is connected to a signal generator which is set to a frequency of 825 Hz. The cone of the loudspeaker oscillates with simple harmonic motion with an amplitude of 0.05 mm. What are (i) the maximum speed and (ii) the maximum acceleration of the cone?

(b) A second, identical loudspeaker, Y, is connected to the output of the same signal generator and placed so that the two loudspeakers face each other and are separated by 2.0 m. A microphone placed at A, midway between X and Y, detects a maximum audible signal.

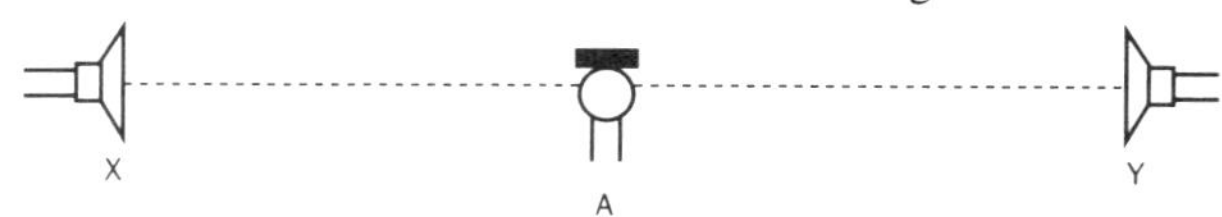

(i) Calculate the wavelength of the sound waves emitted by the loudspeakers.

(ii) Describe and explain what will be heard when the microphone is steadily moved 0.5 m towards Y. (Velocity of sound in air = 350 m s^{-1}.)

(O & C, June 1993)

18 When looking at a neon lamp through a diffraction grating it is noticed that the central maximum is a single image but that the subsidiary maxima are made up of several close but distinct images. From this it can be deduced that neon light consists of several wavelengths of light.

Explain what feature of the observation shows that several different wavelengths are emitted from the lamp.

If the diffraction grating is replaced with one having a smaller distance between the slits what effect would this have on the pattern?

The helium atom has an energy difference between two energy states that is almost equal to one of the energy differences for neon. Explain how you would demonstrate this fact by comparing photographs of the spectra of helium and neon.

(Lond., January 1992)

19 (a) (i) State **two** properties of the light sources necessary to produce an observable interference pattern for two overlapping beams of light.

(ii) Explain what is meant by *path difference* when considering the interference of waves from two sources and its relevance in deducing the positions of maxima and minima of light intensity.

(b) In a Young's slits experiment to determine the wavelength of monochromatic light the fringe separation y is given by

$$y = \frac{DX}{d}$$

(i) State what each of the symbols on the right hand side of the equation represent.

(ii) Suggest suitable values for d and D when light of wavelength 7×10^{-7} m is used.

(iii) Draw a diagram showing the arrangement you would use when carrying out the experiment.

(iv) Explain carefully how you would measure each of the quantities y, D and d. State clearly the instrument you would use in each case and the steps you would take to ensure that the result is as accurate as possible.

(c) (i) Draw a labelled sketch of the interference pattern formed when the monochromatic source used in (b) is replaced by a source which emits radiations of two wavelengths, one in the red and one in the blue region of the spectrum.

(ii) Explain how this experiment shows that the wavelength of red light is greater than that of blue light.

(d) A linear polarizer consists of two pieces of Polaroid, one of which is fixed in position while the other can be rotated relative to it. The polarizer is placed between the monochromatic light source and the slits. Describe the effect of rotating the moveable Polaroid on the interference pattern observed.

(AEB, 1989)

20 (a) Define *simple harmonic motion*.

(b) The displacement of a body undergoing SHM is given by $y = A \sin \omega t$.

(i) Explain what A and ω represent.

(ii) Draw a graph showing how y varies with t.

(iii) Underneath this, and using the same scales for t, sketch graphs showing how the velocity v and the acceleration a vary with t.

(c) A mass m hangs on a string of length l from a rigid support. The mass is pulled aside, so that the string makes an angle θ with the vertical, and then released.

(i) Show that the mass executes SHM, stating any assumptions made.

(ii) Prove that the period T of this SHM is given by $T = 2\pi\sqrt{\frac{l}{g}}$

(iii) A student times a simple pendulum to determine T. Does it matter how many oscillations are counted? Does it matter from where the counts are taken — the end or the middle of the swing? Give reasons.

(d) A piston in a car engine performs SHM
The piston has a mass of 0.50 kg and its amplitude of vibration is 45 mm. The revolution counter in the car reads 750 revolutions per minute. Calculate the maximum force on the piston.

(WJEC, 1990)

21 A bathroom is a cavity which produces resonance of sound with very little damping. This has the effect of improving the sound made by some singers.

(a) What is heard by a singer which suggests that a bathroom gives rise to *resonance*?

(b) (i) State what is meant by *damping*.

(ii) Give **one** reason why there is very little damping of sound waves in a typical bathroom.

(c) A small bathroom has walls which are 3.0 m apart. Standing waves are set up in the air in the bathroom with vibration nodes at these walls.

(i) Sketch diagrams to represent the standing waves produced at the two lowest resonant frequencies.

(ii) Calculate the lowest resonance frequency assuming that the speed of sound in the bathroom is 340 m s^{-1}.

(AEB, 1993)

22 The sketch-graphs at the top of the page show the variation of amplitude with driving frequency for two different systems experiencing forced oscillations.

(a) Explain carefully what properties of the system determine the form of the curve in **each** case.

(b) Give **one** technologically useful or important example of an oscillatory system whose behaviour is represented by the curve in Fig 1 and **one** whose behaviour is represented by the curve in Fig. 2.

(WJEC, 1993)

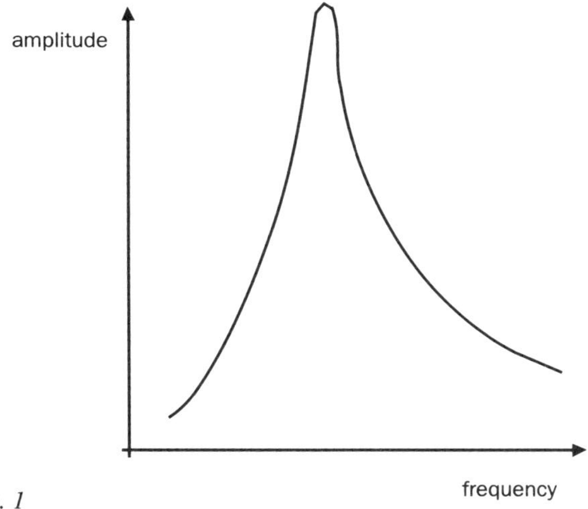

Fig. 1

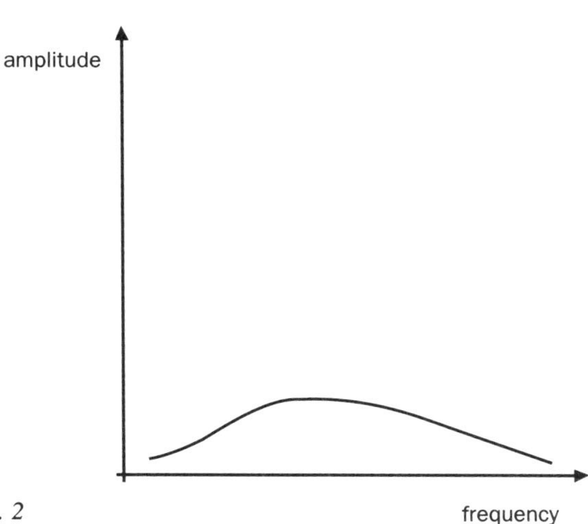

Fig. 2

23 An object moves along a straight line with simple harmonic motion of amplitude 10 cm and frequency $\frac{1}{\pi}$ Hz.

Sketch graphs to show the variation of

(a) its *acceleration* with *displacement*,

(b) its *displacement* and *velocity* with *time*.

(Plot both on the same time axis.)

Insert appropriate numerical values along the axes of each sketch graph.

(WJEC, 1994)

24 Define *simple harmonic motion*.

Explain why a clock can be designed around a system executing simple harmonic motion. Give one example of such an oscillating system.

A mass oscillates at the end of a spring. Describe the energy changes that occur when the mass goes from the lowest point of its motion, through the midpoint, to the highest point.

(Lond., 1994)

25 This question is about some aspects of the production and use of ultrasound. The following figure shows a circuit incorporating a transducer frequently used in the laboratory.

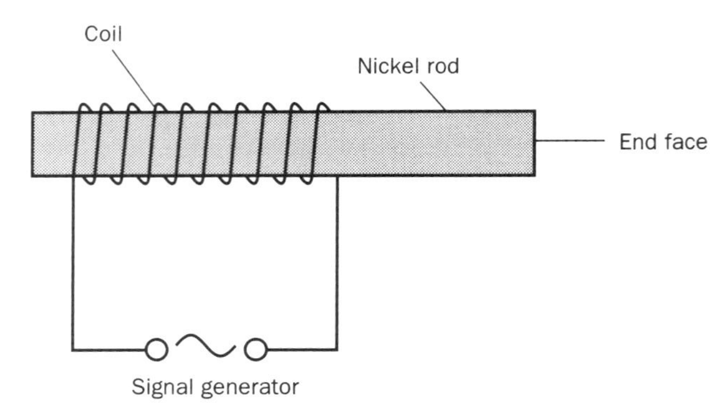

The transducer is a nickel rod which becomes shorter when subjected to a magnetic field. The rod is made to undergo mechanical oscillations, in which it expands and contracts, by the application of a varying magnetic field produced by current in the coil. These oscillations give rise to ultrasonic waves in the surrounding medium.

A certain transducer has a typical *natural frequency* of 40 kHz. A pulse of magnetic flux will cause the rod to undergo *damped longitudinal oscillations* at its natural frequency. Alternatively the rod can undergo *resonant oscillations* by applying a suitable sinusoidal magnetic field variation.

When used as a receiver, the received wave produces mechanical oscillations in the rod which are converted into voltage variations.

(a) (i) Explain the meaning of the terms in italics.
 (ii) Why does a single pulse of magnetic flux make the rod oscillate?
 (iii) Draw a sketch graph to show the oscillations of the 40 kHz transducer when it is subject to pulses at a frequency of 10 kHz. Include a suitable time scale for the time axis.
 (iv) State and explain how the dimensions of the rod will affect the intensity of the emitted ultrasound.

(b) The rod oscillates so that each end face is an antinode with a single node in between. Given that the speed of longitudinal waves in nickel is 480 m s^{-1}, show that the length of rod required is 6.0 cm.

(AEB, 1990)

SECTION

3 Electricity and Magnetism

3.1 Current electricity

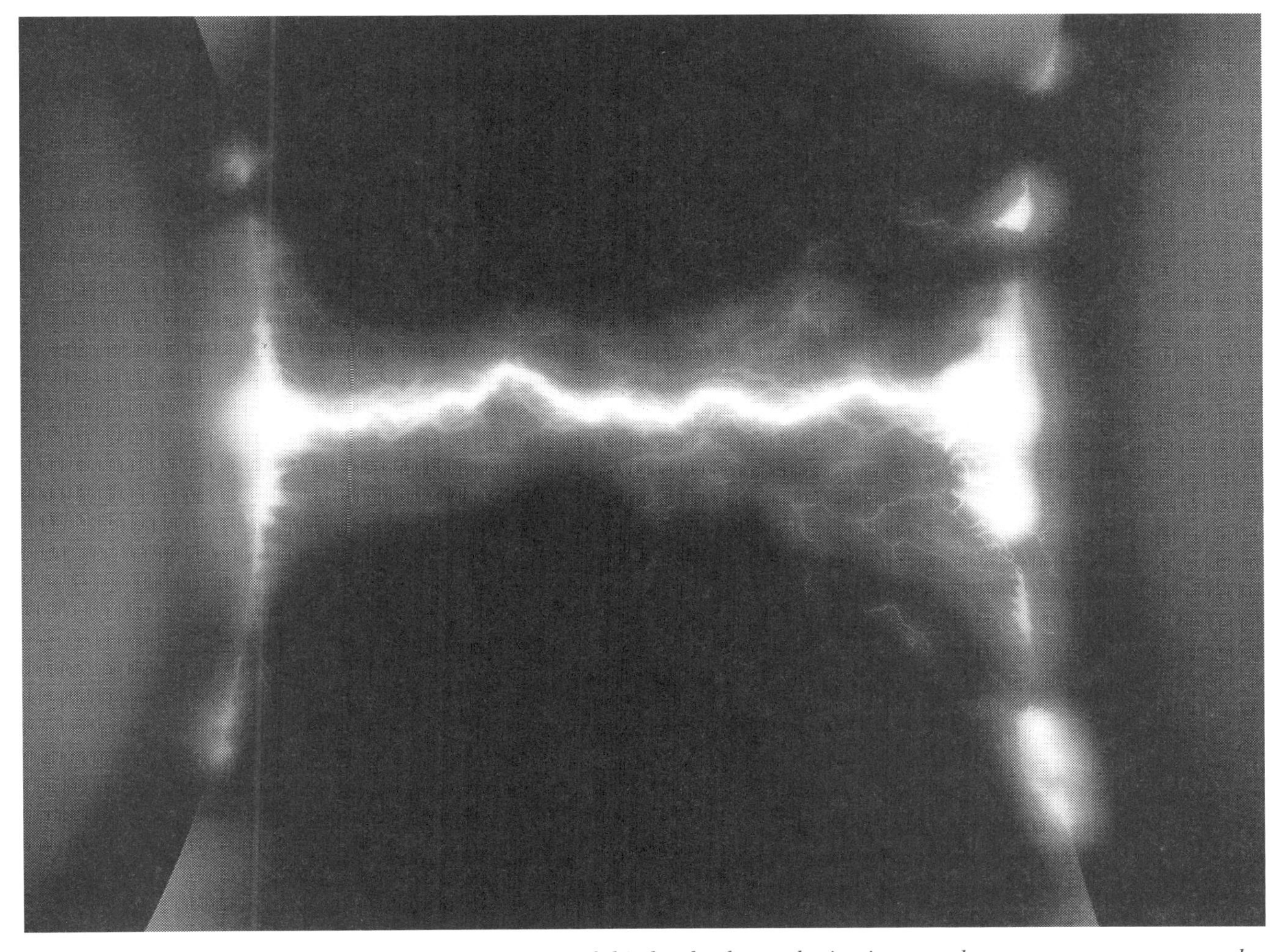

This photograph, which also appears on the cover of this book, shows the intricate and tortuous patterns created by an artificially generated electrical discharge between two metal spheres

Fundamental quantities

Electric charge

Common symbol	*Q*, *q* (special symbol for the charge on 1 electron or proton is *e*)
SI unit	coulomb (C)

Electric charge is a property possessed by protons and electrons which creates an ***electric field***, which is a region of space in which forces are exerted on other bodies with charge. **Like** charges ***repel***; **unlike** charges ***attract*** within this field. The charges carried by the electrons and protons within atoms are equal in magnitude, and opposite in sign. Protons carry **positive** charge, while electrons carry **negative** charge. Neutrons are electrically neutral. In an atom with equal numbers of protons and neutrons, the total positive and negative charges are equal and opposite, so the **resultant** charge is zero. Positive **ions** are atoms which have lost one or more electrons and so have an excess of positive charge. An atom which has gained one or more electrons is therefore called a negative ion. A lone electron removed from an atom is often referred to as a negative ion. **Ionisation** is the name given to the process by which electrons are gained or lost from atoms. It can be caused by collisions with atoms or particles, the effect of radiation, an electric field, chemical action, etc.

Protons are bound within the nucleus by very strong forces, and cannot escape from it, but electrons are relatively weakly attached to atoms by electrostatic forces. Consequently, they can be easily removed from their parent atoms. In conductors, electrons are free to move in the spaces between atoms, and so can carry charge from place to place within solid materials. In gases and liquids, the atoms of the material are free to flow. If these atoms are ionised, the liquid or gas can conduct electricity.

Charged particles such as electrons experience a **force** when in an **electric field**, which may be applied by connection of a cell or battery. Figure 3.1 shows a battery connected across the ends of a metallic conductor.

In the battery, **chemical** energy is used to separate electrons from atoms (work is done against the attractive forces binding the electron to the atom). One terminal becomes deficient in electrons (i.e. becomes **positively** charged) while the other acquires an excess of electrons (the **negative** terminal). You could picture an unconnected battery as having its negative terminal crowded with electrons trying to escape, while the positive terminal is populated by positive ions which are desperately seeking electrons. (The 'voltage' of the battery could be thought of as a measure of the strength of this 'urge'.) When the terminals are connected to the **conductor**, the free electrons in it are **repelled** by the excess negative charge on the negative terminal (and attracted to the positive terminal). They therefore flow slowly round the complete circuit from the negative terminal to the positive. The electrons then 'pair up' with the positive ions concentrated at the positive terminal.

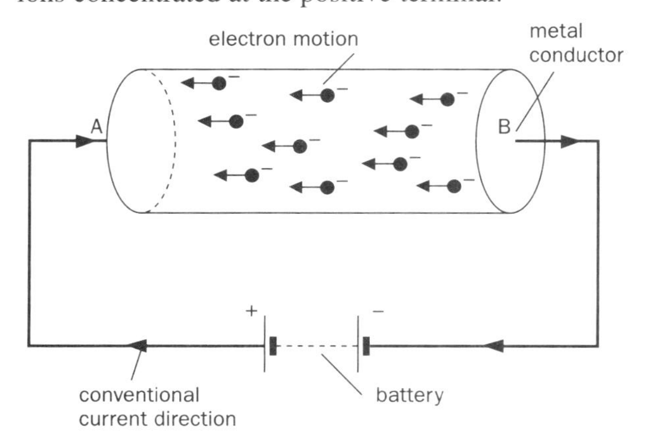

Figure 3.1 Electron flow and current direction

> You should note that the conventional current direction is **opposite** to the direction of electron flow.

If no energy were lost in the circuit, this process could go on indefinitely with no further input of energy, since the energy needed to separate the electrons would be returned to the battery when each electron arrived back at the positive terminal. Currents started in circuits made of superconducting materials continue without any source of energy in the circuit. In practice, the charge carriers (i.e. electrons) lose energy in **collisions** with the atoms in the conductor, whose ***internal energy*** is increased. The energy supplied to the charge carriers by the battery is therefore **dissipated** to the surroundings by the **heating** effect of the current.

The charge carried by one electron = -1.6×10^{-19} C. This quantity of charge (often given the symbol *e*) is believed to be a fundamental property of matter, in that all electrons (and protons) carry this **exact** amount of charge. (Quark theory suggests a smaller amount of charge, but that's another story...)

Electric current

Common symbol	*I*
SI unit	ampere (A)

> **Electric current** is defined as the **rate of flow** of electric **charge** (carried by charged particles).

A current is said to be steady if the quantity of charge crossing a given cross-section per second is constant.

The direction of **conventional** current is taken to be the direction of flow of **positive** charge. In liquids and gases (and some semiconductors) the positive ions physically move in this direction, while any negative charge carriers such as electrons flow the opposite way. In metals, the only charge carrier is the electron, so only negative charge flows. (*You just have to remember that electrons flow in the opposite direction to **conventional** current.*)

Demonstration of flow of charge carriers

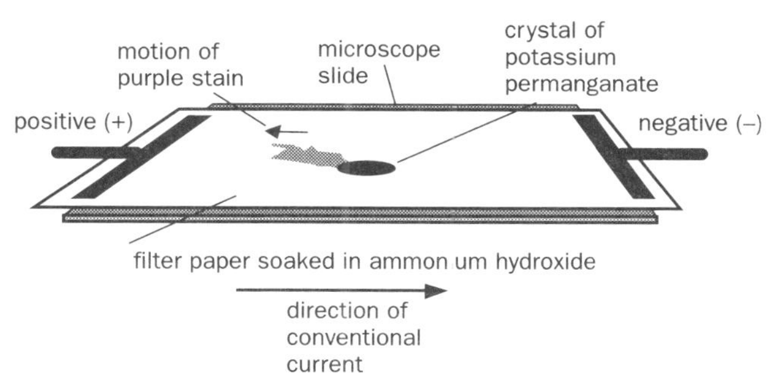

Figure 3.2 Charge flow demonstration

Figure 3.2 shows a simple laboratory demonstration of the movement of charge carriers in a material. It consists of a piece of filter paper soaked in ammonium hydroxide solution, placed on a microscope slide which acts as an insulating support. A small crystal of potassium permanganate is placed on the filter paper where it dissolves. A high-tension power supply is connected using crocodile clips. The purple stain is seen to move very slowly towards the positive terminal. Reversing the direction of the potential difference causes the purple stain to reverse its direction.

The ammonium hydroxide is weakly ionising, so the ions produced carry a small electric current through the filter paper. The potassium permanganate dissociates into positive and negative ions when it dissolves. The purple colour is carried by the **negative** ions, which are attracted to the positive terminal of the supply.

Referring to Figure 3.1, if (Q) is the quantity of charge flowing into (or out of) the conductor in time (t), then if the **current** (I) is **steady** (i.e. its value does not change with time), the magnitude of the current is given by:

$$I = \frac{Q}{t}$$

(A; C; s)

Equation 1

If the current is not steady, the **instantaneous** current is equal to the instantaneous **rate of flow of charge**:

$$I = \frac{dQ}{dt}$$

(A; C; s)

Equation 2

Current density

Common symbol	J
SI unit	ampere per metre^2 (A m^{-2})

This is a useful quantity to electrical engineers, as it allows them to calculate the cross-sectional area of conductor needed to carry a given current. Using too big an area is wasteful of material, while overheating could occur if the area were insufficient.

Current density (J) at any point in a current-carrying conductor is defined as the current (I) per unit cross-sectional area (A) at that point.

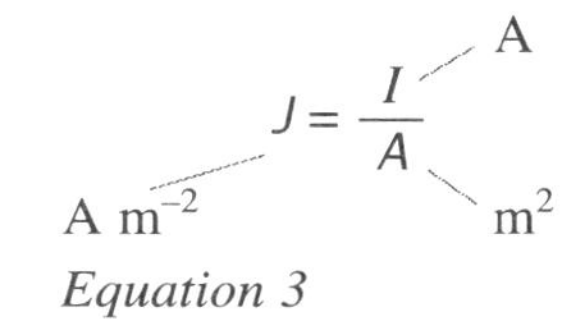

Equation 3

Definition of the coulomb

Since $Q = It$, we can define the **coulomb** as:

the quantity of electric charge that passes a given point in a circuit when a current of 1 A flows for 1 s.

Guided examples (1)

1. A wire which forms part of a circuit has a diameter of 0.50 mm. Calculate
   (a) the steady current through the wire if a charge of 400 μC passes a point in the wire in 8 ms;
   (b) the current density in the wire
   (c) the number of electrons per second passing through the wire (electron charge $= 1.6 \times 10^{-19}$ C).

Guidelines

(a) Use Equation 1 (take care with units
(b) Use Equation 3 after calculating the cross-sectional area of the wire
(c) Calculate the total charge passing per second, then divide this by the charge on one electron.

Guided examples (1) *continued*

2.

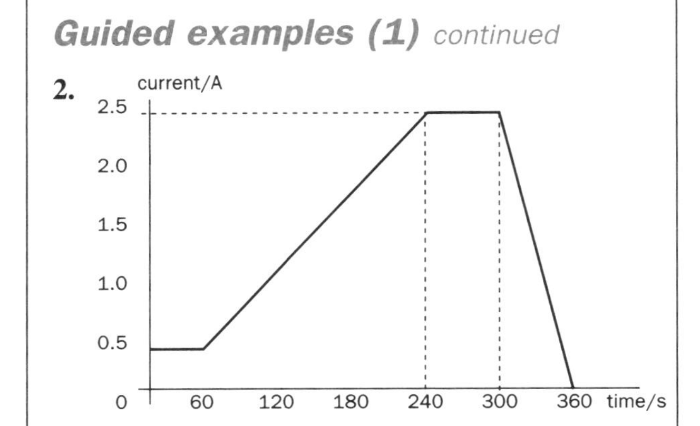

The current supplied by a particular battery varies with time as shown in the above graph. Calculate the total charge delivered.

Guidelines

(charge) = (current) × (time) is true for a **steady** current, or for the **average** current.

(average current) × (time) is equal to the **area** bounded by the graph and the time axis.

Conduction mechanisms

When there is current through a material, charge is transferred from one place to another. The nature of the charge carriers depends on the material.

Metals

In a metal, the atoms exist as positive **ions** surrounded by electrons, some of which are free to move within the crystal structure. At all temperatures above absolute zero, all the particles are in constant motion. The ions, held in place by strong binding forces, simply vibrate about their fixed positions, while those electrons that are free can move from place to place. Without an external electric field, the electrons move rapidly and randomly in all directions. There is no net motion in any particular direction. This is illustrated diagrammatically in Figure 3.3.

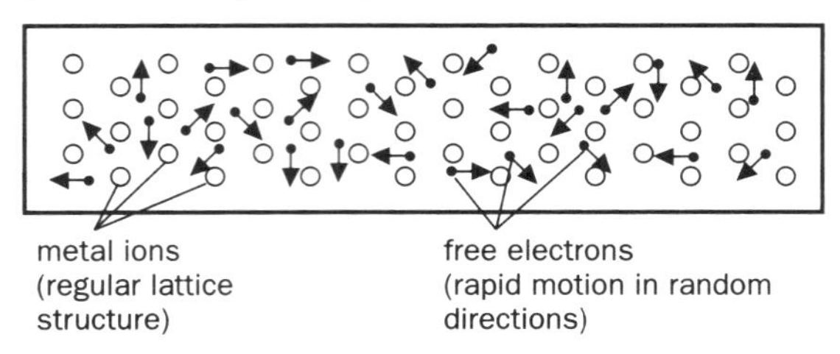

Figure 3.3 Metal sample with no applied field

When an electric field is applied, e.g. by connecting a battery, all the charged bodies (the positive nuclei, orbital electrons and free electrons) experience a **force**. The positive nuclei are attracted towards the more **negative** potential, while the electrons are attracted towards the more **positive** potential. However, only the free electrons can move under the action of this force. These electrons will be accelerated by this force, but their speed is limited by collisions with the metal ions. Consequently the resultant motion of the 'conduction' electrons is best described as a 'drift' towards the more positive potential, superimposed on the random motion which is always present at all temperatures above absolute zero, as shown in Figure 3.4.

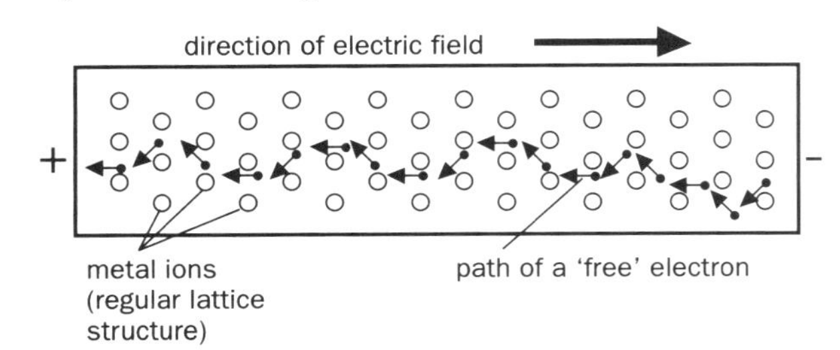

Figure 3.4 Single free electron drifting through a metal under the influence of an electric field

The free electrons are accelerated by the electric field, gaining kinetic energy from the field. In the collisions with the metal ions, this kinetic energy is converted to vibrational kinetic energy of the ions, with a resulting increase in the temperature of the metal. This energy is dissipated to the surroundings as heat energy. This is the heating effect of the electric current. Individual electrons never acquire a steady velocity, but the large numbers involved can be thought of as having an average 'drift velocity'. Electrical insulators are made from materials with very small numbers of charge carriers, which reduces the magnitude of the current to a negligible amount.

Calculation of electron drift velocity

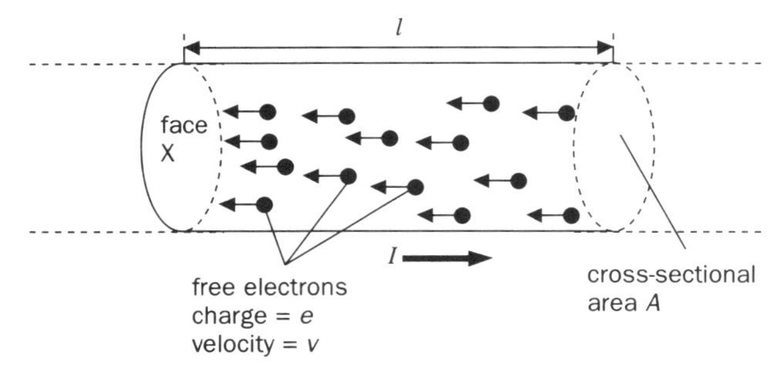

Figure 3.5 Free electron flow through a section of conductor

Figure 3.5 shows a section of a conductor of length (l), cross-sectional area (A) carrying a current (I). The current is carried by free electrons, each possessing charge (e), which can be assumed to have an average drift velocity (v) through the conductor. Assuming that there are (n) free electrons per metre^3 of the material:

volume of section $= Al$
number of free electrons in this section $= nAl$
total charge free to move $= nAle$

time taken for all electrons to emerge from face X is given by: (time) = (distance)/(velocity)

$\therefore$ time taken $= l/v$

since (current) = (charge)/(time):

$$I = \frac{Q}{t} = \frac{nAle}{l/v}$$

$$\therefore \quad I = nAve$$

(units: I: A, n: m^{-3}, A: m^2, v: $m\,s^{-1}$, e: C)

Equation 4

Equation 4 can be re-arranged to give the drift velocity:

$$v = \frac{I}{nAe}$$

showing that the drift velocity is proportional to the current, and inversely proportional to the number of charge carriers per m^3, the cross-sectional area of the conductor and the charge on the charge carriers.

Also, since current density $J = I/A$

$$J = nev$$

(units: J: $A\,m^{-2}$, n: m^{-3}, e: C, v: $m\,s^{-1}$)

Equation 5

Guided example (2)

Calculate the drift velocity of free electrons in a copper wire of cross-sectional area 0.8×10^{-6} m^2 carrying a current of 5.0 A. The number of free electrons per m^3 of pure copper is 1.0×10^{29}, and the charge on each electron is 1.6×10^{-19} C.

Guidelines

Use Equation 4. Simply substitute the given values into the re-arranged equation.

The drift velocity you calculate may seem very low. An electron in this conductor would take over 40 minutes to travel one metre! This speed should not be confused with the very high random velocities of the electrons at normal temperatures, nor with the speed of the electric field through the material. On connection to the battery, the electric field is established in all points of the wire virtually instantaneously, so that all electrons experience the force at virtually the same instant. This is why there is no noticeable delay between operating a light switch, and the lamp coming on, since the current starts almost instantly in every part of the circuit.

Semiconductors

These are materials with the following electrical properties:

- they contain around ten billion (1.0×10^{10}) times fewer charge carriers per m^3 than good conductors
- their resistivities are typically ten billion times greater than good conductors
- they have **negative** temperature coefficients of resistance (their resistance **decreases** as temperature increases) over certain ranges of temperature
- their resistivity may be light-sensitive.

Intrinsic semiconductors

These are **pure** materials whose method of atomic bonding does not liberate free charge carriers, except for the small number 'shaken' free by thermal vibrations at non-zero temperatures. For example, **silicon** has four electrons in its outer shell, and therefore four vacancies or 'holes', since the outer shell can contain eight electrons. When silicon atoms combine, they share their outer electrons, so each nucleus in the body of the material is surrounded by a full complement of eight outer-shell electrons. This arrangement is shown schematically in Figure 3.6.

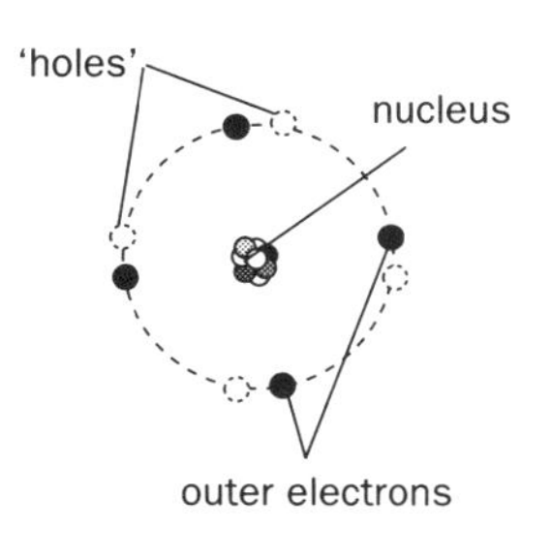

Figure 3.6 The bonding of atoms in pure silicon

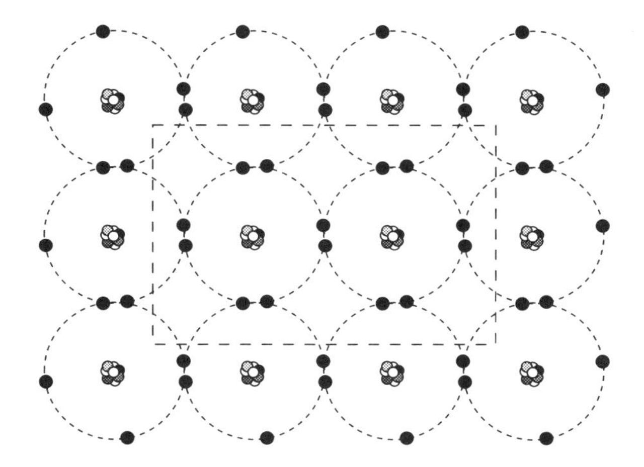

The only free electrons available for conduction are those shaken free by thermal vibrations of the atoms. It follows that conductivity can improve with rising temperature, if the effect of releasing more electrons compensates for the increased resistance associated with the increased vibration of the atoms.

Extrinsic semiconductors

The electrical behaviour of pure semiconductors like silicon and germanium would be only a mild curiosity, if it were not for the fact that adding carefully-controlled numbers of impurity atoms (a few parts per million) enables the characteristics of these materials to be manipulated. This process is called **'doping'**. There are two types of extrinsic semiconductor:

n-type semiconductors

To create an **n-type** semiconductor, **pentavalent** atoms are added to the pure semiconductor. These are atoms with **five** electrons in their outer shell, such as arsenic and phosphorus (group 5 in the periodic table), which therefore have valency 5. Only four of these electrons are needed for bonding the pentavalent atom to the silicon atoms, so each impurity atom contributes a free electron. The impurity atom is referred to as a **donor** atom. This arrangement is shown in Figure 3.7.

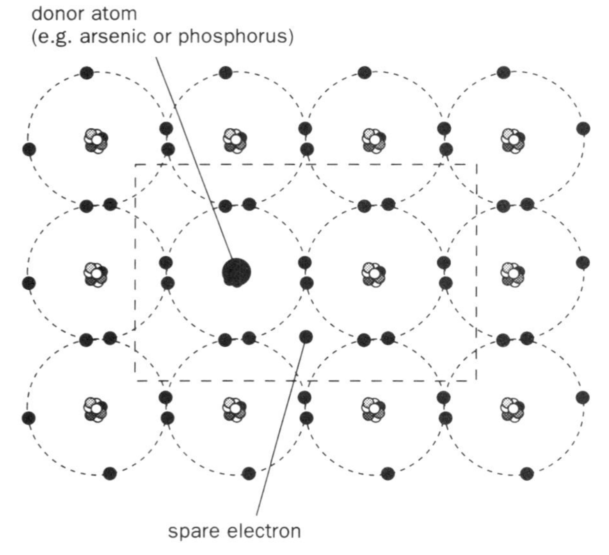

Figure 3.7 n-type semiconductor

In an n-type semiconductor, the majority charge carriers are **negatively** charged electrons, including those liberated by thermal vibrations and those donated by the impurity atoms.

p-type semiconductors

In this case, the doping atom is **trivalent**, coming from group 3 of the periodic table, with three outer-shell electrons. Examples are gallium and indium. Such an atom is unable to supply the necessary four electrons to complete the bonding with the semiconductor atom, so one neighbour-bond is left incomplete. This creates a vacancy, which is usually called a 'hole'.

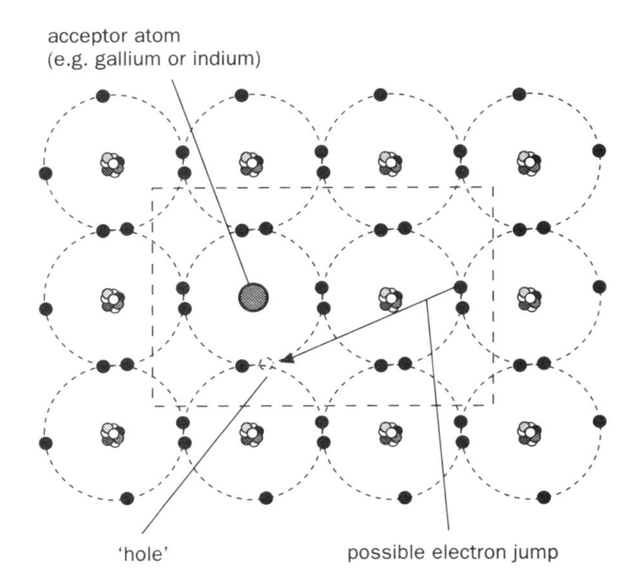

Figure 3.8 p-type semiconductor

Under the influence of an electric field, electrons from neighbouring atoms may jump across to fill the 'hole', leaving behind a hole in their own bond, as shown in Figure 3.8. As this process is repeated, the hole appears to drift through the material, in the direction of the applied electric field. This effect can be seen in a traffic jam – if one car pulls out of the queue it leaves a vacancy; if each car behind moves forward to fill the gap, the 'hole' can be seen to move rapidly backwards along the queue, while there is little real movement of the waiting cars.

The hole thus behaves as a **positive** charge carrier, and p-type semiconductors have the characteristics of materials in which the majority charge carriers are positive.

Potential difference (voltage)

Common symbol	V, v
SI unit	volt (V)

One of the difficulties with the subject of electricity is that we cannot see the charge flowing around a circuit. This is why we have to use **models** to help us understand it. You may have come across the model in which the transfer of energy from a battery to a lamp around a circuit is pictured by imagining each electron to carry, in addition to its charge, a 'bundle' of energy which it collects from the battery and delivers to the lamp. The electrons then return to the battery to collect more energy and repeat the process. Although this model is an adequate aid towards a basic understanding of electricity, it is too basic for our purposes. In particular, it is clearly at variance with our experience that the light comes on almost instantly when the circuit is made, despite the fact that electrons move only very slowly through metals. In order to explain the transfer of energy around the circuit, we should extend this model by adding to it the concept of an **electric field**.

You should be familiar with the idea that a **mass** in a **gravity** field has **potential energy**. If it falls to a lower height, it can do work, thereby reducing its potential energy. In Section 1.11 on gravitational fields, we consider different heights to represent different **potentials**. Moving from one potential to another (i.e. changing height) involves a gain or loss of potential energy. Gravitational potential is difficult to measure directly, but it can be calculated.

In an **electric** field, it is **charge** which possesses the potential energy. An electric circuit channels the electric field along the conductors, so the **potential** changes from place to place in the circuit, and when charge moves from a higher to a lower potential, its potential energy is reduced, and energy is given out to the surroundings. This is what happens in the lamp in our example. In the cell, chemical energy is used to move the charge from a lower to a higher potential. A major difference from gravitational fields is that electrical potential is very easy to measure directly. A **voltmeter** measures the **potential difference** between two points in a circuit simply by connecting its terminals to the two points in question. This topic is explored in more detail, both in Section 1.11, and in Section 3.6 on Electric fields.

Electrical components are designed to convert energy from one form to another. Cells, batteries, generators, solar panels, thermocouples, etc. all convert some form of energy to electrical energy; while lamp and heater filaments, motors, wires, etc. all convert electrical energy to some other form. These energy conversions are the useful features of electrical components, and potential difference and its units are defined in terms of **energy** changes.

The **potential difference** (p.d.) between two points in a circuit is defined as the amount of electrical energy changed to other forms of energy per coulomb of charge flowing between them.

1 volt is the p.d. between two points in a circuit in which 1 joule of energy is converted to other forms when 1 coulomb of charge passes between them.

In general, if an amount of energy W is converted when charge Q flows between two points in a circuit, then the p.d. V between the two points is given by:

$$V = \frac{W}{Q}$$

(V: V; W: J; Q: C)

Equation 6

Since $Q = It$, where Q is the charge flowing through a component when current I passes for time t, then substituting for Q in Equation 6 we obtain:

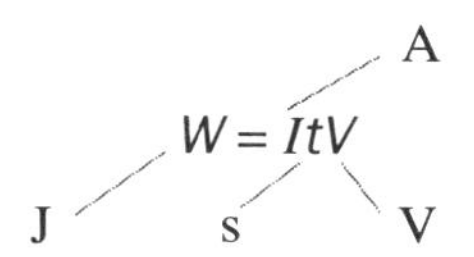

Equation 7

You should note that:

- Current passes between two points as a result of a difference in potential (p.d.) between them (no p.d. – no current).
- Battery terminals have a p.d. between them due to the chemical action of the cells, which forces charge to flow through any conductor connected across the terminals.
- It is nearly always the **difference** of potential between two points in a circuit which is important.
- Sometimes the potential at **a point** may be specified. This means the potential difference between that point and some point specified as having **zero** potential. Any part of a circuit connected to **earth** is considered to be at zero potential.

Electromotive force (e.m.f.)

Cells, batteries, dynamos, solar panels etc. cause the charge passing through them to move from a lower to a higher potential. To achieve this, work must be done on the charge, so these devices have the effect of converting other forms of energy into electrical energy. These components have a potential difference across them, even when there is no current. To distinguish between devices like these, which produce an **increase** of potential, from components which dissipate energy, and so cause a **decrease** of potential, the term **electromotive force** is used for the energy **supplied** per coulomb of charge flowing. This topic will be covered in more detail in a later section, but for now:

The **electromotive force** (e.m.f.) of a cell or other device which **supplies** electrical energy is equal to the potential difference across the terminals of the cell when there is **no current** in the cell.

Energy and power in d.c. circuits

Electrical power

Common symbol	P
SI unit	watt (W)

The electrical **power** of a device is the rate at which it converts electrical **energy** into other forms of energy.

If the p.d. across a device is V and the current through it is I, the electrical energy W converted by it in time (t) is given by:

$$W = ItV$$

Thus the electrical **power** P of the device is given by:

$$P = \frac{W}{t} = \frac{ItV}{t}$$

$$\therefore \quad P = IV$$

W A V

Equation 8

The unit of power is the watt (W) and equals an energy conversion rate of 1 joule per second.

If all the electrical energy is converted to heat by the device, it is called a **passive** resistor, and the rate of production of heat will also be IV.

The resistance (R) of the resistor is given by $R = V/I$. Substituting for (I) in Equation 8 gives:

$$P = \frac{V}{R} \times V$$

$$\therefore \quad P = \frac{V^2}{R}$$

W V Ω

Equation 9

and substituting for (V) in Equation 8 gives:

$$P = I \times IR$$

$$\therefore \quad P = I^2R$$

W A Ω

Equation 10

Electrical resistance R is explained in detail starting on page 199.

You should note that:

- Equation 9 and Equation 10 are valid only when all of the electrical energy is turned into heat (i.e. for a 'pure' resistor in which there is no other form of energy conversion).
- Equation 8 ($P = IV$) gives the rate of production of all forms of energy.

Thus, if a motor takes a current of 5 A when its applied p.d. is 10 V, then electrical power is being supplied to it at a rate of 50 W. However, perhaps only 40 W of mechanical power is obtained, the rest being dissipated as heat.

Electrical energy

Common symbols	E, W
SI unit	joule (J)

energy = power × time

If power is in watts, and time in seconds, then the unit of energy is the joule. This is an inconvenient unit for commercial use, one joule being a very small quantity of energy. Consequently, domestic and commercial electricity supplies are measured in larger units. The domestic unit of electrical energy is the **kilowatt-hour**.

The kilowatt-hour (kWh) is the quantity of energy converted to other forms of energy by a device of power 1 kilowatt in a time of 1 hour.

Thus the energy converted by a device in kWh is calculated by multiplying the power of the device in kW by the time in hours for which it is used.

A 3 kW electrical radiator running for 4 hours consumes 12 kWh of electrical energy – often called 12 'units' (currently priced at around 6 pence per unit).

To calculate the number of joules of energy equal to one kilowatt hour, we use the fact that:

$$\begin{aligned} \text{energy (J)} &= \text{power (W)} \times \text{time (s)} \\ &= 1000 \text{ W} \times 3600 \text{ s} \\ &= 3.6 \times 10^6 \text{ J} \end{aligned}$$

This calculation shows just how small one joule is, and why we have adopted the much larger kilowatt hour as the standard domestic unit of electrical energy.

Electrical resistance

Common symbol R, r
SI unit ohm (Ω)

When a given potential difference exists between two points on a conductor, the current between those points depends on the ease with which the conductor can transport the charge. This in turn depends on the **dimensions** of the conductor and the properties of the material, and, to a lesser extent, on the **temperature** of the conductor.

The **electrical resistance** of a conductor is a measure of its opposition to the flow of charge, i.e. current.

Electrical resistance is defined by the equation:

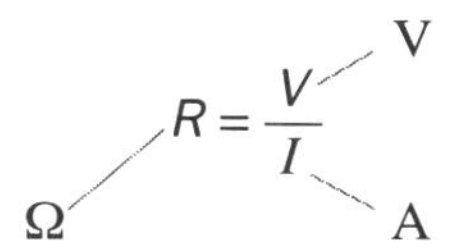

$$R = \frac{V}{I}$$

Equation 11

This leads us to the definition of the **ohm**:

A conductor has a resistance of **one ohm** (1.0 Ω) if there is a current of **one ampere** (1.0 A) through it when the p.d. across it is **one volt** (1.0 V) (see Figure 3.9).

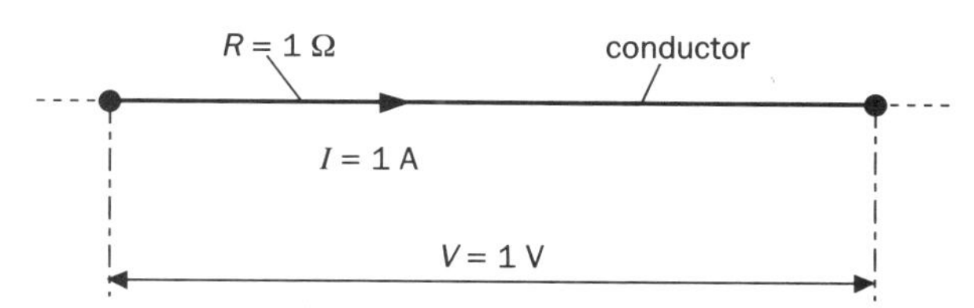

Figure 3.9 Electrical resistance

You should note that:

- The resistance of a conductor is the ratio of p.d. applied to current flowing.
- The greater the p.d. required to 'push' a given current through the conductor, the greater its resistance.
- For **metallic** conductors, the ratio V/I is constant, provided physical conditions such as temperature do not change. Thus a graph of V against I for such a conductor would be a straight line through the origin. This is a special case, illustrated by the graphs in Figure 3.10.

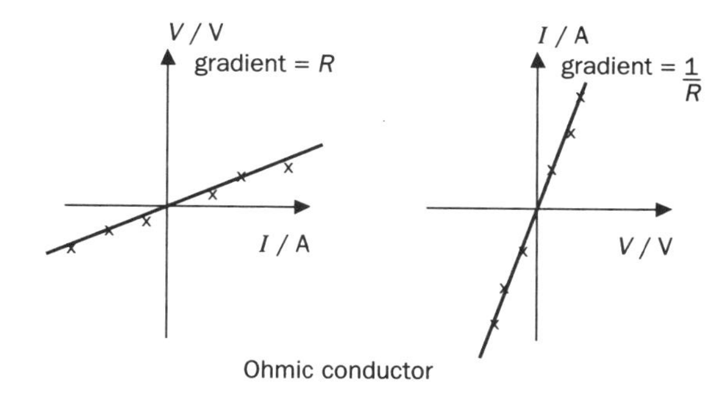

Figure 3.10 V/I and I/V graphs for an ohmic conductor

A conductor whose V/I graph is a straight line passing throught the origin is said to be an **ohmic** conductor as it obeys **Ohm's law**.

Ohm's law states that: The current through a metallic conductor is directly proportional to the p.d. across it, provided the temperature and other physical conditions remain constant.

Remember that this is a special case, applying only under the specified conditions.

Origin of resistance

Resistance arises from the interactions between the **charge carriers** moving through the conductor, and the **atoms** (or **ions**) of the material themselves. In the case of metals, as the electrons 'drift' through the crystal lattice, under the action of an applied p.d., they lose some of their energy in **collisions** with the atoms (strictly speaking - **ions**) of the material. In turn, the crystal lattice gains vibrational kinetic energy, which produces a rise in **temperature** of the conductor – the **heating effect** of the current. Thus electrical energy is dissipated to the surroundings.

A rise in temperature causes increased amplitude of vibration of the atoms (ions) of the crystal lattice. This presents a larger **collision cross-section** ('bigger target') to the drifting electrons. Collisions are thus more frequent, and rate of energy loss **increases**. To maintain the same current, a **larger** p.d. would be needed, i.e. the resistance has **increased**. This is generally true for all metallic conductors.

(**Superconduction** occurs at very low temperatures, when conditions are such that electrons flow through the material with zero loss of energy. Once started, a current in a superconducting circuit needs no p.d. to maintain it. Since no energy is dissipated, such a current will exist indefinitely without decreasing. Most metals exhibit superconductivity at temperatures near absolute zero; the search is on for 'room-temperature' superconductors – Nobel prizes await their discoverers! Towards the end of 1993, a claim was made for a material which became a superconductor at only –23 °C.)

Conductance

Common symbol G
SI unit ampere per volt ($A\ V^{-1}$)
unit name – **siemens** (S)
equivalent to ohm^{-1} (Ω^{-1})

We are often more concerned with how well a piece of wire **conducts** electricity, rather than how well it **resists** it, and the **conductance** of a conductor is a measure of the ease with which an electric current passes through it. If there is an electric current (I) through a conductor when the potential difference across it is (V), the conductance (G) is defined by the equation:

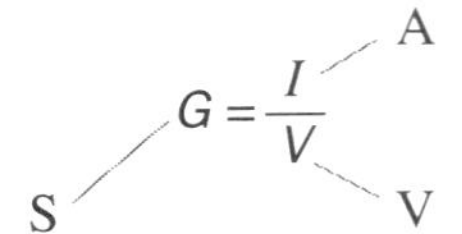
$$G = \frac{I}{V}$$

Equation 12

Since electrical resistance (R) is defined as

$$R = \frac{V}{I}$$

an alternative definition of conductance is:

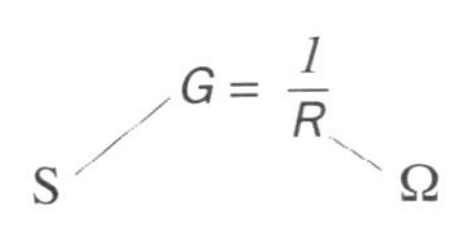
$$G = \frac{1}{R}$$

Equation 13

Resistivity and conductivity

Resistivity

Common symbol ρ (Greek letter 'rho')
SI unit ohm metre (Ω m)

This is a property of a **material** (unlike resistance and conductance which are properties of a **component**). Resistivity does not depend on the **dimensions** of the component or device, only on the materials from which it is made.

Experiments show that the **resistance** (R) of a **uniform** conductor is **directly** proportional to its length (l) and **inversely** proportional to its cross-sectional **area** (A):

$R \propto l$ and also $\propto \frac{1}{A}$

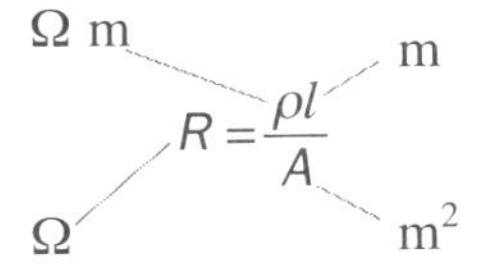
$$R = \frac{\rho l}{A}$$

Equation 14

where ρ is the **resistivity** of the material of the conductor, and is a constant for that material.

> **Resistivity** can be defined as **numerically** equal to the resistance of a sample of the material of unit ***length*** and unit ***cross-sectional area*** at a particular temperature (i.e. the resistance between opposite faces of a cube of material of side 1 m) (see Table 3.1).

In this magnetic levitation demonstration a small magnet floats freely above a nitrogen-cooled specimen of ceramic superconductor

The experimental determination of the resistivity of a material involves making accurate measurements of the **length** (l), cross-sectional **area** (A) and **resistance** (R) of a **uniform** sample of the material. Conveniently, for a good conductor, this could take the form of wire. Length could be measured with a metre rule, while a micrometer screw gauge would be necessary to measure the diameter of the wire with sufficient accuracy. The diameter should be measured at several points along the length of the wire, and in different directions across the wire. An average of these measurements should compensate for variations in thickness along the wire and any ovality. Resistance can be calculated by connecting a d.c. supply across the ends of the length of wire. The current through the wire could be measured with an ammeter, while the p.d. across it should be measured by a 'null' method, e.g. by using a potentiometer as described in the next section.

Material	Class	Resistivity Ω m
copper	good conductor	1.6×10^{-8}
graphite	conductor	8.0×10^{-6}
silicon	semiconductor	2.3×10^{3}
quartz	insulator	5.0×10^{16}

Table 3.1 Typical values of resistivity at room temperature

Conductivity

Common symbol: σ (Greek letter 'sigma')
SI unit: ohm^{-1} metre^{-1} or siemens per metre (Ω^{-1} m^{-1} or S m^{-1})

The electrical **conductivity** of a material is defined by the equation:

$$\sigma = \frac{1}{\rho}$$

S m^{-1} (for σ), Ω m (for ρ)

Equation 15

You should note that, in general:

- words ending in **-or** refer to **devices** (conductor, resistor, inductor etc)
- words ending in **-ance** refer to a property of that **device** (e.g. resist**ance**, conduct**ance**, induct**ance**, etc.)
- words ending in **-ivity** refer to a property of a **material** (e.g. resist**ivity**, conduct**ivity**, etc.).

Guided examples (3)

1. Calculate the electrical energy supplied by a 12 V battery when:

   (a) a charge of 50 C passes through it

   (b) a steady current of 5.0 A flows through it for 10 minutes.

   Guidelines

   (a) Use Equation 6.

   (b) Use Equations 1 and 6. Remember to use SI units.

2. An electric kettle takes 2.5 minutes to boil a quantity of water when it is connected to a 240 V supply. Assuming all the electrical energy is converted into 2.4×10^5 J of heat energy, calculate the current taken from the supply.

   Guidelines

   Use Equations 1 and 6. Remember to use SI units.

3. An electrical heating element is to be designed so that it dissipates energy at the rate of 1.2×10^3 W when it is connected to a 240 V supply. If the element is to be made of nichrome wire of diameter 0.6 mm, and resistivity 1.1×10^{-6} Ω m, calculate the length of wire needed.

   Guidelines

   Use Equations 9 and 14.

4. The live rail of an electric railway is made of steel of cross-sectional area 60 cm^2 and resistivity 1.0×10^{-7} Ω m. Calculate:

   (a) the electrical resistance of 0.75 km of the rail

   (b) the electrical conductivity of the steel

   (c) the conductance of a 5 m section of the rail.

   Guidelines

   (a) Use Equation 14.

   (b) Use Equation 15.

   (c) Remember to re-calculate the resistance of the new length of rail, then use Equation 13.

Self-assessment

SECTION A
Qualitative assessment

1. (a) What is an **electric current**?
   (b) State the relationship between current and **charge**.
   (c) Explain what is meant by a **steady** current.
   (d) Define **current density** at a point in a current-carrying conductor.
2. Describe the mechanism of conduction in metals in terms of **'free'** electrons. Explain why the temperature of a conductor is likely to rise when there is a current through it.
3. There is an electric current (I) through a section of conductor of length (l) and cross-sectional area (A). If there are (n) free electrons carrying charge (e) per m^3 of conductor, show that
   (a) the average drift velocity (v) is given by
   $v = I/nAe$
   (b) the current density (J) is given by
   $J = nve$
4. At room temperature, state the similarities and differences between conductors and semiconductors in terms of
   (a) number of charge carriers per unit volume available for conduction
   (b) resistivity
   (c) effect of temperature on resistivity.
5. What is the difference between **intrinsic** and **extrinsic** semiconductors? Describe the nature of the charge carriers in:
   (a) **intrinsic** semiconductors
   (b) **n-type** semiconductors
   (c) **p-type** semiconductors.
6. Define **potential difference** and its SI **unit**. What analogy can you draw between potential difference and **temperature** difference?
7. (a) Define electrical **power** and its SI **unit**.
   (b) Show that if the current through a component is (I) when the p.d. across it is (V), the power dissipated (P) is given by
   $P = VI.$
   (c) Hence show that if the component has resistance (R), alternative expressions for power are:
   (i) $P = I^2R$
   (ii) $P = V^2/R.$

SECTION B
Quantitative assessment

(***Answers:*** 2.9×10^{-16}; 1.5×10^{-4}; 5.0×10^{-3}; 0.65; 0.72; 0.81; 1.2; 2.0; 2.5; 6.8×10^3; 3.5×10^4; 2.5×10^7; 7.9×10^7.)

(Charge on electron = 1.6×10^{-19} C)

1. Calculate the quantity of charge which flows through a light bulb in 45 minutes if the p.d. across the lamp is 100 V and the resistance of the filament can be assumed to be 40 Ω.
2. If the quantity of charge flowing from a battery every minute is 120 C, calculate the current flowing.
3. A 220 V electric heater is switched on for 10 minutes. Calculate the amount of heat energy produced if 160 C of charge flows through the heater element.
4. If the p.d. through which electrons are accelerated in a television tube is 1800 V, calculate the gain of kinetic energy of each electron. If the mass of an electron is 9.1×10^{-31} kg, calculate the velocity with which an electron hits the screen.
5. Calculate the resistance of 55 cm of constantan wire of diameter 0.75 mm if its resistivity is 5.2×10^{-7} Ω m.
6. A steady current of 7.5 mA exists axially along a cylindrical conductor of cross-sectional area 0.25 mm^2, length 6.0 m and resistivity 4.0×10^{-6} Ω m. Calculate the p.d. across the ends of the cylinder.
7. A p.d. of 2.0 V is applied across the ends of a 1.5 m length of copper wire of diameter 0.20 mm. Calculate
   (a) the resistance and conductance of the wire
   (b) the current in the wire and the current density
   (c) the average drift velocity of the electrons in the wire.

   For copper, number of free electrons per m^3 = 1.0×10^{29}; resistivity = 1.7×10^{-8} Ω m.
8. The heating element in a hair dryer dissipates energy at a rate of 400 W when it is connected to a 240 V supply. If the element is made of nichrome wire of total length 2.4 m, calculate the diameter of the wire used. (Resistivity of nichrome = 1.1×10^{-6} Ω m.)

3.2 Behaviour of d.c. in series and parallel circuits

Series circuit

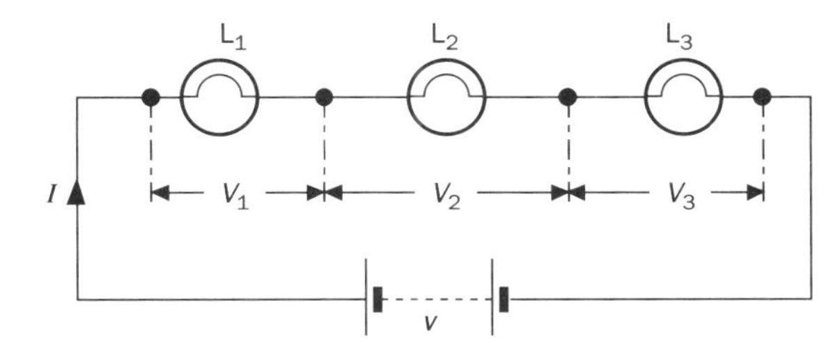

Figure 3.11 Three lamps connected in series

Figure 3.11 shows three lamps (L_1, L_2 and L_3) connected in **series** with a battery, which creates a potential difference (V) across the lamps. In such a circuit:

- **The current is the same (= I) at all points in the circuit** – i.e. the current through each lamp is the same.
- The total p.d. (V) across the three lamps is equal to the **sum** of the individual p.d.s across each lamp (assuming the connecting wires have negligible resistance), i.e.

  $$V = V_1+V_2+V_3$$

  Equation 1

Parallel circuit

Figure 3.12 shows three lamps (L_1, L_2 and L_3) connected in **parallel** with a battery, which creates a potential difference (V) across the lamps. In such a circuit:

- **The potential difference is the same (=V) across each lamp.**

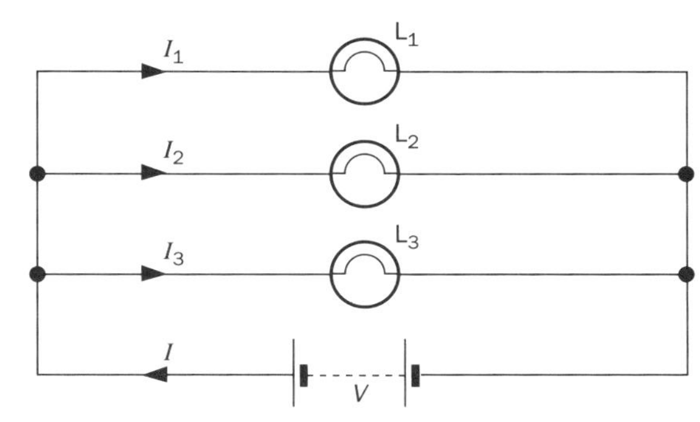

Figure 3.12 Three lamps connected in parallel

- Since charge is conserved, the total current (I) through the battery is equal to the sum of the individual currents through each lamp, i.e.

  $$I = I_1+I_2+I_3$$

  Equation 2

Combination of resistors

Series connection

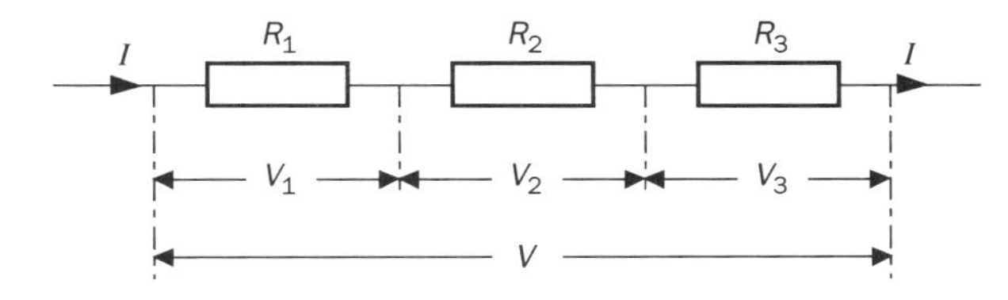

Figure 3.13 Resistors in series

Figure 3.13 shows three resistors R_1, R_2 and R_3 connected in **series**.

V is the **total p.d.** across the resistors.

I is the total current through the resistors.

V_1, V_2 and V_3 are the p.d.s across R_1, R_2 and R_3

R_T is the **total effective resistance** of R_1, R_2 and R_3.

Since energy is conserved:

$$V = V_1 + V_2 + V_3$$

but from the definition of resistance:

$$V = IR$$

$$\therefore \quad IR_T = IR_1 + IR_2 + IR_3$$

Dividing through by I:

$$\therefore \quad R_T = R_1+R_2+R_3$$

Equation 3

Thus when resistors are combined **in series**, the total effective resistance R_T is equal to the **sum** of the individual resistances.

Parallel connection

In this type of connection, alternative routes are provided for the current, which splits so that there is a fraction of the current through each resistor.

- the **lowest** value resistors carry the **greatest** proportion of the current
- the **total** effective resistance is **less** than that of the **smallest** resistor in the combination.

Figure 3.14 shows three resistors R_1, R_2 and R_3 connected in parallel. The potential difference across each resistor is the same, and equal to V.

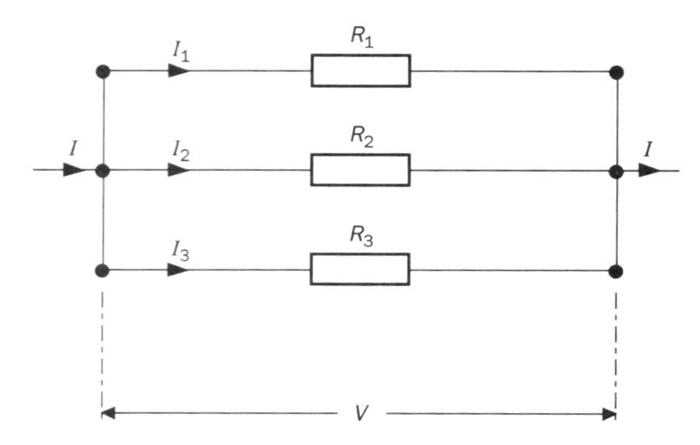

Figure 3.14 Resistors in parallel

Since charge is conserved:

$$I = I_1 + I_2 + I_3$$

From the definition of resistance:

$$I = \frac{V}{R}$$

$$\therefore \quad \frac{V}{R_T} = \frac{V}{R_1} + \frac{V}{R_2} + \frac{V}{R_3}$$

Dividing through by V:

$$\therefore \quad \frac{1}{R_T} = \frac{1}{R_1} + \frac{1}{R_2} + \frac{1}{R_3}$$

Equation 4

When resistors are connected **in parallel**, the total effective resistance of the combination is obtained by using Equation 4.

Special case for identical resistors

If N resistors each of resistance R are connected in parallel, then the **total effective resistance** R_T is given by:

$$R_T = \frac{R}{N}$$

Equation 5

Circuit calculations

The basic equation relating p.d. (V), current (I) and resistance (R)

$V = IR$

is a powerful tool in analysing circuits and solving problems, but some care needs to taken when applying it. The values used in the equation can apply to a single component, group of components or the whole circuit, but you must not mix them. The following example illustrates the use of the equation in a circuit.

Calculate the unknown currents and p.d.s in the circuit shown in Figure 3.15.

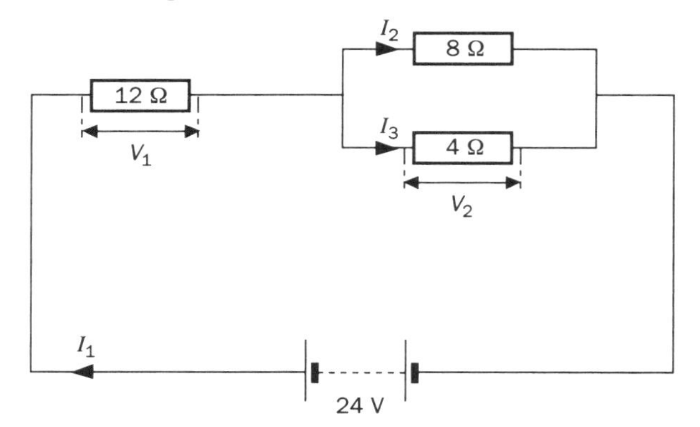

Figure 3.15 Using the equation V = IR in a circuit

The first task is to identify what is happening in the circuit.

The 24 V battery is pushing a current (I_1) through the whole circuit, which splits into I_2 and I_3 to pass through the two parallel resistors, before rejoining to return to the battery. The potential difference across the **whole** circuit is 24 V. As the charge flows through the 12 Ω resistor, energy is dissipated, and the **potential** decreases. So there is a **potential difference** (V_1) between one side of the resistor and the other. (The side nearest the **positive** terminal of the battery is at the higher potential.) Energy is also dissipated in the parallel pair of resistors, and the potential difference across this pair is V_2.

You are given the p.d. across the whole circuit (24 V), and to calculate the current through the circuit (I_1) you must calculate the resistance of the whole circuit:

Parallel pair

$$\frac{1}{R} = \frac{1}{4} + \frac{1}{8} = \frac{3}{8}$$

$$\therefore \quad R = \frac{8}{3} = 2.7\ \Omega$$

Adding this to the 12 Ω resistor in series gives a total resistance of 14.7 Ω.

We use Ohm's law to calculate I_1:

$$I_1 = \frac{24}{14.7}$$
$$= 1.63 \text{ A}$$

This is the current through the battery, and is also the current through the 12 Ω resistor.

Applying

$$V = IR$$
$$V_1 = 1.63 \times 12$$
$$\therefore \quad V_1 = 19.6 \text{ V}$$

Knowing V_1 enables V_2 to be calculated directly, since the sum of the p.d.s in the circuit must add up to the total p.d. of 24 V:

$$V_1 + V_2 = 24$$
$$\therefore \quad V_2 = 24 - 19.6$$
$$\therefore \quad V_3 = 4.4 \text{ V}$$

Since this is the p.d. across both the 4 Ω and the 8 Ω resistors, the currents through them can now be calculated:

$$I_2 = 4.4/8$$
$$\therefore \quad I_2 = 0.55 \text{ A}$$
$$I_3 = 4.4/4$$
$$\therefore \quad I_3 = 1.1 \text{ A}$$

As a final check, you should notice that $(I_2 + I_3)$ is equal to I_1 (within rounding errors).

Guided examples (1)

1. Calculate the total effective resistance of the following circuit:

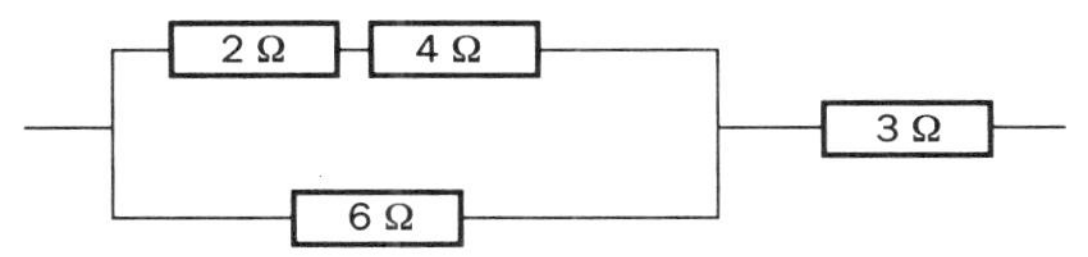

Guidelines

The 2 Ω and 4 Ω resistors are in series. Combine these first, then combine this with the 6 Ω resistor in parallel. Finally combine this result with the 3 Ω resistor in series.

2. For the circuit shown below, calculate the:

(a) effective resistance of the parallel group
(b) total circuit resistance
(c) total current (I)
(d) potential difference (V_1) across the parallel group
(e) potential difference (V_2) across the 3.5 Ω resistor
(f) currents (I_1 and I_2).

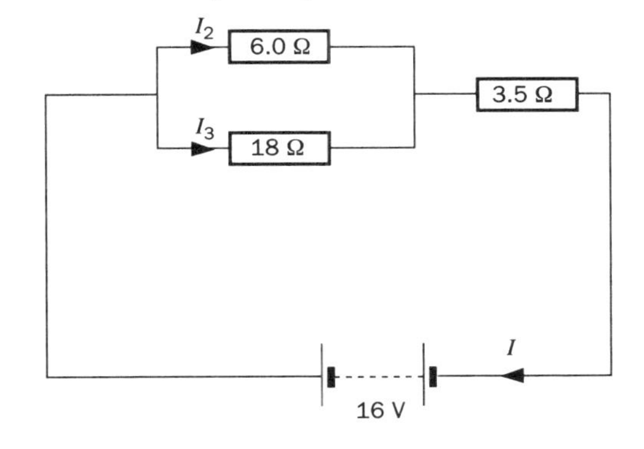

Guidelines

(a) Use Equation 4.
(b) Add the series resistor to answer (a).
(c) Apply $V = IR$ to the whole circuit.
(d) Apply $V = IR$ to the parallel pair, where I is the current **through the pair**, and R is the effective resistance **of the pair**.
(e) Use $V = IR$.
(f) You calculated the p.d. across the pair in (d). Use this in $I = V/R$ where R is the resistance of each resistor to calculate the current in each resistor in turn. Check that the sum of these two currents is equal to the total current calculated in (c).

3. For each of the circuits (a) and (b) below, calculate the unknown currents.

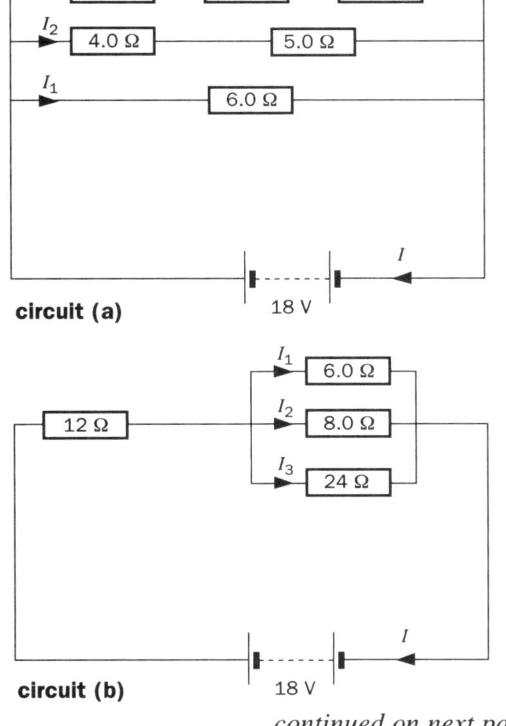

continued on next page

Guided examples (1) continued

Guidelines

Calculate the total resistance, adding the series resistors in each branch first, before combining the branches in parallel. Use this to calculate the total current.

In (a) the p.d. across each branch is the same (18 V) since they are connected in parallel. Use this, and the resistance of each branch, to calculate each of the currents in the branches.

In (b) calculate the p.d. across the 12 Ω resistor, and subtract this from 18 V (the supply p.d.) to give the p.d. across the parallel group. This will be the p.d. across each resistor in the parallel group, so use this to calculate the current in each resistor. (Check that the sum of these currents is equal to the total current in the circuit.)

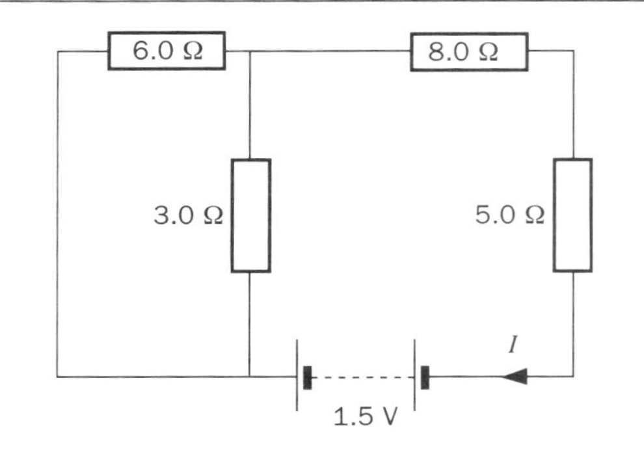

network (a)

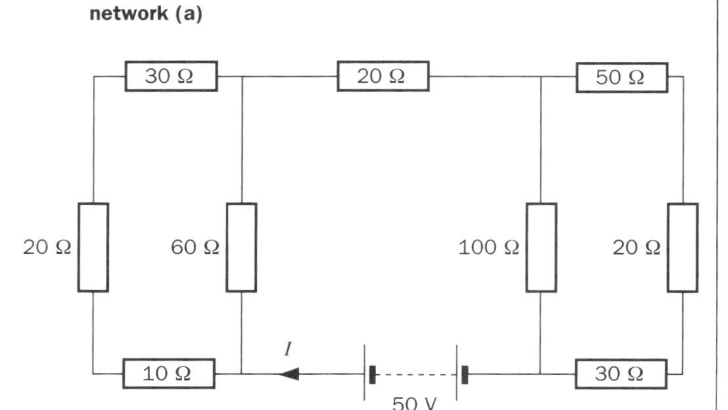

network (b)

4. Calculate the total current flowing in each of the resistor networks shown to the right:

Guidelines

You must identify which components are connected in **parallel** and which are in series. Whenever there is a junction, the branches from the junction are parallel branches. Using these ideas, the two networks can be re-drawn in more familiar form as shown below

Apply the techniques from the earlier examples to find the unknown currents.

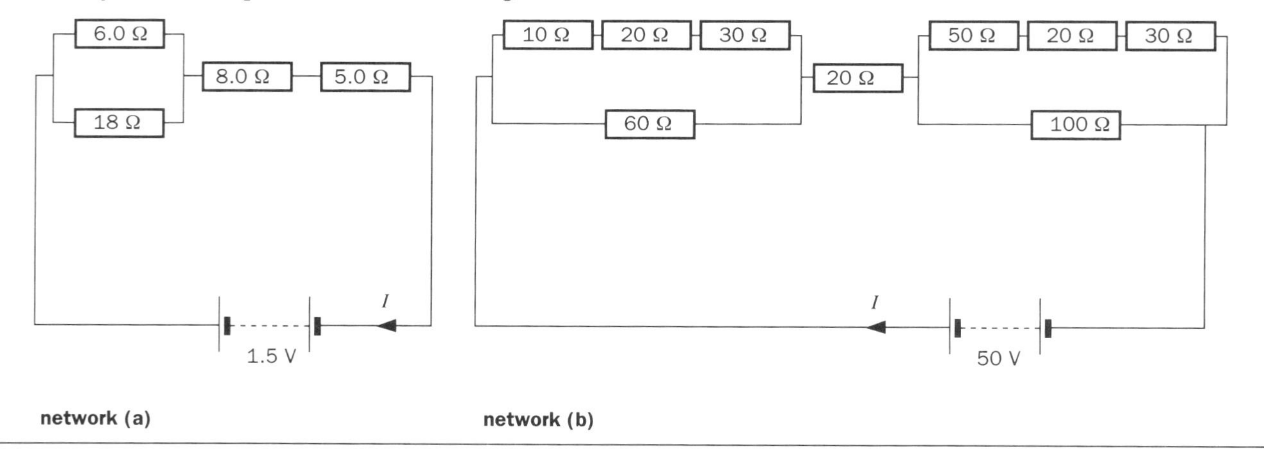

network (a)

network (b)

Current/p.d. characteristics

The **characteristic** of a component is a graph of current against p.d. for the component for a range of values within the component's operating limits. Since some components behave differently when the current through them is in the reverse direction, it is important to measure p.d. and current when the current is in both forward and reverse directions.

A typical circuit for obtaining the characteristic is shown in Figure 3.16 (at top of next page).

The battery and variable resistor together form a **potential divider**. This enables a continuously-variable p.d. (ranging from 0 up to the maximum desired) to be applied to a component connected between points X and Y. The current through the component is measured with the ammeter, while the voltmeter measures the p.d. across the component.

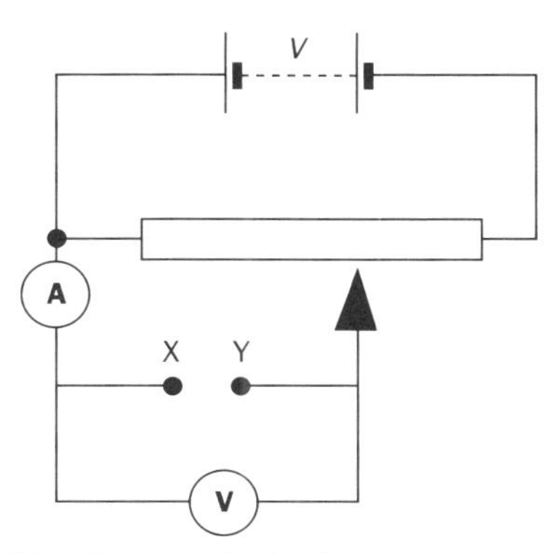

Figure 3.16 Current/p.d. characteristic circuit

A suitable range of p.d.s is applied, and these are recorded, together with the corresponding values of current. To measure the **reverse characteristic**, the component is simply connected to X and Y the opposite way round. Figure 3.17, shows some typical characteristics for some common components.

Ohmic conductor

A straight line through the origin is obtained, showing that Ohm's law is obeyed; since the gradient is constant the resistance is constant, for both directions of current flow.

Filament lamp

The curve shows that resistance increases as current increases. This is due to the rise in temperature of the filament, caused by the heating effect of the current.

Semiconductor diode

The curve shows that the semiconductor diode allows virtually no current when a negative p.d. is applied (***reverse bias***), and that a ***forward-biased*** p.d. of about 0.6 V is needed before the diode will conduct in the forward direction. When the p.d. is greater than this, resistance is very low, allowing a large increase in current for a small increase in p.d.

Light-emitting diode (LED)

These devices have a similar characteristic to a standard semiconductor diode, but visible light is released when current is in the forward direction.

Thermistor

The curve may increase or decrease in gradient. Thermistors may have either a positive or negative temperature coefficient. A negative temperature coefficient means that its resistance **decreases** with temperature, unlike that of most resistors. The normal increase in resistance with temperature is more than offset by the release of extra charge carriers in some materials.

Kirchhoff's laws

These rules summarise the behaviour of currents and p.d.s in circuits with steady currents. They are based on the concepts of conservation of charge and conservation of energy.

When a steady current exists, charge cannot build up in one place, nor can it become depleted. Therefore we arrive at the first of Kirchhoff's laws:

Kirchhoff's first law (Kirchhoff 1)

The algebraic sum of the currents at a junction is zero:

i.e. $\sum I = 0$

Equation 6

Put more simply: at a junction in a circuit, the sum of the currents entering the junction must be equal to the sum of the currents leaving it.

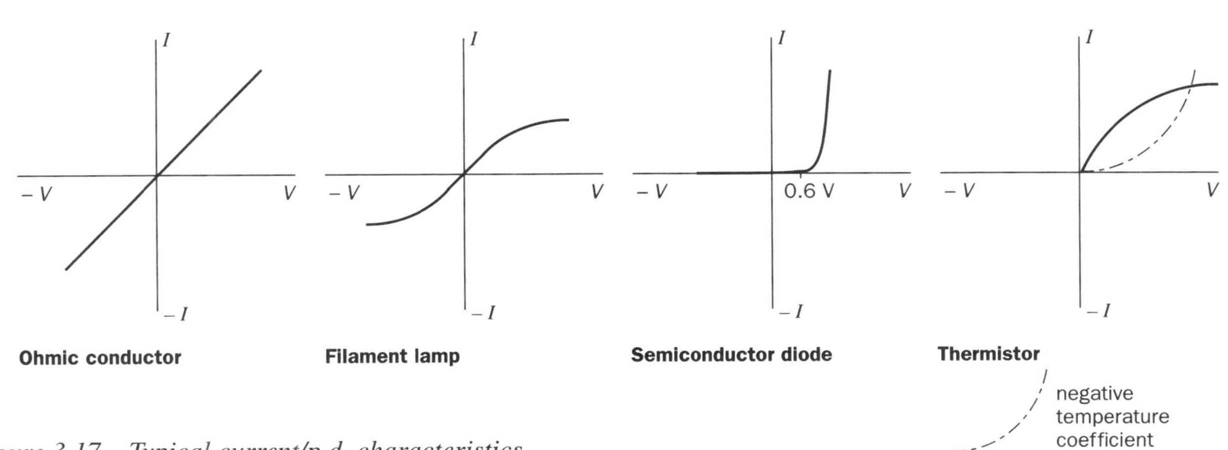

Figure 3.17 Typical current/p.d. characteristics

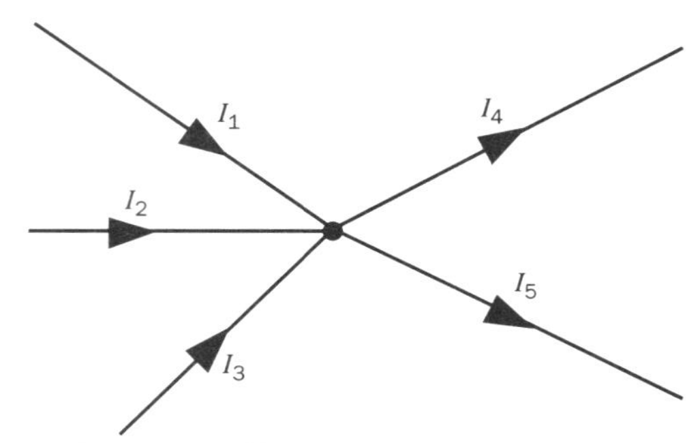

Figure 3.18 Kirchhoff's first law

Figure 3.18 shows 5 conductors meeting at a point. I_1, I_2 and I_3 are currents into the point while I_4 and I_5 are currents away.

$\sum I = 0$

To apply this equation, we identify currents **into** the junction as **positive** and currents **away** as **negative**:

$I_1 + I_2 + I_3 - I_4 - I_5 = 0$

This rule arises since charge cannot accumulate anywhere **while there are steady currents**. It should be applied as a matter of routine wherever currents split up or rejoin at junctions. If you worked through the guided examples on pages 205–6, you will have seen examples of this working in practice.

Kirchhoff's second law (Kirchhoff 2)

When a current exists in a circuit, charge flows through the various components which make up the circuit. If we consider the **energy changes** which occur within each component, energy is either **supplied** by the component or **dissipated** by it. For cells, batteries etc., the energy **supplied** per coulomb of charge is measured as the **electromotive force** (E) of the cell. For components which dissipate energy, the energy **dissipated** per coulomb of charge is the **potential difference** (V) across the component. If a unit of charge follows a **closed path** around a circuit (i.e. it returns to its original point) then the total of all the energy supplied to it must exactly balance the energy dissipated by it. This is a consequence of the principle of conservation of energy.

Kirchhoff's second law expresses this condition mathematically:

In any closed loop of a circuit, the algebraic sum of the **e.m.f.s** is equal to the algebraic sum of the **potential differences** across all the resistances in the loop.

Since the p.d. (V) across a resistance (R) carrying a current (I) is given by $V = IR$, Kirchhoff's second law can be written:

$$\sum E = \sum IR$$

Equation 7

- $\sum E$ means the sum of the e.m.f.s, and is a measure of the total amount of energy **supplied** to each coulomb of charge flowing round the loop.
- $\sum IR$ means the sum of the p.d.s across the resistive components, and is a measure of the total energy **dissipated** in the loop.

Taken together, Kirchhoff's first and second laws provide powerful tools for calculating the currents through, and p.d.s across, components in the most complex of circuits. You will find it sensible to follow a basic structure when solving circuit problems.

The following guided examples will show you a suitable procedure.

Guided examples (2)

1. A battery charger works by passing a current through a rechargeable cell in the reverse direction to normal current flow. The diagram below shows a particular charger being used to recharge a cell of e.m.f. 1.2 V and internal resistance 0.2 Ω. If the e.m.f. of the battery charger is 5.5 V, and its internal resistance is 1.0 Ω, calculate the current through the rechargeable cell.

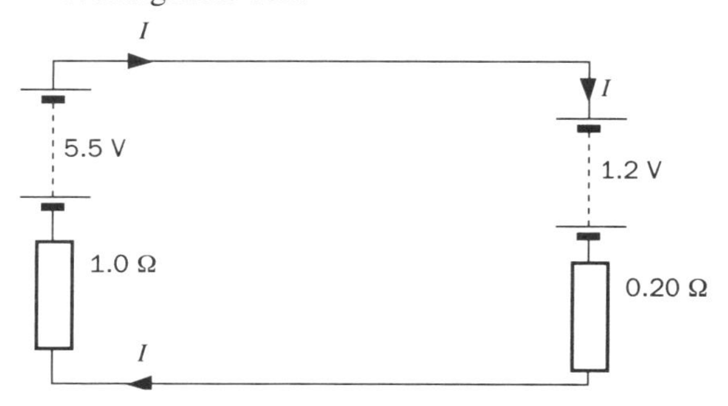

Guidelines

Apply Kirchhoff 2:

Traverse the loop clockwise, and consider the energy changes encountered by a unit charge.

E.m.f.

Traversing clockwise, the charge goes through an **increasing** potential in the 5.5 V battery, so this is a **positive** energy change. In the 1.2 V cell, the charge suffers a **decrease** in potential, so this is a **negative** energy change.

$\therefore \quad \Sigma E = 5.5 - 1.2$

$\therefore \quad \Sigma E = 4.3$

Guided examples (2) *continued*

P.d.

The current (I) is in the direction of travel around all parts of the circuit, so all p.d.s are positive:

$$\sum IR = I \times 0.20 + I \times 1.0$$

$$\therefore \sum IR = I \times 1.20$$

Since

$$\sum E = \sum IR$$

$$\therefore 4.3 = I \times 1.20$$

$$\therefore I = 4.3/1.2$$

$$\therefore I = 3.58 \text{ A}$$

2. The technique can be applied to more complex circuits involving branches.

If the above circuit is modified to include a voltmeter of resistance 1000 Ω in parallel with the battery charger, calculate the:
(a) current through the voltmeter
(b) reading on the voltmeter.

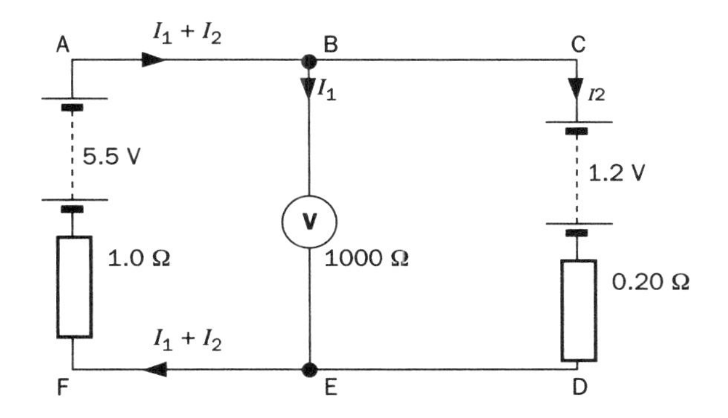

Guidelines

Label the currents I_1 and I_2 as shown. Use Kirchhoff 1 to reduce the number of unknown currents.

Apply Kirchhoff 2 to two closed loops.

For loop ABEF:

Traversing the loop clockwise

$$\sum E = 5.5$$

$$\sum IR = I_1 \times 1000 + (I_1 + I_2) \times 1.0$$

$$\therefore 5.5 = I_1 \times 1000 + (I_1 + I_2) \times 1.0 \quad \text{(i)}$$

For loop BCDE:

$$\sum E = -1.2$$

$$\sum IR = -I_1 \times 1000 + I_2 \times 0.2$$

(direction of I_1 opposes the direction of traverse)

$$\therefore -1.2 = -I_1 \times 1000 + I_2 \times 0.2 \quad \text{(ii)}$$

Equations (i) and (ii) are **simultaneous equations**. Solving them gives:

$I_1 = 0.00192$ A

$I_2 = 3.58$ A

Voltmeter reading $= I_1 \times 1000 = 1.92$ V

The technique for applying Kirchhoff's laws can be summarised:

- **Always** draw a circuit diagram. Label each component with a **unique** symbol (R_1, L_2, etc.).
- Use letters on junctions and points on the circuit so that each loop can be identified.
- Show currents with arrows (guess the direction if necessary). Again use unique labels (I_1, I_2).
- Use Kirchhoff 1 to reduce the number of unknown currents.

Now try one yourself:

Two batteries, of e.m.f. 5.0 V and 3.0 V, and internal resistances 2.0 Ω and 1.0 Ω, are joined in parallel, and the combination is connected to a 10 Ω resistor as shown below.

Calculate:
(a) the currents I_1 and I_2 through each battery
(b) the current through the 10 Ω resistor
(c) the p.d. across the terminals of the batteries.

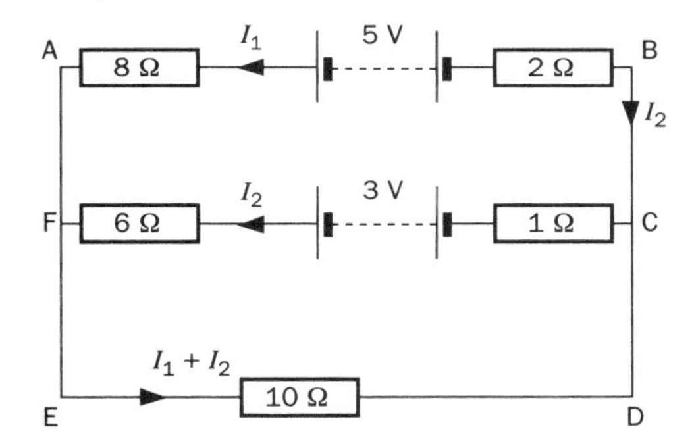

Guidelines

Kirchhoff 1 has already been applied to combine I_1 and I_2.

(a), (b) Apply Kirchhoff 2 to loops ABDEA and FCDEF.

You could also have used loop ABCFA, but generally it is best to choose the loops with the fewest components. Remember to traverse the loops in the directions given, and be very careful to use the correct signs for each current and e.m.f.

(c) Once I_1 and I_2 are known, apply $V = IR$ to calculate the p.d. across the battery terminals.

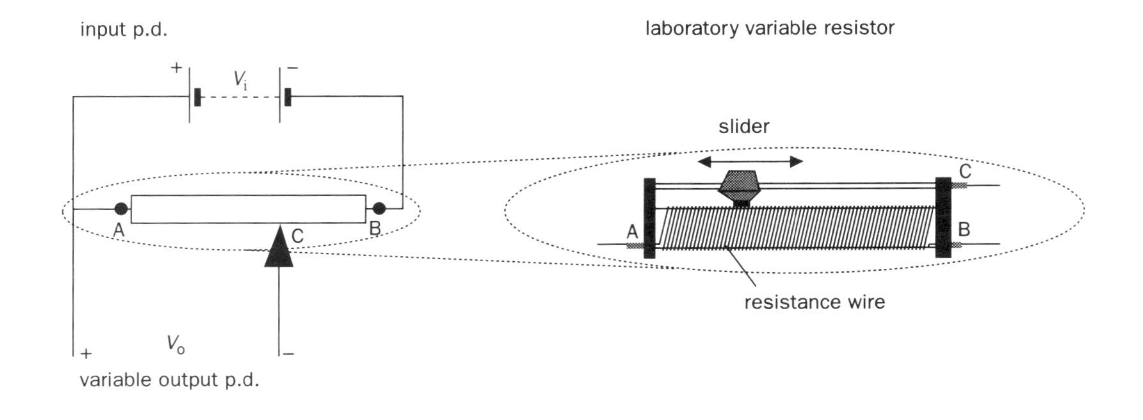

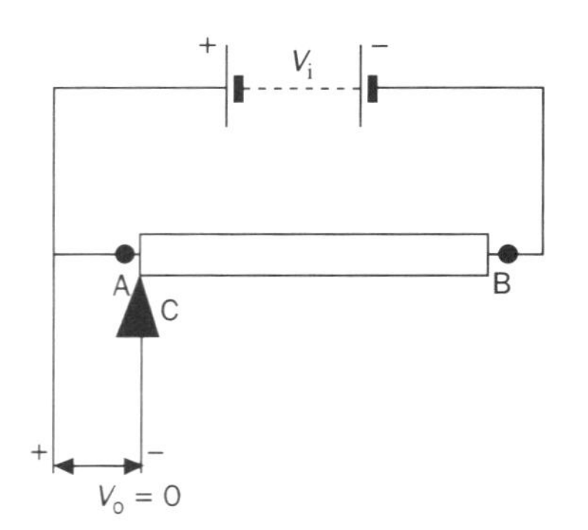

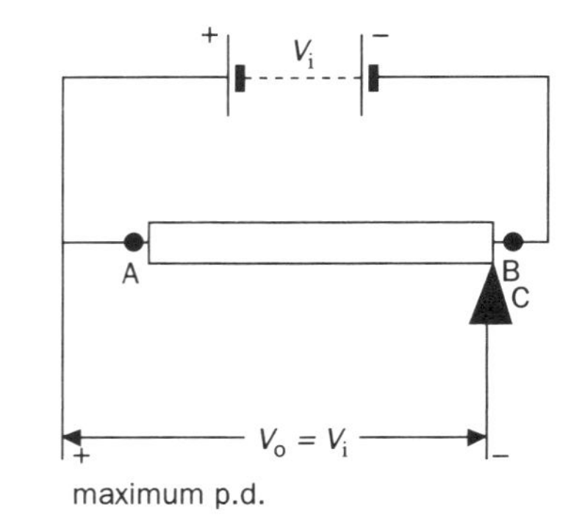

Figure 3.19 Variable resistor used as a potential divider

The potential divider

Used to supply variable p.d.

A **variable** resistor can be connected into a circuit in such a way as to control the p.d. which may be applied to the rest of the circuit, as shown in Figure 3.19.

Used in this way, all three terminals (the two **end** terminals A and B and the **slider** or **wiper** terminal – C) are connected. The input p.d. (V_i) is supplied to the 'fixed' terminals (A and B) of the variable resistor. The output p.d. (V_o) is obtained by connecting to A and the 'slider' terminal (C). Any fraction of the input p.d. may be obtained by varying the position of the slider. With the slider at end A the output p.d. is zero, and with the slider at end B, the maximum output p.d. is obtained (equal to the supply p.d.).

Variable resistors may be the large, linear types used with large currents in the laboratory, or miniature rotary types used in electronic circuits and for volume controls in audio equipment.

Calculation of output p.d.

Figure 3.20(a) shows a variable resistor used as potential divider with the slider in some intermediate position. Figure 3.20(b) shows the equivalent circuit consisting of two separate resistors R_1 and R_2. The output p.d. (V_o) is the p.d. across R_1, and is equal to the p.d. obtained from circuit (a).

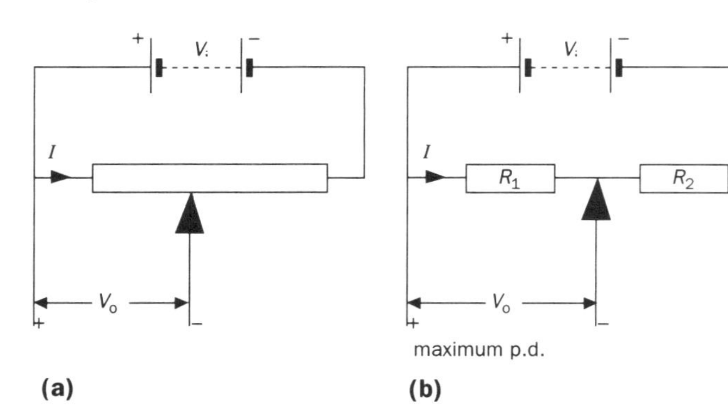

Figure 3.20 Output p.d. from a potential divider

Applying Kirchhoff's laws:

$$V_i = I(R_1 + R_2) \quad \text{and} \quad V_o = IR_1$$

$$\frac{V_o}{V_i} = \frac{IR_1}{I(R_1 + R_2)} = \frac{R_1}{(R_1 + R_2)}$$

$$V_o = V_i \frac{R_1}{(R_1 + R_2)}$$

Equation 8

For example, a p.d. of 12 V can be obtained from a 100 V supply by setting R_1 at 1500 Ω, R_2 at 11000 Ω:

$$V_o = V_i \times \frac{1500}{(1500 \times 11000)} = 12 \text{ V}$$

Light-dependent potential divider

Figure 3.21 shows how a **light-dependent resistor** (LDR) can be used in a potential divider to provide an output p.d. (V_o) which varies with **light intensity**.

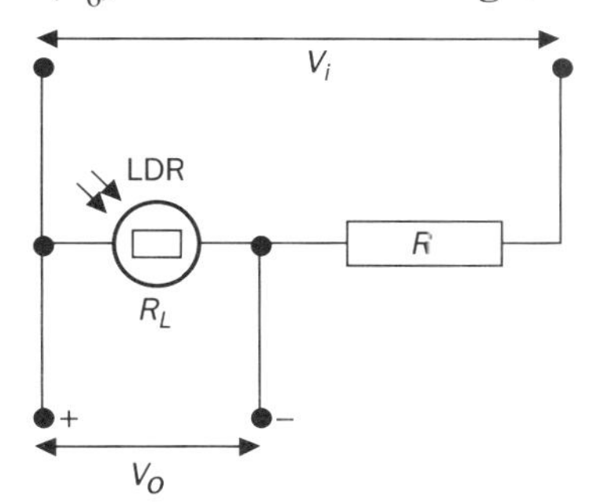

Figure 3.21 Use of a LDR in a light-dependent potential divider

A LDR is a resistor made from a semiconducting material in which electrons are liberated when light shines on the surface of the material. In total darkness, the only free electrons are those 'shaken' free by thermal vibrations of the atoms, so the resistance of the LDR is very high. Light energy has the effect of liberating electrons, reducing the resistance of the device.

The output p.d. (V_o) is given by

$$V_o = V_i \times \frac{R_L}{(R_L + R)} \quad \text{where } R_L = \text{LDR resistance.}$$

In **bright light**, R_L is low (<100 Ω) compared with R, so (V_o) is low. As the light intensity decreases, R_L increases until it reaches its maximum value (around 10 MΩ) in total darkness. V_o will then have its maximum value (approximately equal to V_i).

Since the output p.d. depends on the light intensity, such a circuit could be used to control any process which depends on light level. At the simplest level, this could mean automatically switching on lamps whenever darkness fell. A switching circuit could be set to operate when V_o reached a pre-determined limit, corresponding to a particular level of light intensity. If R were replaced by a **variable resistor**, this would provide some manual adjustment of the value of V_o at a particular light intensity. For example, if the lamps were set to switch on at a p.d. equal to $V_i/2$, R could be adjusted so that this occurred at any desired level of illumination.

If R and R_L were interchanged, the circuit would give an output p.d. which **increased** as light intensity **increased**.

Temperature-dependent potential divider

A thermistor (see Figure 3.22) is a device whose resistance varies markedly with temperature.

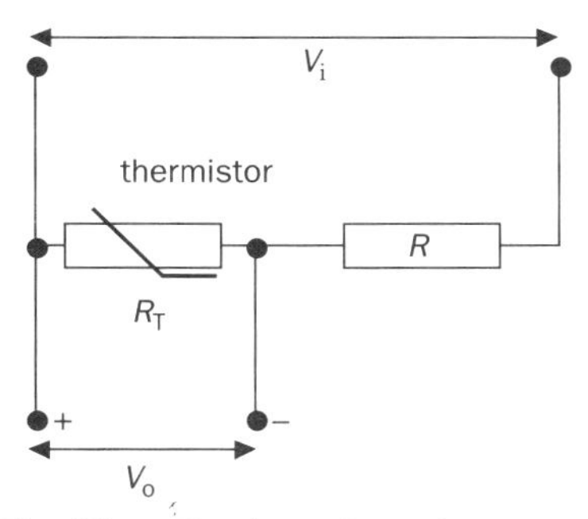

Figure 3.22 Use of a thermistor in a temperature-dependent potential divider

With **increasing temperature**:

The resistance of a **negative** temperature coefficient thermistor **decreases**;

The resistance of a **positive** temperature coefficient thermistor **increases**.

The output p.d. (V_o) is given by:

$$V_o = V_i \times \frac{R_T}{(R_T + R)} \quad \text{where } R_T = \text{thermistor resistance.}$$

For a **negative** temperature coefficient thermistor, R_T will **increase** as temperature **decreases**. This means that the output p.d. will rise as temperature falls. This rising p.d. could be used to trigger a frost alarm, or switch on a heater in order to maintain temperature above a given limit. Replacing R with a variable resistor would allow manual adjustment of this 'trigger temperature'.

If R_T and R were interchanged, the output p.d. would decrease as temperature decreased.

Potential dividers and measurement

The accuracy with which current and p.d. can be measured with analogue (pointer-type) ammeters and voltmeters is limited. A voltmeter connected in parallel with a component inevitably draws some current from the circuit, so it changes the behaviour of the circuit to which it is connected. An ideal voltmeter would draw no current. Modern electronic (digital) voltmeters and cathode ray oscilloscopes come very close to achieving this ideal, as they have very high resistances of many millions of ohms.

A type of potential divider known as a **potentiometer** can also be used as a very accurate voltmeter. The basic circuit is shown in Figure 3.23.

The driver cell drives a current through the uniform slide-wire. As this wire is uniform, the potential decreases uniformly along its length. The potential difference between points A and C (the sliding

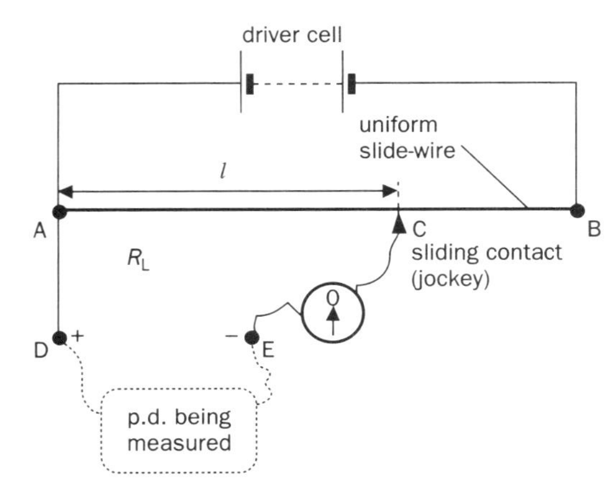

Figure 3.23 Simple potentiometer circuit

contact) is therefore proportional to the length of wire (l) between A and C.

The p.d. to be measured is connected to points D and E in the circuit, with the more positive potential connected to D. There are three possibilities:

(i) If this p.d. is less than the p.d. across AC, there will be a current from D to E (through the external circuit being measured), giving a reading on the galvanometer.
(ii) If the p.d. is greater than the p.d. across AC, the current direction will be reversed (i.e. from E to D).
(iii) If the p.d. is exactly equal to the p.d. across AC, then there will be **no current**.

In condition (iii), the potentiometer is said to be **balanced**, and the length (l_1) is proportional to the p.d. being measured (V_1). Other p.d.s (V_2, V_3, etc.) can be connected to D and E, producing different **balance lengths** (l_2, l_3 etc.).

In general, since balance length is proportional to p.d. applied:

$$\frac{V_1}{V_2} = \frac{l_1}{l_2}$$

Equation 9

If V_1 is known, l_1 and l_2 can be measured, enabling V_2 to be calculated.

You should note that:

- This method is a **null** method, since, at balance, no current flows between D and E. This means that the potentiometer can be used to measure a p.d. without changing it. It can also be used to measure the **true e.m.f.** of a cell.
- The galvanometer reads zero, so any inaccuracy of the galvanometer does not affect the measurement.
- The apparatus is somewhat cumbersome and time-consuming to use.
- It is not a 'direct-reading' instrument – the p.d. must be calculated from the length measurements.
- The driver cell must maintain a constant potential difference during all measurements.
- The final accuracy of the calculation depends on the accuracy with which V_1 is known.

Guided examples (3)

1. The diagram below shows a light-dependent resistor (LDR) connected in series with a 10 kΩ resistor and a 12 V d.c. supply.

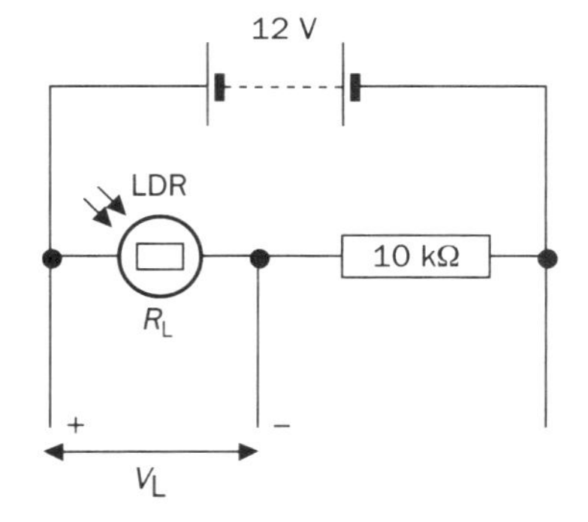

Calculate:

(a) the p.d. across the LDR (V_L) when
  (i) it is in the dark, and has a resistance of 8.0 MΩ, and
  (ii) it is in bright light, and its resistance is 500 Ω
(b) the resistance of the LDR in lighting conditions which make $V_L = 4.0$ V.

Guidelines

Equation 8 can be used for (a) and (b).

2. When a cell of e.m.f. (E) is connected to a slide-wire potentiometer, the balance length is found to be 64.0 cm. This cell is then replaced by a standard cell of e.m.f. 1.02 V, and a new balance length of 44.0 cm is obtained. Calculate the value of E.

Guidelines

Use Equation 9.

3. A potentiometer wire has a driver cell of e.m.f. 1.50 V and negligible internal resistance. If the

Guided examples (3) *continued*

100 cm long slide-wire has a resistance of 2.00 Ω, calculate:

(a) the value of the resistor which must be connected in series so that the p.d. across the wire is 10.0 mV

(b) the e.m.f. (E) of a thermocouple if it is balanced against 65.0 cm of the potentiometer wire

(c) the current through the thermocouple if the jockey is moved to the 50.0 cm mark.

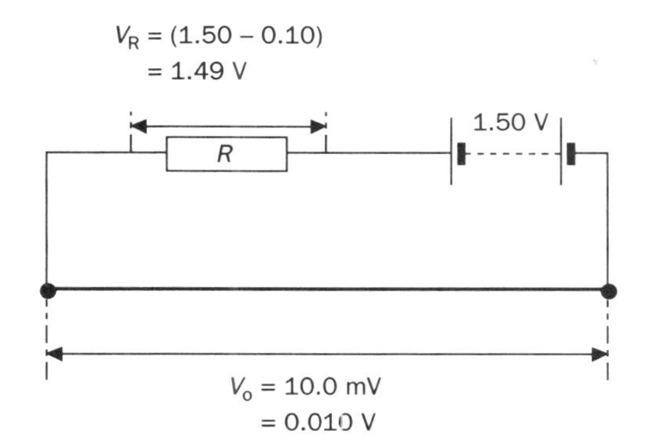

Guidelines

(a) Calculate the current in the slide-wire, then use this to find the value of R so that the p.d. across it is 1.49 V.

(b) Calculate the p.d. across 65.0 cm of the wire. This will be equal to the p.d. of the thermocouple.

(c)

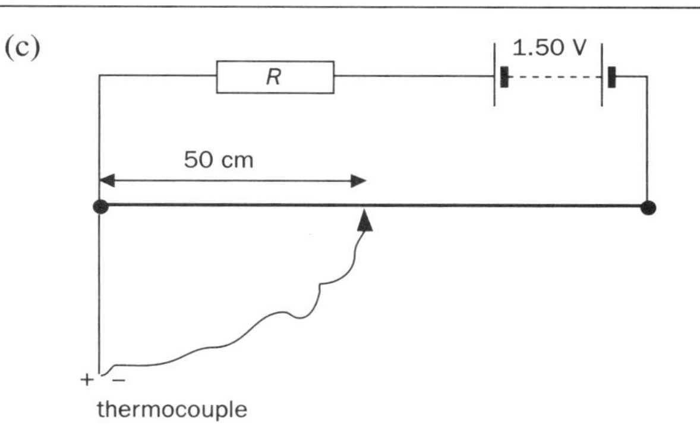

thermocouple circuit

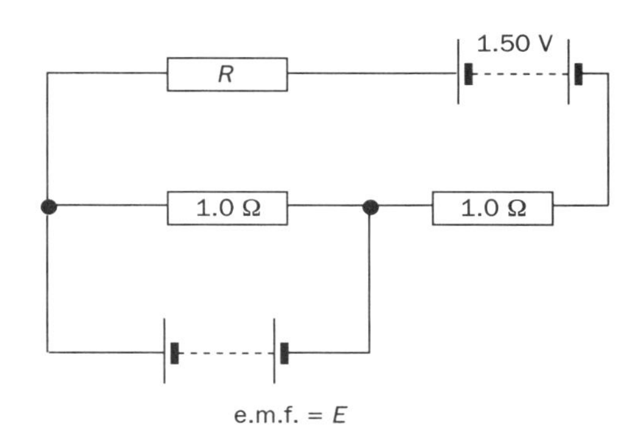

equivalent circuit

The thermocouple acts as a source of e.m.f. (calculated in part (b)). When the jockey is at the mid-point of the wire, the e.m.f. of the thermocouple will not be balanced and there will be a current through the thermocouple. Use Kirchhoff's laws to calculate the magnitude and direction of this current.

Electromotive force and internal resistance

Electromotive force (e.m.f.)

Common symbol	E
SI unit	volt (V)

The electromotive force of a source (battery, generator, thermocouple etc.) is defined as the **energy** (chemical, mechanical, thermal, etc.) converted into electrical energy when **unit charge** (i.e. 1 coulomb) passes through it.

This is similar to the definition of potential difference (p.d.), but they deal with different aspects of an electrical circuit. E.m.f. applies to a source **supplying** electrical energy, while p.d. refers to the conversion of electrical energy **within the external circuit**. The term electromotive force could be misleading if not properly understood, since it has nothing to do with force and is a measure of energy per unit charge.

In general, if a charge Q (in coulombs) passes through a source of e.m.f. E (in volts), then the electrical energy W (in joules) supplied by the source is given by:

$$W = QE$$

(J, C, V)

Equation 10

Heart pacemakers are sometimes permanently implanted to produce and maintain a normal heart rate in persons having a heart block. Advanced technology electric batteries are used to power such devices

Measuring e.m.f.

Ideally, cells and batteries should supply a constant e.m.f., which would be unaffected by the magnitude of the current drawn from them. In practice, this is never the case. If a voltmeter is connected to the battery of a torch, it may read 3.0 V when switched off, but this could drop to 2.5 V or less when switched on. If another lamp were connected in parallel, the voltmeter reading would drop further. This causes problems when trying to measure the e.m.f. of a battery or other source of electrical energy.

A voltmeter measures potential difference, and if it is connected to the terminals of an electrical supply such as a battery, it indicates what is called the **terminal p.d.** (V) of the battery. If the voltmeter is perfect (i.e. has infinite resistance), and the battery is not part of any circuit, then no current will be drawn from the battery, and the indicated reading will be equal to the **e.m.f.** (E) of the battery. If an external circuit is connected to the battery (or the voltmeter is not perfect, and allows a small current), then the reading on the voltmeter will be less than E. The reason for this drop is that when there is a current through the cell, some of its energy is converted into heat by the **internal resistance** of the cell. The decrease is referred to as the **'lost volts'** of the cell, and is proportional to the magnitude of the current. From this we can loosely define e.m.f. as being equal to:

> The terminal p.d. of the source on **open circuit** (i.e. when not maintaining a current).

Internal (or source) resistance

All cells and batteries have resistance, simply because they are made from materials which have some electrical resistance. The magnitude of the internal resistance varies with cell type, but is often very low in rechargeable batteries.

If a perfect voltmeter (i.e. one of infinitely high resistance, through which there could be no current) is connected to the terminals of a cell, it records its e.m.f. In Figure 3.24(a), this is the value (E).

When an external circuit in the form of a resistor (R) is connected to the cell, there will be a steady current (I) through the cell, and the voltmeter will indicate a lower reading, (V) in Figure 3.24(b) *(see bottom of page)*. This value is the **terminal p.d.** of the cell, and also the p.d. across R (assuming the connecting wires have negligible resistance).

Since V is less than E, not all of the energy supplied per coulomb by the cell (i.e. E) is available to the external circuit: some has been 'lost' in driving the current through the cell's internal resistance (r).

Applying the principle of conservation of energy to the complete circuit of Figure 3.24(b):

For each coulomb of charge:
(energy supplied by cell) = (energy dissipated in R) + (energy dissipated in cell)

i.e. (e.m.f.) = (terminal p.d.) + (lost volts)

In symbols:

$E = V + v_r$

where v_r is the p.d. across the internal resistance of the cell, a quantity which cannot be measured directly, only calculated.

From the equation $E = V + v_r$ we see that the sum of the p.d.s across all the resistances in a circuit (external and internal) equals the e.m.f. supplied.

Since $V = IR$ and $v_r = Ir$ we can rewrite the previous equation as:

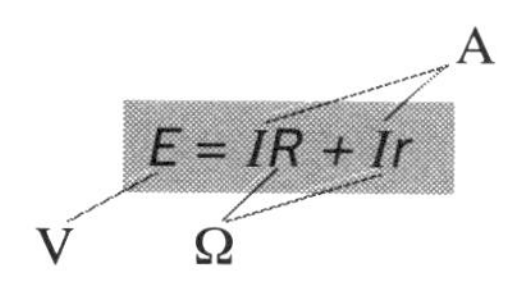

Equation 11

This can be re-arranged into other useful forms:

$E = I(R + r)$

or

$E = V + Ir$

You should note that:

- The magnitude of the internal resistance of a supply depends on several factors, and is not always constant (although it is assumed to be constant for most calculations). However, estimates of its value are useful, and can be obtained from the procedure outlined in the above example.
- Low-voltage supplies from which high currents are taken (e.g. car batteries) must have a low internal resistance. Conversely, high-voltage supplies like the laboratory 5000 V e.h.t. supply must have an internal resistance of millions of ohms to limit the current supplied if they are accidentally short-circuited.

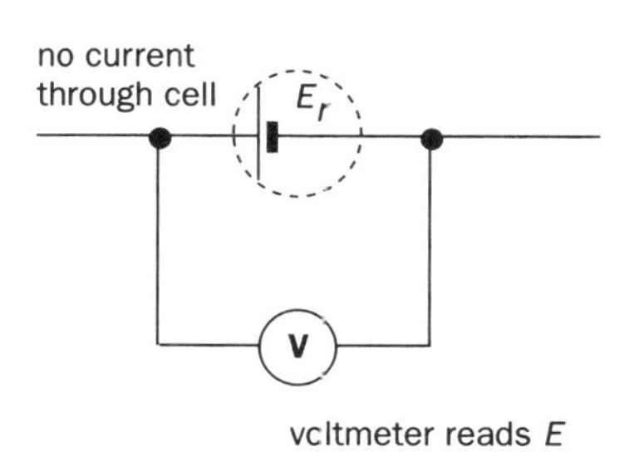

(a) A perfect voltmeter measures the e.m.f. of a cell on open circuit

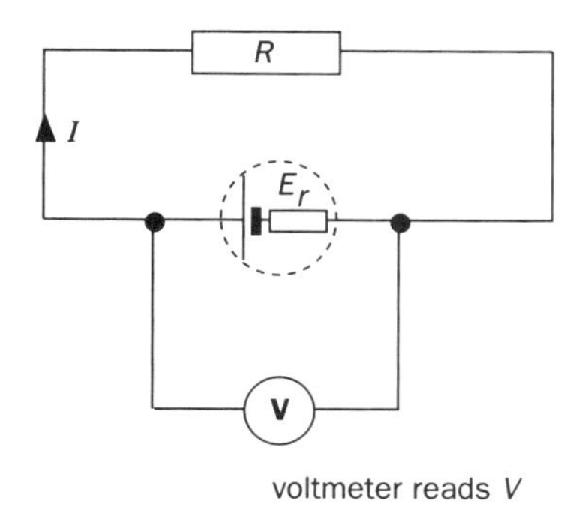

(b) Effects of internal resistance

Figure 3.24 E.m.f. and terminal p.d.

Guided examples (4)

1. A high resistance voltmeter indicates a reading of 1.5 V when connected to a dry cell on open circuit. When the cell is connected to a lamp of resistance R, there is a current of 0.30 A, and the voltmeter reading falls to 1.2 V. What is (a) the e.m.f. of the cell; (b) the internal resistance of the cell; (c) the value of R?

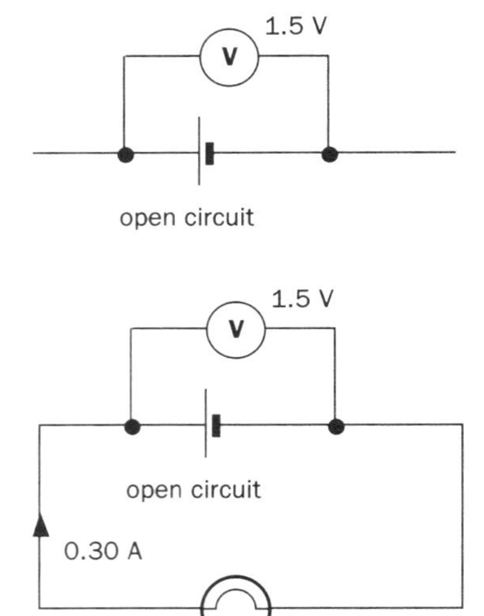

Guidelines

The first clue is the phrase: 'high resistance voltmeter'. This is meant to tell you that it takes such a negligible current that it measures the true e.m.f. of a cell on open circuit.

(b) Use Equation 11. Remember that $IR = V$ (the p.d. across the battery).

(c) Use a variation of Equation 11.

2. A car battery has an e.m.f. of 12 V and an internal resistance of 0.04 Ω. Calculate the terminal p.d. of the battery while the engine is being started, if it supplies a current of 100 A to the starter motor. Describe and explain the appearance of the headlamps if they are switched on while the engine is being started.

Guidelines

Use Equation 11. Remember that $IR = V$ (the p.d. across the battery).
The headlights use bulbs rated at 12 . (i.e. they give normal illumination when the p.d. is 12 V).

Self-assessment

SECTION A
Qualitative assessment

1. State two differences between **series** and **parallel** connection in d.c. circuits.
2. Derive the equations for the effective resistance R_T of three resistors R_1, R_2 and R_3 connected
   (a) in **series**
   (b) in **parallel**.
   (c) What is the effective resistance of n **identical** resistors of resistance R connected in **parallel**?
3. Draw a circuit diagram of an arrangement for determining the **current/p.d. characteristic** of various electrical components, and briefly describe how you would use the circuit.
4. Sketch the graphs you would expect to obtain for
   (a) an ohmic conductor
   (b) a semiconductor diode
   (c) a filament lamp.
   Explain the shape of each graph.
5. State and briefly explain **Kirchhoff's laws**.
6. Using diagrams to illustrate your answer, explain how a variable resistor can be used to supply a continuously variable p.d.
7. Two resistors R_1 and R_2 are connected in series with a d.c. supply of p.d. (V). State an expression for the p.d. across
   (a) R_1
   (b) R_2
   in terms of R_1 , R_2 and V.
8. Explain how
   (a) a **light-dependent resistor** (LDR) can be used in a potential divider circuit to provide a p.d. which increases as light intensity increases
   (b) a **negative temperature coefficient thermistor** can be used in a potential divider circuit to provide a p.d. which increases as temperature increases.
9. Explain the principle of the **potentiometer method** for measuring potential differences. State two advantages and two disadvantages of this method.

Self-assessment *continued*

10. Define **electromotive force** (e.m.f.) of a cell or battery.

11. Explain what is meant by the **internal resistance** and **terminal potential difference** of a cell or battery.

SECTION B
Quantitative assessment

(***Answers:*** 6.7×10^{-5}; 1.0×10^{-4}; 1.7×10^{-4}; 4.2×10^{-3}; 0.24; 0.31; 0.40; 0.48; 0.50; 0.50; 0.50: 0.63; 0.65; 0.70; −0.77; 1.0; 1.0; 1.1; −1.1; 1.3; 1.3; 1.4; 1.4; 1.6; 1.9; 2.0; 2.0; 2.1; 2.5; 2.5; 3.0; 3.1; 3.3; 3.6; 3.9; 4.4; 4.8; 5.0; 5.3; 5.5; 6.0; 6.7; 7.5; 8.0; 12; 12.5; 18; 37; 38; 46; 860; 2.0×10^3; 2.0×10^3; 6.0×10^4.)

1.

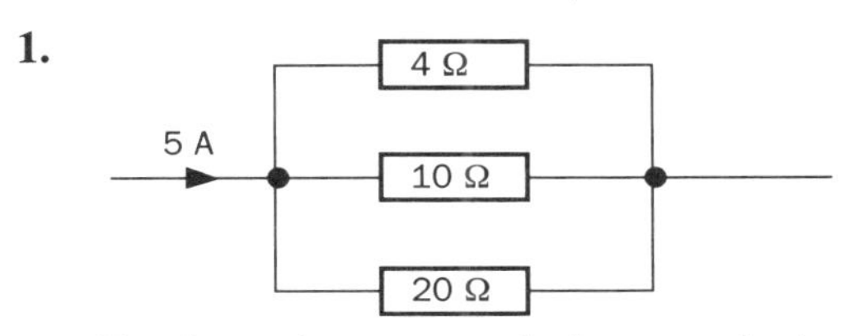

For the resistor network shown, calculate

(a) the total resistance of the combination

(b) the p.d. across each resistor

(c) the current through each resistor.

2.

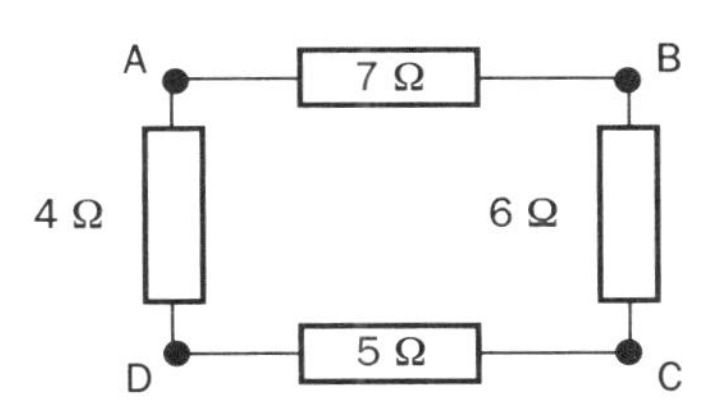

The diagram shows four resistors connected in a closed square. Calculate the total effective resistance of the combination if connection is made between:

(a) A and B (b) B and C

(c) C and D (d) D and A

(e) A and C (f) D and B.

3.

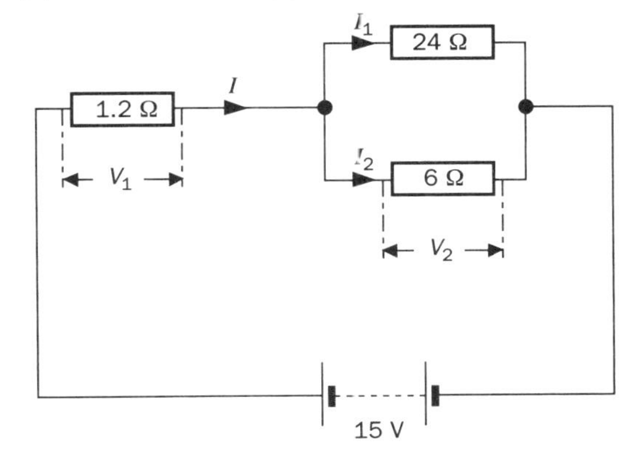

For the circuit shown, calculate

(a) the total resistance of the circuit

(b) the currents and p.d.s shown.

4.

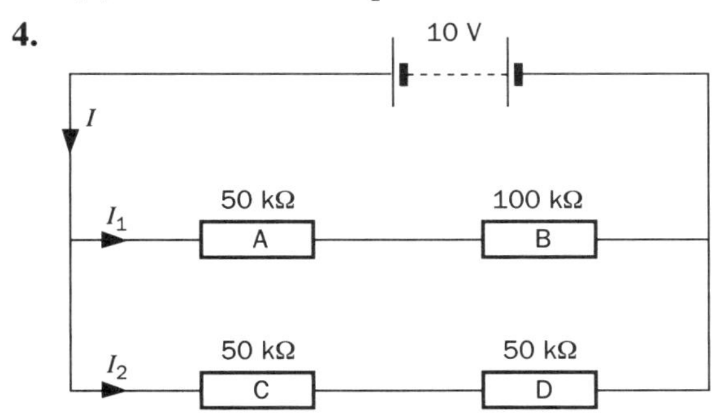

For the circuit shown, calculate;

(a) the total resistance

(b) the total current (I)

(c) the branch currents (I_1 and I_2)

(d) the p.d. across resistor B

(e) the p.d. across resistor D.

5.

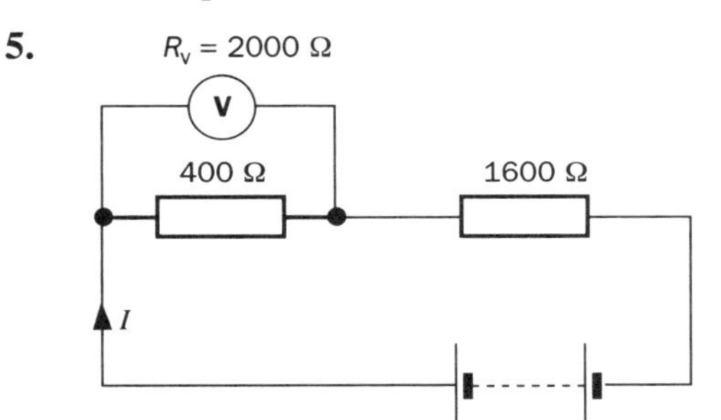

In the circuit shown, a d.c. supply of p.d. V and negligible internal resistance is connected to 400 Ω and 1600 Ω resistors arranged in series. A voltmeter of resistance 2000 Ω connected across the 400 Ω resistor gives a reading of 8.0 V. Calculate:

(a) the magnitude of V and hence the p.d. across the 1600 Ω resistor

(b) the p.d. across the 1600 Ω resistor when the voltmeter is disconnected.

6. Two batteries, each of e.m.f. 10 V and internal resistance 4.0 Ω, are joined in parallel (i.e. like poles are connected together). The combination is then connected to an 8 Ω resistor. Calculate:

(a) the current through each cell

(b) the current through the resistor

(c) the p.d. across the terminals of each cell.

Self-assessment *continued*

7.

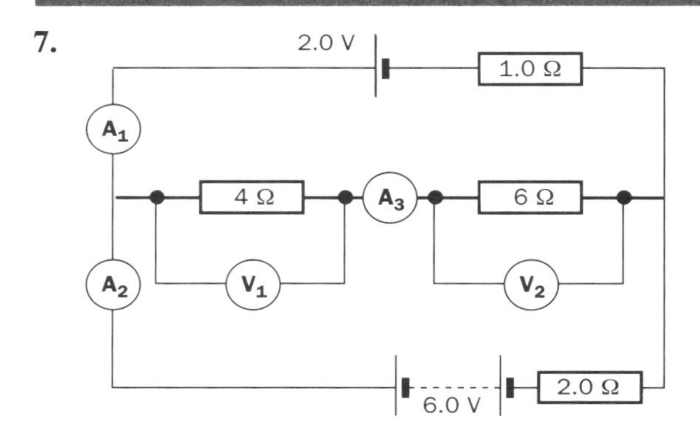

Apply Kirchhoff's laws to the circuit shown above and hence calculate:

(a) the current indicated by the ammeters A_1, A_2 and A_3

(b) the p.d. indicated by the voltmeters V_1 and V_2.

8. A battery of e.m.f. 8.0 V and internal resistance 2.0 Ω is connected in parallel with one of e.m.f. 4.0 V and internal resistance 1.5 Ω. An 8 Ω resistor is then connected between the positive and negative terminals. Use Kirchhoff's laws to calculate:

(a) the current through the 8 V battery

(b) the current through the 4 V battery

(c) the current through the 8 Ω resistor.

9. A 4.0 Ω and a 5.0 Ω resistor are connected in series to a battery of e.m.f. 4.0 V and internal resistance 1.0 Ω. Calculate:

(a) the total current in the circuit

(b) the p.d. across the 4 Ω resistor

(c) the p.d. across the battery terminals.

10. A battery of e.m.f. 6.0 V is connected across a 10 Ω resistor. If the p.d. across the resistor is 5.0 V, calculate:

(a) the current in the circuit

(b) the internal resistance of the battery.

11. A torch bulb is powered by three 1.5 V cells connected in series. If the p.d. across the bulb is 3.75 V and it dissipates heat at the rate of 900 mΩ, calculate:

(a) the current through the bulb

(b) the internal resistance of each cell

(c) the energy dissipated in each cell in 5 minutes.

12. When the battery in a toy car delivers a current of 0.75 A, the p.d. across its terminals is 1.3 V. When the car's motion is reversed, the current delivered by the battery and the p.d. across its terminals become 1.2 A and 0.80 V respectively. Calculate, for the battery,

(a) the e.m.f.

(b) the internal resistance.

13. The thermistor in the potential divider circuit shown has a resistance which varies between 100 Ω at 100 °C and 6.0 kΩ at 0 °C.

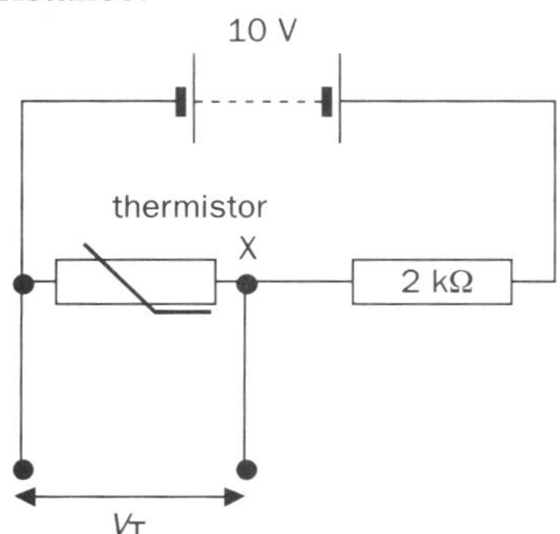

Calculate the p.d. across the thermistor at

(a) 100 °C

(b) 0°C

14. The light-dependent resistor in the potential divider circuit shown below has a resistance which varies between 1000 Ω in bright light, and 5.0 MΩ in darkness.

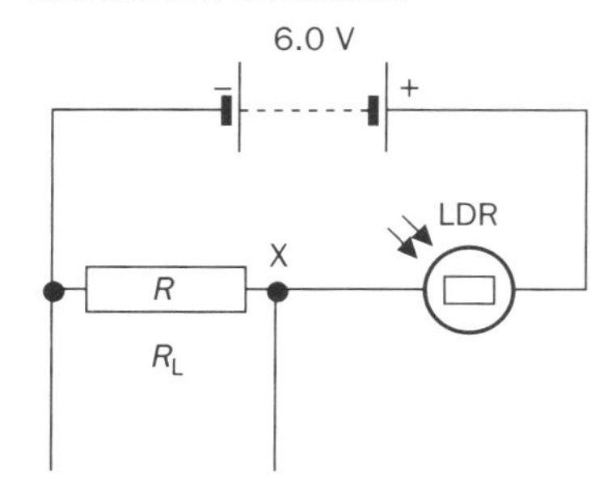

Calculate the value of the series resistor R which is needed so that the potential at point X is

(a) 4.0 V in bright light

(b) 2.4 mV in darkness.

15. A slide-wire potentiometer has a driver cell of e.m.f. 4.0 V and negligible internal resistance. If the slide-wire is 100 cm long and has a resistance of 1.5 Ω, calculate

(a) the length of the potentiometer wire needed to balance a p.d. of 2.8 V

(b) the resistance which must be connected in series with the slide-wire to give a p.d. of 7.0 mV across the whole wire

(c) the e.m.f. (E) of a thermocouple which is balanced by 60 cm of the wire, set up as in part (b).

3.3 The cathode ray oscilloscope (CRO)

This is a very important and versatile electrical instrument which uses an electron beam to display waveforms on a fluorescent screen. The name cathode ray oscilloscope derives from the fact that electron beams were originally called **cathode** rays. The CRO's ability to produce a steady picture of a rapidly varying p.d. is particularly useful because it allows accurate measurements to be made of transient phenomena. At the heart of the oscilloscope is the cathode ray tube shown schematically in Figure 3.25.

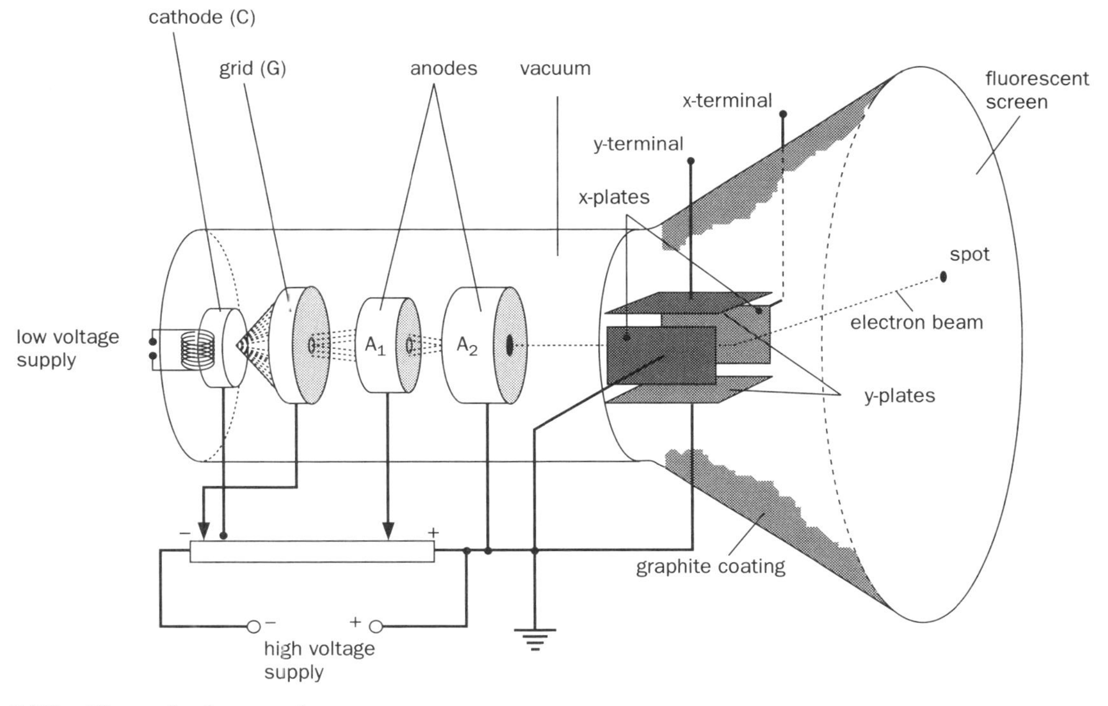

Figure 3.25 The cathode ray tube

The tube can be divided into three main elements.

The electron gun

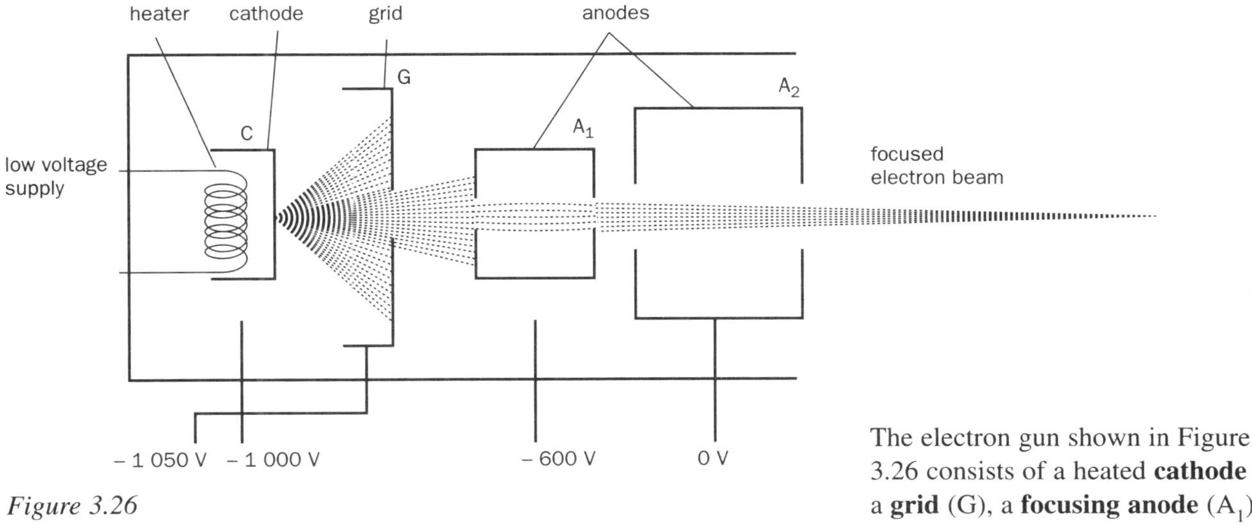

Figure 3.26 Typical electron gun system

The electron gun shown in Figure 3.26 consists of a heated **cathode** (C), a **grid** (G), a **focusing anode** (A_1) and an **accelerating anode** (A_2).

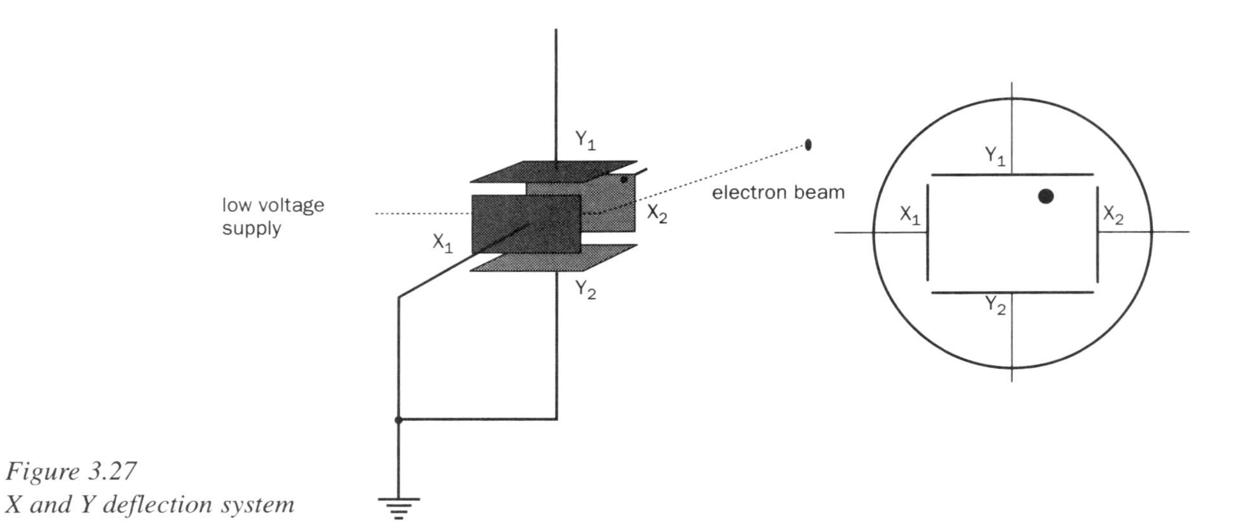

Figure 3.27
X and Y deflection system

When the cathode is heated by the current from a low voltage supply, free electrons are **thermionically** emitted from its surface. These electrons are strongly attracted by the two anodes which are at a positive potential with respect to the cathode.

Beam brightness

The electrons first encounter the **grid**, which is **negative** with respect to the cathode and controls the number of electrons reaching the anodes and screen. Making the grid more negative repels some electrons, and reduces the number of electrons getting through. Making it less negative increases the number. Thus by varying the grid potential we can control the number of electrons which strike the screen and hence the **brightness** of the spot on the screen.

Beam focus

Apart from accelerating the electron beam towards the screen, the anodes also have a **focusing** effect on the beam. The sharpness of the spot produced on the screen is controlled by adjusting the potential of anode (A_1) with respect to the cathode. This alters the angle of the cone of electrons issuing from the grid, so that the apex of the cone falls on the screen to give the finest possible spot.

Anode (A_2) is at zero potential, and its main function is to accelerate the electrons.

The deflection system

The beam deflection system consists of two pairs of parallel plates at right angles to each other, as shown in Figure 3.27 *(see top of page)*. When a p.d. is applied across a pair of plates, the electric field created exerts a force on the electrons. The electric field between the Y-plates causes **vertical** deflection of the beam and that between the X-plates produces **horizontal** deflection.

X and Y shift

If the oscilloscope is switched on (with the time-base off), a stationary spot should be visible on the screen where the electron beam strikes it. By varying the steady p.d. between the X- or Y-plates, the horizontal and vertical position of the spot on the screen can be controlled. On the oscilloscope these are called the X- and Y-shift controls.

X- and Y-inputs

When the oscilloscope is being used for observing and measuring, the voltages being studied are applied to either or both of the X- and Y-inputs. This causes changes in the position of the spot on the screen, and enables measurements to be made.

It is also possible to use Helmholtz coils to provide a deflection system. In this case the electron beam would be deflected by the magnetic field between the coils which would also be positioned at right angles to each other.

The display system

Figure 3.28 *(see top of page 221)* shows a typical **cathode ray tube**. The zinc sulphide gives a glow of light when electrons collide with it. The atoms of the coating become excited by absorbing the energy of the incoming electrons, and this energy is re-released as a burst of light. Thus the motion of the spot, produced by the electron beam on the screen, is seen as a glowing trace. The addition of other chemicals can produce different colours.

There is a graphite coating inside the tube which ends close to the screen. This is at earth potential (0 V) and provides a return path for electrons which strike the screen as well as for secondary electrons produced from the impact of electrons with the zinc sulphide. Otherwise there would be a large build-up of static charge on the tube.

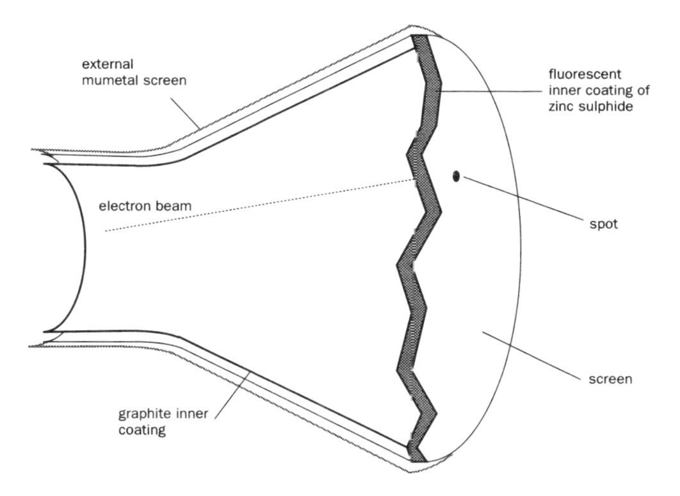

Figure 3.28 CRO display system

The graphite also shields the electron beam from external electric fields which would give unwanted deflections.

A mumetal screen surrounds the tube and shields it from external magnetic fields.

Using the CRO – the time-base

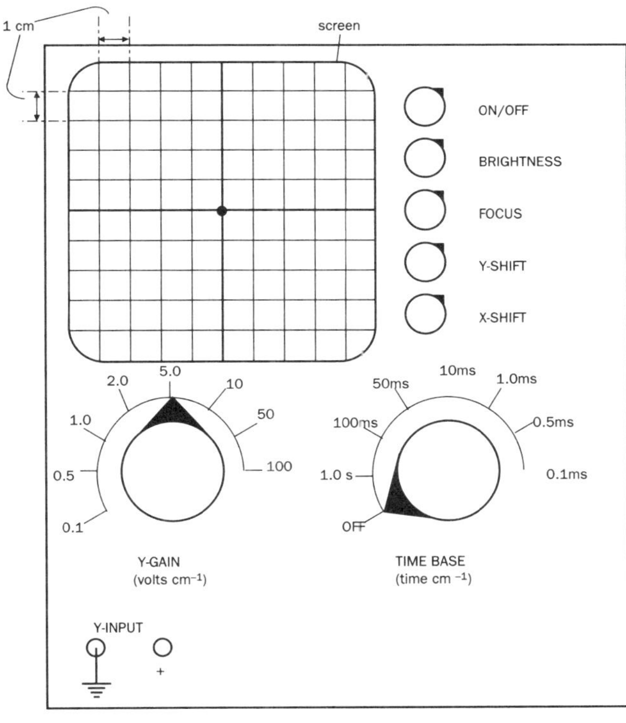

Figure 3.29 Basic controls on a laboratory CRO

Figure 3.29 shows the essential features and controls found on a basic school oscilloscope. Input p.d.s to be studied are applied to the Y-input terminals (one of which is earthed) and go through an amplifier before reaching the Y-plates. The amplification is altered by the Y-GAIN control. This enables a very wide range of potential differences to be measured.

When the oscilloscope is switched on with the time-base off and no p.d. applied to the input, the spot appears in the centre of the screen.

If a steady direct potential difference is applied to the Y-input. (e.g. from a cell) the spot will move up or down to a new stationary position from its central position by an amount depending on the size of the p.d. and on the Y-GAIN setting (see Figure 3.30). If the non-earthed terminal is made **positive**, the spot moves **upwards**.

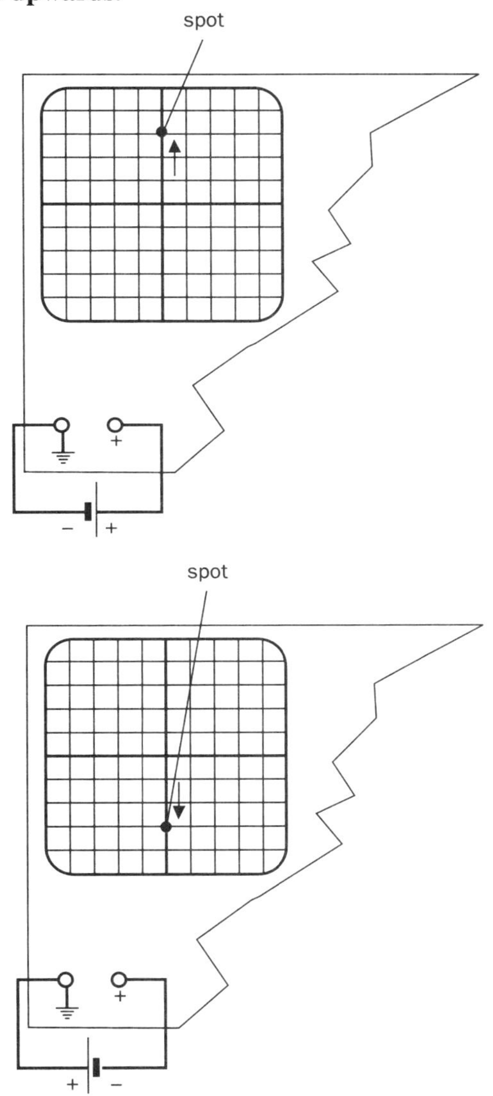

Figure 3.30 Spot deflection produced by steady D.C. *potential difference (time-base off)*

If an alternating p.d. is applied to the input terminals the spot moves vertically up and down at the frequency of the applied p.d. as shown in Figure 3.31. This produces a vertical straight line trace on the screen whose length represents the peak-to-peak value of the p.d. applied.

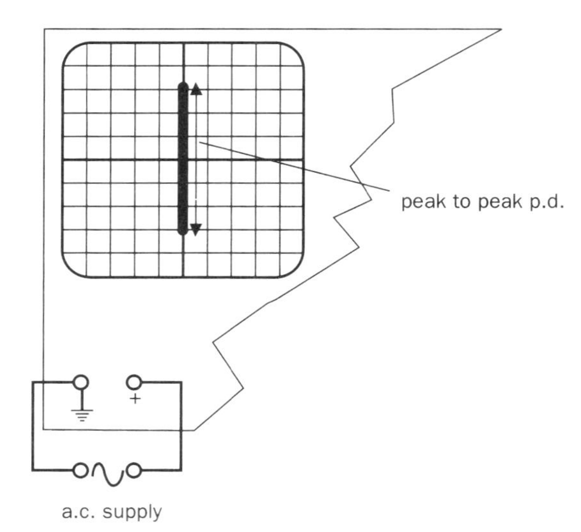

Figure 3.31 Trace produced by an alternating p.d. (time-base off)

In order to see how the alternating p.d. varies with time, the X-plates are connected to an internal circuit called the **time-base** which applies a saw-tooth p.d. as shown in Figure 3.32.

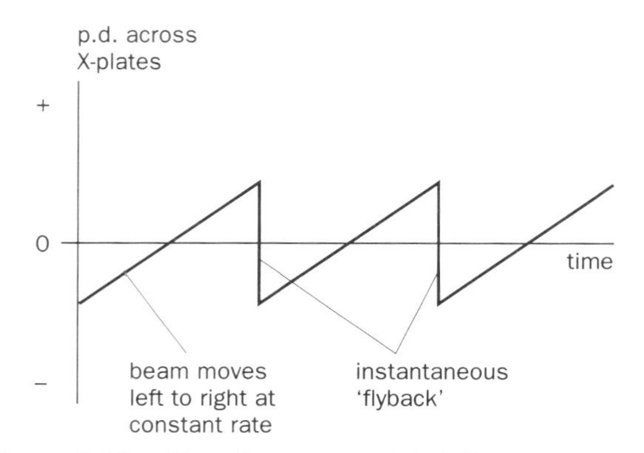

Figure 3.32 Time-base potential difference (saw-tooth waveform)

This moves the spot horizontally across the screen at a **constant** rate which is controlled by the **time-base** setting. The cycle of the time-base p.d. starts off with a negative value, which pulls the spot to the extreme left of the screen. The p.d. becomes positive at a constant rate, which sweeps the spot at a constant speed across the screen. At the end of the cycle, the saw-tooth p.d. changes back to its initial negative value in a very short time (called the **flyback** time) and this returns the spot to its starting position, and the cycle repeats. Time-base settings can range from several seconds per cm, where the spot moves very slowly across the screen, to microseconds per cm. In practice, with time-base settings of shorter than around 10 ms per cm, the trace looks like a continuous line rather than a moving spot. The combined effects of the alternating p.d. applied to the Y-plates and the time-base p.d. applied to the X-plates produces the alternating trace shown in Figure 3.33.

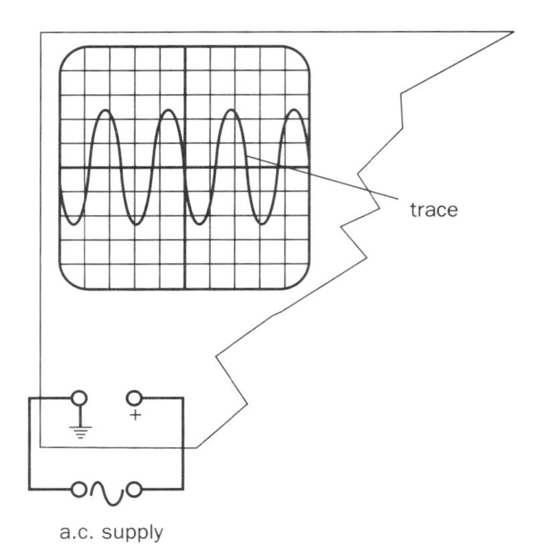

Figure 3.33 Trace obtained with alternating p.d. (time-base on)

Uses of the CRO

Measuring potential difference

Both a.c. and d.c. voltages can be measured by connecting to the Y- input terminals of the CRO.

D.c. potential difference With the time-base switched off the spot is displaced vertically by an amount which depends on the size of the p.d. and the Y-GAIN setting (Figure 3.34 (*see below and top of next page*)).

With no p.d. applied to the Y-input, use the Y-shift control to position the spot on one of the grid lines. Apply the p.d. to be measured, the spot will be deflected vertically. Use the Y-GAIN control to produce the maximum possible deflection (improves accuracy by reducing the percentage error in reading the scale).

Measure the deflection of the spot (in cm) and multiply by the Y-GAIN setting (in V cm^{-1}). This gives the p.d. in volts.

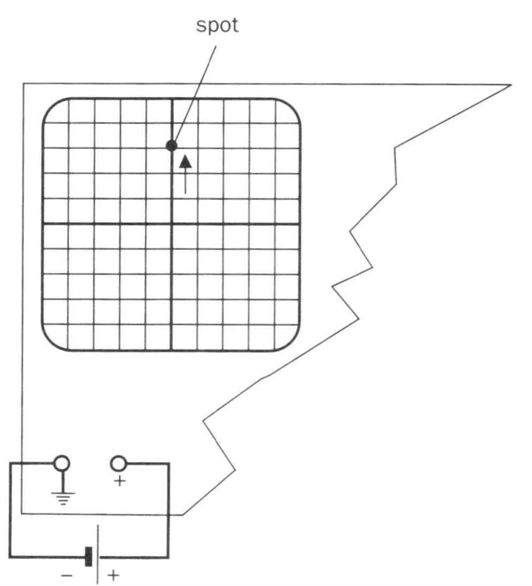

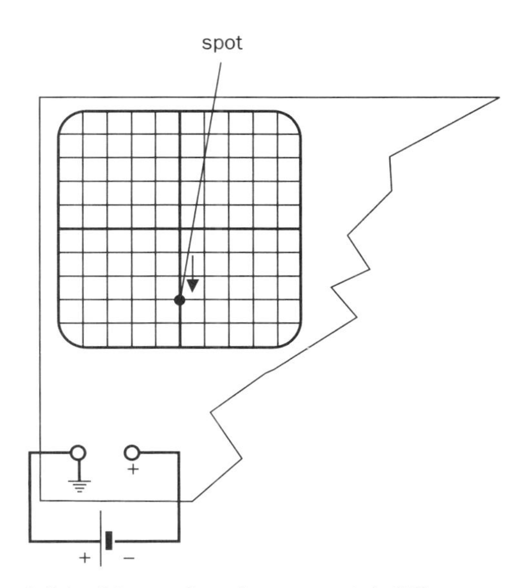

Figure 3.34 Measuring d.c. potential difference

A.c. potential difference With the time-base switched off, the length of the vertical trace obtained represents the **peak-to-peak** value of the applied potential difference (V) as shown in Figure 3.35.

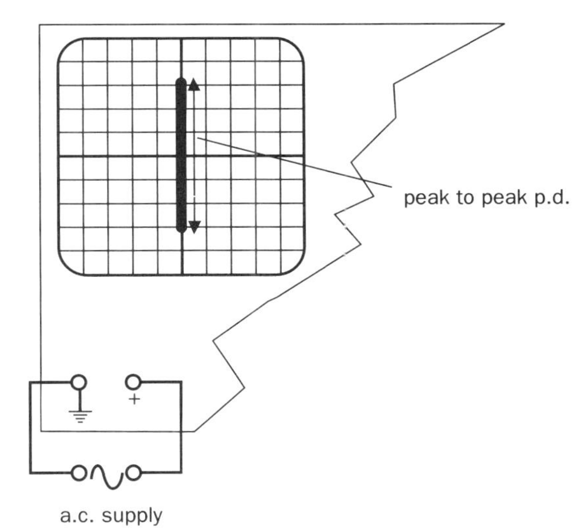

Figure 3.35 Measuring an alternating potential difference

To determine the peak-to-peak potential difference, measure the length of the trace (in cm) and multiply by the Y-GAIN setting (in V cm^{-1}).

To determine the **amplitude** of an alternating p.d., it is usual to have the time-base switched on, when a sinusoidally varying trace is obtained, as shown in Figure 3.36.

Use the Y-shift control to centre the trace on the screen. Measure the height of the peaks above the centre line (in cm). Multiplying by the Y-GAIN setting (in V cm^{-1}) gives the amplitude (or peak value) of the alternating potential difference.

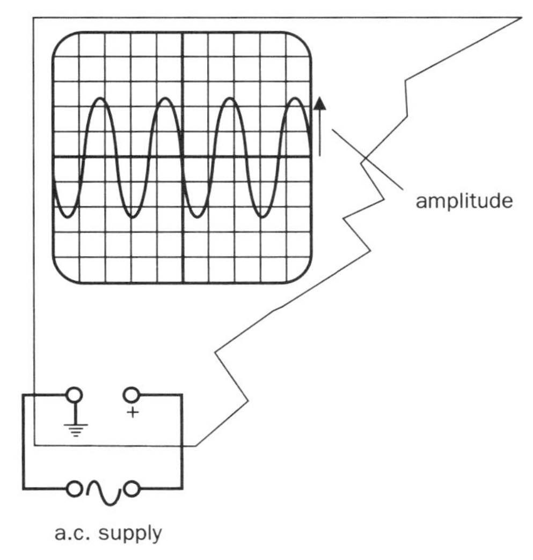

Figure 3.36 Measuring the amplitude of an alternating waveform

The CRO is a very useful form of voltmeter because:

- both d.c. and a.c. voltages can be measured
- the electron beam is a virtually weightless pointer with an almost instantaneous response
- it has a very high resistance to d.c. and a very high impedance to a.c. – hence it draws very little current from the circuit it is connected to.

Measuring time intervals

Since the time-base is marked in (time cm^{-1}), it is possible to determine the **time interval** between two events, as long as the two events produce visible changes in the trace seen on the CRO screen. Two extra controls not shown in Figure 3.29 must be used now:

The **fine** (or variable) time-base control must be set to CAL. This ensures that the trace takes an exact number of milliseconds to cross each cm on the screen.

The X-GAIN must be set to minimum, again to ensure that the time-base setting accurately represents the movement of the spot on the screen.

Example A microphone beside a starting pistol detects the initial sound of the gun and, a short time later, the echo of the sound from a cliff some distance away. These events are seen on the CRO screen as voltage peaks which are separated by a horizontal distance of 6.0 cm: as shown in Figure 3.37.

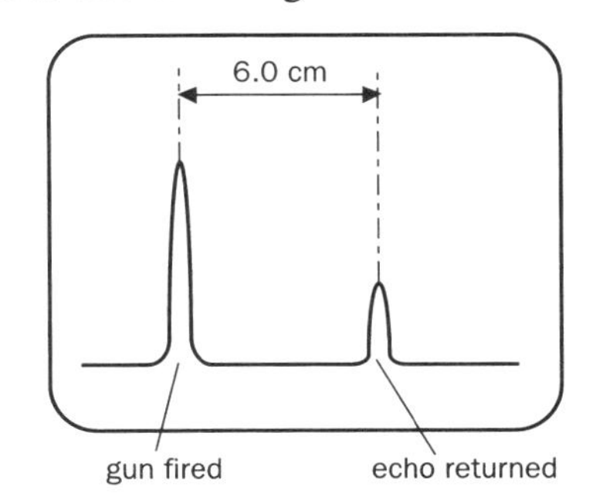

Figure 3.37 Trace obtained on the CRO

If the time-base setting is 100 ms cm^{-1}, then the time interval between making the sound and receiving the echo is:

$6 \times 100 = 600$ ms = 0.6 s

Measuring frequency

Since frequency = 1/(time period), measuring frequency using a CRO involves using the above method to measure the time interval between successive cycles of the alternating waveform, and using this to calculate the frequency. In hospitals, CROs are used to monitor heart beat. The muscles of the heart are triggered to contract by electrical signals carried through the nerve fibres. These electrical signals can be received through electrodes stuck to the skin of the chest, enabling medical staff to study the rhythm and behaviour of the heart's contractions.

Suppose the heart trace shown in Figure 3.38 is obtained with the time-base setting at 0.5 s cm^{-1}.

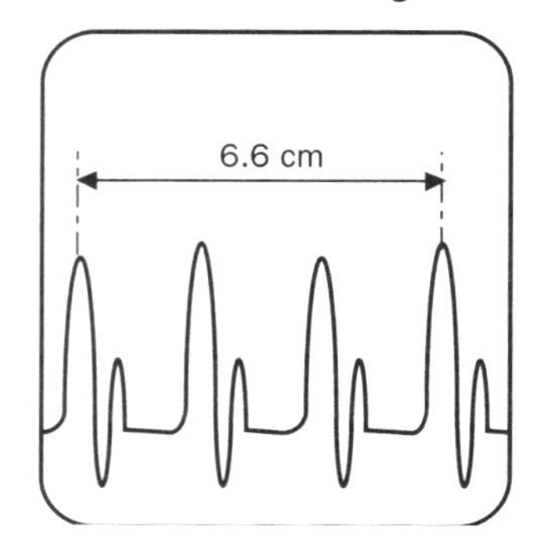

Figure 3.38 Heart trace on CRO screen

Clearly each beat of the heart is slightly irregular, but an average heart rate can be calculated by measuring the time taken for several beats, and calculating an average period.

	Time for three beats	$= 6.6 \times 0.5$	= 3.3 s
	Period, T	= 3.3/3	= 1.1 s
$\therefore$	frequency f	$= 1/T$	= 1/1.1
$\therefore$	f	= 0.909 beats per second (Hz)	
$\therefore$	heart rate	$= 0.909 \times 60$	
		= 55 beats per minute	

Displaying waveforms and phase differences

A double-beam oscilloscope has the facility to display two separate signals on the same screen. The two traces can be synchronised so that they always start together, and in this way, phase differences can be observed and measured. The arrangement shown in Figure 3.39 allows the sound waves emitted by the loudspeaker (L) and those received by the microphone (M) to be simultaneously displayed. One trace (Y_1) on the CRO screen is the signal fed to L and the other trace (Y_2) is the signal from M.

By moving M slowly towards L, a position is found for which traces Y_1 and Y_2 are as shown in Figure 3.40(a). At this position the distance between the loudspeaker and the microphone is equal to a whole number of wavelengths and the waves from L are in phase with those received by M.

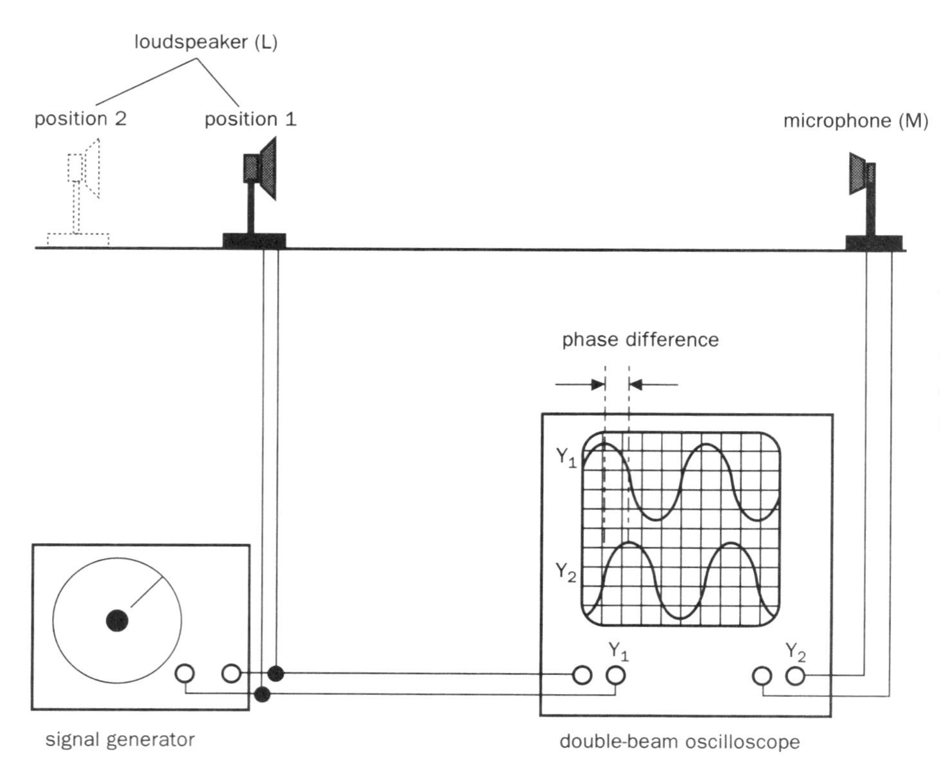

Figure 3.39 Use of double-beam oscilloscope to display and measure phase difference

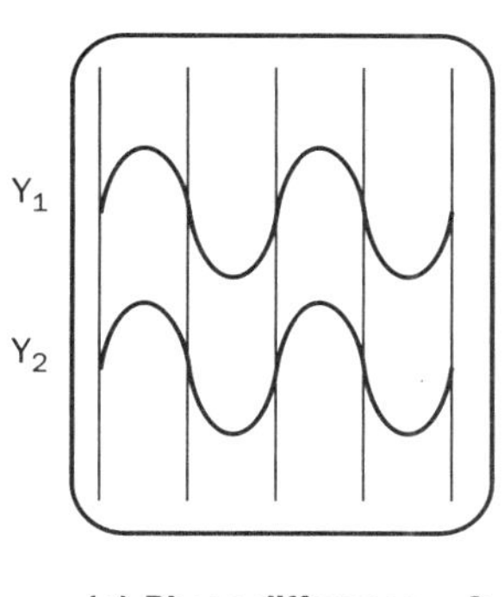

(a) Phase difference = 0

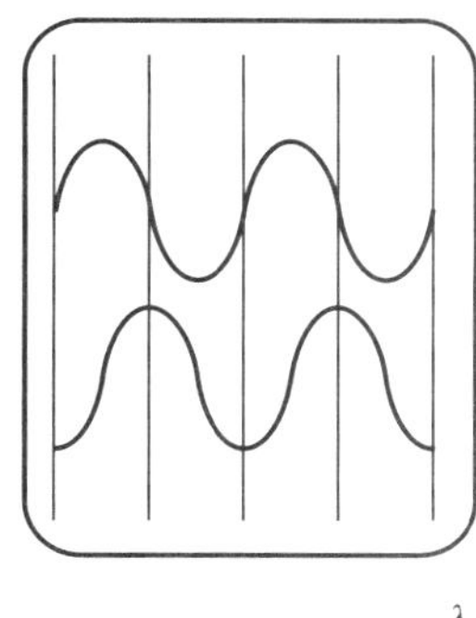

(b) Phase difference = $\frac{\lambda}{4}$

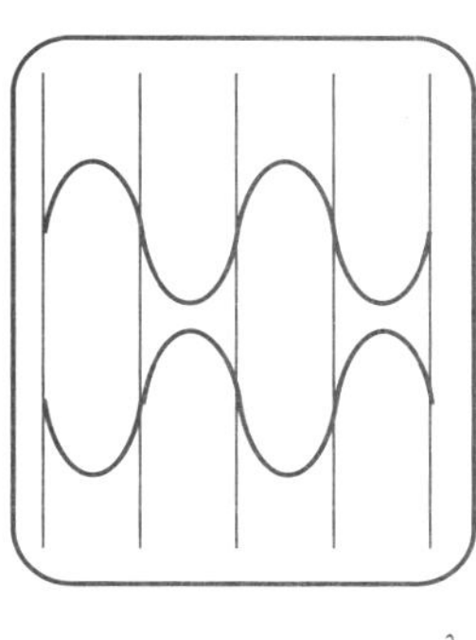

(b) Phase difference = $\frac{\lambda}{2}$

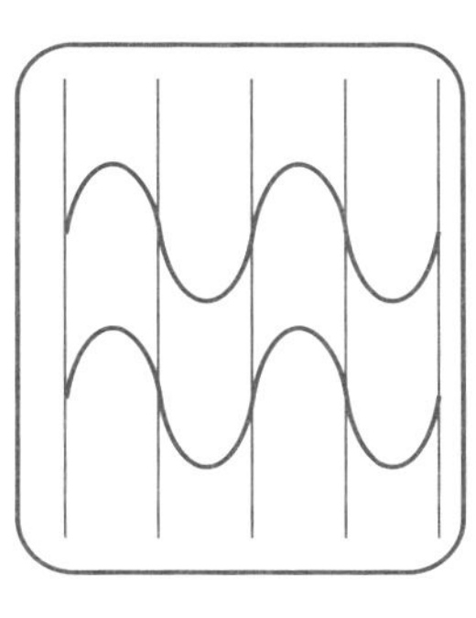

(a) Phase difference = 0

Figure 3.40 Appearance of traces as M is moved towards L

If M is moved a further distance $\lambda/4$ away from L the Y_2 trace changes to the form shown in Figure 3.40(b) where the two signals have a phase difference of $\lambda/4$ (or 90° or $\pi/2$ rad). Continued movement of M away from L causes further increases in the phase difference between Y_1 and Y_2 until a new position is reached for which there is zero phase difference. L has now been moved a distance equal to one whole wavelength. This distance could be measured to enable the wavelength of the sound to be determined.

Guided examples (1)

1.

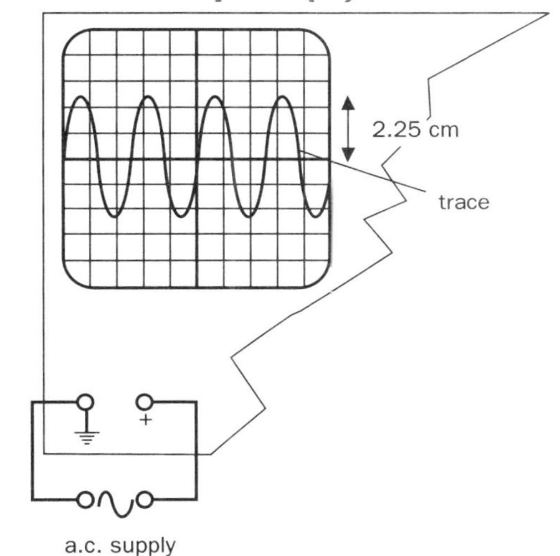

A sinusoidal alternating voltage applied to the Y-input terminals of a CRO produces a trace on the screen as shown in the diagram. If the Y-amplification control and time-base settings are 5 V cm^{-1} and 50 ms cm^{-1} respectively, and the grid lines on the screen are 1 cm apart, calculate:

(a) the frequency of the applied voltage

(b) the peak value of the voltage.

Guidelines

(a) Period of waveform (T) = length of 1 cycle (in cm) × time-base setting (in s cm^{-1}).

Frequency (f) of applied voltage is given by:

$$f = \frac{1}{T}$$

(b) Peak value of voltage, V_p = amplitude of trace (in cm) × Y-gain setting (in V cm^{-1})

2.

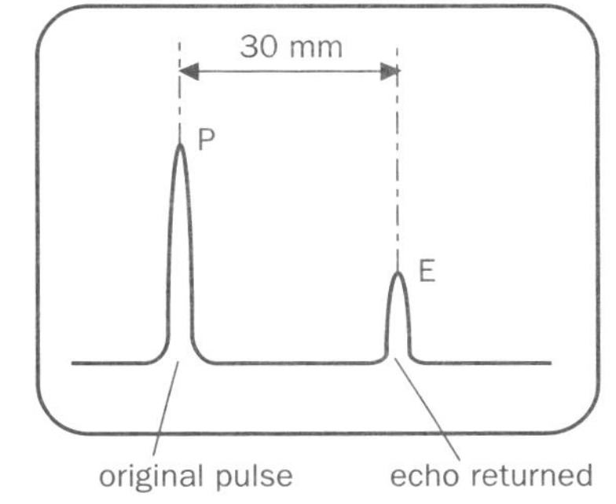

A student uses a CRO to measure the speed of sound in steel. The oscilloscope is used to measure the time taken for a brief pulse of ultrasound to travel through a 3.0 m length of steel and back again. The diagram shows the trace obtained on the CRO screen when the time-base control was set at 0.33 ms cm^{-1}. If the distance on the screen between the original pulse (P) and the reflected pulse (E) is 30 mm, what is the speed of sound in steel?

Guidelines

Pulse travel time (t) is given by:

$$t = \frac{\text{(distance between original and reflected pulses)}}{\text{(time-base setting)}}$$

∴ pulse speed (v) is given by:

$$v = \frac{\text{(total distance travelled by sound waves)}}{t}$$

Self-assessment

SECTION A
Qualitative assessment

1. Draw and label a diagram showing the basic construction of a cathode ray tube. You should identify the **cathode**, **grid**, **focusing** and **accelerating anodes**, **deflection plates** and **screen**.
2. The cathode ray tube can be thought to consist of three main parts:
   (i) the electron gun
   (ii) the deflection system
   (iii) the display system.
   Briefly describe and explain the function of each of these elements.
3. What is the **time-base** in an oscilloscope? Sketch a graph showing how a linear time-base p.d. varies with time and explain its form.
4. Explain how a CRO can be used to measure:
   (a) a **d.c. potential difference**
   (b) the **peak value** and **frequency** of an a.c. potential difference
   (c) the time interval between two events.
5. State the advantages of the CRO as a p.d. measuring instrument.
6. Explain how a double-beam CRO can be used to show phase differences.

SECTION B
Quantitative assessment

(***Answers:*** 0.30; 0.016; 4.0; 20; 400; 1.3×10^3; 1.4×10^3)

1. A CRO is being used to study an alternating potential difference obtained from a transformer connected to the mains supply. The p.d. has a peak-to-peak value of 8.0 V and a frequency of 50 Hz. If the Y-GAIN control is set at 5.0 V cm^{-1} and the time-base setting is 50 ms cm^{-1}, calculate
   (a) the peak-to-peak height of the trace in cm
   (b) the number of complete cycles which will be displayed if the trace is 8.0 cm wide.
2. A CRO screen displays the trace shown when the Y-amplification control and time-base settings are 100 mV cm^{-1} and 0.20 ms cm^{-1} respectively. Obtain values for
   (a) the peak p.d.
   (b) the frequency of the signal.

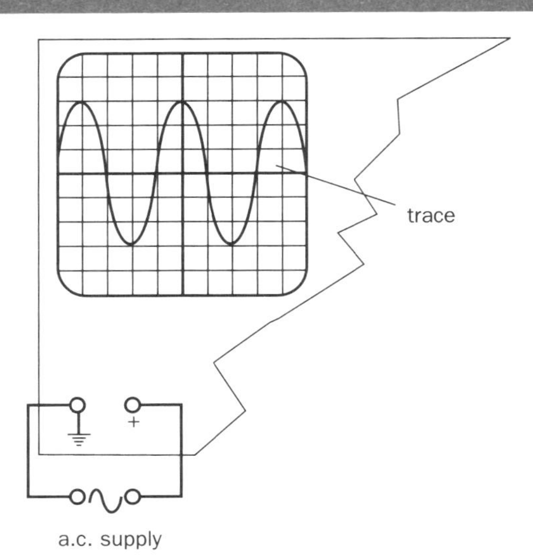

3. 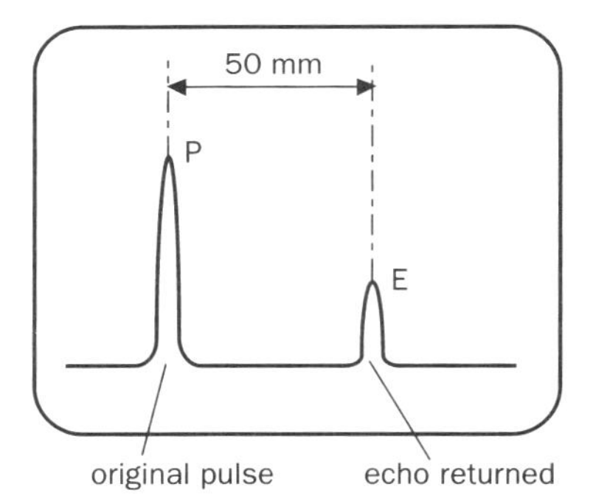

   A CRO is used to determine the speed of sound in water. The experiment involves measurement of the time taken for a sound pulse to travel through a 3.5 m depth of water and back again. A microphone connected to the Y-input terminals of the oscilloscope is used to pick up the emitted and reflected sound pulses. The diagram shows the trace obtained, A being the original pulse and B the final pulse after reflection. If the time-base setting is 1.0 ms cm^{-1}, what is the value for the speed of sound obtained in the experiment.
4. A CRO has its Y-GAIN control set at 2.0 V cm^{-1}. A sinusoidally alternating signal is applied to the Y-input terminals to give a steady trace with the time-base set so that the electron beam takes 10 ms to cross the screen. If the trace obtained has a peak-to-peak height of 4.0 cm and contains 4 complete cycles, calculate
   (a) the peak value, and
   (b) the frequency of the applied potential difference.

3.4 Magnetic effects of electric currents

Magnetism

As we have already stated in Section 1.11 a **gravitational** field exists in the region around any mass and, in this region, other masses experience a **gravitational** force. We also know that an **electric** field exists around charges, and in this region, other charges will be subjected to an **electric** force. **Magnets** give rise to a **magnetic** field in the region around them and in this region other magnets and magnetic materials will experience a **magnetic** force. Magnets have been known on Earth for centuries. The strange attraction between lodestone and ordinary iron was studied by the Greek philosopher Thales as long ago as 600 BC. The Chinese probably had a crude magnetic compass by about 200 AD, but the compass as we know it did not appear in the West until about 1200 AD.

The magnetic field around a magnet may be represented using the idea of **field lines**, in a similar manner to electric and gravitational fields. By convention, the field **direction** is that of the force on a N-pole of a magnet placed at the point. The relative density (lines per unit area) of the field lines is a measure of the magnetic field strength in a particular region. Thus parallel field lines indicate a uniform field, whereas converging field lines tell us that the field is increasing in strength (see Figure 3.41).

The Earth behaves as though it had a bar magnet at its centre. Observation of the direction of the magnetic field at the Earth's surface leads to the pattern shown in Figure 3.42 *(see page 228)*. Near the equator, the lines are nearly parallel to the surface, but nearer the poles, the lines become nearly perpendicular to the surface.

A freely-suspended magnet will always align itself with the direction of the local magnetic field. Since the Earth has a magnetic field, this property is used for navigation, since the alignment of the magnet with the Earth's magnetic field defines the direction we call North–South. Within a magnet, the magnetic effects seem to be concentrated at regions called **poles**. Experiments and observations show that:

- There are two types of pole. Unmagnetised magnetic materials are attracted towards either pole.
- Opposite poles exert attractive forces on each other, while similar poles repel each other.

Figure 3.41 Magnetic field lines

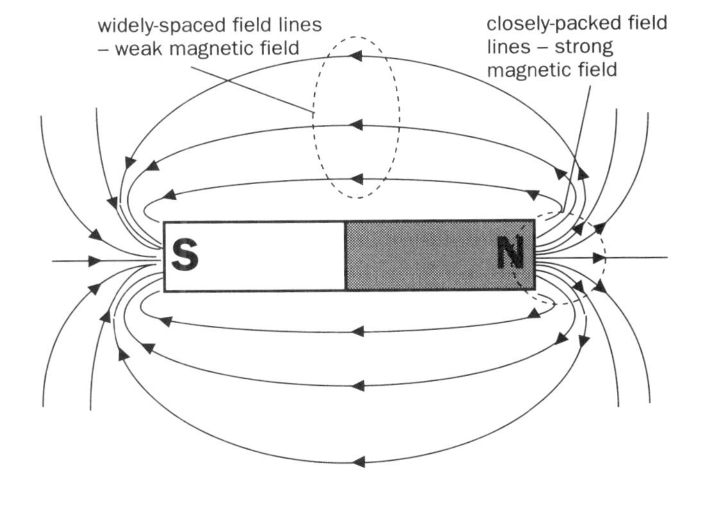

(a) Non-uniform magnetic field around a bar magnet

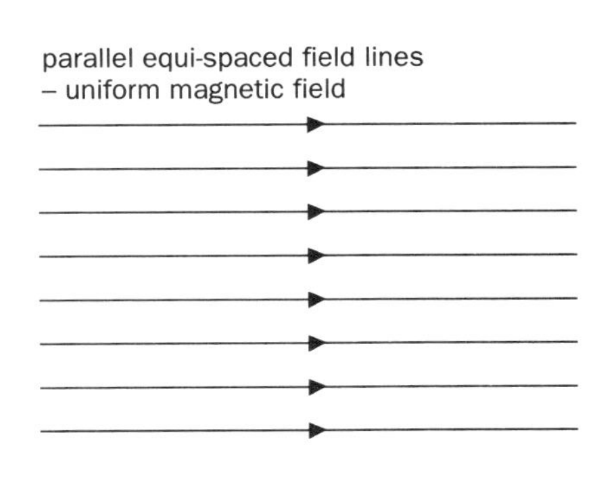

(b) Uniform magnetic field

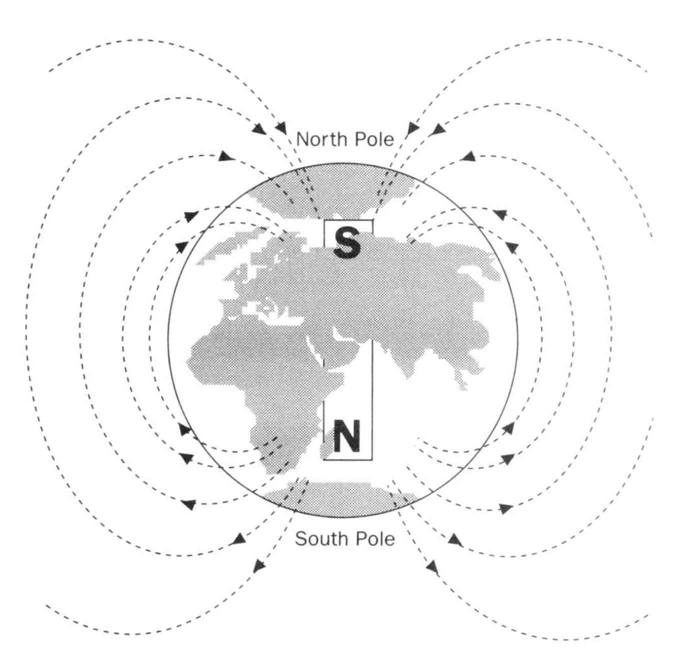

Figure 3.42 Earth's magnetic field

- In the absence of another magnetic field, a freely suspended magnet will align itself so that one pole points towards the Earth's North Pole, while the other points towards the South Pole.
- All magnets have (at least) two poles, and the pole which is attracted towards the Earth's North Pole is called the north-seeking pole (N-pole or north pole) and the other pole the South-seeking pole (S-pole or south pole). This leads to the slightly confusing conclusion that the type of magnetic pole under the Earth's North Pole is in fact a S-pole, since the N-pole of a compass needle is attracted to it. This is why it is better to refer to the poles of magnets as N-poles or S-poles rather than north or south poles.

In 1820, Hans Christian Oersted discovered that electric currents produce magnetic fields, and, shortly after, James Clerk Maxwell proved that all magnetic fields have their origins in the movement of charges. Electrons and protons are the fundamental charged particles of atoms, and it is their motion which is responsible for the magnetic properties of materials. Each orbiting electron represents a circular electric current, and produces a tiny magnetic field. In most materials, the random alignment of atoms means that there is no overall magnetism. In materials which have magnetic properties, these 'atomic magnets' align themselves in small groups, called **domains**, which behave as tiny magnets. Normally, these domains lie in random directions (Figure 3.43(a)), and the material is not magnetised. Under the influence of a weak external magnetic field, some domains can rotate their orientation, making the material into a weak magnet (Figure 3.43(b)).

The more domains become aligned in a common direction, the stronger the magnet. Once all the domains in a piece of iron are aligned in the same direction, the material is said to be **saturated** and the magnetic effect cannot be further increased (Figure 3.43(c)). If the structure of the material is such that these tiny magnets remain in their aligned positions, the material is said to be permanently magnetised, retains its own magnetic field, and is able to exert magnetic forces on other magnets. Very few elements (iron, nickel, cobalt and a few rare earth elements) can be permanently magnetised. These are called **ferromagnetic** materials, and respond even to weak magnetic fields. A ferromagnetic material is described as **hard** if it retains its magnetism, while **soft** magnetic materials are those which can be magnetised easily, but quickly lose their magnetism once the applied field is removed. **Paramagnetic** materials show a weak response to a strong magnetic field, while **diamagnetic** materials become magnetised in opposition to an applied field.

(a) Unmagnetised iron – random orientation no resultant magnetic field

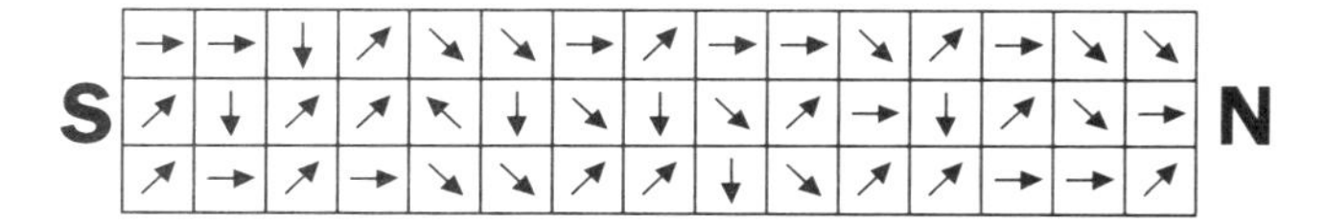

(b) Weak magnetic field – some domains align material becomes a weak magnet

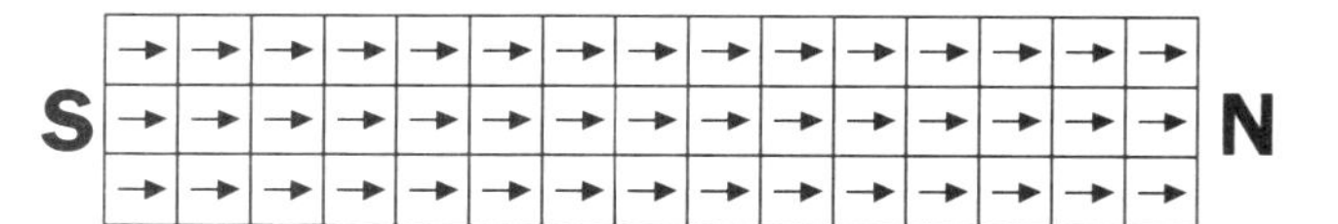

(c) Strong magnetic field – all domains align (saturation) material becomes a strong magnet

Figure 3.43 Magnetic domains

Magnetic fields near current-carrying wires

Since an electric current is a flow of charged particles, it follows that a magnetic field will be produced around any conductor which carries an electric current. The shape and magnitude of the field depends on the current flowing, the arrangement of the conductor and the medium in which it is situated. If you look back to Figure 3.41, you will see that the **strength** of a magnetic field at any point depends on the concentration of the field lines at that point. We model the magnetic field by thinking of it as being filled with **magnetic flux**, which is represented by field lines in diagrams. This concept will be explored further in Section 3.5, but for now it gives us a useful means of measuring magnetic field strength in terms of the concentration of flux. So long as a magnetic field is steady (i.e. unchanging), its strength, as represented by the flux density, can be measured using a **Hall probe**. Such a probe consists of a small slice of semiconducting material which can be placed anywhere in the field. When the probe is supplied with current and inserted into a magnetic field, a voltmeter connected to the probe indicates a p.d. which is proportional to the flux density component at right angles to the probe's flat tip. The probe's action depends on the Hall effect which is dealt with in greater detail at the end of this section.

Straight, current-carrying wire

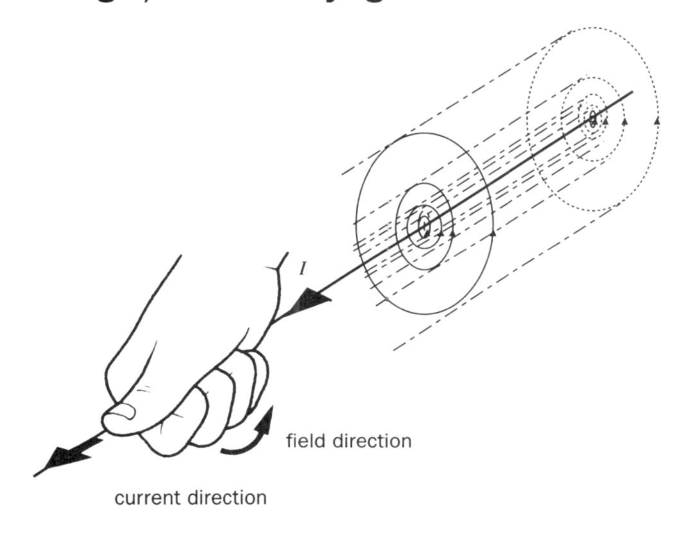

Figure 3.44 Magnetic field around a straight current-carrying conductor – right-hand grip rule

This is the simplest arrangement for producing a magnetic field from an electric current. The field which is generated from a straight, current-carrying wire forms a set of concentric cylinders, centred on the wire, as shown in Figure 3.44. The magnetic field direction is **tangential** to the field lines and it can be predicted using the **right-hand grip rule**. If you imagine gripping the wire in your **right** hand with the thumb pointing in the current direction, the direction of the field is given by the way in which the fingers curl around the wire.

Representation of current and field directions in diagrams

You will often need to represent the directions of currents and fields in diagrams. Since magnetic fields are three dimensional, it is wise to adopt some simple conventions to show these directions in your diagrams. Usually, the directions of the current and the magnetic field are at right angles, so a convenient approach is to draw the magnetic field parallel to the plane of the paper, with the conductors perpendicular to it. This leaves the problem of showing whether the current is flowing into the paper or out from it, which is normally indicated as shown in Figure 3.45.

The direction of the current is shown by either the dot or the cross in the centre of the conductor. Think about the appearance of a dart – going away from you, all you see is the flight at the back, hence the cross; coming towards you, the point comes first, hence the dot.

Measurement of the flux density (B) around a current-carrying wire shows that it is directly proportional to the current ($B \propto I$) and inversely proportional to the distance (r) from the centre of the wire ($B \propto 1/r$). This is represented in the graphs shown in Figure 3.46 *(see p. 230)*.

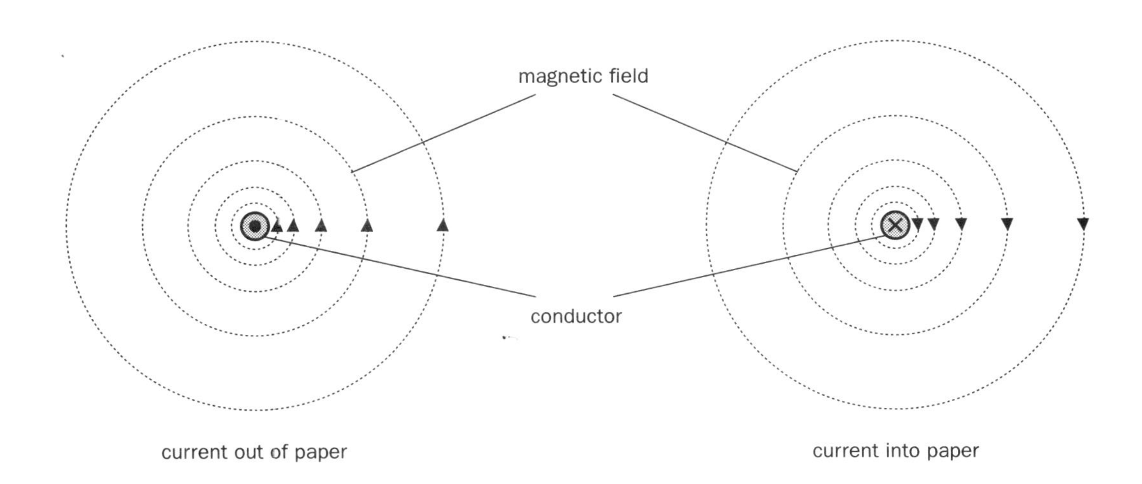

Figure 3.45 Indication of current and field directions

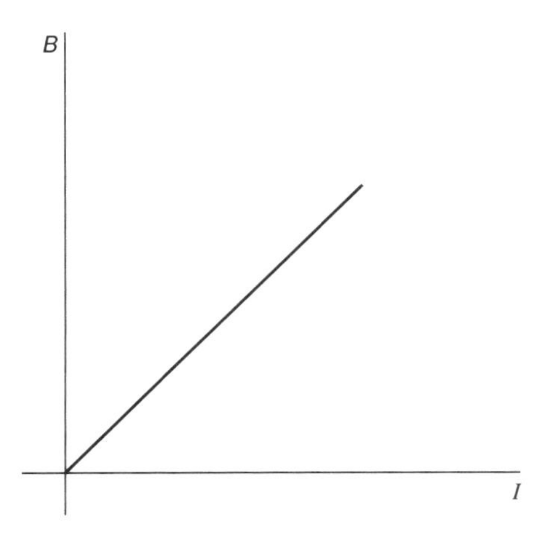

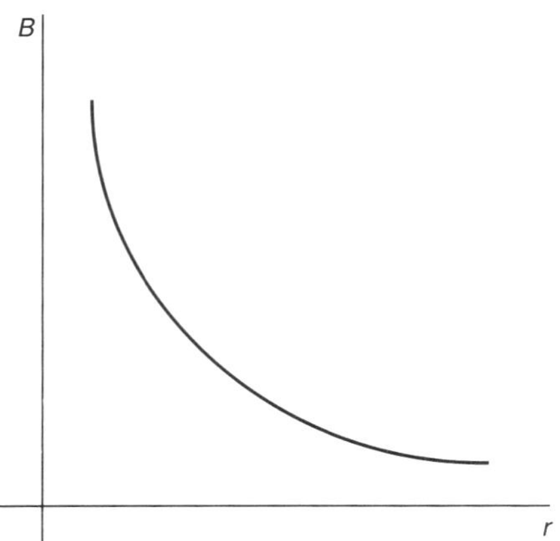

Figure 3.46 Variation of magnetic field strength with current and distance

It can be shown that the flux density (B) at a distance (r) from the centre of a wire carrying a current (I) is given by:

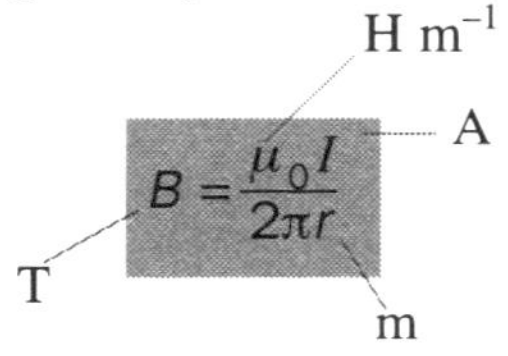

$$B = \frac{\mu_0 I}{2\pi r}$$

Equation 1

μ_0 is a constant of proportionality called the **permeability of free space** (i.e. of a vacuum).

$\mu_0 = 4\pi \times 10^{-7}$ henry per metre (H m^{-1})

(The **henry** (H) is the unit of **inductance**, which will be defined later.)

Strictly speaking we should use the permeability of air (μ_a), since this is the medium in which the wire is situated, but $\mu_0 \approx \mu_a$, so for most purposes using μ_0 is acceptable.

Permeability is a constant which measures a medium's ability to transmit a magnetic field. Each material has its own **absolute permeability** value (μ). It is convenient to refer to the **relative permeability** (μ_r) of a material which is defined as the ratio of its permeability to that of a vacuum.

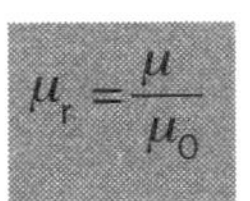

$$\mu_r = \frac{\mu}{\mu_0}$$

Equation 2

The value of μ_r is very large for ferromagnetic materials. Pure iron, for example, has a μ_r value $\approx 5 \times 10^3$ and for some alloys of iron, nickel and cobalt $\mu_r \approx 1 \times 10^4$.

Long solenoid

A solenoid consists of many turns of insulated wire wound on a hollow, insulating tube. Most solenoids are cylindrical in shape, but there are square and flat rectangular prism types. When there is a current (I) through the windings, the field produced inside the solenoid, well away from the ends, is uniform as shown in Figure 3.47 *(at bottom of page).*

Measurement of the flux density B inside the solenoid shows that it is directly proportional to the current (I) and the number of turns per metre (n) of the windings. It can be shown that the magnetic field strength (B) at the centre of a long solenoid of n turns m^{-1}, carrying a current I in a medium of permeability μ, is given by:

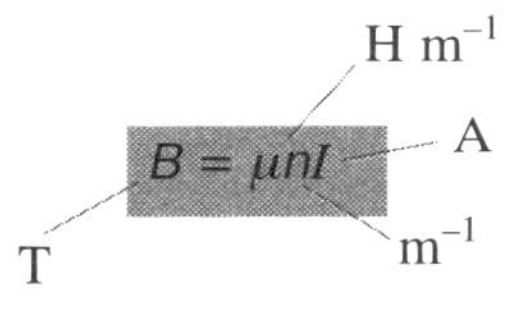

$$B = \mu n I$$

Equation 3

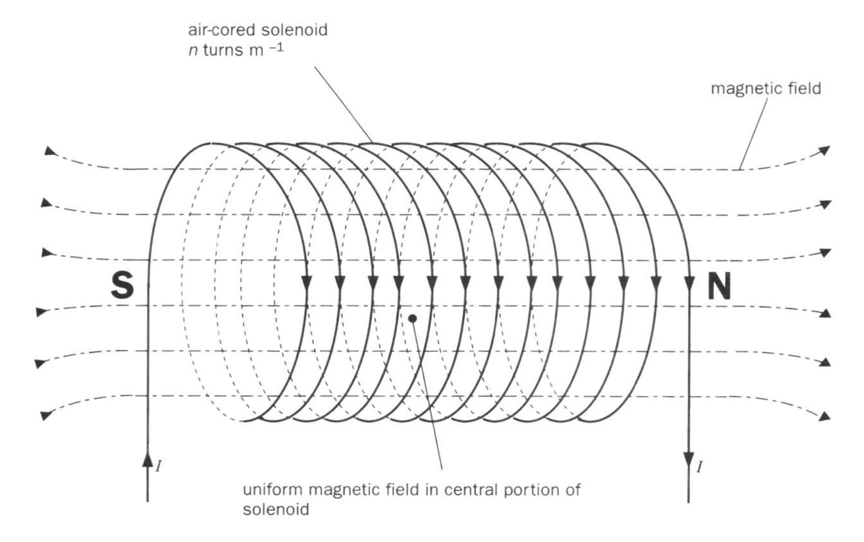

Figure 3.47 Uniform magnetic field inside a long, air-cored solenoid

Placing a ferromagnetic core inside the solenoid increases B to a very high value. As an example, consider a 0.2 m long solenoid of 400 turns carrying a current of 0.5 A.

$$n = 400/0.2 = 2000 \text{ turns m}^{-1}$$

With an air core:

$$\mu = \mu_0 = 4\pi \times 10^{-7} \text{ H m}^{-1}$$

$$\therefore \quad B = 4\pi \times 10^{-7} \times 2000 \times 0.5 = 1.3 \times 10^{-3} \text{ T}$$

With a soft iron core of relative permeability (μ_r) equal to 5000:

$$B = 5000 \times 4\pi \times 10^{-7} \times 2000 \times 0.5 = 6.3 \text{ T}$$

So adding an iron core of relative permeability 5000 increases the field strength at the centre of the solenoid by a factor of 5000. The field strength will be approximately uniform along the length of the solenoid, except approaching the ends. At the ends, the field strength decreases to half of that at the centre.

Predicting the direction of the field

The direction of the field inside a solenoid can be predicted using the **right-hand grip rule** as shown in Figure 3.48.

If you imagine your fingers are curling round so that they point in the direction of the current in the solenoid, then your thumb points in the field direction.

Remember that the end from which the magnetic field lines **leave** the solenoid is the N-pole.

Flat circular coil

Figure 3.49 shows the general shape of the magnetic field around a flat, circular coil.

Measurements of the flux density at the centre of such coils shows that it is directly proportional to the current (I) in the coil as well as the number of turns (N) of the coil. It is inversely proportional to the radius (r) of the coil. Thus for a coil of (N) turns and

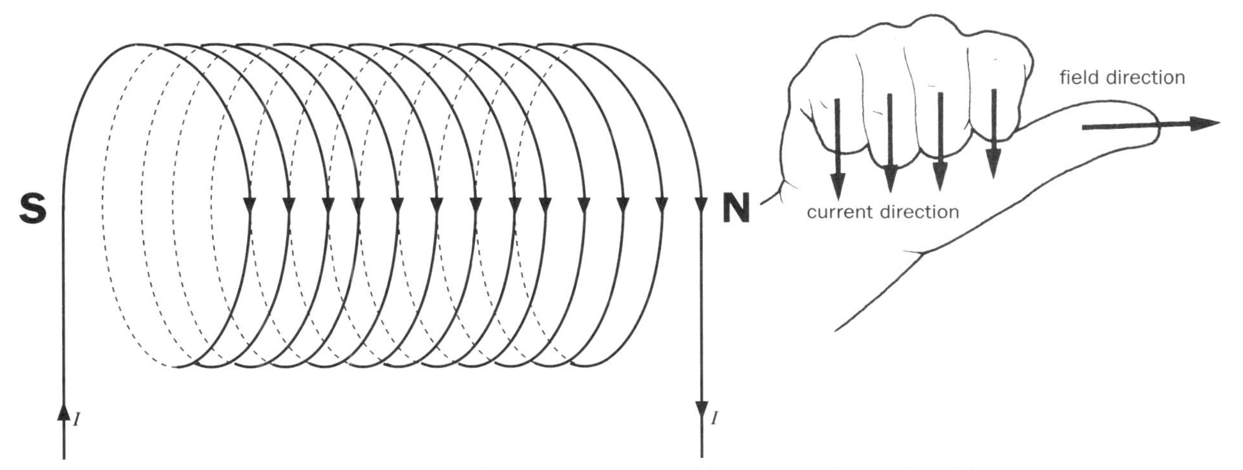

Figure 3.48 Using the right-hand grip rule to predict the field direction in a solenoid

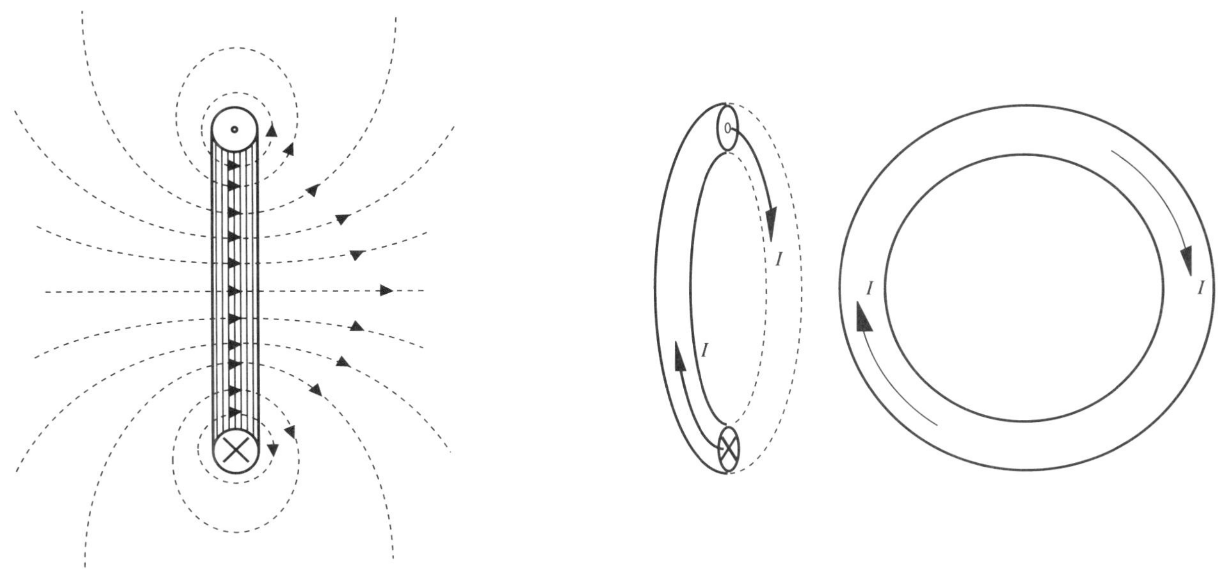

Figure 3.49 Magnetic field around a flat, circular, current-carrying coil

radius (r) carrying a current (I) situated in a medium of permeability (μ), the flux density (B) at the centre is given by:

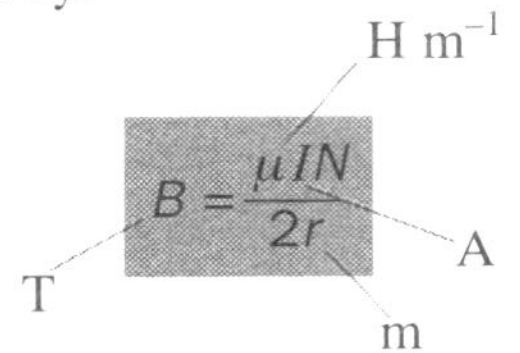

Equation 4

Force on a current-carrying wire in a magnetic field

Whenever two magnetic fields interact, there may be forces exerted. Calculating the magnitude and direction of the force is a complex business, except in a few simple cases which we can consider here. Figure 3.50 shows an arrangement which can be used to demonstrate the fact that a current-carrying wire placed in a magnetic field experiences a force. The strong magnets are attached to the soft iron frame with opposite poles facing each other, so that a strong magnetic field is created in the space between. The top-pan balance is set to read zero when the magnets are placed on the pan. When there is a current (I) through the clamped wire, an electromagnetic force is created which acts on the current in the wire. According to Newton's third law, there will be an equal and opposite force exerted on the magnets, and this will cause a change in the reading of the balance. A downward force on the magnets will cause an increase in the reading, while an upward force will cause a decrease.

The following factors can be varied to investigate their effect on the magnitude and direction of the force:

- magnitude and direction of the current
- strength and direction of the magnetic field (strength can be varied in a crude manner by adding extra magnets to the soft iron frame)
- length of conductor (this can be crudely varied by adding additional iron frames with magnets either side of the original).

The electromagnetic force is produced as a result of the interaction between the magnetic field surrounding the wire when it carries a current, and the magnetic field between the permanent magnets. This is shown in Figure 3.52 on the next page.

Predicting the direction of the force

The direction of the force on the wire depends on the field and current directions and may be predicted using **Fleming's left-hand rule** as shown in Figure 3.51.

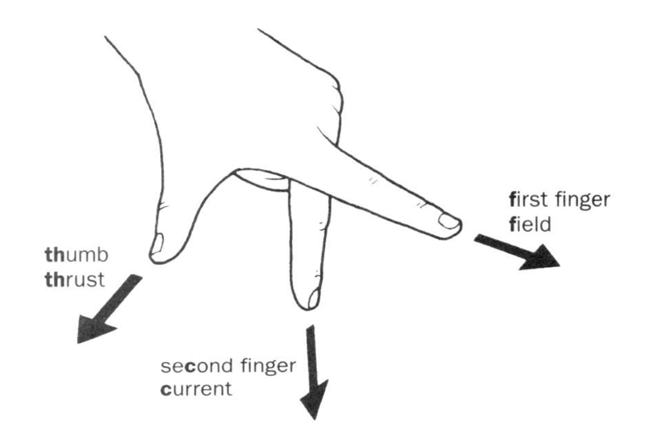

Figure 3.51 Fleming's left-hand rule

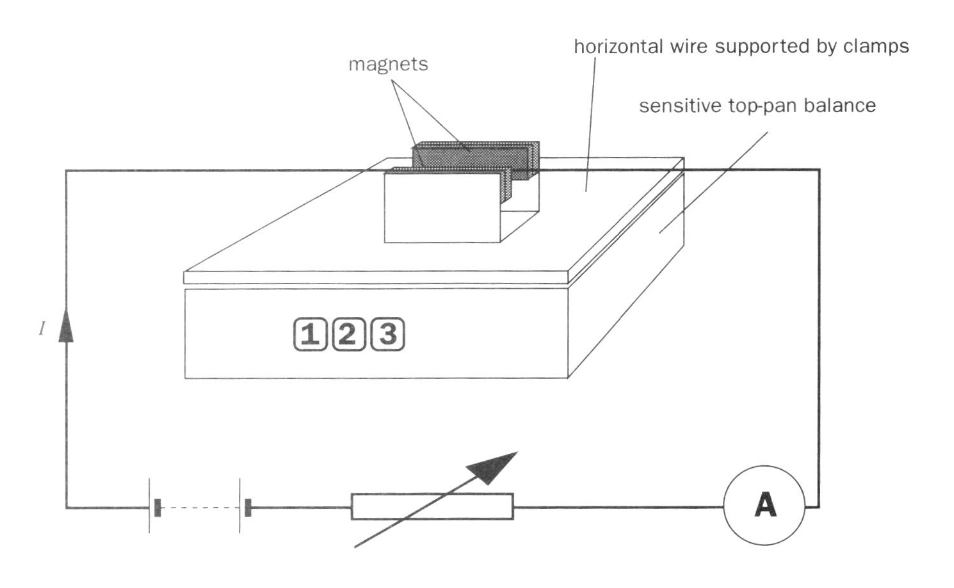

Figure 3.50 Measuring the force on a current-carrying conductor in a magnetic field

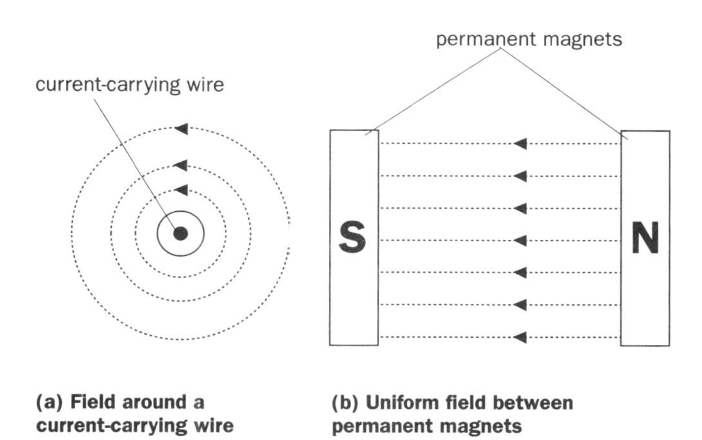

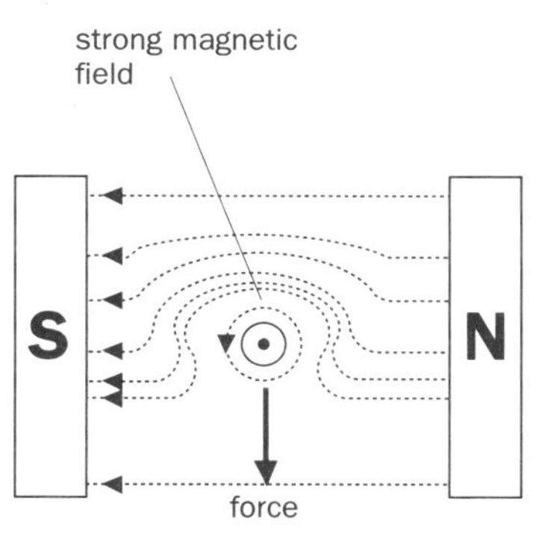

(a) Field around a current-carrying wire

(b) Uniform field between permanent magnets

(c) Resultant field due to interaction of (a) and (b)

Figure 3.52 Force on a current-carrying conductor in a magnetic field

Experiment shows that the size of the force (F) acting on the wire is directly proportional to:

- the strength of the magnetic field (i.e. $F \propto B$)
- the size of the current (i.e. $F \propto I$)
- the length of the wire in the field (i.e. $F \propto l$).

Thus for a wire of length (l) carrying a steady current (I) situated in a magnetic field of strength (B) which is at right angles to the wire, the force (F) acting on the wire is given by:

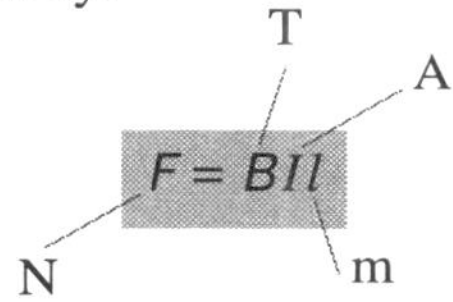

Equation 5

You should note that:

- Equation 5 may be re-arranged as $B = \dfrac{F}{Il}$ from which:

Magnetic field strength (B) is defined as the force acting per unit current in a wire of unit length which is perpendicular to the field.

and

A magnetic field has a flux density of **1 tesla** if a wire of length **1 metre** carrying a current of **1 ampere** and perpendicular to the field experiences a force of **1 newton** in a direction which is perpendicular to both the field and the current.

- If the wire is at some angle (θ) to the magnetic field, the force (F) is given by:

$$F = BIl \sin \theta$$

Equation 6

Then:

(i) with the wire perpendicular to the field, $\theta = 90°$, $\sin 90° = 1$ and $F = BIl$.

(ii) with the wire parallel to the field, $\theta = 0°$, $\sin 0° = 0$, and $F = 0$.

Parallel current-carrying wires

Figure 3.53 on page 234 shows the shape of the resultant magnetic field for two straight, parallel wires which are placed side by side and carry currents (a) in the same direction and (b) in opposite directions.

Each of the wires is in the magnetic field of the other and so they experience a force which is at right angles to both the current and the field. For currents in the **same** direction the force is **attractive** and for currents in **opposite** directions the force is **repulsive**.

Note that in the case of the attractive force (currents in the same direction) there is a **neutral point** between the wires where the magnetic fields are equal in magnitude and opposite in direction. If the two currents are equal in magnitude, the neutral point would be mid-way between the wires (assuming the medium has uniform permeability). Otherwise the neutral point would be closer to the wire with the smaller current.

Magnitude of the force

Consider two wires X and Y of length (l) and carrying currents (I_x) and (I_y) which are positioned a distance (r) apart in air as shown in Figure 3.54 on page 234.

Each wire is in the magnetic field produced by the other wire. Consider a length (l) of wire Y, which is a distance (r) from wire X. At this distance from wire X, the magnetic field due to X is equal to (B_x), and the force (F_y) acting on Y is given by:

$$F_y = B_x I_y l$$

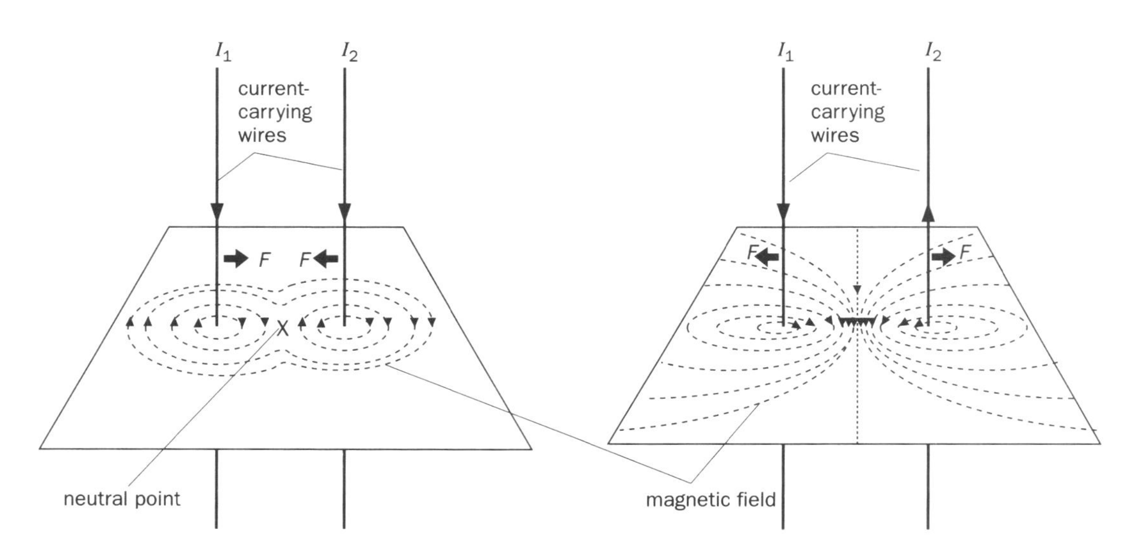

Figure 3.53 Fields around parallel current-carrying wires

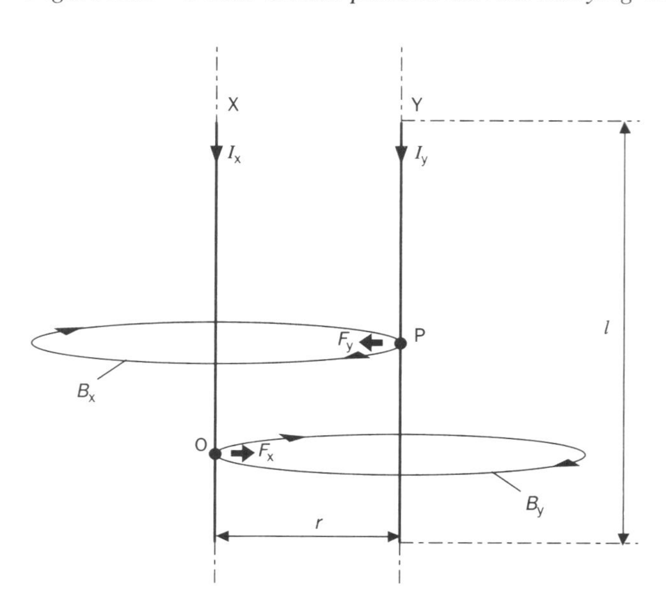

Figure 3.54 Force between two long, parallel, current-carrying wires

$$F_y = \left(\frac{\mu_0 I_x}{2\pi r}\right) I_y l$$

$$F_y = \frac{\mu_0 I_x I_y}{2\pi r} l$$

The direction of F_y is predicted using the left-hand rule at point P.

Current I_y is downwards, field B_x is coming out of the paper and so F_y acts to the left.

Similarly the force F_x acting on wire X is given by:

$$F_x = \frac{\mu_0 I_x I_y}{2\pi r} l$$

i.e. the forces on the wires are equal in magnitude, and opposite in direction.

Thus the **force per unit length** (F/l) acting on either wire is given by:

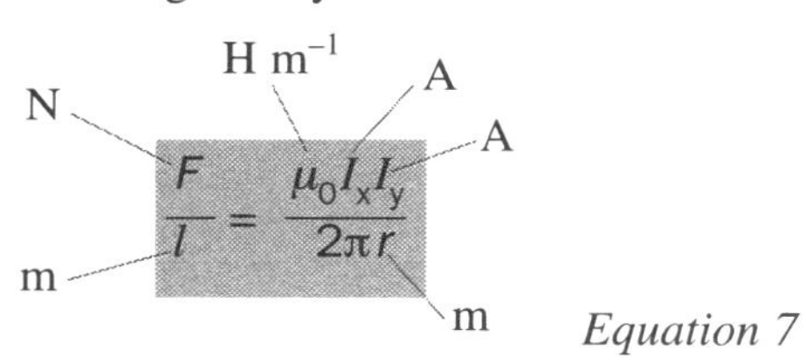

$$\frac{F}{l} = \frac{\mu_0 I_x I_y}{2\pi r}$$

Equation 7

The ampere is defined in terms of the force between two current-carrying wires.

One **ampere** is that steady current which when flowing in each of two infinitely long, straight, parallel wires of negligible cross-section placed one metre apart in a vacuum causes each wire to exert a **force** of 2×10^{-7} newton on each **metre length** of the other wire.

This defines the ampere, which is one of the base units, in terms of the base units of mass, length and time. It also fixes the value of the constant μ_0.

From Equation 7,

$$\mu_0 = \frac{2\pi r F}{I_x I_y l}$$

And from the ampere definition:

$r = 1$ m
$F = 2 \times 10^{-7}$ N
$I_x, I_y = 1$ A
and $l = 1$ m

$$\therefore \quad \mu_0 = \frac{2\pi \times 1(\text{m}) \times 2\pi \times 10^{-7}\ (\text{N})}{1(\text{A}) \times 1(\text{A}) \times 1(\text{m})}$$

$$= 4\pi \times 10^{-7}\ \text{N A}^{-2}\ (\text{H m}^{-1})$$

Guided examples (1)

Permeability of free space, $\mu_0 = 4\pi \times 10^{-7}\ \text{H m}^{-1}$

1. Two long, parallel wires are placed 2.0 cm apart and carry equal currents of 5.0 A in opposite directions. Calculate the magnetic flux density at a point P which is:

(a) midway between the two wires

(b) 2.0 cm from one wire and 4.0 cm from the other.

Guidelines

Use the left-hand rule to find out the direction of B for each wire and then draw a sketch to help you to see whether the B-values for the two wires should be added or subtracted to obtain the resultant magnetic field (B_P) at point P. (Use $B_P = B_x + B_y$ where B_x, B_y are the flux densities due to the currents in each wire.)
Use Equation 6.
Drawing a sketch will help you to see where point P is in relation to the wires. The left-hand rule should give magnetic field directions at P which show that $B_P = B_x - B_y$.

2. The diagram below shows two long, parallel wires X and Y in which steady currents of 10 A and 20 A flow in the same direction. The two wires are laid horizontally at a distance of 0.1 m from each other.

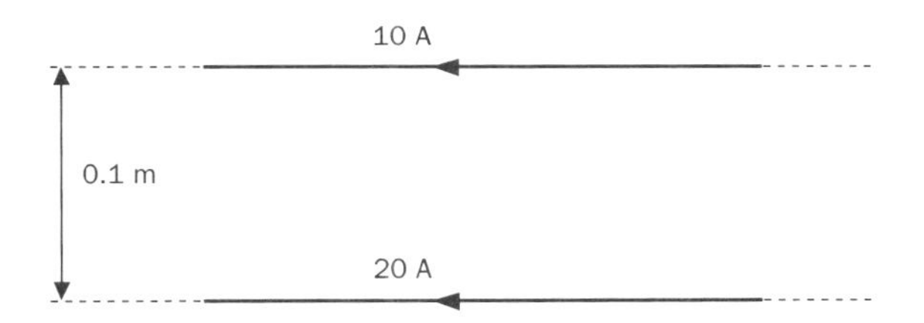

(a) what is the force acting on a 1.0 m length of each wire?

(b) If a uniform magnetic field of flux density 4.0×10^{-5} T acts vertically downwards, what is the new resultant force acting on one metre of each wire?

Guidelines

(a) Use Equation 6.

(b) Use the left-hand rule to obtain the directions of the forces on X and Y due to the vertical magnetic field.

Resultant force on X = (F due to magnetic field of X) + (F due to vertical magnetic field)

3. At some point A there is a magnetic field strength of 2.0×10^{-5} T acting horizontally towards the right.

What is the size and direction of the current in:

(a) a flat circular coil of 200 turns and radius 4.0 cm if it is to be arranged so that the resultant magnetic field strength at A is 4.0×10^{-5} T acting towards the left?

(b) a 0.2 m long solenoid consisting of 500 turns of wire so as to produce the same resultant magnetic field strength of 4.0×10^{-5} T at point A?

Guidelines

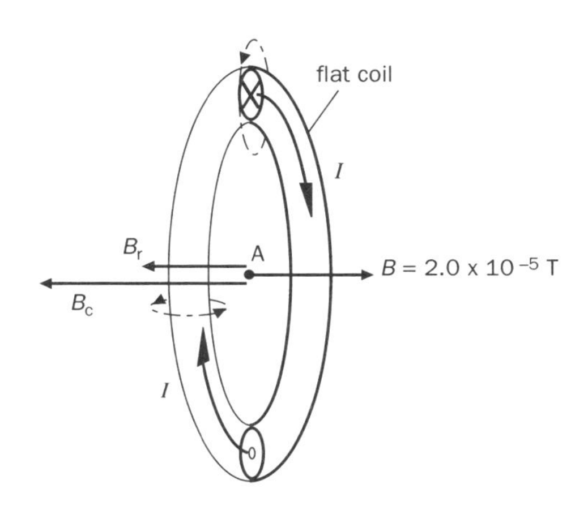

(a) The coil is arranged so that its centre is at point A as shown in the diagram.

I = the current in the coil

B_c = the flux density produced by the coil

and B_r = the resultant flux density

Then: $B_r = B_c - B$

$\therefore B_c = B_r + B$

Use Equation 7 to calculate the magnitude and direction of the current required to produce B_c.

(b) Draw a sketch of the arrangement and use the same thinking as in (a). This time use Equation 5 for the magnetic field strength (B_s) produced by the solenoid.

Rectangular coil in a magnetic field

Figure 3.55 shows a rectangular coil of (N) turns having dimensions ($a \times b$) and carrying a current (I). The coil is pivoted so that it can rotate about its central axis.

Figure 3.56 shows an end view of the coil. Diagram (a) shows the coil parallel to the field (i.e. the **normal** to the plane of the coil is at 90° to the field). Diagram (b) shows the coil at a particular instant when the angle between the normal and the magnetic field direction is $\theta°$.

Each side of the coil experiences a force whose direction is given by the left-hand rule. With the current as shown, there are **no** forces acting on sides WX and ZY. Sides XY and WZ experience vertical forces (i.e. in a direction at 90° to both the side of the coil and the field direction). These forces produce a couple of torque T which rotates the coil about its central axis. Referring to Figure 3.56(b) we see that:

magnitude of force (F) is given by:

$F = BIaN$

torque (T) = force (F) × perpendicular distance between the forces:

$T = BIaN \times b \sin \theta$

But: $a \times b$ is equal to the **area** (A) of the coil.

∴

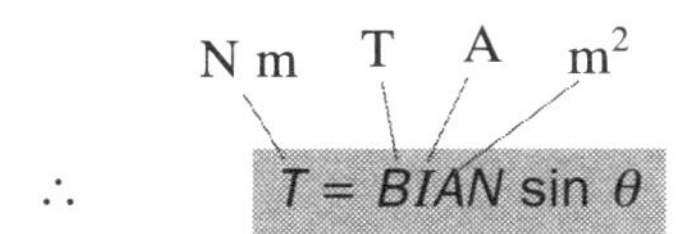

Equation 8

T has its maximum value (= $BIAN$) when $\theta = 90°$ (i.e. when the coil plane is **parallel** to the field). As the coil rotates T decreases until it becomes zero when $\theta = 0°$ (i.e. when the coil plane is **perpendicular** to

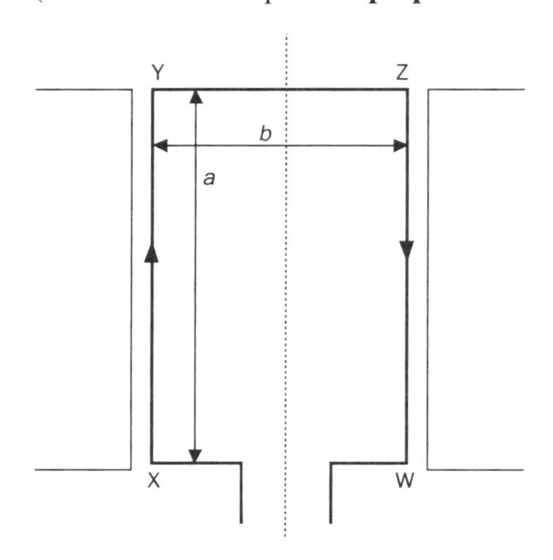

Figure 3.55 Force on a coil in a uniform magnetic field

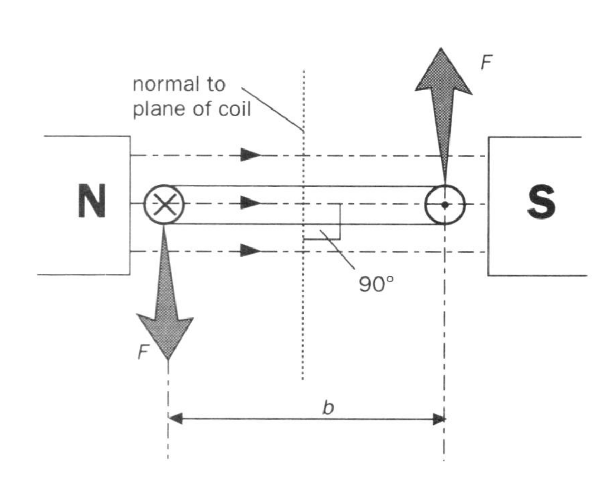

(a) Normal to plane of coil perpendicular to field

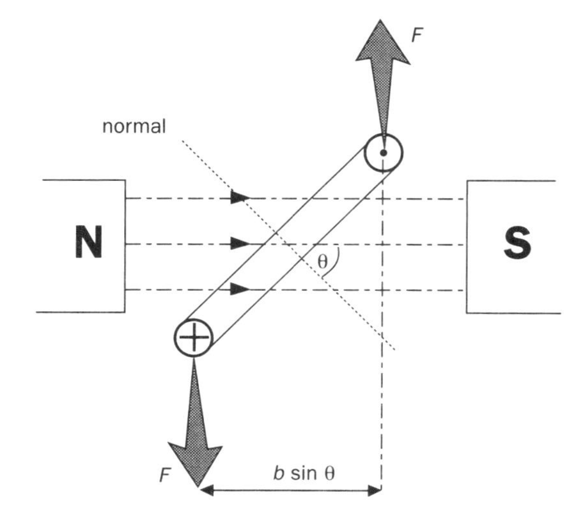

(b) Normal to plane of coil at angle to field

Figure 3.56 End view of coil

the field). Remember that θ is the angle between the **normal** to the coil plane and the field.

Applications – galvanometer and motor

Moving-coil galvanometers and electric motors make use of the fact that a current-carrying coil in a magnetic field experiences a torque.

The moving-coil galvanometer

Figure 3.57 shows the basic construction of one type of moving-coil instrument.

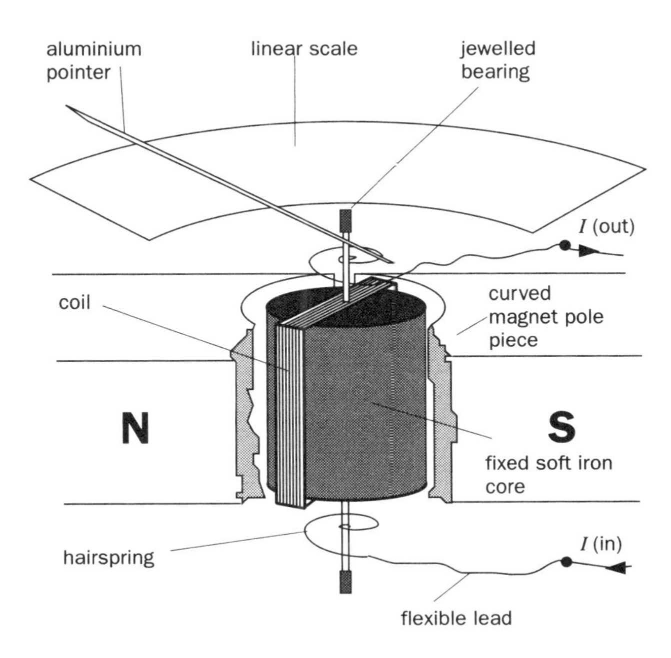

Figure 3.57 The moving-coil galvanometer

A small, rectangular coil of many turns of fine wire is pivoted in the magnetic field between the **curved** poles of a horse-shoe magnet. The coil moves in the intense, **radial** field which exists in the air gap between the pole pieces and a fixed soft-iron cylinder, as shown in Figure 3.58. The radial field ensures a **constant torque** for all positions of the coil and this means that the instrument can have a linear scale.

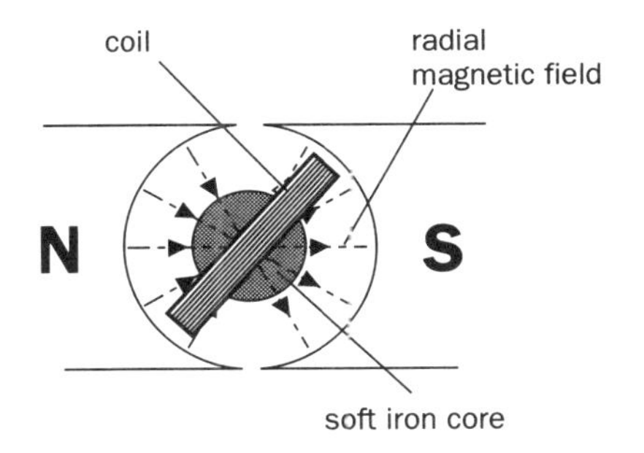

Figure 3.58 Radial magnetic field

The current being measured is led in and out of the coil by the hairsprings which also provide a torque to oppose the coil's motion. When there is a current (I) through the coil it turns until the torque on it due to the field–current interaction is exactly balanced by the opposing torque provided by the hairsprings.

You should note that:

- The springs, bearing, pointer, scale, etc. are made of non-magnetic metals such as aluminium so as not to affect the instrument's magnetic field.
- The coil is wound on an aluminium frame. This material is chosen for its combination of low density, high conductivity and the fact that it is not ferromagnetic. As the coil rotates in the magnetic field, **eddy currents** are set up in the aluminium frame by electromagnetic induction (see Section 3.5). These eddy currents create a magnetic field, and the interaction of this eddy current field with the field of the permanent magnets creates a couple which **opposes** the direction of rotation. The magnitude of this couple is proportional to the **speed of rotation** of the coil, so the effect of this is to slow the movement of the coil. This is called **damping**, and is desirable in such instruments as it prevents the needle oscillating for some time before coming to rest on the true reading. Careful design of the aluminium frame means that the coil can be **critically damped**, which means that the coil comes to rest at the true reading as quickly as possible, without overshooting.
- The small size and lightness of the coil makes the meter very compact and sensitive, but it also means that only very small currents can be tolerated. To measure larger currents the meter is fitted with a resistor connected **in parallel** (called **a shunt**) whose resistance is much smaller than that of the coil and so diverts the majority of the current.
- The meter can be used as a sensitive **voltmeter** to measure small potential differences. A larger potential difference can be measured by the addition of a large value resistor **in series** with the meter (called a **multiplier**).

The simple d.c. motor

The electric motor is very similar to the moving-coil galvanometer, but in this case the coil must be made to rotate continuously. The basic principle of a simple d.c. motor is shown in Figure 3.59 on page 238. It consists of a rectangular coil which is free to rotate about a central axis and situated between permanent **magnet poles**.

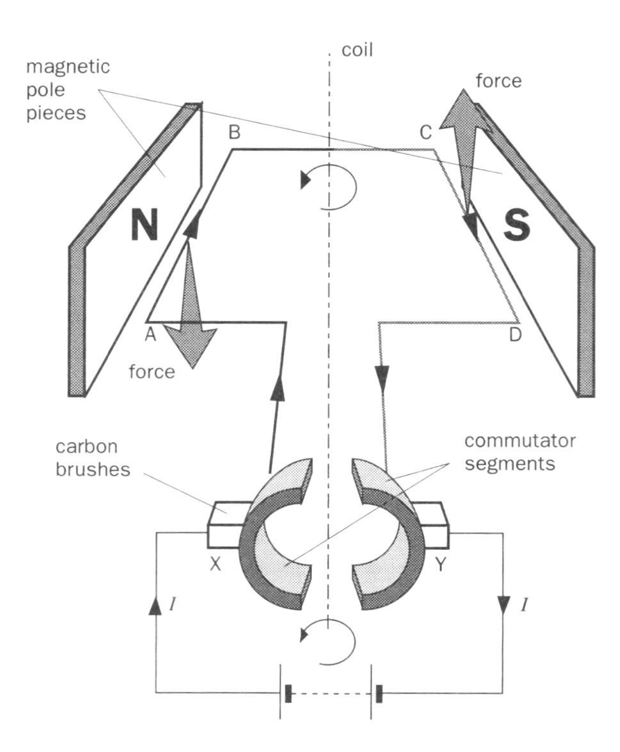

Figure 3.59 The simple d.c. motor

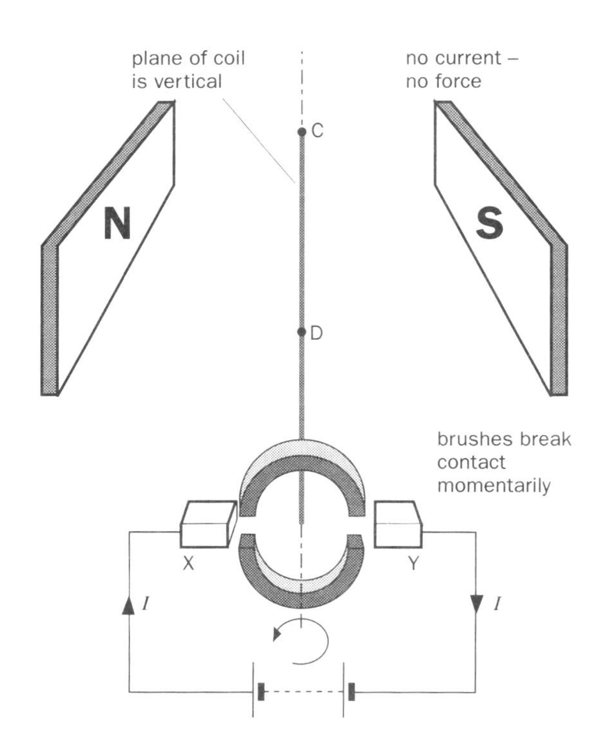

Figure 3.60 Action of the commutator

The current is fed into the coil through carbon **brushes** which press against the **split-ring commutator**. At the instant shown, application of the left-hand rule gives a **downward** force on side AB and an **upward** force on side CD. This produces an anti-clockwise couple (viewed from the commutator end) which turns the coil until it reaches the vertical position, as shown in Figure 3.60.

Despite the fact that there is no current in the coil when it is vertical (since the brushes are then lined up with the commutator gaps) the coil's momentum carries it past this position. Side CD now comes into contact with the positive brush X and BA is in contact with brush Y. Thus the commutator ensures that the current is always in the same direction and this means that the coil continues its anti-clockwise rotation. We shall consider the motor in greater detail when we deal with electromagnetic induction in Section 3.5.

The electric starter motor in a car has to provide a very high torque in order to start the engine

Guided examples (2)

1. A superconducting wire of length 2.0 cm carries a current of 1200 A. It is situated in a magnetic field of flux density 2.5×10^{-2} T. Calculate the force acting on the wire if it is (a) parallel, (b) perpendicular and (c) at an angle of 30°, to the field direction.

Guidelines

Use $F = BIl \sin \theta$ making sure that all units are SI and noting the value of θ in each case.

2. The coil in a simple electric motor has dimensions of 30 mm × 15 mm and consists of 200 turns of fine copper wire wound on a light insulating frame. It is situated in a uniform magnetic field of flux density 0.8 T. Calculate the torque acting on the coil when there is a current of 50 mA through it and its plane is: (a) parallel, (b) perpendicular and (c) at 60° to the magnetic field.

Guidelines

Use Equation 8 (SI units throughout).

Force on individual moving charges in a magnetic field

You have already seen how a wire placed in a magnetic field experiences a force when there is a current through it. Since we regard the current in the wire as a flow of electrons it seems reasonable to conclude that the force exerted on the wire in the field is the resultant of the forces acting on each of the moving electrons (or other charged particles).

Figure 3.61 shows a length (l) of wire containing (n) electrons of charge (q) and average drift velocity (v) which is situated in a magnetic field of flux density (B) directed at right angles to the wire.

The average time (t) taken by each electron to travel distance (l) is given by: $t = l/v$

And the current (I) in the wire is equal to the total charge (nq) passing a section per unit time (t),

i.e. $I = \dfrac{nq}{t} = \dfrac{nqv}{l}$

Then the total force acting on the wire is given by:

$$force = BIl = Bnqvl/l = Bnqv$$

This expression is the resultant of the forces on the n

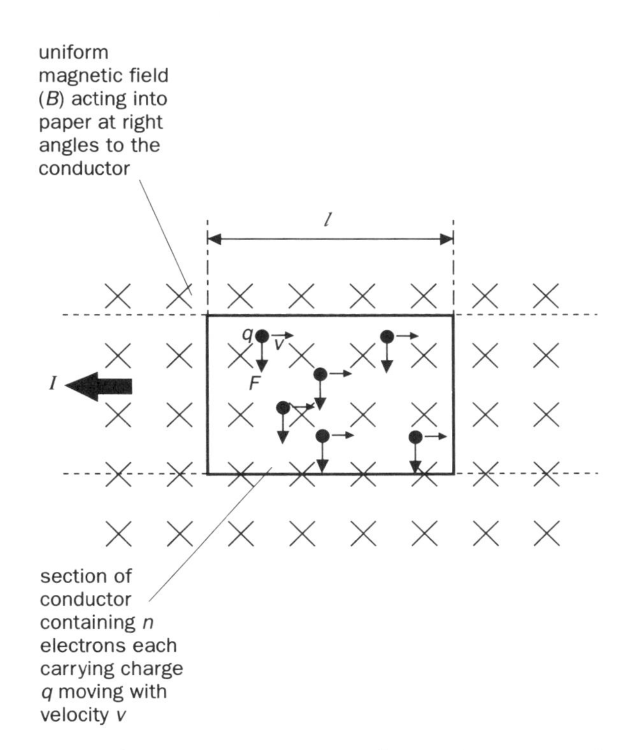

Figure 3.61 Force on moving charges in a magnetic field

electrons contained in the wire. Thus the force (F) acting on each individual electron is given by:

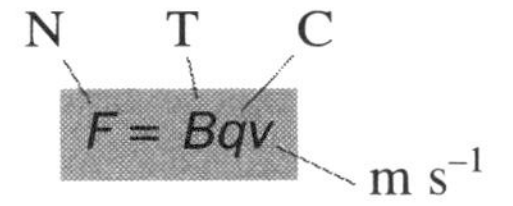

Equation 9

You should note that:

- This equation applies for any moving, charged particle in a magnetic field.
- If the charged particle moves at an angle (θ) to the field direction, the force (F) exerted on it is given by:

$$F = Bqv \sin \theta$$

Equation 10

- There is no force on the charged particle when:
  (i) it is stationary (i.e. $F = 0$ when $v = 0$)
  (ii) it moves parallel to the field (i.e. $F = 0$ when $\theta = 0°$).
- The direction of the force can be obtained using the **left-hand rule** but you must bear in mind that in this rule the current direction is that of **conventional** current (which is opposite to the direction of flow of electrons or other negative charges).
- The force causes the moving charged particles to be deflected towards one side of the conductor. This is called the **Hall effect**, and is used to measure the strength of magnetic fields (see later).

Charged particle path in a magnetic field

When a charged particle moves at right angles to a magnetic field the constant force ($F = Bqv$) exerted on it is perpendicular to both the particle velocity and the field direction. Thus although the force continually changes the particle's direction of motion, it has no effect on its speed (and hence on its kinetic energy). The result is that the charged particle moves in a circular path when it is in the field (see Figure 3.62).

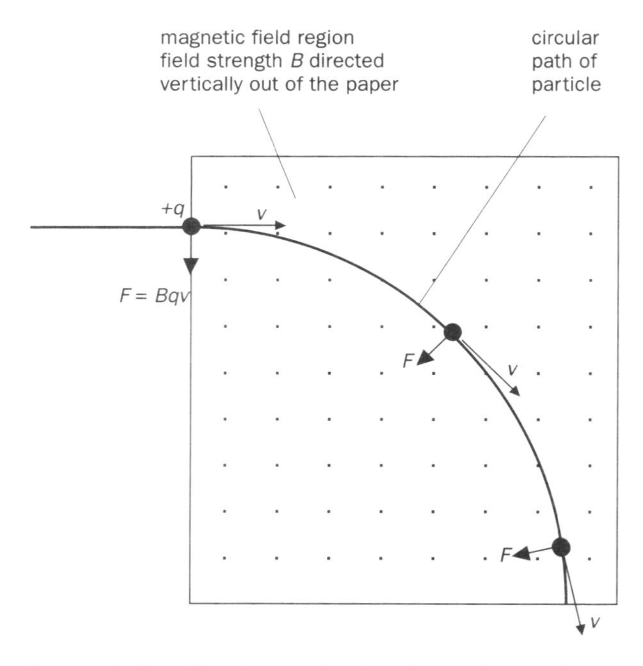

Figure 3.62 Circular path of a charged particle in a perpendicular magnetic field

As you will remember from Section 1.8 on Motion in a circle, a body can only move in a circular path if a centripetal force acts on it. In the case of a charged particle moving in a magnetic field it is the force due to the field ($F = Bqv$) which provides the necessary centripetal force. Thus for a particle of mass (m), charge (q) which moves with speed (v) at right angles to a magnetic field of density (B), the radius (r) of the circular path which it follows is given by:

$$Bqv = \frac{mv^2}{r}$$

$$\therefore \quad r = \frac{mv}{Bq}$$

(units: r — m; m — kg; v — m s^{-1}; B — T; q — C)

Equation 11

You should note that:

- From Equation 11 a smaller radius (i.e. a tighter turn) is produced by a smaller mass or lower velocity of particle, or by a stronger applied magnetic field or greater charge on the particle.
- If the charge moves through the field at any angle other than 90° the path it follows is a **helix**.
- Because of the way they move, charged particles can become 'trapped' within magnetic fields. This is the case in certain regions around the Earth, called the **Van Allen radiation belts**, where the Earth's field causes large numbers of charged particles from space to become trapped in a great spiralling motion from one pole to the other. The spectacular **Aurora Borealis** (Northern Lights) glow in the Northern Hemisphere is produced when charged particles from the radiation belts enter the Earth's atmosphere.

The luminous glow of the Aurora Borealis or Northern Lights is a spectacular example of the motion of charged particles in a magnetic field

- If the path of the particle can be made visible, or be detected in some way, measurements of the radius of curvature of the path provide information about the mass, charge, and velocity of the particle. This is particularly useful for identification of particles from bubble chamber and cloud chamber photographs. In bubble chambers and cloud chambers, the velocity of the particle is not constant, since it is causing ionisation of the particles of the medium, and hence losing kinetic energy. As the velocity decreases, so does the radius of the path, and the particle follows a spiral path. This gives further insights into the nature of the particle, as some are more intensely ionising than others.

The Hall effect

When a section of conducting material carries a current in a magnetic field, each of the charged particles carrying the current experiences a force at right angles to their direction of motion and the direction of the magnetic field. This causes the charged particles to accumulate on one side of the conductor, which produces a small potential difference across the sides of the conductor. This phenomenon is known as the Hall effect.

In Figure 3.63, a metallic conductor of rectangular cross-section carries a current (I) from right to left and at right angles to a magnetic field of flux density (B).

Since we are dealing with a metal, the current (I) is due to a drift of negatively-charged electrons from left to right. We consider one such electron of charge (q) and average drift velocity (v).

Using the left-hand rule and remembering that this uses conventional current direction you can see that a force $F = Bqv$ will act downwards on the electron. Thus electrons are forced to move towards face X causing a build-up of negative charge on X and a consequent build-up of positive charge on Y. This continues until the p.d. between X and Y becomes large enough to prevent any further movement of charge. This maximum p.d. between X and Y is called the **Hall voltage** V_H and it is established when:

> The force on each electron due to the magnetic field is balanced by the oppositely directed force due to the electric field between X and Y.

If (V_H) is the Hall p.d. across the width (d) of the slab and (E) is the final value of the electric field strength then since field strength = potential gradient (see later, page 273) we see that:

$$E = \frac{V_H}{d}$$

Then:

$$qE = Bqv$$

$$\frac{V_H}{d} = Bv$$

$$\therefore \quad V_H = Bvd \qquad \text{(i)}$$

Also, the current (I) through the slab is given by:

$I = nAvq$

where n is the number of charge carriers (electrons) per unit volumeand A = cross-sectional area of the slab = zd.

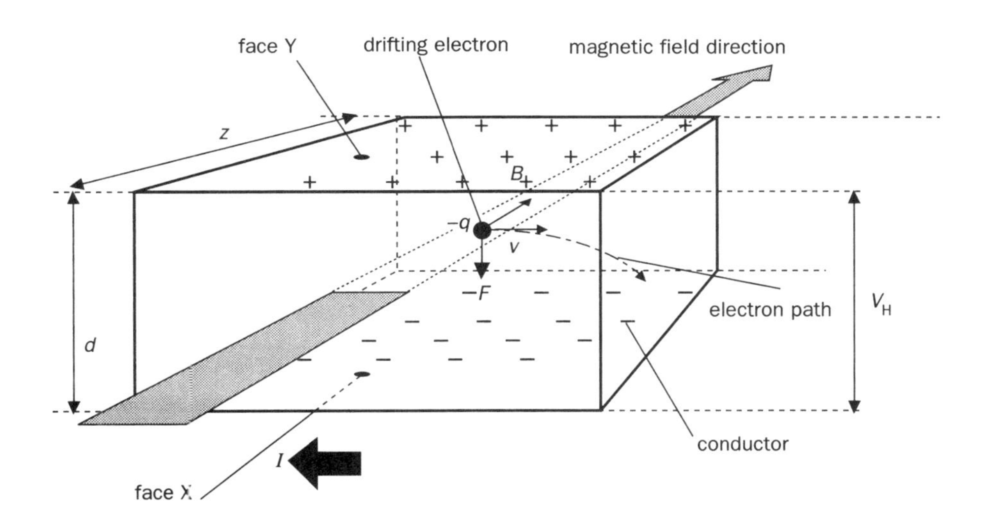

Figure 3.63 Origin of the Hall effect in a metal conductor

$$\therefore \quad v = \frac{I}{nAq} = \frac{I}{nzdq}$$

Substituting for v in Equation (i) gives:

$$V_H = B \frac{I}{nzdq} d$$

$\therefore$

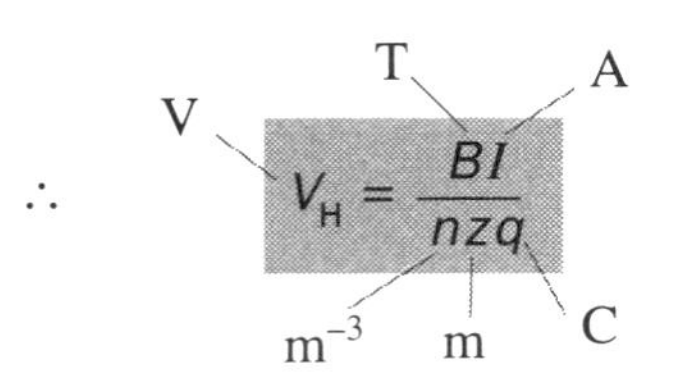

Equation 12

You should note that:

- Equation 12 shows that the Hall voltage $V_H \propto 1/n$, and so for a given current, field strength and thickness of conductor, V_H is largest for materials with the lowest values of n. Metals have large n-values and so V_H is very small for metals. Semiconductors have much lower values of n, so a much larger Hall potential is produced under the same conditions. If the value of n is too low, as in the case of bad conductors, it is difficult to get a current to pass through the material. For practical use, the choice of material is a compromise between a value of n sufficiently large to allow a significant current to pass without overheating, and low enough to develop a measurable Hall potential.
- V_H should be measured with a high-resistance voltmeter, to prevent any significant current.

Using the Hall effect

Identification of majority charge carriers in semiconductors Since the direction of the electromagnetic force on the charged particles depends only on the **direction** of the current and not on the **sign** of the charge carrier, the polarity of the Hall potential can be used to identify the sign of the charge carriers in a material. Figure 3.64 below shows a sample of semiconductor in which the current is due to a drift of **positive** charges from right to left.

With the current direction from right to left, and the magnetic field directed into the paper as shown, the force exerted on each charge due to the magnetic field acts downwards (predicted using the left-hand rule). The charges accumulate on face X (the lower face), thus face X becomes positively charged and Y negatively charged. The polarity of the faces can be determined by connecting a voltmeter to face X and face Y. If the **positive** terminal of the voltmeter is connected to face X, a **positive** p.d. will be measured. If the p-type semiconductor in Figure 3.64 is replaced by a material in which the charge carriers are negative, the direction of the electromagnetic force is still downwards, but this time, the charge accumulating on the lower face (face X) will be negative, as shown in Figure 3.65.

The voltmeter indicates a **negative** p.d., showing that face X is negative compared to face Y. If the directions of the current and magnetic field are known, the polarity of the charge carriers can be identified.

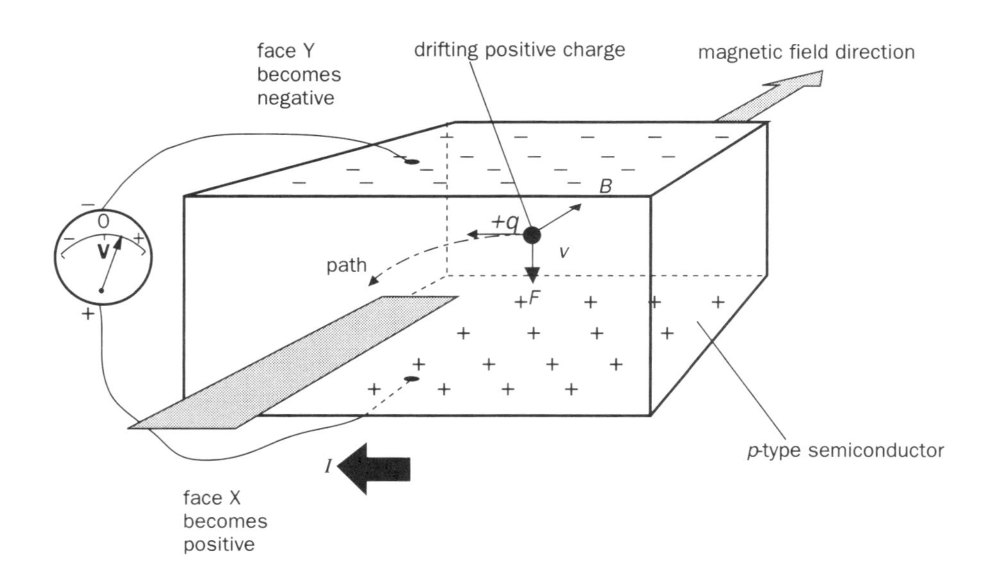

Figure 3.64 p-type semiconductor slab carrying a current in a magnetic field

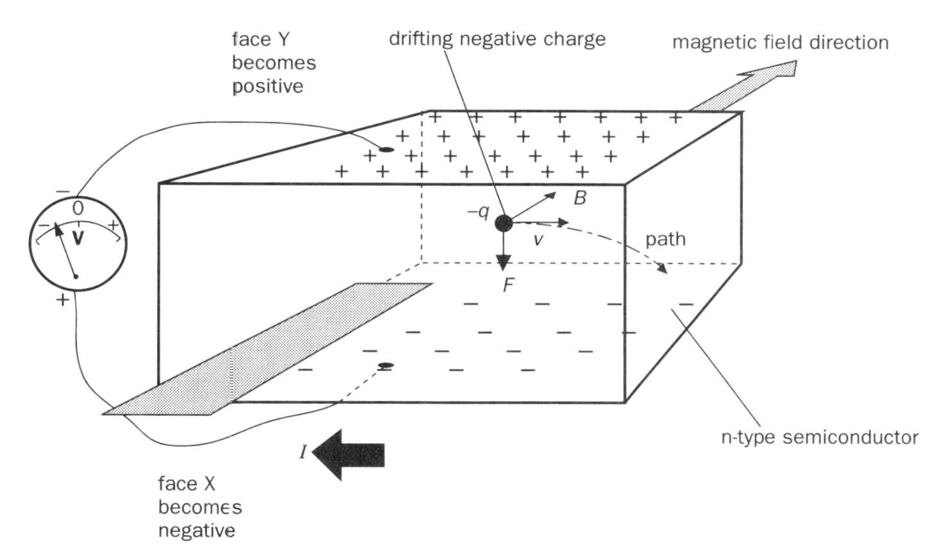

Figure 3.65 n-type semiconductor slab carrying a current in a magnetic field

Measurement of *B* – Hall probe

Figure 3.66 shows the principle of operation of a Hall probe. The semiconductor slice is positioned with its largest face perpendicular to the magnetic field whose flux density (B) is to be measured. The current (I) through the semiconductor is controlled by the rheostat (R) and it is measured by the milliammeter. A high resistance millivoltmeter is used to measure the Hall p.d. (V_H) generated across the slice. A practical Hall probe would consist of such a semiconductor mounted on the end of a hollow rod containing all the wiring (shielded from the magnetic field) so that the slice could be inserted into a desired position in the field. Additional circuitry would include a potential divider to ensure that the millivoltmeter reads zero when the probe is not in a magnetic field. The device could be adapted to measure fluctuating magnetic fields by replacing the millivoltmeter with an oscilloscope or datalogger. The Hall p.d. (V_H) is given by:

$$V_H = \frac{BI}{nzq}$$

From which:

$$B = \frac{V_H}{I}(nzq)$$

(nzq) is a constant for a given probe and so B can be calculated from the V_H and I readings. In a practical arrangement the millivoltmeter scale could be calibrated in tesla to give direct measurement of magnetic field strength.

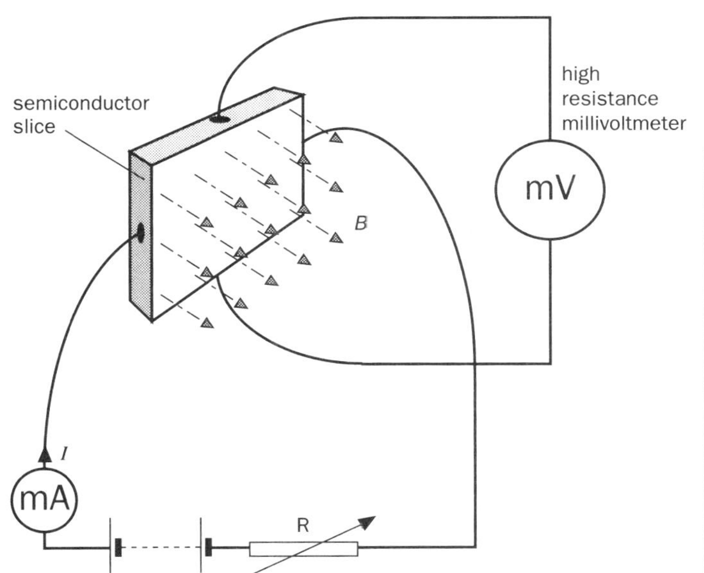

Figure 3.66 Principle of the Hall probe to measure magnetic field strength

Guided examples (3)

1. Calculate the magnetic flux density directed at right angles to a beam of electrons travelling at a speed of 2.0×10^6 m s^{-1} which will deflect the beam into a circular path of radius 20 cm.
(Charge on an electron, $e = -1.6 \times 10^{-19}$ C; electron mass, $m_e = 9.1 \times 10^{-31}$ kg.)

Guidelines

Use Equation 11 – (Remember SI units throughout.)

Guided example continued on next page

Guided examples (3) *continued*

2. In a cloud chamber high energy particles leave a track of tiny droplets as they travel though the ethanol-saturated air. A positively charged particle travels in a cloud chamber with a speed of 1.2×10^6 m s^{-1} at right angles to a magnetic field of flux density 0.05 T. If the particle moves in a circular path of radius 50 mm, calculate the charge to mass ratio of the particle.

Guidelines

Use the fact that the centripetal force needed for the particle's circular motion is provided by the force due to the magnetic field:

i.e. $$Bqv = \frac{mv^2}{r}$$

charge to mass ratio is q/m.

3. A 0.50 mm thick slice of n-type germanium is placed with its plane perpendicular to a magnetic field of flux density 0.40 T. If a Hall p.d. of 10 mV is generated across the slice when the current through it is 20 mA, calculate the number of free electrons per unit volume in germanium (charge on an electron, $e = -1.6 \times 10^{-19}$ C).

Guidelines

Use Equation 12 – (Remember SI units throughout.)

Self-assessment

SECTION A
Qualitative assessment

1. Magnetic fields are represented using **field lines**.

(a) What does the direction of the field lines indicate?

(b) What does the relative density of the field lines tell you about the field?

(c) How are the field lines arranged so as to show
 (i) a **uniform** field
 (ii) a field of **increasing** strength?

2. Sketch the magnetic field around the Earth, indicating the direction of the field lines. Discuss the extent to which the Earth's magnetic field may be assumed constant.

3. (a) Sketch an arrangement which could be used to show that a current-carrying conductor placed in a magnetic field experiences a force.

(b) State the factors which determine the magnitude of the force.

(c) Use **Fleming's left-hand rule** to determine the force direction in each of the cases shown in the diagram below.

4. (a) Define **magnetic field strength** (B) in terms of the force exerted on a current-carrying wire in a magnetic field.

(b) Define the **tesla** (T).

(c) A wire which is carrying a current (I) at right angles to a magnetic field of flux density (B)

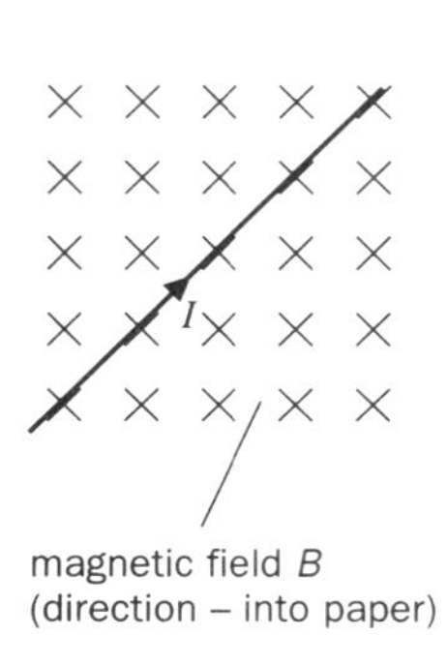

(i)

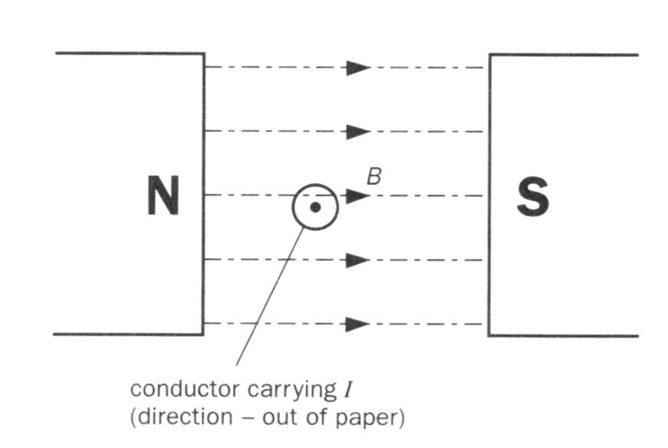

(ii)

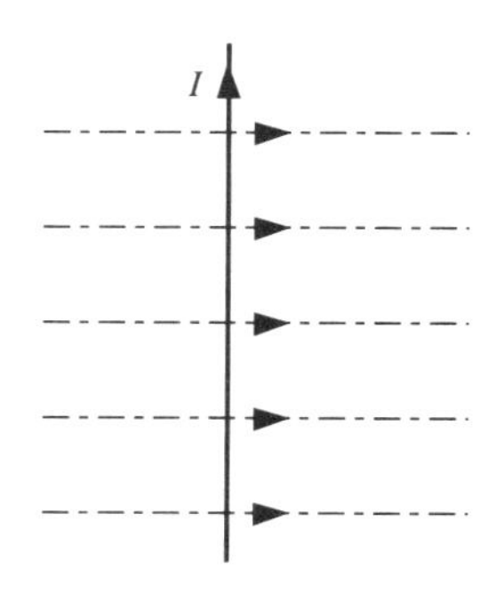

(iii)

Self-assessment *continued*

experiences a force (F). What is the angle (θ) between the wire and the field for which the force is

(i) $F/2$ and

(ii) 0.

5. (a) A coil of (N) turns each having an area (A) is placed in the uniform magnetic field between permanent magnet poles. If the field has a flux density (B) and the coil carries a current (I), show that the torque (T) acting on the coil at any instant is given by:

$$T = BIAN \sin \theta$$

where θ is the angle between the normal to the plane of the coil and the field.

(b) What is the value of (T) when the coil plane is:

(i) at **right angles** and

(ii) **parallel** to the direction of the magnetic field?

6. A moving-coil galvanometer has a rectangular coil of (N) turns, each of area (A), in a radial magnetic field of magnetic field strength (B).

(a) What is the torque (T) exerted on the coil when there is a current (I) through it?

(b) The coil rotation is opposed by springs which provide a restoring torque of ($k\theta$), where k is a constant for the springs and θ is the coil deflection in radians. Derive a relationship between I and θ.

(c) What is the purpose of the fixed, soft-iron cylinder in such a meter?

(d) The coil in most moving-coil meters is small and light. Give one advantage and one disadvantage of this.

7. (a) Draw a labelled diagram of a **simple d.c. motor**.

(b) Explain the purpose of the **split-ring commutator** in such a motor.

8. Sketch the shape of the magnetic field in each of the following cases:

(a) Straight, current-carrying wire.

(b) Two long, parallel wires placed side by side and carrying currents in the same direction.

(c) Long solenoid.

9. How does the value of flux density near a long, straight wire vary with

(a) the current (I) through it and

(b) the distance (r) from the centre of the wire?

10. (a) What is meant by the **permeability** (μ) of a medium?

(b) Define the **relative permeability** (μ_r) of a medium.

11. Derive an equation for the force acting on unit length of a long straight wire carrying a current (I_1), when it is placed at a distance (r) from a parallel long, straight wire carrying a current (I_2).

12. (a) Define the **ampere**.

(b) Show how the definition of the ampere fixes the value of the permeability of free space.

13. (a) Solenoid D has **twice** the number of turns as solenoid P, and is **four** times as long. For the same current, how does the flux density in D compare with that in P?

(b) A solenoid which is in the form of a spring is stretched to 1.5 times its original length. If the current in the solenoid remains constant, how is the flux density affected?

(c) An air-cored solenoid has a flux density (B) at its centre when there is a current (I) through it. What is the new flux density when a ferromagnetic alloy core of relative permeability (μ_r) = 6000 is placed inside the solenoid and the current is reduced to $I/1000$?

14. A particle having a charge (q) is moving with a velocity (v) in a magnetic field where the magnetic field strength (B) is perpendicular to the particle's path.

(a) Write an expression for the distance moved by the particle in a time t.

(b) Since current is charge moved per unit time, calculate the current due to the charged particle in time t.

(c) The force (F) exerted on a current (I) flowing through a conductor of length (l) in a magnetic field strength (B) is given by $F = BIl$.
If you consider the answer to (a) to be the effective length of the current element, derive an expression for the force acting on the particle in the magnetic field.

Self-assessment *continued*

15. The force acting on the charged particle in Question 14 causes it to move in a circular path of radius (r). If the particle has a mass (m), derive an expression for r in terms of q, B, v and m.

16. (a) What is the **Hall effect**?

(b) The diagram shows a slab of conductor having (n) charge carriers per unit volume and carrying a current (I). The slab is placed with its largest faces perpendicular to a magnetic field of flux density (B). Derive an expression for the **Hall potential** (V_H) which is developed across the shaded faces of the slab.

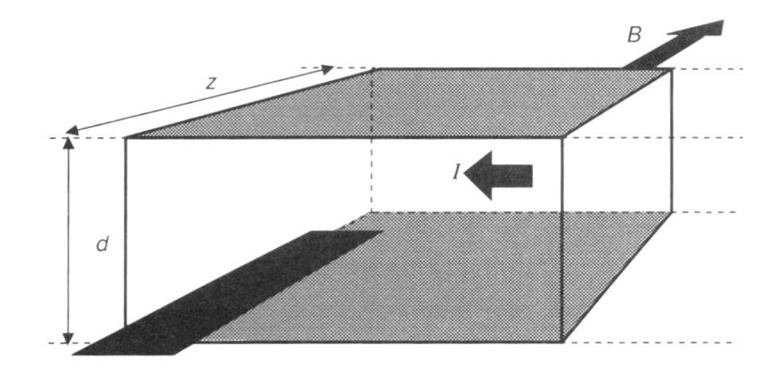

17. Explain how the Hall effect is used in

(a) identifying the **sign** of the majority charge carriers in semiconductors

(b) the measurement of magnetic flux density using a Hall probe.

SECTION B
Quantitative assessment

(***Answers:*** 0; 4.5×10^{-6}; 2.0×10^{-5}; 3.8×10^{-5}; 1.6×10^{-3}; 4.0×10^{-3}; 0.010; 0.012; 0.020; 0.020; 0.040; 0.20; 0.71; 1.3; 270; 1.2×10^{3}; 8.7×10^{28}.)

(Electronic charge	$e = 1.6 \times 10^{-19}$ C
Electronic mass	$m_e = 9.11 \times 10^{-31}$ kg
Gravitational field strength	$g = 10$ N kg^{-1}
Permeability of free space	$\mu_0 = 4\pi \times 10^{-7}$ H m^{-1})

1. A wire carries a current of 2.5 A and 85 mm of its length lies in a magnetic field of flux density 5.5×10^{-2} T. Calculate the force acting on the wire when it is: (i) parallel, (ii) perpendicular and (iii) at 60° to the field direction.

2.

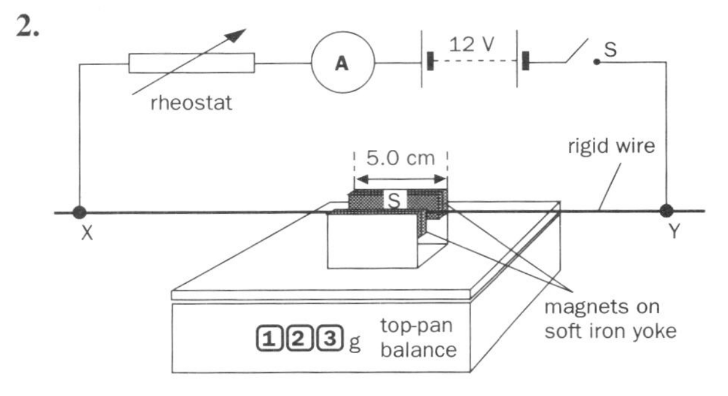

The arrangement shown in the diagram is used to demonstrate and measure the force exerted on a current-carrying wire XY in a magnetic field. The magnets on the soft iron yoke stand on a sensitive top-pan balance and provide a uniform horizontal magnetic field at right angles to XY. The reading on the balance is 272.0 g when switch S is open. When S is closed and the rheostat is adjusted so as to give a total circuit resistance of 6.0 Ω, the balance reads 274.0 g.

(a) Briefly explain why the reading on the balance changes when there is a current through XY.

(b) If the length of XY in the magnetic field is 5.0 cm, calculate:

(i) the extra force acting on the balance pan when S is closed

(ii) the magnetic flux density acting at right angles to XY

(iii) the new balance reading obtained (in g) when the connections to the 12.0 V supply are reversed ($g = 10$ N kg^{-1}).

3. A superconducting wire has a diameter of 3 mm and a density of 8600 kg m^{-3}. Calculate the current through the wire which causes it to 'float' without any visible support when it is placed at right angles to a horizontal magnetic field of flux density 5×10^{-4} T.

4. Two long, parallel wires which are 5.0 mm apart carry currents of 10 A in opposite directions. Calculate:

(a) the magnetic flux density midway between the two wires

(b) the force per unit length acting on each wire.

5. (a) At what perpendicular distance from the axis of a long wire carrying a current of 2.0 A is the flux density 2.0×10^{-5} T?

Self-assessment *continued*

(b) What is the flux density at the centre of a flat circular coil of 100 turns each of radius 50 mm with a current of 30 mA through it?

(c) A solenoid of 400 turns has a length of 100 mm.

(i) What is the current through it if the flux density at its centre is 2.0×10^{-4} T?

(ii) What is the new current needed to produce the same flux density when an iron cylinder of relative permeability 2000 is inserted into the solenoid?

6.

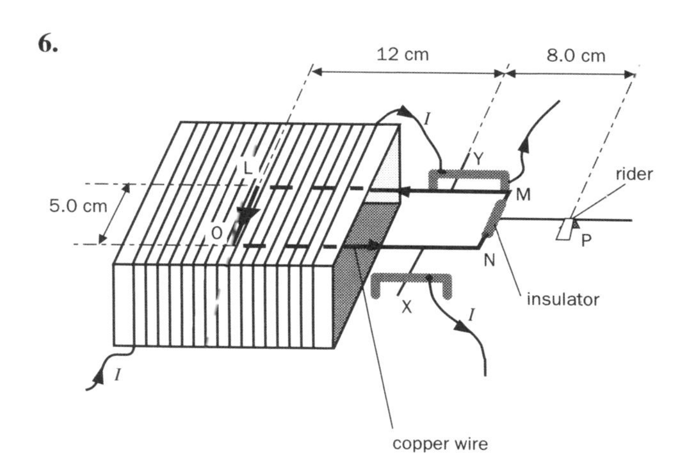

The diagram shows a simple form of current balance being used to measure the flux density inside a flat, rectangular solenoid having 3000 turns per metre. The solenoid is connected in series with the horizontal stiff copper wire LMNO, so that the solenoid current is also in the copper wire. The stiff copper wire is freely pivoted on the axis XY, and is in equilibrium when no current flows. LO is at right angles to the solenoid's central axis. The magnetic field of the solenoid can be assumed to be uniform within the coil, and parallel to the axis of the solenoid.

When a current (I) is switched on, there is an electromagnetic force on LO directed vertically downwards. A rider of mass 0.05 g has to be placed 8.0 cm from XY to restore equilibrium.

What is the value of the current I?
($\mu_0 = 4\pi \times 10^{-7}$ H m^{-1}; $g = 10$ N kg^{-1})

7. An electron travelling with a velocity of 5.0×10^7 m s^{-1} enters a magnetic field of flux density 4.0×10^{-4} T at right angles to its direction of motion. Calculate the radius of curvature of the electron path while it travels in the field.

8.

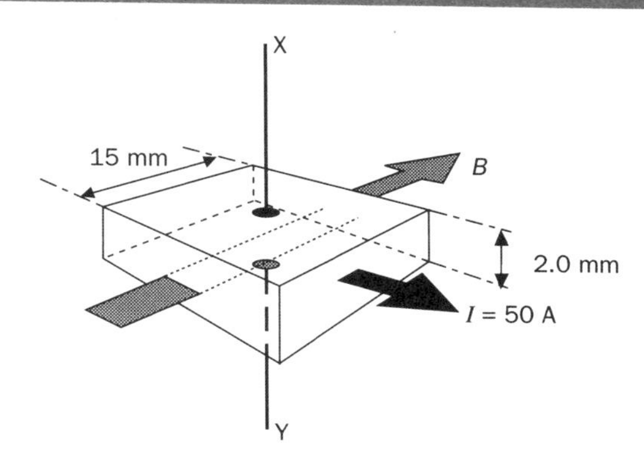

(a)

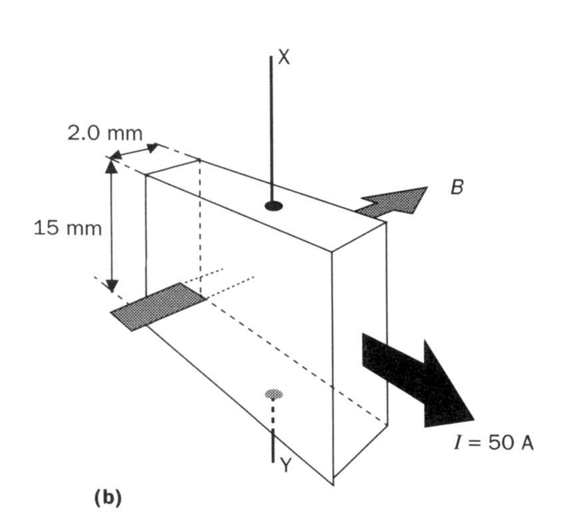

(b)

Part (a) shows a rectangular copper strip in which a current of 50 A flows at right angles to a magnetic field of flux density 2.5 T. A Hall p.d. of 0.6 μV develops across the terminals XY.

(i) What is the number of free electrons per m^3 for copper?

(ii) Calculate the Hall p.d. obtained when the strip is arranged as shown in (b) with the I and B values remaining the same.

(iii) How would the Hall p.d. obtained change if the copper strip was replaced with a semiconductor strip of the same dimensions? Explain your answer.

3.5 Electromagnetic induction

As we have already seen, when a conductor carries a current in a magnetic field, it experiences a force. This momentous discovery, made by Ampere in 1820, eventually led to the discovery of **electromagnetic induction**, which is the principle behind the whole electricity generating industry. This discovery is attributed to Faraday in England, and Henry in the USA at about the same time, 1831.

Magnetic flux and field strength

The magnetic field strength B has previously been defined in terms of the force acting on a unit magnetic pole. A magnetic field is the region around a magnet in which magnetic forces are exerted. It was once thought that there must be something flowing through this region from N-pole to S-pole. This was given the name **magnetic flux**, a name which has stuck. The **strength** of the field can be thought of as the amount of flux passing through **unit cross-sectional area** – i.e. a measure of the concentration of the flux. There is, of course, no such substance as flux, but the concept is a useful one when we come to consider electromagnetic induction.

Magnetic flux

Common symbol Φ (Greek upper case 'phi')
SI unit weber (Wb)

If a uniform magnetic field of strength (B) acts perpendicular to an area (A), the magnetic flux (Φ) passing through the area is given by:

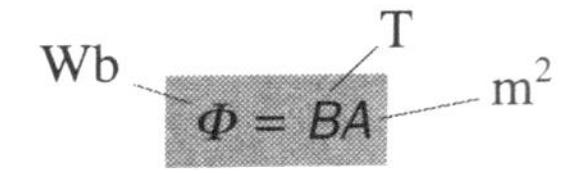

Equation 1

Since $B = \Phi/A$, magnetic field strength can be re-defined as **magnetic flux density**:

> Magnetic flux density B is defined as the **magnetic flux** Φ perpendicular to **unit area** A.

The unit of B is therefore the **weber per metre**^{2} (Wb m^{-2}). This is identical to the **tesla** (T), and the two units and definitions are completely interchangeable.

If the plane of the area is not perpendicular to the field, but is inclined at some angle θ to the field, then Equation 1 is modified to become:

$$\Phi = BA \sin \theta$$

Simple experimental phenomena

Relative motion between field and conductor (dynamo effect)

Stationary field – moving conductor

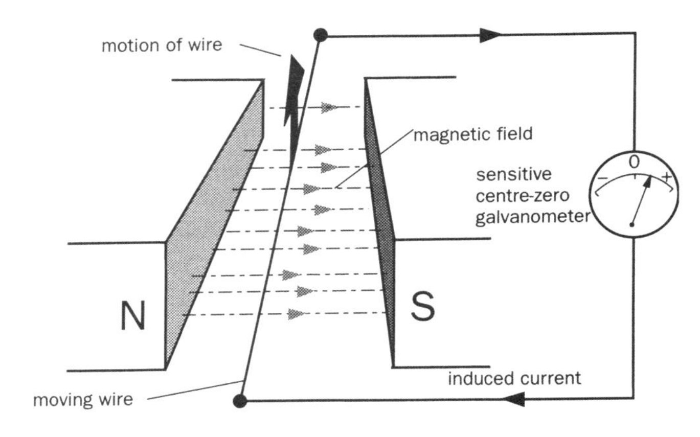

Figure 3.67 Electromagnetic induction produced by movement of a conductor in a stationary magnetic field (dynamo effect)

When a wire is moved vertically in a horizontal magnetic field as shown in Figure 3.67, a sensitive galvanometer connected to its ends shows a deflection. We say that an **electromotive force (e.m.f.)** has been **induced** in the wire. This e.m.f. causes an **induced current** to flow, as indicated by the galvanometer. Careful observation reveals the following.

The **magnitude** of the induced e.m.f. is proportional to:

- the **speed** of movement of the wire
- the **strength** of the magnetic field.

The **direction** of the induced e.m.f. is reversed by:

- reversing the direction of **motion of the wire**
- reversing the direction of **the magnetic field**.

NB: The induced e.m.f. is zero if the wire is held stationary relative to the field.

Stationary conductor – moving field

The above effect can also be produced by holding the wire stationary, and moving the magnet. Winding the wire into a coil allows the field to intersect with additional **turns** of wire:

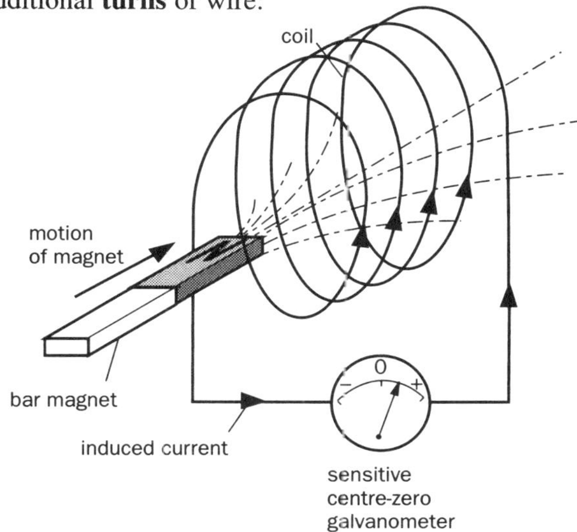

Figure 3.68 Electromagnetic induction produced by movement of a magnetic field relative to a stationary conductor

Figure 3.68 shows the effect of moving a bar magnet into a coil of wire. The **magnitude** and **direction** of the induced e.m.f. are controlled by the **speed** and **direction** of motion as above. In addition, the **magnitude** of the induced e.m.f. is also proportional to:

- the **number** of turns of wire in the coil.

What causes the e.m.f.?

This is a complex process that demands careful study. Figure 3.69 shows a straight conductor rolling across a table top. A magnetic field acts vertically downwards, so the conductor 'cuts' through the magnetic flux.

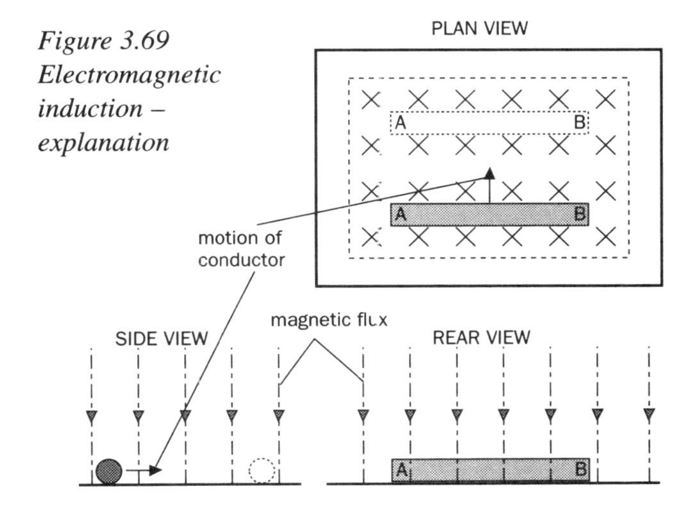

Figure 3.69 Electromagnetic induction – explanation

The charged particles (protons, electrons) which are a part of the atoms of the conductor are moving with it, so they are effectively an **electric current** flowing in the direction of motion of the conductor. As we have seen previously, a current in a magnetic field experiences a **force**, and the direction of this force is given by Fleming's left-hand rule. Both the protons and electrons feel this force, but only the electrons are free to move. Figure 3.70 shows the behaviour of the **electrons** in the conductor:

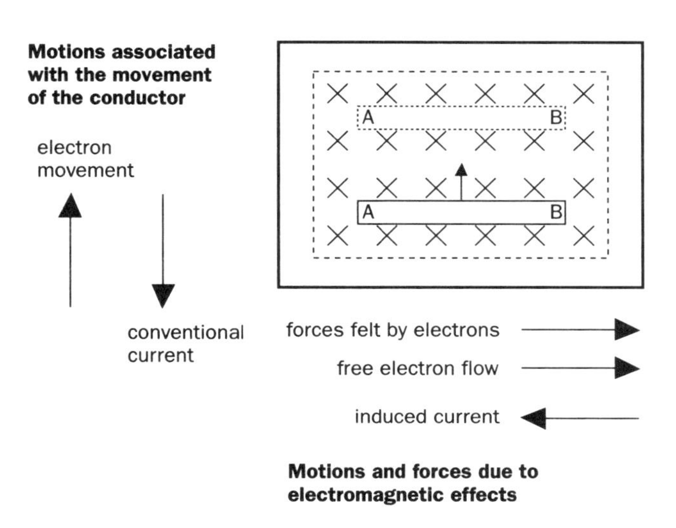

Figure 3.70 Forces on electrons in a conductor moving in a magnetic field

As a result of this force, electrons flow from A to B within the conductor. This electron flow means that the direction of the **induced conventional current** is from B to A. The direction of the **conventional e.m.f.** is also in this direction. The direction of this induced e.m.f. (and any current) can be predicted using **Fleming's right-hand rule:**

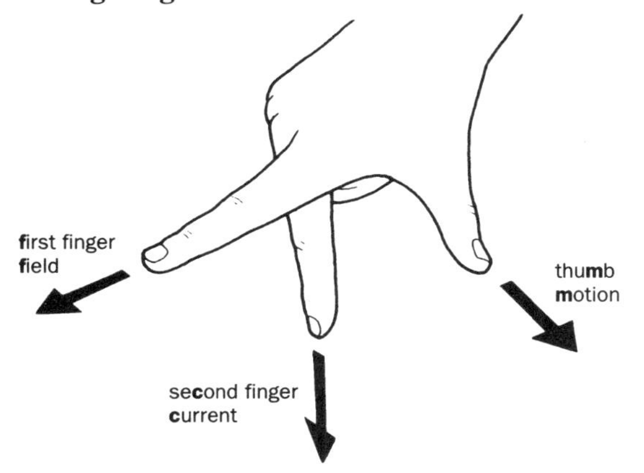

Figure 3.71 Fleming's right-hand rule relating the directions of magnetic field, motion and induced current

The thumb, first and second fingers of the right hand are held at right angles to each other as shown in Figure 3.71. If the thumb indicates the direction of **motion**, and the first finger the **field** direction, then the second finger indicates the direction of the induced **current** (and e.m.f.).

It is important to think clearly about the directions of the conventional e.m.f. and the **internal electric field** created by the redistribution of the electrons in the conductor. If the conductor is not connected to any external circuit, electrons will accumulate at end B. This creates an **internal electric field**, as shown in Figure 3.72. The electric force of this field acts to oppose the motion of the electrons. The electrons will stop flowing when the electric force balances the electromagnetic force.

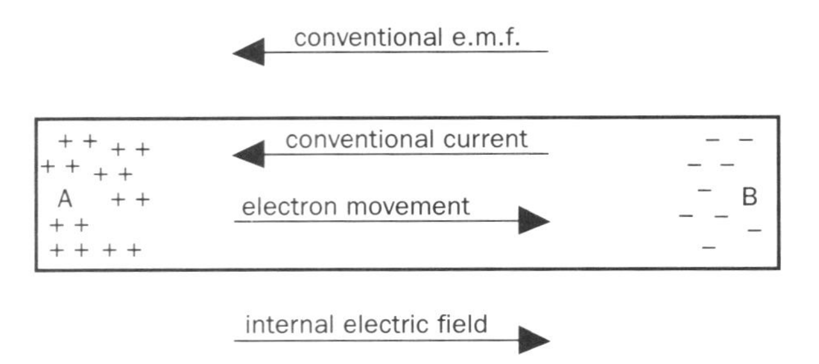

Figure 3.72 Creation of the internal electric field

The internal electric field is caused by the accumulation of electrons at one end, i.e. end B becomes more negative. This effect is a **result** of the electron movement, not the cause of it. If the conductor is connected to an external circuit, there will be no accumulation of electrons, and the electrons will flow continuously due to induction.

Changing magnetic field (the transformer effect)

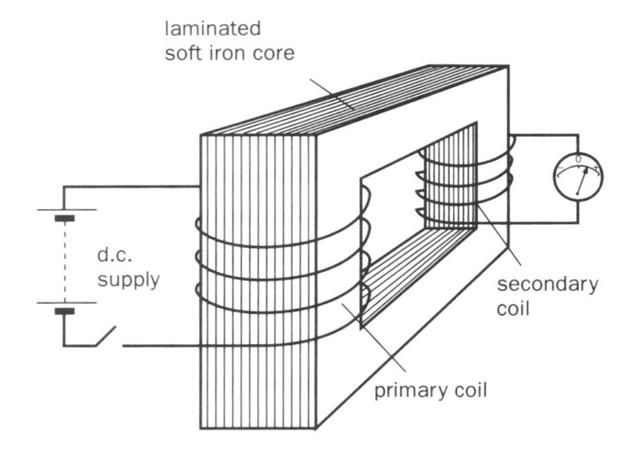

Figure 3.73 Electromagnetic induction by varying magnetic field (transformer effect)

In this case the magnetic field is created by the current flowing in a coil of wire (the **primary** coil) and the e.m.f. is induced in another separate coil (the **secondary** coil). The two coils are not connected electrically, but are **linked** by the magnetic flux. The iron core concentrates the flux produced so improving the linkage.

With the set-up shown in Figure 3.73, the galvanometer across the secondary coil shows a deflection only at the instant the primary current is either **switched on** or **off**. It registers **zero** whenever the primary current is **zero** or **constant**. Whenever the primary current is changing (growing or decaying) the magnetic field linking the secondary coil will also be changing. It is this changing magnetic field that induces the secondary e.m.f.

Laws of electromagnetic induction

Faraday's law

This law relates the **magnitude** of the induced e.m.f. to the **speed** at which the magnetic flux is changing.

Faraday's law states that:

> The magnitude of the **induced e.m.f.** is proportional to the **rate of change of flux cutting**, or to the **rate of change of flux linkage.**

The first statement is useful in situations where there is relative motion between conductor and field, the second applies to effects caused by changing magnetic fields.

Flux cutting and **flux linkage** are terms used to describe the interaction between the conductor and the magnetic field. As a conductor moves through a magnetic field, you can imagine it 'cutting' through the field, rather like a wire through cheese. The changing **linkage** between a varying magnetic field and a conductor is less easy to visualise, but produces the same result.

Faraday's law can be expressed by the equation:

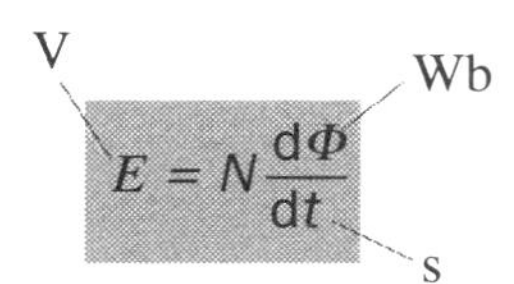

$$E = N\frac{d\Phi}{dt}$$

Equation 2

where N = number of turns of wire linking the flux

$\frac{d\Phi}{dt}$ = rate of change of flux linkage.

Guided example (1)

Flux cutting

An aircraft wing can be considered as a straight conductor of length 30 m. If the aircraft flies at a steady velocity of 100 m s^{-1} through a magnetic field of strength 2.0×10^{-7} T acting vertically downwards, calculate the magnitude and direction of the e.m.f. induced between its wing tips.

Guidelines

Consider distance moved in one second.

Calculate the area swept out by wing per second.

Calculate flux cut by wing ($\Phi = BA$) in one second.

Calculate rate of flux cutting.

Consider the direction of the electromagnetic force on the electrons. On which wing tip will they accumulate?

Another way in which a change in flux linkage can be produced is by rotating a coil in a magnetic field. The maximum flux linkage occurs when the plane of the coil is perpendicular to the field. If the field and the plane of the coil are parallel, the flux linkage is zero (Figure 3.74).

Guided example (2)

Coil rotation

A circular coil of 20 turns of radius 10 cm is situated with its plane perpendicular to a magnetic field of strength 0.5 T. Calculate the e.m.f. induced if the coil is rotated through 90° in a time of 50 ms.

Guided example (2) *continued*

Guidelines

Calculate the maximum flux linking each turn of the coil, and hence calculate the total flux linkage for N turns.

Minimum flux linkage is zero (plane of coil parallel to field).

The rate of change of flux linkage is then given by: *(change of flux linkage)/(time taken)*

Lenz's law

At this point, the alert reader may be asking 'where does the electrical energy come from?' As if by magic, an e.m.f. appears in the conductor, and if its ends are connected, a current results. This current represents electrical energy which will in turn be dissipated to the surroundings by the heating effect of the current through the conductor.

Of course it is not magic. There has been an input of energy to generate this current. Lenz's law is simply a statement of the principle of conservation of energy, as applied to electromagnetic induction.

Lenz's law states that the **direction** of the induced e.m.f. is such that it tends to **oppose** the flux change which causes it, and does oppose it if there is an induced current.

Mathematically, this statement is incorporated into Faraday's law by the inclusion of a **minus** sign in Equation 2:

$$E = -N\frac{d\Phi}{dt}$$ *Equation 2(a)*

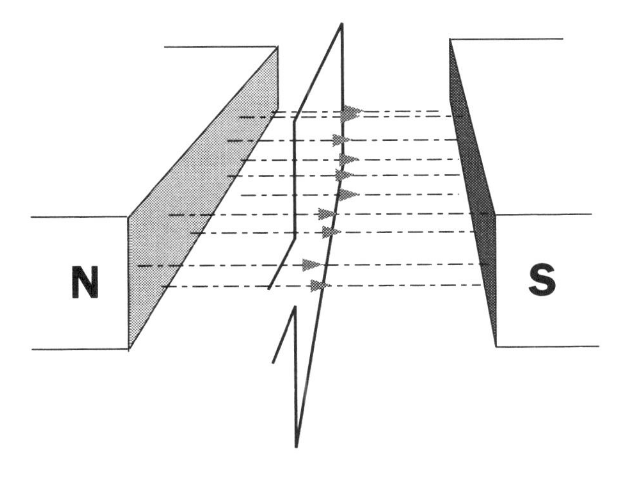

(a) Maximum flux linkage

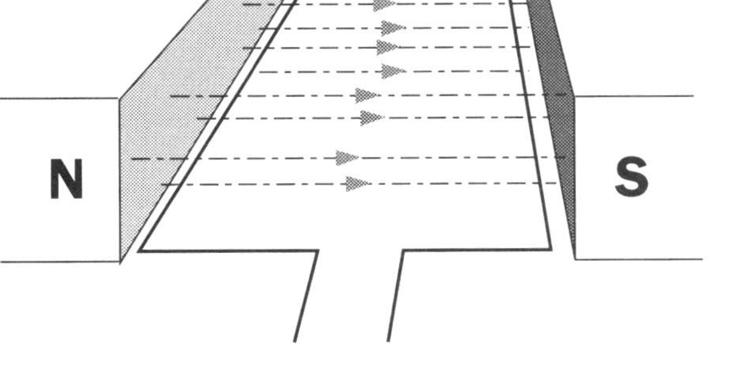

(b) Zero flux linkage

Figure 3.74 Change in flux linkage produced by a coil rotating in a uniform magnetic field

When the induction is caused by relative **motion** between magnetic field and conductor, and an induced current results, this current produces an induced magnetic field. This field exerts a force which **opposes** the motion. Thus work must be done in moving against this force, which requires an input of **energy.**

Consider what happens when a bar magnet is pushed into a coil of wire. Figure 3.75 shows the bar magnet's field 'cutting' (or linking with) the coil, and so inducing a current I in it. The induced current must produce a magnetic field which opposes this change. The right-hand-grip rule then shows that the current direction to give this field is anti-clockwise as shown. i.e. this face of the coil acts as a magnetic N-pole.

Thus whatever is pushing the magnet has to do work against the **repulsive** force generated between like magnetic poles. It is the energy required to do this work that is converted into electrical energy.

If the magnet is pulled out from the coil, there must be a force of attraction to resist this motion. Consequently the direction of the induced current is opposite to that in Figure 3.75. This time, work must be done against the **attractive** force which is acting to prevent the magnet being removed.

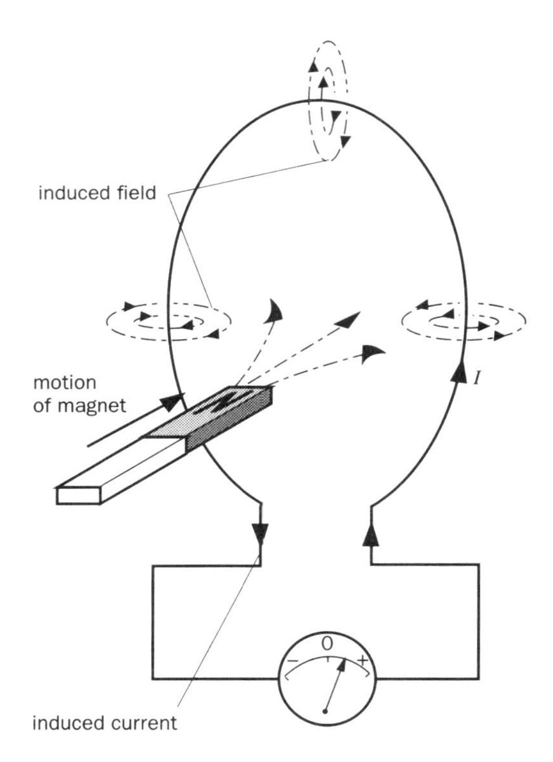

Figure 3.75 Lenz's law – induced field in coil exerts force to oppose motion of magnet

Guided examples (3)

1. A flat, square coil of 250 turns, each of area 1.6×10^{-3} m^2, is connected to a galvanometer and placed with its plane perpendicular to a magnetic field of flux density 0.35 T. If this flux density is reduced to zero in 0.050 s, and the total resistance of the coil and galvanometer is 100 Ω, calculate:

(a) the e.m.f. induced in the coil

(b) the induced current.

Guidelines

(a) Calculate the change in magnetic flux linking the coil (use $\Phi = BAN$) then apply Faraday's law to calculate the induced e.m.f.

(b) Use *(current) = (e.m.f.)/(resistance)* to calculate the current.

2. A coil of 200 turns, each of area 6.0 cm^2, has its plane perpendicular to a magnetic field whose flux density varies with time as shown by the following graph:

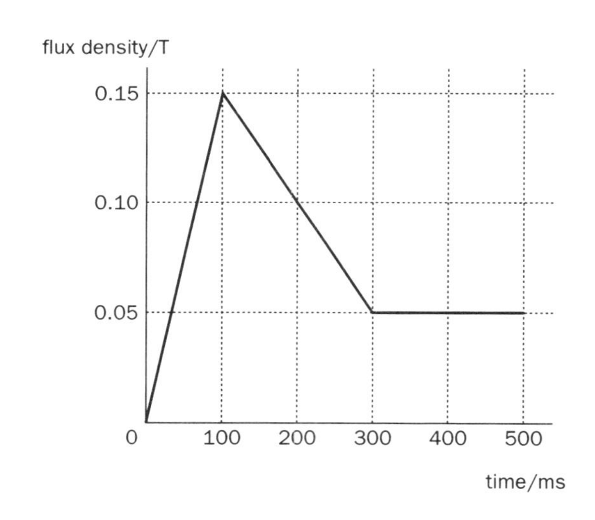

(a) Calculate the total magnetic flux linkage (i) during the first 100 ms; (ii) between 100 ms and 300 ms; (iii) between 300 ms and 500 ms.

(b) Sketch a graph of induced e.m.f. against time between 0 and 500 ms. Show numerical values on each axis.

Guided examples (3) continued

Guidelines

(a) The change in magnetic flux density is the difference between the initial and final values. Don't forget to use SI units, and that the coil has more than one turn.

(b) Calculate the e.m.f. for each time period, using Faraday's law. If the rate of change of flux linkage is constant for a time, then the e.m.f. induced will be constant for that time.

3. A closed wire loop has an electrical resistance of 4.0×10^{-3} Ω and is in the shape of an equilateral triangle of side 4.0 cm. The loop is situated in a uniform magnetic field of flux density 0.4 T acting in a direction perpendicular to the plane of the coil. If the wire loop is rotated through 90° to a new position in which its plane is parallel to the field in a time of 5.0 ms, calculate:

(a) the e.m.f. induced

(b) the current in the loop

(c) the energy dissipated by the current.

Guidelines

(a) Calculate the area of the loop and hence the change in flux density when the loop is rotated (flux density is zero when the plane of the loop is parallel to the field). Then use Faraday's law to calculate the e.m.f.

(b) Use $I = E/R$.

(c) Energy dissipated = I^2Rt.

E.m.f. induced in a rotating coil

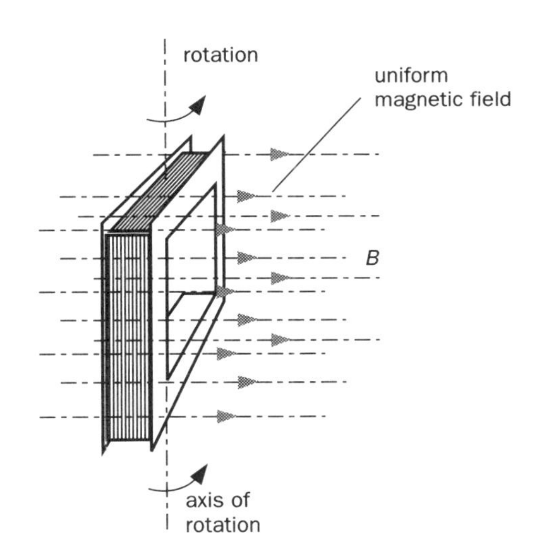

Figure 3.76 Pivoted coil in a uniform magnetic field

Figure 3.76 shows a coil constructed by wrapping N turns of wire round a rectangular frame of area A. The coil is pivoted about a vertical axis through its centre and rotates about this axis with a uniform angular velocity ω. A uniform, horizontal magnetic field of flux density B passes through the coil.

θ = angle between normal to the plane of the coil and the field (rad)

t = time after $\theta = 0$ (s)

Φ = flux through coil (Wb)

In Figure 3.77(a) (when $t = 0$), the normal to the plane of the coil is parallel to the field, so the flux linking each turn of the coil is a maximum (= BA).

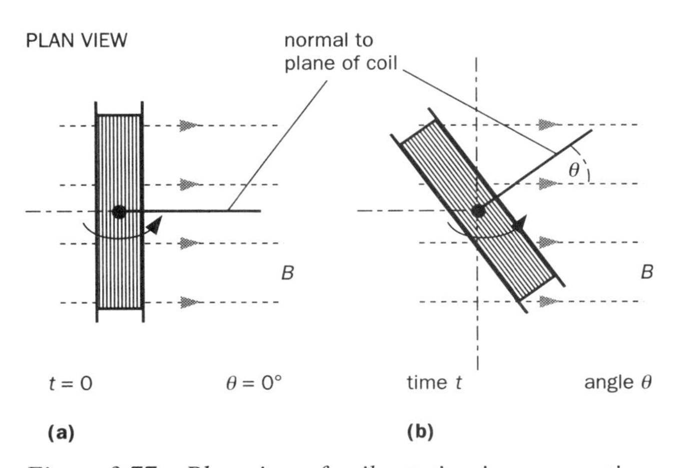

Figure 3.77 Plan view of coil rotating in a magnetic field

Figure 3.77(b) shows the coil rotated through an angle θ at some later time t.

The flux Φ linking each turn is given by:

$$\Phi = BA \cos \theta \quad \text{(by definition of } \Phi\text{)}$$

but $\theta = \omega t$ (by definition of angular velocity)

$$\therefore \quad \Phi = BA \cos \omega t$$

To calculate the e.m.f. E induced in the coil, we use:

$$E = -N \frac{d\Phi}{dt} \quad \text{(Faraday's and Lenz's laws)}$$

$$\therefore \quad E = -N \frac{d}{dt}(BA \cos \omega t)$$

$$\therefore \quad E = -BAN \frac{d}{dt}(\cos \omega t)$$

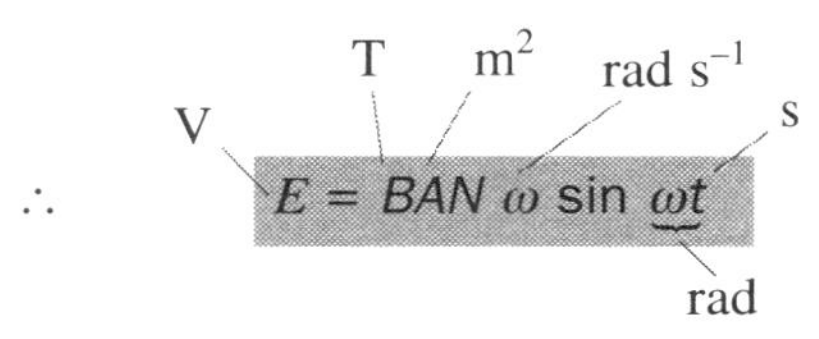

Equation 3

Maths window

Standard differentiation of trig. functions:

If $y = \cos t$

then $\dfrac{dy}{dt} = -\sin t$

If $y = \cos At$

then $\dfrac{dy}{dt} = -A \sin At$

$\therefore \quad \dfrac{d}{dt}(\cos \omega t) = -\omega \sin \omega t$

Figure 3.78 shows how the flux linking the coil and the e.m.f. induced in it vary with time. The e.m.f. is alternating and sinusoidal.

NB: The e.m.f. is a maximum when the flux linking the coil is zero. However, the **rate of change** of flux linkage is a maximum (the gradient of the t/Φ graph is at its steepest) and hence the e.m.f. is a maximum.

At (a):

$\theta = 0°$ $\sin\theta = 0$ $\cos\theta = 1$

$\therefore \quad \Phi = BAN \cos \omega t = BAN$

$E = BAN\omega \sin \omega t = 0$

At (b):

$\theta = 90°$ $\sin\theta = \pm 1$ $\cos\theta = 0$

$\therefore \quad \Phi = BAN \cos \omega t = 0$

$E = BAN\omega \sin \omega t = BAN\omega$

At (c):

$\theta = 180°$ $\sin\theta = 0$ $\cos\theta = -1$

$\therefore \quad \Phi = BAN \cos \omega t = -BAN$

$E = BAN\omega \sin \omega t = 0$

At (d):

$\theta = 270°$ $\sin\theta = -1$ $\cos\theta = 0$

$\therefore \quad \Phi = BAN \cos \omega t = 0$

$E = BAN\omega \sin \omega t = -BAN\omega$

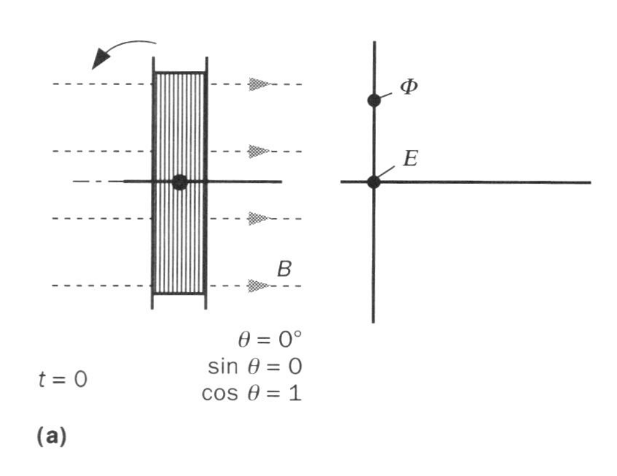

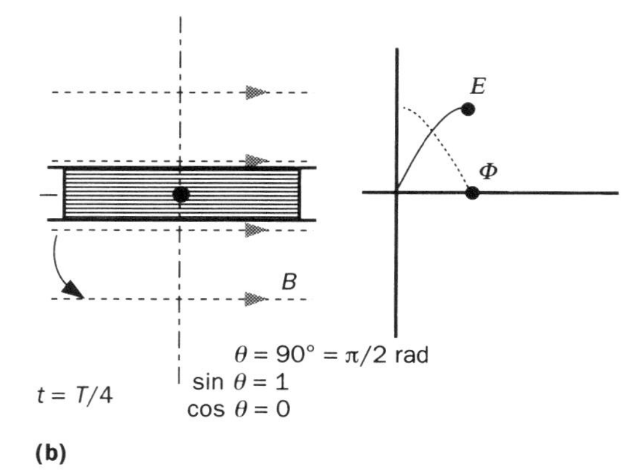

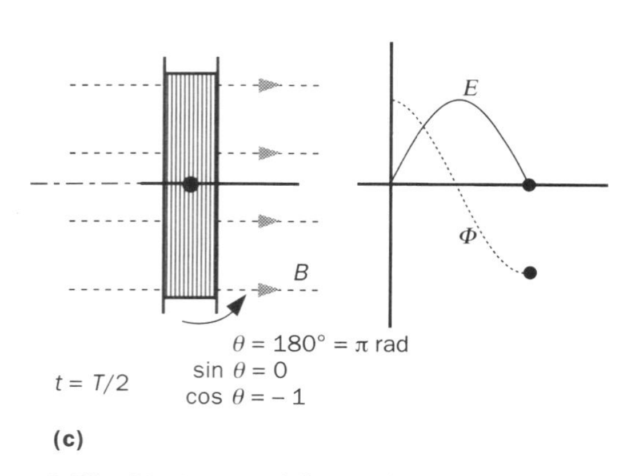

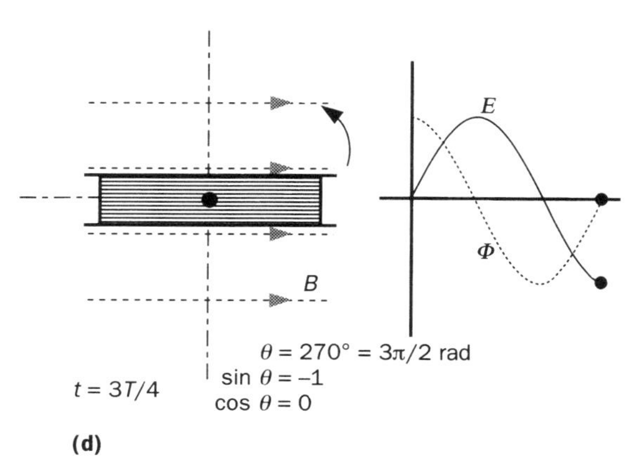

Figure 3.78 Variation of flux linkage and induced e.m.f. as the coil rotates

Guided example (4)

A flat, circular, 250 turn coil of radius 2.5 cm rotates with a uniform angular velocity of 1200 r.p.m. in a uniform magnetic field of flux density 0.09 T directed at 90° to the axis of rotation.

(a) Calculate the magnitude of the induced e.m.f. when the plane of the coil is

(i) perpendicular to the field

(ii) parallel to the field

(iii) at 60° to the field.

(b) If the combined resistance of the coil and a galvanometer connected to its ends is 120 Ω, calculate the maximum current in the coil. State the position of the coil when the current is a maximum.

Guidelines

(a) Calculate the coil area and convert r.p.m. to rad s^{-1}. Then use Equation 3, remembering that θ in the equation is the angle between the **normal** to the plane of the coil and the field direction.

(b) Maximum current will occur when the e.m.f. is a maximum.

Self-inductance

Common symbol L
SI unit henry (H)

When there is a current in a wire, a magnetic field is produced. If the magnitude (or direction) of the current changes, then so does the strength (or direction) of the magnetic field. This changing magnetic field links the wire producing it, so a current-carrying conductor induces an e.m.f. in itself.

According to Faraday's law, the magnitude of the induced e.m.f. will be proportional to the rate of change of flux linkage, which is itself proportional to the rate of change of current in the wire. Thus the induced e.m.f. (E) is proportional to the rate of change of current (I):

i.e. $E \propto \dfrac{dI}{dt}$

$$\therefore \quad E = L\frac{dI}{dt}$$

(V, H, A s^{-1})

L is the constant of proportionality in the above equation, and is called the **self-inductance** (or just **inductance**) of the wire. This only assumes significance if the wire is wound in a coil.

A coil is said to have an inductance of **1 henry** if an e.m.f. of **1 volt** is induced in the coil when the current in the coil changes at the rate of **1 ampere per second**.

Adding a core of magnetic material greatly increases the magnetic field and therefore inductance of a coil. Such a coil is often called a choke coil (or simply a choke).

The direction of the induced e.m.f. is of course given by Lenz's law – i.e. it acts so as to oppose the change causing it. Thus if the current is increasing, then the e.m.f. induced acts so as to send a current back in the opposite direction. In other words, self-inductance inhibits the rate at which a current can increase:

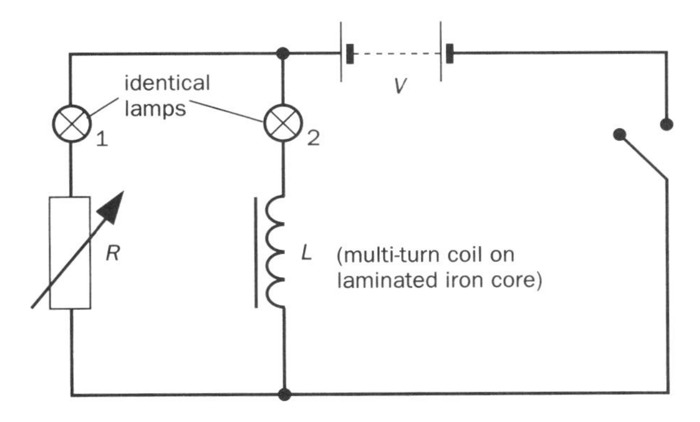

Figure 3.79 Circuit for demonstrating the effect of inductance on current growth

In the circuit shown in Figure 3.79, the variable resistor is adjusted until the two bulbs are glowing with equal brightness with a steady current: i.e. the resistor and the coil have equal resistance. Then the current is switched off.

On switching on, it is noticed that lamp 1 reaches full brightness almost instantly, while lamp 2 takes a few seconds to become fully lit. Figure 3.80 on page 256 shows how the currents in the two lamps grow with time.

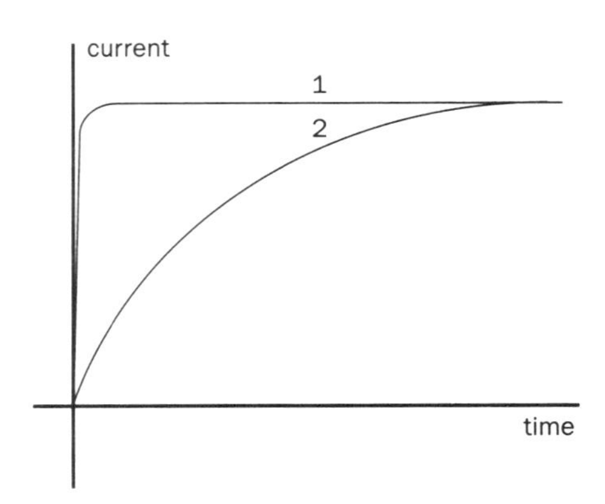

Figure 3.80 Graph of current against time for an inductive circuit

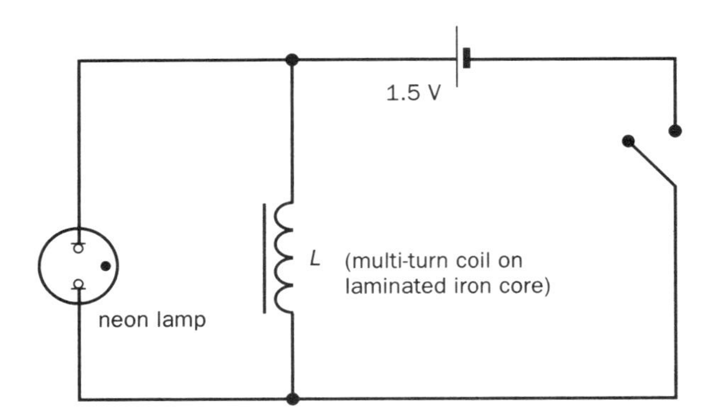

Figure 3.81 Lighting a neon bulb by self-inductance

The resistor causes negligible induction, so there is virtually no delay in the growth of current through it (graph 1). However, the coil possesses considerable self-inductance. As the current through it starts to increase from zero, the magnetic field starts to grow. The large self-inductance effect produced by this growing field causes a **back e.m.f.**, which tries to oppose the current through the coil. As a result, the current in the coil can only increase slowly, hence the delay before the current has become large enough to light the lamp.

Combining Faraday's and Lenz's laws gives the final equation:

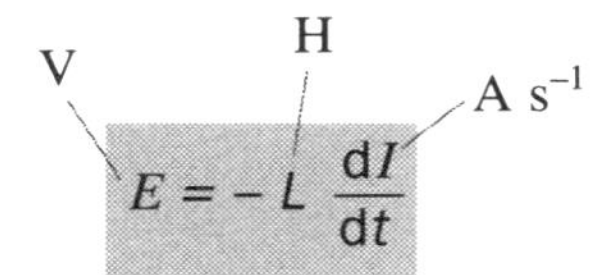

$$E = -L\frac{dI}{dt}$$

Equation 4

The minus sign reminds us that the self-induced e.m.f. **opposes** the change of current producing it.

The circuit shown in Figure 3.81 illustrates the effect of self-inductance in a more dramatic way. When the switch is closed, there is current in the coil, but the neon lamp fails to light, since the p.d. is too low to cause conduction in the neon gas. (Several hundred volts are needed.)

When the switch is opened, the lamp flashes briefly, showing the presence of a brief 'spike' of high voltage. Switching off the current in the coil causes its magnetic field to decay rapidly. This sudden change induces an e.m.f. in the coil large enough to light the lamp. According to Lenz's law, the direction of this e.m.f. acts so as to oppose the change producing it. So the induced e.m.f. tries to sustain the current through the coil in the same direction as it was flowing before switching off.

The effect of the sudden, rapid decrease in current is to cause an induced e.m.f. which acts to try to maintain the current. When there is a current in the coil, energy is stored in the magnetic field. This is the energy released in the high-voltage 'spike'.

Since the rate of change of current is high, the induced e.m.f. is also high. This effect is dangerous, since the high e.m.f. can cause sparking to occur as switches open. To prevent the spark, the whole switch may be immersed in oil, or a capacitor may be connected in parallel with the switch, to absorb the energy.

Inductors have great importance in circuits where the current is changing rapidly, i.e. alternating currents (see page 303).

Mutual inductance

Common symbol	M
SI unit	henry (H)

This involves two coils, as shown in Figure 3.82. A current I flowing in the first coil (the primary)

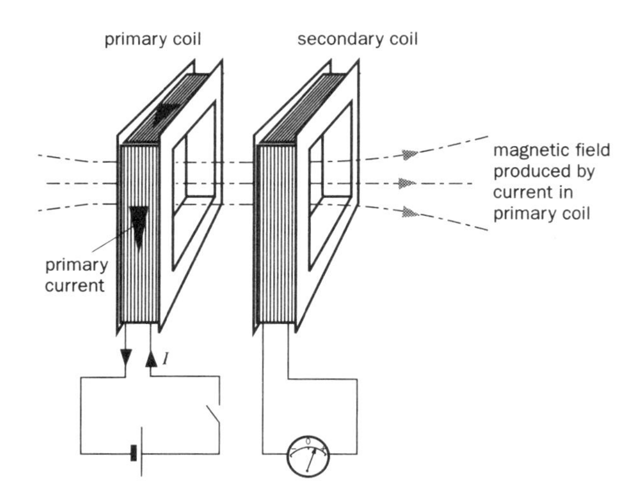

Figure 3.82 Mutual inductance

produces a magnetic field which **links** the secondary. If the current in the primary **changes**, the flux linking the secondary will change. Thus an e.m.f. E will be **induced** in the secondary. This effect is called **mutual inductance**, and the equation is as follows:

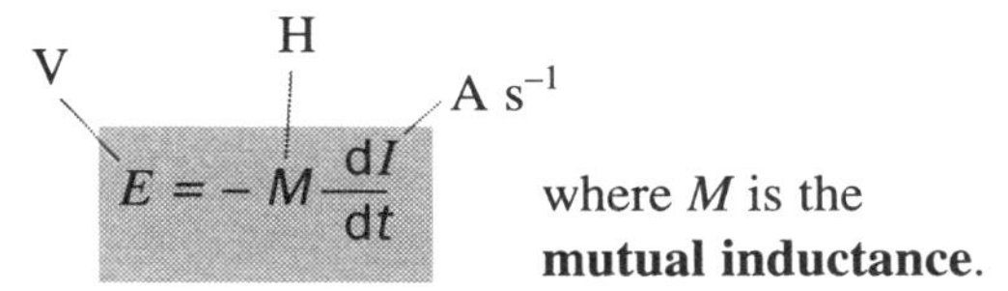

where M is the **mutual inductance**.

Equation 5

There is only an induced e.m.f. in the secondary coil if the current in the primary is **changing**. This is the basis of the **transformer** for stepping up or down alternating voltages.

Two coils have a mutual inductance of **1 henry** if an e.m.f. of **1 volt** is induced in either coil when the current in the other changes at the rate of **1 ampere per second**.

Energy stored in an inductor

When the current in an inductor is suddenly switched off, a spark may be produced across the switch gap as it opens. As seen before, the induction effect tries to sustain the current as the magnetic field collapses, causing the spark. The energy stored in the magnetic field is released in the spark.

It can be shown that the energy W stored in an inductor of inductance L carrying a current I is given by:

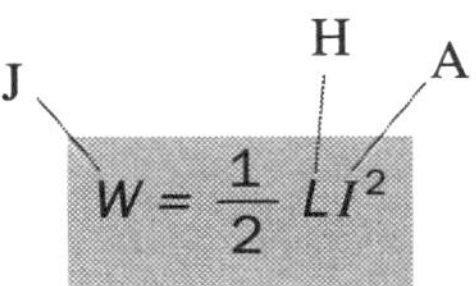

Equation 6

Since every current produces a field, energy must be supplied to establish the field, so no current can ever change instantaneously.

Guided example (5)

A back e.m.f. of 2000 V is induced in a coil when the current in it decreases from 5.0 A to 0 in 2.0 ms.

Calculate

(a) the self-inductance of a coil

(b) the energy stored in this coil when it is carrying a current of 5.0 A.

Guided example (5) *continued*

Guidelines

(a) Standard application of Equation 4.

(b) Standard application of Equation 6.

The electric motor

There are two main parts to an electric motor – the rotating part or **armature**, and the stationary part or **stator**.

- The armature consists of several equally spaced coils wound on a laminated soft iron core. Each end of a coil is connected to a section of the **commutator**, which rotates with the armature.
- The stator carries the magnetic pole-pieces, shaped to create a radial field in which the armature rotates.
- The electromagnetic force, and hence the torque, on the armature is nearly constant at all positions of the armature due to the use of
  (i) many coils
  (ii) curved pole pieces
  (iii) the soft iron core.
- The laminations of the core reduce the circulation of **induced eddy currents within the core**. These reduce the efficiency of the motor due to their heating effect, and the generation of opposing forces (Lenz's law).

The rotation of the coils of the motor in the magnetic field causes an induced e.m.f. (the back e.m.f. E) which opposes the applied p.d. (V). If (I) is the current in the coils, which have resistance (R), then:

$$V - E = IR$$

Equation 7

The value of E is proportional to the rate of change of flux cutting (Faraday's law), so E is proportional to the speed of rotation of the motor.

As the speed increases, E increases, so I decreases.

When a motor performs work on a load, its speed falls to a steady value.

Multiplying Equation 7 by I and re-arranging:

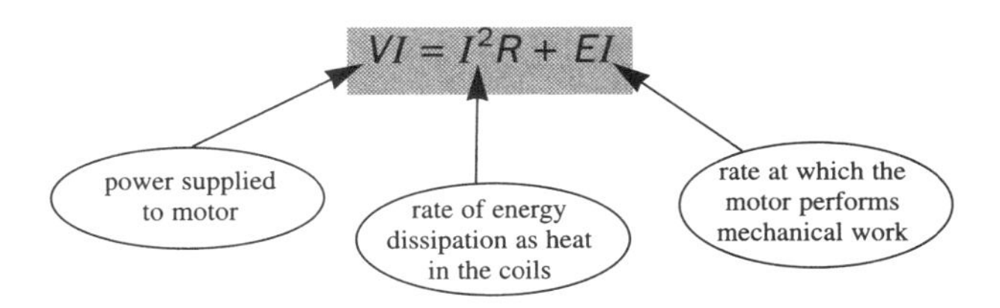

Equation 8

Efficiency of a motor

The percentage efficiency of an electric motor is defined by the equation:

$$efficiency = \frac{mechanical\ power\ output}{electrical\ power\ input} \times 100\%$$

from which:

$$efficiency = \frac{EI}{VI} \times 100\% = \frac{E}{V} \times 100\%$$

The efficiency is high when EI is high, and this happens when I^2R is low (i.e. when R is small).

Eddy currents

When the magnetic field passing through a conductor is changed in any way, e.m.f.s are induced in the conductor. The change of flux may be caused by relative motion between the conductor and the field or by variation of the strength of the field. These induced e.m.f.s cause currents, called **eddy currents**, to circulate within the conductor, even if it is isolated. Because these currents can follow short, low-resistance paths in the conductor, they can reach very high values, producing strong heating, magnetic and other effects.

The magnetic effect

According to Lenz's law, any magnetic field created by the induced eddy currents will **oppose** the motion or flux change which produced them. Any metal object moving through a magnetic field will therefore experience forces resisting its motion – a **braking** effect. This can be easily demonstrated using a pendulum consisting of a copper or aluminium (both good conductors) plate swinging between the poles of a strong magnet.

Figure 3.83(a) shows the copper plate oscillating freely with no magnetic field present. As soon as the electromagnet is switched ON (Figure 3.83(b)), the oscillations die away rapidly, and the pendulum quickly comes to rest. The damping effect is caused by the eddy currents induced in the copper plate as it cuts through the flux created by the electromagnet. The eddy currents create a magnetic field, which interacts with the field of the electromagnet to produce a force which resists the motion. Since the magnitude of the eddy currents is proportional to the rate of change of flux, the force created will be proportional to the velocity of the pendulum. The greatest braking force will be experienced as the pendulum passes through the mid-point of its swing.

If the solid plate is replaced by a **slotted** plate (Figure 3.83(c)), the damping effect is much reduced. This is because the slots constrain the eddy currents to much shorter paths, reducing the magnetic field they can produce.

The Eurostar train is propelled by powerful electric motors

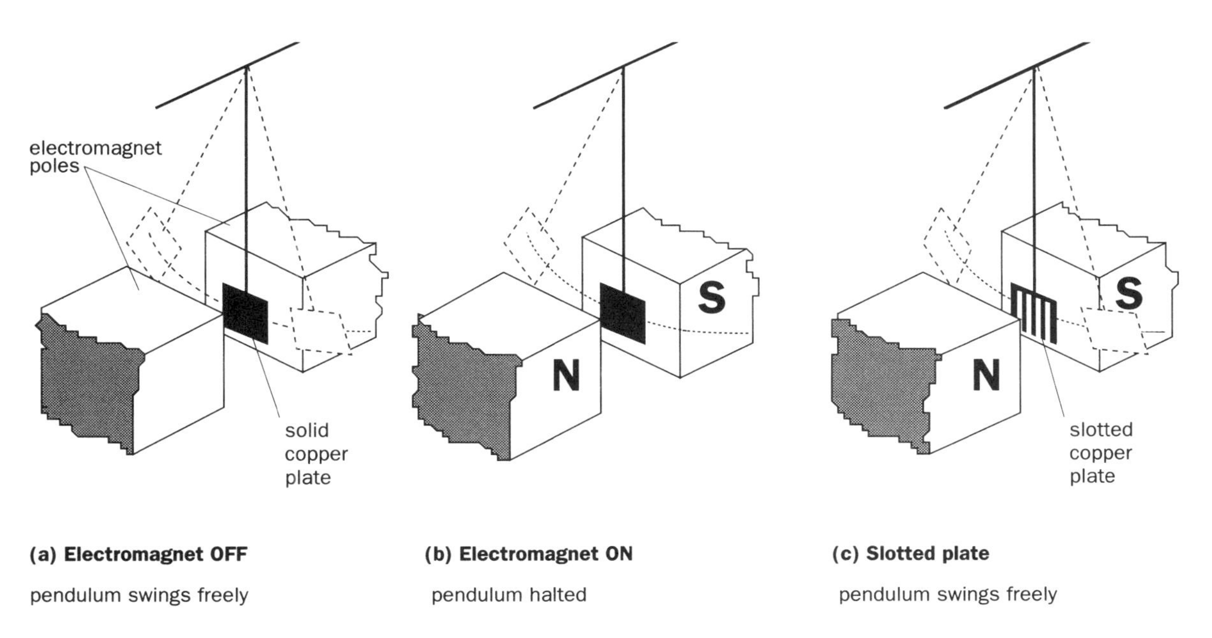

(a) Electromagnet OFF
pendulum swings freely

(b) Electromagnet ON
pendulum halted

(c) Slotted plate
pendulum swings freely

Figure 3.83 Eddy current damping

This magnetic damping is used in moving-coil meters. The coil is wound on a light non-magnetic metal frame (aluminium is ideal). When a current flows in the coil, the frame rotates between the magnetic poles. Eddy currents induced in the frame cause an opposing magnetic field, so the frame experiences a resisting force, proportional to its angular velocity. Consequently, the frame and the pointer come to rest rapidly. When the frame stops, there is no opposing force, so the pointer stays at rest.

The heating effect

Eddy currents can reach very high values; and so cause heating of the conductor. The **induction furnace** uses this effect to cause localised heating of metals. A water-cooled coil carrying a high-frequency alternating current surrounds the metal to be heated. The rapidly-changing magnetic field produced by the alternating current in the coil induces enormous eddy currents in the metal, causing rapid heating. The localised, rapid heating means that adjoining parts, outside this coil, remain cool.

Eddy current losses

The **armature** in an electric motor and the **rotor** in a generator are both pieces of iron moving in a magnetic field. Similarly, the iron **core** of a transformer is situated in the rapidly-changing field produced by the currents in the coils. In each case, eddy currents are induced, causing heating effects and a dissipation of energy. The energy losses are minimised by making the iron parts from thin **laminations** (sheets) which are insulated from one another. The eddy currents are then restricted to very small regions, and their effect is minimised.

Self-assessment

SECTION A Qualitative assessment

1. Briefly describe two methods for demonstrating electromagnetic induction, one involving a permanent magnet, and the other a current-carrying coil. In each case, explain in terms of flux linkage how an e.m.f. is induced.

2. (a) When a conductor moves through a magnetic field, state the factors influencing the **magnitude** of the induced e.m.f.

(b) Explain how to predict the **direction** of the induced e.m.f.

(c) Explain, in terms of the forces on the electrons, how the e.m.f. is produced.

3. (a) State **Faraday's** law and **Lenz's** law.

(b) When a bar magnet is withdrawn from a single loop of wire, an e.m.f. is induced which causes a current to flow in the wire. Explain, using a diagram, how the induced current direction is predicted by Lenz's law.

Self-assessment *continued*

(c) Which fundamental principle of physics is incorporated in Lenz's law?

4. (a) A coil has (N) turns of area (A). If the coil is situated in a magnetic field of flux density (B) so that the normal to the plane of the coil makes an angle (θ) with the direction of the field, state the magnitude of the flux linking the coil.

(b) If the coil in part (a) rotates with angular frequency (ω), derive an equation for the magnitude of the e.m.f. induced in the coil.

(c) When is the e.m.f. (i) a **maximum**? (ii) **zero**?

(d) Sketch a graph showing how the induced e.m.f. varies with time.

5. (a) Explain how **eddy currents** are produced.

(b) Give an example of energy loss due to eddy currents, and explain how this loss can be minimised.

(c) Briefly explain the use of eddy current braking in moving-coil meters.

6. (a) Explain what is meant by the term **self-inductance** as applied to a coil.

(b) State and define the unit of self-inductance.

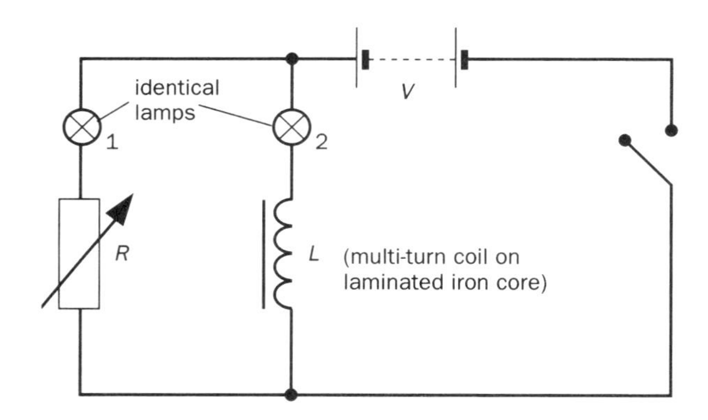

(c) In the circuit shown above, explain carefully why lamp 1 lights immediately the switch is closed, while lamp 2 takes several seconds to reach the same brightness.

7. Define **mutual inductance**. State and define its unit.

8. (a) State the equation for the energy stored in a coil of inductance (L) carrying a current (I).

(b) Where is the energy stored?

9. When the armature coils of an electric motor rotate, they cut through the magnetic field of the magnets. This causes a **back e.m.f.** (E) to be induced, which opposes the applied p.d. (V).

(a) What is the relationship between V, E and the current (I) in the coils, of total resistance (R)?

(b) Derive an expression for the mechanical power output of the motor.

SECTION B
Quantitative assessment

(***Answers:*** 2.0×10^{-5}; 3.9×10^{-4}; 7.8×10^{-4}; 7.1×10^{-3}; 8.0×10^{-3}; 0.014; 0.020; 0.098; –3.3; 8.0; 10; 43; 45.)

1. A flat, circular, 50-turn coil of diameter 5.0 cm having a resistance of 25 Ω is situated with its plane at right angles to a uniform magnetic field of flux density 0.20 T. If the flux density is steadily reduced to zero in 0.20 s, calculate

(a) the initial flux linking one turn of the coil

(b) the e.m.f. induced in the coil as the field changes

(c) the charge which flows through the coil.

2. A coil of 500 turns of fine wire, each of area 4.0×10^{-4} m^2, is placed with its plane normal to a magnetic field. The ends of the coil are connected together. The magnitude of the magnetic flux density varies with time as shown in the graph below.

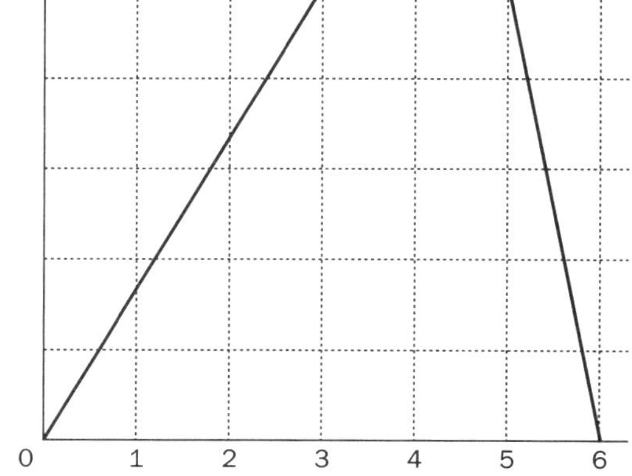

Self-assessment *continued*

(a) Calculate

(i) the maximum flux linking one turn of the coil

(ii) the e.m.f. induced in the coil during the first 3 ms.

(b) Sketch a graph showing how the e.m.f. induced varies with time over the 6 ms time interval shown. Show numerical values on your graph.

3. A closed, rectangular wire loop has dimensions of 5.0 cm × 4.0 cm and an electrical resistance of 50 mΩ. The loop is situated in a uniform magnetic field of flux density 0.95 T which is directed at right angles to the plane of the coil. If the flux density decreases steadily to 0.45 T in 2.5 ms, calculate:

(a) the magnitude of the current induced

(b) the energy dissipated in the loop.

4. A 400 turn coil of area 15 cm^2 rotates with a constant angular velocity of 30 rev s^{-1} in a uniform magnetic field of flux density 0.40 T. Calculate the induced e.m.f. at the instant when the plane of the coil is:

(a) parallel to the magnetic field

(b) at 20° to the magnetic field.

5. (a) A back e.m.f. of 10 mV is induced in a coil when the current flowing in the coil grows at a rate of 0.50 A s^{-1}. Calculate the self-inductance of the coil.

(b) The coil is placed close to a second coil. If an e.m.f. of 5.0 mV is induced in this second coil when the current in the first changes at the rate of 0.70 A s^{-1}, calculate:

(i) the mutual inductance of the pair of coils

(ii) the e.m.f. that would be induced in the first coil if a current in the second coil falls at the rate of 2.0 A s^{-1}.

3.6 Electrostatics

When we studied current electricity in Section 3.1 we were dealing with moving electric charges and the effects which they produce. In **electrostatics** we are dealing with electric charges which are at rest (or only flowing for brief periods). Many of the effects produced by static electricity are probably well known to you. Some clothes are made of material that becomes highly charged by friction and when you take them off you can often hear and see the crackle of sparks caused by the sudden release of the charge. A plastic pen rubbed on your coat-sleeve will attract small pieces of paper and if you rub a balloon on your jumper it can stick to a wall. Some of the clothes which have been rubbing together in a tumble drier will often stick to each other and dust collects on TV screens and records due to the attractive forces created. Lightning is an awesome and spectacular example of electricity in action. Friction within clouds can cause them to become very highly charged. Eventually the charge becomes so great that the insulation of the air breaks down, causing the electric charge to flow to the ground as an enormous spark. The energy of the discharge is so great that it produces an intense trail of light, heat and sound. All the effects we have mentioned are due to the accumulation of charge on objects (and can be explained by considering the transfer of negatively charged electrons in atoms).

The discovery of static electricity is attributed to the Greek scientist Thales (c.625–547 BC) who is thought to have carried out some of the earliest known scientific experiments. He noticed that a piece of amber which had been rubbed briskly with fur attracted light objects such as feathers. In fact our words electron and electricity are derived from **elektron**, the Greek word for amber.

Electric charges

In general two different materials which are rubbed together can become electrically charged by friction and exert **attractive** or **repulsive** forces on each other. Early experimental investigations showed that there were two kinds of electric charge called **positive** and **negative** which are opposite and so cancelled each other. The charge sign allocation which was originally made on a purely arbitrary basis makes **polythene** rubbed with wool **negatively** charged and **cellulose acetate** rubbed with silk **positively** charged. Figure 3.84 shows how the basic law of force between electric charges can be deduced by freely suspending a strip of charged material and bringing up to it another charged strip. We find that **like** charges **repel** and **unlike** charges **attract** each other.

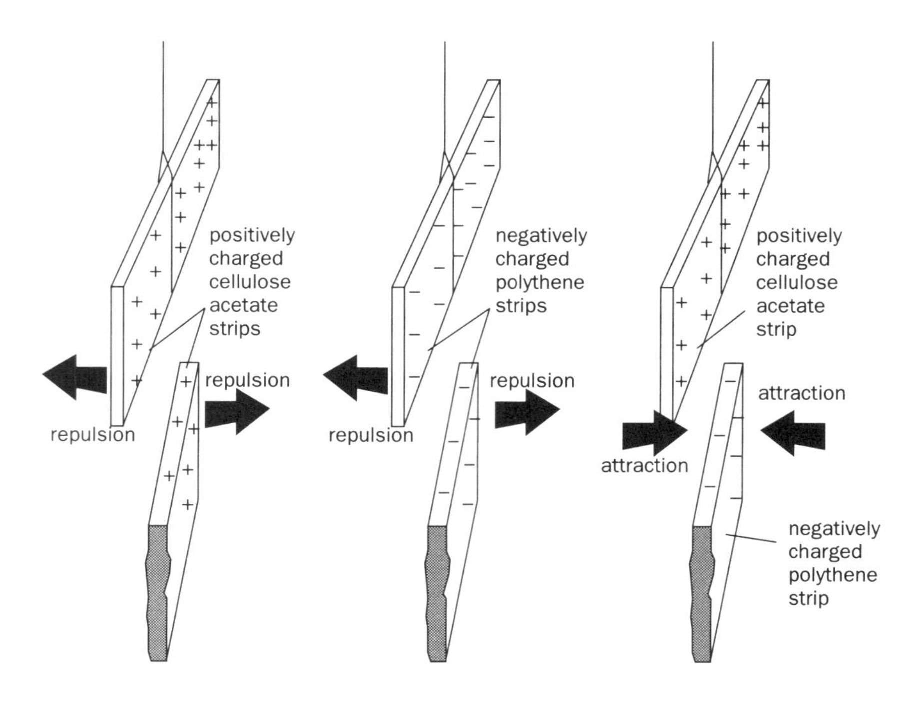

Figure 3.84 Basic law of force between charges

Charging by friction

We can explain charging by friction by looking at what happens to the surface atoms of the rubbed materials. The atoms basically consist of a small, positively-charged nucleus containing protons and neutrons around which there is a 'cloud' of negatively-charged electrons. Normally, atoms have the same number of protons and electrons and their equal, but opposite, charges cancel each other out. The result is that we cannot detect any charge at all and the material is said to be **uncharged**. The action of rubbing causes a transfer of **electrons** from the surface atoms of one of the rubbed materials to the other. The material which **gains** electrons acquires a **negative** charge and that which **loses** electrons becomes **positively** charged. Electric charge is always conserved and so the amount of negative charge gained by one material is equal to the positive charge gained by the other.

Thus when, for example, a polythene strip is rubbed with a woollen cloth, electrons are transferred from the surface of the cloth onto the polythene. This leaves the wool deficient of electrons and, therefore, positively charged. The polythene gains the electrons and, therefore, becomes negatively charged (see Figure 3.85).

Conductors and insulators

As we have already explained, electrical **conductors** (e.g. metals) contain electrons which are free to move within the material. Thus electrons gained by a conducting material when it is rubbed are able to move throughout the material and if electrons are lost those which remain can redistribute themselves in the material. Thus the charge acquired by a conductor spreads itself over its whole surface. If the conductor is connected to **earth**, the charge acquired by the conductor can flow to the Earth. So, if a conductor is to be given a charge, it must be **insulated** from the Earth (or from any other large conducting body which could absorb its charge).

Insulators (e.g. polythene, glass, etc.) are very different in that there are no free electrons to move through the material. Thus if one region of an insulating object becomes charged by addition or removal of electrons, it does not cause a movement of electrons within the body. The charge then stays in the region in which it has been placed (i.e. it is **static**).

The human body and the Earth are relatively good conductors and so when you try to charge a metal rod by rubbing, the charge produced flows along the rod and through your body to earth. This does not happen in the case of insulators such as polythene, for example, because even though you may be holding the material, the charge which it is given does not move. The Earth can be regarded as having an infinite capacity for charge. Electrons from a negatively charged body will flow to the Earth, while a positively charged body connected to earth will be neutralised by electrons flowing from the Earth into the body.

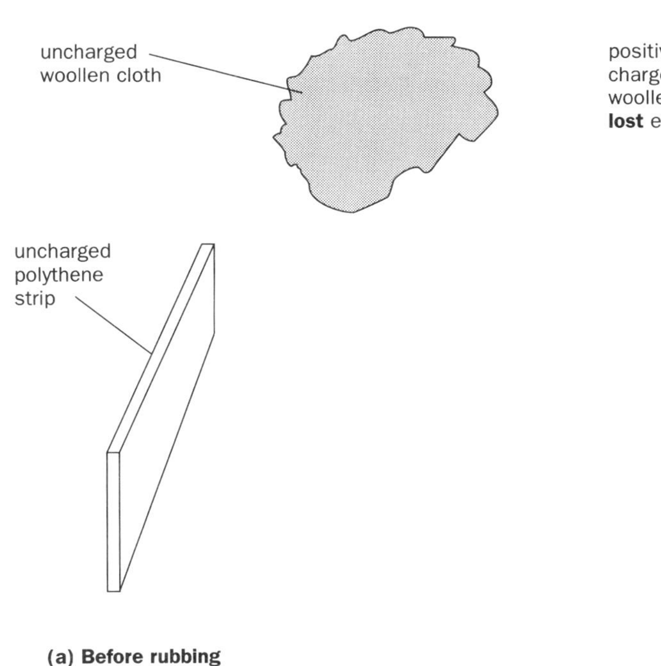

(a) Before rubbing

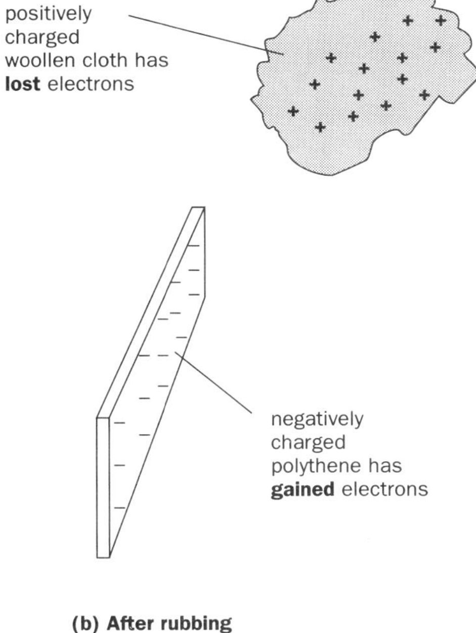

(b) After rubbing

Figure 3.85 Charging by friction

Charging by induction

A body which has been charged by friction or any other means can then be used to charge an uncharged body by the process of electrostatic **induction**. The induced charge is always opposite in sign to that of the inducing charge used. Figure 3.86 shows one method of charge transfer by induction.

A negatively-charged rod is brought close to two metallic spheres (A, B) in contact. This causes some electrons on A to be repelled into B. The charges have been separated by **induction**. When the spheres are separated, still holding the rod near to A, the charges are trapped on the spheres. Each sphere then retains its own induced charge when the rod is removed.

Figure 3.87 shows the charge-inducing procedure adopted for a single metallic sphere. A positively-charged rod is used to illustrate the method in this case.

The positive charge on the rod induces a movement of electrons in the conducting sphere which causes negative charge to accumulate on one side and positive charge on the other. When the sphere is touched with a finger, it becomes connected to the Earth through the human body, and electrons are attracted to the sphere, where they neutralise the positive charges. (We say that the positive charge is caused to leak away to earth.) The earth connection must be removed before the charging rod is removed. When the rod is finally removed the induced negative

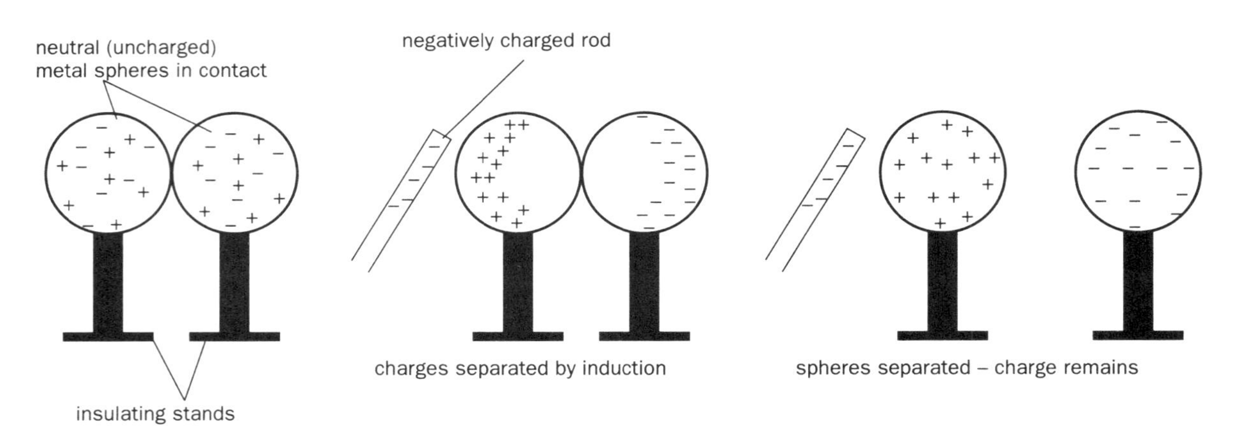

Figure 3.86 Charging by induction – two bodies

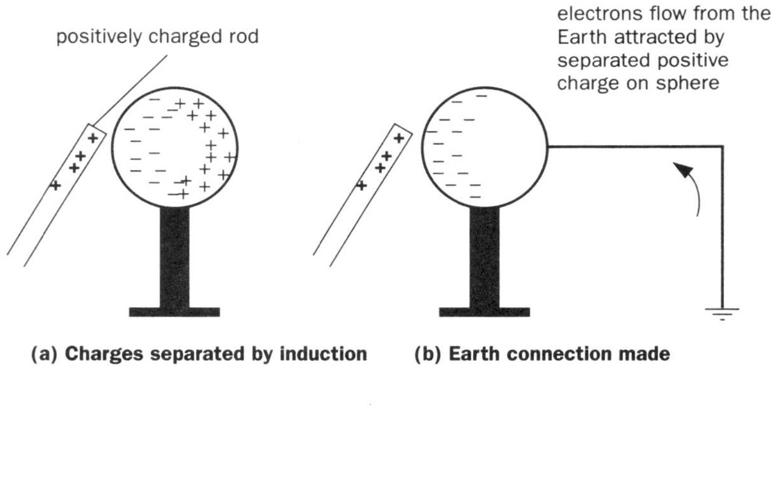

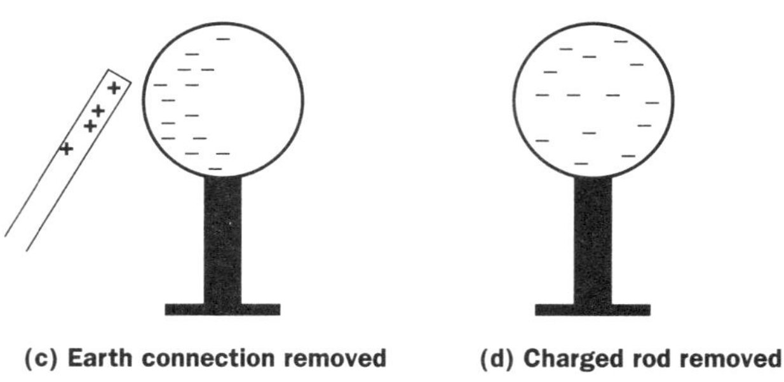

Figure 3.87 Charging by induction – single body

charge distributes itself uniformly throughout the sphere's surface which now has an overall negative charge.

The attraction of a small piece of paper by a plastic pen which has been charged by rubbing can be explained in terms of induction. As the charged pen is brought close to the paper it separates the charges on the paper (which is a conductor) by induction (see Figure 3.88).

The electrons in the paper are repelled as far from the pen as possible, leaving the closer side of the piece of paper positively charged. The **attractive** force between the opposite charges (negative on the pen and positive on the adjacent side of the paper) is greater than the **repulsion** between the more distant like charges and so the resultant effect is that the paper is attracted to the pen. It may even fly up and stick to the pen. It may then fall off when all its charge has been lost and the process repeats itself. Thus the paper can sometimes be seen to fly up and drop from the pen several times in quick succession.

Applications of electrostatics

Although electrostatics was the first branch of electricity to be studied, it was for many years thought to be more of a curiosity than an item of practical value. This is certainly not the case nowadays when most electrostatic phenomena have been put to such good use in industry. The phenomenon of electrostatic forces, where there is attraction by unlike charges and repulsion by like charges, is put to work in a wide variety of modern machines and processes.

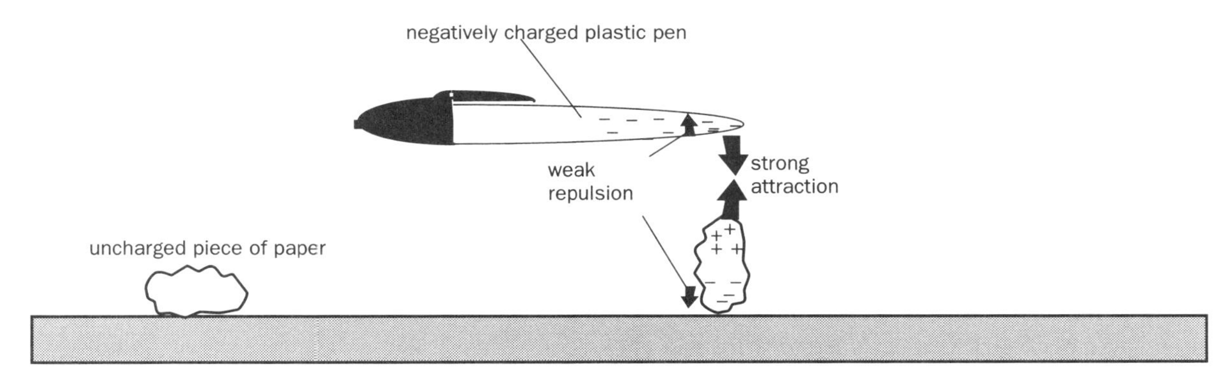

Figure 3.88 Attraction of an uncharged body

Nature's electrostatics extravaganza – lightning is an electrical discharge process of awesome proportions

For instance, in the painting of car bodies in an assembly-line, tiny droplets of spray paint are all given the same electrical charge. As a result they repel each other, and are attracted towards the car body, settling on it as a more even coating. A similar process is used in electrostatic powder-coating systems where the individual particles with which a metal object is to be coated are all given the same electric charge. This results in a very even coating which is then baked on to produce a hard, lasting finish. Electrostatic precipitators are devices which are used to extract flue-ash or other polluting particles from a system before they are discharged into the atmosphere. This is a very important application of electrostatics since it almost eliminates what might otherwise be very heavy atmospheric pollution. Precipitators of the type used to remove flue-ash in modern coal-fired power stations consist of a number of electrically charged wire grids and plates. The grids give a negative charge to each of the ash particles which are then attracted and adhere to the positively charged plates. These are mechanically shaken to remove the ash. Electrostatic microphones, loudspeakers and photocopiers also depend on electrostatic phenomena for their operation. The Van de Graaff generator essentially consists of a moving belt of insulating material which transfers charge to a metallic dome and can produce potential differences as high as 14 MV. It is used to test insulation and other equipment designed to withstand high voltages. In nuclear research the high voltage is used to accelerate charged subatomic particles to very high speeds (see Section 5.5 for a more detailed account of this use of a Van de Graaff generator).

There are also potential hazards associated with some electrostatic phenomena. As a result of friction with the air, electric charge can build up on an aircraft during flight and its discharge could cause an explosion if, for example, the aircraft was re-fuelled. For this reason the rubber tyres are made slightly conducting and any charge accumulated leaks to the Earth on landing. Buildings have to be protected from the potentially destructive effects of lightning bolts by the lightning conductors which are fixed to them. These direct the discharge through a copper strip to the Earth and so leave the building safe. In addition, since charge concentration increases on sharply-curved surfaces, especially points, the sharp tip of a lightning conductor acts to produce a very intense electric field. This helps to dissipate the charge induced in the building by the presence of a highly charged thundercloud, and so can prevent a lightning strike.

Coulomb's law of force

In the 1780s the French physicist Charles de Coulomb performed precise experiments on the magnitude of the forces which exist between charged spheres. His results are summarised in his law which may be stated as follows:

> The magnitude of the force between two point charges (Q_1 and Q_2) which are a distance (r) apart is directly proportional to the product of the charges and inversely proportional to the square of their distance apart (see Figure 3.89) – this is called **Coulomb's law.**

Strictly speaking the law only applies for point charges, but subatomic particles such as protons and electrons may be treated as approximate point charges. Charged spheres only approximate to point charges when their separation is large.

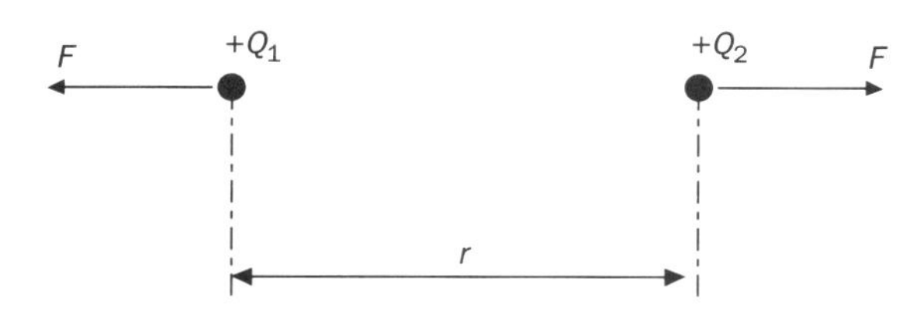

(a) Like charges repel

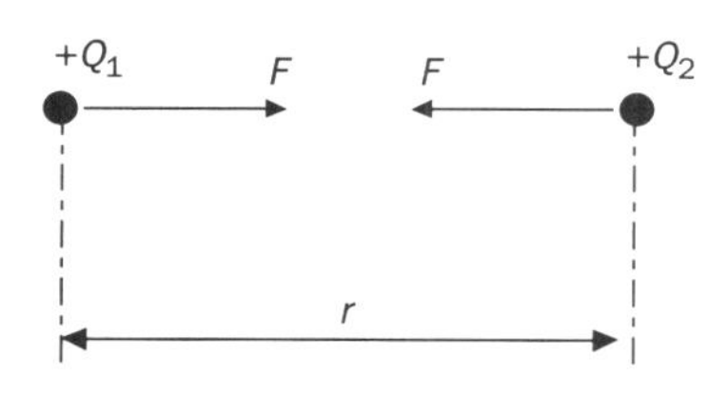

(b) Opposite charges attract

Figure 3.89 Coulomb's law of force between point electric charges

Coulomb's law can be expressed mathematically as:

$$F \propto \frac{Q_1 Q_2}{r^2}$$

From which:

$$F = k\,\frac{Q_1 Q_2}{r^2}$$

where k is a constant of proportionality whose value depends on the insulating medium in which the charges are located. The property which determines the value of k is called the **permittivity** ε of the medium. A material which reduces the force considerably compared with the value when the charges are in air or a vacuum is said to have a high permittivity.

Conventionally, this constant is expressed as:

$$k = \frac{1}{4\pi\varepsilon_0}$$

where ε_0 is the permittivity of free space (i.e. a vacuum) (= 8.85×10^{-12} farad metre $^{-1}$ (F m^{-1}))

$$\text{Then, } k = \frac{1}{4\pi \times 8.85 \times 10^{-12}} = 8.99 \times 10^9 \text{ N m}^2 \text{ C}^{-2}$$

Thus for two point charges (Q_1) and (Q_2) which are situated a distance (r) apart in a vacuum, the force (F) which they exert on each other is given by:

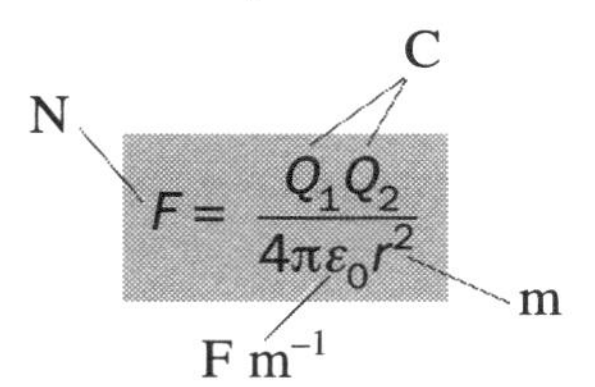

Equation 1

You should note that:

- Since the permittivity of air (ε_{air}) at normal pressure = $1.0005 \times \varepsilon_0$, we generally assume that $\varepsilon_{air} = \varepsilon_0$.
- The fact that ε for water $\approx 80\varepsilon_0$ means that the force between charges in water is 1/80th of the value in air. This is why salt dissolves in water. Salt's crystalline structure is provided by the forces of attraction which exist between the positively charged sodium ions and the negatively charged chlorine ions. When salt is put into water the forces are so reduced that the crystal structure collapses.
- The **relative permittivity** ε_r of a medium is simply the ratio of its permittivity to that of a vacuum and it is defined by the equation:

$$\varepsilon_r = \frac{\varepsilon}{\varepsilon_0}$$

Equation 2

(ε_r is dimensionless)

- Coulomb's law of force between electric charges is analogous to Newton's law of gravitational force between masses. The two equations are shown in Table 3.2 for comparison.

Table 3.2 Comparison of electric and gravitational forces

	Electric fields	**Gravitational fields**
Quantity which 'feels' the force	charge Q	mass m
Constant of proportionality	$\frac{1}{4\pi\varepsilon}$ where ε is the **permittivity** of the medium, whose value depends on the nature of the medium	G the **universal gravitational constant**, the same value for all media, including vacuum
Relationship with distance r	inversely proportional to r^2	inversely proportional to r^2
Force equation	$F = \frac{1}{4\pi\varepsilon}\frac{Q_1Q_2}{r^2}$	$F = G\frac{m_1m_2}{r^2}$
Direction of force	**like** charges **repel** **opposite** charges **attract**	**all** masses **attract**
Relative strength	strong at close range – responsible for chemical bonding	weak except for massive bodies – responsible for motion of planets

Guided examples (1)

1. Use the data given below to calculate the magnitude of the **electric** and **gravitational** forces acting between the proton and the electron in an atom of hydrogen:

   Mean distance between the proton and electron, $r = 5 \times 10^{-11}$ m

   Electron (and hence proton) charge $e = 1.6 \times 10^{-19}$C

   Electron mass $m_e = 9.1 \times 10^{-31}$ kg

   Proton mass $m_p = 1.7 \times 10^{-27}$kg

   Assume that, $G = 6.67 \times 10^{-11}$N m^2 kg^{-2}

   and $\frac{1}{4\pi\varepsilon_0} = 8.99 \times 10^9$ N m^2 C^{-2}

   Guidelines

   Simply apply Equation 1 and its gravitational equivalent (see Table 3.2 on page 267).

2. Three point charges A, B, and C are situated in air as shown in the diagram below.

   Charge A is 3.0 cm from charge B which is 2.0 cm from charge C. Assuming that $1/4\pi\varepsilon_0 \approx 9.0 \times 10^9$ N m^2 C^{-2}, calculate the resultant force acting on C.

   Guidelines

   The force on C due to A (F_1) is **repulsive** (acts to the **right**) and that due to B (F_2) is **attractive** (acts to the **left**). The resultant force (F) on C is then given by: $F = F_2 - F_1$.

   Equation 1 is used to calculate forces F_1 and F_2.

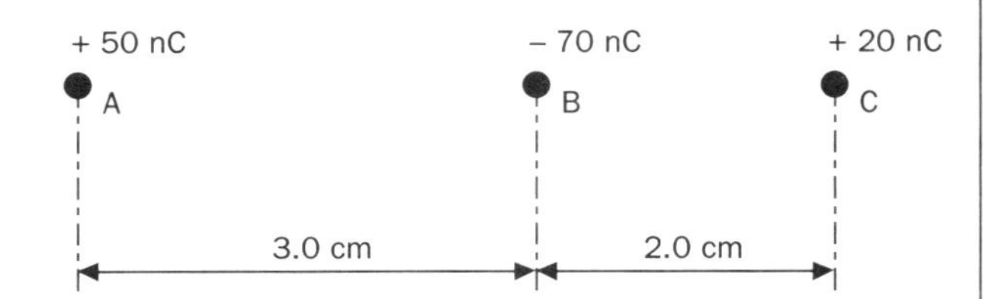

Electric fields

The concept of an electric field is used in physics to explain how separated charges can affect one another. Electric fields are regions which exist around particles and objects which have an electric charge. The charged polythene or cellulose acetate strips which we mentioned earlier have an electric field around them and it is this field which exerts forces on the electrons in an insulated conductor and so separates charges.

> An **electric field** is a region in which an **electric charge** experiences a force.

This is analogous to a gravitational field in which a mass experiences a force, or a magnetic field in which a magnetic pole feels a force.

The direction of the field at any point is defined as that direction in which a small, positive charge would move if it were placed at the point. For example, the direction of the electric field between the two charged plates A and B shown in Figure 3.90 is from B to A (i.e. from the more positive plate, B, to the more negative plate, A). A small, positive charge placed between the plates moves as shown.

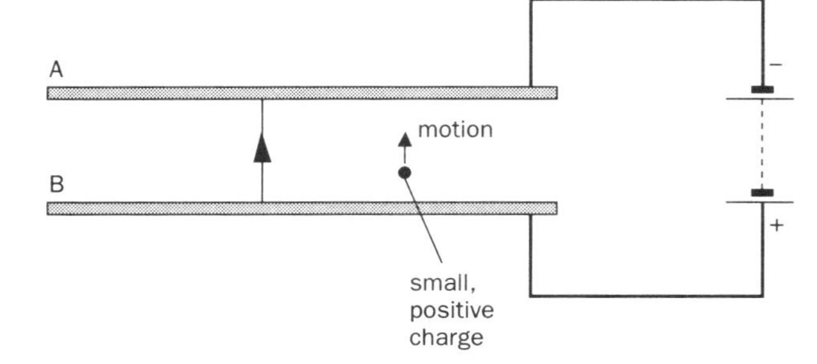

Figure 3.90 Direction of electric field

As in gravitation and magnetism, we make use of **field lines** when describing electric fields. When the lines are **parallel** the field is **uniform** (i.e. its strength and direction are constant – see Figure 3.91(a), where the lines are parallel and equally-spaced near the centre of the region between the plates). In Figure 3.91(b), the

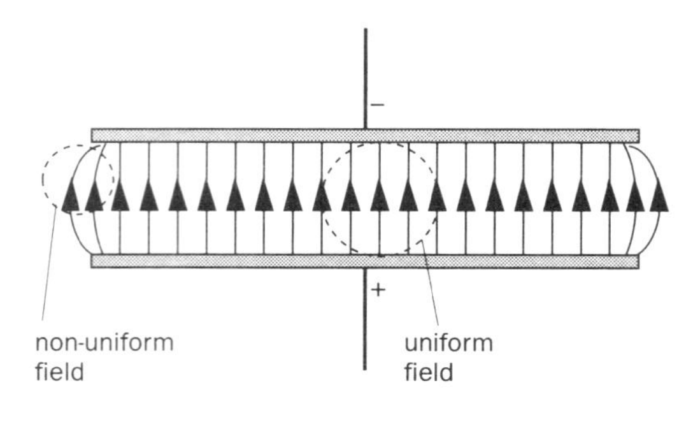

(a) Electric field between parallel charged plates

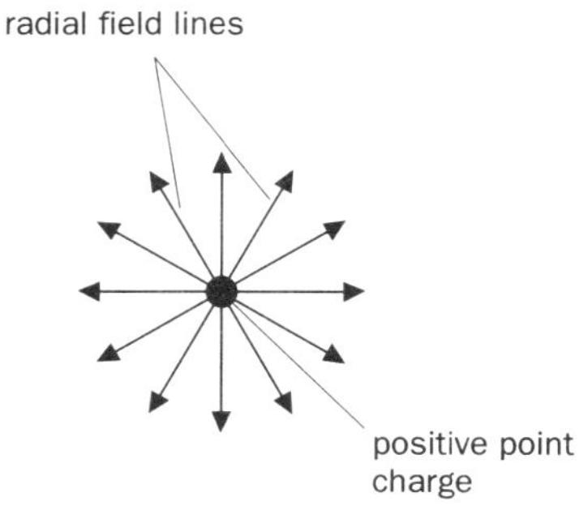

(b) Electric field around a point charge

Figure 3.91 Electric field shapes

field pattern indicates that the strength of the field decreases with distance from the charge. The direction of the arrows indicates the direction of the field at any point.

Electric field strength

Common symbol E

SI unit newton per coulomb ($N\ C^{-1}$) or volt per metre ($V\ m^{-1}$)

> The **field strength** at a point in an electric field is the **force** per **unit charge** exerted by the field at that point.

In Figure 3.92, a small charge ($+q$) experiences a force (F) at a particular point P in an electric field.

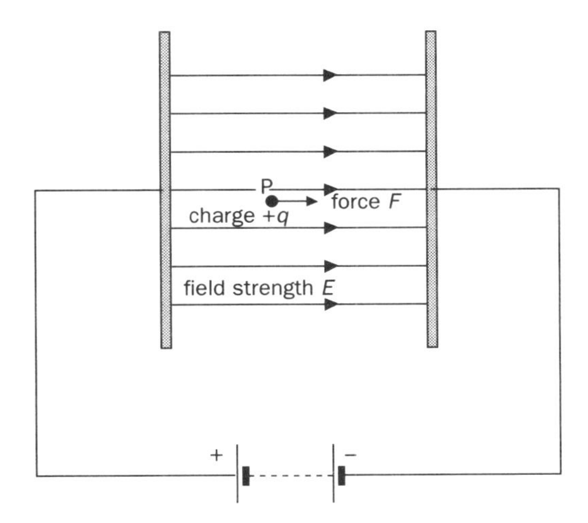

Figure 3.92 Force on a point charge in an electric field

The field strength E at point P is given by:

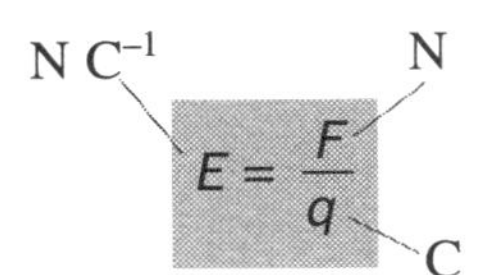

$$E = \frac{F}{q}$$

Equation 3

You should note that:

- If the field is uniform, then E is the magnitude of the field strength at all points in the field.
- Electric field strength is defined in terms of force per **unit** charge. In practical measurements of field strength, a unit charge (1 coulomb) would be so large that it would have a profound effect on the field being measured. This is why a **small** charge is considered. **Small** means of such low magnitude that it has negligible effect on the field being measured.
- From Equation 3 the force F exerted on a charge q at a point where the field strength is E is given by:

$$F = qE$$

- The fact that field strength E is a vector quantity, having direction as well as magnitude, should be borne in mind when calculating the resultant field strength at a point due to a number of charges.
- Electric field strength ($E = F/q$) is analogous to gravitational field strength ($g = F/m$).
- Also expressing forces in terms of fields we obtain:

  Electric force (F) on a charge (q) is: $F = qE$

  Gravitational force (F) on a mass (m) is: $F = mg$

Field strength due to a point charge

As we saw earlier, the electric field around a point charge is **radial**. Figure 3.93 shows a small test charge ($+q$) placed in air at a point A which is distance (r) from a point charge ($+Q$).

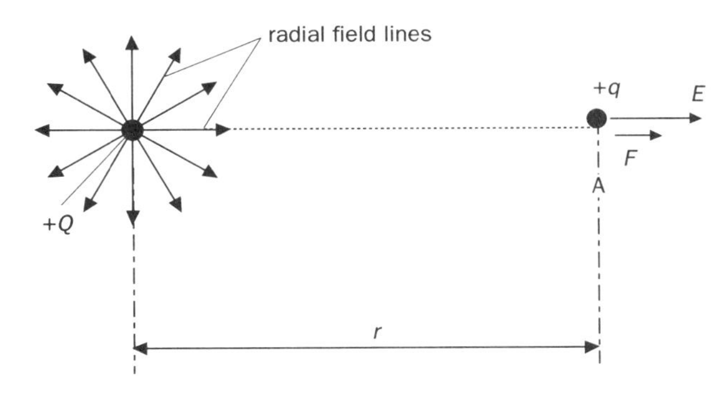

Figure 3.93 Field strength due to a radial field

Using Coulomb's law the force (F) acting on the small charge ($+q$) due to the charge ($+Q$) is given by:

$$F = \frac{1}{4\pi\varepsilon_0}\frac{Qq}{r^2}$$

(acting radially outwards from the centre of $+Q$)

Then from the definition of electric field strength we have that the field strength E at point A is given by:

$$E = \frac{F}{q} \quad \text{(since } q \text{ is the small test charge)}$$

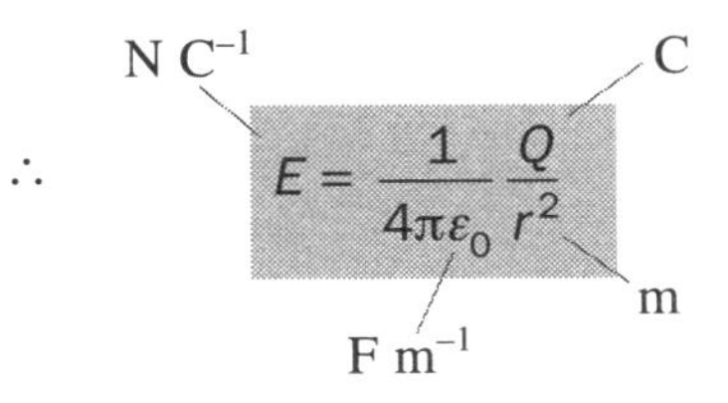

$$\therefore \quad E = \frac{1}{4\pi\varepsilon_0}\frac{Q}{r^2}$$

Equation 4

You should note that:

- The direction of E is radially outwards from the centre of $+Q$.
- E around a point charge is inversely proportional to the square of the distance from the charge (i.e. it follows an inverse square law).

- When calculating the resultant E at a point due to a number of charges, the problem must be solved vectorially.
- Electric field strength and gravitational field strength are directly analogous:

  Electric field strength (E) at distance (r) from point charge (Q)

$$E = \frac{1}{4\pi\varepsilon_0}\frac{Q}{r^2}$$

  Gravitational field strength (g) at distance (r) from point mass (M)

$$g = G\frac{M}{r^2}$$

Electric potential

Common symbol V
SI unit volt (V)

An electric force acts on a charge situated in an electric field. If the charge is moved over a distance against this force, work is done against the electric force and the system gains electric potential energy. The amount of potential energy gained is equal to the work done in moving the charge. Figure 3.94 shows the behaviour of an alpha-particle, which approximates to a point charge, in the electric field surrounding a large nucleus. (It was observations of the behaviour of alpha-particles passing through thin metal films that led Rutherford to identify the nucleus as possessing all of the positive charge in the atom.) An alpha-particle, being positively charged, experiences a repulsive force everywhere in the field due to the nucleus. If the alpha-particle approaches the nucleus head-on, the repulsive force causes the alpha-particle to lose kinetic energy, and the system gains an equal amount of electric potential energy.

The alpha-particle will be brought to rest when the gain of electric potential energy of the system is equal to the initial kinetic energy of the particle. The particle will then reverse its direction and be accelerated away from the nucleus, as the system returns its electric potential energy and gains kinetic energy, as shown in Figure 3.94(b). The repulsive force is felt by both bodies, but, because the alpha-particle has much less mass than the nucleus, it will gain the vast majority of the kinetic energy.

The amount of electric potential energy which is associated with a charge at some point in an electric field depends on:

- the strength (E) of the field at that point
- the magnitude of the charge.

Thus the electric potential energy associated with a unit positive charge at different points in a field can be used as a measure of the field strength at these points.

The change in electric potential energy which occurs when a **unit charge** is moved between two points in a field is interpreted as the change in **potential** (V) of the field. In the case shown in Figure 3.94, points close to the nucleus are at a higher electric potential than points further away. The zero of electric potential is defined as the potential at an **infinite** distance from any electric charge. You should note that infinite is a relative term. It might mean several metres in the case of a charge of, say, 100 μC or millimetres when we are dealing with the field due to subatomic particles such as the electron. With this in mind, we can now define electric potential as follows:

> The **electric potential** (V) at any point in an electric field is defined as the work done (W) per unit positive charge ($+Q$) moved from **infinity** to the point.

α-particle approaching nucleus

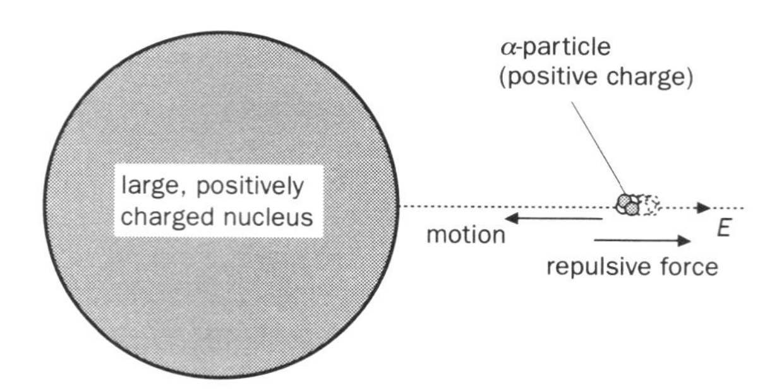

(a) In moving against the field, the charge–field system gains electric potential energy

α-particle moving away from nucleus

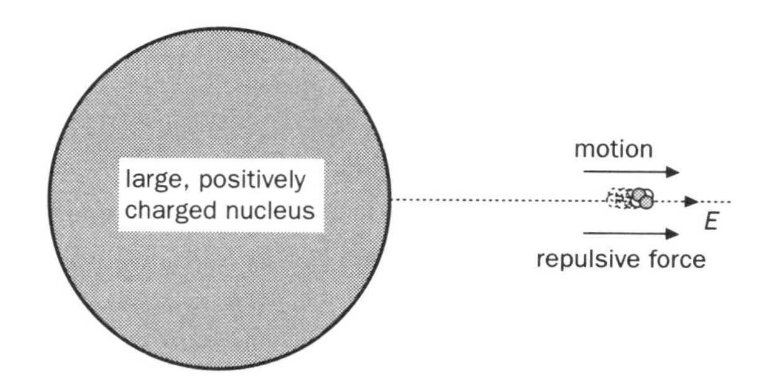

(b) When the charged particle moves in the direction of the electric field, the system loses its electric potential energy and gains kinetic energy

Figure 3.94 Behaviour of an alpha-particle in a repulsive electric field

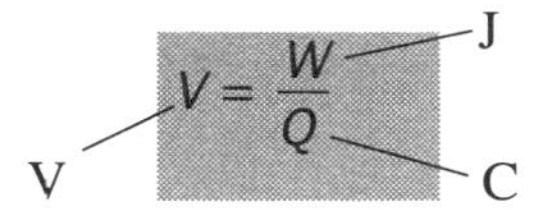

Equation 5

You should note that:

- **Electric potential** (V) is a property of a point in a field, so it only depends on the nature of the charge producing the field, whereas **electric potential energy** is a property of a charge–field system and depends on both the strength of the field and the size of the charge placed in the field.
- From Equation 5: 1 volt = 1 joule coulomb^{-1}.
- As with gravitational fields, the true zero of electric potential is at an infinite distance from a charge. For most practical purposes, however, we assume that the zero of potential is that of the Earth. This does not present a problem since the potential of the Earth is constant and we are normally dealing with differences of potential rather than absolute values.
- From Equation 5, it follows that the electric potential energy (W) of a charge (Q) at a point in an electric field where the potential is (V) is given by: $W = QV$.
- Electric potential is a scalar quantity which can be added directly when more than one field is involved whereas electric field strength is a vector quantity, and calculation of the resultant electric field strength involves more complex vectorial addition. This is one of the reasons why potential is an easier quantity to use than field strength when describing electric fields. A second and equally important reason is that the energy changes which result from the movement of a charge in an electric field are easy to calculate if the potentials are known.

- Electric potential $V = \frac{W}{Q}$ joule coulomb^{-1}
- Gravitational potential $V = \frac{W}{m}$ joule kilogram^{-1}

Electric potential due to a point charge

Figure 3.95 shows an isolated point charge ($+Q$) situated at a point A in a medium of permittivity ε.

At an infinite distance from the point charge, the electric potential (V) is zero.

At B, distance (r) from the point charge, the electric potential (V) is given by:

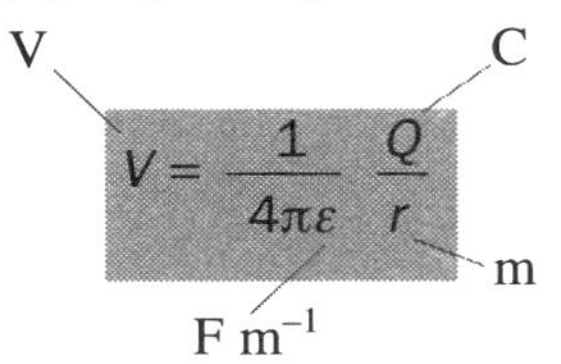

Equation 6

Maths window

Referring to Figure 3.95:

The variation with distance from Q of electric force F on a positive test charge q is shown below:

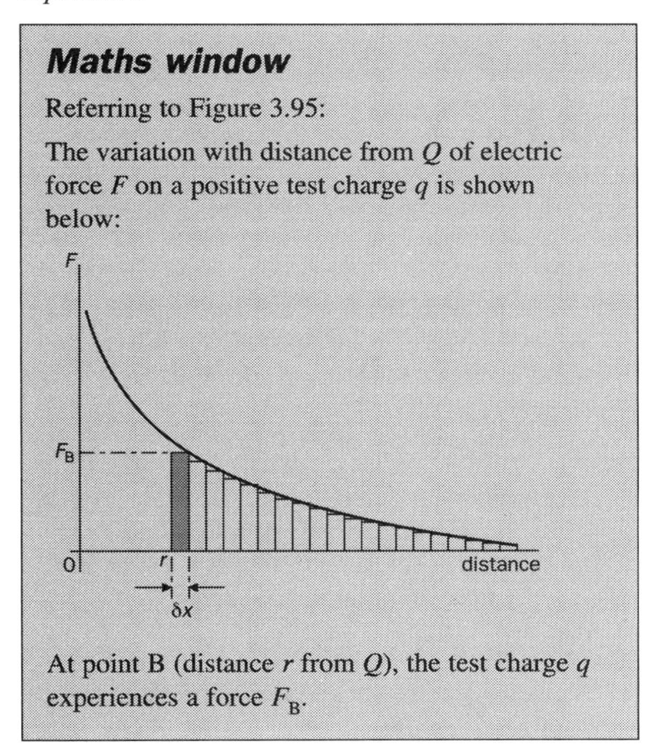

At point B (distance r from Q), the test charge q experiences a force F_B.

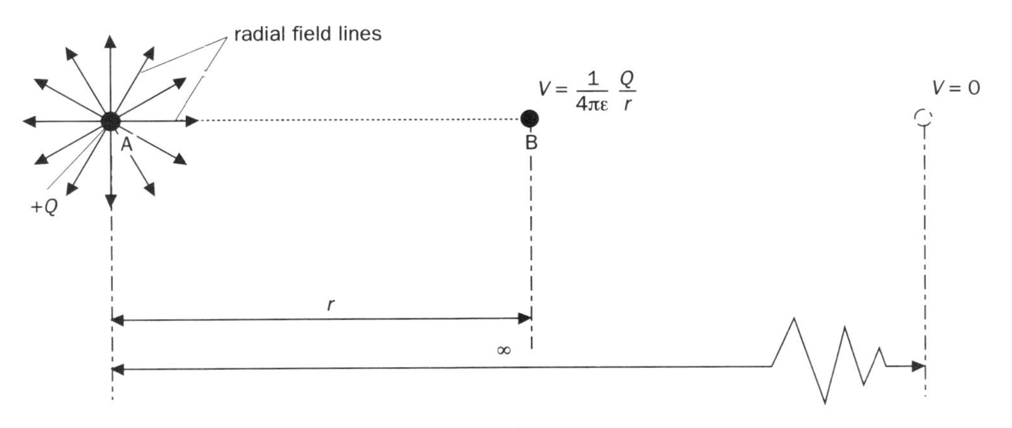

Figure 3.95 Electric potential due to an isolated point charge

Maths window *continued*

The work done (δW) in moving the test charge a small distance (δx) is given by:

$$\delta W = F\,\delta x$$

This is equivalent to the **area** of the shaded strip in the diagram, as long as δx is small enough to make F constant over this distance.

Therefore the total work done (W) in bringing the test charge from infinity to point B will be equal to the total area of all the rectangular strips. This area can be found by mathematical **integration**.

$$W = -\int_{\infty}^{r} F\,dx$$

The minus sign indicates that the force decreases as distance (x) increases.

$$W = -\int_{\infty}^{r} \frac{1}{4\pi\varepsilon}\frac{Qq}{x^2}\,dx = -\frac{Qq}{4\pi\varepsilon}\int_{\infty}^{r}\frac{1}{x^2}\,dx = \frac{Qq}{4\pi\varepsilon}\frac{1}{r}$$

Since $V = W/q$:

$$V = \frac{1}{4\pi\varepsilon}\frac{Q}{r}$$

You should note that:

- V decreases with distance r from an isolated point charge. Points which are equal distances from the centre of a radial field, in any direction, have equal potential.
- The resultant potential at a point due to a number of point charges is equal to the algebraic sum of the potentials due to each of the charges.
- The electric potential energy (W) of a test charge (q) at a distance (r) from a point charge (Q) is given by:

J, C, m, $F\,m^{-1}$

$$W = \frac{1}{4\pi\varepsilon}\frac{Qq}{r}$$

Equation 7

- Electric potential (V) at some distance (r) from a point charge (Q) is

$$V = \frac{1}{4\pi\varepsilon}\frac{Q}{r}$$

- Gravitational potential (V) at some distance (r) from a point mass (M) is

$$V = G\frac{M}{r}$$

Equipotentials

An **equipotential** surface is one over which the potential in an electric field remains constant. No energy change occurs and no work is done when a charge moves along such a surface. It follows from Equation 7 that all points which are equidistant from a point charge have the same potential. Thus the equipotential surfaces around an isolated point charge are concentric spheres (which appear as circles if we consider only two dimensions) centred on the charge as shown in Figure 3.96.

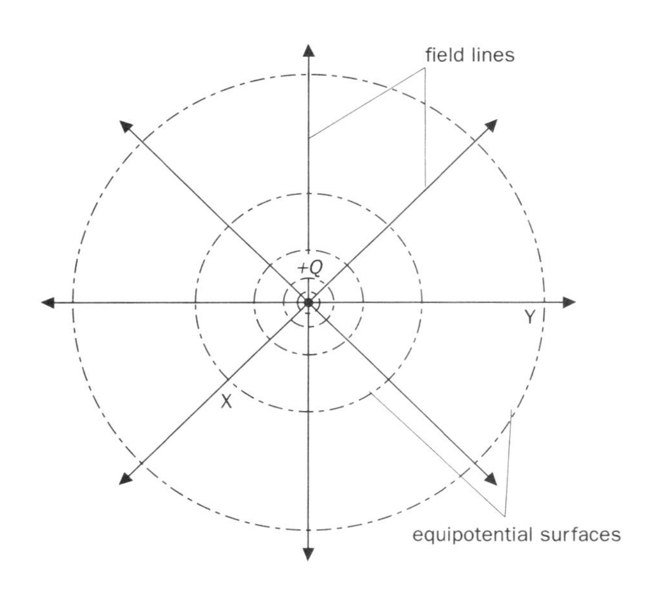

Figure 3.96 Equipotentials and field lines around an isolated point charge

You should note that:

- In Figure 3.96, the p.d. between adjacent equipotentials is constant. Close to the charge, the equipotentials are close together, showing a large potential gradient. Further away, the spacing increases, showing a smaller potential gradient.
- At any point, the electric field direction is perpendicular to the equipotential surface.

Electric potential difference (p.d.)

Common symbol	V
SI unit	volt (V) or joule per coulomb ($J\,C^{-1}$)

The concept of electric potential is equally applicable to the field due to static charge and that produced in a wire in which an electric current is flowing.

In practice, we are usually concerned with **potential differences** (p.d.s) between two points rather than with absolute values of potential.

From the definition of potential it follows that:

> The **potential difference** (p.d.) between two points in an electric field is the **work done** (or **energy changed**) per **unit charge** moved from one point to the other.

i.e. $V = \frac{W}{Q}$

From which the unit of potential difference can be defined as follows:

> The p.d. between two points is said to be 1 **volt** if the work done (or energy change) is 1 **joule** per **coulomb** of charge moved between them.

Consider an electron moving in a vacuum between two points which have a p.d. of 1 volt between them. The electron accelerates under the influence of electric field force.

Energy gained by the electron

= work done by the field force.

$$W = QV$$

Since the charge on an electron

$= 1.6 \times 10^{-19}\,\text{C}$

$$W = (1.6 \times 10^{-19}\,\text{C}) \times (1.0\,\text{J C}^{-1}) = 1.6 \times 10^{-19}\,\text{J}$$

This small amount of energy which an electron acquires when it accelerates through a p.d. of 1 volt is called the **electronvolt** (eV). It is a commonly used unit in atomic and nuclear physics.

Relationship between field strength and potential

Figure 3.97 shows the uniform electric field of strength (E) between two metal plates (A and B) which are separated by a distance (d) and have a p.d. (V) between them.

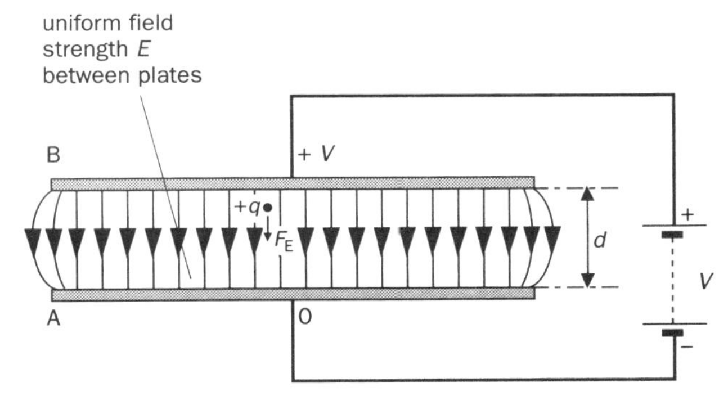

Figure 3.97 Moving a charge in the uniform field between two charged plates

If a small charge ($+q$) is moved from A to B, the work done (W) is given by:

$$W = F_E d = Eqd \qquad \text{(i)}$$

(Since the electric force $F_E = Eq$)

Also, from the definition of p.d.:

$$W = qV \qquad \text{(ii)}$$

$\therefore$ combining (i) and (ii):

$$Eqd = qV$$

$\therefore$

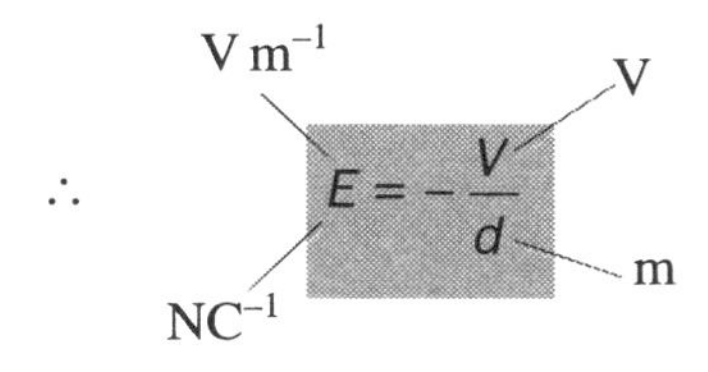

$$E = -\frac{V}{d}$$

Equation 8

Equation 3 and Equation 8 show that E can be expressed in the units: **newton per coulomb** (N C^{-1}) or **volt per metre** (V m^{-1}).

You should note that:

- Equation 8 applies if the electric field is uniform (i.e. constant in magnitude and direction at all points).
- The negative sign indicates that the direction of the field is the direction of decreasing potential.
- If the field is not uniform, calculus notation is used and it can be shown that:

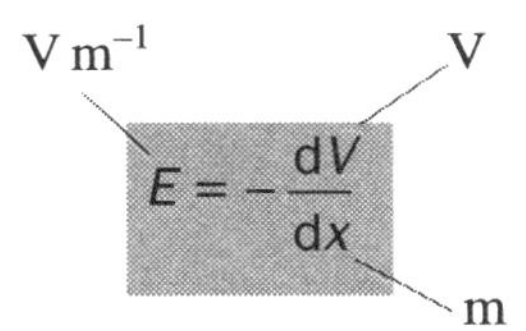

$$E = -\frac{dV}{dx}$$

Equation 9

i.e. the **field strength** at a point is equal to the **potential gradient**.

Guided examples (2)

1. Calculate (a) the field strength and (b) the potential at P, midway between two small spheres A and B which are 40 cm apart in air and carry charges of –4.0 nC and –6.0 nC respectively. You may assume that the permittivity of air = ε_0 and that $\frac{1}{4\pi\varepsilon_0} = 9 \times 10^{+9}\,\text{m F}^{-1}$ (1.0 nC = 1.0×10^{-9} C)

Guidelines

(a) You must remember that field strength (E) is a vector quantity and that you must assign signs (+ or –) to denote field direction.

Guided examples (2) continued

Then resultant field strength at P (E_p) is given by:

E_p = (field strength due to the –4 nC charge) + (field strength due to the –6 nC charge)

$$E_p = -E_A + E_B$$

Use Equation 4 to calculate the field strength due to each charge.

(b) Potential (V) is a scalar quantity

Resultant potential at P (V_p) is given by:

V_p = (potential due to the –4 nC charge) + (potential due to the –6 nC charge)

$$V_p = V_A + V_B$$

Use Equation 6 to calculate the potential due to each charge. Since both charges are negative, V_A and V_B are also negative.

2. Four charges of +5.0 nC, +8.0 nC, –2.0 nC and –4.0 nC are positioned in air at the corners A, B, C and D of a square of side 10 cm as shown in the diagram.

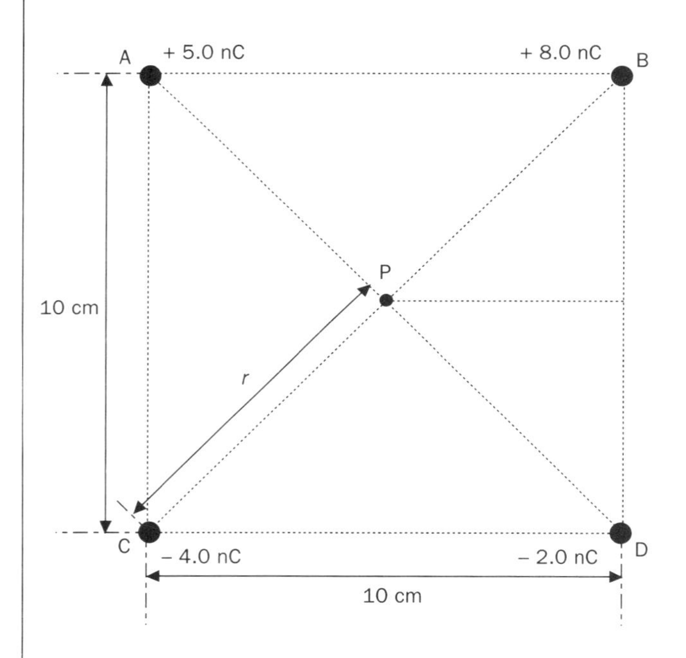

Calculate (a) the field strength and (b) the potential at the centre (P) of the square.

(Assume that the permittivity of air = ε_0 and that $\frac{1}{4\pi\varepsilon_0} = 9{\cdot}0 \times 10^9$ m F^{-1})

Guidelines

(a) Use simple trigonometry to calculate the distance (r) from each of the charges to point P.

Then use Equation 4 to work out the value of E at P due to each charge (E_A, E_B, E_C and E_D).

Once again remember that field strength (E) is a vector quantity. The field strength due to the positive charges acts in a direction away from the charges and that due to the negative charges acts towards the charges.

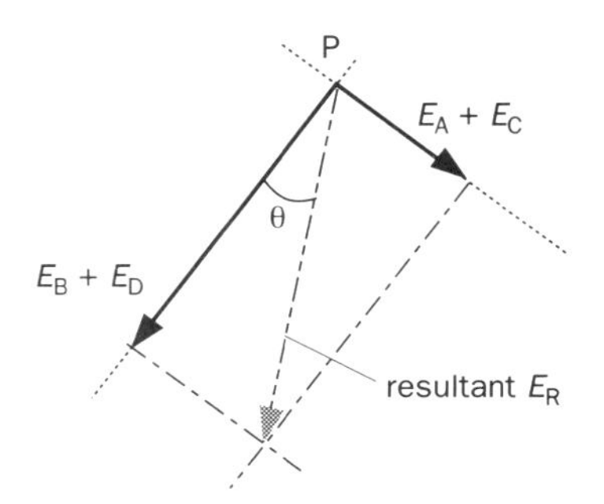

Thus E_B and E_D act in the same direction along a diagonal and E_A and E_C also act in the same direction along the other diagonal. The magnitude of the resultant E_R is then calculated using Pythagoras's theorem, and its direction (θ) is given by:

$$\theta = \tan^{-1} \frac{E_A + E_C}{E_B + E_D}$$

Note that this is the angle measured from a diagonal.

(b) Potential is a scalar quantity. Use Equation 6 to work out the potential due to each charge and then add them algebraically to obtain the resultant potential at P. Remember V_A and V_B are positive and V_C and V_D are negative.

3. Two parallel, horizontal metal plates are placed 4.0 cm apart in an evacuated chamber. If the upper plate is kept at a positive potential of 1.0×10^4 V relative to the lower plate, calculate:

(a) the strength of the uniform electric field between the plates, and

(b) the speed which an electron acquires if it moves from rest from the positive to the negative plate under the influence of the field.

(electron mass, $m_e = 9.1 \times 10^{-31}$ kg; electron charge, $e = 1.6 \times 10^{-19}$ C)

Guided examples (2) *continued*

Guidelines

(a) Since the field between the plates is assumed to be uniform, Equation 8 can be used to calculate the field strength (E).

(b) The energy change (W) which occurs when a charge (Q) moves through a p.d. (V) is given by: $W = QV$. In this case the electron gains kinetic energy at the expense of the potential energy lost by the charge–field system.

$$\therefore \frac{1}{2} m_e v^2 = QV$$

Hence, the final speed (v) of the electron can be found.

Self-assessment

SECTION A Qualitative assessment

1. A cellulose acetate strip which has been briskly rubbed with a piece of silk will pick up small pieces of paper. Explain:

(a) what is happening when the strip is being rubbed with the silk

(b) how the silk is able to pick up the pieces of paper.

2. A piece of polythene held in the hand and rubbed with a cloth becomes electrically charged whereas a metallic rod held in the hand cannot be charged by rubbing. Explain this in terms of the difference between an electrical **conductor** and an **insulator**.

3. Briefly describe

(a) a useful application of electrostatics

(b) a potential hazard resulting from the build-up of static charges.

4. (a) State **Coulomb's** law of force between point charges.

(b) Explain what is meant by the **permittivity** (ε) of a medium.

(c) Define **relative permittivity** (ε_r) of a medium.

(d) Explain the analogy between Coulomb's law of force between electric charges and Newton's law of gravitational force between masses.

5. (a) Define **electric field strength** (E) and state its units.

(b) How is the field strength direction specified?

(c) Explain the analogy between **electric** and **gravitational** field strength.

6. Sketch the electric field patterns for:

(a) a point negative charge and

(b) a pair of parallel, metal plates which have a p.d. between them.

7. Show that the **electric field strength** (E) at a distance (r) from an isolated point charge ($+Q$) in a medium of permittivity (ε) is given by:

$$E = \frac{1}{4\pi\varepsilon} \frac{Q}{r^2}$$

8. Explain the energy changes which occur when:

(a) a charge is moved a distance against an electric field

(b) a charge is allowed to move under the influence of an electric field.

9. (a) Define **electric potential** (V) at a point in an electric field and state its unit.

(b) Although the theoretical zero of potential is at infinity, the Earth is taken as the practical zero of potential. Why is this not a problem in most calculations?

(c) Give two reasons why potential is an easier quantity to use than field strength when describing electric fields.

(d) Is potential a vector or a scalar quantity?

10. (a) What is an **equipotential** surface?

(b) Is work done when a charge is moved along an equipotential surface?

(c) Sketch the form of the electric field lines and the equipotentials for a point charge, making the p.d. between consecutive equipotentials the same.

Self-assessment *continued*

11. Define

(a) **electric potential difference** (p.d.) between two points in an electric field

(b) the unit of p.d. (i.e. the **volt**).

12. Show that for the uniform field between two charged, parallel metal plates the field strength is given by $E = -V/d$ where d = the distance between the plates and V = the p.d. between the plates.

SECTION B
Quantitative assessment

(***Answers:*** 1.3×10^{-13}; 9.4×10^{-10}; 2.3×10^{-8}; 3.8×10^{-8}; 4.0×10^{-8}; 4.0×10^{-7}; 7.6×10^{-7}; 5.0×10^{-6}; 2.1×10^{-3}; 0.012; 2.3; 60; 750; 2.4×10^{3}; 4.0×10^{3}; 1.5×10^{7}; 1.6×10^{21}.)

(You may assume that $\frac{1}{4\pi\varepsilon_0} = 9.0 \times 10^9$ m F^{-1})

1. Calculate the electrostatic force in each of the following cases:

(a) two electrons having a charge of -1.6×10^{-19} C which are 1.0×10^{-10} m apart in a vacuum

(b) two hairspray droplets of diameter 5.0×10^{-2} mm which have each been given a charge of 1.0×10^{-9} C and whose surfaces are 2.0 mm apart in air.

2. Two small conducting spheres are suspended from the same point by fine, insulating threads of length 80 cm. When the spheres are each given a charge of $+Q$ coulomb, they repel each other until they are in equilibrium at a distance of 4.0 cm as shown in the diagram. If the mass of each sphere is 20 mg and the gravitational field strength, (g) = 10 N kg^{-1}, calculate:

(a) the repulsive force acting and

(b) the charge, on each sphere.

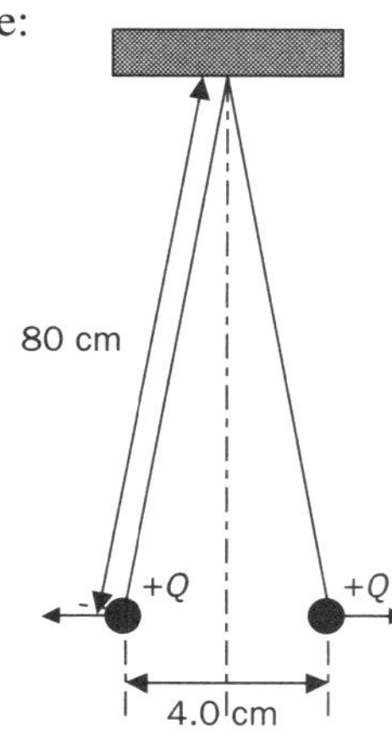

3. A nucleus of uranium-235 has a radius of 9.0×10^{-15} m and contains 92 protons and 143 neutrons. Assuming that the nucleus can be treated as a point charge, find (a) the electric field strength and (b) the potential at the nuclear surface (proton charge = $+1.6 \times 10^{-19}$ C).

4. (a) Calculate the work done in moving a point charge of 4.0×10^{-9} C through a distance of 25 mm in the opposite direction to a uniform electric field of strength 4.0×10^{3} N C^{-1}.

(b) What is the p.d. between two points in an electric field if a charge of +2.0 μC gains 4.8×10^{-3} J of kinetic energy when it is allowed to move in the direction of the field in a vacuum?

5. In an alpha-scattering experiment, alpha-particles (i.e. helium nuclei containing 2 protons) are directed at a thin gold foil in an evacuated chamber. An alpha-particle with an initial kinetic energy of 2.9×10^{-13} J has a head-on collision with a stationary gold nucleus (containing 79 protons) and rebounds along its original path. At the closest distance of approach all the alpha-particle's kinetic energy has been converted into electric potential energy and it momentarily comes to rest before moving back along its incident path. If the charge on a proton is $+1.6 \times 10^{-19}$ C, calculate:

(a) the closest distance of approach of the alpha-particle to the nucleus

(b) the repulsive force acting on the alpha-particle at this distance.

6.

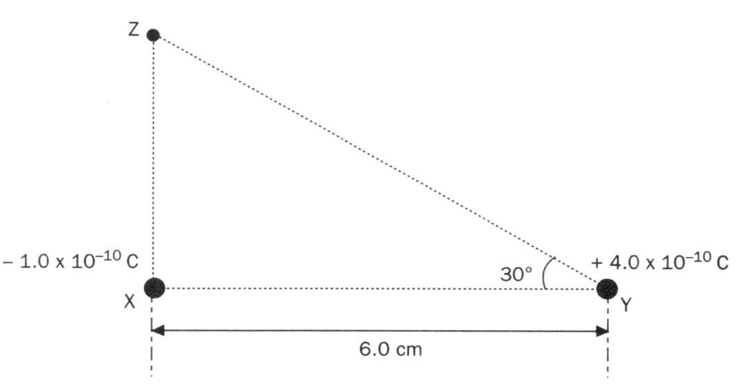

Two point charges of -1.0×10^{-10} C and $+4.0 \times 10^{-10}$ C are positioned at points X and Y which are 6.0 cm apart as shown in the diagram. A point Z is positioned vertically above A as shown.

Self-assessment *continued*

Calculate:

(a) the resultant field strength at Z (magnitude, and direction measured from the vertical)

(b) the distance from X of the point along the straight line joining the charges at which the resultant electric potential is zero.

7. The diagram shows two parallel metal plates A and B which are 5.0 cm apart in a vacuum and have a p.d. of 200 V between them. An oil droplet of mass 4.0×10^{-9} kg which carries a charge of -2.0×10^{-10} C is released close to the lower plate A.

Calculate:

(a) the electric field strength between the plates (assumed uniform)

(b) the **resultant** force acting on the droplet

(c) the droplet's kinetic energy on reaching plate B

(d) the work done by the electric field force in moving the droplet from A to B.

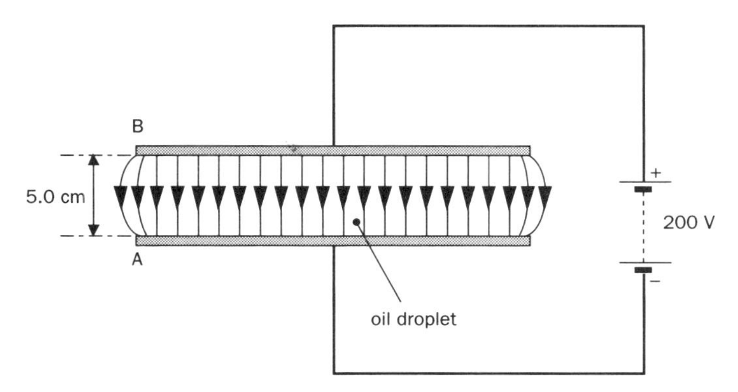

3.7 Storing charge

Electric current is the flow of charge, usually carried by electrons. We think of electric current in a circuit so that the electrons are in continuous motion through the conductors of the circuit. But what happens when a switch is closed in an incomplete circuit containing a battery? We usually say that there is no current, but is this the whole truth?

In the incomplete circuit shown in Figure 3.98, the cell creates an excess of electrons on one side of the gap, and removes them from the other side. This creates a **potential difference** across the gap. When the wires were first connected to the cell, electrons (negative charge) were made to flow briefly in a clockwise direction. The flow quickly stops when the p.d. across the gap (due to the separated charges) becomes equal to V, the supply p.d. We can think of the wires on either side of the gap as 'storing' a tiny quantity of charge. The amount of charge which can be stored depends on:

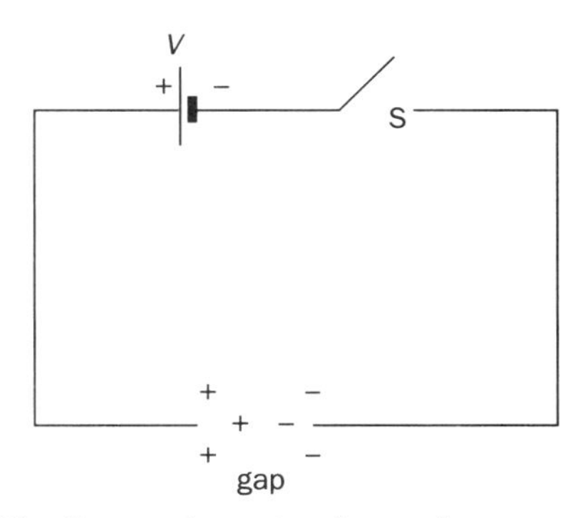

Figure 3.98 Incomplete circuit – only momentary current flow

- The p.d. – higher p.d. means more charge can flow before equilibrium is reached.

- The material between the gap – this also sets a limit on the p.d., since even air can be made to conduct if the electric field becomes high enough for sparking to occur.
- The size of the gap – this controls the strength of the electric field.
- The shape and size of the conducting wires at the gap – very simply, more conducting material means more charge stored.

There does not have to be a 'gap'. An isolated conductor can store charge.

Capacitance

Common symbol	C
SI unit	farad (F)

> Capacitance (C) of an isolated conductor is defined as the ratio of charge stored to the change of potential.

If a charge (Q) causes a change of potential (V), then the capacitance (C) is defined by the equation:

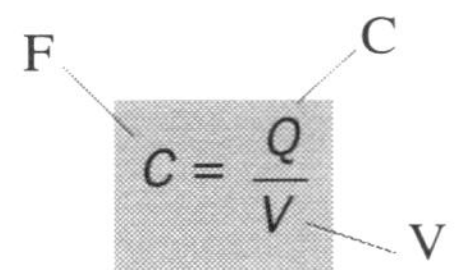

$$C = \frac{Q}{V}$$

Equation 1

The unit of capacitance is the coulomb volt^{-1} (C V^{-1}). The unit name for the coulomb volt^{-1} is the **farad** (F).

1 F is an enormous capacitance, so much smaller sub-multiples are often used:

1 microfarad (1 μF) = 1×10^{-6} F

1 picofarad (1 pF) = 1×10^{-12} F

NB: Take care not to confuse the symbol for the *quantity* capacitance (C) with the abbreviation for the *unit* of charge – the coulomb (C).

Analogy between storing charge and storing liquid

Adding a given quantity of charge (or liquid) to two different conductors (or containers) produces different changes in potential (or liquid level). Figure 3.99 on page 279 shows this effect.

Adding equal quantities of liquid to the two cans causes a much smaller change in **depth** in the wider can: we say it has a greater **capacity**. The greater the change in depth, the greater the **pressure** on the bottom of the container.

Adding equal quantities of charge to the two conductors causes a much smaller increase in **potential** in the larger one: we say it has a greater **capacitance**. As we have seen, physical size is important to capacitance, but other factors such as the nature of the material surrounding the conductor, and the proximity of other conductors are also important.

Capacitors

Adding charge to an isolated conductor increases its potential (V). If another conductor is brought closer to the charged conductor, the potential of the charged conductor will decrease. Since, according to the principle of conservation of charge, the charge (Q) cannot change, it follows from Equation 1 that the capacitance must have increased.

A **capacitor** is an arrangement of conductors and insulators designed to store large amounts of charge. The most common design of capacitor consists of a pair of conducting plates separated by an insulating material called the **dielectric**. The dielectric could be air, paper, oil, mica, etc.

Figure 3.100 shows the simplest such arrangement, consisting of two flat conducting plates separated by an insulator:

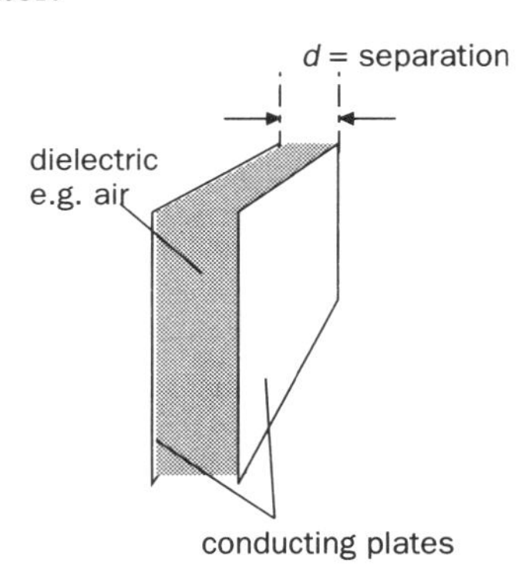

Figure 3.100 Simple parallel-plate capacitor

The capacitance of such a parallel-plate capacitor depends on:

- the distance between the plates (d)
- the area of overlap of the plates (A)
- the material of the dielectric between the plates.

Thus the capacitance (C) of a simple parallel-plate capacitor is given by:

$$C \propto \frac{A}{d}$$

$$\therefore C = \frac{\varepsilon A}{d}$$

F, F m^{-1}, m^2, m

Equation 2

where ε is a constant called the **absolute permittivity** of the dielectric.

Permittivity is a property of the material of the dielectric.

The permittivity of free space (a vacuum) is given the symbol ε_0 and has the value 8.9×10^{-12} F m^{-1}. (For all practical purposes, the permittivity of air can be assumed to have this value.)

It is found experimentally (see page 282) that more charge can be stored for a given potential difference when an insulator (the dielectric) fills the space between the plates – in other words, the capacitance of the arrangement is increased by adding a dielectric.

The **relative permittivity** (ε_r) of the dielectric is given by:

$$\varepsilon_r = \frac{C}{C_o}$$

where C_0 = capacitance of the capacitor with a vacuum between the plates

C = the new capacitance with space between the plates filled by the dielectric.

So for a parallel-plate capacitor we have (from Equation 2):

$$\varepsilon_r = \frac{\varepsilon A/d}{\varepsilon_0 A/d}$$

$\therefore$ 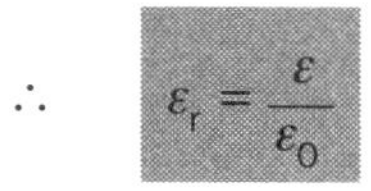

$$\varepsilon_r = \frac{\varepsilon}{\varepsilon_0}$$

Equation 3

You should note that:

- Unlike ε and ε_0 which have units of F m^{-1}, ε_r is simply a ratio (i.e it is a pure number, with no unit).

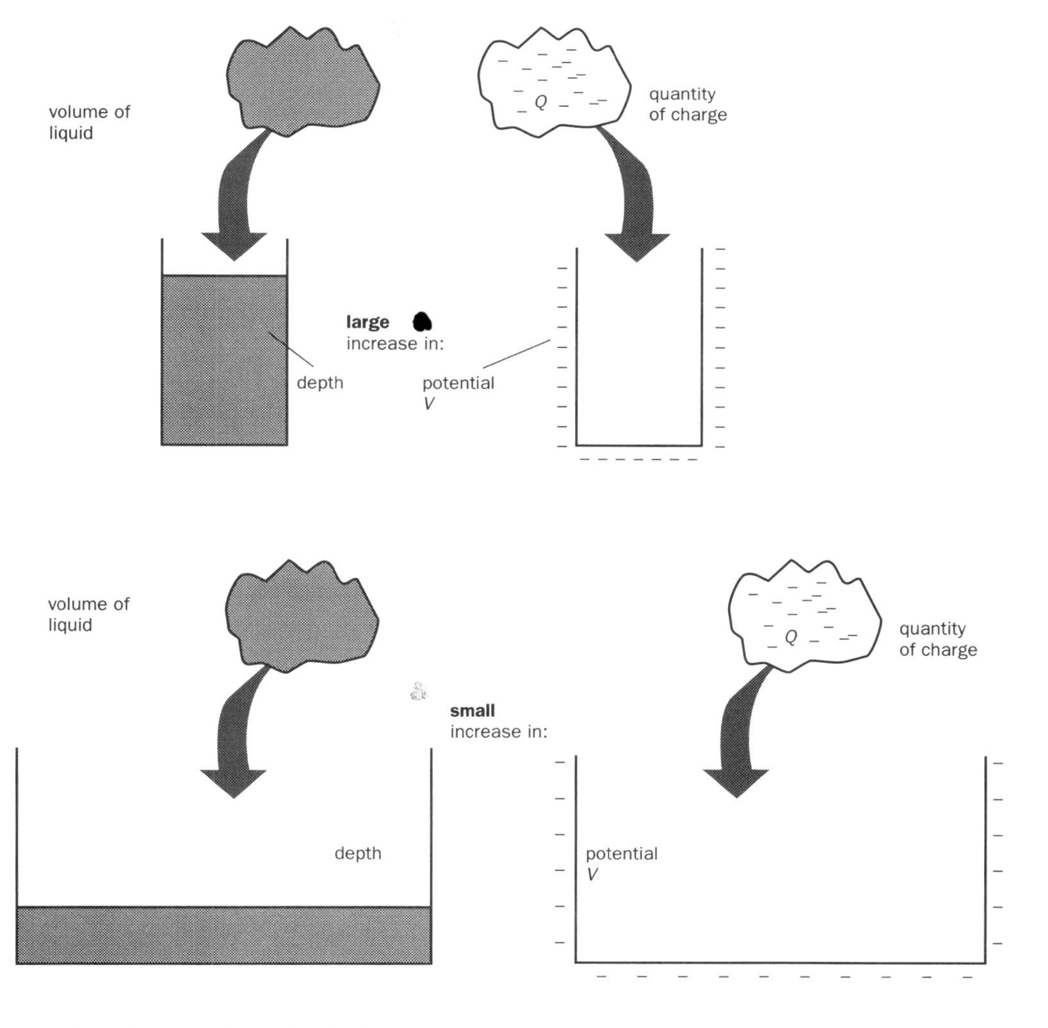

Figure 3.99 Analogy between liquid and charge storage

So, for a parallel-plate capacitor with a dielectric of relative permittivity ε_r between its plates, Equation 2 becomes:

$$C = \frac{\varepsilon_r \varepsilon_0 A}{d}$$

Equation 4

Action of the dielectric in a capacitor

A dielectric material is an electrical insulator, and is used between the plates of practical capacitors to:

- increase the capacitance obtained from a given size of capacitor
- separate the plates of the capacitor, and prevent them touching
- enable high potential differences to be used without electrical breakdown.

The dielectric acts by reducing the strength of the electric field between the plates. When a dielectric is present between the plates of a charged capacitor, its molecules are affected by the electric field between the plates. The positively-charged nuclei of the dielectric atoms are pushed slightly in the direction of the electric field, while the negatively-charged electrons feel a push in the opposite direction. This disturbs the normally uniform charge distribution of the molecules, distorting their shape so that they become electric **dipoles** with an excess of positive charge at one side and negative charge at the other, as shown in Figure 3.101. The dielectric is then said to be **polarised**. (In some materials, the molecules are already dipoles. The action of the applied electric field simply aligns them.)

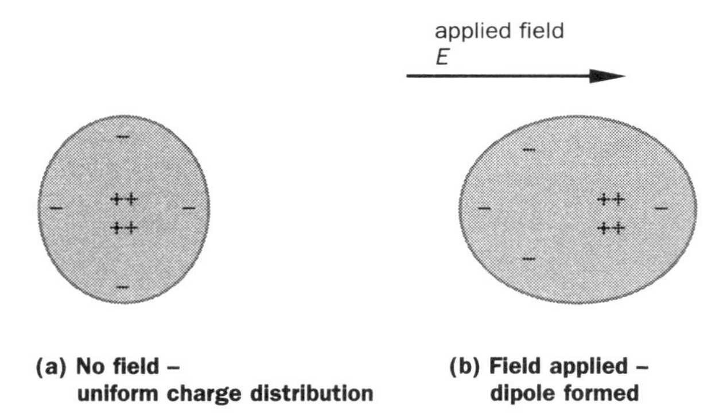

Figure 3.101 Dipole formation

Note that the internal field (E_d) created by the dipole **opposes** the externally applied field (E_0). Between the plates of a capacitor, the internal electric field of the dielectric opposes the applied electric field due to the charge on the capacitor plates. The **resultant** electric field (E_r) is therefore equal to ($E_0 - E_d$), as shown in Figure 3.102.

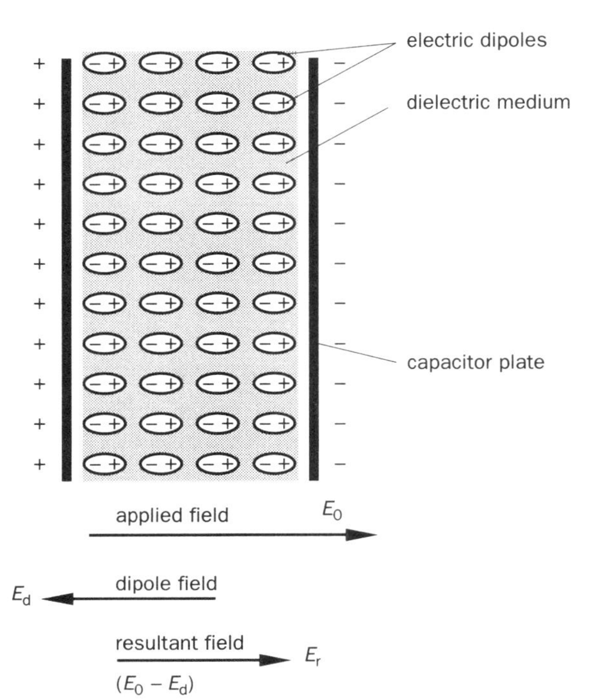

Figure 3.102 Dielectric between charged capacitor plates

You should note that:

- the overall charge of the dielectric is zero
- charges on adjacent dipoles cancel each other
- at the surface of the dielectric, the charges adjacent to the capacitor plates tend to reduce the potential difference between the plates.

If the capacitor is isolated (i.e. it is not connected to a battery), then the total charge stored (Q) cannot change, and the effect of adding the dielectric is to decrease the p.d. (V) across the capacitor. Since capacitance (C) is given by $C = Q/V$, this has the effect of increasing the capacitance.

If the capacitor is connected to a battery, adding a dielectric causes extra charge to flow from the battery to maintain the p.d. across the plates. Thus extra charge is stored for the same p.d., i.e. the capacitance is increased.

Investigation of factors affecting the capacitance of a parallel-plate capacitor

The parallel-plate capacitor used in this experiment is made from two large metal plates with an air gap between them. The thickness of the air gap is controlled by small polythene spacers placed at the corners as shown in Figure 3.103.

The apparatus is assembled as shown in Figure 3.104. The lower plate is connected to Earth, so is held at zero potential (0 V), while the other plate can be raised to a potential V by touching it with a 'flying lead' from the unearthed terminal of an HT (high tension or high voltage) supply, which is then removed. In order to create this potential difference, charge must flow briefly, and this charge is stored by the capacitor.

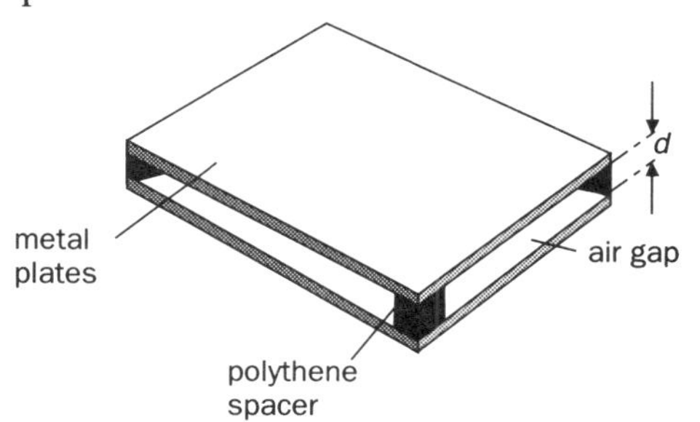

Figure 3.103 Form of parallel-plate capacitor used for laboratory investigation of characteristics

To measure the charge on the capacitor, its charge is removed by touching the top plate with the screened lead from the d.c. amplifier. This device is set up so that the charge flowing into it is proportional to the reading on the microammeter.

Since $C = \frac{Q}{V}$, measuring the charge and p.d. enables the capacitance to be calculated. If the experiments are carried out at a constant value of V, then the charge measured (Q) is directly proportional to the capacitance (C).

Variation of capacitance with plate separation *d*

The plate separation is varied by stacking the polythene spacers as shown in Figure 3.105:

The charge stored (at constant potential) is measured for various values of d, and a graph of Q against $1/d$ gives a straight line through the origin, as shown in Figure 3.106.

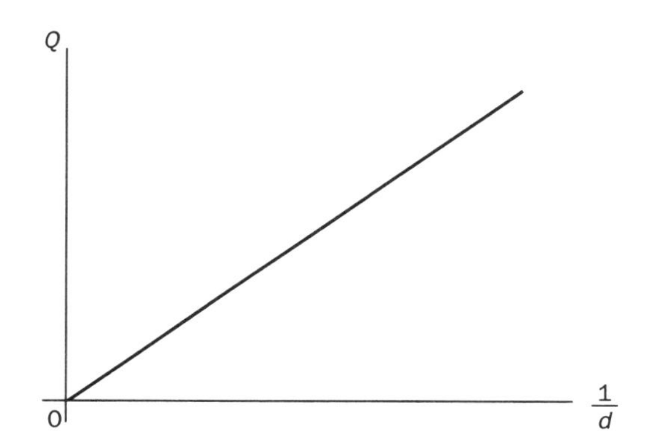

Figure 3.106 Graph of Q against 1/d for a parallel-plate capacitor

Since $C \propto Q$ and $Q \propto 1/d$

$$\therefore \quad C \propto \frac{1}{d}$$

i.e. the capacitance of a parallel-plate capacitor is **inversely proportional** to the separation distance between the plates.

Variation of capacitance with area of overlap *A*

For this experiment the plate separation d is kept constant, and the area of overlap is varied by sliding one plate aside as shown in Figure 3.107 on page 282. The non-overlapping parts of the plates can be considered to be so remote from other conductors as

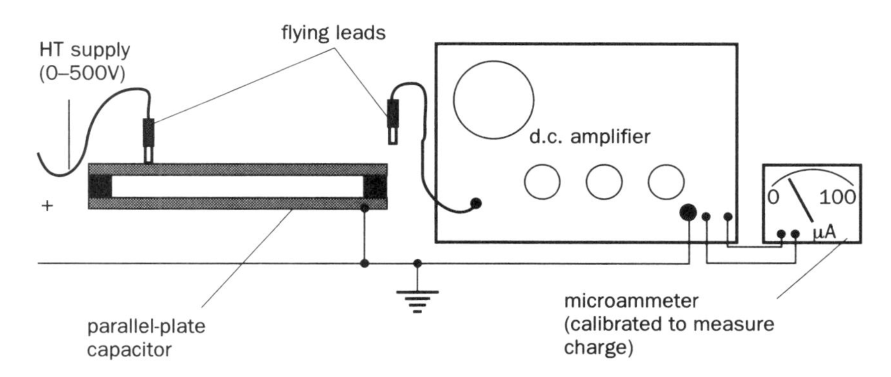

Figure 3.104 Apparatus for investigating the factors affecting capacitance

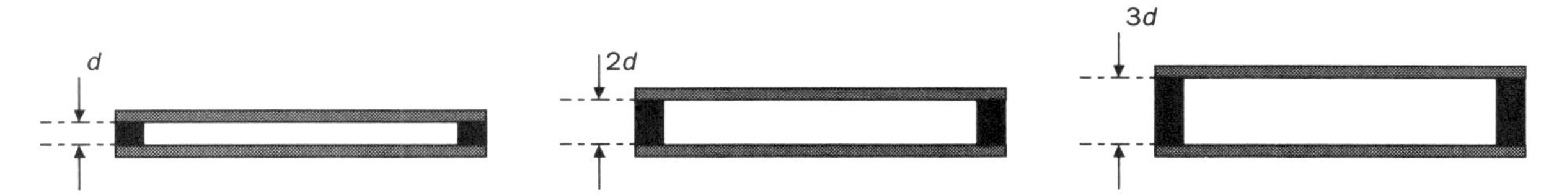

Figure 3.105 Variation of plate separation

to have no effect on the capacitance of the overlapping portion.

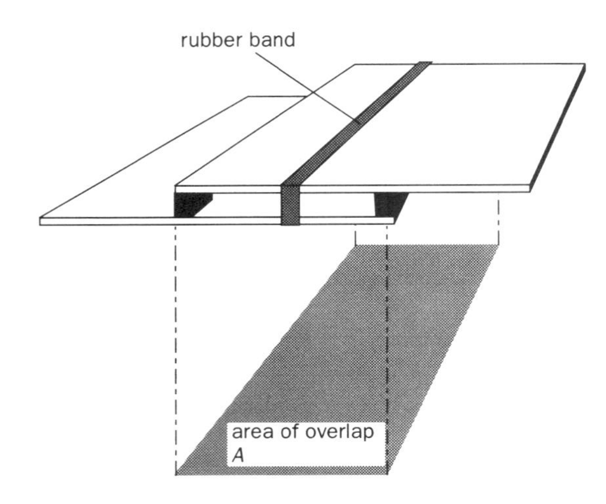

Figure 3.107 Variation of area of overlap A

Charge stored (Q) is measured for various values of area (A), and a graph of Q against A gives a straight line through the origin, as shown in Figure 3.108.

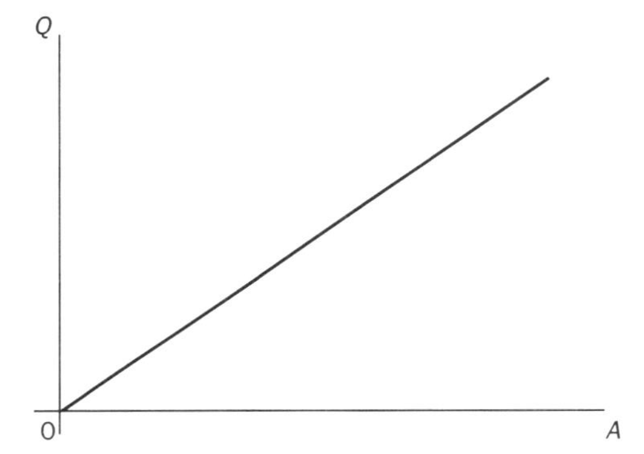

Figure 3.108 Graph of Q against A for a parallel-plate capacitor

Thus charge stored (Q) is directly proportional to area of overlap (A) of the plates.

Since $C \propto Q$ and $Q \propto A$

$\therefore$ $C \propto A$

i.e. the capacitance of a parallel-plate capacitor is **directly proportional** to the area of overlap of the plates.

Variation of capacitance with the permittivity (ε) of the dielectric

In this case, both the area of overlap (A) and the separation distance (d) are kept constant; only the material separating the plates is changed (see Figure 3.109). Different materials have different effects on the capacitance of the capacitor, and the **permittivity** (ε) of the materials is known.

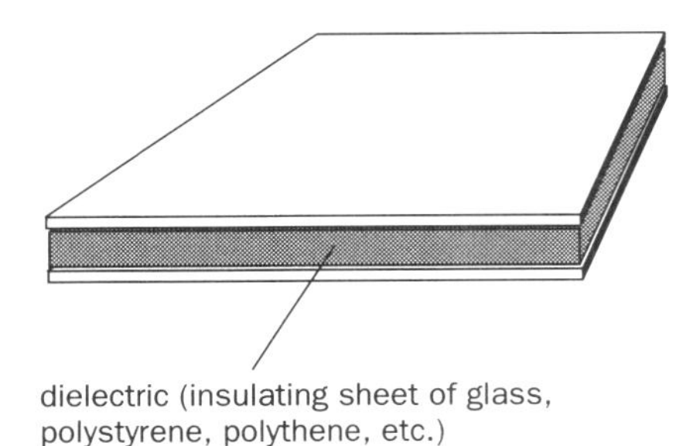

Figure 3.109 Investigating the effect of dielectric on capacitance

A graph of charge against permittivity gives a straight line which tends towards the origin, as shown in Figure 3.110.

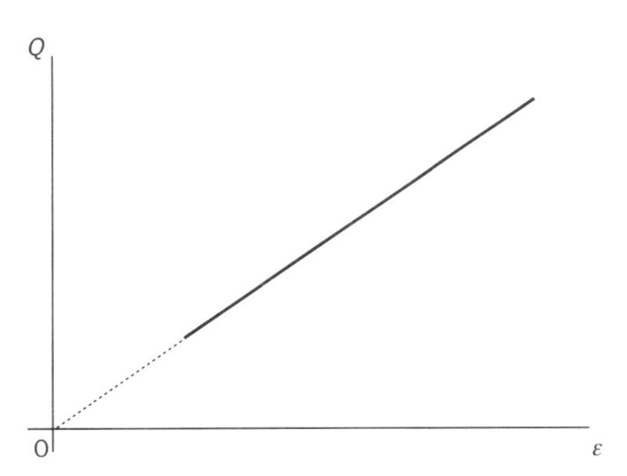

Figure 3.110 Graph of Q against ε for a parallel-plate capacitor

Since $C \propto Q$ and $Q \propto \varepsilon$

$\therefore$ $C \propto \varepsilon$

i.e. the capacitance of a parallel-plate capacitor is **directly proportional** to the permittivity of the material between the plates.

Action of a capacitor

Charging

In the circuit shown in Figure 3.111, closing the switch (S) allows electrons to be drawn from plate X (leaving it positively charged) and deposited on plate Y (which becomes negatively charged). This does not happen instantaneously:

- When the switch is first closed, there is no opposing potential, so there is a relatively large current, and charge rapidly builds up on the capacitor.
- The charge on the capacitor causes its potential to rise; this p.d. opposes the cell p.d. (V_0), and the current decreases as the charge increases.
- Eventually the p.d. across the capacitor becomes equal to the supply p.d. (V_0). No further charge can flow, ($I = 0$) and the capacitor is now fully charged.

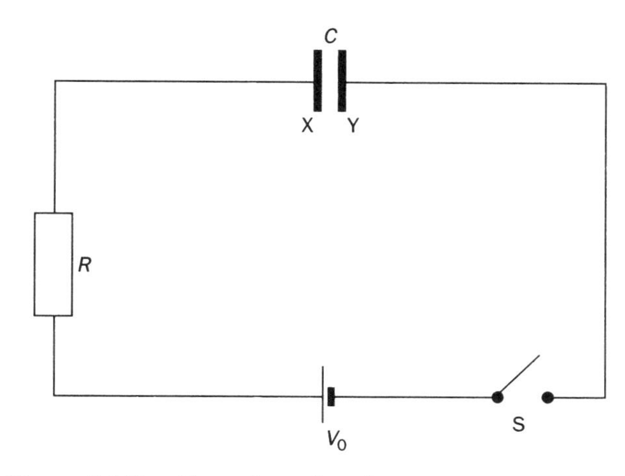

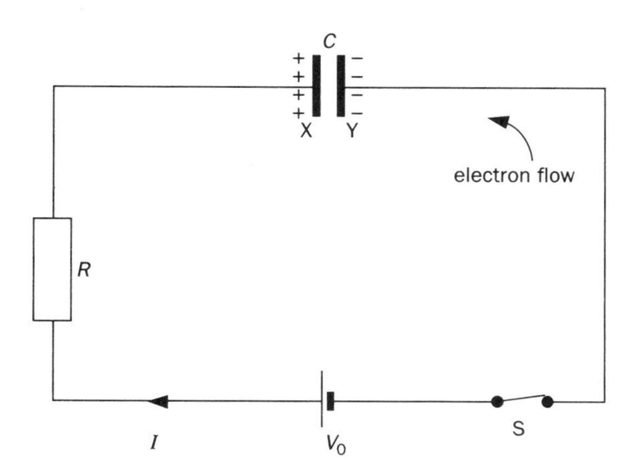

Figure 3.111 Capacitor charging

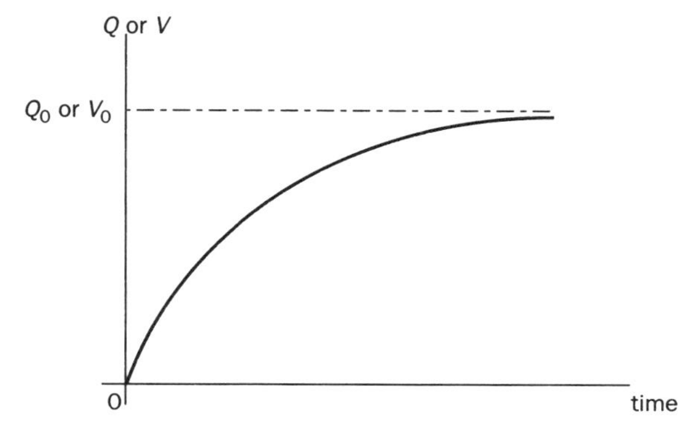

Figure 3.112 Growth of charge (or p.d.) on a capacitor

Figure 3.112 is a typical graph showing the growth of charge (and hence p.d. across the plates) on a capacitor. The rate of growth of charge (and hence time taken for the capacitor to become fully charged) is controlled by the resistance R and the capacitance C. Increasing either slows the process: a larger resistance reduces the current (i.e. the rate of flow of charge), and a larger capacitance means more charge must flow to achieve the same change in potential.

Some of the energy provided by the cell is stored by the electric field of the capacitor and some is dissipated to the surroundings by the heating effect of the current through the resistor.

Discharging

In the circuit shown in Figure 3.113, when switch S is closed, electrons flow round the circuit from the negative plate Y to the positive plate X. (Equivalent to a conventional current from X to Y.)

Again the process is not instantaneous:

- When the switch is first closed, the charge on the plates (and therefore the p.d. across them) is a maximum, so the current in the wires is a maximum.
- As the charge decays, the p.d. decreases, causing the current to decrease.
- Eventually all the charge will have left the capacitor, the p.d. will be zero, hence the current will be zero.
- The energy stored in the capacitor has all been dissipated as heat energy.

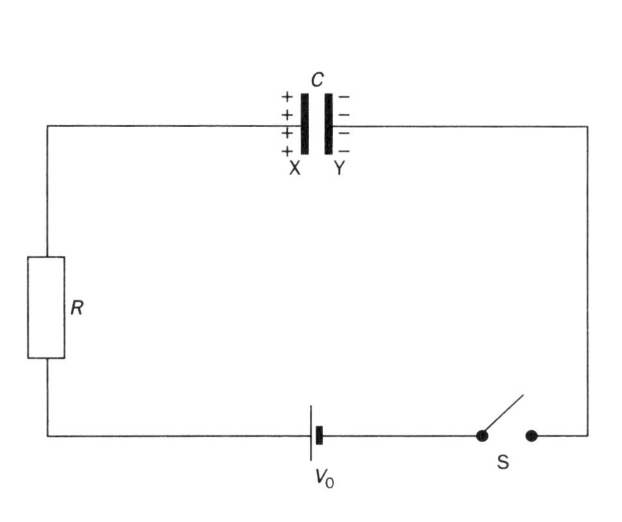

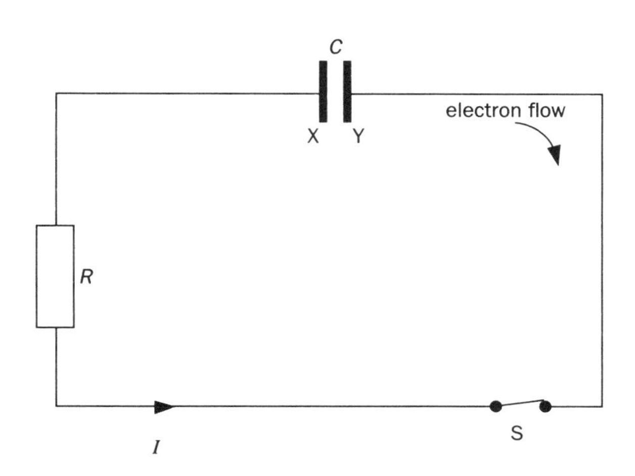

Figure 3.113 Capacitor discharging

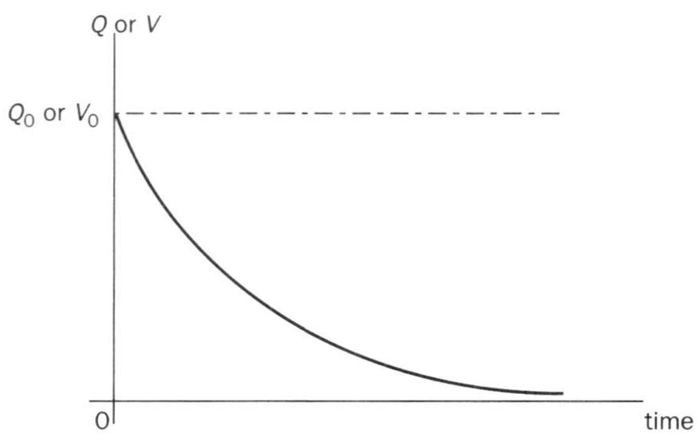

Figure 3.114 Decay of charge (or p.d.) on a capacitor

Figure 3.114 shows a typical **exponential decay** curve for a capacitor discharging.

It is worth noting that since current is equal to the *rate of flow of charge*,

i.e. $I = \frac{dQ}{dt}$

then the current flowing into or out of the capacitor is equal to the **gradient** of the graph of charge against time.

Effect of direct current (d.c.) on a capacitor

Since the plates are separated by an insulating medium (the dielectric) it is impossible for there to be a steady direct current 'through' a fully charged capacitor. Thus a lamp connected in series with a capacitor and a d.c. supply will not light. However, if the capacitor was initially uncharged, the brief flow of charge onto the capacitor when the circuit is first switched on could be enough to produce a brief flash of light from the lamp. This is illustrated in Figure 3.115(a) where a d.c. supply is connected in series with a switch, a lamp and a capacitor.

In practice, the capacitance must be large to give an observable flash of light. Figure 3.115(b) shows the steady-state condition when the capacitor is fully charged and there is no current.

Effect of alternating current (a.c.) on a capacitor

An alternating current reverses its direction every half-cycle. The a.c. mains operates at a frequency of 50 Hz, so every 1/100th of a second, it reverses direction. When an a.c. supply is connected to a capacitor, the effect is very different to that of d.c. During the first half of a cycle, electrons flow one way round the circuit, and one plate of the capacitor becomes positively charged while the other becomes more negative. When the current reverses during the next half of the cycle, the electrons are removed from the negative plate and pass to the positive plate, reversing the previous situation. Thus there is a continuous alternating current in the circuit connected to the capacitor, even though no charge crosses the gap between its plates. So a lamp connected in series with a capacitor and an a.c. supply could light. The size of the current will depend on the capacitance and on the frequency of the supply. If the capacitance is small, or the frequency low, only a small amount of charge needs to flow onto the capacitor plates for the p.d. across the capacitor to equal the supply p.d. In other words, the current will be very low. High capacitance and high frequency give high current.

Figure 3.116 shows an a.c. supply connected in series with a lamp, a switch and a capacitor. The lamp lights whatever the direction of the current, and because of the eye's persistence of vision, it will appear to give continuous illumination even though in fact it is cycling from full to near-zero brightness twice during every cycle of the alternating supply.

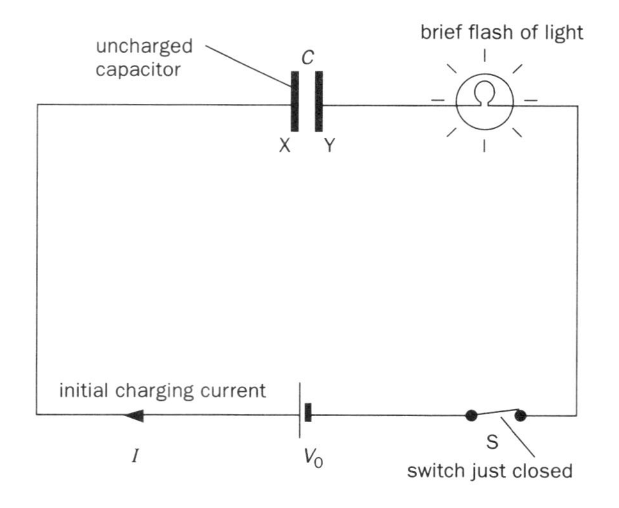

(a) Initial flow of current on first closing switch

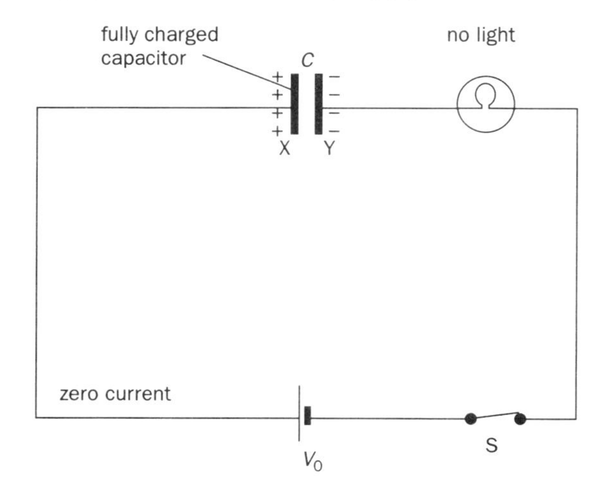

(b) Steady-state condition

Figure 3.115 Effect of d.c. on a capacitor

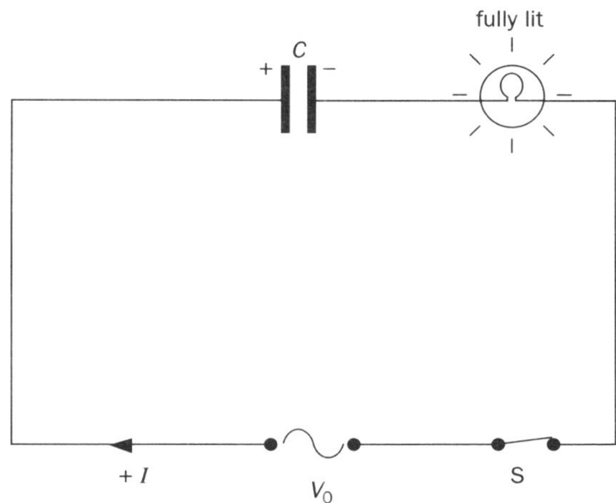

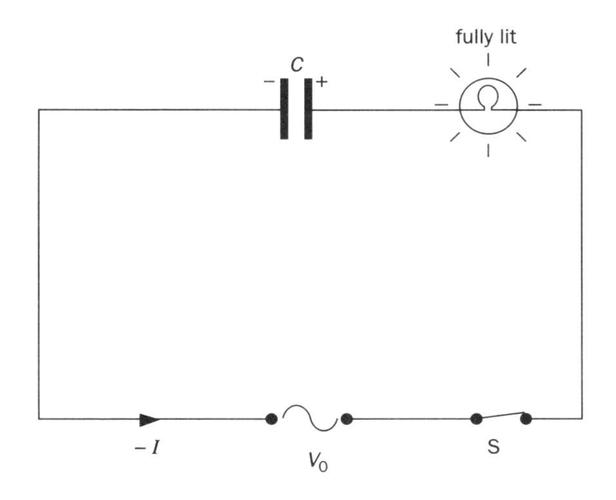

Figure 3.116 Effect of a.c. on a capacitor

Energy of a charged capacitor

The charge (Q) on a capacitor is directly proportional to the p.d. (V) across it since $Q = CV$. Thus a graph of p.d. against charge is a straight line through the origin, as shown in Figure 3.117.

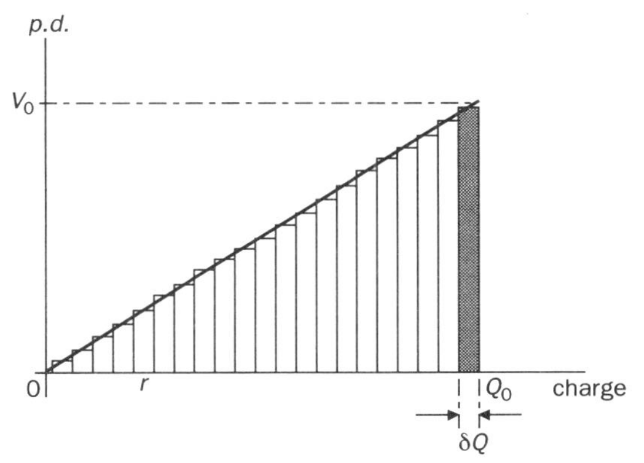

Figure 3.117 Graph of p.d. (V) against charge (Q) for a capacitor

Consider a capacitor of capacitance (C) whose p.d. is V_0 when the charge on its plates is Q_0. If the capacitor is allowed to discharge, its p.d. will fall. However, if a tiny amount of charge δQ flows out, so that the change in p.d. is negligible, then the energy lost (W) is given by:

$W = V_0\,\delta Q$ (from the definition of p.d.)

From the graph: $W = V_0\,\delta Q$ gives the area of the dark-shaded strip. If all of the charge (Q) were to flow out of the capacitor, reducing the p.d. to zero, then the energy released would be given by the sum of the areas of all the shaded strips, i.e. the area bounded by the graph and the charge axis.

Thus the energy released (W) = *area under graph*

i.e. $$W = \frac{1}{2}\,Q_0V_0$$

In general, if the p.d. of a capacitor changes by an amount (V) when a quantity of charge (Q) flows in (or out), then the energy gained (or lost) by the capacitor is given by:

$$W = \frac{1}{2}QV$$

J, C, V

Equation 5(a)

Also, since $Q = CV$, the energy stored in a charged capacitor can be written:

$$W = \frac{1}{2}CV^2$$

J, F, V

Equation 5(b)

and

$$W = \frac{1}{2}\frac{Q^2}{C}$$

J, C, F

Equation 5(c)

You should note that:

- The energy of a charged capacitor is stored in the electric field between the plates.
- If the capacitor is charged from a battery (or similar source), the charge (Q) flows at constant p.d. (V). In this case, the energy drawn from the cell is equal to QV (i.e. twice the energy stored in the capacitor). The 'missing' energy is converted into heat in the connecting wires.

Combination of capacitors

As in the case of resistors, capacitors can be connected in **series** or in **parallel**.

Capacitors in parallel

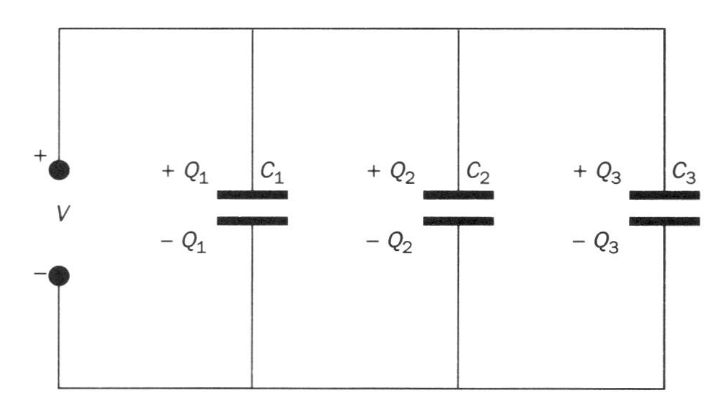

Figure 3.118 Parallel connection of capacitors

Consider three capacitors of capacitance C_1, C_2 and C_3 connected in parallel with a potential difference V as shown in Figure 3.118. The p.d. across each capacitor will be the same, and equal to V. However the **charge** stored on each capacitor will be different, depending on the value of its capacitance.

i.e. $Q_1 = C_1V \quad Q_2 = C_2V \quad$ and $Q_3 = C_3V$

the total charge stored (Q_T) is equal to the sum of the three charges:

$$Q_T = Q_1 + Q_2 + Q_3$$

i.e. $Q_T = C_1V + C_2V + C_3V$

so $C_TV = C_1V + C_2V + C_3V$

where C_T is the total capacitance of the combination.

$$\therefore \quad C_T = C_1 + C_2 + C_3$$

Equation 6

To find the total capacitance of several capacitors in parallel, simply add their individual capacitances.

You should note that:

- The charges stored on capacitors connected in parallel are in the same ratio as their capacitances

  i.e. $Q_1{:}Q_2{:}Q_3 = C_1{:}C_2{:}C_3$

- The expression for capacitors in parallel is similar to that for resistors in series.

Capacitors in series

Consider three capacitors of capacitance C_1, C_2, and C_3, connected in series as shown in Figure 3.119. If a p.d. (V) is applied, charge will flow as follows: electrons will flow from the negative terminal of the cell to a plate of C_3, making it negatively charged. Simultaneously electrons will flow from a plate of C_1

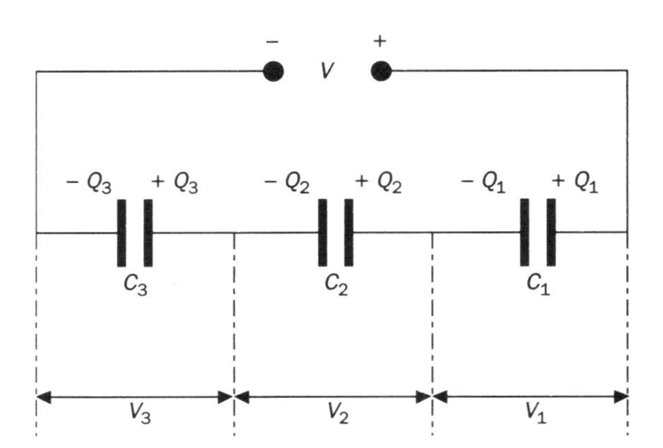

Figure 3.119 Series combination of capacitors

leaving it positively charged. Electrons will flow between the plates of the capacitors, leaving them charged as shown. Equal charge (Q) is stored on each capacitor. Charge flow will stop when the total p.d. across the capacitors is equal to the supply p.d. (V).

i.e. $V = V_1 + V_2 + V_3$

$$\therefore \quad \frac{Q}{C_T} = \frac{Q}{C_1} + \frac{Q}{C_2} + \frac{Q}{C_3}$$

where C_T is the total capacitance of the combination.

Dividing by Q we obtain:

$$\frac{1}{C_T} = \frac{1}{C_1} + \frac{1}{C_2} + \frac{1}{C_3}$$

Equation 7

You should note that

- The expression for capacitors in series is similar to that for resistors in parallel.
- All the capacitors store equal charge.
- The p.d.s across the capacitors are different, but add up to the total p.d.

Joining charged capacitors

Consider two charged capacitors C_1 and C_2, carrying charges Q_1 and Q_2 respectively, which are connected as shown in Figure 3.120.

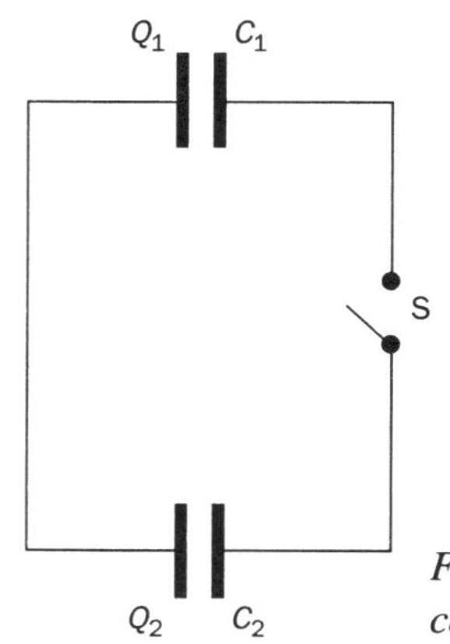

Figure 3.120 Joining charged capacitors

When the switch is closed, charge will flow until the p.d. across the capacitors is equal.

The final charge and energy stored by each capacitor, the total charge remaining, and the final p.d. can be calculated using the following facts:

- the total charge will remain constant (since charge must be conserved)
- both capacitors end up with the same p.d. (since they are in parallel)
- the capacitors are connected in parallel, so the total capacitance of the combination (C_T) is given by:

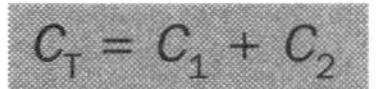

$$C_T = C_1 + C_2$$

If the two capacitors have equal p.d.s before connection, no charge flows on connection, and none of the stored energy is dissipated to the surroundings. If the p.d.s are different, the final energy stored after connection is less than the sum of the energies stored by the two capacitors before connection. The 'missing' energy is dissipated as heat when the charge does work against the resistance of the connecting wires.

Charging and discharging of capacitors

Capacitor discharge (or decay of charge)

Consider a capacitor of capacitance (C) with an initial charge (Q_0) at time (t) = 0 connected in series with a resistor (R) and a switch as shown in Figure 3.121.

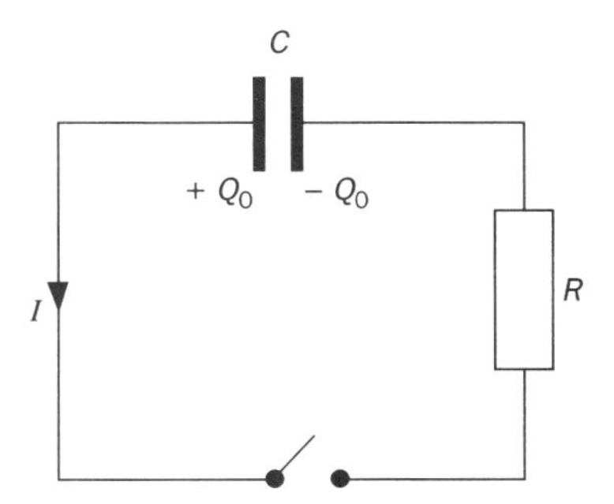

Figure 3.121 Capacitor discharge

When the switch is closed, a current flows in the circuit as shown and the capacitor is said to **discharge**. Energy is dissipated to the surroundings by the heating effect of the current through the resistive parts of the circuit.

Banks of capacitors forming part of the 20 kV, 2 MJ pulsed power supply for a high power Nd–glass laser

Variation of charge with time

Figure 3.114 shows how the charge decays with time. The shape of the graph suggests **exponential** decay, and it can be shown that the equation describing this decay with time is:

$$Q = Q_0 e^{-\frac{t}{CR}}$$

Equation 8(a)

where Q = charge on the capacitor at any time (t)

Variation of p.d. with time

Since the p.d. (V) is proportional to the charge (Q), the relationship of p.d. with time is given by:

$$V = V_0 e^{-\frac{t}{CR}}$$

Equation 8(b)

where V = p.d. across the capacitor at any time (t)

Variation of current with time

Since the current in the circuit will be proportional to the p.d. across the capacitor, the relationship of current with time is given by:

$$I = I_0 e^{-\frac{t}{CR}}$$

Equation 8(c)

where I = current flowing at any time t.

Half-life ($T_{1/2}$) of a *CR* circuit

The idea of half-life should be familiar to you from elementary work on radioactivity. The charge (and hence p.d. and discharge current) decreases by equal fractions in equal time intervals. Figure 3.122 shows a typical graph of charge (Q) against time (t) for a discharging capacitor. An identical shape of graph is obtained if either the p.d. (V) across the capacitor or the discharge current (I) is plotted against t.

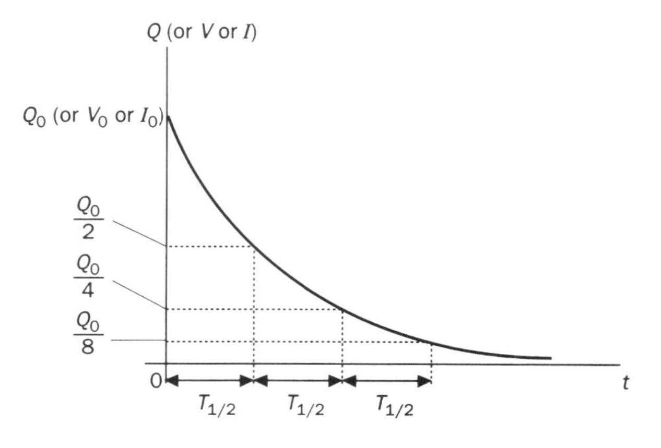

Figure 3.122 Q/t graph for a discharging capacitor, showing the meaning of half-life

The **half-life ($T_{1/2}$)** of a *CR* circuit is defined as: the **time taken** for the charge on the capacitor (or the p.d. across it, or the discharge current through it) to decrease to **half** of any initial value.

Time constant (T) of a *CR* circuit

The **time constant (T)** of a *CR* circuit is defined as: the **time taken** for the charge on the capacitor (or the p.d. across it, or the discharge current through it) to decrease to **$1/e$** of any initial value.

The time constant of a combination of a capacitor and a resistor is easy to calculate, and is used as a measure of the discharge time, especially for timing applications.

The value of T is given by:

$$T = CR$$

(units: T in s, C in F, R in Ω)

Equation 9

If *CR* is large, the decay will be slow and vice versa as shown in Figure 3.123.

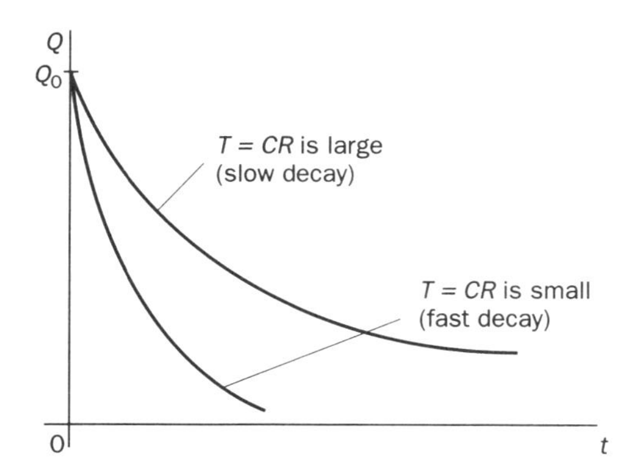

Figure 3.123 Effect of time constant on the speed of decay of charge in a CR circuit

Maths window

If $t = CR$, then Equation (7) becomes:

$$Q = Q_0 e^{-\frac{CR}{CR}} \quad \text{i.e. } Q = Q_0 e^{-1}$$

$$\therefore \quad Q = \frac{Q_0}{e}$$

where $e = 2.718$ (base of natural logarithms)

Growth of charge

If a source of p.d. is added to the circuit shown in Figure 3.121, we obtain the circuit shown in Figure 3.124, which enables the charging behaviour of a capacitor to be studied. When the switch is closed, there is a large initial charging current, which decreases with time. This causes the charge on the capacitor to increase from zero, rapidly at first, until it reaches a final value when the charging current will stop, and the p.d. across the capacitor will have risen to equal the supply p.d.

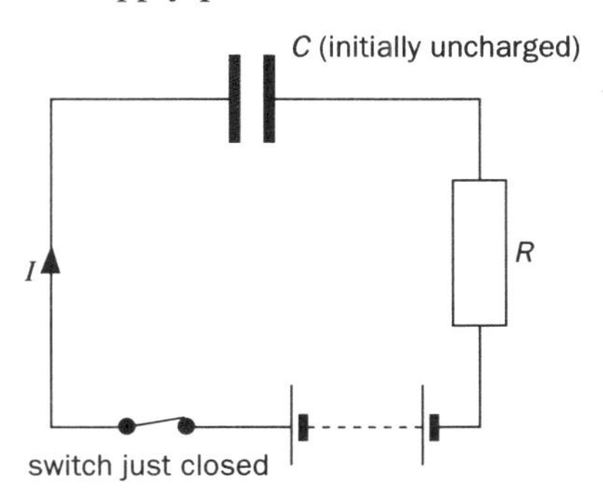

Figure 3.124 Capacitor charging circuit

Variation of charge with time

Figure 3.112 shows how the charge grows with time. The shape of the graph suggests **exponential** growth, and it can be shown that the equation describing this growth with time is:

$$Q = Q_0\left(1 - e^{-\frac{t}{CR}}\right)$$

Equation 10

where Q = charge on the capacitor at any time (t)
and Q_0 = charge on the capacitor when $t = \infty$

Variation of p.d. with time

As before, the equation for the variation of p.d. with time has the same form, since the p.d. across the capacitor is proportional to the charge stored.

$$V = V_0\left(1 - e^{-\frac{t}{CR}}\right)$$

Equation 11

where V = p.d. across the capacitor at any time (t)
and V_0 = p.d. across the capacitor when $t = \infty$

Variation of current with time

The charging current decreases with time, as the p.d. across the capacitor rises. Therefore the charging current **decays** exponentially with time, and the equation is:

$$I = I_0 e^{-\frac{t}{CR}}$$

Equation 12

Guided examples (1)

1. Three capacitors, of capacitance 2 μF, 6 μF and 12 μF are connected in series with a 12 V d.c. supply. Calculate
(a) the total capacitance
(b) the charge on each capacitor
(c) the total energy stored.

Guidelines

(a) Use Equation 7 to calculate the total capacitance.
(b) Use Equation 1.
(c) Use Equation 5(b).

2. A 10 μF capacitor is charged by a 50 V supply, and then connected across an uncharged 20 μF capacitor. Calculate:
(a) the initial charge on the 10 μF capacitor
(b) the final p.d. across the capacitors
(c) the final charge on each capacitor
(d) the initial energy stored
(e) the final energy stored.

Guidelines

(a) Use Equation 1 to calculate the initial charge.
(b) Use Equation 6 to calculate C_T for the parallel combination, then calculate V, since the total charge is constant.
(c) Calculate the charge on each capacitor (same p.d.).
(d) Initial energy stored (use Equation 5(b)).
(e) Final energy stored is sum of energies on capacitors.

3. A washing machine timer uses a capacitor and a resistor in series to switch off the heater after a pre-set time. A 1000 μF capacitor is charged to a p.d. of 25 V, and allowed to discharge through a resistor of 470 kΩ. If the switch is triggered when the p.d. falls to 5 V, calculate
(a) the time for which the heater is on
(b) the new value of resistor required to operate the heater for 25 min.

Guidelines

(a) Use Equation 11, take natural logs of both sides, and solve for t.
(b) As for (a), but solving for R.

Self-assessment

SECTION A
Qualitative assessment

1. (a) Write down an equation which relates **electric charge, potential difference** and **capacitance**.
(b) State the SI units for the quantities in (a).
(c) Define the term **capacitance**.

2. A parallel-plate capacitor of capacitance (C) has plates of area (A), which are a distance (d) apart in air. What is the capacitance (in terms of C) of a similar capacitor:
(a) having plates of area $0.50A$ which are a distance $2d$ apart in air
(b) having plates of area A which are separated a distance $3.5d$ by a dielectric whose relative permittivity (ε_r) has a value of 7.0.

3. Explain how the insertion of a dielectric between the plates of a charged parallel-plate capacitor increases the capacitance when the capacitor is:
(a) isolated
(b) connected to a battery.

4.

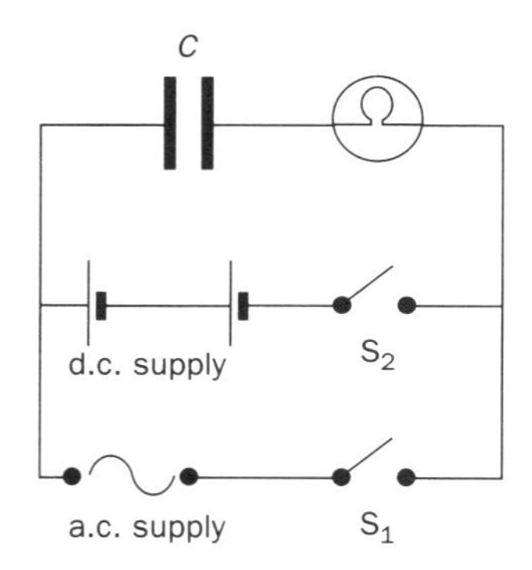

In the circuit shown, C is a capacitor of high capacitance, initially uncharged. State and explain what happens in the circuit when:
(a) S_2 is closed and S_1 is left open
(b) S_1 is closed and S_2 is left open.

5. When a capacitor of capacitance C carries a charge Q, the p.d. across it is V. Derive from first principles an expression for the energy stored in the capacitor in terms of
(a) $\boldsymbol{Q}$ and $\boldsymbol{V}$
(b) $\boldsymbol{C}$ and $\boldsymbol{V}$
(c) $\boldsymbol{Q}$ and $\boldsymbol{C}$.

6. Derive an expression for the total effective capacitance of three capacitors connected
(a) in **parallel**
(b) in **series**.

7. Two charged capacitors are charged to different p.d.s and then connected in parallel. State and explain what happens to the total energy stored **before** and **after** connection.

8.

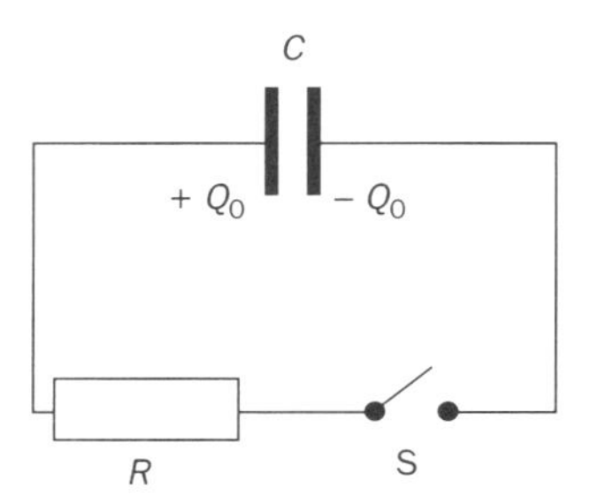

The circuit shows a charged capacitor C connected in series with a resistor R and a switch S.
(a) Sketch a graph of **charge** against **time** from the instant when the switch is closed.
(b) Show on the graph the effect of **increasing** C or R.
(c) State an equation representing this decay, stating the meaning of each term used.
(d) Use the graph to explain what is meant by the '**half-life**' of a CR circuit.
(e) Define the term '**time constant**' as applied to a CR circuit.

9.

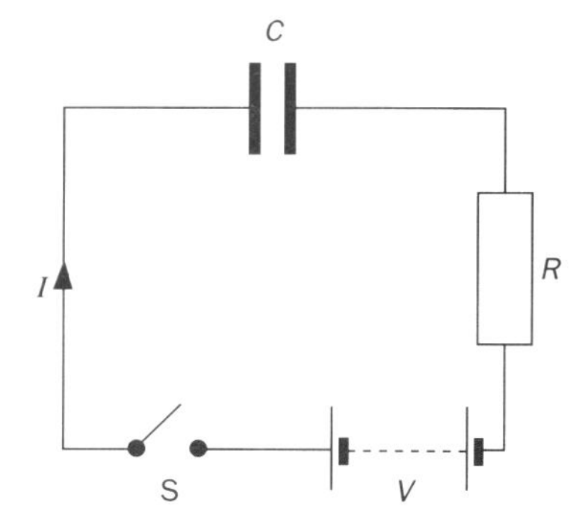

The circuit shows an uncharged capacitor of capacitance (C) connected in series with a resistor of resistance (R), a battery of p.d. (V) and a switch S.
(a) From the instant that the switch is closed, sketch a graph of
(i) p.d. across the capacitor against time
(ii) current in the circuit against time.

Self-assessment *continued*

(b) State the equations representing
   (i) the **growth of p.d.** across the capacitor with time
   (ii) the **decay of current** in the circuit with time.

SECTION B
Quantitative assessment

(***Answers:*** 3.7×10^{-10}; 8.9×10^{-10}; 1.0×10^{-9}; 4.0×10^{-9}; 1.0×10^{-5}; 1.4×10^{-5}; 1.8×10^{-5}; 2.5×10^{-5}; 4.0×10^{-5}; 4.8×10^{-5}; 4.5×10^{-4}; 9.0×10^{-4}; 1.0×10^{-3}; 0.010; 0.023; 0.20; 2.0; 2.5; 3.4; 5.3; 9.4; 11; 50; 3.1×10^{10}.)

(Electronic charge $e = 1.6 \times 10^{-19}$ C.

$\varepsilon_{air} = \varepsilon_0 = 8.9 \times 10^{-12}$ F m^{-1}.)

1. A capacitor stores 5.0×10^{-9} C of electric charge. Calculate
   (a) the number of electrons which must have flowed into the capacitor
   (b) the capacitance of the capacitor if the p.d. across it is 5.0 V
   (c) the amount of charge which must be removed from the capacitor to reduce its p.d. to 1.0 V.

2. A parallel-plate capacitor has square plates whose sides are 250 mm long, and which are situated 1.5 mm apart. Calculate the capacitance of this capacitor when the material between the plates is:
   (a) air
   (b) a dielectric of relative permittivity 2.4.

3. Two 15 μF capacitors are connected in parallel, and the pair are connected in series with a 45 μF capacitor and a battery. If the p.d. across the parallel pair is 30 V, calculate:
   (a) the charge on each capacitor
   (b) the e.m.f. of the battery
   (c) the total capacitance of the circuit
   (d) the total energy stored in the capacitors.

4. A 500 W flash lamp fitted to a camera operates when a 5000 μF capacitor discharges through the lamp. The capacitor charges from a 9 V battery. Calculate the energy stored in the capacitor.

5. If all the energy stored by a 250 μF capacitor charged to a p.d. of 7500 V could be used to lift a man of mass 75 kg, what would be the greatest vertical height through which he could be raised? (Assume gravitational field strength (g) to be 10 N kg^{-1}.)

6. A 5.0 μF capacitor is charged to a p.d. of 10 V. It is then removed from the supply and connected in parallel with an uncharged 20 μF capacitor. Calculate:
   (a) the charge on each capacitor after connection
   (b) the total capacitance of the combination
   (c) the p.d. across the capacitors.

7. Calculate the energy stored in a 20 μF capacitor when charged to a p.d. of 500 V. If a second, initially uncharged, 30 μF capacitor is then connected in parallel with it, calculate the new value of the energy stored by the combination. Account for this change.

8. A 4.0 μF capacitor is charged from a 12 V battery and then discharged through a 0.4 MΩ resistor. Calculate
   (a) the initial charge on the capacitor
   (b) the charge on and p.d. across the capacitor 2.0 s after it starts to discharge.

9. A voltmeter of resistance 150 kΩ is connected across a 50 μF capacitor. If at time $t = 0$ the voltmeter reading is 20 V, calculate
   (a) the charge on the capacitor at time $t = 0$
   (b) the voltmeter reading after 10 s
   (c) the time which must elapse after $t = 0$ before 95% of the energy stored has been dissipated.

3.8 Alternating current

An electric current is the flow of charge carriers through a conductor under the influence of an electric field. In the case of electrons flowing through a metal, this flow is superimposed on the random intermolecular motion of the electrons at all temperatures above absolute zero. It is convenient to ignore this random motion, and concentrate on the motion caused by the applied field.

In the case of direct current, e.g. due to a battery, the electrons flow in one direction only. In a lamp, emission of energy is due to collisions between the moving electrons and atoms of the filament. The direction of motion is controlled by the polarity of the battery, but has no effect on the emission of energy (see Figure 3.125).

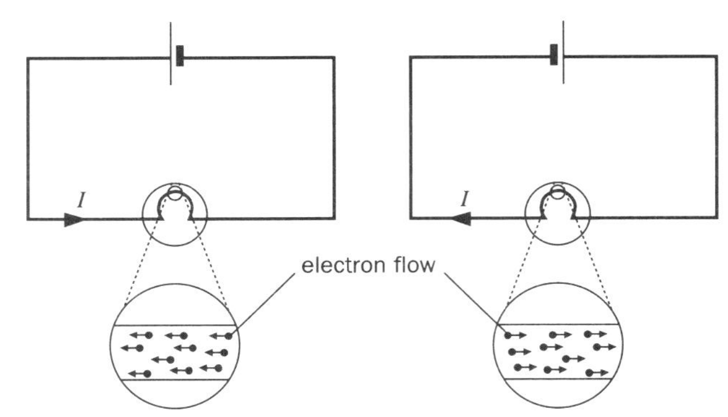

Figure 3.125 Direct current – electrons flow in one direction only

A source of alternating p.d. reverses its polarity regularly, which has the effect of causing the electrons to reverse their direction of motion with the same frequency as the source, as shown in Figure 3.126.

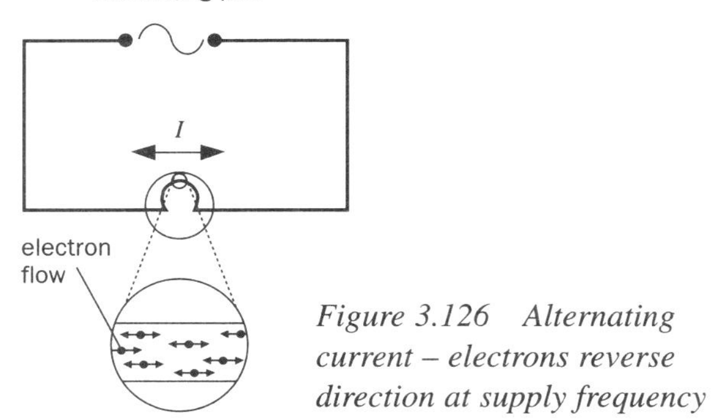

Figure 3.126 Alternating current – electrons reverse direction at supply frequency

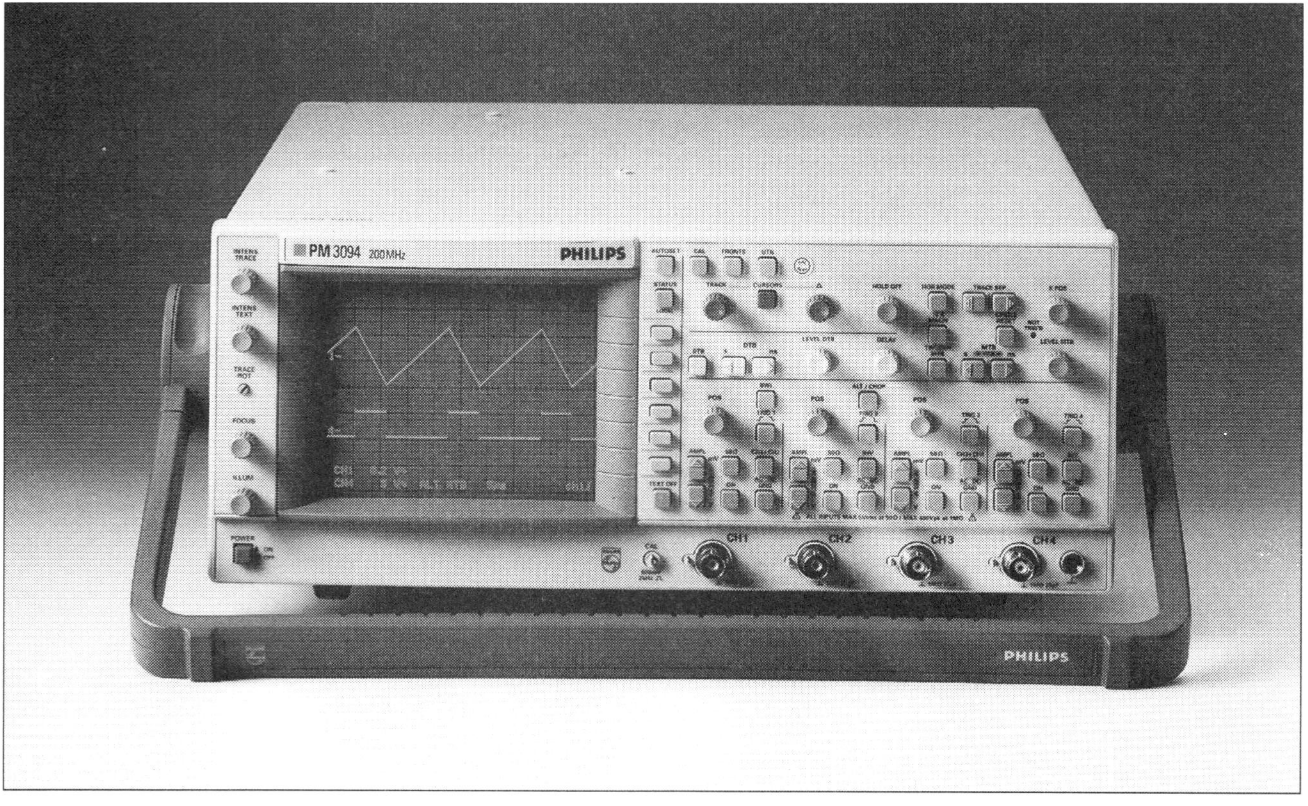

Triangular and square waveforms displayed on a sophisticated oscilloscope.

Of course, at any instant, the electrons are only flowing one way! The flow of electrons through the filament releases energy as before, but there are instants, while the direction reverses, when the electrons are stationary, so the lamp flashes on and off. However, at frequencies greater than about 20 Hz, the human eye and brain will interpret this as continuous illumination, due to the 'persistence of vision'.

The manner in which the motion of the electrons varies depends on the supply. A coil rotating with uniform velocity in a magnetic field gives rise to a p.d. which varies sinusoidally with time. Figure 3.127 shows this and other possible types of a.c.

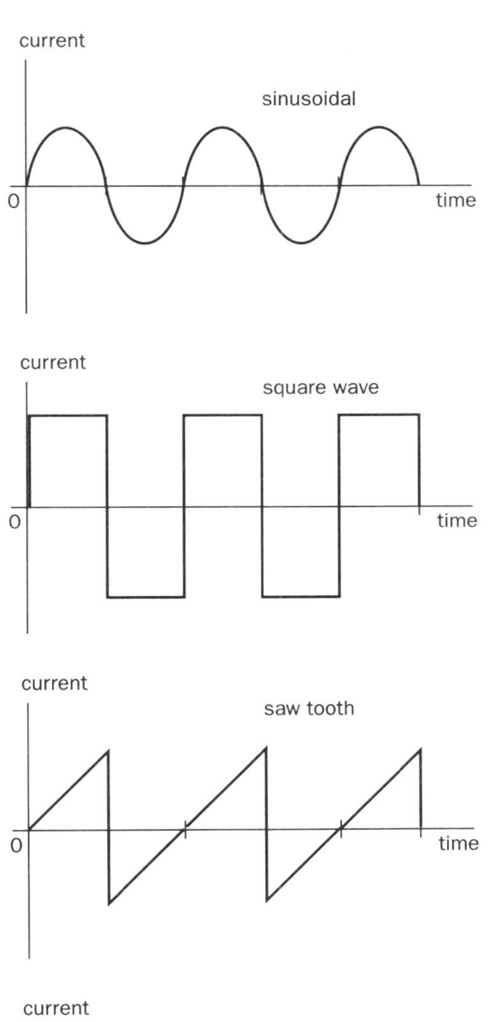

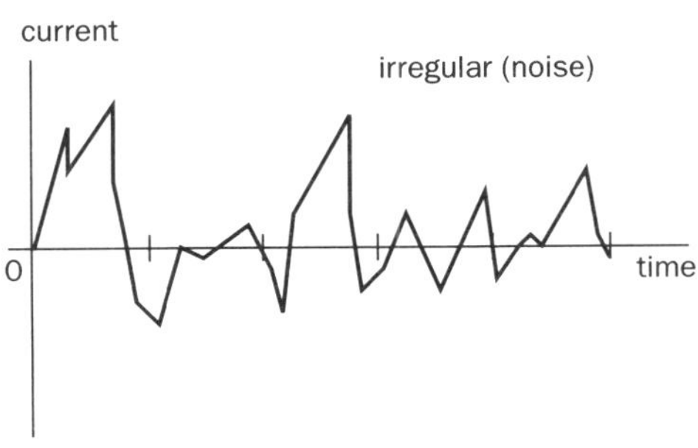

Figure 3.127 Examples of alternating current waveforms

For sinusoidal waveforms:
The p.d. and current can be represented by equations of the form:

$$V = V_0 \sin \omega t$$

Equation 1

$$I = I_0 \sin \omega t$$

Equation 2

where V, I = p.d. or current at time t
V_0, I_0 = peak values of the p.d. and current
ω = angular frequency of the supply

Terms and definitions

Figure 3.128 shows two alternating current waveforms with the same frequency. The upper diagram shows a sinusoidal waveform, while the lower diagram shows a square wave.

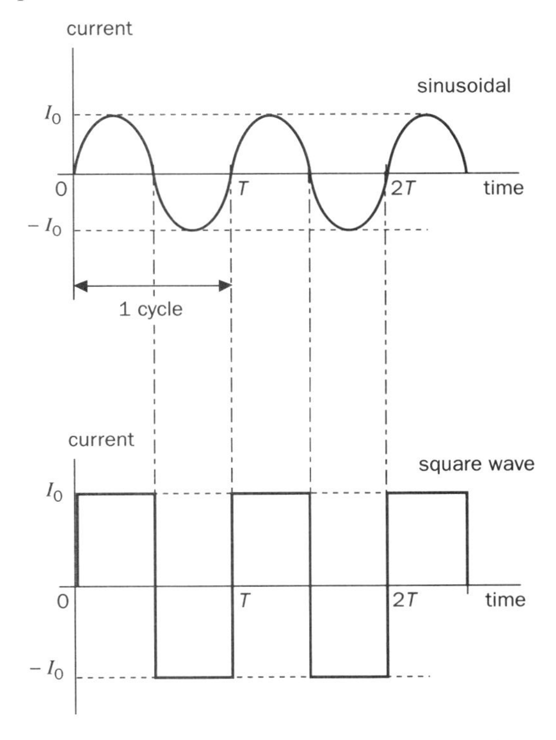

Figure 3.128 Terms and definitions in a.c. theory

- In both cases, the direction of the current changes periodically.
- In the case of sinusoidal a.c. the magnitude of the current shows continuous variation.
- One **cycle** is defined as one complete alternation or oscillation.
- The **period** T is defined as the time (in seconds) taken for one complete cycle.
- The **frequency** f (in Hz) is defined as the number of cycles per second.
- As with waves and SHM:

$$f = \frac{1}{T}$$

and $\omega = \frac{2\pi}{T} = 2\pi f$

- Hence Equation 1 may be written as:

$$V = V_0 \sin 2\pi ft$$

and Equation 2 as:

$$I = I_0 \sin 2\pi ft$$

- The peak value of p.d. or current is the maximum instantaneous value reached (V_0 or I_0). It is analogous to the amplitude of a wave.
- The peak-to-peak value of p.d. or current is a measure of the difference between $+V_0$ and $-V_0$ (or $+I_0$ and $-I_0$).

Root-mean-square (r.m.s.) values

As p.d. and current are continuously varying, we need some way of representing an 'average' value. However, the average value over a complete cycle is clearly zero. The peak value could be used, but is not representative of the full cycle.

In Britain, the a.c. mains voltage is often quoted as 240 V r.m.s., and frequency 50 Hz. The figure of 240 V is not the **peak** value (V_0), but is a measure of the **effective** value of the p.d. It is related to the peak value by:

$$V_0 = \sqrt{2} \times V_{rms} = 339.4 \text{ V}$$

Figure 3.129 illustrates the difference between the peak value and the r.m.s. value of the UK mains electrical p.d.

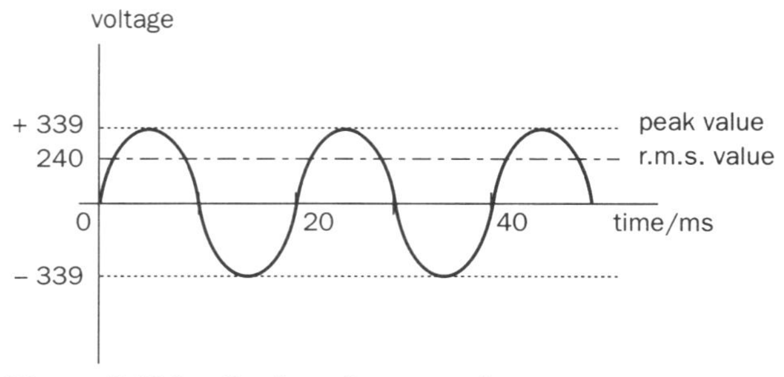

Figure 3.129 Peak and r.m.s. values

For an a.c. waveform, the **r.m.s.** or **effective** value of current (or voltage) is defined as the steady, direct value of the current (or voltage) which converts electrical energy to other forms **at the same rate** as the alternating current (or voltage).

Thus if a lamp is connected to the mains, or to a d.c. supply of 240 V, the power dissipation is the same, i.e. the brightness will be identical.

The following worked example should further clarify the meaning of the term **root-mean-square**.

Consider a sinusoidal alternating current of **peak value** 2.0 A through a lamp of resistance 3.0 Ω as shown in the diagram:

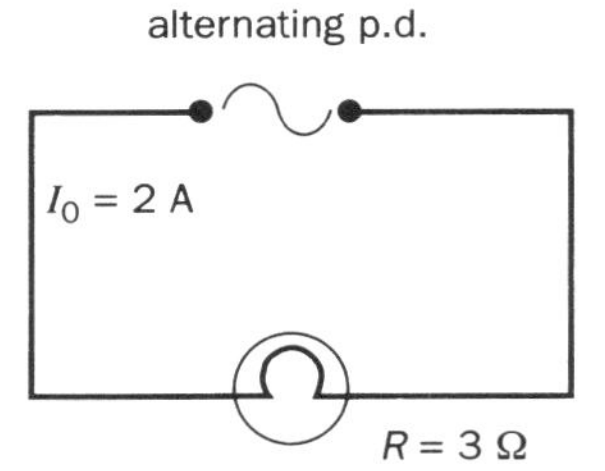

Peak value of power P_0 is given by:

$$P_0 = I_0^2 R = 2^2 \times 3 = 12 \text{ W}$$

(**Minimum** value of power = 0, when the current is momentarily zero)

∴ average value of *power* ($\bar{P}$) is given by:

$$\bar{P} = \frac{0 + 12}{2} = 6 \text{ W}$$

Now consider the same lamp being lit to the **same brightness** by a d.c. supply:

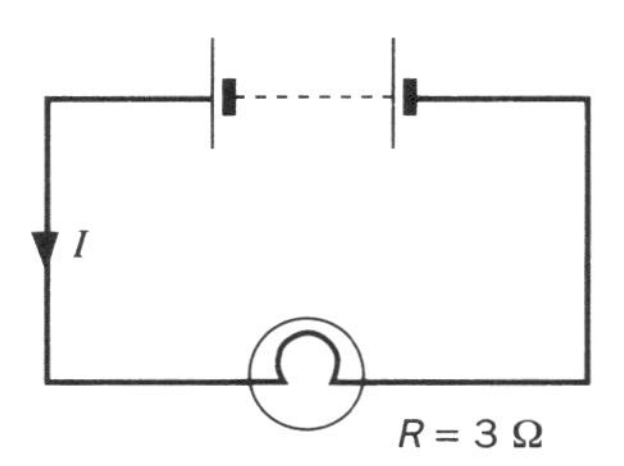

The power dissipated must also be 6 W (if the brightness is the same) and since

$$P = I^2 R$$

$$\therefore \quad I = \sqrt{\frac{P}{R}}$$

$$\therefore \quad I = \sqrt{\frac{6}{3}} = \sqrt{2} = 1.41 \text{ A}$$

Since the **effective** value of the alternating current is that direct current which dissipates energy at the same rate, 1.41 A is the **effective** value (or **root-mean-square** value) of an alternating current of **peak** value 2.0 A.

In general, for a **sinusoidal** alternating current:

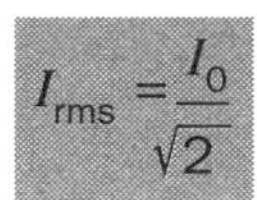

$$I_{rms} = \frac{I_0}{\sqrt{2}}$$

Equation 3

and by the same analysis:

$$V_{rms} = \frac{V_0}{\sqrt{2}}$$

Equation 4

Maths window

For a sinusoidally alternating current, the instantaneous value (I) is given by:

$$I = I_0 \sin \omega t$$

and the average power is given by:

$$P = (\text{average value of } I^2) \times R$$

$$\therefore \quad P = (\text{average value of } (I_0 \sin \omega t)^2) \times R$$

$$\therefore \quad P = I_0^2 R \times (\text{average value of } (\sin^2 \omega t))$$

The following diagram shows $\sin \omega t$ and $\sin^2 \omega t$ plotted against time:

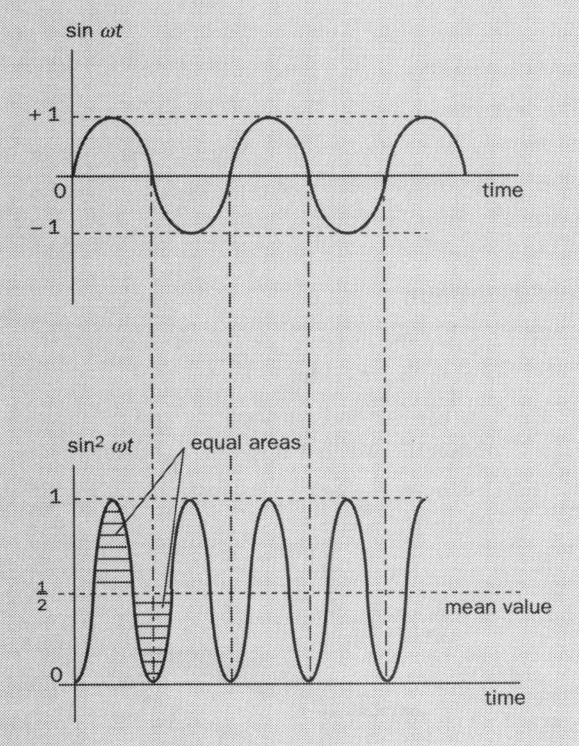

The average value of $\sin^2 \omega t$ is $1/2$,

$$\text{so} \quad P = I_0^2 R \times \frac{1}{2}$$

$$\therefore \quad P = \left(\frac{I_0}{\sqrt{2}}\right)^2 R$$

Maths window *continued*

$$\therefore \quad I_{rms} = \frac{I_0}{\sqrt{2}}$$

The effective current $= \sqrt{(\text{mean value of } I^2)}$

i.e. the root-mean-square value.

R.m.s. values for rectangular waves

For a square wave the $\sqrt{(\text{mean value of } I^2)}$ is obtained as follows.

Consider a square wave alternating current with peak value I_0 as shown in Figure 3.130(a). The **mean** value of the current² is shown in Figure 3.130(b).

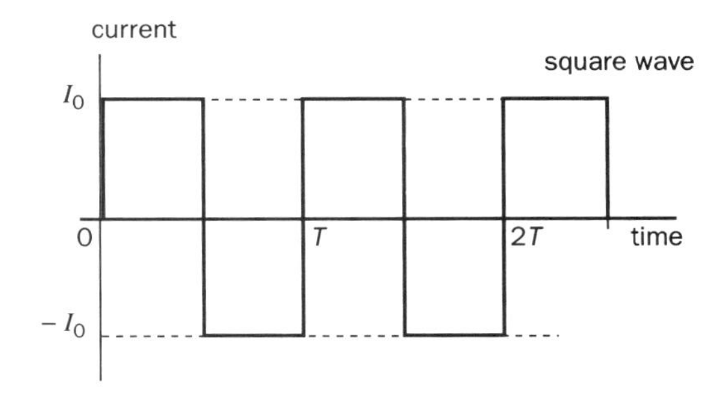

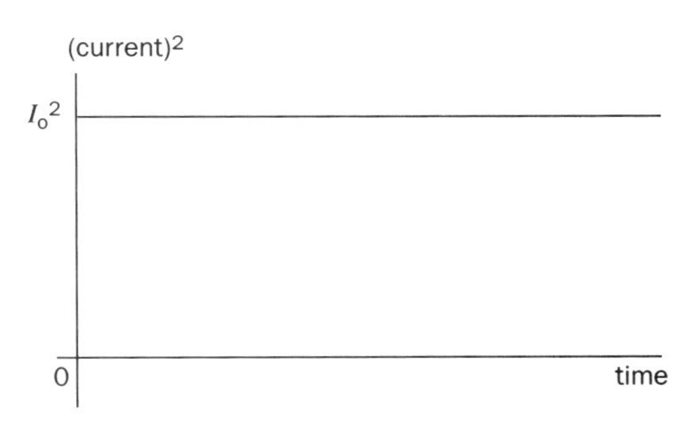

Figure 3.130 Obtaining the r.m.s. values for a square wave

$$\therefore \quad I_{rms} = I_0$$

For a **square** waveform which is **symmetrical about the time axis**, the r.m.s. value is **equal** to the peak value.

This is not the case for waveforms which do not have this symmetry. The following examples illustrate one method for calculating the r.m.s. value:

Guided examples (1)

1. Calculate the r.m.s. (effective) value of the alternating p.d. shown:

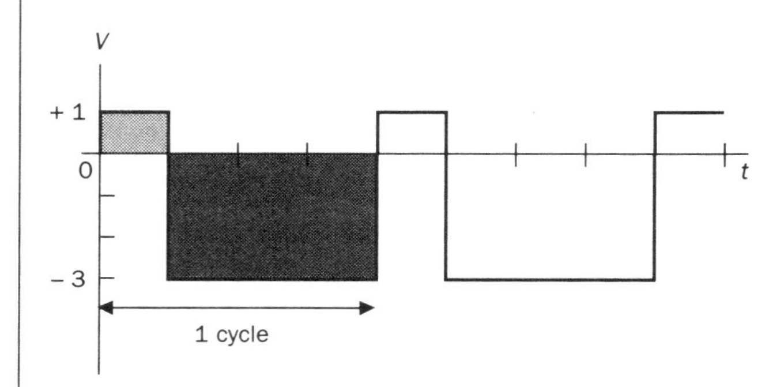

(i) Squaring produces the following waveform:

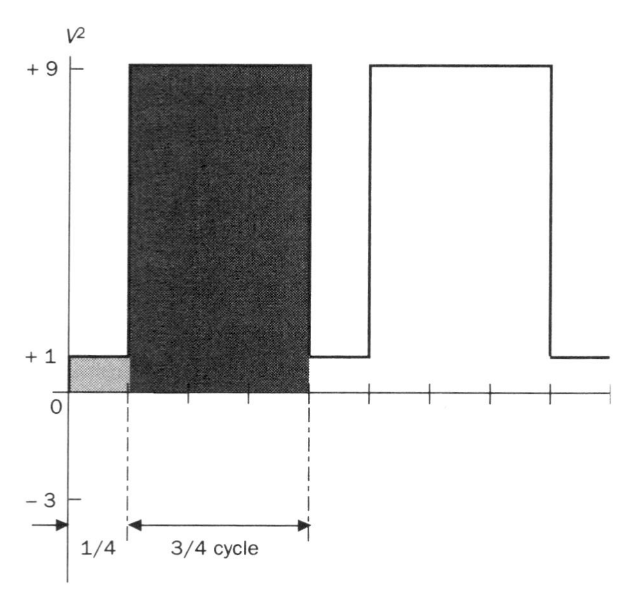

(ii) The **mean** value of V^2 is given by:

$$\overline{V^2} = \left(1 \times \frac{1}{4}\right) + \left(9 \times \frac{3}{4}\right) = 7$$

and

$$V_{rms} = \sqrt{\overline{V^2}} = \sqrt{7} = 2.65 \text{ V}$$

Now try this example for yourself:

2. Calculate the r.m.s. value of the alternating current shown:

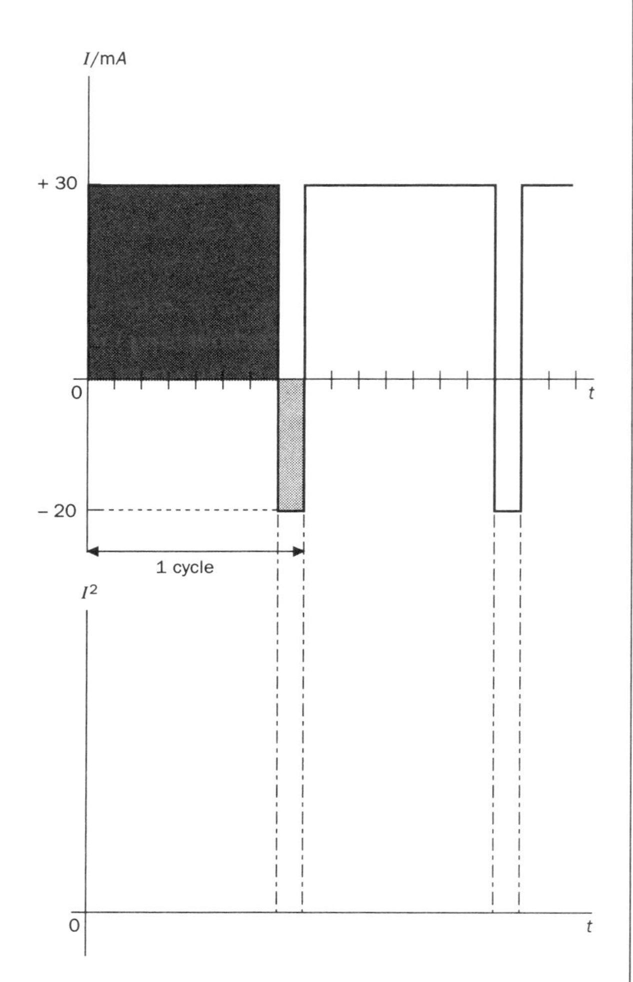

Guidelines

(i) Square the current over one cycle, and sketch the resulting waveform of I^2.

(ii) Use the method outlined in example 1 to obtain the **mean** value of I^2 over one cycle.

(iii) Calculate the square root of this value to obtain I_{rms}.

The transformer

Figure 3.131 shows a simple transformer, which consists of two separate coils of insulated wire wound on a laminated core made from sheets of soft iron separated by an insulating varnish. The sheets are glued together so that they are electrically insulated from each other. The two coils are not linked electrically, the only connection between them being any magnetic field in the core.

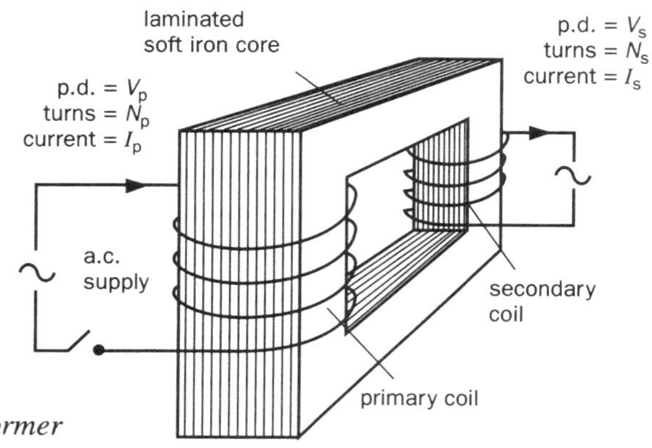

Figure 3.131 Simple transformer

One of the coils is referred to as the **primary coil**, and the other is called the **secondary coil**. The number of turns on each (N_p and N_s) is usually different.

In Section 3.5 on Electromagnetic induction, it was shown that a **changing** current in the primary coil **induces** an e.m.f. in the secondary coil. An **alternating current** in the primary will therefore create an **alternating magnetic field** in the core, which **links** with the turns of the secondary coil. This alternating magnetic field therefore induces an e.m.f. in the secondary, alternating with the **same frequency** as that in the primary. This e.m.f. will cause an alternating current to flow in any external circuit connected to the coil.

Assuming 100% efficiency, it can be shown that the p.d.s across the primary and secondary coils (V_p and V_s respectively) are related to the number of turns of wire (N_p and N_s respectively) as follows:

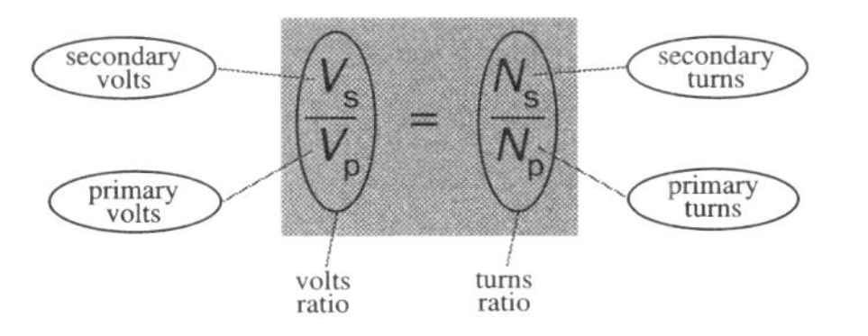

Equation 5

If $N_s > N_p$ then $V_s > V_p$ – a **step-up** transformer
If $N_s < N_p$ then $V_s < V_p$ – a **step-down** transformer

Guided examples (2)

1. The primary coil of a transformer has 500 turns and the secondary coil 2000 turns. A 5 V alternating supply is connected to the primary. Calculate the voltage across the secondary.

2. In order to light a 12 V lamp, a transformer with 2400 turns in the secondary coil is used to step-down the voltage of the a.c. mains (240 V). Calculate the number of turns in the primary coil.

Guidelines

Use Equation 5 in each case.

Power and current in transformers

In a perfectly efficient transformer, there would be no energy losses; i.e. the power delivered to the primary would be equal to the power delivered by the secondary.

800 kV transformers at a large substation in South America

i.e. $P_p = P_s$

$\therefore \quad I_pV_p = I_sV_s$

$$\therefore \quad \frac{I_s}{I_p} = \frac{V_p}{V_s}$$

Equation 6

Combining Equation 5 and Equation 6:

$$\frac{I_s}{I_p} = \frac{V_p}{V_s} = \frac{N_p}{N_s}$$

You should note that:

- N_p, V_p and I_p are the number of turns, p.d. and current respectively in the primary coil and N_s, V_s and I_s are the number of turns, p.d. and current respectively in the secondary coil.
- In practice, the secondary current in the transformer depends on the resistance of the load connected to it, not the turns ratio. The primary current drawn from the supply is controlled by the turns ratio.

Energy losses

The above equations assume 100% efficiency, i.e. all the electrical energy supplied to the primary is converted to electrical energy in the secondary.
In practice, a small amount of energy is dissipated to the surroundings:

- **Coil resistance** causes energy to be dissipated due to the heating effect of current in the coils. This is minimised by using high conductivity copper wire.
- **Eddy currents** are circulating electric currents **within** the iron core of the transformer. These are **induced** in the iron core by the alternating magnetic field. As iron has considerable resistance, energy losses due to heating of the core would be significant if these currents were not minimised. The laminated construction of the core cuts down possible paths for the flow of eddy currents (see Figure 3.132).

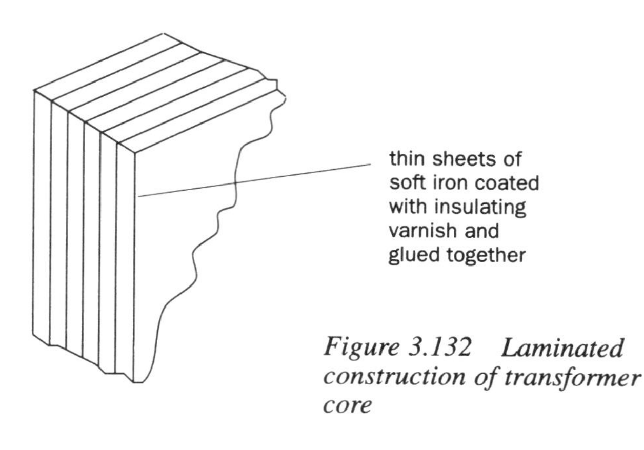

Figure 3.132 Laminated construction of transformer core

- **Magnetic flux leakage**. Some of the changing magnetic flux produced in the primary may not actually link with the secondary coil. The energy loss due to this effect is minimised by appropriate design – the primary and secondary coils are often overlapping, for optimum flux linkage.
- **Hysteresis** is the name given to the reluctance of a material to undergo changes in magnetisation. During each cycle of a.c., the core reverses the polarity of its magnetisation, which requires energy. This is minimised by careful choice of material (with **soft** magnetic properties).

A well-designed transformer should suffer energy losses of only 1% or so.

Transmission of electrical energy

Since *power = p.d. × current*, it is possible to transmit electrical energy at, for example, a rate of 1 million watts using an infinite number of combinations of p.d. and current, some of which are shown in the following table:

P.d. /V		*Current* /A		*Power* /W
1	×	1 000 000	=	1 000 000
10	×	100 000	=	1 000 000
100	×	10 000	=	1 000 000
1 000	×	1 000	=	1 000 000
10 000	×	100	=	1 000 000
100 000	×	10	=	1 000 000
1 000 000	×	1	=	1 000 000

Using transformers, the p.d. of an alternating supply can be stepped-up or stepped-down at will with negligible cost in terms of energy wasted.

The choice of which combination of p.d. and current is best is a compromise based on the following factors.

If the conducting wires have resistance R, and carry a current I, the rate of energy loss due to heating of the wires is equal to I^2R. In other words, the rate of loss of energy is proportional to the **square** of the current (since R is constant). So reducing the current by a half brings a fourfold saving in energy.

However, reducing the current means a corresponding increase in p.d. (to transmit the same power). This increases the cost and complexity of the switch gear used to control the distribution of the supply around the country. It also increases the danger to the public since it is much more difficult to insulate 1 000 000 V than 100 V.

In the UK, voltages up to 400 kV are used in the 'supergrid' while most electricity is transmitted at 132 kV nationally. This is stepped-down progressively to 33 kV and 11 kV for distribution to industry and domestic substations, where it is finally stepped-down to 240 V for homes.

Thus electrical power is transmitted at very high voltage (and hence low current) to minimise power losses due to heating of the transmission cables. Electricity generating stations therefore produce alternating current due to the ease with which its voltage can be stepped-up and stepped-down using transformers.

Guided example (3)

1. Calculate the power loss in transmitting 2 MW of electrical power through cables of total resistance 2 Ω, if the voltage is
   (a) 4000 V
   (b) 400 kV.

Guidelines

Calculate the current:

current = power/voltage

Then calculate the power loss:

$power\ loss = current^2 \times resistance$

Compare the two answers.

Rectification

The major advantage of a.c. is the ease with which its voltage can be changed using transformers. However, many components require a current in one direction only – a **direct** current. Rectification is the process by which a direct current is produced from an alternating current.

Half-wave rectification

Figure 3.133 shows a simple arrangement for rectifying an alternating current. A diode only passes current when it is '*forward-biased*'. With a single diode in series with the load as shown, current can only pass in the direction A to B. This occurs during the half of the cycle when A is more positive than B. When the supply p.d. reverses, the secondary p.d. also reverses. However the diode is no longer forward-biased, and does not conduct. So, for this half of the cycle, there is no current. The current supplied is **direct** current, since its direction does not alter, but the magnitude shows considerable variation, and for half the time, there is no current.

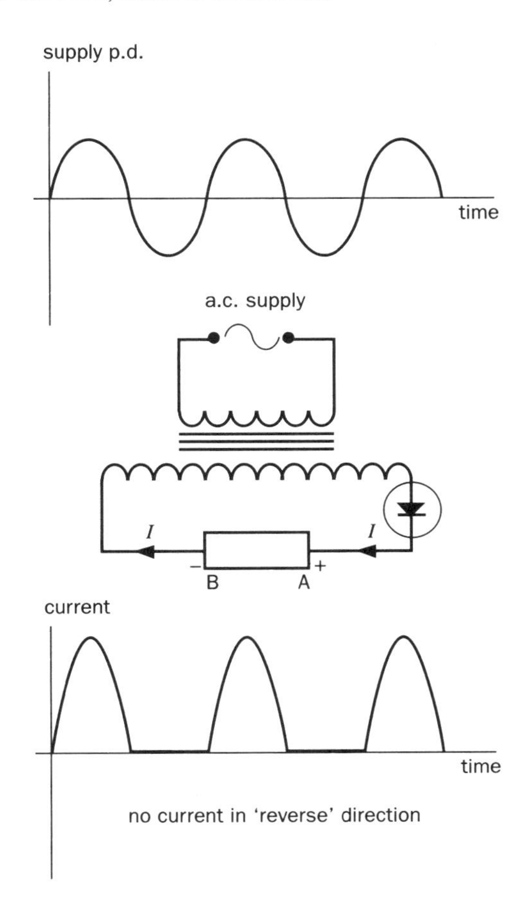

Figure 3.133 Half-wave rectification

A better approximation to a steady direct current is obtained using two diodes.

Full-wave rectification

Two diodes are arranged as shown in Figure 3.134 on page 300 with a centre-tap transformer. Effectively the load is supplied by two separate half-wave rectifiers. During one half of the cycle, only rectifier 1 is forward-biased so it is the only one to conduct, allowing current from A to B through the load. During the second half of the cycle, only diode 2 conducts, but there is still a current through the load from A to B. Thus the output is a fluctuating direct current through the load during *both* halves of each cycle of the a.c. supply.

The full-wave bridge rectifier

Four diode rectifiers 1, 2, 3 and 4 are arranged in a square as shown in Figure 3.135 on page 300. During the first half of a cycle, if U is the more positive terminal, there is a current in the direction U – V – W since diode 1 is forward-biased. Then through the load from A to B, returning to the supply

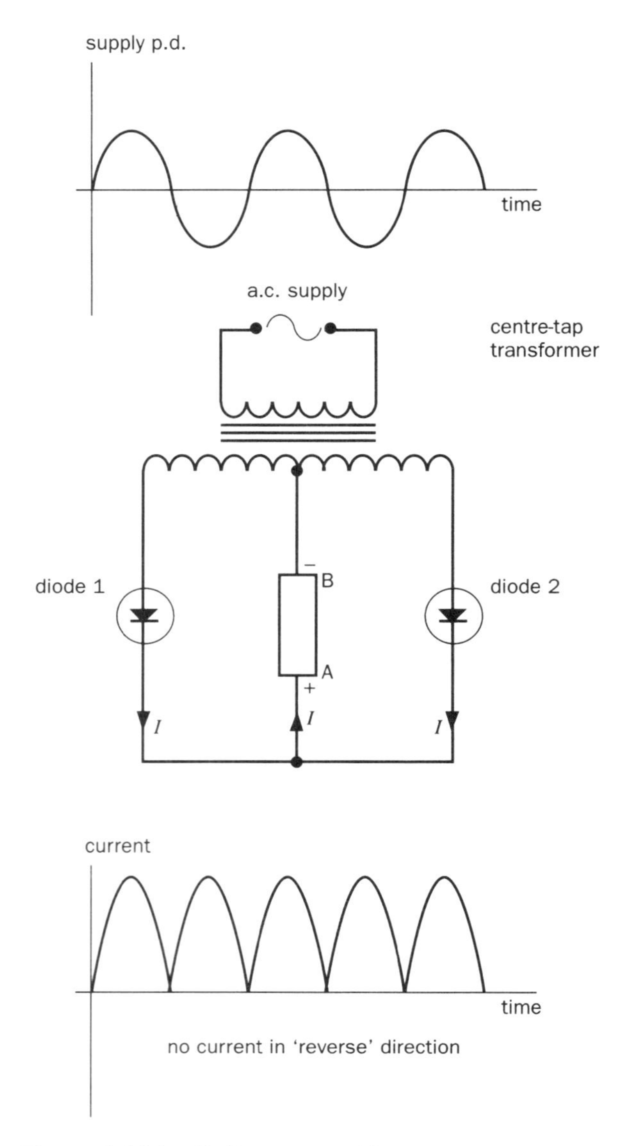

Figure 3.134 Full-wave rectification

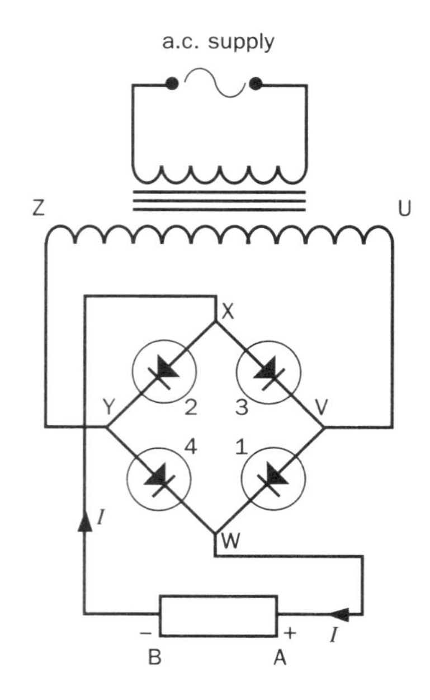

Figure 3.135 Full-wave bridge rectifier

along the path X – Y – Z since diode 2 is also forward-biased. During this half-cycle, diodes 3 and 4 are reverse-biased, and do not conduct.

During the second half, Z becomes more positive, so diodes 3 and 4 are now forward-biased. So the current path becomes Z – Y – W then through the load from A to B as before, returning along the path X – V – U.

Smoothing

A rectified alternating current still fluctuates between zero and a peak value when the above methods are used. In order to minimise the fluctuations, a *smoothing* capacitor (sometimes called a *reservoir* capacitor) can be connected in parallel with the load, as shown in Figure 3.136.

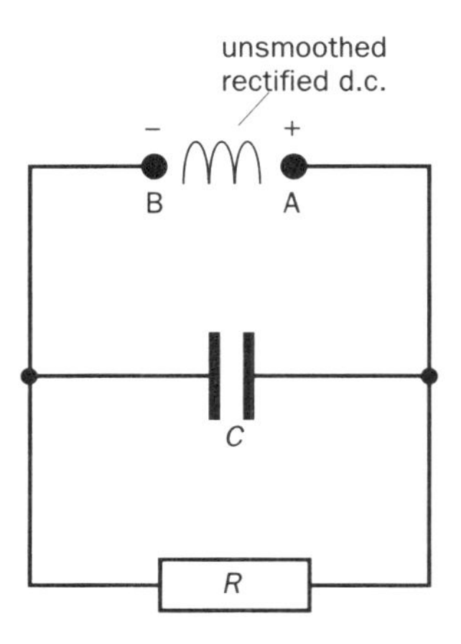

Figure 3.136 Use of capacitor for smoothing

The action of the capacitor can be readily explained by considering Figure 3.137 on page 301.

If the capacitor is sufficiently large, then the time constant of the capacitor and the load resistor (CR) will be large. Consider the sequence of events shown in Figure 3.137. The dashed line is the unsmoothed rectified p.d. The line AB shows the rising p.d. at switch on. The heavy line is the p.d. across the capacitor (and hence the p.d. across the load).

Initially the rising supply p.d. sends a current through R in Figure 3.136, and causes the p.d. across the capacitor to increase to the peak value (almost). This is shown by the line AB in Figure 3.137. As the supply p.d. decreases from its peak, the capacitor cannot discharge through the supply (wrong polarity), only through the load R. The p.d. can only fall slowly (due to the large time constant, CR) shown by the line BC. Meanwhile the supply p.d. has fallen to zero and started to increase again. At point C, the supply p.d. becomes greater than the p.d. across the capacitor, and the capacitor starts to recharge up to the peak p.d. (line

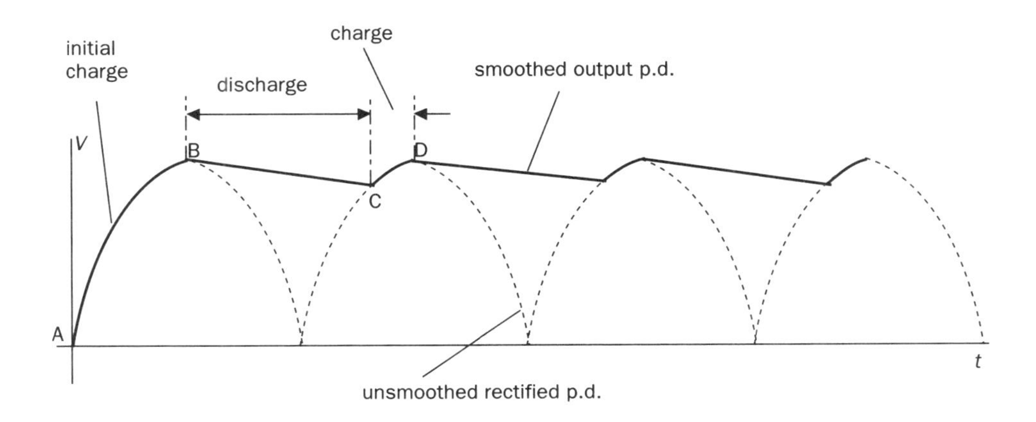

Figure 3.137 Smoothing of the output p.d. from a full-wave rectifier

CD). The process repeats, giving rise to an almost steady p.d., with only a *ripple* of small amplitude. The capacitor must be matched to the output load R to give a suitable time constant. If C is too small, the time constant will be too short, allowing the capacitor to discharge too much between peaks of p.d. If C is too large, it could draw too much current on its first charging cycle, possibly damaging the rectifier.

3.9 A.c. circuits

A.c. in resistive, capacitive and inductive circuits

Alternating current has very different behaviour in circuits containing **inductors** and **capacitors**. There may be a **phase difference** between the p.d. across some components and the current passing through them. It is quite possible that at some instants, the p.d. across a component may be zero when the current is a maximum, and vice versa! To see how this strange state of affairs can be explained, consider an alternating p.d. $V = V_0 \sin \omega t$ applied in turn across a resistor, a capacitor and an inductor.

Pure resistor of resistance *R*

This is the simplest of the three cases, and the relationship between the p.d. across the component and the current through it is straightforward.

At any time (t) the p.d. (V) across the resistor is given by: $V = V_0 \sin \omega t$

And the current I in the circuit is given

by: $I = \frac{V}{R} = \frac{V_0 \sin \omega t}{R}$

i.e. $I = I_0 \sin \omega t$ where $I_0 = \frac{V_0}{R}$

Thus both I and V are **sine** functions which vary with time as shown in Figure 3.138.

- In a purely resistive a.c. circuit the p.d. (V) and the current (I) are **in phase**.
- The opposition to a.c. which this circuit presents is the **resistance**, $R = \frac{V}{I}$.
- The resistance of such a circuit is not affected by the **frequency** of the supply.

Pure capacitor of capacitance *C*

This situation is less simple. As you already know, a capacitor does not conduct electricity. The two plates are separated by an insulator. However, an alternating current does exist in the circuit to which it is connected, as the charge alternately flows on and off each plate in turn. The p.d. across the capacitor depends on the amount of charge present on it – more charge, more p.d. But the current is greatest when there is no charge – i.e. zero p.d.! To examine this in more detail, consider an alternating p.d. applied across a capacitor, as shown in Figure 3.139.

At any time (t) the p.d. (V) across the capacitor is given by:

$$V = V_0 \sin \omega t$$

and the charge (Q) at this instant is given by:

$$Q = CV = CV_0 \sin \omega t$$

therefore the current (I) at this instant is given by:

$$I = \frac{dQ}{dt}$$

$$= \frac{d}{dt}(CV_0 \sin \omega t)$$

$$= CV_0 \frac{d}{dt}(\sin \omega t)$$

$$= \omega CV_0 \cos \omega t$$

$$\therefore \quad I = I_0 \cos \omega t \quad (\text{where } I_0 = \omega CV_0)$$

This shows that V is a **sine** function while I is a **cosine** function, and they vary with time as shown in Figure 3.139(b). So we see that the p.d. is a maximum when the current is a minimum. This can be understood if we remember that the charge flowing into a capacitor is only opposed by the p.d. across it. When the capacitor is uncharged, there is no p.d. and therefore no opposition to current, so the current is maximum. When the capacitor is fully charged, the p.d. across it reduces the current to zero.

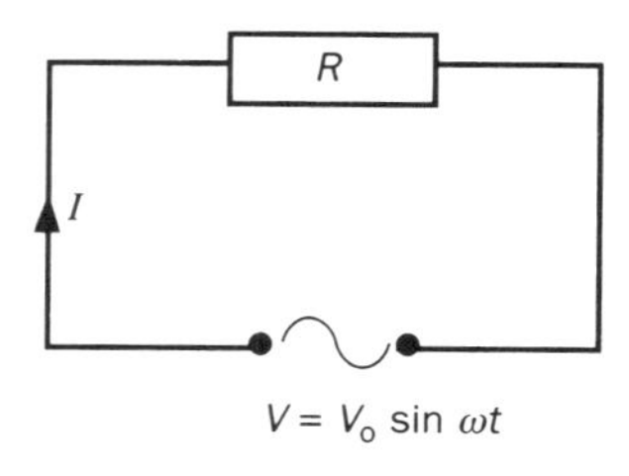

(a) Circuit

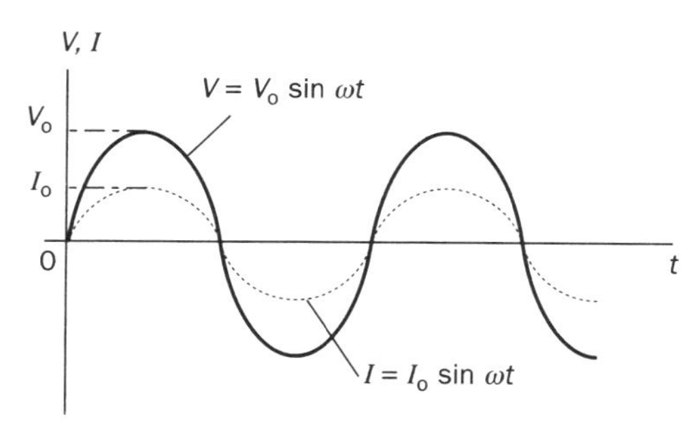

(b) Variation of p.d. and current with time

Figure 3.138 Alternating p.d. applied across a pure resistor

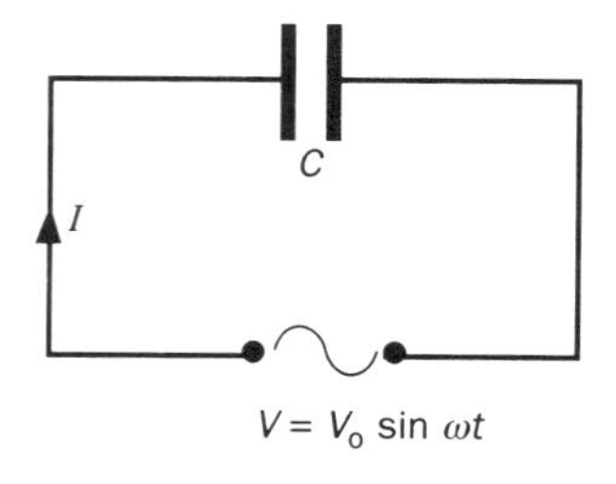

(a) Circuit

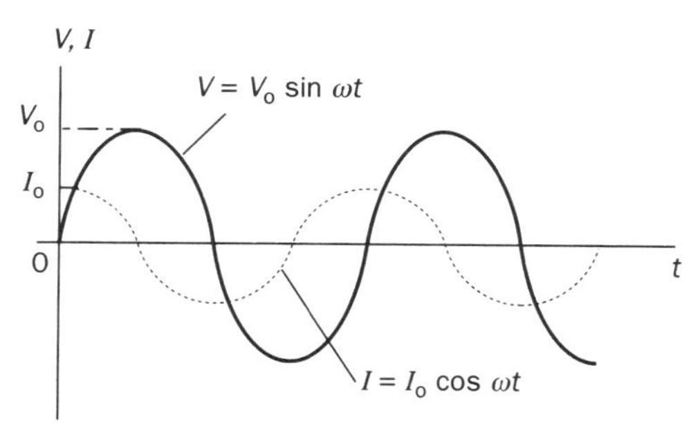

(b) Variation of p.d. and current with time

Figure 3.139 Alternating p.d. applied across a pure capacitor

- In a purely capacitive circuit, the current (I) **leads** the p.d. by 90° (π/2 rad) – current must flow into the capacitor before a p.d. can be developed across it.
- The opposition to flow of a.c. produced by a pure capacitance is known as the **capacitive reactance** (X_C).
- $X_C = \frac{V_{rms}}{I_{rms}} = \frac{V_0}{I_0} = \frac{V_0}{\omega C V_0} = \frac{1}{\omega C}$

  and since $\omega = 2\pi f$, where f is the a.c. frequency:

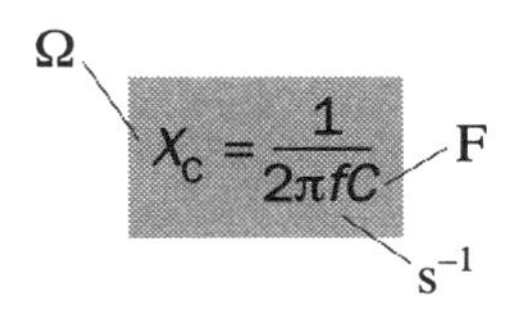

Equation 1

You should note that:

- X_C decreases with increasing frequency and capacitance. At high frequencies, there is little time for a large p.d. to build up on the capacitor. Similarly, larger capacitors can store more charge for a given rise of p.d., so they present less opposition to the current than smaller ones.
- $X_C = \frac{1}{2\pi f C} = \frac{1}{2\pi f Q/V}$

  units: $\frac{1}{s^{-1} C V^{-1}} = \frac{V}{A} = \Omega$

Pure inductor of inductance *L*

A **pure** inductor is one which has zero resistance to direct current – i.e. no p.d. is developed across it due to a direct current. An alternating current, however, causes an alternating magnetic field, which induces a **back e.m.f.**, which presents opposition to the current. The back e.m.f. will be a maximum when the **rate of change** of current is a maximum. The rate of change of current is a maximum at the instant that it is changing from positive to negative, i.e. when it is **zero**. Therefore we have another situation in which current and p.d. are out of phase, as shown in Figure 3.140.

The alternating current $I = I_0 \sin \omega t$ through the inductor sets up a varying magnetic flux which links up with the coil and induces a back e.m.f. in it whose value is given by

$$E = -L\frac{dI}{dt}$$

If V = applied p.d. at any time t then applying Kirchhoff's second law, we have that:

$$V - L\frac{dI}{dt} = 0$$

$$V - L\frac{d}{dt}(I_0 \sin \omega t) = 0$$

$$V - \omega L I_0 \cos \omega t = 0$$

$$\therefore \quad V = \omega L I_0 \cos \omega t$$

$$\therefore \quad V = V_0 \cos \omega t \quad (\text{where } \omega L I_0 = V_0)$$

In this case V is a **cosine** function and I is a **sine** function. They vary with time as shown in Figure 3.140(b).

- In a purely inductive a.c. circuit the p.d. (V) **leads** the current (I) by 90° (π/2 rad).
- The opposition to a.c. presented by a pure inductance is known as the **inductive reactance** X_L.
- $X_L = \frac{V_{rms}}{I_{rms}} = \frac{V_0}{I_0} = \frac{\omega L I_0}{I_0} = \omega L$

  (and since $\omega = 2\pi f$)

$\therefore$ 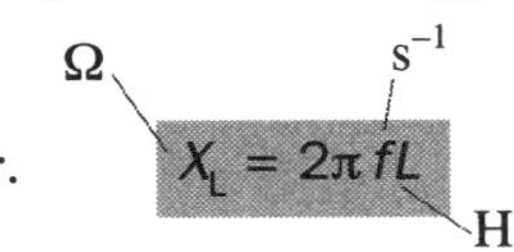

Equation 2

You should note that:

- X_L **increases** with increasing frequency and inductance. Higher frequencies mean a higher rate of change of current, and therefore higher back e.m.f. Higher inductance also causes a higher back e.m.f., which leads to a higher opposition to current – higher **reactance**.

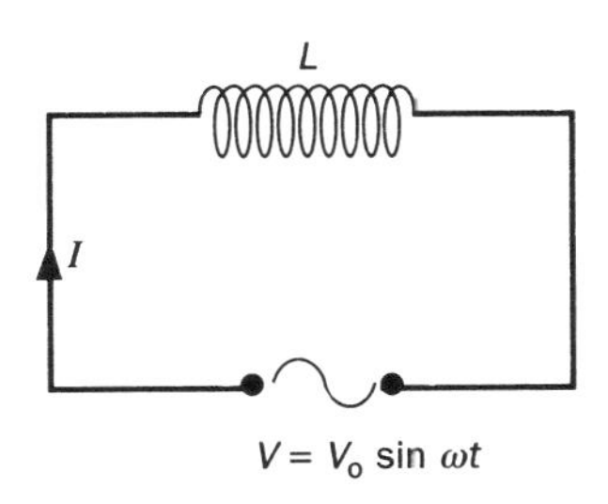

(a) Circuit

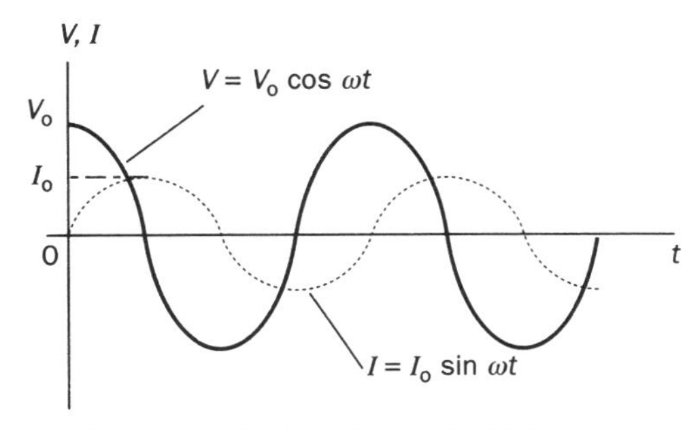

(b) Variation of p.d. and current with time

Figure 3.140 Alternating p.d. across a pure inductor

- $X_L = 2\pi f L = 2\pi f \dfrac{E}{dI/dt}$

units: $\dfrac{s^{-1}V}{A\,s^{-1}} = \dfrac{V}{A} = \Omega$

Maths window

If $y = \sin x$

then $\dfrac{dy}{dx} = \cos x$

If $y = \sin Ax$ (where A is a constant)

then $\dfrac{dy}{dx} = A \cos Ax$

Phasor diagrams

In this type of diagram the alternating current in an a.c. circuit and the p.d. across the circuit components are represented by **vectors** (or **phasors**) which show the phase relationship between the two quantities. One of the quantities (usually the current) is drawn as the **reference** vector. The other quantities (the p.d.s) are drawn as vectors at angles representing their **phase difference** from the reference vector.

Figure 3.141 shows the phasor diagram for a.c. circuits which are: (a) purely resistive, (b) purely capacitive and (c) purely inductive. In (b) and (c), notice that the current and p.d. vectors are drawn at 90°, since the **phase difference** between p.d. and current is 90°.

R–C series circuit

Consider an alternating p.d. (V) of frequency (f) applied across a resistor of resistance (R) and a capacitor of capacitance (C) connected in series as shown in Figure 3.142(a). The phasor diagram for this circuit is shown in Figure 3.142(b). In any **series** circuit the current (I) in the circuit is the same through each component and it is therefore used as the **reference phasor**.

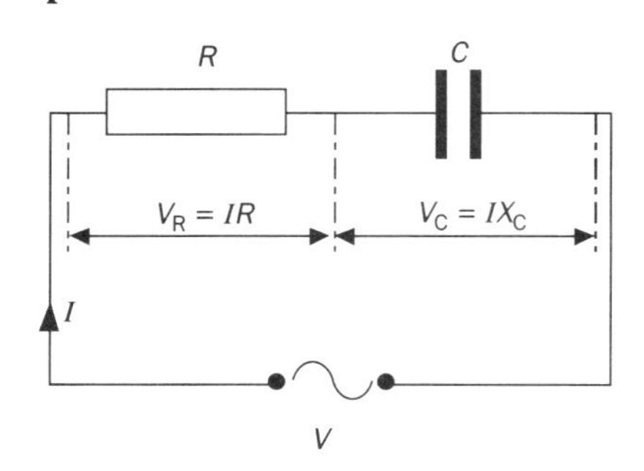

(a) R–C series circuit

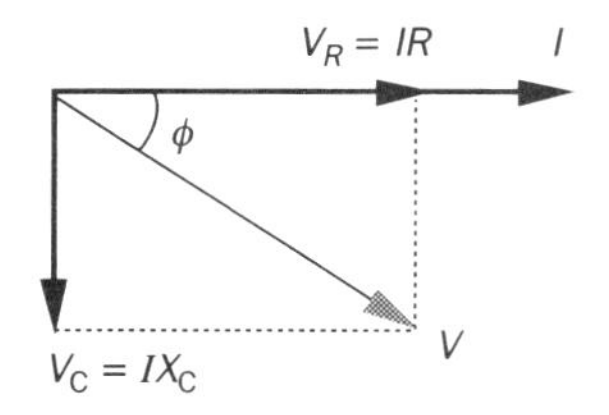

(b) Phasor diagram

Figure 3.142 R–C series circuit

Applying Pythagoras's theorem to the phasor diagram we have:

$$V^2 = V_R^2 + V_C^2 = I^2R^2 + I^2X_C^2 = I^2(R^2 + X_C^2)$$

$$\therefore \quad V = I\sqrt{R^2 + X_C^2}$$

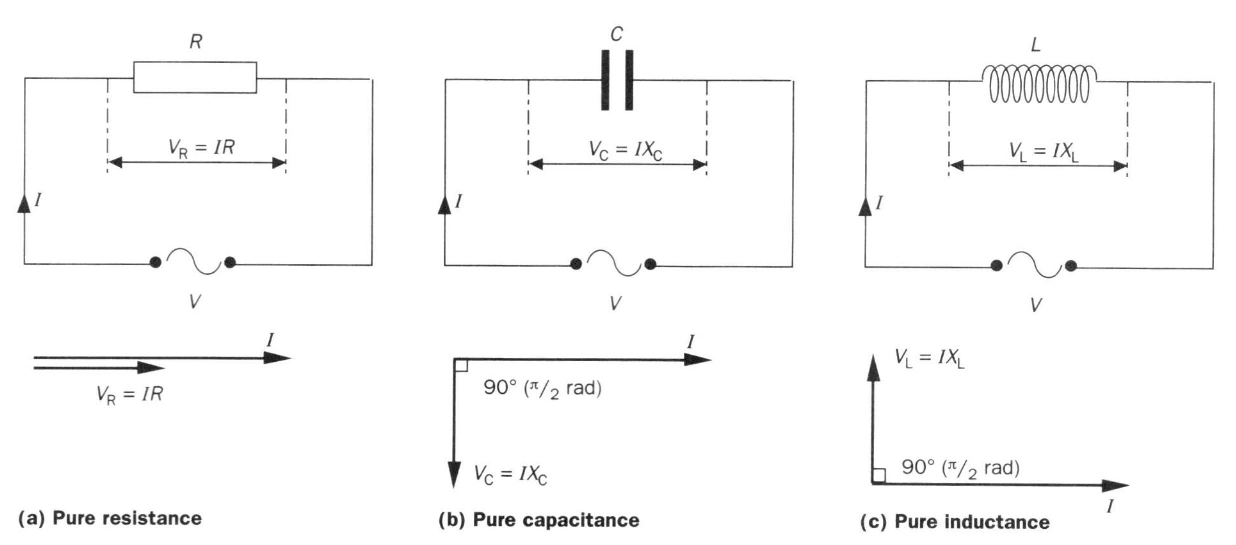

(a) Pure resistance **(b) Pure capacitance** **(c) Pure inductance**

Figure 3.141 Phasor diagrams

from which we can define the **impedance** (Z) of the R–C circuit as:

$$Z = \frac{V}{I} = \sqrt{R^2 + X_C^2}$$

Equation 3

You should note that:

- $X_C = \dfrac{1}{2\pi fC}$
- The impedance, (Z) is the total **opposition** to a.c. flow due to **resistance** (R) and **capacitive reactance** (X_C) which the circuit presents – Z is in **ohms** (Ω).
- The current (I) in this circuit **leads** the applied p.d. (V) by a phase angle (ϕ) given by:

$$\tan \phi = \frac{V_C}{V_R} = \frac{IX_C}{IR}$$

$$\therefore \quad \phi = \tan^{-1} \frac{X_C}{R}$$

Equation 4

R–L series circuit

In this case an alternating p.d. (V) of frequency (f) is applied across a resistor of resistance (R) and an inductor of inductance (L) connected in series. Figure 3.143 shows the circuit and its phasor diagram.

From the phasor diagram we have:

$$V^2 = V_R^2 + V_L^2 = I^2R^2 + I^2X_L^2 = I^2 (R^2 + X_L^2)$$

$$\therefore \quad V = I\sqrt{R^2 + X_L^2}$$

from which we can define the **impedance** (Z) of the R–L circuit as:

$$Z = \frac{V}{I} = \sqrt{R^2 + X_L^2}$$

Equation 5

You should note that:

- $X_L = 2\pi fL$
- The impedance (Z) in this case is the total opposition to a.c. due to resistance and inductive reactance.

The current (I) in this circuit lags behind the applied p.d. (V) by a phase angle (ϕ) given by:

$$\tan \phi = \frac{V_L}{V_R} = \frac{IX_L}{IR}$$

$$\therefore \quad \phi = \tan^{-1} \frac{X_L}{R}$$

Equation 6

R–L–C series circuit

The alternating p.d. V is now applied to a resistor, a capacitor and an inductor connected in series as shown in Figure 3.144 on page 306.

From the phasor diagram we have:

$$V^2 = V_R^2 + (V_L - V_C)^2 = I^2R^2 + I^2 (X_L - X_C)^2$$
$$= I^2\left[R^2 + (X_L - X_C)^2\right]$$

$$\therefore \quad V = I\sqrt{R^2 + (X_L - X_C)^2}$$

from which we can define the **impedance** (Z) of the R–L–C circuit as:

$$Z = \frac{V}{I} = \sqrt{R^2 + (X_L - X_C)^2}$$

Equation 7

The impedance (Z) in this case is the total opposition to a.c. due to resistance, inductive reactance and capacitive reactance.

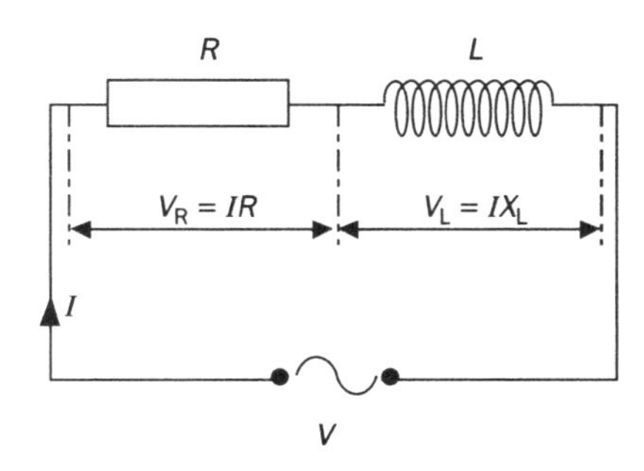

(a) R–L series circuit

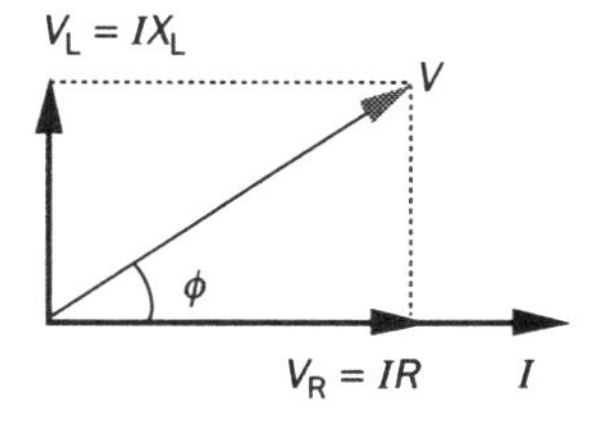

(b) Phasor diagram

Figure 3.143 R–L series circuit

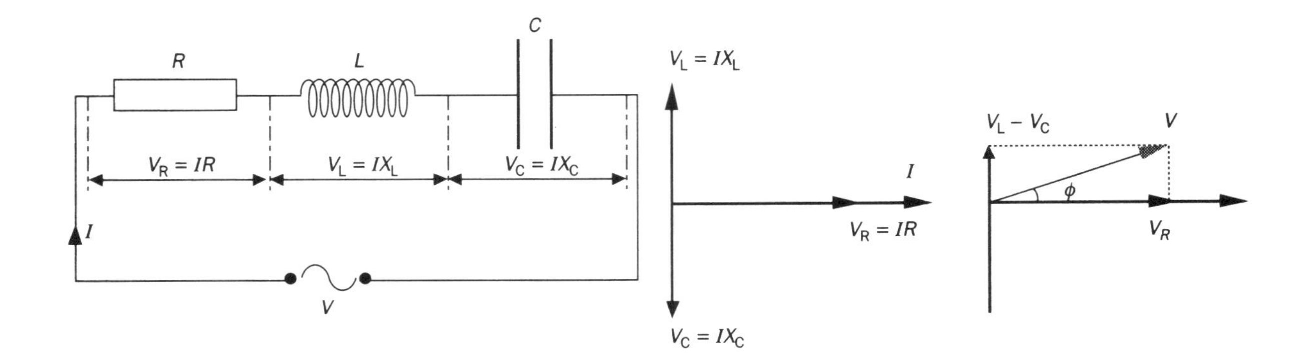

Figure 3.144 R–L–C series circuit

The current (I) in this circuit lags behind the applied p.d. (V) by a phase angle (ϕ) given by:

$$\tan \phi = \frac{V_L - V_C}{V_R} = \frac{I(X_L - X_C)}{IR}$$

$$\therefore \quad \phi = \tan^{-1} \frac{(X_L - X_C)}{R}$$

Equation 8

You should note that:

- If $V_C > V_L$, the current (I) **leads** the applied p.d. (V) by angle (ϕ).
- If $V_C < V_L$, the current (I) **lags behind** the applied p.d. (V) by angle (ϕ).
- In order to take account of the fact that in some cases $V_L > V_C$ and in others $V_C > V_L$, Equation 7 and Equation 8 are written as:

$$Z = \frac{V}{I} = \sqrt{R^2 + |X_L - X_C|^2}$$

$$\phi = \tan^{-1} \frac{|X_L - X_C|}{R}$$

where $|X_L - X_C|$ means 'the difference between' X_L and X_C.

Resonance of R–L–C series circuit

For an R-L-C series circuit, the total impedance (Z) is given by:

$$Z = \sqrt{R^2 + |X_L - X_C|^2}$$

We want to consider how R, X_L, X_C and Z vary with the frequency (f) of the applied p.d.

Inductive reactance, $X_L = 2\pi f L$ i.e. $X_L \propto f$

Capacitive reactance, $X_C = \dfrac{1}{2\pi f C}$ i.e. $X_C \propto \dfrac{1}{f}$

Resistance (R) is independent of f.

Impedance (Z) varies with the frequency f since it involves both X_L and X_C.

Z has its minimum value when $X_L = X_C$.

Then: $Z = R$

The current (I) in the circuit then has its maximum value:

$$I = \frac{V}{Z} = \frac{V}{R} \quad (\text{when } X_L = X_C)$$

Figure 3.145 shows the variation of R, X_L, X_C and Z with frequency f.

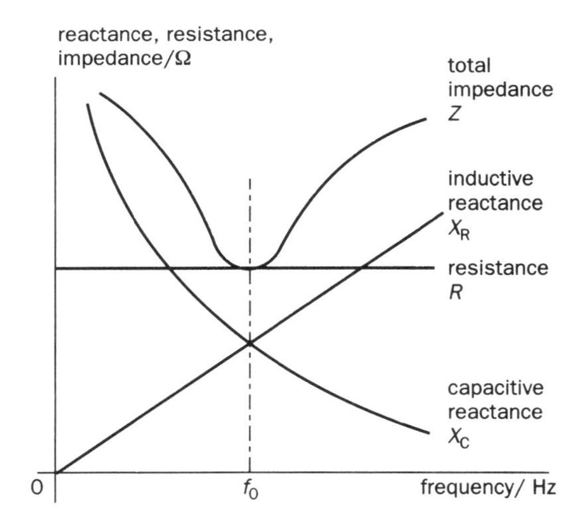

Figure 3.145 Variation of R, X_L, X_C and Z with frequency of applied p.d.

The minimum value of Z (= R) occurs at a certain value of frequency called the **resonant frequency** (f_0). The circuit then behaves as if it were purely **resistive** and the current flowing in it has its **maximum** value (see Figure 3.146).

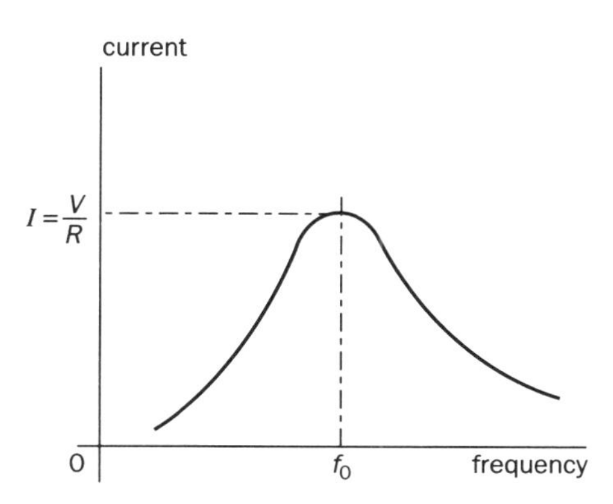

Figure 3.146 Variation of current with frequency in an R-L-C series circuit

The phasor diagram for a series resonant circuit is shown in Figure 3.147.

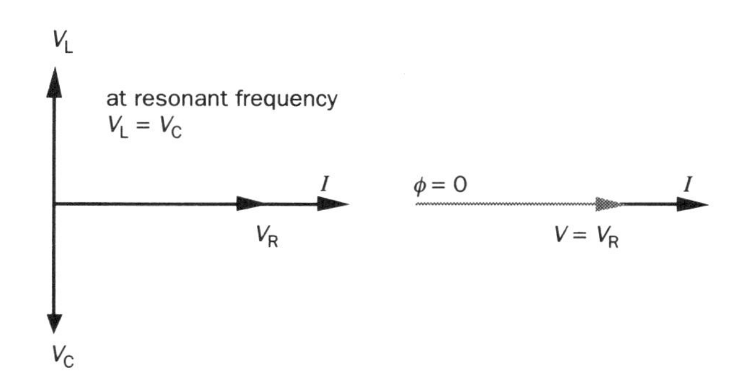

Figure 3.147 R–L–C series circuit phasor diagram

You should note that when $f = f_0$:

$X_L = X_C$ and hence $V_L = V_C$

From which, supply p.d. $V = V_R$

Phase angle $\phi = 0$ (i.e. the current (I) and the applied p.d. (V) are **in phase** with each other)

The p.d. across the inductor (V_L) and that across the capacitor (V_C) may be much greater than the applied p.d. (V).

$$I = \frac{V}{R}$$

Then $V_L = IX_L = \frac{V}{R} X_L$

and $V_C = IX_C = \frac{V}{R} X_C$

Since R is usually small compared with X_L (or X_C), V_L (or V_C) can be very much greater than V.

The resonance frequency f_0 is given by:

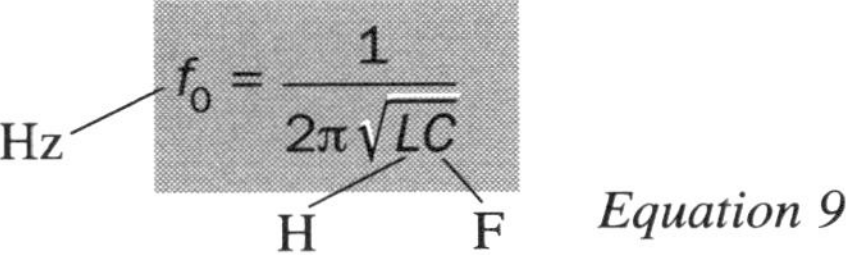

$$f_0 = \frac{1}{2\pi\sqrt{LC}}$$ *Equation 9*

Equation 9 is derived as follows:

At the resonance frequency (f_0):

$$X_L = X_C$$

$$2\pi f_0 L = \frac{1}{2\pi f_0 C}$$

$$f_0^2 = \frac{1}{4\pi^2 LC}$$

$$\therefore \quad f_0 = \frac{1}{2\pi\sqrt{LC}}$$

Guided examples (1)

1. A 200 Ω resistor is connected in series with a 0.20 H inductor. If an alternating p.d of 48 V and frequency 50 Hz is applied across the arrangement, calculate:

(a) (i) the inductive reactance of the inductor

(ii) the impedance of the circuit

(iii) the current in the circuit

(iv) the p.d. across the resistor and the inductor.

(b) Draw a phasor diagram for the circuit and hence calculate the phase angle between the current and the applied p.d.

Guidelines

(a) (i) Use Equation 2 to calculate X_L.

(ii) Use Equation 5 to calculate Z.

(iii) Use Equation 5 to calculate I.

(iv) $V_R = IR$ and $V_L = IX_L$.

(b) Draw the phasor diagram – remember that V_L leads I by 90° and V_R is in phase with I. Then use Equation 6 to find ϕ.

2.

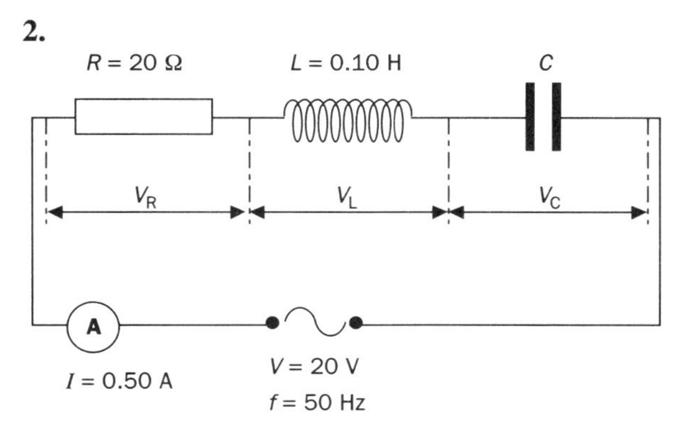

Guided example continued on next page

Guided examples (1) *continued*

The a.c. ammeter in the circuit shown above reads 0.50 A.

(a) Calculate:

(i) the impedance

(ii) the inductive reactance

(iii) the capacitive reactance

(iv) the value of C in μF

(v) the p.d. across each component (i.e. V_R, V_L and V_C).

(b) Draw a phasor diagram for the circuit and use it to calculate the phase angle between the current and the applied p.d.

Guidelines

(a) (i) Calculate Z directly from the p.d. and current values given (Ohm's law!).

(ii) Use Equation 2 to calculate X_L.

(iii) Use Equation 7 to calculate X_C as it is the only unknown.

(iv) Use the value of X_C you have just obtained in (iii) to find C.

(v) Use the fact that $V_R = IR$, $V_L = IX_L$ and $V_C = IX_C$.

(c) Remember that V_R is in phase with I, V_L leads I by 90° and V_C lags behind I by 90°.

3. A circuit consists of a 150 Ω resistor, a 4.0 μF capacitor and a 0.50 H inductor connected in series.

(a) At what frequency will the circuit impedance be a minimum?

(b) If the p.d. applied to the arrangement is 20 V r.m.s. at this resonant frequency, calculate:

(i) the current in the circuit

(ii) the p.d. across each circuit component.

(c) Draw a phasor diagram for the circuit and show numerical p.d. values on it. What is the phase angle between the current and the applied p.d.?

Guidelines

(a) The impedance of the circuit is a minimum at the resonant frequency. Use Equation 9.

(b) (i) At the resonance frequency, the impedance is a minimum ($Z = R$) and the current has its maximum value.

(ii) $V_R = IR$; $V_L = V_C = IX_L = IX_C$.

(c) Remember that V_R is **in phase** with the current, V_L **leads** the current by 90° and V_C **lags** the current by 90°.

Self-assessment

SECTION A
Qualitative assessment

1. Distinguish between **direct** and **alternating** current.

2. A sinusoidal, alternating current can be represented by the equation

$$I = I_0 \sin \omega t$$

What does each term in this equation represent?

3. (a) Explain what is meant by the **root-mean-square** (r.m.s.) value of an alternating current.

(b) What is the relationship between the **r.m.s.** and **peak** values of

(i) a sinusoidal alternating current, and

(ii) a square wave alternating current?

4. (a) An alternating e.m.f. is induced in the secondary coil of a transformer when an alternating p.d. is applied to the primary coil. Explain how this happens.

(b) Why are the coils wound on a soft iron core?

(c) What is the relationship between the number of turns on each coil and

(i) the primary and secondary **voltages**

(ii) the primary and secondary **currents**?

5. State the main causes of energy loss in a transformer and explain how they can be minimised.

6. Why is electrical power transmitted at very high voltage?

Self-assessment *continued*

7. (a) What is **rectification** ?

(b)

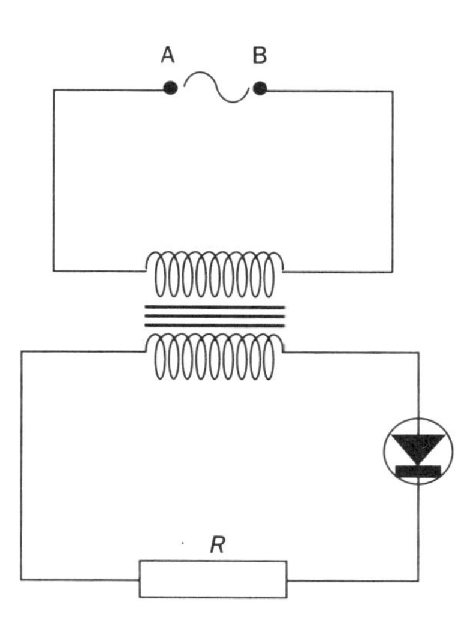

For the circuit shown above describe how the p.d. across R will vary with time when an alternating p.d. is applied between A and B.

(c) The arrangement in (b) gives **half-wave** rectification. Explain how two diodes can be used to produce **full-wave** rectification of an alternating p.d.

(d) The diagram below shows the circuit of a full-wave bridge rectifier. Explain how it works.

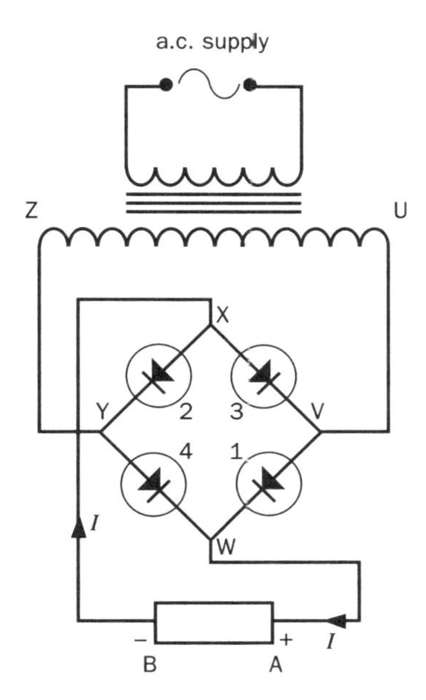

(e) Explain how a capacitor can be used to smooth the output from the full-wave bridge rectifier.

8. (a) What is the phase relationship between applied p.d. and current in an a.c. circuit which is purely:

(i) **resistive**

(ii) **capacitive**

(iii) **inductive**.

(b) Use phasor diagrams to show the phase relationships in each of the cases in (a).

9. (a) Define: (i) **capacitive reactance X_C** and

(ii) **inductive reactance X_L**.

(b) What are the units of X_C and X_L?

(c) How does (i) X_C, and (ii) X_L vary with the **frequency** of the applied p.d.?

10.

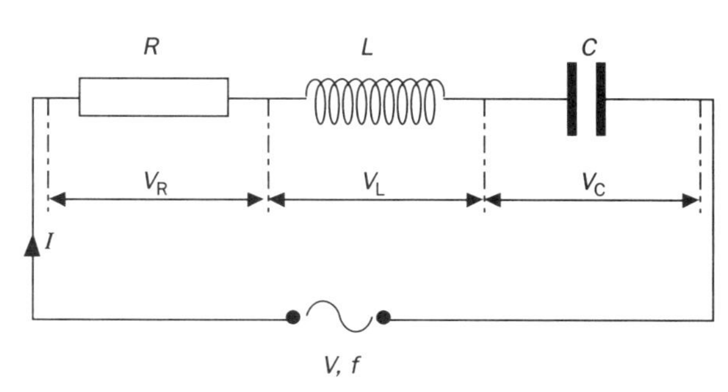

The R–L–C series circuit shown above is supplied by an alternating p.d. of r.m.s. value V and frequency f.

(a) Taking the direction of the current phasor as being the reference direction, draw phasors for V_R, V_L and V_C and hence obtain a value for their resultant.

(b) Explain what is meant by the **impedance** of an a.c. circuit and give an expression for Z in terms of R, X_L and X_C.

(c) What is the **phase angle** between the resultant p.d. and the current?

11. An a.c. circuit consists of a resistor R, an inductor L and a capacitor C connected in series with an alternating p.d. of variable frequency applied across the arrangement.

(a) How does

(i) the resistance

(ii) the inductive reactance X_L and

(iii) the capacitive reactance X_C

of the circuit vary with the **frequency** of the applied p.d.?

Self-assessment *continued*

(b) What is the relation between X_L and X_C when the impedance Z of the circuit is a **minimum**?

(c) At what value of frequency does the current in the circuit have its **maximum** value? Derive an expression for this frequency in terms of L and C.

(d) Draw a phasor diagram for the circuit when the current is a **maximum**.

SECTION B
Quantitative assessment

(***Answers:*** 0; 0; 0.016; 0.045; 0.050; 0.15; 0.25; 0.32; 2.5; 4.4; 6.3; 7.9; 8.5; 8.8; 10; –12; 16; 16; 30; 35; 50; 58; 73; 78; 95; 96; 99; 130; 250; 250; 340; 420; 500; 640; 670; 750; 780; 1.5×10^3; 3.0×10^3; 7.8×10^3; 1.2×10^4; 4.2×10^5; 7.8×10^5; 1.2×10^6.)

1. An electric toaster is rated at 240 V, 1500 W (r.m.s. values). Calculate the:

(a) r.m.s. current
(b) peak current
(c) peak p.d.
(d) peak power
(e) average power.

2. A coil is rotated at constant angular velocity in the uniform magnetic field between two permanent magnet poles. The peak-to-peak value of the e.m.f. induced in the coil is 24 V. Calculate:

(a) the r.m.s. value of the e.m.f.
(b) the instantaneous value of the e.m.f.
 (i) 1/2 a period
 (ii) 3/4 of a period
 after the e.m.f. is a maximum.

3. Calculate the r.m.s. value of the alternating quantities whose variation with time is as shown in each of the diagrams below.

4. A generating station supplies electrical power to an industrial complex through thick cables having a total resistance of 0.78 Ω. The station has an output p.d. of 12 kV and is situated 20 km from the complex. If the power is supplied by a current of 100 A, calculate:

(a) the p.d. across the connecting cables
(b) the p.d. at the complex
(c) the power loss incurred in transmission
(d) the power delivered to the complex
(e) the percentage of the station's power output which is delivered to the complex.

Repeat the question assuming that the station's output p.d. is 1.2 kV and the current is 1000 A.

5. (a) An ideal transformer has a 5000 turn primary winding and is used to provide 24 V r.m.s. when connected to the 240 V mains supply How many turns are there in the secondary winding?

(b) If the output terminals of the transformer in (a) are connected to a device rated at 24 V, 60 W, calculate:
 (i) the current in the secondary circuit
 (ii) the current in the primary circuit.

(c) In practice it is found that the secondary p.d. and current provided by the transformer are

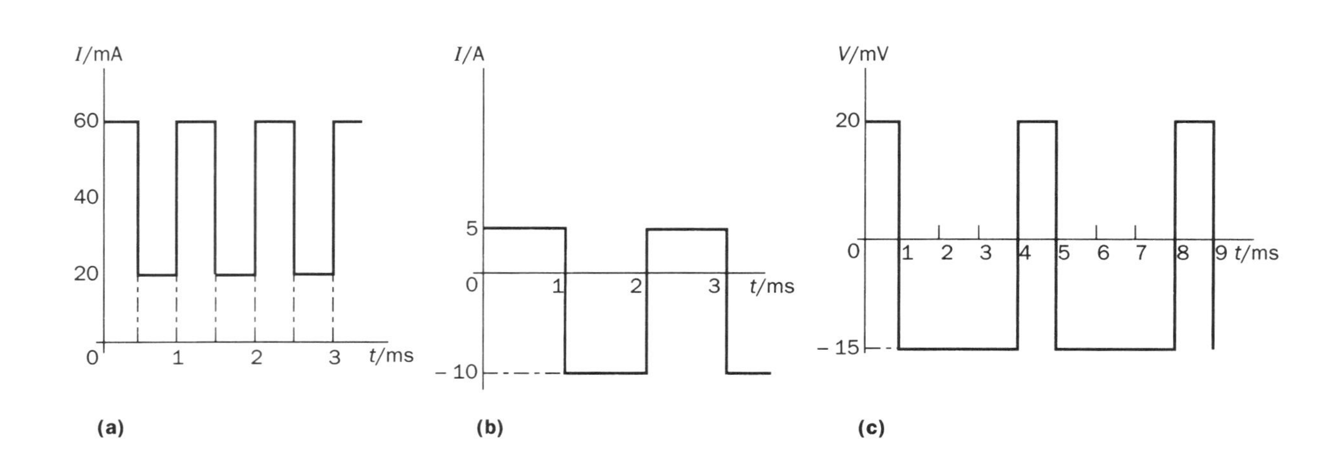

(a) (b) (c)

Self-assessment *continued*

24 V and 2.4 A r.m.s. when the primary current is 0.25 A. Calculate the percentage efficiency of the transformer.

6.

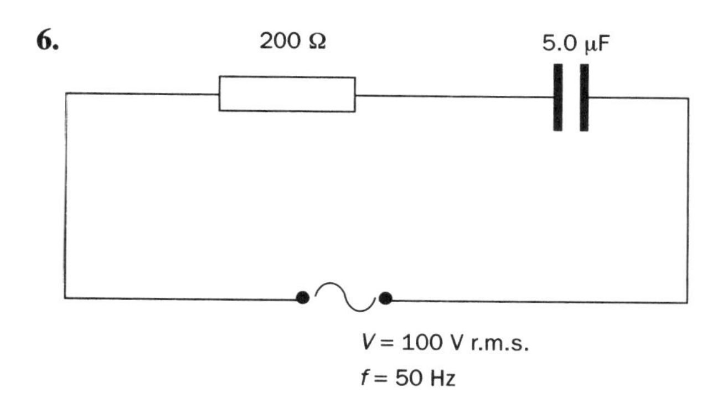

The diagram shows a 200 Ω resistor and a 5.0 μF capacitor connected in series with an alternating voltage source of negligible impedance. Calculate:

(a) the reactance of the capacitor
(b) the impedance of the circuit
(c) the r.m.s. value of the current
(d) the p.d. across each component
(e) the mean rate at which energy is supplied by the source.

Draw a phasor diagram for the circuit and use it to find the phase difference between the supply p.d. and the current in the circuit.

7.

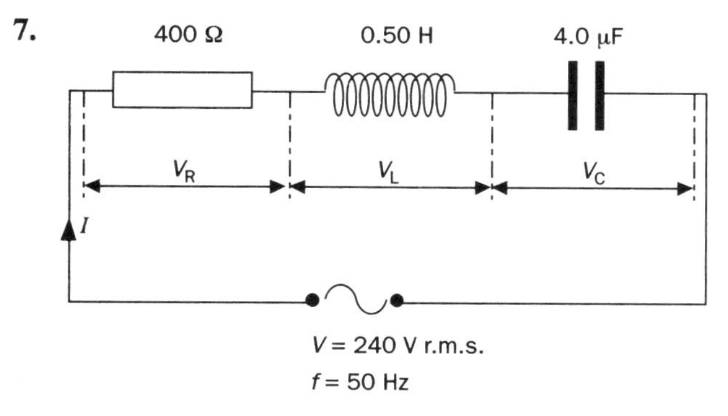

For this R–L–C series circuit, calculate:

(a) the impedance
(b) the r.m.s. value of the current
(c) the p.d. across each component (V_R, V_L and V_C).

Draw a phasor diagram and use it to find the phase difference between the supply p.d. and the current in the circuit.

8.

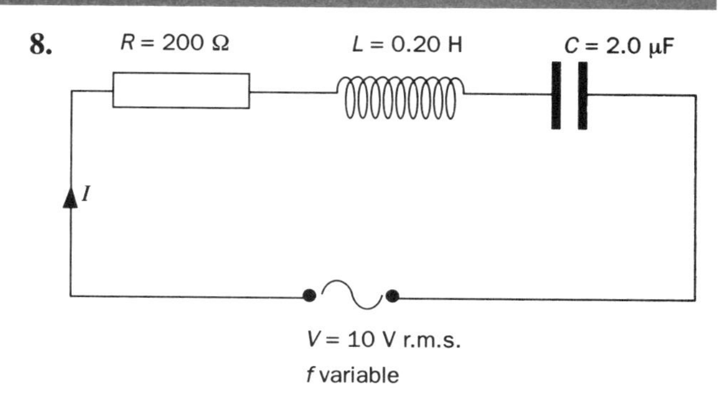

A 200 Ω resistor of negligible inductance, a 0.20 H inductor having a resistance of 5.0 Ω and a 2.0 μF capacitor are connected in series as shown.

The a.c. output of a signal generator is set at 10 V r.m.s. and connected across the combination as shown.

(a) Starting with the frequency at a low value, it is increased until the current in the circuit is a maximum. Calculate:
   (i) the frequency at which the current is a maximum
   (ii) the maximum r.m.s. current in the circuit.

(b) (i) Draw a phasor diagram for the circuit when the current is a maximum and calculate the phase difference between the supply p.d. and the current in the circuit.
   (ii) Calculate the p.d. across each component when the current is a maximum.

Past-paper questions

1 A car driver parks his car with two 60 W headlamps and four 6 W sidelamps still switched on. The six lamps are connected in parallel and powered by a 12 V battery.

Explain the phrase *connected in parallel.*

Calculate the current in the battery.

(Lond., June 1994)

2 (a) State Ohm's law.

(b) The following are four electrical components:

A a component which obeys Ohm's law

B another component which obeys Ohm's law but which has a higher resistance than A

C a filament lamp

D a component, other than a filament lamp, which does not obey Ohm's law.

(i) For each of these components, sketch current–voltage characteristics, plotting current on a **vertical** axis, and showing both positive and negative values. Use one set of axes for A and B, and separate sets of axes for C and D. Label your graphs clearly.

(ii) Explain the shape of the characteristic for C.

(iii) Name the component you have chosen for D.

(NEAB, 1992)

3 (a) (i) State Ohm's law.

(ii) Define resistance.

(b) (i) Sketch graphs showing the variation of resistance with temperature for (I) a metallic conductor and (II) an intrinsic (pure) semiconductor.

(ii) *Briefly* describe the mechanisms of conduction in the components in (b) (i).

(WJEC, 1993)

4 Find the current I in the circuit shown below.

(WJEC, 1991)

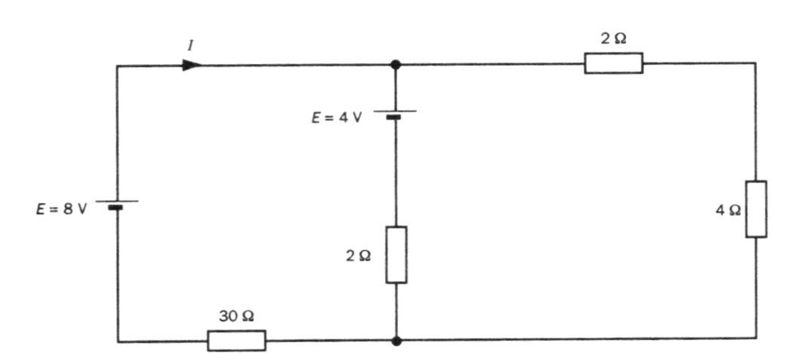

5 In the following circuit, C is a cell of e.m.f. 2 V and internal resistance 1 Ω. Calculate

(a) the current through A,

(b) the current through B,

(c) the p.d. between the terminals of the cell.

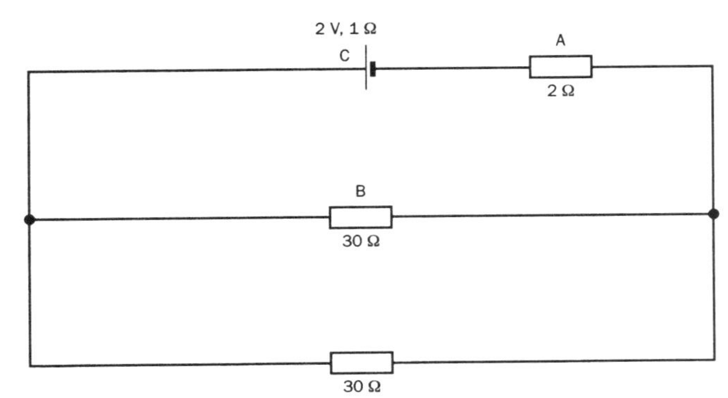

(WJEC, 1994)

6 (a) Define the *electromotive force* of, and explain the meaning of the *internal resistance* of, a d.c. supply.

A high resistance voltmeter reads 1.5 V when attached to the terminals of a dry battery. When the battery is supplying a current of 0.30 A through an external resistance R the voltmeter reads 1.2 V.

Calculate (i) the e.m.f. of the battery

(ii) the value of R

(iii) the internal resistance of the battery

and (iv) the energy converted from chemical to electrical energy by the battery in 2.0 seconds.

(b) State Kirchhoff's first law and explain how it is accounted for in terms of conservation.

(O & C, 1994)

7 In order to supply sufficient current to light a resistance of 0.50 Ω a student uses two dry batteries, each with e.m.f. 1.5 V and internal resistance 1.0 Ω. They are connected in parallel as shown in the diagram.

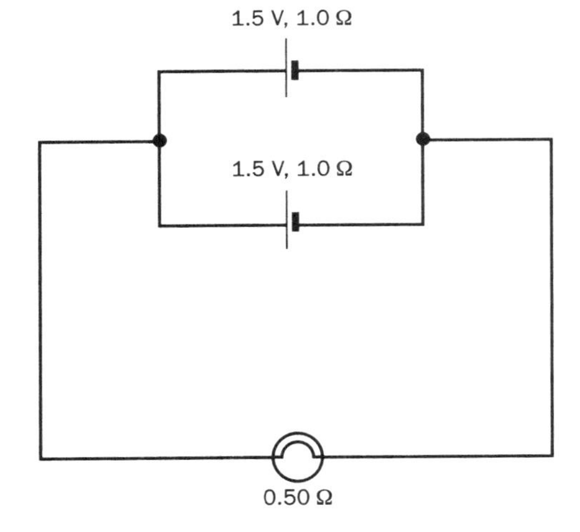

Calculate the power dissipated by the lamp.

(O & C, 1994)

8 The battery in the circuit shown at the top of the following page has e.m.f. 5.4 V and drives a current of 0.30 A through a lamp. The voltmeter reading is 4.8 V.

Explain why the voltmeter reading is less than the e.m.f. of the cell.

Calculate values for

(a) the internal resistance of the battery, and

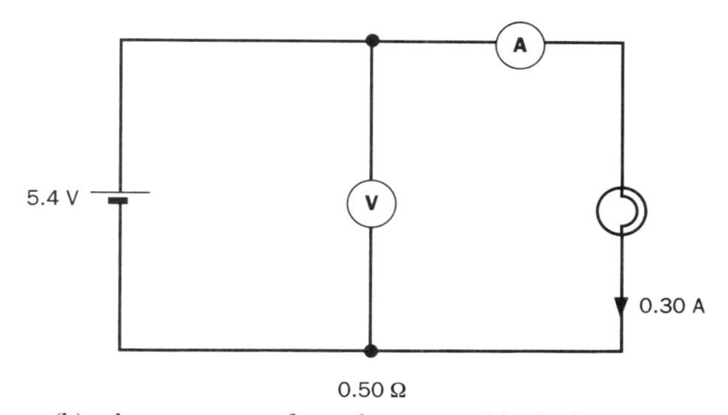

(b) the energy transformed per second in the lamp.

State *two* assumptions you made in order to complete these calculations.

(Lond., June 1992)

9

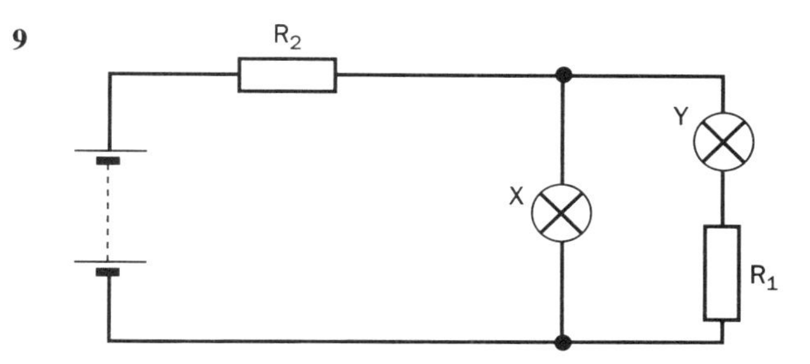

In the circuit shown above, X is a 12 V, 12 W lamp and Y is a 6 V, 12 W lamp. The cell has e.m.f. 24 V and internal resistance 2.0 Ω . The resistors R_1 and R_2 are such that each lamp operates at its rated voltage.

Calculate

(i) the current in X and the current in Y
(ii) the p.d. across the terminals of the cell
(iii) the values of resistors R_1 and R_2.

(NEAB, 1994)

10 Explain in terms of electrons, why the electrical resistance of a metallic conductor increases with increasing temperature.

(WJEC, 1990)

11 There are *n* free electrons in unit volume of a wire and each electron carries a charge *e*.

(a) Show that when a current flows, the current density *J* is given by

$J = nev$

where *v* is the drift velocity of the electrons.

(b) A p.d. of 4.5 V is applied to the ends of a 0.69 m length of manganin wire of cross-sectional area 6.6×10^{-7} m^2. Calculate the drift velocity of the electrons along the manganin wire.

The resistivity of manganin is 4.3×10^{-7} Ω m and *n* for manganin is 10^{28} m^{-3}.

($e = 1.6 \times 10^{-19}$C)

(WJEC, 1992)

12 (a) An electric fire is operated from a 240 V mains supply. The element is made of wire which has a resistance per metre of 4.5 Ω at 0 °C. The temperature coefficient of resistance of the material of the wire is 1.0×10^{-4} K^{-1}. The temperature of the wire is 1200 °C when the current is 12 A.

(i) Determine the resistance of the element at 1200 °C.
(ii) Calculate the length of the wire used.

(b) Explain why the resistance of a metallic conductor increases as the temperature increases.

(c) Explain why the element reaches an equilibrium temperature.

(d) A designer decides to reduce the power output to half by using the same wire but halving its length. Discuss whether this change has the desired effect.

(AEB)

13 (a) Define *temperature coefficient of resistance.*

(b) A 100 W, 240 V electric light bulb has a tungsten filament.

(i) Calculate the resistance of the filament when operating at its rated value.
(ii) If the filament resistance at 0 °C is 51 Ω, calculate the temperature of the filament in (b) (i).
(The temperature coefficient of resistance of tungsten = 0.0058 K^{-1}.)

(WJEC)

14 The graph below shows how the current in a light emitting diode (LED) varies with the potential difference across its ends. A student is asked to build the circuit shown in the diagram in which a current of 15 mA in the LED is driven by a 5 V supply of negligible internal resistance.
What is the voltage across the LED at the stated current?
Calculate

(i) the resistance of the resistor, R,
(ii) the power dissipated in the LED.

(Lond., June 1993)

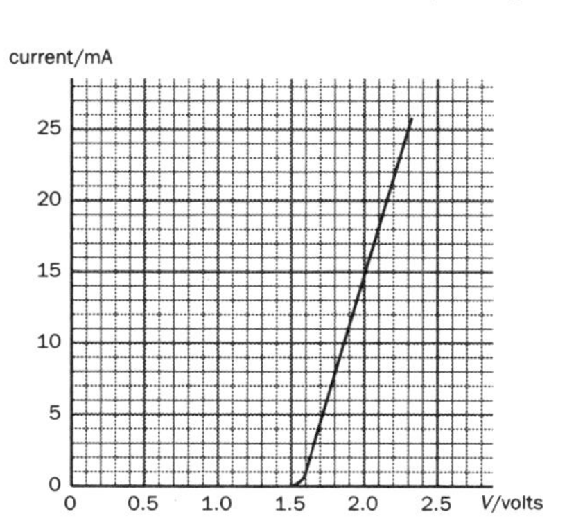

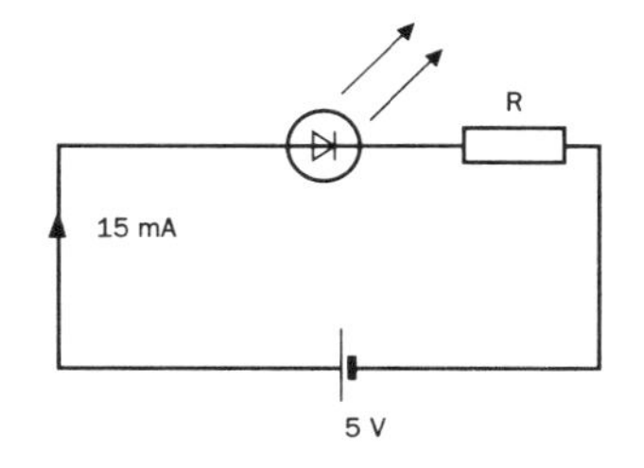

15 (a) The diagram shows the circuit of a slide-wire potentiometer.

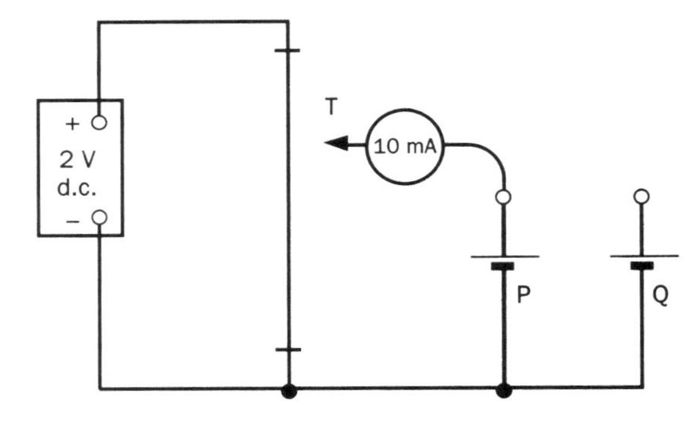

Describe the procedure you would follow and the measurements you would make to compare the e.m.f.s of the two cells P and Q.

The 10 mA meter is converted to a voltmeter and used to compare the terminal potential differences of cells P and Q. The resistance of the 10 mA meter is 8 Ω. Draw a diagram to show how you would convert the 10 mA meter to measure up to 2 V full scale deflection. Calculate the value(s) of any component(s) you use.

In what circumstances is the potential difference of a cell equal to its e.m.f.?
Explain your answer.

(Lond., June 1993)

16 Two parallel plates have a potential difference of 200 V across them. Draw a diagram to show the shape of the electric field between the plates. Add to the diagram equipotentials at 50 V and 100 V.

(Lond., 1994)

17 (a) The word 'field' is used in physics in connection with electrostatics and gravitation. For each of these explain what is meant by *field* and *field strength.*

An electrostatic field may be produced by a small charge and a gravitational field by a small mass. Explain clearly

(i) *two* ways in which these fields behave similarly, and

(ii) *two* ways in which these fields behave differently.

(Lond., January 1993)

18 Outside the sphere, a charged conducting sphere behaves as if the charge were concentrated at its centre. The electric field inside the sphere is zero.

One sphere of radius 5.0 cm carries a positive charge of 6.7 nC.

(a) (i) Show that the potential at the surface of the sphere is about 1200 V.

(ii) Calculate the capacitance of the sphere.

(Permittivity of free space, $\varepsilon_0 = 8.9 \times 10^{-12}$ F m^{-1}.)

(b) (i) Calculate the electric field strength just outside the surface of the sphere.

(ii) Sketch a labelled graph to show how the electric field strength varies with distance from the centre of the sphere to a distance 20 cm from the centre. Include a suitable scale on your graph.

(AEB)

19 (a) For each of the following, state whether it is a scalar or a vector and give an appropriate unit.

(i) Electric potential.

(ii) Electric field strength.

(b) Points A and B are 0.10 m apart. A point charge of $+3.0 \times 10^{-9}$ C is placed at A and a point charge of -1.0×10^{-9} C is placed at B.

(i) X is the point on the straight line through A and B, between A and B, where the electric potential is zero. Calculate the distance AX.

(ii) Show on a diagram the approximate position of a point, Y, on the straight line through A and B where the electrical field strength is zero. Explain your reasoning, but no calculation is expected.

(NEAB, 1992)

20 (a) A point charge of +5 nC is situated in air.

(i) Sketch to scale the equipotential surfaces corresponding to 45 V, 15 V, and 5 V respectively.

(ii) A point charge of –5 nC is placed at 8 cm from the +5 nC charge. Show to scale the equipotential surface corresponding to 0 V.

(iii) Sketch in (not necessarily to scale) some of the lines of force in the vicinity of the two charges of +5 nC and –5 nC.

$$\left(\frac{1}{4\pi\varepsilon_0} = 9 \times 10^9 \text{ F}^{-1}\text{ m}\right)$$

(WJEC, 1989)

21

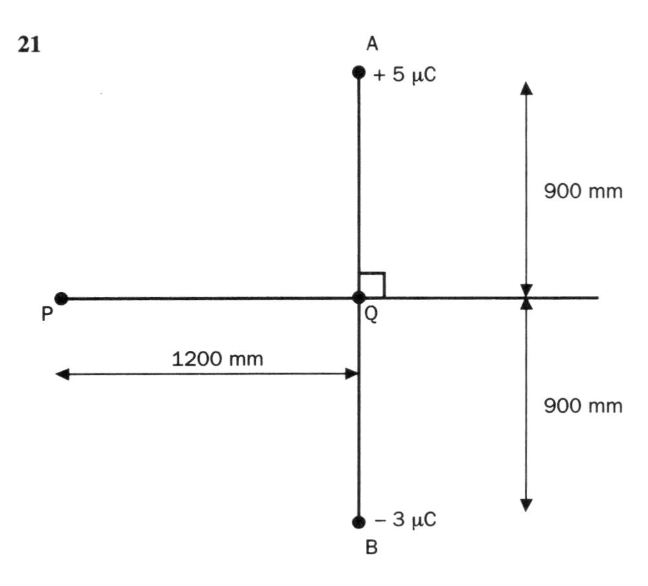

Two point charges of +5 μC and –3 μC are placed at points A and B as shown in the diagram above.

Calculate the work done in moving a charge of –3 μC from P to Q.

$$\left(\text{Take } \frac{1}{4\pi\varepsilon_0} = 9.0 \times 10^9 \text{ F}^{-1}\text{ m}\right)$$

(WJEC, 1991)

22 (a) (i) Define *electric potential*

(ii) The electric potential V at a distance r from a point charge Q *in vacuo* (free space) is given by

$$V = \frac{Q}{4\pi\varepsilon_0 r}$$

Write down an expression for the electric potential energy of a point charge of q when at a distance r from a point charge of Q *in vacuo* (free space).

(b) An α-particle of charge $+3.2 \times 10^{-19}$ C and mass 6.8×10^{-27} kg is travelling at a speed of 1.2×10^{-7} m s^{-1} directly towards a fixed nitrogen nucleus of charge $+11.2 \times 10^{-19}$ C.

Assuming that initially they are far apart calculate the closest distance of approach.

$$\left(\text{Take } \frac{1}{4\pi\varepsilon_0} = 9.0 \times 10^9 \text{ F}^{-1}\text{ m}\right)$$

(WJEC, 1994)

23 Explain the meaning of *capacitance*.

Two capacitors C_1 and C_2 are connected in series and then charged with a battery. The battery is disconnected and C_1 and C_2, still in series, are discharged through an 80 kΩ resistor. The time constant for the discharge is found to be 4.8 seconds. Calculate

(a) the capacitance of C_1 and C_2 in series, and

(b) the capacitance of C_1 if C_2 has a capacitance of 100 μF.

(Lond., January 1993)

24 (a) Two capacitors of capacitance C_1 and C_2 are connected in series. Derive an expression for the capacitance of a single equivalent capacitor.

(b)

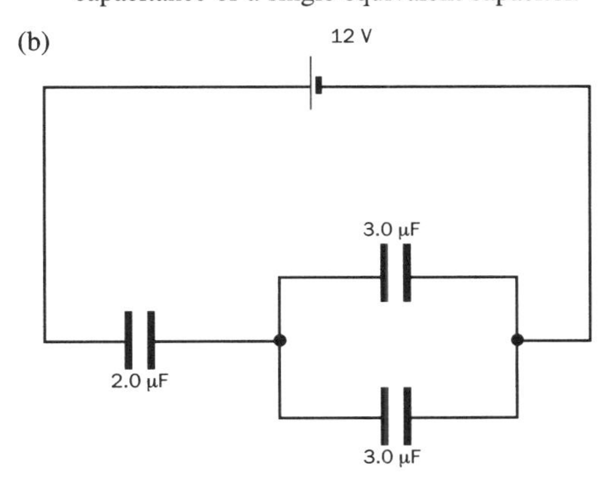

For the circuit shown, calculate

(i) the capacitance of this combination of capacitors

(ii) the total energy stored in the capacitors

(iii) the p.d. across the 2.0 μF capacitor.

(NEAB, 1992)

25 (a) (i) Define *capacitance*.

(ii) Write down the expression for the capacitance of a parallel-plate capacitor giving the meanings of the symbols involved.

(b) (i) A 12 pF parallel-plate air capacitor is connected to a 10 V battery. Find the charge on each plate and the energy stored in the capacitor when it is fully charged.

(ii) The battery is now disconnected and the capacitor isolated with the charges remaining on the plates. The separation of the plates is now doubled. Find the new value for the energy stored in the capacitor.

(iii) Account for any difference between your answers to (b)(i) and (b)(ii).

(WJEC, 1993)

26 The circuit for an electronic flash gun for a camera is shown below.

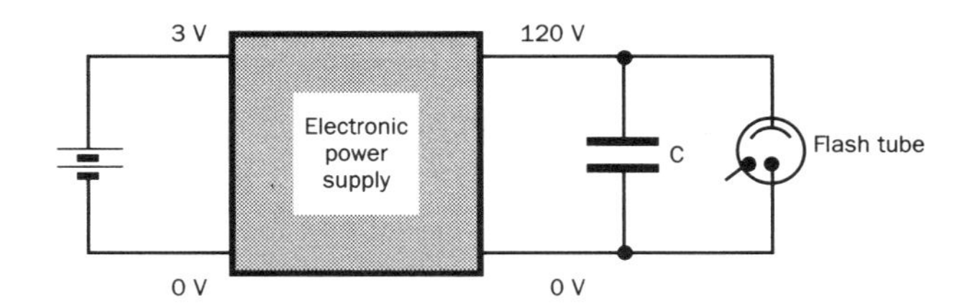

(a) To operate the flash tube C must store 1.0 J at a p.d. of 120 V. Calculate the capacitance of C.

(b) When the flash tube is triggered, the capacitor discharges completely in 1.5 ms converting 70% of the stored energy into light. Calculate the mean light power output during the flash.

The electronic power supply takes 20 s to charge the capacitor during which the battery supplies 4.0 J. Assuming that the battery has negligible internal resistance, calculate

(i) the average charging current delivered by the 3.0 V battery, and

(ii) the average load resistance presented to the 3.0 V battery by the electronic power supply.

(O & C, 1993)

27 (a) In the circuit shown in Figure 1, both switches are open and both capacitors discharged. The capacitor C_1, of value 2 μF, is charged by closing switch S_1.

(i) Calculate the charge on C_1 and the energy stored on it.

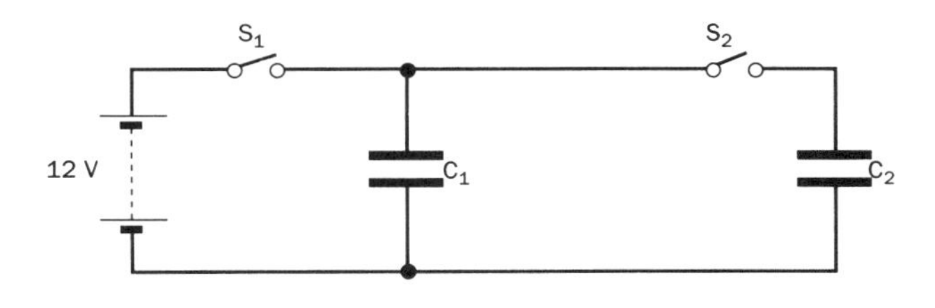

Fig. 1

S_1 is opened after which S_2 is closed so connecting capacitor C_2, of value 3 μF, across C_1.

(ii) Calculate the final voltage across C_2 and the total energy stored on the two capacitors. Account for the difference between this energy and the energy calculated in (i).

(b) In another experiment using capacitor C_1, a resistor R is placed in series, Figure 2.

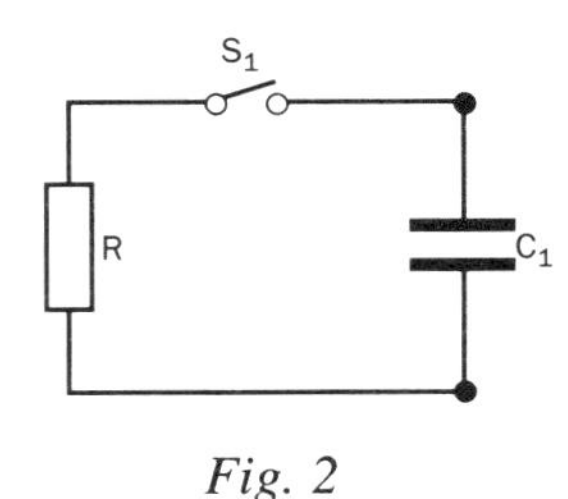

Fig. 2

The voltage, V_c, across C_1, is plotted against time from the closing of the switch, Figure 3.

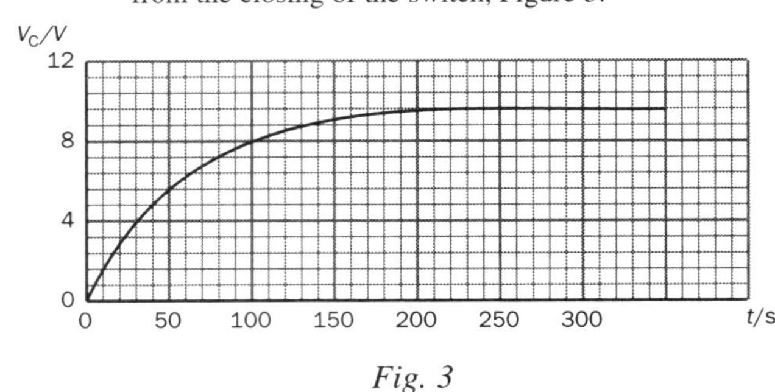

Fig. 3

(i) Sketch a graph of the voltage V_R across R, over the same time interval. Estimate, from the graph, Figure 3, the time at which $V_C = V_R$.

(ii) What is the significance of the quantity RC in this circuit, Figure 2? Explain the shape of the V_C against time graph, shown in Figure 3, using the quantity RC in your explanation.

(iii) Find the value of R.

(O & C, 1992)

28 The graph shows the charge on a capacitor plotted against the potential difference across it. Calculate

(i) the capacitance of the capacitor

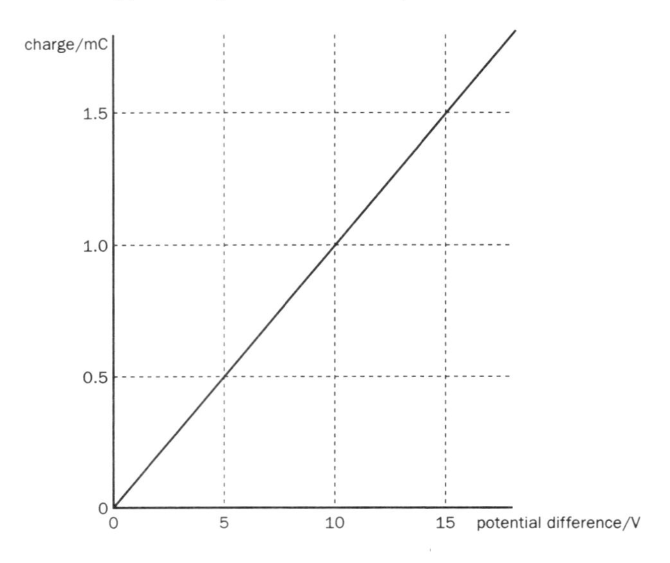

(ii) the energy stored by the capacitor when charged to a potential difference of 9.0 V.

(b) The capacitor in (a) is connected in series with a 10 kΩ resistor, a switch and a 9.0 V battery of negligible internal capacitance. The capacitor is initially uncharged.

Calculate

(i) the current flowing through the battery immediately after the switch is closed

(ii) the total energy supplied by the battery in charging the capacitor to a potential difference of 9.0 V. State why this value is different from your answer to (a)(i).

(NEAB, 1993)

29 The diagram shows a circuit used to measure the capacitance of a capacitor. A student opens switch S and plots a current–time graph for the capacitor circuit. The diagram also shows the graph obtained.

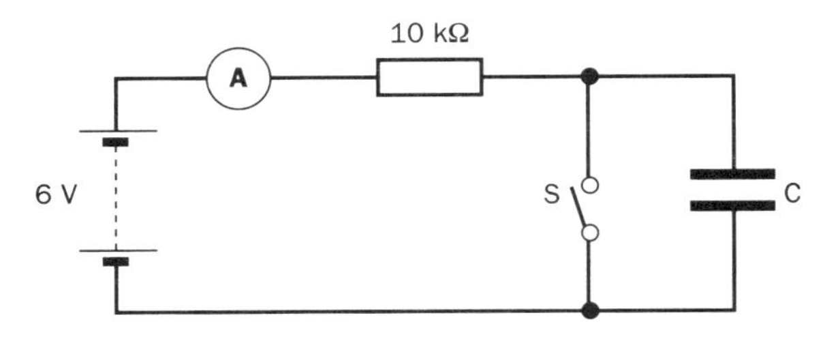

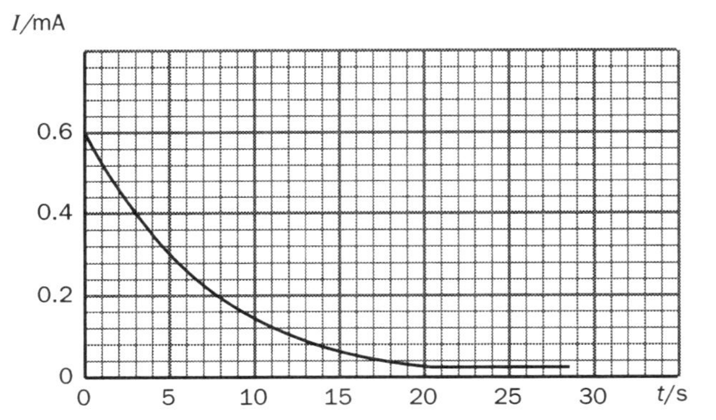

State the readings that the student would take in order to plot the graph.

Use the graph to find the final charge on the capacitor. What is the capacitance of the capacitor if the e.m.f. of the battery is 6 V?

The diagram at the top of page 317 shows a circuit used to measure the capacitance of a capacitor. The signal generator supplies a square wave to the circuit and the oscilloscope measures the voltage across the resistor. The diagram also shows the graph obtained.

The Y-sensitivity is 0.5 V cm^{-1} and the time–base speed is 20 ms cm^{-1}. Estimate the time constant for the circuit. What is the capacitance of the capacitor if the resistance of the resistor is 15 kΩ?

A student replaces the above capacitor with one about 100 times smaller and attempts to measure its capacitance without any other changes to the circuit.

Sketch the display the student would obtain. With

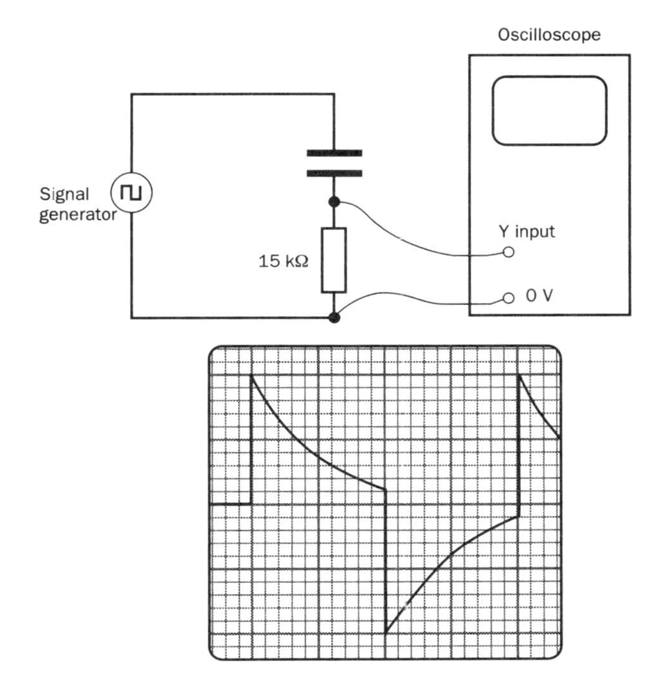

reference to your sketch state why it would not be possible to measure the capacitance without modifying the circuit.

State *one* modification that could be made to enable the smaller capacitance to be measured.

(Lond., June 1994)

30 (a) Define the term *magnetic flux*.

A solenoid of resistance 0.20 Ω has area of cross-section 1.00 cm^2, length 15.0 cm and 300 turns. (It is long enough to assume that the flux density is constant throughout the solenoid when it carries a constant current.)
A cell of e.m.f. 1.50 V and internal resistance 0.10 Ω is connected in series with the solenoid and a switch.
Calculate
(i) the flux density at the centre of the solenoid, and
(ii) the flux linking the solenoid after the switch is closed and the current is steady.
Explain why the current takes time to grow to its steady value.
(Permeability of vacuum $\mu_0 = 4\pi \times 10^{-7}$ H m^{-1}.)

(Lond., January 1992)

31 (a) Sketch the magnetic field lines produced by a straight wire carrying an electric current. On your sketch show the directions of the field and current.

(b) A long straight conductor P carrying a current of 2 A is placed parallel to a short straight conductor Q of length 0.12 m carrying a current of 3 A, the directions of the currents being the same. The conductors are 0.20 m apart in air.
(i) What is the value of the force experienced by Q?
(ii) What is the direction of this force?

(c) State the magnitude and direction of the force experienced by P and briefly justify your statement.

$$\left(B = \frac{\mu_0 I}{2\pi a}\right)$$

(WJEC, 1993)

32 The magnetic flux density, B, near a long straight current-carrying conductor depends on the distance, r, from the conductor. Describe how you would investigate experimentally the relationship between B and r.

State the other factors which affect the value of B.

A long straight wire produces a field of 6.0 μT at a distance of 30 cm from it in vacuum.

Calculate the current in the wire.

A second similar wire, carrying a current of 8.0 A, is now placed parallel to and 10 cm from the first wire. Calculate the force per unit length on the wire. The currents are in opposite directions. Is the force attractive or repulsive?

(Permeability of free space $\mu_0 = 4\pi \times 10^{-7}$ H m^{-1}.)

(Lond., January 1993)

33 (a)

The diagrams above show a long straight vertical wire and a solenoid. The current directions are indicated by arrows. Copy these diagrams and sketch the magnetic fields generated by each of these conductors alone.

(b)

The diagram above shows two long straight parallel conductors, K and L. P is a point midway between the wires and PQ is a line perpendicular to the plane containing the wires. The direction of the current through wire K is shown.

The resultant field at P is zero. Explain, with the aid of a diagram, what you can deduce about the current through the wire L.

What is the direction of the resultant field along the line from P towards Q? Sketch a graph to show how you would expect the magnitude of the field to vary with distance along PQ.

(Lond., June 1992)

34 (a) Figure 1 shows three long, straight, parallel wires, P, M and Q, which carry equal currents I in the same direction in a plane perpendicular to the paper. The direction of I is **into** the paper.

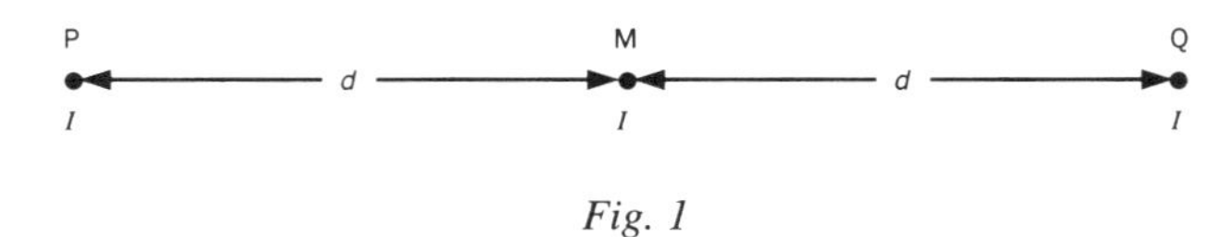

Fig. 1

(i) Draw an arrow on the diagram to represent the direction of the resultant magnetic field at wire Q due to the currents in P and M. Label this B.

(ii) Draw a second arrow to show the force on Q due to the currents in P and M. Label this F.

(b) A uniform copper rod XY, 0.50 m long, lies on the surface of a wooden laboratory table in an East–West direction, as shown in Figure 2. The Earth's magnetic field has the same magnitude and direction at all points along the rod.

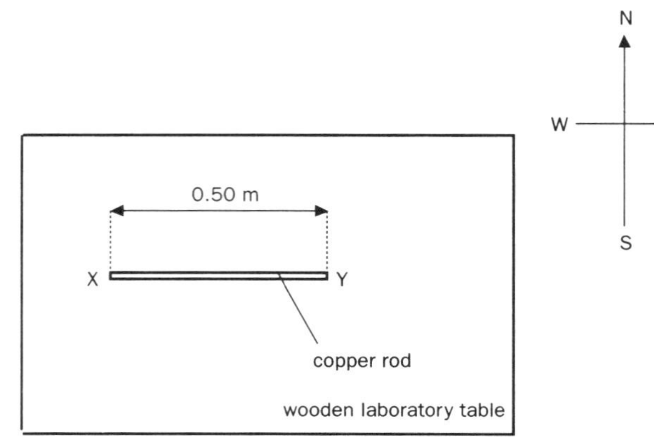

Figure 2

When a potential difference is applied to the ends of the rod, in the correct direction and of sufficient magnitude, the rod will rise from the table. assume that the electrical connections to the rod cause no significant restraint.

(i) Mark on the diagram at X and Y using + and – signs, the direction in which the potential difference should be applied.

(ii) Calculate the minimum potential difference which will cause the rod to rise from the table.
(Density of copper $= 8.9 \times 10^3$ kg m^{-3}
Resistivity of copper $= 1.7 \times 10^{-8}$ Ω m
Horizontal component of the Earth's magnetic field $= 1.8 \times 10^{-5}$ T)

(O & C, 1994)

35 Describe the *Hall effect*. Use a simple theory for a single carrier material to derive an expression for the Hall voltage. How is the Hall *coefficient* defined?

(a) Explain how measurements of Hall voltages can determine whether charge carriers in the material under examination are negative (electrons) or positive (holes).

(b) Explain why, for the same value of current through a specimen and the same value of B-field applied, the Hall voltage is likely to be much larger in a single-carrier semiconductor than in a metal.

(c)

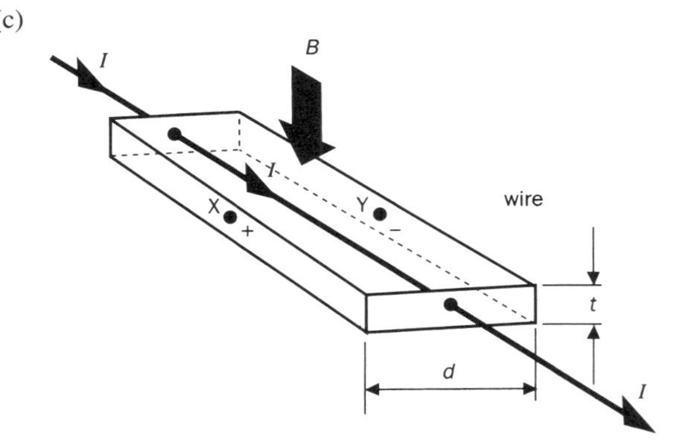

A rectangular specimen of width d (10 mm) and thickness t (2 mm) carries a current I (30 mA) in the direction shown. A uniform magnetic field B (1.8 T) is applied in the direction shown. The Hall voltage is measured between X and Y. X is positive with respect to Y. Calculate the Hall voltage and Hall coefficient, given that there are 10^{22} charge carriers per m^3. What is the sign of the carriers?

(O & C, 1993)

36 (a) (i) Show that a particle of mass m and charge Q moving with speed v at right angles to a magnetic field of flux density B travels in a circular arc of radius r given by

$$r = \frac{mv}{QB}.$$

(ii) Calculate this radius for an α particle of energy 6.4×10^{-13} J travelling at right angles to a magnetic field of flux density 0.50 T.
mass of α particle $= 6.7 \times 10^{-27}$ kg
magnitude of charge of electron $= 1.6 \times 10^{-19}$ C

(b) The deflection of β particles in a magnetic field can be readily demonstrated in the laboratory using a permanent magnet and a Geiger–Muller tube and counter.

Discuss the difficulties involved in demonstrating the deflection of α particles in a magnetic field.

(NEAB, 1993)

37 (a) Give the equation for the force on a particle carrying a charge e^-

(i) in an electric field of intensity E

(ii) while moving with velocity v at right angles to a magnetic force of flux density B.

(iii) Draw diagrams to indicate the directions of the forces. (Make sure that you show the direction of the fields in relation to + and –, and N and S.)

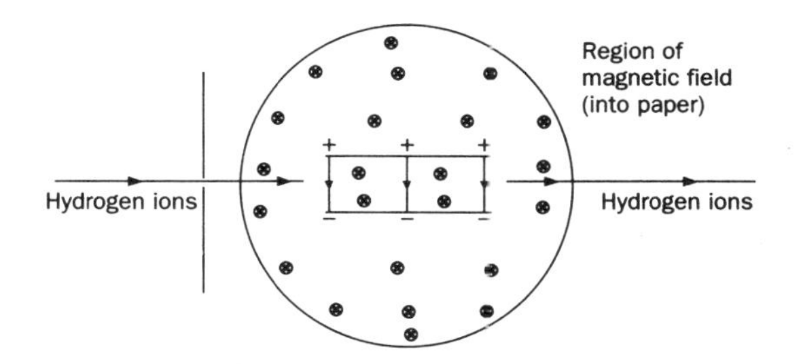

(b) A beam of electrons is directed into a region of uniform magnetic field of flux density *B* along a path at right angles to the direction of *B*.
   (i) What is the path followed by the electrons when they enter the field?
   (ii) Explain why this path is followed.

(c) Hydrogen ions moving at various speeds are directed at a region of combined electric and magnetic fields as shown in the diagram above.

The electric field is between two parallel plates 10 mm apart with a potential difference *V* across them, while the magnetic field of flux density 0.1 T is at right angles to the electric field.
   (i) Calculate the value of *V* required so that ions of speed 100 m s^{-1} pass through the region of the two fields without being deviated.
   (ii) What is the kinetic energy per ion as they leave the combined fields?

$(M_u = 1.67 \times 10^{-27}$ kg)

(WJEC, 1990)

38 (a) Electrons are accelerated from rest through a p.d. of 4000 V in an evacuated tube, then they enter a uniform magnetic field *B* of flux density 10^{-3} T which is at right angles to the electron beam as shown in the diagram below.

   (i) Calculate the speed *v* of the electrons on entering the magnetic field.
   (ii) Calculate the magnitude of the force experienced by an electron in the magnetic field.
   (iii) Show the direction of this force by an arrow on the above diagram.

(b) Explain why the electrons move in a circular path and calculate the radius of this path.

(WJEC, 1993)

39 State the laws of magnetic induction. Describe how you would demonstrate experimentally the relationship between the magnitude of the change of flux and the magnitude of the induced e.m.f.

A long magnet is removed from the centre of a coil of 20 turns. The speed of the magnet is controlled to maintain an induced e.m.f. of 60 μV across the coil. Removing the magnet in this way takes two minutes. Calculate the change of flux through the coil.

Explain with the aid of a diagram, how you would determine the polarity of the magnet from this experiment.

(Lond., January 1993)

40 (a) Explain how the laws of electromagnetic induction predict the magnitude and direction of induced e.m.f. s. Describe two simple experiments which you could carry out in the laboratory to illustrate the laws.

(b) A powerful electromagnet produces a uniform field in the gap between its poles, each of which measures 0.10 m × 0.08 m. There is no field outside the gap. A circular coil of 80 turns and radius 0.09 m is placed so that it encloses all the flux of the magnetic field (Figure 1).

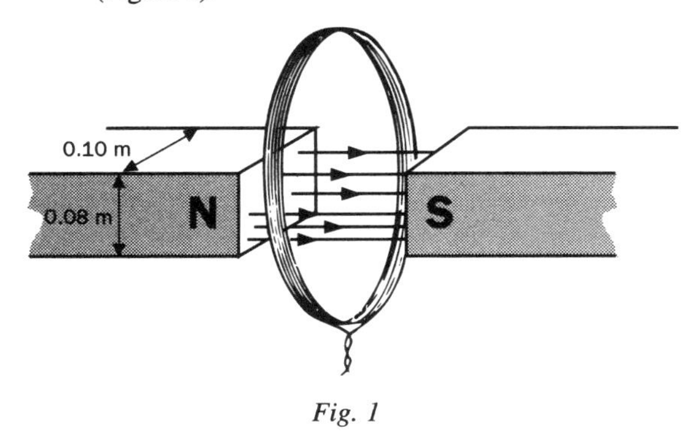

Fig. 1

   (i) The current in the electromagnet is reduced so that the field falls linearly from 0.20 T to zero in 5.0 s. Calculate the initial total flux in the gap and hence the e.m.f. generated in the coil during this time.
   (ii) The coil is part of a circuit in which the total resistance is 24 Ω. Calculate the current in the circuit while the field is collapsing and the magnetic field which this current produces at the centre of the coil.
   (iii) Figure 2 shows the poles and coil. Copy this diagram and show on your copy the direction of the induced current in the coil and the direction of the magnetic field which this current produces. Explain how your application of the laws led you to these deductions about the directions of field and current.

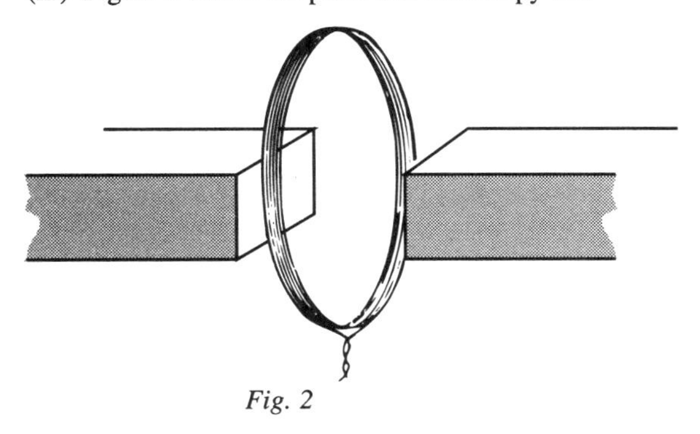

Fig. 2

(O & C, 1992)

41

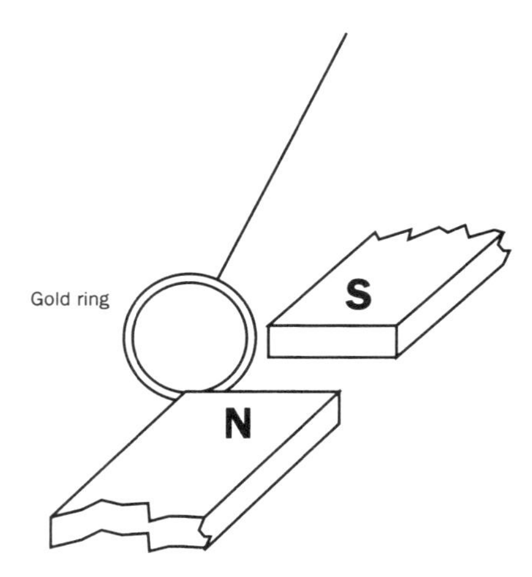

The diagram shows a gold ring on a silk thread about to swing through a magnetic field. Two students predict differently the influence of the magnetic field on the motion of the ring.

Student A says that the effect of the field is to increase the speed of the ring as it enters the field and to slow it down as it leaves: the net effect is zero.

Student B says there is only one effect: to slow the ring down.

State which student is correct. Explain why the ring behaves as this student describes.

(Lond., June 1992)

42 State Lenz's law

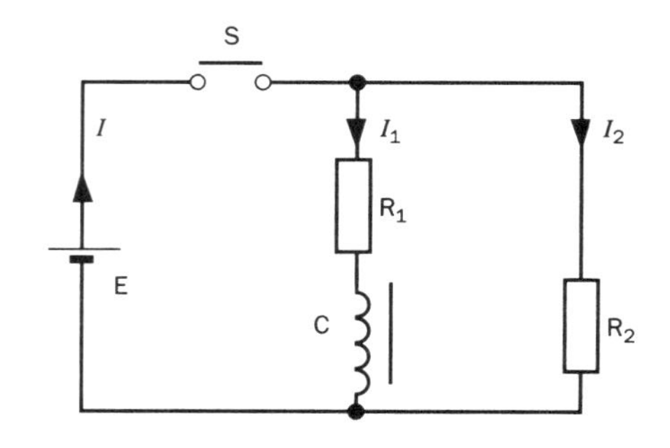

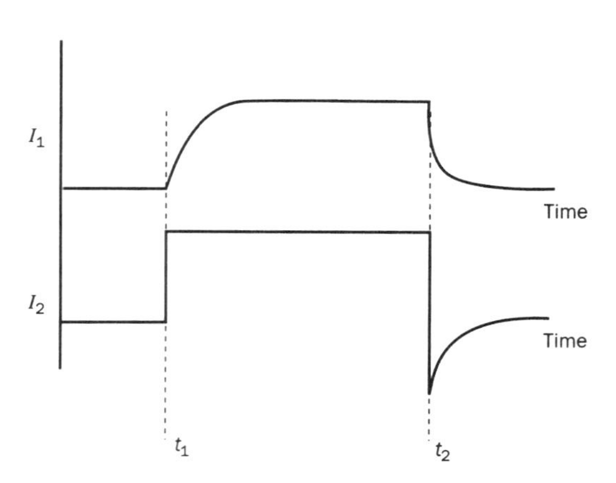

The coil C in the circuit shown above consists of several hundreds of turns of fine wire wound on a soft iron bar.

The graphs show how the currents I_1 and I_2 vary with time from just before the moment t_1 when the switch is closed to just after the moment t_2 when the switch is opened again.

Explain

(i) what inhibits the initial growth of I_1, and

(ii) why the current I_2 reverses at time t_2.

Sketch a graph showing how the current I in the battery varies over the same time interval.

(Lond., January 1994)

43 The graph below shows the variation of current, I, with time, t, in a circuit consisting of a coil connected in series with a switch and a 4.0 V cell of zero internal resistance. The switch is closed at $t = 0$.

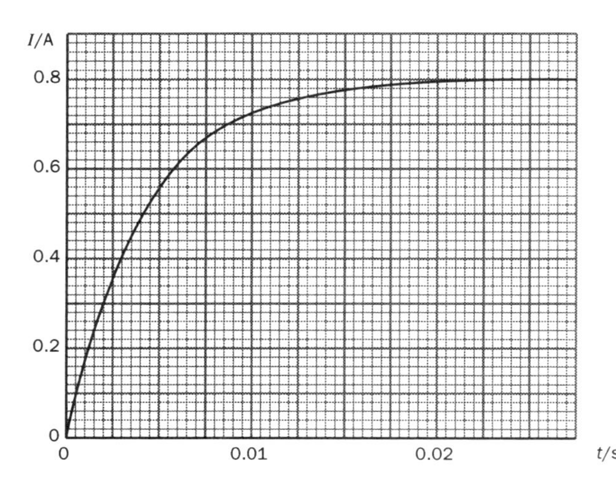

Use the graph to determine the initial rate of change of current with time and the final steady current. Hence calculate

(i) the resistance of the coil

(ii) the inductance of the coil

(iii) the energy stored in the magnetic field of the coil when the current is constant.

(NEAB, 1994)

44

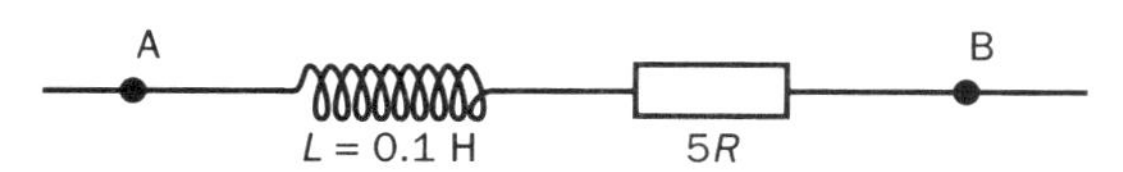

The current through AB (from A to B) increases uniformly from zero to 12 A in 6 s and then remains constant at 12 A. What is the p.d. across AB

(a) after 3 s

(b) after 7 s?

(WJEC, 1990)

45 (a) Figure 1 shows a copper disc of radius 120 mm situated in a uniform magnetic field B of 1.5×10^{-2} T. The plane of the disc is perpendicular to B which points into the plane of the diagram. The disc is rotated about an axis through its centre O at 2500 revolutions per minute.

(i) Calculate the magnetic flux cut by a radius OP in one second.

(ii) A stationary connection is made from a sliding contact touching the rim at M through an ammeter A to a similar contact at the centre of the disc. This is indicated by the dotted lines on

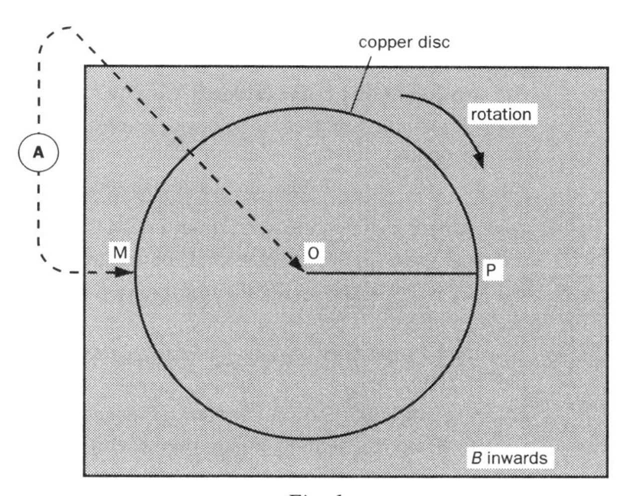

Fig. 1

the diagram. The resistance of the connection MAO is 0.01 Ω and the resistance between the rim and the centre of the disc is negligible. Calculate the magnitude of the current through A.

(iii) Explain whether a conduction electron in the copper disc will experience a force of magnetic origin towards the centre of the disc or towards the rim as a result of its motion in the magnetic field. Deduce the direction of the current through A.

(O & C, 1994)

46 (a) State the laws of electromagnetic induction.

(b)

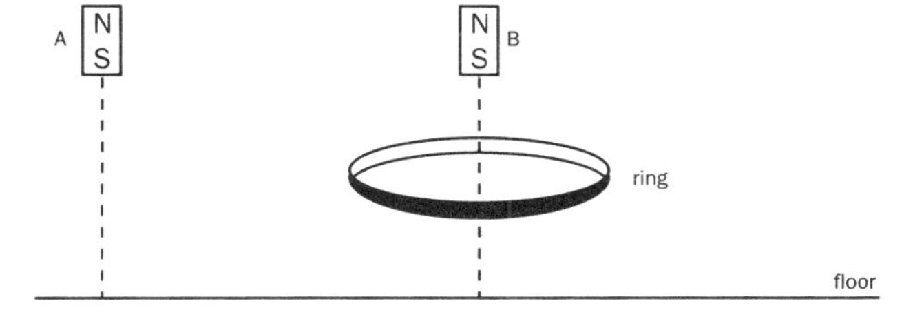

Two identical bar magnets A and B are held at the same height above the floor and sufficiently far apart as not to interact with one another. A copper ring is placed vertically below B so that magnet B can move vertically through this ring without touching it.

(i) Describe, and *briefly* explain, what happens when the two magnets are simultaneously released.

(ii) The experiment is repeated but this time B has its south pole uppermost. What happens now when the magnets are simultaneously released? Explain your answer.
Ignore the effects of the Earth's field.

(WJEC, 1994)

47 The diagram shows a simple electromagnetic brake consisting of a copper disc attached behind the road wheel of a vehicle so that it rotates with it. The magnetic field B across part of the disc is provided by an electromagnet whose current is regulated by a 'brake pedal'. With no

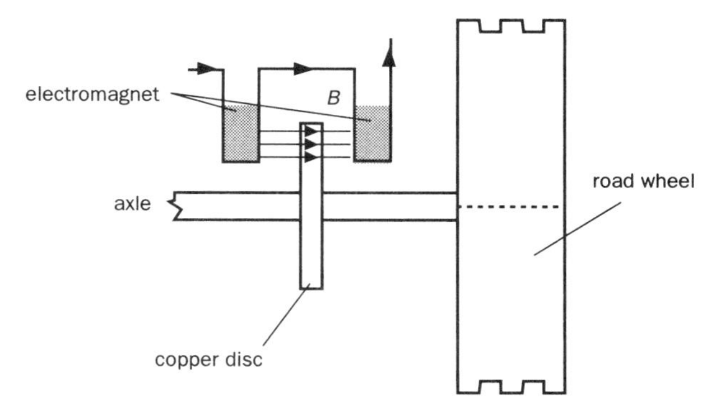

pressure on the pedal, B is zero and it is increased progressively as the pedal is pressed.

(i) Explain in terms of Lenz's law how braking is obtained.

(ii) Explain why such a system would be good at motorway speeds but should not be the **only** braking system on a vehicle.

(Lond., June 1992)

48

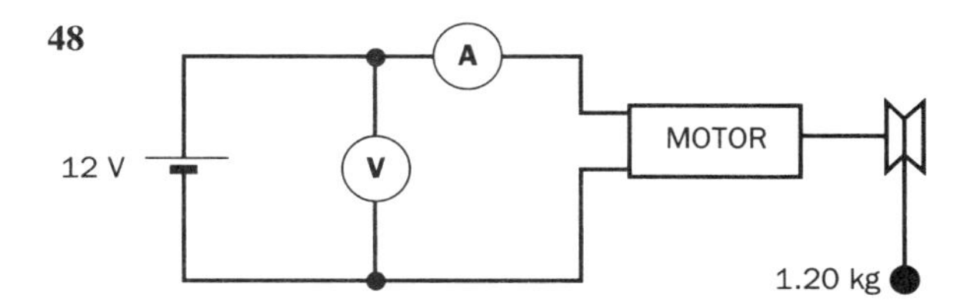

The motor shown in the diagram above lifts a 1.20 kg mass through a vertical distance of 2.60 m in 4.8 s. The ammeter and voltmeter readings are 0.92 A and 10.6 V respectively.

Calculate

(i) the energy transferred to the mass from the motor

(ii) the energy transferred from the battery to the motor and

(iii) the efficiency $\left[\dfrac{\text{Power}_{\text{out}}}{\text{Power}_{\text{in}}}\right]$ of the motor.

Explain why, if the motor is to be used to raise a heavier mass, the ammeter current must increase, and the voltmeter reading will then decrease.

Why would you expect the motor efficiency to be low when the load to be raised is heavy?

(Lond., June 1992)

49 (a) (i) Explain what is meant by root mean square voltage.

(ii) Write down an expression relating r.m.s. and peak voltage for a sinusoidal electricity supply.

(b) A voltage given by $V = 339.4 \sin(100\pi t)$ is applied in turn across the terminals of the devices shown in diagrams A (below) and B (on page 322).

(i) In **each** case, what is the total energy dissipated by the device over a period of 100 s?

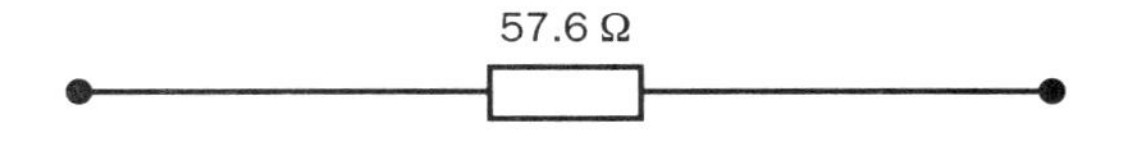

Diagram A

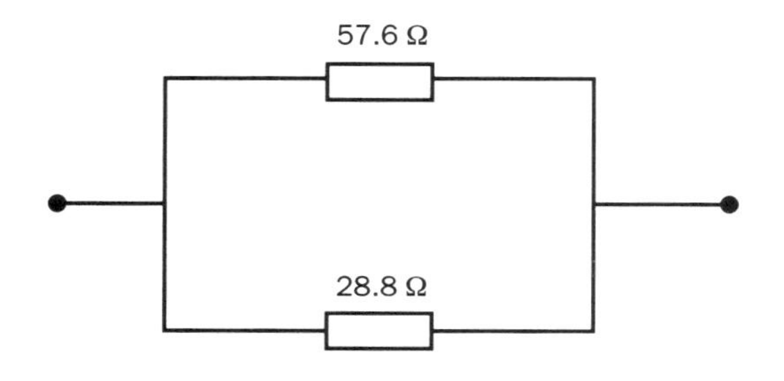

Diagram B

(ii) Suggest a practical device whose operation may be described by the application of the above voltage to A or B.

(WJEC, 1991)

50 Graph A shows how the voltage across a coil varied with the current flowing through it. One graph was obtained using d.c. and the other using a.c. of frequency 200 Hz.

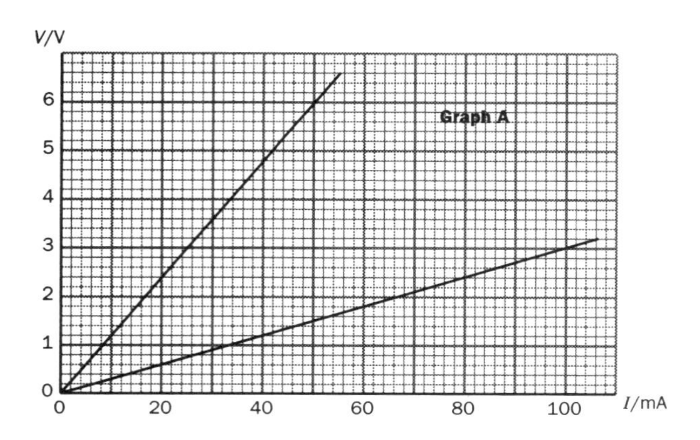

(a) Explain why graph A must be the one obtained when a.c. was used.

(b) Determine the resistance of the coil.

(c) Determine the inductance of the coil.

(AEB, 1990)

51 A farmer installs a private hydroelectric generator to provide power for equipment rated at 120 kW 240 V a.c. The generator is connected to the equipment by two conductors which have a total resistance of 0.20 Ω. The system is shown schematically in Figure 1 below.

(a) The equipment is operating at its rated power. Calculate.
- (i) the power loss in the cables
- (ii) the voltage which must be developed by the generator
- (iii) the efficiency of the transmission system.

(b) An engineer suggests that the farmer uses a transformer to convert the generator output to give a p.d. of 2400 V at the end of the transmission line as shown in Figure 2 below.
A second transformer is to be used to step down this p.d. to 240 V.
- (i) Explain briefly how a transformer makes use of electromagnetic induction to produce an output voltage several times bigger than the input voltage.
- (ii) The transformers are 100% efficient. Calculate the power loss in the new transmission system.

(AEB, 1992)

52 Figure 1 on page 323 shows a circuit to rectify 50 Hz alternating current using a single diode to supply a load resistance R. Figure 2 shows a similar circuit with the addition of a smoothing capacitor C.

(a) For each of the circuits, draw a sketch graph showing both the a.c. supply waveform (use a dotted line) and the d.c. voltage across R (use a solid line). Put numerical scales on both the voltage and time axes.

(b) In circuit 4.2 the capacitor and resistor have values of 1000 μF and 200 Ω respectively.
- (i) When the supply is switched on the capacitor charges fully during the first quarter cycle. Calculate the mean current in the diode during charging.
- (ii) In later cycles, the capacitor is topped up on each peak of the cycle and supplies energy to the resistor R for the remainder of the cycle. Estimate by how much the voltage across the capacitor changes during the cycle. State any assumptions you make in your estimate.

(c) Draw a circuit diagram for full-wave rectification using four diodes to provide the supply across R and C. State two ways in which the ripple waveform will be different in the full-wave rectifier circuit.

(O & C, 1993)

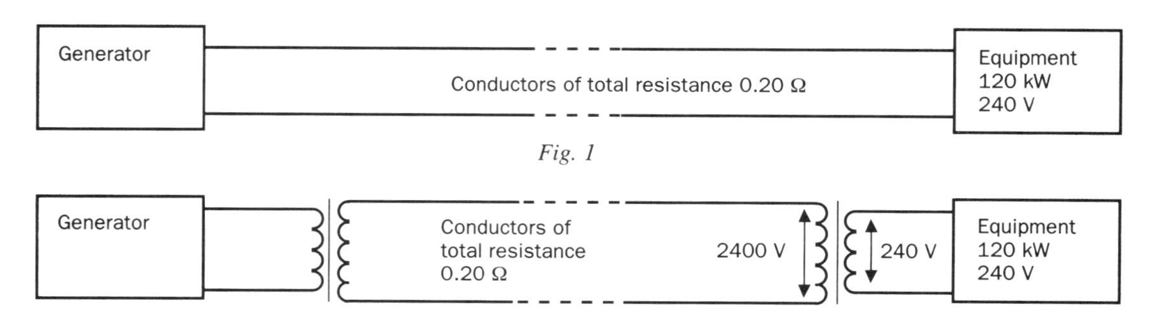

Fig. 1

Fig. 2

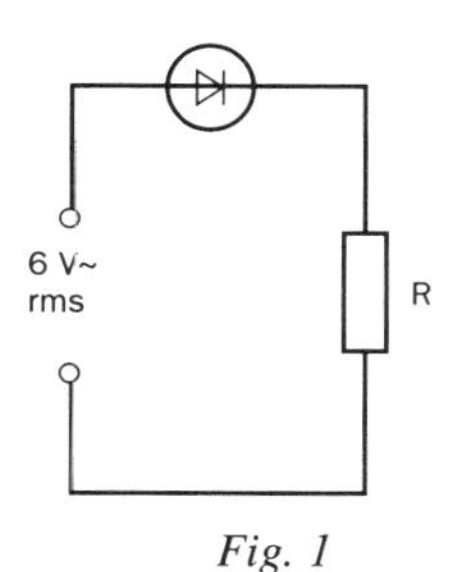

Fig. 1

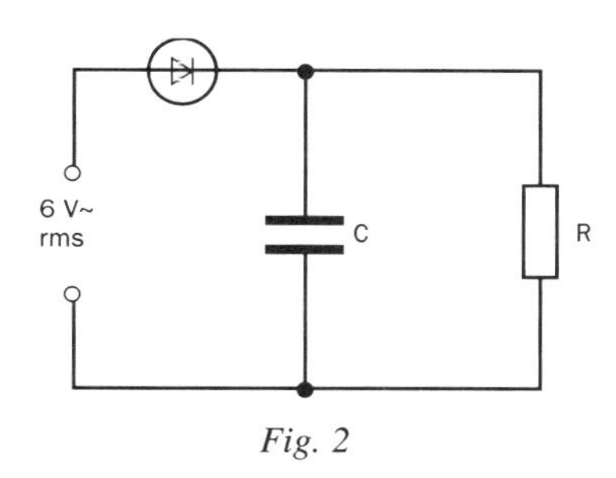

Fig. 2

53 The graph shows the variation of reactance with frequency for a capacitor.

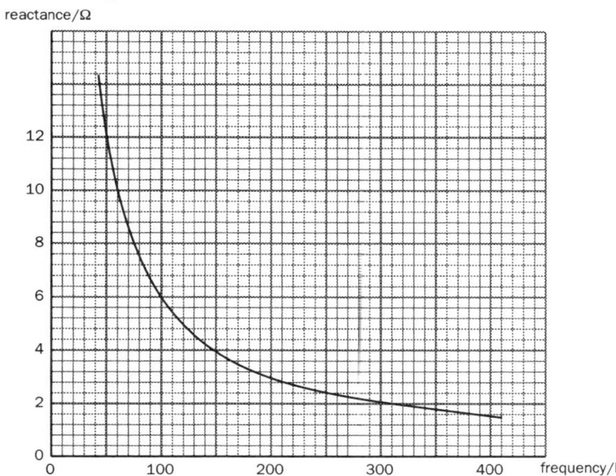

(a) (i) Write down an expression for the reactance of a capacitor in terms of frequency.

(ii) Use the graph to determine the value for the capacitance of the capacitor which was used to produce data for the graph.

(iii) Draw a clearly labelled diagram to show the phase relationship between the alternating current and the voltage waveforms when a sinusoidal supply is connected across the capacitor.

(iv) A calibrated signal generator (oscillator), which has an internal resistance of 6 Ω and provides an e.m.f. of 4 V r.m.s at all frequencies, is available. Describe how you would proceed to obtain data which would enable you to investigate the variation of reactance with frequency. Include in your description:

— a circuit diagram

— the measurements you would make and how you would use them

— the instruments you would use and their ranges.

(b) (i) Sketch a graph showing the variation of reactance with frequency for a **pure** inductor which has the same reactance as the capacitor at 100 Hz.

(ii) Calculate the inductance of the inductor.

(c) (i) What would be the circuit impedance at 100 Hz if the capacitor and inductor were in series with the supply of internal resistance 6 Ω? Show your reasoning clearly.

(ii) Sketch a graph showing how the impedance would vary with frequency for this circuit.

(AEB, 1994)

54 An inductor of inductance L and a capacitor of capacitance C are connected in series across a sinusoidal supply. The circuit has a resonant frequency of 600 Hz.

(a) Sketch a graph to show how the impedance of the circuit varies with frequency when the frequency of the supply is increased from a low frequency to about 1 kHz.

(b) By considering the reactances in the circuit, show that the resonant frequency, f, is given by

$$f = \frac{1}{2\pi\sqrt{LC}}$$

(c) When the a.c. supply is at the resonant frequency, state the phase difference between

(i) the current in the circuit and the applied p.d.

(ii) the p.d. across the capacitor and the p.d. across the inductor.

(NEAB, 1991)

55 A sine-wave signal generator is connected in series with a capacitor and a resistor. The output of the generator measured by a digital voltmeter is 4.0 V r.m.s. The voltage across the resistor is found to be 3.0 V r.m.s. and across the capacitor to be 2.7 V r.m.s.

(a) Explain the meaning of the term *r.m.s.* and why it is useful.

(b) Explain qualitatively why the voltage measured across the generator terminals does not equal the voltages measured across the resistor and the capacitor.

(c) The value of the resistor is 1.0 kΩ and of the capacitor is 1 μF. Find the frequency setting of the signal generator.

(d) Draw graphs of the voltage against time across (i) the resistor and (ii) the capacitor. Use the same time axis for the two graphs. Label each graph clearly, and the axes with appropriate scales and units.

(O & C, 1994)

56 (a) (i) Explain why transformers are important in the transmission of electrical energy by the National Grid from the power station to the domestic consumer.

(ii) State and explain **two** features of the design of the transformer core which help to minimise energy loss.

(b) A transformer has 600 turns on the primary and 120 turns on the secondary. An alternating p.d. of 20 V r.m.s. is applied to the primary and a resistance of 4.0 Ω is connected across the secondary. If it is assumed that the efficiency of the transformer is 100% and the resistance of the coils negligible, calculate

(i) the secondary r.m.s. potential difference

(ii) the secondary r.m.s. current

(iii) the primary r.m.s. current.

(c) You are supplied with a coil which has resistance and inductance, a variable capacitor and a source of sinusoidal e.m.f. giving a constant voltage at a fixed frequency. The usual apparatus of a physics laboratory is also available.

(i) What do you understand by resonance in a series circuit?

(ii) Explain how you would demonstrate series resonance using the fixed frequency supply.

(d) A series circuit contains a coil of resistance R and inductance L, a capacitor and a sinusoidal source. Resonance occurs at 80 Hz when the capacitance is 5.00 μF. The supply p.d. is 120 V r.m.s. and the current in the circuit at resonance is 0.60 A r.m.s. Calculate, at 80 Hz,

(i) the reactance of the capacitor
(ii) the inductance, L, of the coil
(iii) the resistance, R, of the coil
(iv) the impedance of the coil.

(NEAB, 1994)

57 (a) (i) Explain why power is transmitted by the National Grid at high voltages and why alternating currents are used.

(ii) The power output of a power station is 500 MW at an r.m.s. p.d. of 132 kV. If the total resistance of the power line is 2.0 Ω, calculate the r.m.s. current in the line and the percentage power loss in transmission due to heat produced in the power line.

(b)

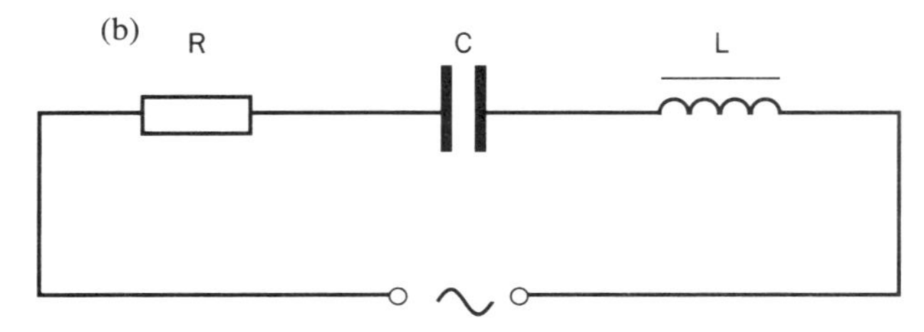

In the circuit shown in the diagram resistor R has resistance 100 Ω, capacitor C has capacitance 20 μF and the pure inductor L has inductance 0.20 H.

The supply p.d. is sinusoidal. R.m.s. potential differences and currents are referred to throughout.

When the current in the circuit is 0.50 A, V_L, the p.d. measured across the inductor, is 25 V.

(i) Calculate the reactance of L and the frequency of the supply.
(ii) Calculate V_R, the p.d. across R, and V_C, the p.d. across C.
(iii) Draw a phasor diagram showing V_R, V_L, V_C and V_S, the p.d. of the supply.
(iv) Calculate V_S and the phase difference between the current and the supply voltage.

(NEAB, 1992)

58 A series *LCR* circuit is set up as shown. The frequency of the source S (of zero impedance) is $\frac{1000}{2\pi}$ Hz.

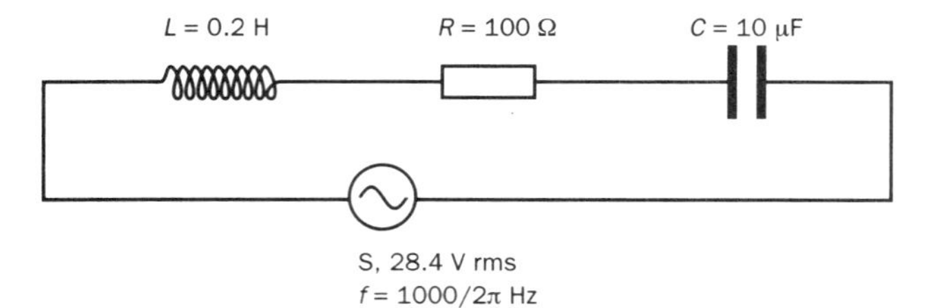

Calculate for the circuit

(a) the impedance
(b) the r.m.s. current
(c) the power dissipated.

(WJEC, 1992)

59 Figure 1 shows how the impedance of the *LCR* circuit shown in Figure 2 varies with frequency. The supply provides a voltage of 5.0 V r.m.s., which is constant for all frequencies and loads.

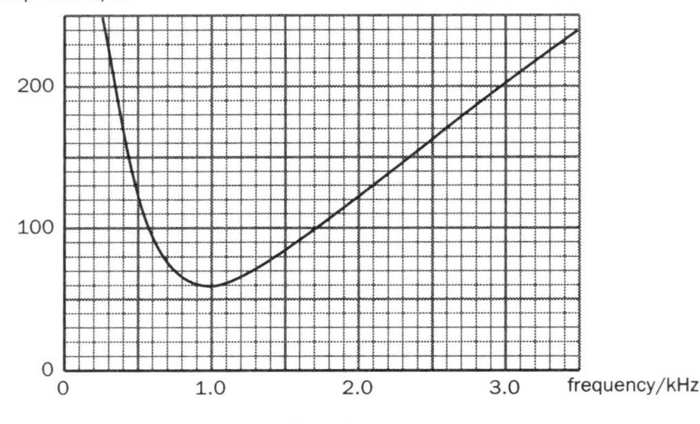

Fig. 1

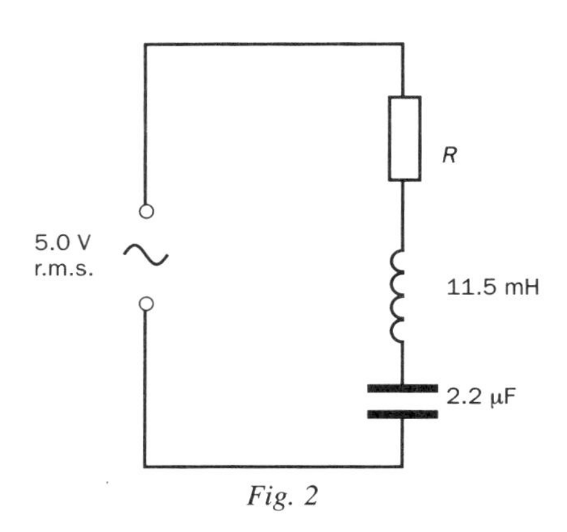

Fig. 2

(a) Determine
(i) the value of the resistance in the circuit
(ii) the r.m.s. current in the circuit at resonance.

(b) Determine
(i) the reactance of the capacitor at resonance
(ii) the r.m.s. potential difference across the capacitor at resonance
(iii) the r.m.s. potential difference across the inductor at resonance.

(AEB, 1993)

60 (a) How would you use the CRO to measure the phase difference across a resistor and an inductor connected in series with an a.c. supply? Describe how the circuit being investigated is connected to the oscilloscope, draw a diagram of what is seen on the screen and explain how the conclusions are arrived at.

(WJEC, 1989)

SECTION

Matter

4.1 Solids

Atomic size, mass and density

Radius

The **radius** of an atom of the natural elements varies between 0.04 nm (i.e. 4×10^{-11} m) for hydrogen and 0.22 nm (i.e. 2.2×10^{-10} m) for rubidium. As a comparison, the nucleus of an atom is about 100 000 times smaller and a human blood corpuscle is about 100 000 times larger.

Mass

Atomic **masses** can be accurately measured using a **mass spectrometer** (see Section 5.4 on Atomic and nuclear masses for more detail). It is found that they range from 1.7×10^{-27} kg for hydrogen to 4×10^{-25} kg for uranium.

On the carbon scale, the **atomic mass** of a particular atom is the mass of that atom compared to the mass of a carbon-12 ($^{12}_{6}C$) atom. The atomic mass of carbon is taken as exactly 12 which gives a value for the atomic mass of hydrogen of 1.008. In general, the atomic mass in g of any element contains the same number of atoms as 12 g of carbon-12. This number is called the **Avogadro constant** (N_A) and has the value 6.02×10^{23} .

Density

Common symbol	ρ (Greek letter 'rho')
SI unit	kilogram per metre^3 (kg m^{-3})

The **density** of a substance is defined as the **mass** per **unit volume** of the substance.

It is calculated from the formula:

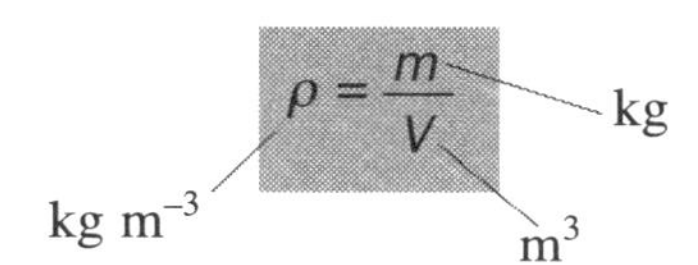

Equation 1

Scientists and engineers need to know the density of the different materials which they might use in a manufacturing or construction project. From the known density values and the volume of material used they can calculate the mass and hence the weight of a particular structure. The density of all the materials used in the construction of a bridge or an aircraft must be accurately known.

The density of an element depends on :

- the mass of each of its atoms, and
- the way the atoms are packed together.

The fact that osmium, at a lowly 76th place in the Periodic Table, is the densest of all the elements clearly shows that density is not solely determined by the atomic mass. There are many elements with more massive atoms, and if density only depended on this, all these elements would have higher densities. So osmium's atoms must be packed in an arrangement which is particularly efficient to achieve such high density from comparatively light atoms. Of course, each atom of osmium is mostly empty space, and while its density (2.25×10^4 kg m^{-3}) may be high by everyday standards, it is definitely 'small potatoes' compared to the density of nuclei (around 4×10^{17} kg m^{-3}!)

Guided examples (1)

1. Given that the Earth is a sphere of mean radius 6.4×10^6 m and of mass 6×10^{24} kg, estimate the mean density of the Earth.

Guidelines

Simply apply Equation 1.

2. Since virtually all the mass of the atom is concentrated in the nucleus and since this is about $10^5 \times$ smaller in size than the atom, it can be said that matter mainly consists of empty space. Can you imagine the dramatic volume reduction which would occur if the atoms which constitute our planet were to lose all their electrons and so eliminate matter's empty space? Assuming that all the atoms which constitute planet Earth have been stripped of their electrons and that nuclear density = 4×10^{17} kg m^{-3} and Earth mass = 6×10^{24} kg, calculate
   (a) the volume of the Earth and
   (b) the radius of the Earth under these circumstances.

Guidelines

(a) Simply apply Equation 1 to calculate the new volume.

(b) Volume of a sphere (V) is given by:

$V = \frac{4}{3}\pi R^3$ where R is the radius.

Solve for R.

Interatomic force and potential energy

Solid substances consist of atoms which are held together by the **attractive** forces which they exert on each other. The forces are electrical and give rise to interatomic bonds of various types (i.e. ionic, covalent, metallic and Van der Waals). There are also interatomic **repulsions**, otherwise matter would collapse into an infinitely small volume. The fact that solid matter exists without falling apart or collapsing suggests that these two forces (attraction and repulsion) are in **equilibrium** when solids are in an unstressed state. Theoretical and experimental evidence suggests that for atoms separated by less than one atomic diameter the resultant force is repulsive, while attractive forces predominate for greater atomic separations. In Figure 4.1, the magnitudes of the attractive and repulsive forces between two neighbouring atoms are shown by the relative sizes of the arrows:

F_a – attractive force
F_r – repulsive force

You should note that:

- In Figure 4.1(a), when the atoms are relatively far apart, attraction is greater than repulsion and the resultant of the two forces is **attractive**.
- There is a certain separation (the **equilibrium separation** r_0) shown in Figure 4.1(b) when attraction is equal to repulsion. At this equilibrium separation the resultant of the two forces is **zero**.
- In Figure 4.1(c), when the atoms are closer together than the equilibrium separation distance, repulsion exceeds attraction, and their resultant is **repulsive**.

It is the size of these forces between all atoms which is responsible for the tensile and compressive strength of solids; in other words these forces stop the world either falling apart or collapsing into a tiny, dense solid.

Interatomic force–separation graph

Figure 4.2 shows how the short-range attraction force (given a negative sign by convention) and the very short-range repulsion force between two atoms vary with the separation (r) between the atoms (shown by the dotted graphs). The variation of the resultant force

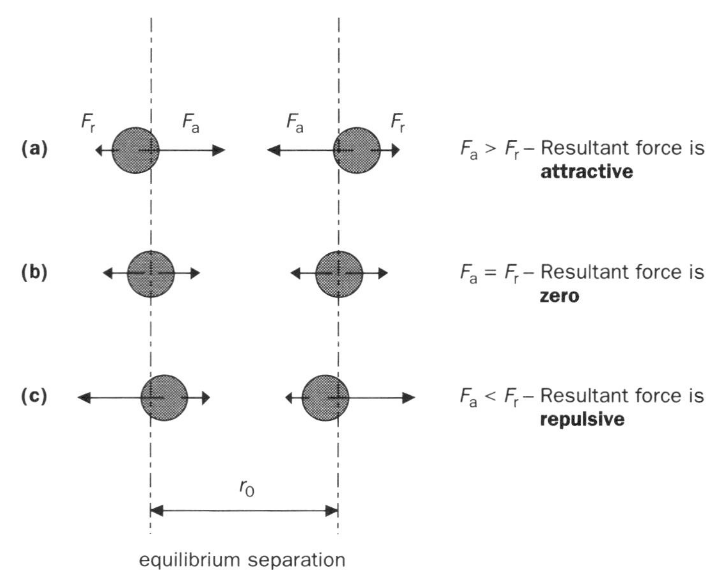

Figure 4.1 Variation of interatomic force between neighbouring atoms with separation distance

(F) with separation (r) is the vector sum of the attractive and repulsive forces, and is shown by the continuous graph.

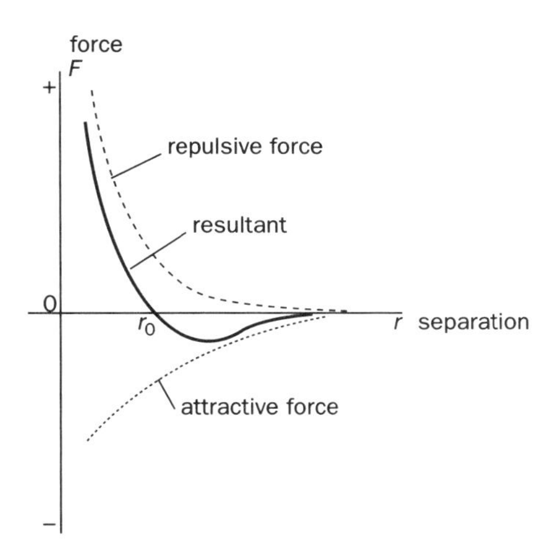

Figure 4.2 Graph of interatomic forces against atomic separation

You should note that:

- At one particular separation called the **equilibrium separation**, r_0, the repulsive force is balanced by the attractive force and so there is **zero resultant force** between the atoms. This is the normal atomic separation distance in a solid. The value of r_0 (typically around 3×10^{-10} m) depends on the atomic size and arrangement of the solid.
- If a compressive force is applied (i.e. we try to push the atoms closer together) the resultant force becomes **repulsive** and work has to be done against this repulsive force in order to decrease r. Solids are very difficult to compress because the repulsive force becomes very large for very small decreases in r from the equilibrium separation r_0.
- When a tensile force is applied (i.e. we try to pull the atoms further apart) work has to be done against a resultant **attractive** force in order to increase r from the equilibrium separation r_0.
- As a result of their thermal energy the atoms in a solid vibrate about their equilibrium separation (r_0), alternately attracting and repelling each other and exchanging kinetic and potential energy. For small amplitude vibrations, the curve is approximately linear near the equilibrium separation position, so the motion of the atoms approximates to SHM. This linear portion of the graph also explains **Hooke's law** for elastic solids: i.e. that change in length is proportional to the force applied (up to the elastic limit).
- Raising the temperature of the material increases the amplitude of vibration of the atoms. When this occurs, the equilibrium separation r_0 increases slightly – this is the phenomenon of **expansion**.
- If the solid is heated sufficiently the vibrational amplitude of the atoms may increase enough so as to overcome the attractive forces and so break the bonds – **melting** occurs. This requires an input of energy sufficient to break the bonds of all the atoms in the solid.

Variation of potential energy with separation

Figure 4.3 shows how the interatomic **potential energy** of a pair of atoms varies with the distance between them. (The corresponding interatomic force–separation graph is reproduced above it for comparison purposes.)

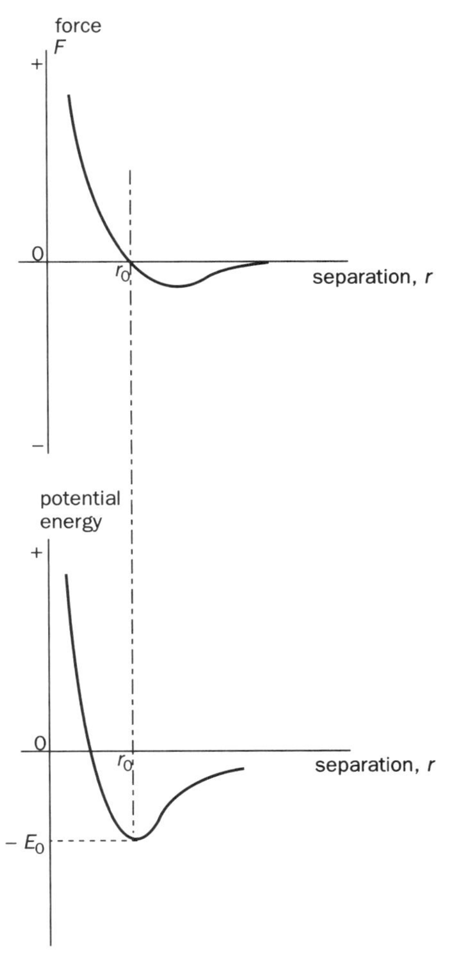

Figure 4.3 Comparison of interatomic force–separation and potential energy–separation graphs

You should note that

- The potential energy has its **minimum** value ($-E_0$) at the **equilibrium** separation distance (r_0). At this separation, the **resultant force** between the atoms is **zero**.
- To increase or decrease the separation distance from r_0, work has to be done against an **opposing** force. If the atoms are to be pushed closer together, work is done against the **repulsive** resultant force, while work is done against the **attractive** force if

the molecules are pulled further apart. In either case, the input of work results in an increase of potential energy.

- E_0 is sometimes called the **bonding energy**. It is the amount of work which would be required to separate the two atoms to an infinite distance apart. At this distance, the potential energy would be zero, and the resultant force between them would also be zero.
- It is generally true that the **area** bounded by a force–distance graph and the distance axis is equal to the **work done** by the force. The shaded area in Figure 4.4 is therefore equal to the work done in moving the atoms from their equilibrium separation to infinity, i.e. the **bonding energy** (E_0). (See Elastic energy on page 338.)

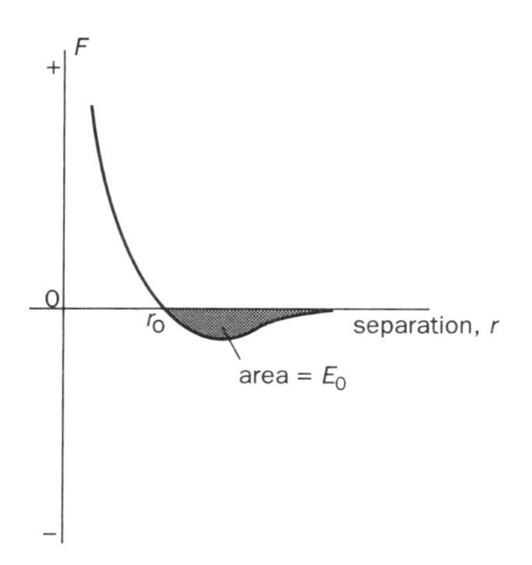

Figure 4.4 Bonding energy (E_0) obtained from the force–separation graph

- If we look at the potential energy–separation graph, the **force** at any separation can be calculated by measuring the **gradient** of the graph.

Mathematically:

$$F = -\frac{dE}{dr}$$

Look carefully at Figure 4.3, and compare the gradient of the energy–separation curve on either side of the equilibrium position:

- when $r < r_0$ the gradient is **negative**, so the resultant force is **positive** (i.e. **repulsive**)
- when $r > r_0$ the gradient is **positive**, so the resultant force is **negative** (i.e. **attractive**)
- when $r = r_0$ the gradient is **zero**, so the resultant force is **zero**.

Solid structures

In a liquid or a gas the atoms (or molecules or ions), which we shall refer to simply as the 'particles', are able to move around quite freely. Thus a substance in either of these states has no fixed shape. In solids, however, the particles can only vibrate about mean, fixed positions and it is this lack of mobility which gives solids the very characteristic property of a definite fixed shape. The behaviour of a solid depends not only on the nature of its constituent particles, but also on the way in which they are arranged. Solids may be classified as being either crystalline, amorphous or polymeric and these groupings are related to the internal arrangement of the particles.

Crystalline solids

All the metals (with the possible exception of mercury, which is in the liquid state at room temperature) and many minerals are classified as crystalline. In this type of solid the particles are arranged in a regular

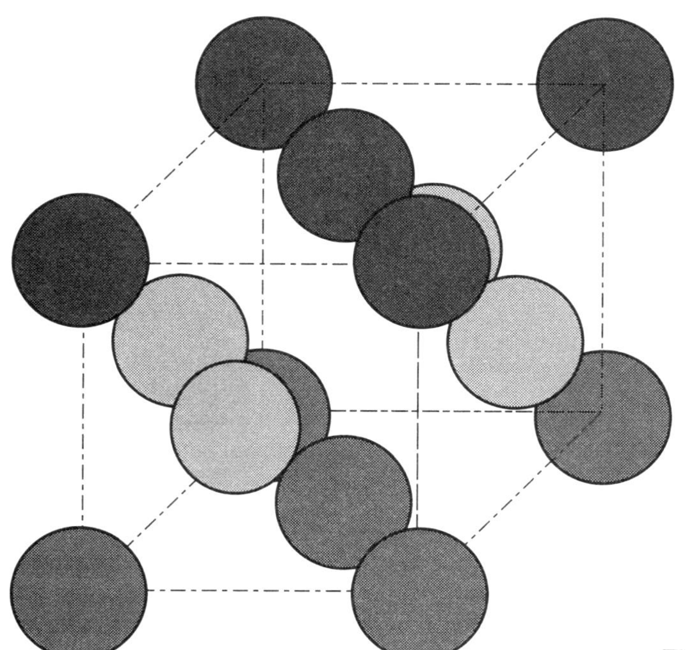

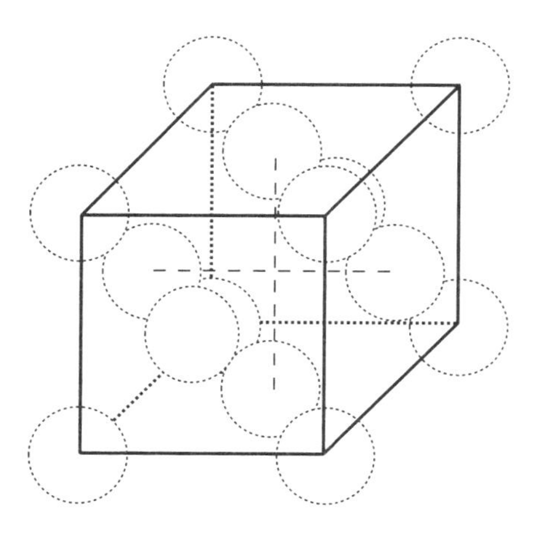

Figure 4.5 Face-centred cubic (FCC) structure

pattern which is repeated in three dimensions throughout the material. The basic repeating unit from which the solid is formed is called a **unit cell**, some examples of which are shown.

In the face-centred cubic (FCC) structure shown in Figure 4.5, the particles are closely packed with one at the centre of each of the six cube faces as well as the eight at the corners. Gold, aluminium and copper crystals are built up from this type of unit cell and salt crystals may be regarded as two interpenetrating FCC structures.

Magnesium and zinc form hexagonal close-packed (HCP) crystals which consist of particle layers in which each particle is surrounded by a hexagonal ring of six others which are in contact with it (see Figure 4.6). The FCC and HCP structures give the closest possible particle packing, and 60% of all metals form crystals of this type.

The less stable metals (e.g. sodium and potassium) form crystals having the more open body-centred cubic (BCC) structure shown in Figure 4.7 in which there is a particle at the centre of the cube in addition to one at each corner.

You should note that:

- With some substances, such as salt or sugar for example, the crystalline form is readily visible. These substances are actually in the form of small, single crystals. The crystals can grow to various sizes, but the shape of the crystal remains clearly visible.

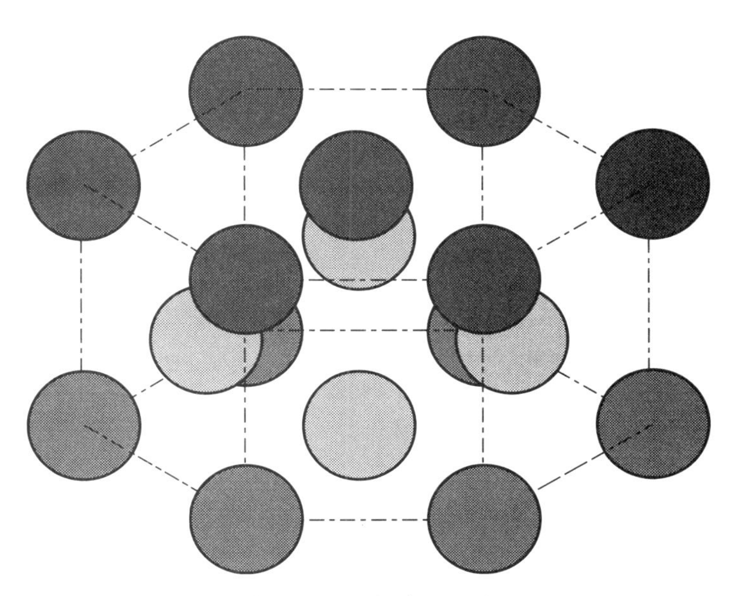
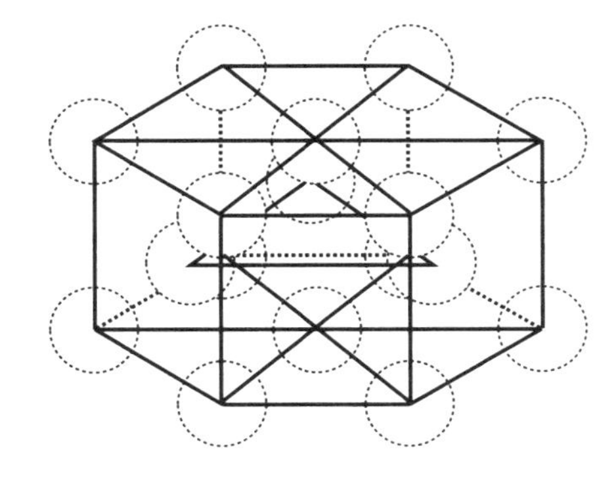

Figure 4.6 Hexagonal close-packed (HCP) structure

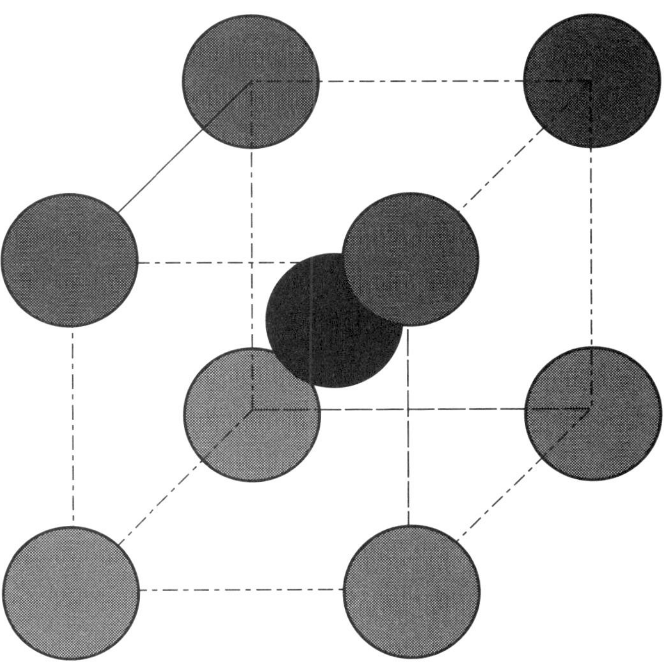
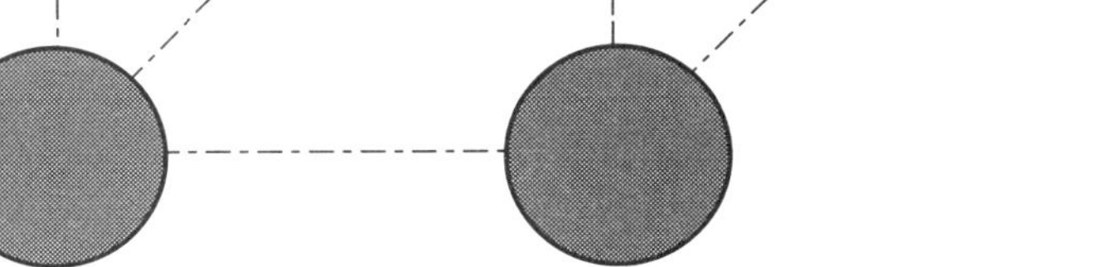
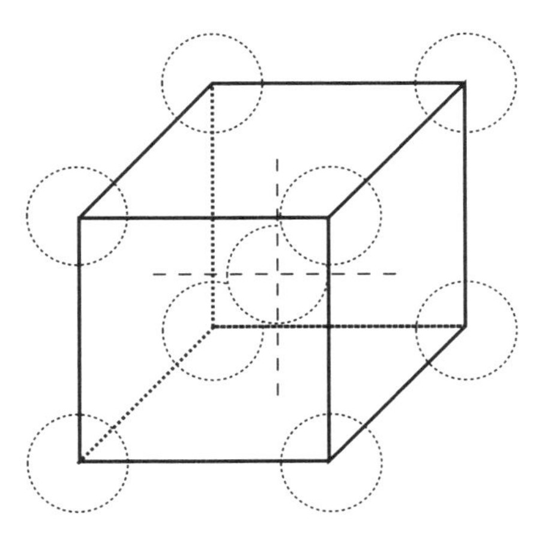

Figure 4.7 Body-centred cubic (BCC) structure

- In other cases the crystalline structure is not quite so obvious. For example in the case of most metals the crystalline structure can only be seen under a microscope after the metal surface has been polished and etched with a dilute acid. Closely-packed, tiny crystals which are known as **grains** are then seen.
- Many crystalline substances, and the metals in particular, are **polycrystalline**. This means that they consist of a large collection of tiny crystals (grains) which are all pointing in different directions. The crystals are joined together at the grain boundaries and this adds strength to the material.
- The particles in a crystalline solid are regularly arranged so as to give a structure having the lowest possible total potential energy and hence the greatest stability. This kind of particle arrangement means that all parts of a crystalline solid need the same amount of energy to cause melting and so such solids have definite melting points. The physical properties of a crystalline solid are very dependent on the particle arrangement within the solid. The two very different forms of carbon, diamond and graphite, whose crystal structures are shown in Figure 4.8, are a vivid illustration of this point.

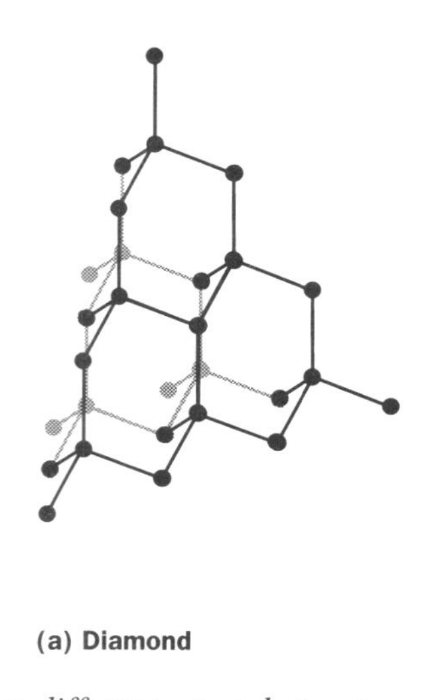

(a) Diamond

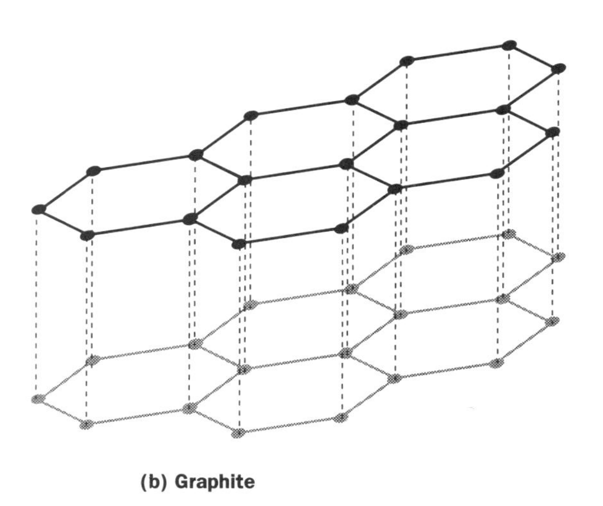

(b) Graphite

Figure 4.8 Two different crystal structures of carbon

Scanning electron micrograph of common table salt (99% sodium chloride). In its pure state the salt crystals are near perfect cubes, but the edges of the grains in this micrograph have been slightly rounded during manufacture

- In contrast to the extreme hardness of diamond, graphite is the soft, flaky material which is used as the 'lead' in pencils. Diamond is transparent and may be classed as an electrical insulator whereas graphite is opaque and a good electrical conductor. Diamond's hardness is partly due to its strongly-bonded tetrahedral structure, a particle arrangement which gives the material great strength and rigidity. In graphite the fact that the carbon atoms are formed as hexagonal layers weakly bound to each other means that they can slide over each other, and this explains its intrinsic softness and flakiness.
- In a single crystal, equivalent atomic planes are parallel to each other, and this gives the crystal its characteristic shape. Single crystals have physical properties whose magnitude depends on the direction in which they are measured relative to the crystal axis. Crystalline solids in which the physical properties such as tensile strength are directional are called **anisotropic** and those in which they are non-directional are called **isotropic**.
- Particular crystalline substances have crystals of a certain general shape. For example, salt crystals are cubic and those of quartz are hexagonal prisms capped by pyramids. This does not mean that all the crystals of a substance have exactly the same shape. Different growth rates of the crystal faces produce slightly different versions of the ideal shape. Thus salt crystals might well be rectangular prisms rather than perfect cubes and in quartz the ideal hexagonal shape might appear slightly distorted. Nevertheless, the general shape is kept since the angles between the faces are still the same (see Figure 4.9).
- **Cleavage** planes are certain well-defined planes, which are generally parallel to the crystal faces, along which fracture will occur when the crystal is struck. They are the result of weaker atomic bonding along certain crystal planes. Cutting a diamond to produce its many facets is a delicate and skilful process which requires accurate knowledge of the material's cleavage planes.

Amorphous solids

The term **amorphous** (meaning 'without shape or form') is used to describe solid substances in which there is no long-range order in the arrangement of the particles. It can be likened to an instantaneous or 'frozen' picture of the internal structure of a liquid. Unlike crystalline materials, amorphous solids such as glass and wax have no regular shape, no cleavage planes and no directional properties. The lack of structural regularity also means that different parts of such a solid may need different amounts of energy to cause melting and so amorphous solids do not have definite melting points. Despite the fact that they exhibit definite solid properties of fixed shape and physical strength, some amorphous materials can be regarded as highly viscous liquids. Glass, for example, will flow very slowly under the influence of its own weight and a window pane which has been in place for a couple of centuries will actually be thicker towards the bottom as the molecules flow downwards under the influence of gravity.

Polymeric solids

Polymers are organic materials which consist of very long **chains** of carbon atoms bonded to hydrogen and other atoms. The identical molecular units which are joined together to form the polymeric chains are called **monomers**. Polymers can be **natural** (e.g. cellulose, protein, rubber, etc.) or **synthetic** (e.g. polythene, polystyrene, nylon, etc.) From a biological viewpoint natural polymers are an extremely important group of materials. Without the strength and rigidity provided by cellulose, plants and trees would be unable to stand and grow. Our own survival is strongly linked to the intake of the very complex

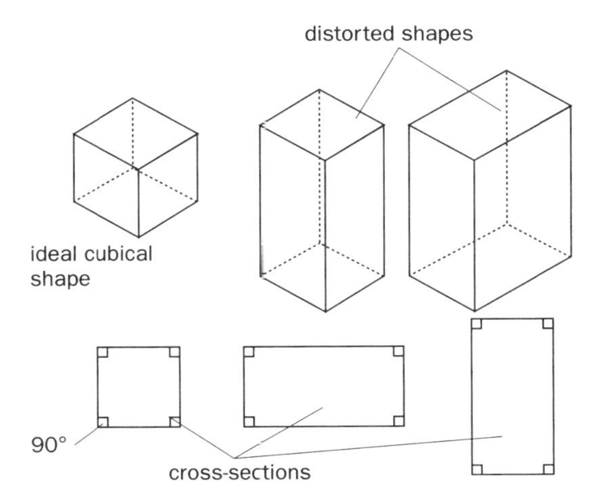

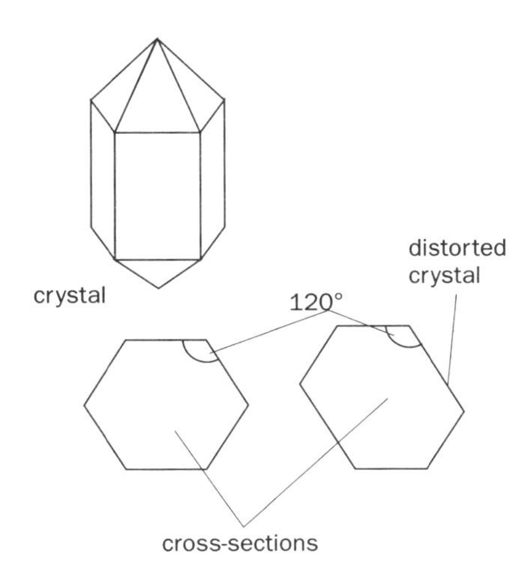

Figure 4.9 General crystal shapes of common crystalline solids

amino acid chains which we call proteins. Synthetic or man-made polymers form the basis of an enormous and ever-expanding plastics industry. The number of everyday articles which are manufactured from artificially created polymeric materials is vast, and constantly growing. The properties of each material are delicately engineered to suit their own specific application. Your own toothbrush is an example of this. The nylon of which the handle is made is very strong and rigid whereas that used for the bristles must combine longitudinal rigidity with some sideways flexibility. In fact, toothbrushes are graded as hard, medium or soft depending on how flexible the nylon bristles happen to be. Modern homes are crammed full of synthetic polymers or 'plastics'.

Look in any kitchen or bathroom and you will see just how many plastic articles we use; the list is endless. How many containers on supermarket shelves are made of plastic? What are the carrier bags made of? Most children's toys are made of one kind of plastic or another.

Synthetic Polymers **Polymerisation** is the production of synthetic polymers by a process which involves a chemical reaction in which thousands of small molecules (the monomers) are linked together to form very large molecular chains. Polyethylene (commonly known as polythene), for example, is produced by the polymerisation of ethylene (C_2H_4), a gas obtained from petroleum. In this process, which is shown in Figure 4.10, the double bonds of a large number of ethylene molecules (the monomers in this case) are broken and this allows them to form the giant polyethylene molecules.

In the case of polyvinyl chloride (PVC) the monomer is vinyl chloride, while for polystyrene, benzene rings are the basic building unit.

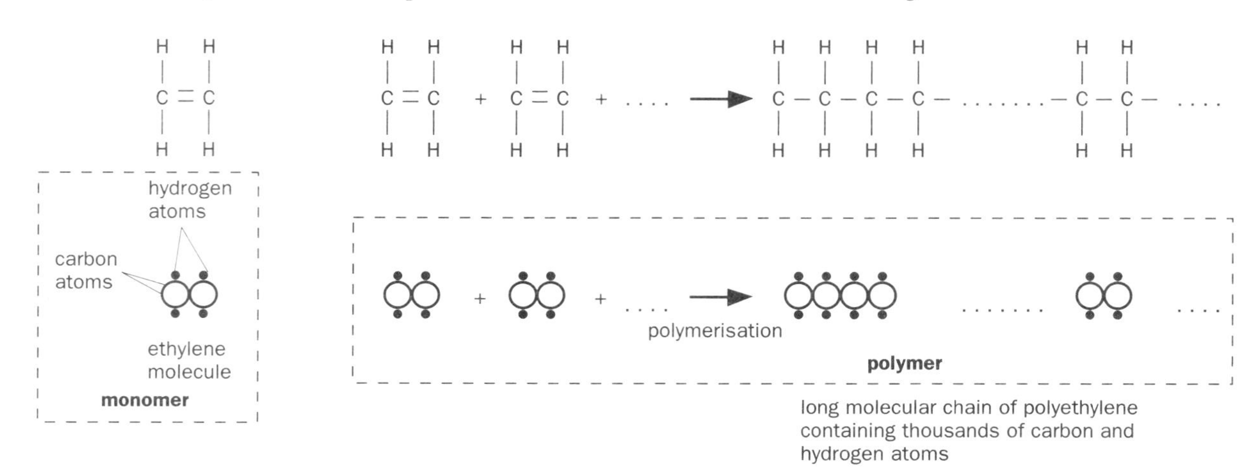

Figure 4.10 Polymerisation of ethylene to produce polyethylene ('Polythene')

These specifically-shaped flexible food and drink containers are an excellent example of the ever-growing number of plastic utensils which have become an integral part of our lives

Branching and cross-linking in polymer molecules

Polymer molecules can be **linear, branched** or **cross-linked** as shown diagrammatically in Figure 4.11 below.

Branching and cross-linking can occur during the chemical polymerisation process. Both of these effects cause the molecules to become more entangled and reduce their ability to move past each other. Thus physical properties such as flexibility, hardness and strength are to a large extent determined by the amount of branching and cross-linking between the polymer molecules. Polythene's flexibility, for example, can be partly attributed to the ease with which its linear molecules can slide over each other. Heavily cross-linked polymers such as ebonite are very rigid because large forces are required to make the molecules move past each other.

Cross-linking is used in the manufacture of vulcanised rubber from its raw form (latex). Latex has very few cross-links between its molecules and so stretches beyond its elastic limit quite easily. Its use is therefore fairly limited. By heating it with a small amount of sulphur the number of cross-links is much increased. The resulting polymer, called vulcanite, is a much more solid and useful material. If the amount of sulphur used is increased to about 4% the very heavy cross-linking created produces a very hard, brittle material called ebonite.

If the molecular chains in a polymer are close and parallel to each other (like the strands in a rope) the structure is termed **crystalline** because it has a certain degree of order. **Amorphous** polymers, on the other hand, have an internal structure which has no order. The molecular chains are completely tangled like the metal strands in wire wool. Linear chain polymers may be crystalline or amorphous; highly branched polymers tend towards being amorphous and highly cross-linked polymers are completely amorphous. Some polymers have both crystalline and amorphous regions. Figure 4.12 shows the different types of internal structure.

Crystallinity in polymers means strength and rigidity whereas amorphous polymers are soft and flexible. This is because although the forces between the molecular chains are weak, in a crystalline structure the chains are close together over relatively large distances and this results in greater stiffness. Thus producing extra crystalline regions provides a second method (cross-linking being the other) for increasing the intermolecular bonding and hence the rigidity of a polymer. The degree of crystallinity also affects both the melting point and density of a polymer. The general effects of crystallinity can be seen by considering the differences in the properties of low-density and high-density polythene. High-density polythene, which is used for 'hard' plastic articles, is more crystalline than the low-density type, which is used for making carrier bags and other flexible items. The greater crystallinity of the high-density type means higher strength, rigidity and melting point.

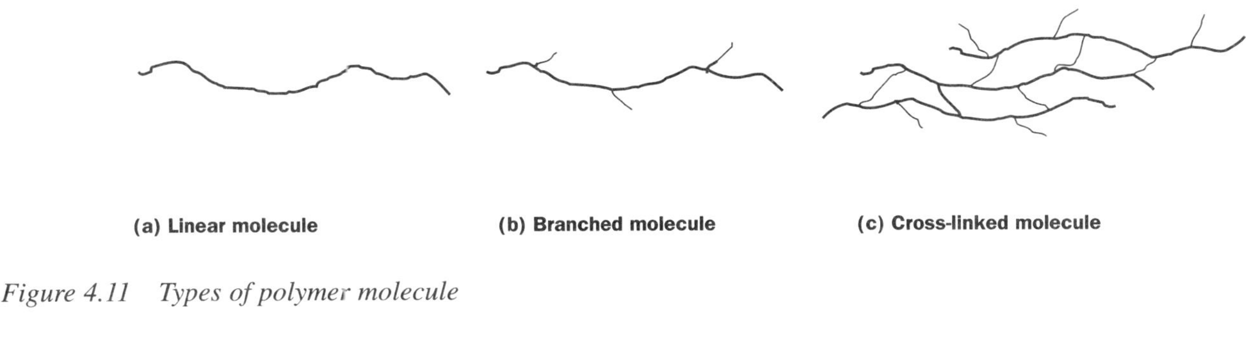

Figure 4.11 Types of polymer molecule

Figure 4.12 Types of polymer structure

General polymer properties

Polymers are:

- good thermal and electrical insulators
- less dense than metals or ceramics
- water and (usually) acid resistant
- attacked by some organic solvents
- more easily worked than metals
- cheaper to produce than metals.

Most or all of the above properties, as well as colour, can be altered by the incorporation of other materials into the polymer.

Effect of external forces on elastic solids

In our study of Newton's laws (Section 1.6) we found that external forces can change the motion of a body. Here we will look at how external forces can affect the shape and size of a body. Changes in shape and size are controlled by a very important material property called **elasticity**. Engineers need to know about the elasticity of the materials they are using, be it in the construction of a bridge or the design of an aircraft. And elasticity is important to us in much less dramatic, but more commonplace objects. The mattress you sleep on and the clothes you wear must retain their elasticity if your comfort is to be ensured. Some materials behave quite normally under the action of external forces. A climber's rope will extend considerably under the action of his weight and a diving board bends when you jump on it. Both of these return to the normal size and shape on removal of the external force. But there are materials whose behaviour can only be described as odd. A perfect example is the substance called 'silly putty'. It stretches under its own weight and so appears to flow when you suspend it, but it can also bounce like a rubber ball when you throw it against a wall!

Engineers and scientists differentiate one material from another by describing mechanical properties such as:

- **Strength**: A material is said to be strong if it can withstand a large force without breaking. Some materials, like stone and concrete, are very useful as load-bearing walls in buildings because of their enormous strength under the action of compressive forces.
- **Stiffness**: This tells us about a material's opposition to changes in shape and size. **Flexibility** is the opposite description. For example, the clothing you wear is comfortable because it is made of materials with very little stiffness, whereas the buttons, zips, etc. which hold the garments together are made of rigid, unyielding materials.
- **Ductility**: This describes a material's ability to suffer permanent deformation without fracture. Ductile materials (e.g. copper) can be permanently deformed or 'worked' (i.e. hammered, pressed, drawn, cut, etc.).
- **Brittleness**: Brittle materials (e.g. cast iron) crack easily. Such materials are very stiff but not very strong. A good example of a brittle object is a boiled sweet which can suddenly snap when you bite it with the rather painful side effect of biting your tongue!
- **Hardness**: This is a measure of the difficulty of marking or scratching a material. Stone is harder than wood and so a piece of stone will scratch a wooden table. The hardest material known to man is diamond in which the carbon atoms are bonded together in a tetrahedral structure which makes them difficult to dislodge.

Much information about the mechanical properties of materials can be obtained by applying increasing tensile forces to samples and measuring the corresponding extensions produced. In order to eliminate the effects of shape and size of the samples tested, force and extension are replaced by **stress** and **strain**. These two quantities are fully defined below.

Stress

Common symbol	σ (Greek 'sigma')
SI unit:	newton metre^{-2} (N m^{-2})
	unit name – pascal (Pa)

> The **stress** on a material is defined as the **force** acting per unit cross-sectional **area** of the material.

A **tensile** stress is one which causes (or tends to cause) an **increase** in length and a **compressive** stress is one which results in a **decrease** in length. Figure 4.13 shows a wire of length (l) and cross-sectional area (A) which has a tensile force (F) applied to it.

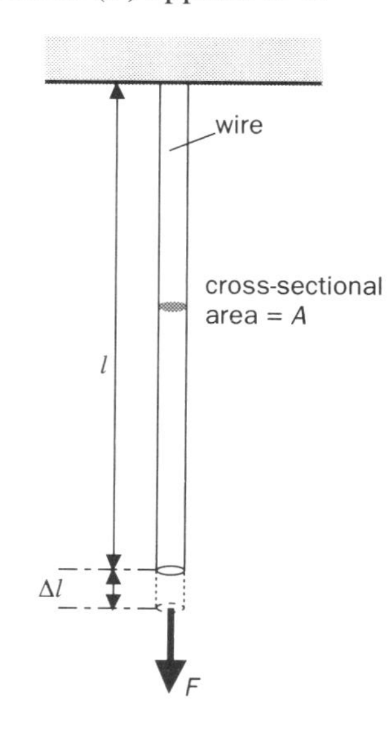

Figure 4.13 Tensile force stretching a wire

By definition, the **tensile stress** in the wire is given by:

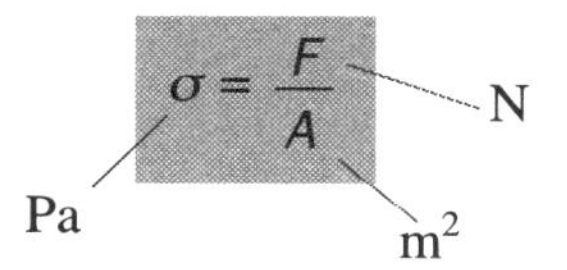

Equation 2
(Strictly speaking, when the wire stretches, the cross-sectional area will decrease slightly. This effect can usually be ignored for small strains, but becomes significant when materials are subjected to large deformations.)

The **breaking stress** of a material sample is the **maximum** stress it can withstand without fracture.

Strain

Common symbol e
No unit – strain is a dimensionless quantity (i.e. a pure number)

The **strain** *(e)* of a material sample is defined as the **extension** produced per **unit length.**

Strain is a measure of the extent of the deformation produced by an applied stress. For the wire shown in Figure 4.13:

the tensile strain (e) is given by:

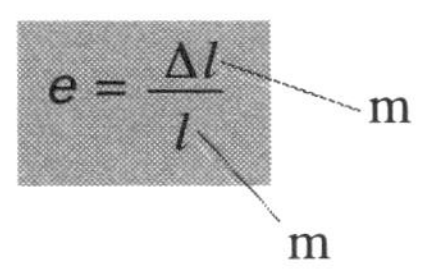

Equation 3

Since both Δl and l are in metres, e has no unit.

Stress–strain graph

When a ductile material (e.g. copper) in the form of a wire is progressively loaded until it breaks, the stress–strain graph obtained from the results is as shown in Figure 4.14.

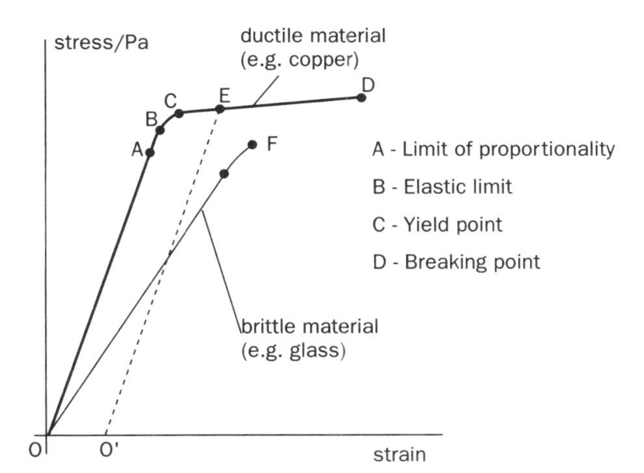

Figure 4.14 Stress–strain graphs for ductile and brittle materials

You should note that:

- Section OA is called the **'proportionality region'**. The strain of the wire is directly proportional to the applied stress, which implies that **Hooke's law** is obeyed (i.e. that the extension of an elastic body is directly proportional to the applied force, so long as the extension is small). In the proportionality region the wire suffers **elastic** deformation (i.e. it returns to its original length when the deforming stress is removed). Referring to the interatomic force–separation graph shown in Figure 4.2, it can be seen that for values of r close to the equilibrium separation (r_0) the graph is approximately linear. This shows that for small displacements from r_0, extension is directly proportional to force (i.e. Hooke's law applies).
- Beyond the **limit of proportionality** (A) Hooke's law no longer applies (i.e. strain is no longer proportional to stress). However, the material still behaves **elastically** – it will return to its original size and shape when the stress is removed.
- The **elastic limit** (B) is the maximum stress which the wire can experience and still regain its original dimensions once the stress is removed. With some materials points (A) and (B) coincide.
- The deformation continues up to a stress value corresponding to point (C), called the **yield point,** where the deformation shows a marked increase for only small increases in load.
- Plastic behaviour occurs for stresses beyond the elastic limit. If stress greater than this value were applied, the wire would retain permanently some of its extension after the stress is removed. For example if the stress corresponding to point (E) is applied to the wire, a permanent strain = OO' will remain after the stress is removed.
- At point (D) – called the **breaking stress** – the wire develops a 'waist' – this is often referred to as 'necking' – and fracture occurs.
- The stress–strain graph OF is obtained for a brittle material such as glass. In this case fracture follows very closely on the elastic limit with no significant plastic deformation occurring.

The Young modulus

Common symbol E
SI unit newton metre^{-2} (N m^{-2})
unit name – pascal (Pa)

As we noted from the stress–strain graph, so long as the limit of proportionality is not exceeded stress is proportional to strain.

The **modulus of elasticity**, also called the **Young modulus** (E), is defined by:

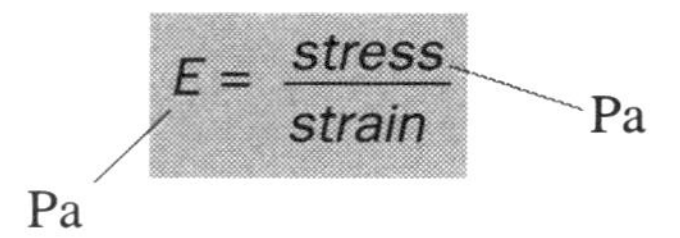

Equation 4

The modulus of elasticity is constant for any particular material, and is a measure of a material's resistance to changes in length. (i.e. it measures elastic **stiffness**). The higher the value of E, the greater stress is needed to produce a given extension.

If we consider a wire of length (l) and cross-sectional area (A) in which a force (F) produces an extension (Δl), then,

$$E = \frac{stress}{strain}$$

$$E = \frac{F/A}{\Delta l/l}$$

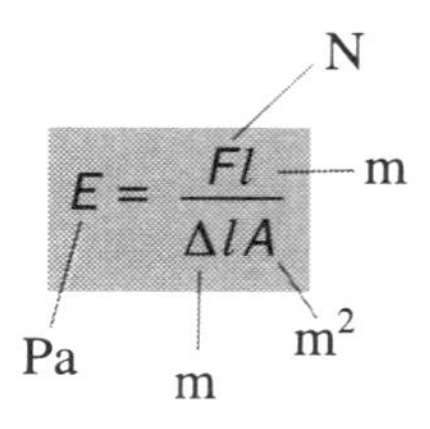

Equation 5

Deductions from stress–strain graphs The general shape, gradient and extent of the stress–strain curve for a material contains a great deal of information about the material's behaviour under stress. From studying such graphs, engineers can select the most suitable material for an application. Figure 4.15 is a typical stress–strain graph for a pure metal.

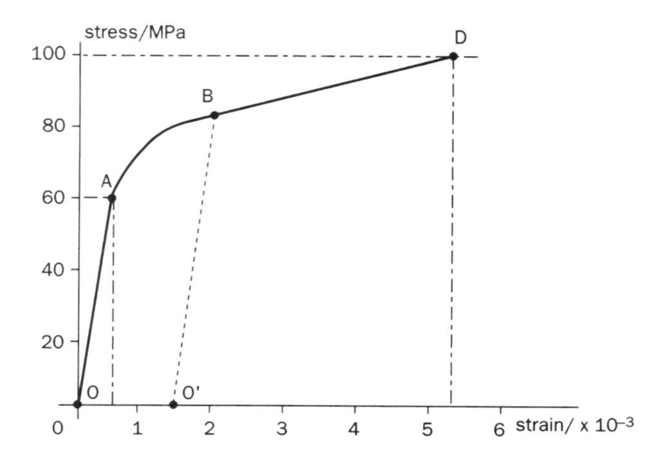

Figure 4.15 Stress–strain graph for a pure metal

Deductions from the graph:

- The **gradient** of the graph in the proportional region (OA) is steep, indicating that this is a relatively **stiff** material with a high value of E. The actual value for this material is approximately:

$$E = \frac{60 \times 10^6}{0.5 \times 10^{-3}} = 1.2 \times 10^{11} \text{ Pa (120 GPa)}$$

- The **breaking stress** of this material is around 100 MPa. The strain at this point is around 5.3×10^{-3} (0.53%).
- Beyond the elastic limit (A) the sample exhibits **plastic** behaviour. If the sample is stretched to point B, removal of the stress means that the sample will suffer a **permanent strain**, indicated by OO' on the graph. In this case, the sample would have a permanent strain of 1.5×10^{-3} or 0.15%. The extent of the **plastic region** indicates relatively high **ductility** – this material could be readily worked into new shapes.

The stress–strain graph of a typical brittle material such as electrical porcelain is shown in Figure 4.16.

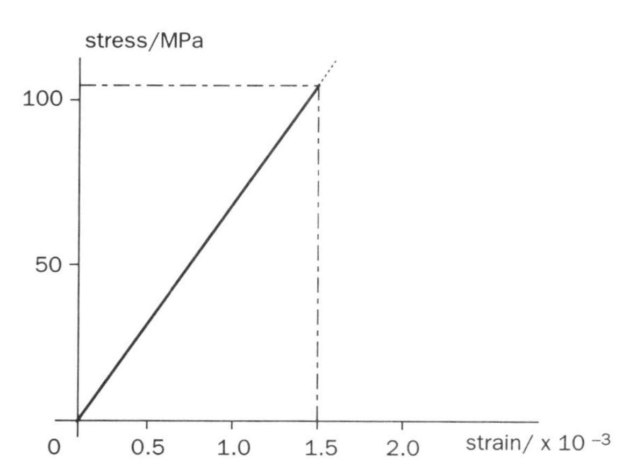

Figure 4.16 Stress–strain graph for electrical porcelain

Deductions from the graph:

- The gradient is fairly steep indicating a fairly high stiffness (note that the scales are different from the previous graph). The value of E is approximately:

$$E = \frac{105 \times 10^6}{1.5 \times 10^{-3}} = 7.0 \times 10^{10} \text{ Pa (70 GPa)}$$

- The breaking stress (105×10^6 Pa) occurs at a strain of 1.5×10^{-3} (0.15%).
- The material is **brittle** – it breaks easily and cannot be permanently stretched.

Guided examples (2)

1. A nylon string of diameter 0.40 mm and length 2.5 m is suspended vertically from a fixed point and its free end is loaded with a mass of 0.50 kg. Given that the Young modulus for the nylon is 3×10^9 Pa, calculate the amount by which the string extends.

Guidelines

Calculate the cross-sectional area of the nylon string, and use Equation 4 to calculate the extension of the string (Δl).

2. A mass of 0.50 kg is attached to a long steel wire of diameter 0.50 mm and whirled in a horizontal circle of radius 1.0 m. Calculate

(a) the angular velocity in rad s^{-1} of the mass at which the wire snaps

(b) the extension produced in the wire just before it snaps.

(For steel, the Young modulus $E = 2.1 \times 10^{11}$ Pa and breaking stress = 6.0×10^8 Pa.)

Guidelines

(a) Calculate the cross-sectional area of the wire. When the wire just snaps, the stress in the wire will be equal to the breaking stress. Use Equation 2 to calculate the tension in the wire. This tension will be the **centripetal force** on the mass.

Use $F = m\omega^2 r$ to calculate the value of ω (the angular velocity) at this tension.

(b) Re-arrange Equation 4 to calculate Δl. Remember to use SI units throughout.

Measurement of the Young modulus

Two wires P and Q of the same material, length and diameter are hung from a common support, as shown in Figure 4.17.

Wire Q is the wire under test and wire P is the comparison wire. The use of a comparison wire as a reference standard avoids errors due to (i) expansion as a result of temperature changes, (ii) sagging of the support. Any small increases in length of wire P are accurately measured by the vernier arrangement between P and Q. The original length (l) of Q is measured using a metal tape measure, and the cross-sectional area (A) is obtained by measuring the diameter of Q using a micrometer at several points and in at least two directions across the wire. These precautions are necessary to minimise errors due to ovality and unevenness in the cross-section of the wire. The mean diameter is then calculated and used to calculate the cross-sectional area.

The test wire is then incrementally loaded and the corresponding extensions are measured and noted. The results are used to plot a graph of load (F) against

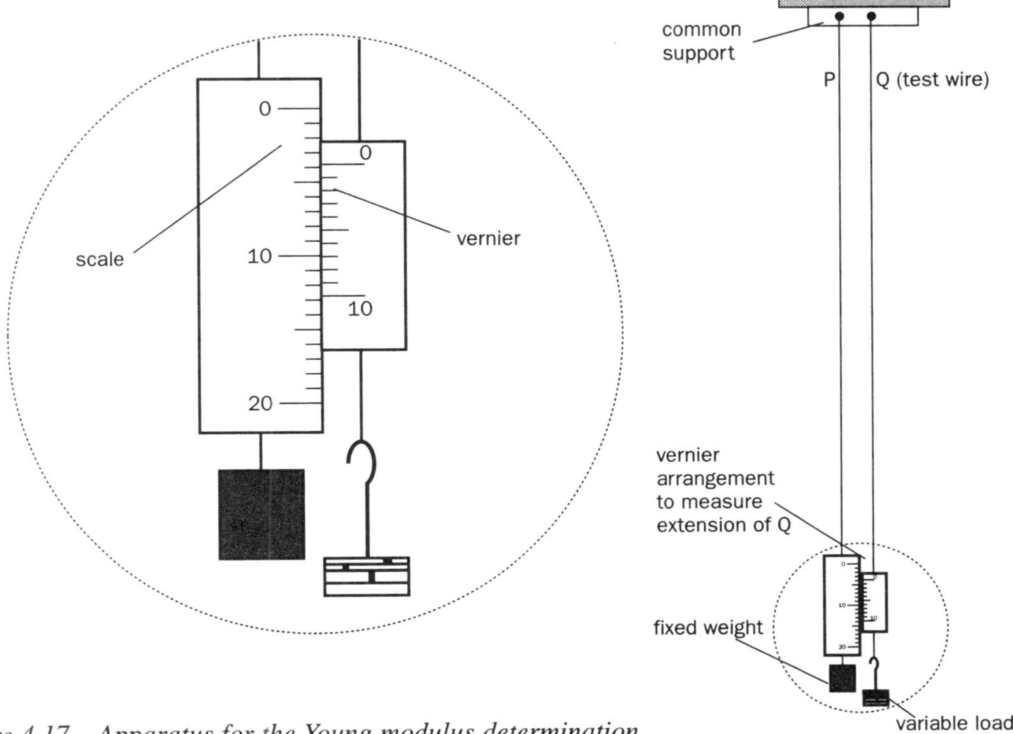

Figure 4.17 Apparatus for the Young modulus determination

extension (Δl). So long as the limit of proportionality is not exceeded, a straight line graph as shown in Figure 4.18 is obtained.

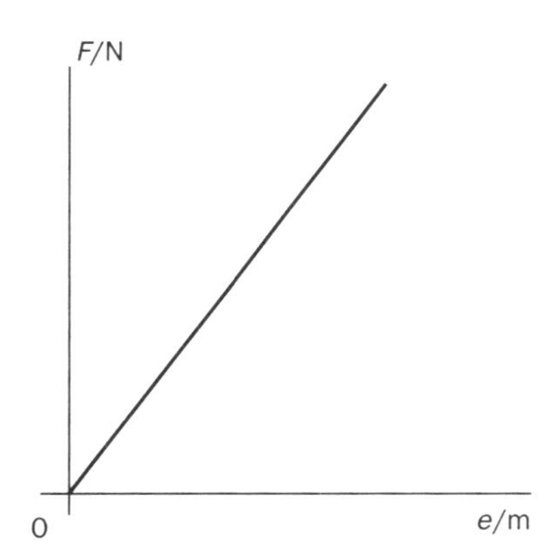

Figure 4.18 Graph of force against extension for a steel wire

Using Equation 5:

$$E = \frac{F/A}{\Delta l/l} = \left(\frac{F}{\Delta l}\right)\frac{l}{A}$$

but $F/\Delta l$ is the **gradient** of the graph of force against extension.

$$\therefore \quad E = \text{gradient of graph} \times \frac{l}{A}.$$

* For increased accuracy the test-wire is gradually **unloaded** and a new set of corresponding extension measurements is obtained at each tension. The **mean** extension at each load would then be used to plot the force–extension graph.

Elastic (strain) energy

When a wire is strained by fixing at one end and attaching a mass to the free end, the gravitational potential energy lost by the mass is stored as elastic energy in the wire.

Consider a wire of unstretched length (l) which extends by an amount (Δl) under the action of a force (F) whose value does not exceed the limit of proportionality (i.e. Hooke's law is obeyed) (Figure 4.13). A graph of force against extension is as shown in Figure 4.18.

The work done by F is equal to the elastic energy stored in the wire.

elastic energy = work done
= average force × extension
(average force must be used since the force is not constant)

$$= \frac{0 + F}{2} \times \Delta l$$

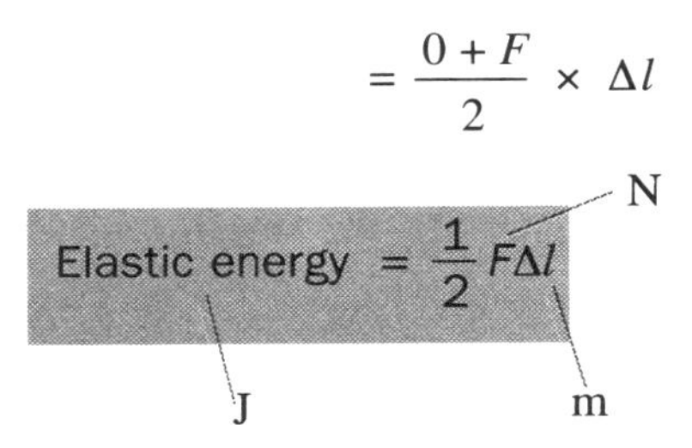

Equation 6

NB: This is the **area** enclosed by the **force–extension** graph.

Up to the elastic limit, the **gradient** of the graph is constant (= $F/\Delta l$)

$$\therefore \quad F = k\Delta l$$

where k is the gradient of the force–extension graph. (This is sometimes referred to as the **elastic constant**.)

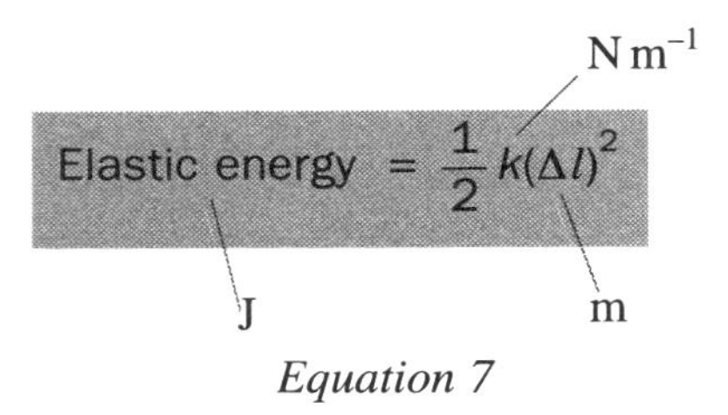

Equation 7

In general, even if the relationship between force and extension is non-linear, the energy stored can be obtained by calculating the area between the curve and the extension axis.

SECTION A

Self-assessment

Qualitative assessment

1. Differentiate between a **crystalline** and an **amorphous** solid. Give one example of each.

2. What does the term **unit cell** mean when used with reference to a crystalline substance?

3. Most metals are **polycrystalline**. Explain what is meant by this term.

4. Explain why crystalline solids have a definite melting point and amorphous solids do not.

5. Distinguish between **isotropic** and **anisotropic** crystalline substances.

6. Explain why different crystals of a particular crystalline substance can have variations on the 'ideal' shape.

7. What are **cleavage planes**? Explain their importance in the production of a finished cut diamond.

8. (a) What is a polymer?
   (b) Give an example of
      (i) a **natural** polymer
      (ii) a **synthetic** polymer.
   (c) Briefly explain how polythene is produced by polymerisation.

9. (a) Distinguish between **crystalline** and **amorphous** polymers.
   (b) State the general effects of increased crystallinity in polymers.

10. The graph ABCD shows how the **resultant force** between a pair of atoms varies with the **distance** between the atoms.
   (a) Over what section of the graph is the resultant force
      (i) repulsive
      (ii) attractive?
   (b) What distance represents the **equilibrium separation** of the atoms? What is the resultant force acting at this distance?
   (c) Explain what is meant by stating that 'the equilibrium separation is stable'.

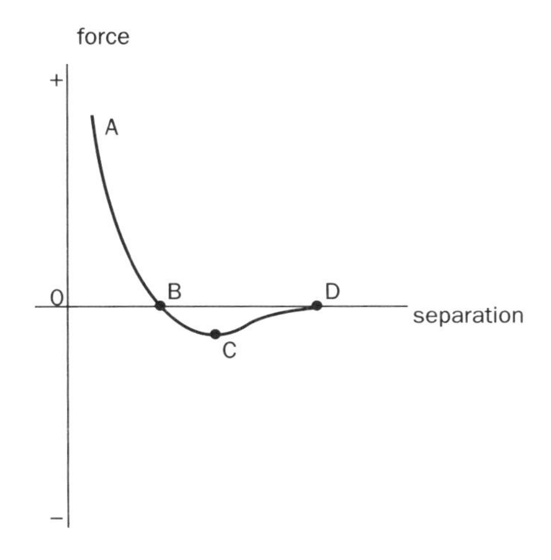

11. (a) Copy the force–separation graph of Question 10 and draw a corresponding graph of **potential energy** against separation directly beneath it.
   (b) Mark the position of the **equilibrium separation** (r_0) on both graphs.
   (c) At what distance does the system of the two atoms have **minimum** potential energy? What is this minimum potential energy called, and how is it defined?
   (d) Explain what happens in energy terms if the two atoms are:
      (i) pulled further apart than (r_0)
      (ii) pushed closer together than (r_0).
   (e) What is the physical significance of the **gradient** of the potential energy–separation curve?
   (f) Explain what is meant by **bonding energy**.

12. The terms **strength, hardness, ductility, brittleness** and **stiffness** are used to describe the mechanical properties of materials. Briefly explain what is meant by each of these terms and give examples of materials displaying these properties.

13. Define and state the SI unit for:
   (a) tensile **stress**
   (b) tensile **strain**
   (c) the **Young modulus** of elasticity.

14. For a metal wire which obeys Hooke's law, the extension produced by a force F is x.
   (a) State the extension when the force is
      (i) $2F$
      (ii) $3F$.
   (b) State the extension produced in a wire of the same material having **twice** the diameter when the force is
      (i) F
      (ii) $4F$.

Self-assessment *continued*

15. Sketch a stress-strain graph for a ductile material. Mark on the graph, and explain, the following:

(a) proportionality region

(b) limit of proportionality

(c) elastic limit

(d) yield point

(e) breaking stress.

16. A length of copper wire is said to show **elastic** behaviour when it is strained within its elastic limit, and **plastic** behaviour when it is strained beyond this value. Explain these terms and sketch a stress–strain graph to show the effect of removing a load from a copper wire which has strained it beyond its elastic limit.

17. Compare **brittle** and **ductile** materials by sketching stress–strain graphs for each.

18. Describe an experiment to measure the Young modulus for steel in the form of a thin wire. State the measurements to be made, and the precautions taken to minimise experimental errors. Describe how the measurements made are used to calculate the Young modulus.

19. A wire of length l and elastic constant k is found to extend by an amount Δl under the action of a force F, whose value does not exceed the limit of proportionality.

(a) Derive an expression for the elastic energy stored in the wire.

(b) Describe how the elastic energy stored in the wire can be obtained from the graph of force against extension.

20.

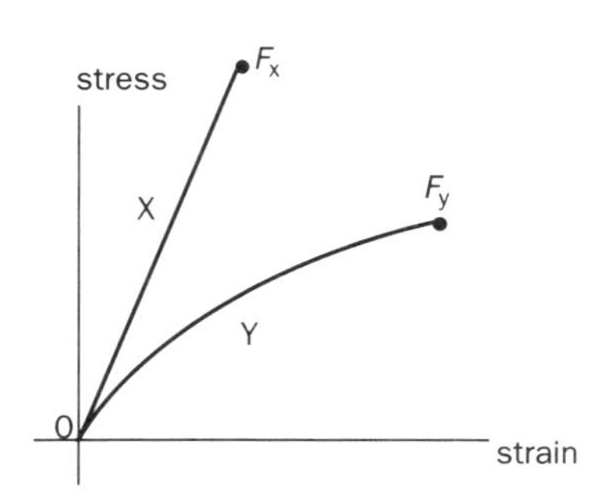

In the stress–strain graphs for two different materials X and Y, F_x and F_y are the points at which fracture occurs for each material. State which of the two materials

(a) obeys Hooke's law up to the point of fracture

(b) has the larger Young modulus

(c) is the stronger.

Give reasons for your answers in each case.

SECTION B
Quantitative assessment

(***Answers:*** 1.0×10^{-3}; 0.014; 0.050; 0.40; 0.80; 63; 1.7×10^{3}; 1.3×10^{8}; 2.0×10^{8}; 5.1×10^{8}; 2.0×10^{9}; 2.0×10^{11}; 2.5×10^{11}.)

1. When a force of 400 N is applied to a steel wire of length 2.0 m and cross-sectional area 2.0 mm^2 it is found to extend by 2.0 mm. Calculate

(a) the tensile stress

(b) the tensile strain

(c) the Young modulus for the wire.

2. A rod of cross-sectional area 8.0×10^{-6} m^2 is made of a material for which the Young modulus is 2.1×10^{11} Pa and whose breaking strain is 1.0×10^{-3}. Assuming Hooke's law is obeyed until the rod fractures, calculate the tensile force needed to cause fracture.

3. An elastic string of cross-sectional area 3 mm^2 and length 2.5 m stretches by 2.5 cm when a force of 4 N is applied to it. Find

(a) the Young modulus for the material of the string

(b) the elastic energy stored in the string when it is stretched.

4.

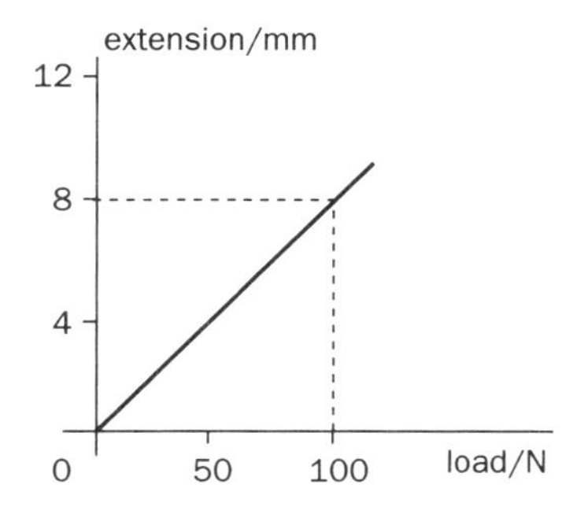

The graph shows the variation of extension with applied load for a wire of length 4 m and radius 0.25 mm.

(a) What is the stress value when the load is 100 N?

Self-assessment *continued*

(b) What is the elastic energy stored in the wire when the load is 100 N?

(c) What is the decrease in gravitational potential energy of the mass which gives the 100 N load?

(d) Why are answers (b) and (c) different?

(e) Calculate the Young modulus for the wire material.

5. A copper wire and a steel wire each of diameter 1.0 mm and length 0.75 m are joined end to end. The composite wire is then suspended from a rigid point and loaded so as to produce an extension of 0.8 mm. If the Young modulus for steel and copper is 2.0×10^{11} Pa and 1.2×10^{11} Pa respectively, calculate the load on the wire.

6. A nylon guitar string of length 0.7 m and cross-sectional area 0.2 mm^2 requires a force of 4.0 N to increase its length by 1.0%. Calculate

(a) the Young modulus for nylon

(b) the energy stored in the string when stretched by this amount.

4.2 Thermal properties

Temperature

Common symbol	T or θ
SI unit	kelvin (K)

Temperature is a term which we commonly use to describe how hot or cold an object is. It is sometimes defined as the **degree of hotness** (or **coldness**) of a system. However, using the physiological sensation of hotness is not the best way to define temperature. Whether an object feels hot, warm or cold when touched depends not only on the individual's sense of touch, but also on the material of the object. The metal parts of a school desk, for example, feel much cooler than the wooden lid and yet they are at the same temperature. Temperature, like mass, length and time, is one of the fundamental quantities and as such it cannot be defined in terms of other quantities.

The best way to define temperature is to use the idea of heat flow resulting from a temperature difference in much the same way as liquid flow is caused by a pressure difference. If two bodies A and B, initially at different temperatures T_1 and T_2, are placed in thermal contact, heat flows from the higher to the lower temperature body until the temperatures equalise. There is then no further transfer of heat and the two bodies are said to be in **thermal equilibrium**. Figure 4.19 illustrates the idea of heat flow and thermal equilibrium between two bodies in contact.

Diagram (a) in Figure 4.19 (page 342) shows a pan of water on a heating plate. Heat is flowing from the hotplate to the pan, which becomes hotter, so the temperature of the hotplate (T_1) must be greater than the temperature of the pan (T_2). In (b), the pan has been placed on a cool surface. T_1 is now less than T_2, so heat flows out of the pan into the surface and the pan becomes colder. If the two bodies are at the same temperature, no heat flows (i.e. heat flow from plate to pan is equal to that from the pan to the plate) and there is no change of temperature. The two bodies are in thermal equilibrium.

We can now define temperature as:

> that property which determines whether or not one body is in thermal equilibrium with another.

Thermal equilibrium is linked with a thermodynamic principle called the **zeroth law** which states that:

> If two bodies A and B are each in thermal equilibrium with a third body C, then A and B must be in thermal equilibrium with each other – **zeroth law**.

This means that if a thermometer (body C) gives the same reading when it is placed in contact with each of two bodies (A and B) we can conclude that A and B are in fact at the same temperature. The definition of temperature as that property which determines thermal equilibrium is therefore a consequence of the zeroth law.

NB: The **zeroth** law was given its rather unusual name because it was proposed some time after the first and second laws of thermodynamics had become accepted. It logically precedes the other two laws and is in fact assumed in them. (Early mountaineers thought there were only five climbable gullies on Ben Nevis, so they named them One to Five, from left to right across the face. Later, a gully to the left of Number One was climbed, so it was named Zero Gully. Now there are Minus One, Minus Two, Point Five, etc.)

The temperature of a body depends on its nature, its mass and the amount of heat energy which has flowed in or out of it. It is measured with a **thermometer** calibrated on a suitable **temperature scale**. You should be careful not to confuse the temperature of a body with the flow of energy as heat. Heat can be thought of as energy on the move. When heat energy flows into a body, its **internal energy** (kinetic and potential energy of its molecules) increases. Heat is energy which flows for reasons other than the application of forces.

Temperature scales

All thermometers measure temperature by measuring a particular **thermometric property** of a substance whose value changes with temperature. A mercury-in-glass thermometer, for example, uses the variation in the length of a mercury column contained in a capillary tube; a thermistor makes use of the change in electrical resistance of a semiconductor with temperature. Having selected a particular thermometric property, a temperature scale is then defined by means of two **fixed points**. These are accurately reproducible single temperatures at which a specific physical event always occurs. The interval between the two fixed points, called the **fundamental interval**, then defines temperature on the scale. Three examples of fixed points are defined below:

> The **ice point** – That temperature at which pure ice can exist in equilibrium with water at **standard atmospheric pressure.***
>
> The **steam point** – That temperature at which steam and pure boiling water are in equilibrium at standard atmospheric pressure.
>
> The **triple point** – That temperature at which the solid, liquid and vapour phases of a particular substance exist in equilibrium. In the case of water we are defining the temperature at which pure ice, water and water vapour can exist in equilibrium.
>
> * Standard atmospheric pressure $= 1.013 \times 10^5$ Pa (N m^{-2})
> It is the pressure exerted at the base of a column of mercury 0.76 m tall.

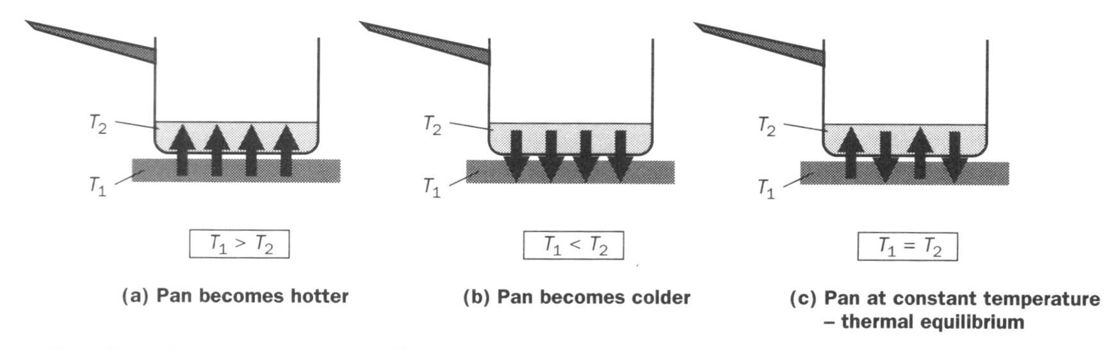

Figure 4.19 Heat flow, temperature difference and thermal equilibrium between two bodies in contact

On the Celsius (C) temperature scale the ice point is taken as zero degrees Celsius (0 °C) and the steam point is 100 degrees Celsius (100 °C) and this gives a fundamental interval of 100 °C. If X_0, X_{100} and X_θ are the measured values of a particular thermometric property at 0 °C, 100 °C and at some unknown temperature θ respectively, then the value of θ in °C is given by:

$$\theta = \frac{(X_\theta - X_0)}{(X_{100} - X_0)} \times 100$$

Equation 1

NB: Common temperature is denoted by θ and is usually measured in °C.

Equation 1 is derived as follows. Figure 4.20 shows the variation of X with θ.

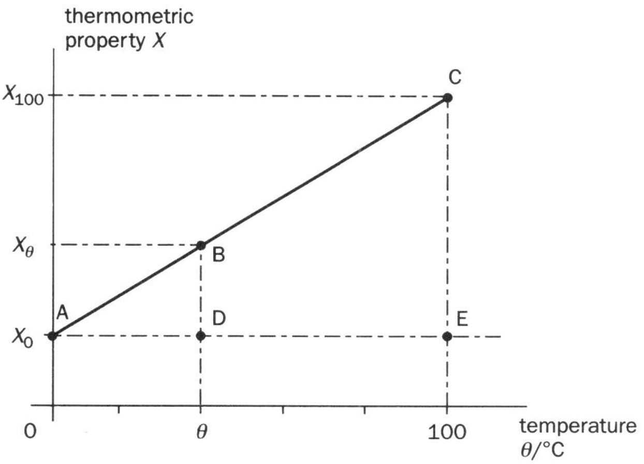

Figure 4.20 Graph of thermometric property against temperature

Triangles ABD and ACE are similar and so:

$$\frac{AD}{AE} = \frac{BD}{CE}$$

$$\frac{(\theta - 0)}{(100 - 0)} = \frac{(X_\theta - X_0)}{(X_{100} - X_0)}$$

$$\frac{\theta}{100} = \frac{(X_\theta - X_0)}{(X_{100} - X_0)}$$

$$\therefore \quad \theta = \frac{(X_\theta - X_0)}{(X_{100} - X_0)} \times 100$$

You should note that:

- The graph is drawn as a straight line.
- A straight line graph **defines** a linear relationship between this thermometric property and temperature measured **with this thermometer**.
- This procedure defines the temperature scale **for this particular property**, and no other.
- When a temperature is measured with a thermometer, the scale used should be quoted (e.g. the temperature of a cup of tea is 78 °C on the mercury-in-glass scale).
- This procedure ensures that the temperature divisions on the thermometer are uniform, producing an easy-to-read scale.

Some common thermometers

When deciding which thermometer is most suitable for measuring a particular temperature many factors, such as accuracy, sensitivity, range, speed of response, size, etc. need to be considered. In each of the outlines which follow, the order of description is:

1. Thermometric property and temperature range
2. Construction/method of use
3. Advantages, disadvantages and typical uses.

Liquid-in-glass thermometer

Thermometric property	Length of mercury (or ethanol) column in a glass tube.
Range	From about – 20 °C to + 350 °C for mercury and about – 100 °C to + 50 °C for ethanol.

Figure 4.21 shows the general construction of a liquid-in-glass thermometer of the type used in the school laboratory.

The thermometer is calibrated by marking the length of the mercury column when the thermometer is measuring the two fixed points (0 °C and 100 °C) and dividing the interval between the marks on the scale into 100 equal divisions. Temperature readings are given directly by the position of the mercury thread along the bore. The tube is graduated in °C or K and it is shaped so as to magnify the thread and so make it clearly visible.

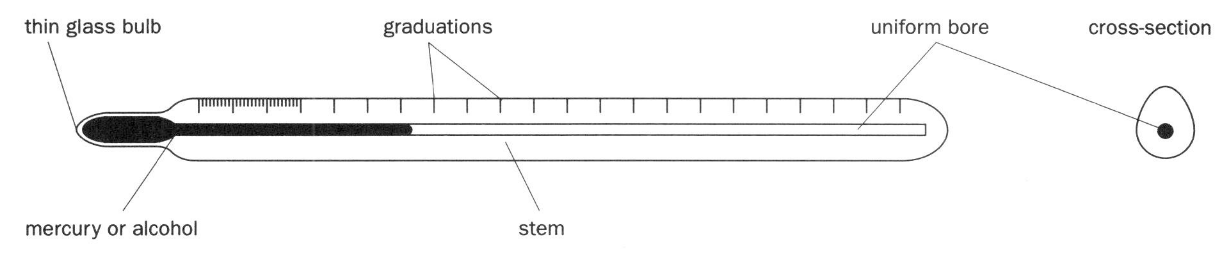

Figure 4.21 Liquid-in-glass thermometer

Advantages:

Simple to use; direct reading; compact; cheap.

Disadvantages:

Limited temperature range; fragile; relatively large heat capacity (compared with thermocouple) which means that it has an effect on the temperature being measured; relatively slow response means that it is less suitable for rapidly-varying temperatures.

Errors:

(i) parallax error in reading
(ii) non-uniformity of capillary bore
(iii) expansion/contraction of glass bulb and stem.

Best suited to measuring fairly steady temperatures (e.g. weather recording), in clinical temperature measurement and laboratory use where temperatures are steady or changing slowly.

Constant-volume gas thermometer

Thermometric property	The pressure of a fixed mass of gas at constant volume.
Range	About –250 °C to +1500 °C.

Figure 4.22 shows a simple version of a constant-volume gas thermometer. The bulb (and hence the air it contains) is placed in the fluid whose temperature is being measured. An increase in temperature causes the pressure of the gas to increase and push the mercury down in tube X and up in tube Y. The height of tube Y is then adjusted so as to bring the mercury level in X back to its original position at the reference mark A. This means that the gas volume has been restored to its original value. The gas pressure (p) is therefore equal to the atmospheric pressure plus the pressure due to the difference in height (h) of the mercury in tubes X and Y.

Using mm of mercury (mm Hg) as the unit of pressure:

$p_\theta = h + p_A$, where p_A is atmospheric pressure (mm Hg).

Equation 1 can be applied, where the thermometric property is the gas pressure, and where p_0, p_{100} and p_θ are the gas pressures at 0 °C, 100 °C and θ °C.

$$\theta = \frac{(p_\theta - p_0)}{(p_{100} - p_0)} \times 100$$

This equation defines temperature on the **constant-volume gas thermometer** scale.

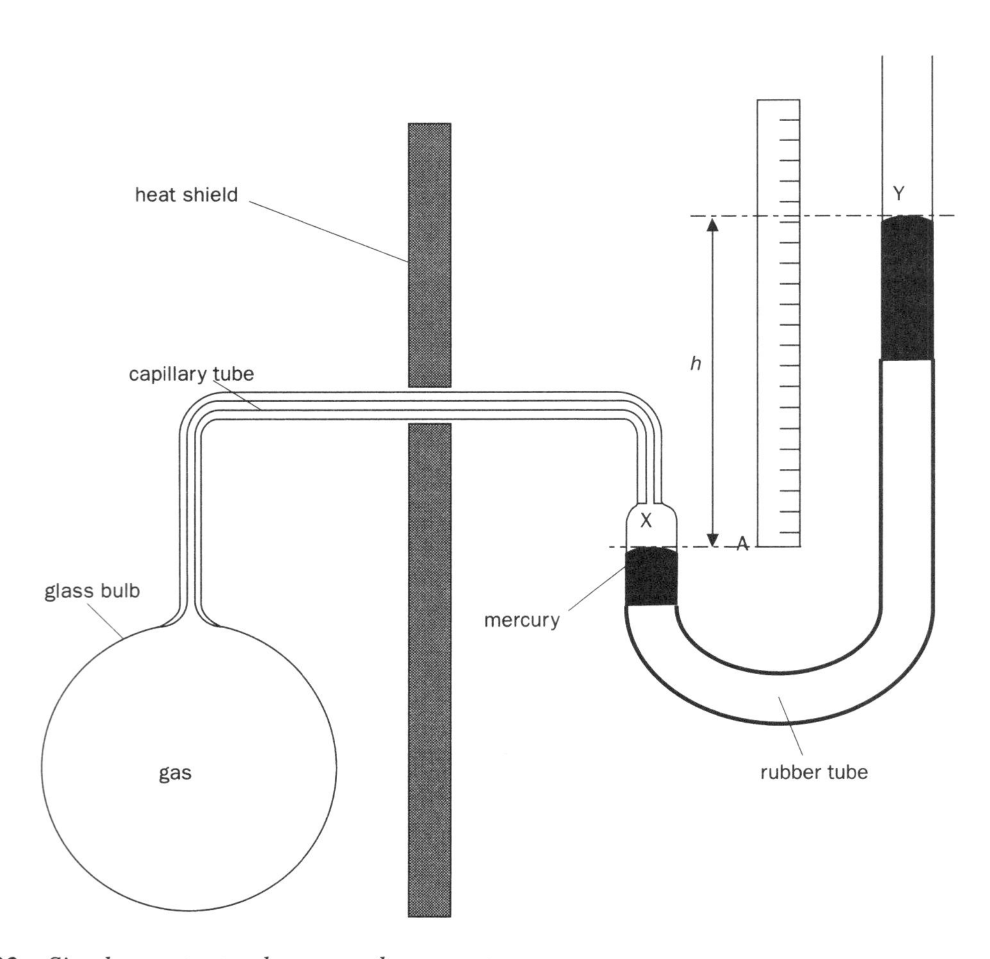

Figure 4.22 Simple constant-volume gas thermometer

Advantages:
  Extremely accurate; very wide temperature range; very sensitive.

Disadvantages:
  Bulky; not direct reading (temperature must be calculated); large heat capacity which means that it is less suitable for rapidly changing temperatures and it affects the temperature being measured.

Errors:
- (i) the temperature of the gas in the capillary tubing is different to that in the bulb
- (ii) expansion of the bulb means that the volume of the gas is not constant
- (iii) air is not an ideal gas
- (iv) parallax error in reading liquid levels on the scale.

Formerly used as a standard to calibrate other practical thermometers.

Platinum-resistance thermometer

Thermometric property	Electrical resistance of a platinum coil.
Range	About – 200 °C to 1200 °C.

This type of thermometer makes use of the fact that the electrical resistance of a pure metal increases with temperature. A measurement of resistance can therefore be used as a measurement of temperature. The metal in this case is in the form of a fine platinum coil wound on a strip of insulating material (e.g. mica) and contained inside a porcelain tube. Platinum is used because of its high temperature coefficient of resistance (small changes in temperature cause a significant change in resistance), its resistance to corrosion and high melting point (1773 °C). In use the thermometer forms one arm of a Wheatstone bridge, as shown in Figure 4.23.

With the porcelain tube in the body whose temperature θ is required, the variable resistor R_V is adjusted until a zero reading on the meter indicates that the bridge is balanced. The standard Wheatstone bridge formula is applied to calculate the resistance of the platinum coil (R_θ):

$$R_\theta = R_V \frac{R_1}{R_2}$$

The unknown temperature θ is defined on the platinum resistance scale as:

$$\theta = \frac{(R_\theta - R_0)}{(R_{100} - R_0)} \times 100$$

where R_0, R_{100} and R_θ are the resistance values at 0 °C, 100 °C and θ °C.

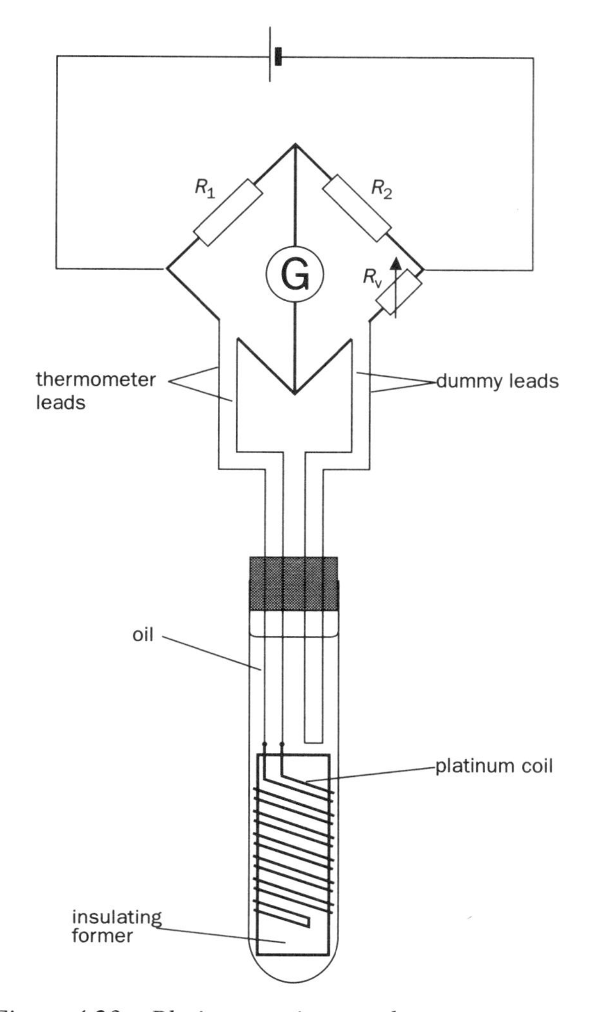

Figure 4.23 Platinum-resistance thermometer

Advantages:
  Very accurate; sensitive to small temperature changes; wide temperature range.

Disadvantages:
  Unsuitable for rapidly varying temperatures because of its large heat capacity; system is bulky and inconvenient when used with a Wheatstone bridge.

Errors:
  Changes in the resistance of the thermometer leads due to temperature changes. This is compensated by the addition of identical 'dummy' leads connected in one arm of the Wheatstone bridge. The change of resistance suffered by the 'live' leads and the dummy leads should be equal, and cancel out. Best used to measure small, steady temperature differences.

Thermistor thermometer

Thermometric property	Electrical resistance of a semiconductor.
Range	About – 50 °C to about +300 °C.

Thermistors are semiconductor devices whose resistance changes markedly with temperature. The type used in thermometry generally has a **negative** temperature coefficient of resistance (i.e. their resistance decreases with increasing temperature). By connecting the thermistor into one arm of a Wheatstone bridge circuit as shown in Figure 2.24 small changes in its resistance can be detected.

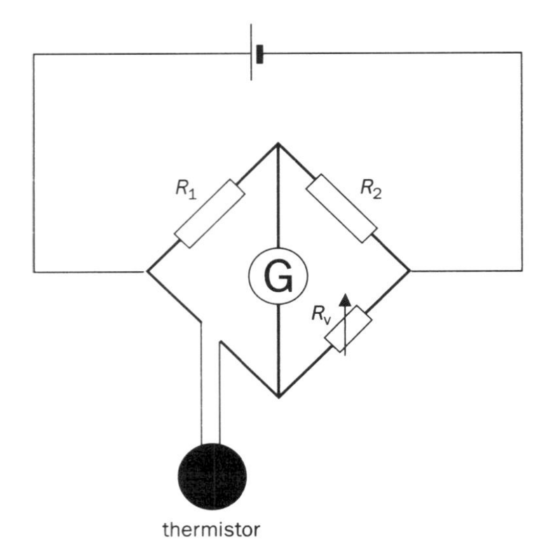

Figure 4.24 Thermistor thermometer used with a Wheatstone bridge

Calculating the unknown temperature follows an identical procedure to that for the platinum-resistance thermometer. Alternatively, the thermistor can be connected directly to an 'ohmmeter' type of meter, calibrated directly in degrees to give a direct reading. Many digital thermometers are of this type.

Advantages:

Very sensitive; cheap; robust; small heat capacity means that they can follow rapidly changing temperatures.

Disadvantages:

Although thermistors are compact, the arrangement with a Wheatstone bridge is bulky; less stable and therefore less accurate than resistance thermometers; non-linear.

Uses:

(i) in bolometers to detect changes in infra-red radiation
(ii) industrial monitoring and temperature control.

Thermocouple thermometer

Thermometric property	The electrical potential difference between two different metals in contact.
Range	Depends on the metals used. For example with a chromel–alumel combination the upper limit is 1100 °C whereas a platinum/platinum–rhodium thermocouple can read temperatures up to 1700 °C.

When two different metals are in good electrical contact free electrons flow in both directions across the junction. Because the electron density is different in each metal the electron flow in one direction exceeds that in the other causing an e.m.f. to develop

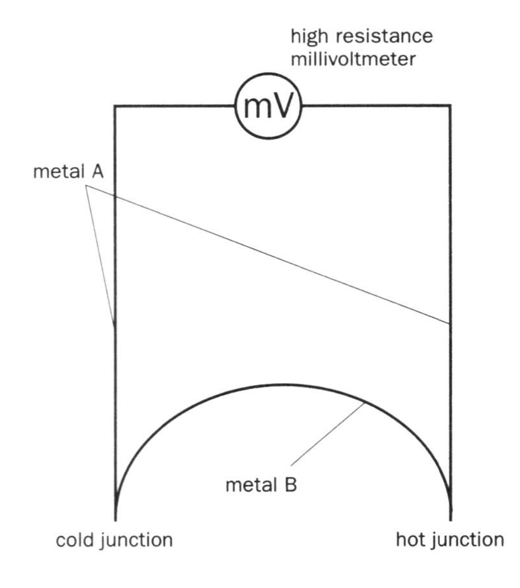

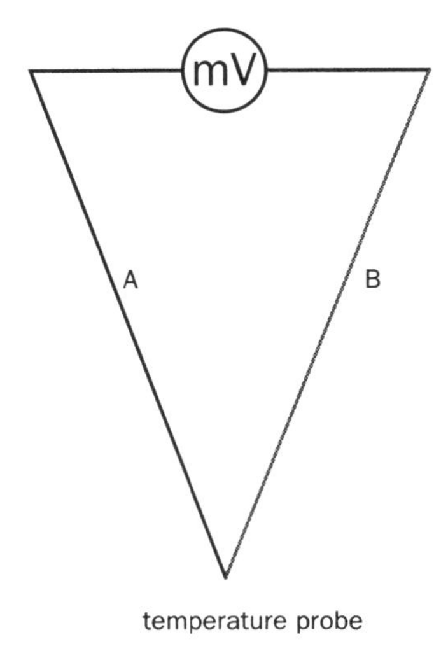

Figure 4.25 Typical thermocouple arrangements

across the junction. This **thermoelectric e.m.f.** which is always present at a junction of different metals is also found to vary with temperature. A typical arrangement for using this effect for temperature measurement is shown in Figure 4.25(a).

One of the junctions is used as the temperature probe and the other is kept at a known temperature (e.g. 0 °C) so as to act as the reference junction. The reading on the millivoltmeter depends on the temperature difference between the two junctions and so by suitable calibration the meter can give direct temperature readings. Figure 4.25(b) shows the arrangement which can be used when high accuracy is not essential. For example, when used to measure high temperatures (about 1000 °C) this arrangement gives an uncertainty of several degrees. On the thermocouple scale, the unknown temperature θ is defined by:

$$\theta = \frac{(E_\theta - E_0)}{(E_{100} - E_0)} \times 100$$

where E_0 E_{100}, E_θ are the e.m.f. values at 0 °C, 100 °C and θ °C.

Advantages:

Wide temperature range; quite accurate; compact and robust; can measure the temperature of small objects and its small heat capacity gives it the ability to follow rapidly changing temperatures; very useful as a temperature probe sensor for connections to computer via interface.

Disadvantages:

Because the e.m.f. produced is small, sensitive electrical equipment has to be used to ensure accurate measurement.

Main uses are for temperature measurement at points and to follow rapidly varying temperatures in datalogging.

Comparison of temperature scales

Each of these thermometers defines its own unique temperature scale according to its thermometric property. Except at the fixed points where, by definition, they agree, thermometers using different thermometric properties may give different measurements of temperature when placed in the same environment. This is because different thermometric properties vary differently with temperature. For example, if the end of the mercury column in a liquid-in-glass thermometer is exactly half-way between the fixed points, then the temperature indicated is 50 °C (i.e. exactly half of the interval between 0 °C and 100 °C). However, the pressure recorded by a gas thermometer for the same temperature may not be exactly half way between the pressures at the fixed points, so would indicate a different temperature. In practice, the two temperatures would be very close. Both thermometers are correct **according to their own scale**, but the disagreement means that we should state which scale is involved whenever a temperature is quoted. In order to avoid such an inconvenience we need a standard temperature scale which is independent of any thermometric property.

The kelvin absolute (thermodynamic) temperature scale

This is a theoretical temperature scale based on the efficiency of an ideal heat engine. It is **absolute** in the sense that it is completely independent of any thermometric properties and it can be shown to be identical with the **ideal gas scale** on which the constant-volume gas thermometer is based. The **kelvin**, which was defined on page 2, is the unit used on both scales and the kelvin temperature T is given by:

$$T = \frac{P_T}{P_{tr}} \times 273.16$$

Equation 2

where P_T is the ideal gas pressure at temperature T
and P_{tr} is the ideal gas pressure at the triple point of water (the fixed point of both the absolute scale and the ideal gas scale).

On the kelvin scale the lower fixed point is zero (i.e. 0 K) – the lowest temperature that in theory can be reached. It is called the **absolute zero** of temperature and although physicists have come very close to achieving it in practice (to within less than a millionth of a degree), it still appears impossible to achieve.

The SI unit of temperature is the kelvin (K):

A temperature of 1 K is 1/273.16 of the temperature of the triple point of water as measured on the thermodynamic or absolute scale of temperature.

The other unit of temperature which is commonly used is the degree Celsius (°C) which is the unit of the Celsius scale and is defined by:

$$\theta = T - 273.15$$

Equation 3

where θ is the temperature in °C and T is the absolute temperature in K.

You should note that:

- Using Equation 2 the ice point (0 °C) is 273.15 K and the steam point (100 °C) is 373.15 K.
- The triple point of water (273.16 K) is 0.01 °C.
- A temperature change of 1 °C = a temperature change of 1 K.

Guided examples (1)

1. A certain platinum resistance thermometer is found to give resistance values of 3.5 Ω at the ice point, 4.80 Ω at the steam point and 4.15 Ω at an unknown temperature. Calculate:
   (a) the value of this temperature on the platinum resistance scale
   (b) the resistance value obtained at a temperature of –10 °C on this scale.

   Guidelines

   (a) Use Equation 1.
   (b) Use Equation 1. Take care with the negative temperature.

2. The resistance (R_T) of a certain wire at a temperature θ °C measured by a constant-volume gas thermometer is given by:

   $R_T = R_0 (1 + 4.0 \times 10^{-3}T + 2.0 \times 10^{-6}T^2)$

   where R_0 (the resistance at 0 °C) = 40 Ω

   Calculate the temperature as measured **on the resistance scale** which corresponds to a temperature of 80 °C as measured by the gas thermometer.

   Guidelines

   Calculate R_{80} and R_{100} and then use Equation 1.

Heat and internal energy

According to the kinetic theory of matter a solid consists of particles (atoms or molecules) which alternately attract and repel each other as they continually vibrate about fixed equilibrium positions. The internal energy of the solid is then the total energy of all its particles. This energy is partly kinetic and partly potential. The kinetic part is due to the vibratory motion of the particles and it depends on the temperature. The potential energy component is stored in the bonds which hold the particles together and it depends on the size of the bonding forces as well as the particle separation. In a solid the kinetic and potential energy components are present in approximately equal proportions and there is a continuous interchange between the two forms. This is not the case in a gas, where the inter-particle forces are weak and the internal energy is almost entirely kinetic.

Heat can be defined as the energy transferred between two bodies due to their different temperatures. It is closely linked to internal energy but they do not mean the same thing. The internal energy of a body can be changed by doing work or by transferring heat. For example, when you rub your hands together on a cold day, the work done against friction causes an increase in the internal energy (and so the temperature) of the hands. The transfer of heat energy from the hands to the surroundings occurs as a result of a temperature difference, and this removal of heat energy causes a decrease in the internal energy of the hands. The relationship between heat energy, internal energy and work is based on the principle of conservation of energy, and states that:

The increase of **internal energy** of a body (ΔU) is equal to the sum of the **heat flowing** into the body (ΔQ) and the **work done** (ΔW) on the body.

(Heat flowing out of or work done by the body is negative, and contributes to a decrease of internal energy.)
This is the **first law of thermodynamics**.

Heat capacity

Common symbol	C
SI unit	Joule per kelvin (or degree Celsius) ($J\ K^{-1}$ or $J\ °C^{-1}$)

When a body is heated, its temperature rise depends on its **mass** and the **material** of which it is composed. Figure 4.26 shows how the temperature (T) of a body of mass (m) changes by an amount (ΔT) when an amount of heat energy (ΔQ) flows into or out of the body.

Temperature rise

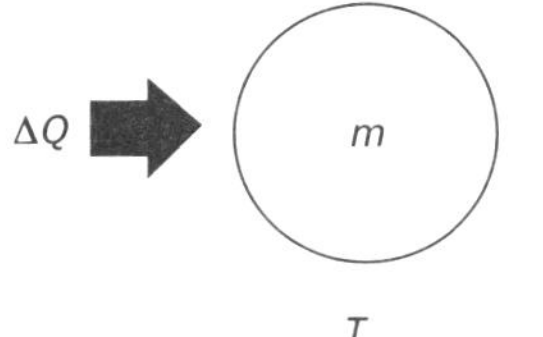

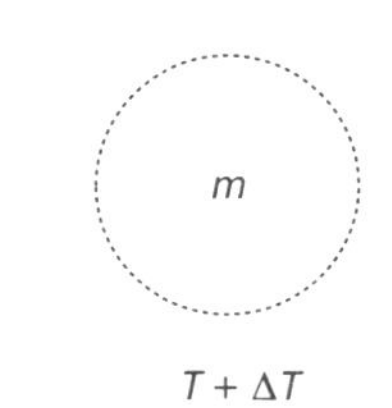

Temperature fall

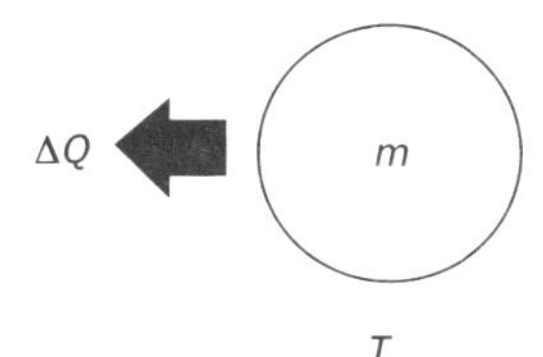

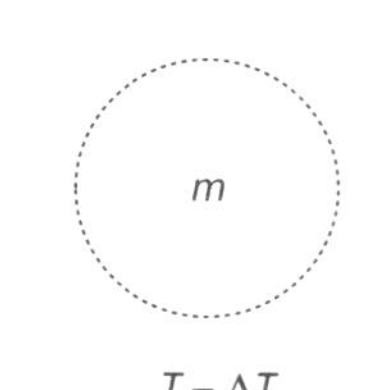

Figure 4.26 Temperature change due to heat flow

The temperature change (ΔT) experienced depends on:

- The mass (m) of the body – the smaller the mass, the larger is the temperature rise,

  i.e. $\Delta T \propto \dfrac{1}{m}$

- Material of which the body is made – objects of the same mass, but different material, experience different temperature rises when they are supplied with the same amount of heat. This fact is taken into account by a quantity called the heat capacity C of the body.

The **heat capacity** of a body is the **amount of heat energy** (ΔQ) which must flow into the body to produce **unit temperature rise** (1 K) in the body.

This may be expressed mathematically as:

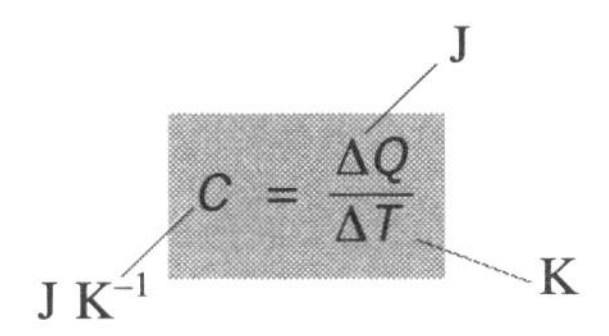

Equation 4

Thus the greater the heat capacity of a body is, the smaller the temperature rise experienced for a given amount of heat supplied.

Specific heat capacity

Common symbol	c
SI unit	joule per kilogram per kelvin (or degree Celsius) (J kg^{-1} K^{-1} or J kg^{-1} °C^{-1})

The term **specific** heat capacity refers to the heat capacity **per kilogram** of a substance. It is a property of a **material**, and is a much more useful quantity than heat capacity, which relates to only one particular body.

The **specific heat capacity** (c) of a **substance** is the **amount of heat energy** needed to produce a temperature rise of **one degree** in **one kilogram** of a substance.

This may be expressed mathematically as:

$$c = \frac{\Delta Q}{m\Delta T} \quad \text{or} \quad \Delta Q = mc\Delta T$$

(c: J kg^{-1} K^{-1}; ΔQ: J; m: kg; ΔT: K)

Equation 5

i.e. if an amount of heat (ΔQ) is supplied to (or removed from) a body of mass (m) and specific heat capacity (c) its temperature changes by an amount ΔT.

You should note that:

- Although the value of c for a material varies with temperature, the variation is only significant at very low temperature or over wide temperature ranges. Thus for normal purposes c is treated as constant.
- Since the equations for C and c involve **changes** in temperature both K and °C can be used as units of ΔT.

Specific heat capacity determinations

From Equation 5, the experimental determination of specific heat capacity requires accurate measurement of the mass and temperature change of the substance, as well as the amount of heat energy which is supplied to it.

Metals The metal whose specific heat capacity (c) is required is in the form of a cylindrical block of known mass (m) with holes provided for the thermometer and the electric immersion heater. The block is well insulated with expanded polystyrene so as to minimise heat loss to the surroundings and oil is used in the holes to give good thermal contact between the heater and thermometer and the metal of the block. The apparatus used is shown in Figure 4.27.

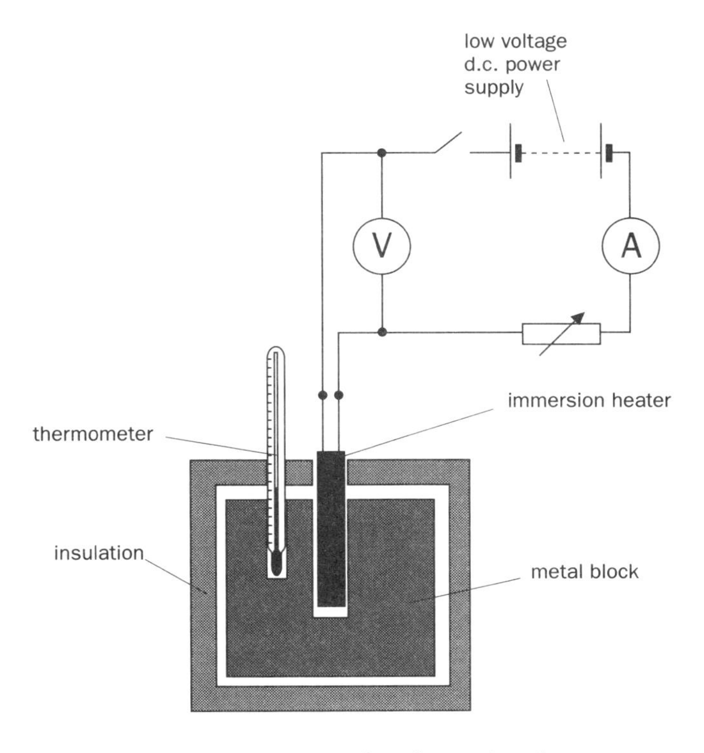

Figure 4.27 Apparatus used to determine the specific heat capacity of a metal block

After the initial temperature (T_1) of the block has been measured and recorded, the switch is closed and the heater current (I_1) and p.d. (V_1) are also measured and recorded. Heating is continued for a measured time (t) such that a substantial temperature rise has been produced. The final temperature (T_2) is then measured and recorded. Then, assuming that the heat lost to the surroundings and given to the thermometer and heater is negligible, we can say that:

(electrical energy supplied by the heater) = (heat gained by the block)

$\therefore \quad V_1 I_1 t = mc\,(T_2 - T_1)$

from which c can be calculated

In practice, the heat lost to the surroundings (H) may be considerable. The accuracy of the determination can be significantly improved by repeating the procedure for the same time interval, but with different current and p.d. values (I_2 and V_2) to obtain a different temperature rise ($T_4 - T_3$). Then for each experiment:

(electrical energy supplied by the heater) = (heat gained by the block) + (heat loss)

For experiment 1:

$$V_1 I_1 t = mc\,(T_2 - T_1) + H \qquad \text{(i)}$$

For experiment 2:

$$V_2 I_2 t = mc\,(T_4 - T_3) + H \qquad \text{(ii)}$$

Subtracting Equation (i) from Equation (ii) then eliminates H and gives:

$$(V_2 I_2 - V_1 I_1)\,t = mc\,[(T_4 - T_3) - (T_2 - T_1)]$$

from which c can be calculated.

$$c = \frac{(V_2 I_2 - V_1 I_1)t}{m[(T_4 - T_3) - (T_2 - T_1)]}$$

Liquids The apparatus used to determine the specific heat capacity of a liquid is shown in Figure 4.28. The liquid is contained in a polished copper or aluminium can called a calorimeter which is suitably insulated to minimise heat losses. The heating is provided by a heating coil which is totally immersed in the liquid under test.

The procedure is very similar to that for the solid, except that the liquid must be stirred continuously during the experiment. A temperature rise ΔT is produced in the liquid by heating it for a time t.

If I, V = heater current and p.d.

m = mass of the liquid

m_c = mass of the calorimeter and stirrer

and c_c = specific heat capacity of the calorimeter and stirrer material

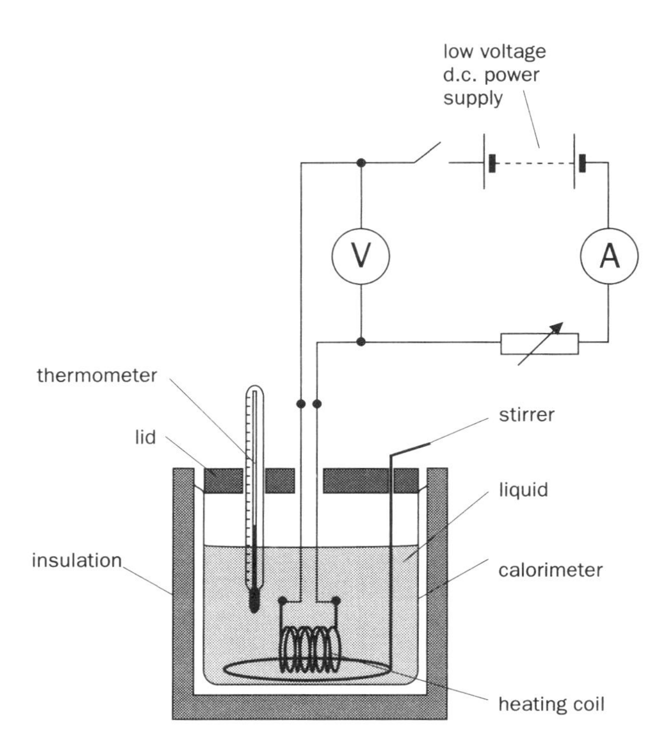

Figure 4.28 Apparatus used to determine the specific heat capacity of a liquid

then, assuming negligible heat loss to the surroundings:

(energy supplied by the heater) = (energy gained by liquid) + (energy gained by calorimeter and stirrer)

$\therefore \quad It = mc\Delta T + m_c c_c \Delta T$

From which c can be found if c_c is known.

Heat losses – cooling correction

In any procedure involving heating or cooling, there is always some heat exchange with the surroundings. Conduction and convection can be minimised by wrapping the apparatus with insulating material or a layer of still air or, in the extreme, evacuating the region around the substance under test. Heat radiation can be reduced by using highly polished surfaces. However, despite all these efforts some heat will still be exchanged.

The simplest way to reduce the effect of heat losses is to conduct experiments so that half the time the apparatus is cooler than the surroundings and half the time it is warmer, thus producing zero overall heat exchange. Thus if room temperature is say 18 °C, then the apparatus could be cooled so as to start the experiment at say 8 °C and the final temperature is set at 28 °C. In this way the apparatus would have temperatures equally above and below room temperature and for equal times. If this is not possible and the experiment begins at room temperature (T_R), the temperature of the apparatus will at all times exceed (T_R). There is then a heat loss to the surroundings which causes a reduced temperature rise. Heat losses will also cause a temperature fall when the heating is stopped. Figure 4.29 shows the temperature variation in an experiment in which the temperature is recorded during heating and after heating has stopped (solid line).

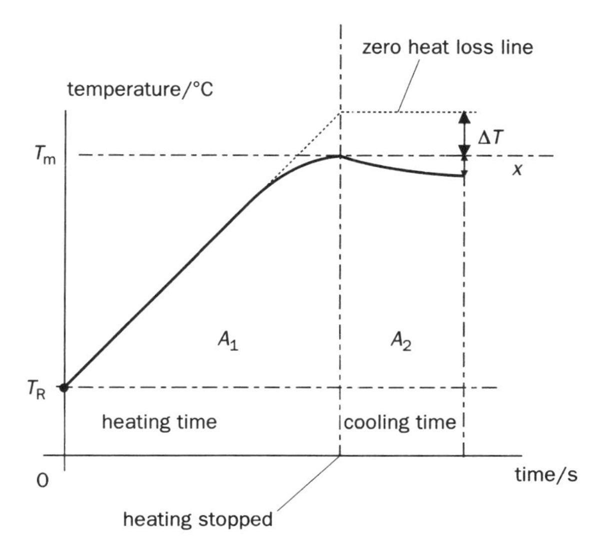

Figure 4.29 Consideration of heat loss to the surroundings – cooling correction

If there had been no heat losses, the maximum temperature would have been ($T_m + \Delta T$) instead of T_m. To apply a cooling correction, the apparatus is allowed to cool for some time, while the temperature is monitored. At the end of this time, the temperature will have fallen by an amount (x).

It can be shown that the cooling correction ΔT is given by:

$$\Delta T = \frac{A_1}{A_2} \times x$$

Thus measuring x and areas A_1 and A_2 enables ΔT to be found.

Guided examples (2)

1. A solid aluminium block of mass 1.5 kg is heated for 7.5 minutes by an electric immersion heater embedded in the block. The mean values of the current through the heater and the p.d. across it are 2.5 A and 12.0 V respectively. Assuming that the heat losses and the heat capacity of the heater are negligible, calculate the temperature rise of the block. (Mean specific heat capacity of aluminium = 9.1×10^2 J kg^{-1} K^{-1})

Guided examples (2) *continued*

Guidelines

Since heat losses are negligible and the heat capacity of the heater is also negligible:

(electrical energy supplied by heater) = (heat energy gained by block)

Remember to use SI units throughout.

2. A copper cube of mass 110 g is heated to a temperature of 100 °C and then rapidly transferred to a well insulated aluminium can of mass 80 g containing 200 g of water at 10 °C. If the final temperature of the cube and water (after stirring) is 14 °C, calculate the specific heat capacity of the copper. You may assume that the heat loss to the surroundings is negligible.

   (Mean specific heat capacities for water and aluminium are 4.2×10^3 J kg^{-1} K^{-1} and 9.1×10^2 J kg^{-1} K^{-1} respectively)

Guidelines

When a hot solid (or liquid) is introduced into a cold liquid, the heat lost by the hot material in cooling is equal to the heat gained by the cold liquid and its calorimeter plus any heat lost to the surroundings (in this case this is negligible).

Remember to use SI units throughout.

3. A small electric immersion heater is used to heat up 200 g of milk in a baby's feeding bottle. The heater operates at a p.d. of 240 V and takes a current of 1.0 A. If the bottle is wrapped in a thick towel during heating so that heat loss to the surroundings may be taken as negligible, how long does it take for the temperature of the milk to rise from 18 °C to 38 °C. The bottle has a heat capacity of 24 J K^{-1} and the specific heat capacity of the milk used is 3.9×10^3 J kg^{-1}K^{-1}.

Guidelines

Since heat loss to the surroundings may be neglected:

(electrical energy supplied by the heater) = (heat gained by milk) + (heat gained by bottle)

Latent heat

We have so far looked at situations in which the temperature of a body changes whenever there is a flow of heat energy in or out of the body. However, there are many circumstances in which transfer of heat does not produce a change of temperature. If the temperature of the water in a kettle is measured while it is switched on, it will be seen that the temperature stops rising when the water starts to boil, and will remain at this temperature despite the fact that the heating element is still supplying heat energy at the same rate. In fact, whenever there is a **change of state** of a material, the change takes place at a constant temperature, which is why the **fixed points** of temperature scales are based on materials melting or boiling.

Latent heat is the heat energy which a body will **absorb** during melting, evaporation or sublimation and which it **gives out** during freezing or condensation. It is called latent (meaning hidden) because it produces a change of state but no temperature change. Figure 4.30 illustrates what happens when a solid is continuously supplied with heat energy, or heat energy is continuously extracted from a gas.

If the diagram in Figure 4.30 is read from left to right, it shows the sequence of events when heat energy is continuously added to a solid, while reading it from right to left illustrates the reverse process, of extracting heat energy from a gas. Starting at the left of the diagram, with a solid at a temperature well below its melting point, we see that the heat supplied causes the temperature of the solid to increase until the **melting point** (temperature T_m) is reached.

> *In terms of kinetic theory we say that the heat energy supplied increases the **kinetic** energy component of the internal energy. The mean molecular separation is also increased slightly (i.e. the solid expands) and this means a small increase in the molecular potential energy.*

Once the melting point has been reached continued heating does not cause any further increase in temperature until **all** the solid has melted. The heat

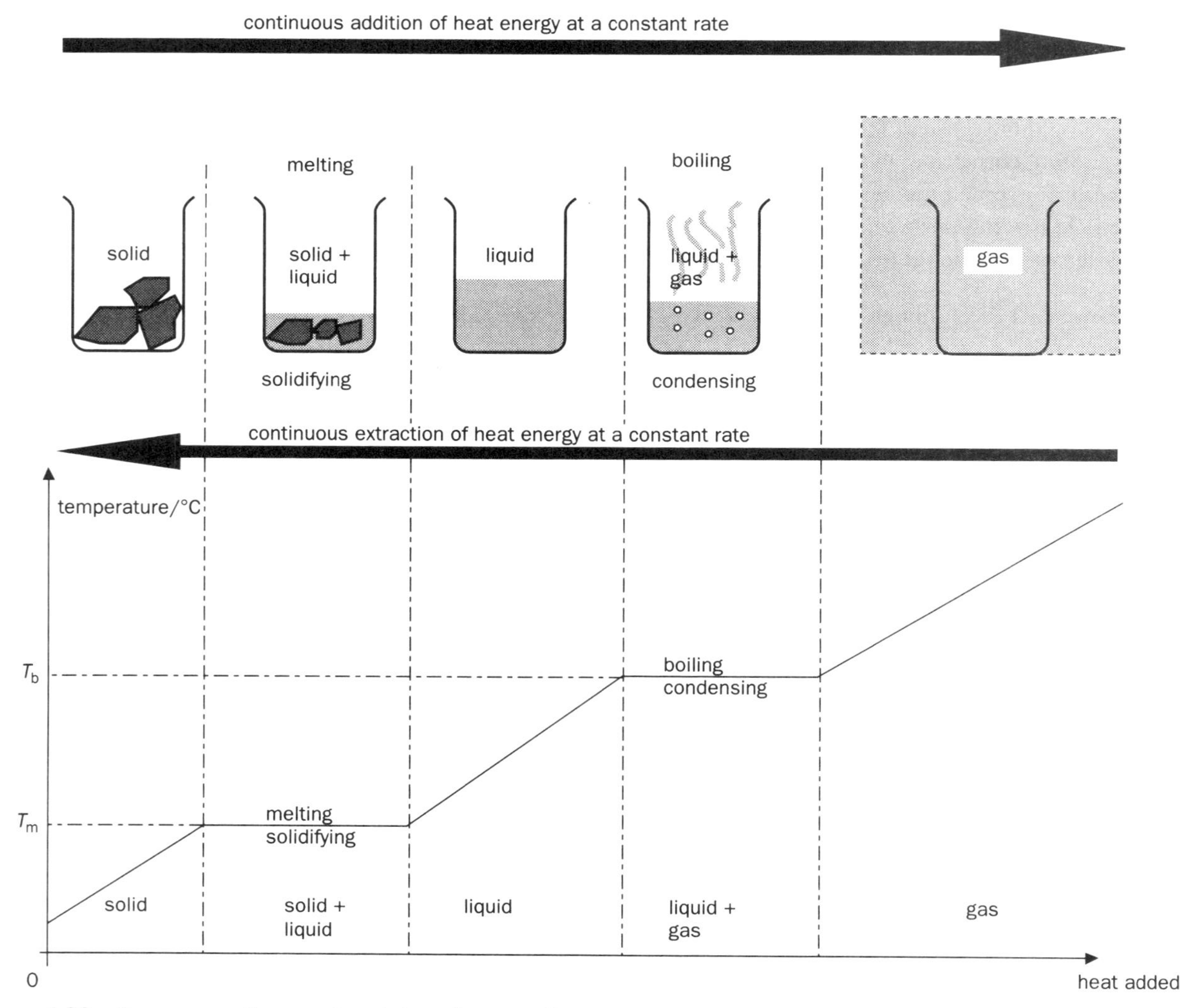

Figure 4.30 Sequence of events involving change of state

supplied to cause the phase change from solid to liquid is called the **latent heat of fusion**.

In microscopic terms we say that the supply of latent heat increases the ***potential*** *energy component of the internal energy by significantly increasing the mean separation of the molecules. Latent heat enables the molecules to overcome the forces which hold them together in a regular crystalline lattice. The solid structure then breaks down to give the greater molecular mobility and disorder of the liquid state.*

When all the solid has melted, the temperature of the resulting liquid can again increase until the **boiling point** (temperature T_b) is reached and **vaporisation** occurs throughout the body of the liquid, not just from the surface. Continued heating from this point on does not cause any further increase in temperature. The **latent heat of vaporisation** is the heat which is supplied at constant temperature to cause the phase change from liquid to vapour.

The latent heat of vaporisation supplies the energy needed (i) to separate the molecules and so allow them to move around independently as gas molecules and (ii) to enable the vapour formed to expand against the pressure of the atmosphere.

Continued heating of the gas, once all the liquid has evaporated, will again produce a rise in temperature.

You should note that:

- Latent heat is distinct from the heat which may have been supplied to a substance to raise its temperature to the melting (or boiling) point.
- Changes of state in the reverse direction (i.e. condensation and fusion or freezing) involve the extraction or release of latent heat from a substance.
- All changes of state occur at constant temperature.
- If the change of state involves an expansion some of the energy supplied to the substance is used to do external work. This can usually be neglected in the case of melting, but it is significant for vaporisation or sublimation in which a large expansion occurs.

Specific latent heat

Common symbol	l
SI unit	joule per kilogram (J kg^{-1})

The **specific latent heat** (l) of fusion (or vaporisation or sublimation) of a substance is the **energy** required to change **1 kg** of the substance from solid to liquid (or liquid to vapour, or solid to vapour) at **constant temperature**.

Thus if an amount of heat energy (ΔQ) is needed to cause a change of state in a mass (m) of a substance, then the specific latent heat (l) of the substance is given by:

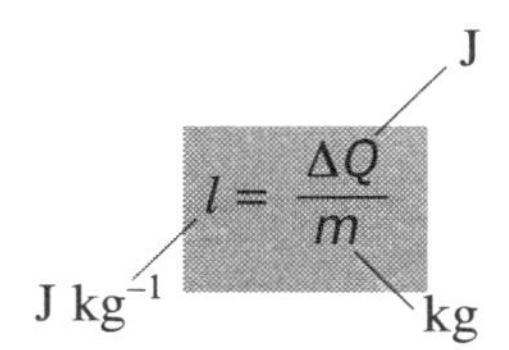

Equation 6

Thus the heat energy ΔQ which must be supplied to cause a change of state of a mass m of a substance is given by:

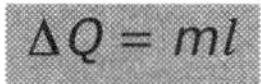

For changes of state in the reverse direction (i.e. liquid to solid, vapour to liquid, and vapour to solid) ΔQ is the amount of heat energy extracted from the substance.

Cooling produced by evaporation If a volatile liquid (i.e. one which evaporates readily at room temperature) such as petrol or methylated spirit is spilt on your hand, it will evaporate rapidly making your hand feel cold. The explanation for this cooling which results from evaporation is that the liquid requires the latent heat of vaporisation in order to change its state from liquid to vapour. The necessary heat is taken from the hand which therefore loses heat and so its temperature decreases. Water also causes the same effect, but it is not so noticeable as methylated spirit or petrol because these liquids have a lower boiling point and so evaporate much more rapidly at the temperature of your hand. Sweating is the body's cooling mechanism. When you are hot you sweat and the liquid produced takes the latent heat it requires in order to evaporate from your body.

We can also explain cooling due to evaporation in terms of kinetic theory. The molecules of a liquid have an average kinetic energy which increases with temperature. Those near the liquid surface which may be moving faster than average can escape from the attractive forces of neighbouring molecules and 'jump' out of the liquid. The most energetic molecules will escape from the liquid altogether (i.e. evaporation occurs) leaving behind a liquid which contains less energetic molecules. Thus the average kinetic energy of the remaining molecules is reduced and this means a decrease in temperature.

Evaporation and boiling Evaporation, or vaporisation, is the name of the process by which a liquid changes state into a gas. This can happen at any temperature, but only at the surface of the liquid, and the rate of evaporation is influenced by the temperature of the liquid and the concentration of vapour molecules near the surface of the liquid. When washing is hung out to dry, we are relying on natural evaporation to remove the excess water from the clothes. The hotter the day, the faster this will happen, but the process will be slower if the air already contains a high proportion of water vapour (i.e. the humidity is high). The wind is also important: newly-evaporated molecules will increase the humidity near the clothing, hindering further evaporation. A good breeze will disperse these molecules, ensuring a plentiful supply of dry air to the washing. The combination of wet clothing and high wind is particularly dangerous to mountaineers and hill-walkers: the high rates of evaporation produced extract latent heat of vaporisation from the body very rapidly. If this rate of extraction of energy exceeds the body's ability to replace it, maybe due to illness, hunger or injury, then the net result is a gradual lowering of body temperature, known as **hypothermia**, which can often be fatal.

Boiling is a special case of evaporation, and takes place only at a specific temperature – the boiling point. Bubbles of vapour form **throughout the body of the liquid**, rise to the surface and burst, releasing the vapour to the atmosphere. This is why boiling liquids appear so disturbed.

Experimental determination of the specific latent heat of vaporisation (l_v)

The value of l_v for a liquid can be determined using the apparatus shown in Figure 4.31.

The liquid under test is electrically heated by a coil carrying a steady current (I_1), and having a p.d. (V_1) across it. When the liquid boils the vapour which is produced passes through holes in the inner wall of the container and into the condenser. Boiling is continued until the temperatures of all parts of the apparatus have steadied. The condensed vapour is then collected over a measured time interval (t) and its mass (m_1) is measured. Then, since steady state conditions prevail, it can be said that in time t:

(electrical energy supplied) = (heat energy to vaporise liquid) + (heat loss to the surroundings)

$$\therefore \quad V_1 I_1 t = m_1 l_v + H \qquad \text{(i)}$$

Using a new heater current and p.d. (I_2,V_2) the new mass (m_2) of vapour which condenses in the **same** time (t) is measured. Then:

$$\therefore \quad V_2 I_2 t = m_2 l_v + H \qquad \text{(ii)}$$

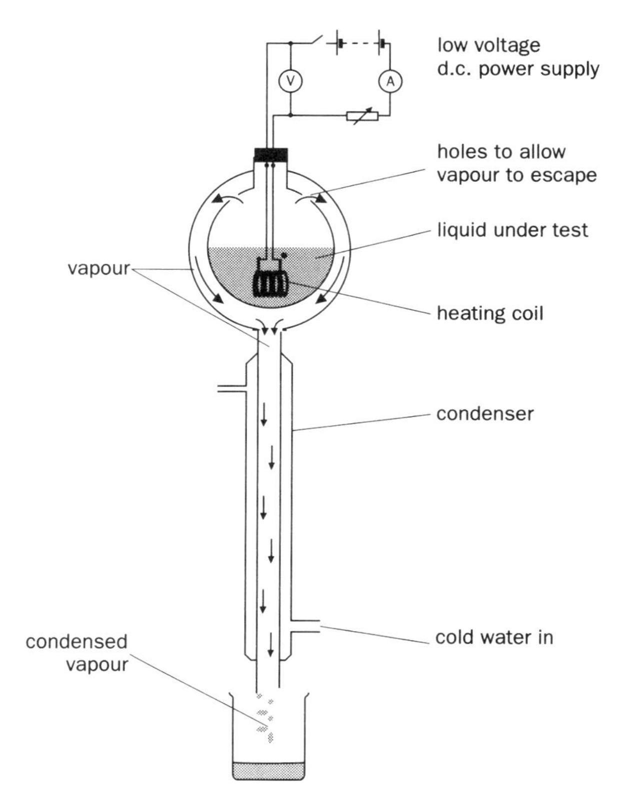

Figure 4.31 Determination of the specific latent heat of vaporisation (l_v) of a liquid

Equation (ii) – Equation (i) gives:

$(V_2 I_2 - V_1 I_1)t = (m_2 - m_1)l_v$

from which l_v can be determined.

Heat losses are kept to a minimum in this experiment because the liquid is in a constant temperature enclosure since it is surrounded by vapour which is at the same temperature as the liquid.

Guided examples (3)

1. Calculate the mass of ice which melts when 2.5 kg of cabbage at a temperature of 20 °C is mixed with ice and water at 0 °C in a well-insulated container. You may assume that heat loss to the surroundings is negligible and that all the cabbage is cooled to 0 °C.
(Specific heat capacity of cabbage = 2.0×10^3 J kg K^{-1}; specific latent heat of fusion of ice = 3.4×10^5 J kg^{-1}.)

Guidelines

Since the temperature remains constant at 0 °C until all the ice has melted, the heat gained by the ice must equal the heat lost by the cabbage.

Use Equation 5 for the heat lost by the cabbage and Equation 6 for the heat gained by the ice.

Guided examples (3) *continued*

2. Calculate the total amount of heat which is given out when 200 g of steam condenses to water at 40 °C. (Specific heat capacity of water = 4.2×10^3 J kg^{-1} K^{-1} and specific latent heat of steam = 2.26×10^6 J kg^{-1}.)

Guidelines

Total amount of heat given out = latent heat given out as steam condenses to water at 100 °C (use Equation 6) + heat given out as the water cools from 100 °C to 40 °C (use Equation 5). Don't forget to use SI units throughout.

3. To ensure that there is no shortage of ice at a party a girl pours 3×10^3 ml of water at 24 °C into a large plastic bag which she then ties securely and places in the freezer. If it can be assumed that the heat capacity of the bag is negligible and it takes 50 minutes for all the water to freeze, calculate the **rate** at which the freezer extracts heat from the water. (Density of water = 1.0×10^3 kg m^{-3}; 1 m^3 = 10^6 ml; specific heat capacity of water = 4.2×10^3 J kg^{-1} K^{-1}; specific latent heat of fusion of ice = 3.37×10^5 J kg^{-1}.)

Guidelines

Total amount of heat extracted = heat extracted in cooling the water from 24 °C to 0 °C (use Equation 5) + latent heat extracted in changing the water into ice at 0 °C (use Equation 6). Then:

$$\textit{rate of heat extraction} = \frac{\textit{total amount of heat extracted}}{\textit{time taken}}$$

4. In an experiment to find the specific latent heat of vaporisation (l_v) of ethanol a student collects 4.8 g of condensed ethanol in 60 s when electric heater current and p.d. were 3.0 A and 24 V respectively.
   (a) Calculate the value of l_v for ethanol from these results.
   (b) The student feels that the value obtained in (a) is too high and decides to obtain a more accurate value by repeating the experiment using new current and p.d. values of 2.5 A and 20.0 V. He then collects 3.25 g of condensed ethanol in 60 s. Use both sets of results to obtain a new value for l_v. Why is this value closer to the true value?

Guidelines

(a) Assuming no heat loss to the surroundings

(electrical energy supplied by heater) = (latent heat needed to vaporise ethanol)

(b) Follow the method used in the experimental determination described on page 354 using both sets of results so as to eliminate heat loss to the surroundings.

Self-assessment

SECTION A
Qualitative assessment

1. Explain how temperature is related to the idea of thermal equilibrium.
2. State the **zeroth law** of thermodynamics.
3. (a) What is a **thermometric** property? What are the characteristics which make a property useful as a thermometric property? Give five properties that are used as thermometric properties.
   (b) Explain what is meant by the terms:
   (i) **fixed point** and
   (ii) **fundamental interval**
   when used with reference to a temperature scale.
   (c) Define the **ice point**, the **steam point** and the **triple point** of water.
4. If X_0, X_{100} and X_θ are the measured values of a thermometric property at 0 °C, 100 °C and θ °C, the value of the unknown temperature θ is given by:

$$\theta = \frac{(X_T - X_0)}{(X_{100} - X_0)} \times 100$$

Derive this equation.

5. Define the **kelvin**. State the relationship between this unit of temperature and the degree Celsius.
6. Complete a summary table as shown below for each of the following types of thermometer: mercury-in-glass; constant-volume gas; platinum-resistance; thermistor; thermocouple.

Self-assessment *continued*

Type	Temp. range	Accuracy	Advantages	Use

7. Why might scales derived on the basis of different thermometric properties differ in the value they give to a particular temperature, even when using the same fixed points?

8. (a) Why is the **thermodynamic** scale of temperature an absolute scale?
   (b) What is the fixed point of this scale?
   (c) What are the values of the ice point and steam point on the thermodynamic scale?
   (d) What is the **absolute zero** of temperature?

9. (a) Explain what is meant by the **internal energy** of a body.
   (b) Why is the internal energy of a gas almost entirely kinetic energy?

10. Define **heat capacity (*C*)** of a body and **specific heat capacity (*c*)** of a substance.
    State the units for each of these quantities.

11. Briefly describe an electrical method for the determination of the specific heat capacity of
    (a) a solid and
    (b) a liquid.

12. (a) What is meant by a change of state or phase?
    (b) What is **latent heat** and how is it related to a change of state?
    (c) Define **specific latent heat of fusion** (or vaporisation) and state its unit.

13. Sketch a typical graph of temperature against time for a solid substance which is supplied with heat so that it undergoes **melting** and **vaporisation.** Show on your graph:
    (a) the **melting point** and the **boiling point**
    (b) the sections of the graph in which the substance is **solid, liquid** or **vapour**, and
    (c) the sections of the graph at which the substance is **melting** and **vaporising**.

14. Explain the cooling which accompanies evaporation in terms of:
    (a) latent heat and
    (b) the escape of high energy molecules.

15. Briefly describe an electrical method for finding the specific latent heat of vaporisation of a liquid.

SECTION B
Quantitative assessment

(***Answers:*** 7.3×10^{-3}; 0.089; 1.9; 17; 29; 55; 57; 380; 400; 600; 2.7×10^{3}; 9.3×10^{5}.)

1. The length of the mercury thread in an uncalibrated mercury-in-glass thermometer was found to be 1.5 cm from the top of the bulb when the thermometer was placed in pure melting ice and a further 21.0 cm when it was placed in pure boiling water at 100 °C. Calculate:
   (a) the length of the mercury thread when the thermometer is placed in a liquid whose temperature is 35 °C on the mercury-in-glass scale
   (b) the temperature on this scale for which the length of the mercury thread would be 5.0 cm.

2. The temperature of a particular liquid is being measured using a resistance thermometer and a constant-volume gas thermometer. The readings which are tabulated below refer to measurements made with the two thermometers.
   Calculate the temperature of the liquid on the:
   (a) resistance thermometer scale
   (b) gas thermometer scale.

	At 0 °C	At 100 °C	At liquid temp. θ °C
Constant-volume gas thermometer pressure readings / 10^5 Pa	1.34	1.85	1.62
Resistance thermometer readings / Ω	30.00	44.40	38.20

Self-assessment *continued*

3. A metal cube of heat capacity 48.0 J K^{-1} is heated to a temperature of 150 °C. It is then rapidly transferred into an insulated beaker of heat capacity 14.0 J K^{-1} containing 150 g of water at 20 °C. Assuming that no heat is lost to the surroundings, calculate the final steady temperature of the cube and water. (Specific heat capacity of water = 4.2×10^3 J kg^{-1} K^{-1}.)

4. A 4.0 kW electric immersion heater is used to heat 30.0 kg of water in a well-insulated domestic hot water tank of heat capacity 800 J K^{-1}. If heat is lost to the surroundings at the rate of 300 J s^{-1}, how long does it take to heat the water from 10 °C to 90 °C. (Specific heat capacity of water = 4.2×10^3 J kg^{-1} K^{-1}.)

5. In an experiment to determine the specific heat capacity of copper, a well-insulated copper block of mass 1.5 kg is heated for 15 minutes by an electric heater which is embedded in it. The temperature of the block is found to rise by 36 °C when a p.d. of 12.0 V is applied across the heater and the current is recorded as 2.0 A. The experiment is repeated with p.d. and current values of 10.0 V and 1.9 A respectively and the heating is again carried out for a time of 15 minutes. If the new observed temperature rise is 28.5 °C, use the two sets of results to calculate the specific heat capacity of copper. You may assume that the heat lost to the surroundings is the same for each experiment.

6. In an experiment to determine the specific latent heat of vaporisation of a particular alcohol a vaporising unit of the type described in the text was used. The results obtained are shown in the table below.

Use these results to calculate:

(a) the specific latent heat of vaporisation of the alcohol

(b) the average rate of heat loss to the surroundings.

7. A kettle having a capacity of 1.6×10^3 cm^3 is marked 240 V, 1.5 kW.

(a) Calculate the time taken to bring a full kettle of water to the boiling point from a starting temperature of 15 °C. What simplifying assumption have you made?

(b) If the automatic cut-off switch fails to work, calculate the time taken, from the moment the boiling point is reached, for 400 cm^3 of water to be vaporised. (Density of water = 10^3 kg m^{-3}; specific heat capacity of water = 4.2×10^3 J kg^{-1} K^{-1}; specific latent heat of vaporisation of water = 2.25×10^6 J kg^{-1}.)

8. Steam at 100 °C is passed into 300 g of water initially at 20 °C in a container of negligible heat capacity which is perfectly lagged. What mass of steam will have been passed into, and condensed in, the water when the water temperature has risen to 35 °C?

(Specific heat capacity of water = 4.2×10^3 J kg^{-1} K^{-1}; specific latent heat of vaporisation of water = 2.3×10^6 J kg^{-1}.)

	P.d. across coil (V)	**Current through coil (A)**	**Mass of liquid collected in 300 s (g)**
Expt. 1	7.4	2.6	5.6
Expt. 2	10.0	3.6	11.0

4.3 Ideal gases

The gas laws

When discussing the thermal behaviour of a gas, there are three inter-dependent variables which need to be considered. These are the volume (V), temperature (T), and pressure (p) of the gas. Experiments performed by Boyle, Amontons and Charles in the 17th and 18th centuries resulted in three **gas laws** which relate these three variables.

Boyle's law

Boyle's law relates the **pressure** and **volume** of a gas at **constant temperature** and states that:

> The **pressure** (p) of a fixed mass of gas at constant temperature is **inversely** proportional to its **volume** (V).

This law may be stated mathematically as

$p \propto \frac{1}{V}$ (at constant temperature)

or $pV = \text{constant}$

Equation 1

(i.e. for a fixed mass of gas at constant temperature, the product of pressure and volume is **constant**). This equation can be used to calculate pressure or volume changes whenever a fixed mass of gas is either **compressed** into a smaller volume (at higher pressure), or allowed to **expand** into a greater volume by reducing the pressure, provided the temperature remains constant throughout the change.

Thus if (p_1) and (V_1) are respectively the pressure and volume of a fixed mass of gas, at some initial stage, and (p_2) and (V_2) are the values measured after expansion or compression at constant temperature:

$$p_1V_1 = p_2V_2$$

A graphical view of the relationship between the pressure p and the volume V of a fixed mass of gas at constant temperature is shown in Figure 4.32 below.

In Figure 4.32(a), when (p) is plotted against (V) a curve called a **rectangular hyperbola** is obtained. In Figure 4.32(b) (p) is plotted against ($1/V$) to give a straight line graph and in Figure 4.32(c) (pV) against (p) produces a straight line parallel to the p-axis. By investigating a fixed mass of gas at different temperatures, a series of graphs can be obtained in each. Each of the curves or lines is called an **isothermal** (since the values plotted were obtained at constant temperature).

Charles' law

Charles' law relates the **volume** (V) and **temperature** (T) of a gas at **constant pressure** and states that:

> The **volume** (V) of a fixed mass of gas at constant pressure is **directly** proportional to its absolute (or kelvin) **temperature** (T).

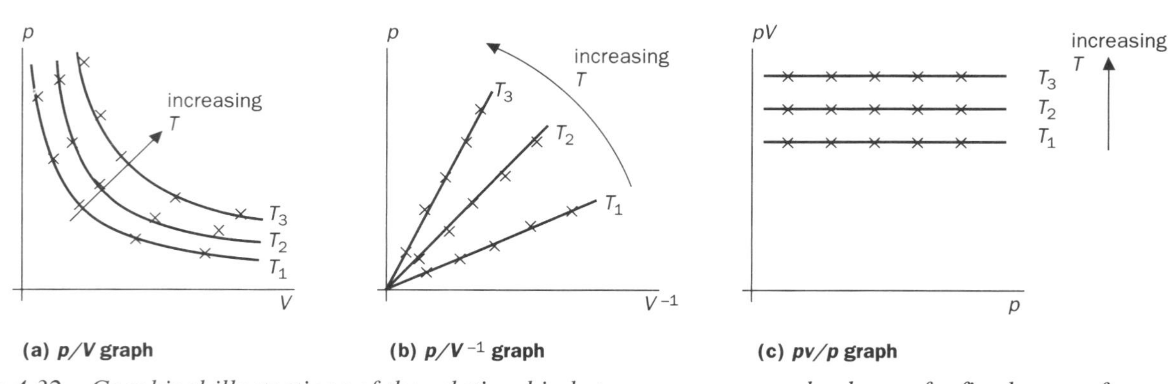

Figure 4.32 Graphical illustrations of the relationship between pressure and volume of a fixed mass of gas – Boyle's law

This may be stated mathematically as:

$V \propto T$ (at constant pressure)

or $\frac{V}{T} = \text{constant}$

Equation 2

This equation is used to calculate changes in volume or temperature whenever a gas expands or contracts at **constant pressure**. Thus if (V_1) and (T_1) are respectively the volume and temperature of a fixed mass of gas at the start of some process, and (V_2) and (T_2) are the values at the end, then if the pressure before and after is constant:

$$\frac{V_1}{T_1} = \frac{V_2}{T_2}$$

(i.e. for a fixed mass of gas at constant pressure the ratio of the volume of the gas to its absolute temperature is constant).

Figure 4.33 below shows the relationship between the volume and the temperature of a fixed mass of gas.

In Figure 4.33(a) volume (V) is plotted against the absolute temperature (T) in kelvins. This yields a straight line graph which passes through the origin when it is extended backwards. In Figure 4.33(b) V is plotted against **Celsius** temperature. This line cuts the temperature axis at approximately –273 °C (the absolute zero of temperature).

Pressure law

The pressure law relates the **pressure** and **absolute temperature** of a gas and states that:

The **pressure** (p) of a fixed mass of gas at constant volume is **directly** proportional to its absolute (or kelvin) **temperature** (*T*).

This may be stated mathematically as:

$p \propto T$ (at constant pressure)

or $\frac{p}{T} = \text{constant}$

Equation 3

(i.e. for a fixed mass of gas at constant volume the ratio of the pressure of the gas to its absolute temperature is a constant).

This equation is used to calculate changes in **pressure** or **temperature** whenever a gas is compressed or rarefied at constant volume. Thus if (p_1) and (T_1) are respectively the pressure and temperature of a fixed mass of gas at the start of some process, and (p_2) and (T_2) are the values at the end, then if the volume before and after is constant:

$$\frac{p_1}{T_1} = \frac{p_2}{T_2}$$

The graph of (p) against (T) (°C) yields a straight line which cuts the temperature axis at approximately – 273 °C as before (see Figure 4.34).

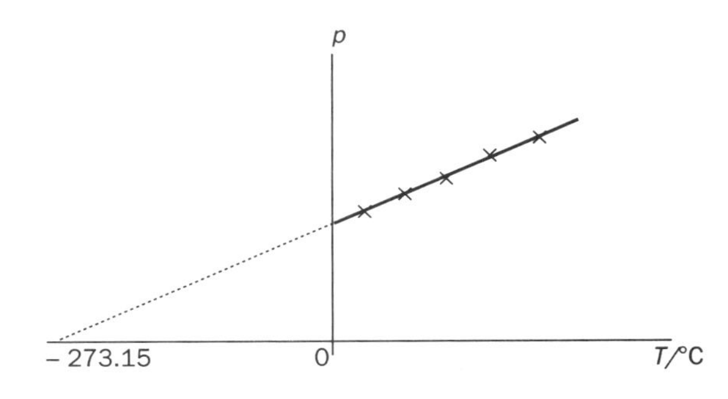

Figure 4.34 Graphical representation of the relationship between pressure and temperature of a fixed mass of gas – pressure law.

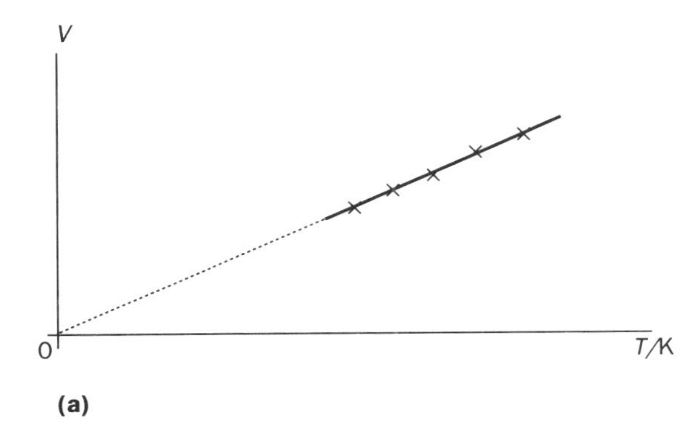

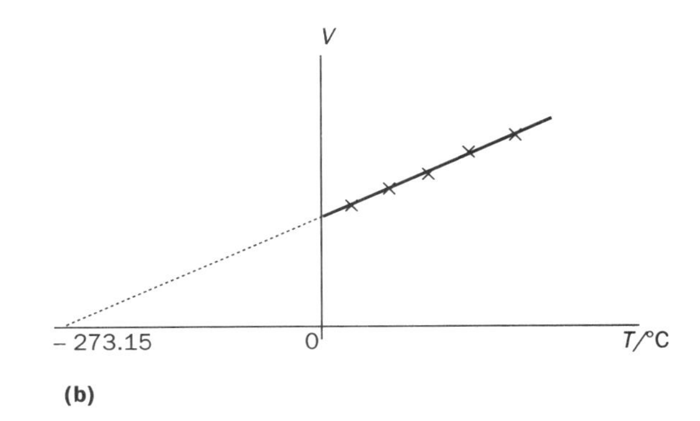

Figure 4.33 Graphical illustrations of the relation between volume and temperature of a fixed mass of gas – Charles' law

The ideal gas equation

An **ideal** or perfect gas is one which obeys the gas laws exactly. Real gas behaviour can be classed as very nearly ideal so long as we are considering gases at low pressures and at temperatures which are significantly higher than those at which liquefaction occurs. Thus, for an ideal gas each of the following applies.

- pV = constant (at constant T)
- V/T = constant (at constant p)
- p/T = constant (at constant V)

From which we have that:

$$\frac{pV}{T} = \text{constant}$$

The magnitude of the constant depends on the mass of gas being considered. If, for example, one mole (defined on page 361) of gas is considered, the constant is the **universal molar gas constant** (R) ($R = 8.31$ J mol^{-1} K^{-1}). Thus for one mole of gas:

$$\frac{pV}{T} = R$$

$$\therefore \quad pV = RT$$

So for n moles of gas:

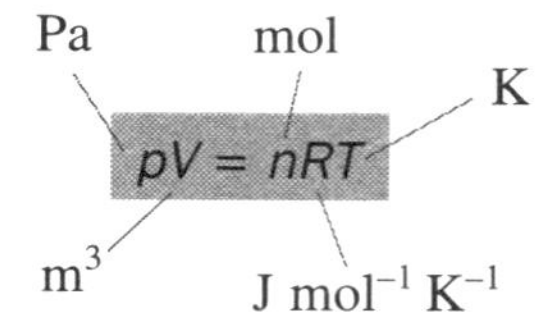

Equation 4

This is the **equation of state of an ideal gas**; sometimes simply called the **ideal gas equation**. You should note that:

- A gas which obeys $pV = nRT$ exactly is an **ideal** gas which is subject to the assumptions made in the **kinetic theory of gases** (see page 362). One of these assumptions is that there are no forces between the gas molecules and this means that the internal energy of an ideal gas is totally kinetic and depends only on gas temperature.
- For a gas of volume (V_1) at a pressure (p_1) and a temperature (T_1) Equation 4 gives:

$$\frac{p_1V_1}{T_1} = nR \qquad \text{(i)}$$

For the **same mass of gas** (i.e. same number of moles of gas) whose volume has become (V_2) at a new pressure (p_2) and temperature (T_2) Equation 4 gives:

$$\frac{p_2V_2}{T_2} = nR \qquad \text{(ii)}$$

Combining (i) and (ii) gives:

$$\frac{p_1V_1}{T_1} = \frac{p_2V_2}{T_2}$$

Equation 5

This is called the **combined gas equation**. It is particularly useful for solving problems in which, volume, pressure and temperature vary simultaneously. When using Equation 5 it does not matter what units are used for p and V so long as they are the same on both sides of the equation, but ***T* must be in kelvins**.

Guided examples (1)

1. A diver swims at a depth of 40 m where the temperature of the water is 4.0 °C.

He inhales 1.2×10^{-5} m^3 of compressed air at a pressure of 7.0×10^5 Pa and suddenly he sees something that panics him into rising to the surface very rapidly without exhaling. Calculate the new volume of the air which he inhaled at 40 m, if the surface temperature and pressure are 20 °C and 1.01×10^5 Pa respectively.

Guidelines

The initial pressure, volume and temperature are given at a depth of 40 m. Use Equation 5 to calculate the new volume, given the new pressure and temperature at the surface, but remember that temperatures must be expressed in kelvins.

2. A gas cylinder of volume 6.0×10^3 cm^3 contains oxygen at a temperature of 20 °C and a pressure of 3.0×10^6 Pa. Calculate:

(a) the equivalent volume of oxygen at standard temperature and pressure (i.e. 0 °C and 1.0×10^5 Pa).

(b) the mass of oxygen in the cylinder if the density of oxygen at standard temperature and pressure (i.e. 0 °C and 1.0×10^5 Pa) is 1.4 kg m^{-3}.

Guidelines

(a) Use Equation 5 – temperatures in kelvins;

(b) *mass = density × volume* – You must use the volume at s.t.p. obtained in (a) since the density value is at s.t.p.

The mole and associated terms

One **mole** is the amount of substance which contains the same number of elementary units (i.e. atoms or molecules) as there are atoms in 12 g of carbon-12 (^{12}C).

The number of atoms in 12 g of carbon-12 (which is equal to the number of elementary units per mole) is called the **Avogadro constant** N_A:

$$N_A = 6.02 \times 10^{23}\ \text{mol}^{-1}$$

The **relative molecular mass** (M_r) of a substance is defined by:

$$M_r = \frac{\text{mass of a molecule of the substance}}{\text{mass of the carbon-12 atom}} \times 12$$

Equation 6

The **molar mass** (M_m) of a substance is the **mass per mole** of the substance:

$$M_m = M_r \text{ (in gram)} = M_r \times 10^{-3} \text{ (in kg)}$$

Equation 7

We can use $pV = nRT$ to calculate the volume of 1 mole (i.e. the **molar volume** V_m) of a gas at standard temperature and pressure (i.e. $T = 0\ °C = 273$ K and $p = 1.013 \times 10^5$ Pa):

$$V_m = \frac{nRT}{p} = \frac{(1\ \text{mol}) \times (8.31\ \text{J mol}^{-1}\ \text{K}^{-1}) \times (273\ \text{K})}{(1.013 \times 10^5\ \text{Pa})}$$

$$= 22.4 \times 10^{-3}\ \text{m}^3$$

The number of moles n in a given mass of gas can be calculated from:

$$n = \frac{M\ (\text{kg})}{M_m\ (\text{kg mol}^{-1})}$$

Equation 8

where M = mass of gas
and M_m = molar mass of the gas.

Guided examples (2)

1. A sealed container of volume 0.8×10^{-3} m^3 contains a gas at a temperature of 320 K and a pressure of 1.5×10^6 Pa. Calculate:

(a) the number of moles and molecules of the gas

Guided examples (2) *continued*

(b) the mass of gas if its molar mass is 32.0×10^{-3} kg

(c) the mass of a single molecule of the gas.

(Given that the universal molar gas constant, $R = 8.31$ J mol^{-1} K^{-1} and the Avogadro constant, $N_A = 6.02 \times 10^{23}$ mol^{-1}.)

Guidelines

(a) Use Equation 4 and the fact that the number of molecules in 1 mole of substance = N_A.

(b) Use Equation 8.

(c) Mass per molecule, $m = \dfrac{M_m}{N_A}$

2. An isotope of lead has a relative atomic mass of 208 and a density of 11.3×10^3 kg m^{-3}. Calculate:

(a) the number of atoms in a lead sample of volume 0.25 m^3

(b) the distance between the centres of two adjacent lead atoms if it can be assumed that they are perfect spheres in contact.

(Given that the Avogadro constant, $N_A = 6.02 \times 10^{23}$ mol^{-1}.)

Guidelines

(a) Use Equation 7 to obtain the mass per mole, M_m.

Then number of moles per m^3

$$= \frac{\text{mass per m}^3}{\text{mass per mole}} = \frac{\text{density}}{M_m}$$

Then multiply by the Avogadro constant (N_A) to obtain the number of atoms per m^3.

Hence obtain the number of atoms in the sample (N).

(b) Volume of atom (V) = volume of sample divided by the number of atoms (N).

Calculate the radius of the atom, assuming it to be a perfect sphere.

Hence the distance between centres of adjacent atoms is calculated.

Kinetic theory of gases

Up to now we have described gas behaviour mainly in terms of large-scale, measurable quantities such as mass, volume, pressure and temperature. The kinetic theory tries to improve our understanding by explaining the quantitative behaviour of gases in terms of a model in which a gas is seen as a large collection of **particles** in continuous motion. Probably the first attempts to give a microscopic explanation of gas behaviour came from **Boyle** and **Hooke** who thought the pressure exerted by a gas was due to the collisions of gas particles with the container walls. **Newton's** laws of motion, which came a short time later, gave the idea some strong mathematical explanation. It was the work of **Boltzmann** and **Maxwell** on the statistical treatment of the distribution of energy among a large sample of particles, however, which was to give the kinetic theory its finishing touch.

The first step in this theory is to visualise a gas as a very large number of molecules in a state of random and perpetual motion. The diffusion of gases and Brownian motion are two effects which give support to this idea. The Brownian motion is the observed continual, haphazard motion of any small particles which are suspended in a fluid. The original observation was made by a Scottish botanist called Robert Brown who noticed that tiny pollen grains suspended in water were subject to continual, jerky movements. The theory was developed by Albert Einstein in 1905 and it gave one of the first methods of measuring N_A which convinced many sceptics of the correctness of kinetic theory. The effect can be clearly observed and studied using the arrangement shown in Figure 4.35.

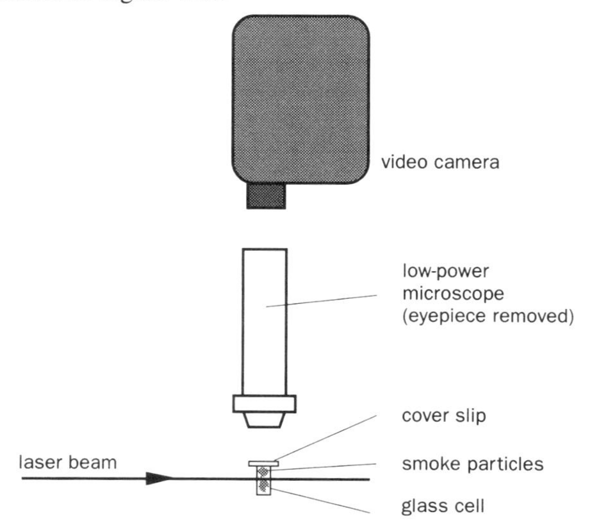

Figure 4.35 Apparatus used for observing Brownian motion

The glass cell is filled with smoke and quickly covered with a cover slip. It is then brightly illuminated as shown and a low-power microscope is used to view the motion of the smoke particles. These appear on the TV screen as tiny light spots which are continually jiggling about.

The observations can be explained by considering what happens to a single smoke particle. The particle is quite large compared to the air molecules which are continually bombarding it from every direction. At any given moment, the smoke particle is forced to move in a particular direction because the number of molecular impacts is greater in that direction than in any other. A moment later the particle may be moving in a new direction as the balance of air molecule impacts changes. Thus the smoke particle is pushed around haphazardly as shown in Figure 4.36.

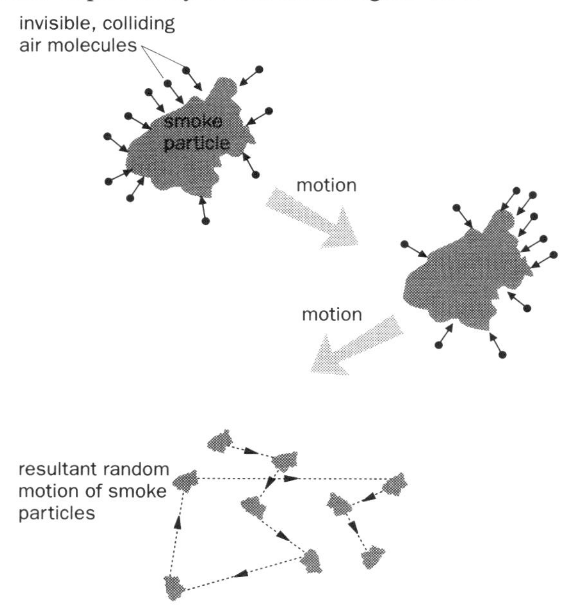

Figure 4.36 Diagrammatic illustration of the random motion of a smoke particle in air

Pressure exerted by an ideal gas

According to kinetic theory the pressure exerted by a gas results from the impact of its molecules with the walls of the vessel which contains the gas. We will now use this idea to derive an expression for gas pressure. The following assumptions are made:

- A gas consists of a large number of identical molecules.
- Intermolecular collisions and those between molecules and the container walls are perfectly elastic (i.e. there is no loss of kinetic energy).
- Intermolecular forces are negligible (except during collisions) and the effect of gravity is ignored. As a result the molecules move in straight lines at constant speed between collisions and their direction of motion is random.

- The collision time is negligible compared with the time spent by a molecule between collisions.
- The volume of the molecules is negligible compared with that of the gas.
- The number of molecules is large enough to allow a meaningful statistical treatment.
- Newton's laws of motion can be applied.

A gas consisting of a large number of **identical** molecules in rapid, random motion is contained in a cubic box of side l (see Figure 4.37).

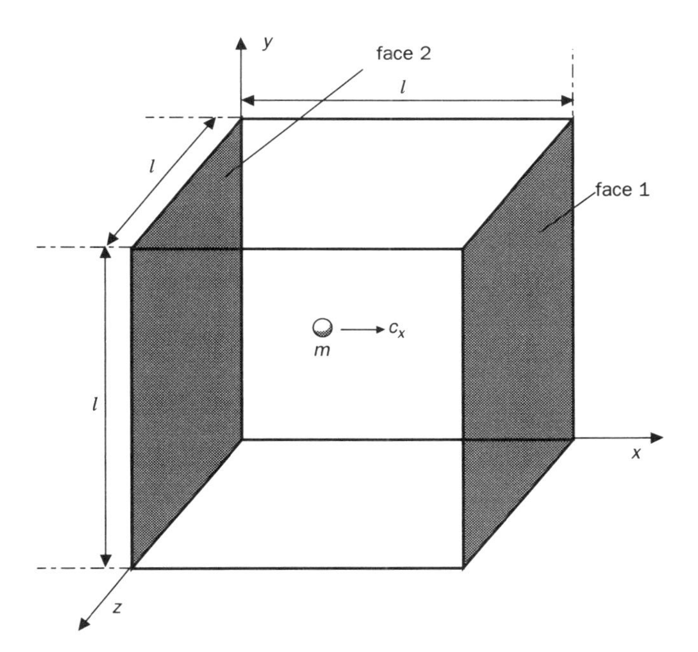

Figure 4.37 Derivation of the expression for gas pressure

Consider a single molecule of mass (m) moving towards face 1 with an x-component of velocity (c_x).

$\therefore$ momentum in the x-direction $= mc_x$

At face 1 the molecule has a **perfectly elastic** collision and its momentum is reversed so that after collision:

new momentum in the x-direction $= -mc_x$

$\therefore$ momentum **change** in the x-direction

$$= mc_x - (-mc_x) = 2mc_x$$

After colliding with face 1 the molecule travels a distance ($2l$) before it again collides with face 1:

time taken (t) to travel distance $2l$ is given by:

$$t = \frac{2l}{c_x}$$

We are trying to derive an expression for the pressure caused by these molecular collisions, so we need to look at the **force** exerted by this collision.

By Newton's second law the force exerted **on** the molecule **by** the wall = the **rate of change of momentum** of the molecule.

Rate of change of momentum of molecule

$$= \frac{\textit{momentum change}}{\textit{time taken}}$$

$\therefore$ Rate of change of momentum of molecule

$$= \frac{2mc_x}{2l/c_x} = \frac{mc_x^2}{l}$$

Also, according to Newton's third law, the force exerted **by** the wall **on** the molecule is equal but oppositely directed to the force exerted **by** the molecule **on** the wall.

$\therefore$ Force (F) exerted by the molecule on the wall is given by:

$$F = \frac{mc_x^2}{l}$$

Since **pressure** = **force/area**, the pressure, (p), which then acts on face 1 is given by:

$$p = \frac{\textit{force}}{\textit{area}} = \frac{mc_x^2/l}{l^2} = \frac{mc_x^2}{l^3}$$

If the gas contains (N) molecules having x-components of velocity $c_{x1}, c_{x2}, \ldots, c_{xN}$, the total pressure ($p$) on face 1 is given by:

$$p = \frac{m}{l^3}(c_{x1}^2 + c_{x2}^2 + c_{x3}^2 + \ldots c_{xN}^2) = \frac{m}{l^3}\left(N\overline{c_x^2}\right)$$

$$\therefore \quad p = \frac{Nm}{V}\overline{c_x^2} \qquad \text{(i)}$$

where $\overline{c_x^2}$ = the **mean square speed** of the gas molecules in the x-direction (i.e. the **mean** of the **squares** of the **speeds** of all the molecules travelling in the x-direction)

and $V = l^3$ = the **volume** of the gas.

Of course, the molecules in the container will be moving in all possible directions. Very few will be moving exactly parallel to the x-axis. However, we can consider each molecule as having **components** of velocity c_x, c_y and c_z in each of the three co-ordinate directions x, y and z. As can be done with two dimensions, the three components can be combined into a **resultant** velocity (c), given by:

$c^2 = c_x^2 + c_y^2 + c_z^2$ (see **Maths window** on page 364)

Similarly, the components of the **mean square velocities** can be combined:

$$\overline{c^2} = \overline{c_x^2} + \overline{c_y^2} + \overline{c_z^2}$$

Also, since there are a large number of molecules in random motion we can assume that there will be equal numbers moving in each of the three co-ordinate directions:

$$\overline{c_x^2} = \overline{c_y^2} = \overline{c_z^2}$$

$$\therefore \quad \overline{c^2} = 3\overline{c_x^2} \text{ from which: } \overline{c_x^2} = \frac{1}{3}\overline{c^2}$$

Substituting for $\overline{c_x^2}$ in Equation (i) we have:

$$p = \frac{1}{3}\frac{Nm}{V}\overline{c^2}$$

from which:

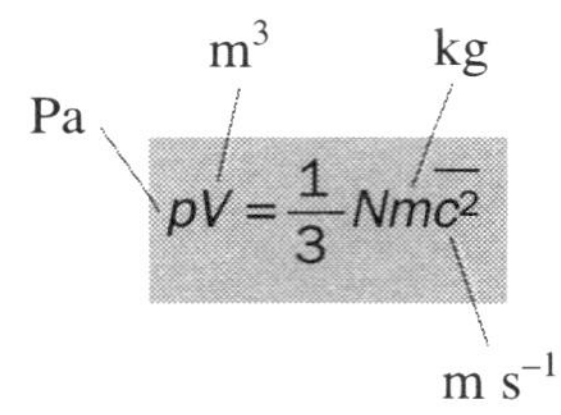

$$pV = \frac{1}{3}Nm\overline{c^2}$$

Equation 9

You should note that:

- Equation 9 links the **macroscopic** properties of gas pressure and volume with the **microscopic** properties of number, mass and speed of the **individual** molecules.
- Since Nm = total mass of gas
  and V = total volume of gas

  the **density** (ρ) of the gas is given by:

$$\rho = \frac{\textit{mass of gas}}{\textit{volume of gas}} = \frac{Nm}{V}$$

Then Equation 9 can be expressed as:

$$p = \frac{1}{3}\rho\overline{c^2}$$

(Pa; kg m^{-3}; m s^{-1})

Equation 10

Maths window

The direction of motion of the molecules is assumed to be totally random.

Consider a molecule whose speed (c) is resolved into x, y and z components c_x, c_y and c_z respectively as shown below:

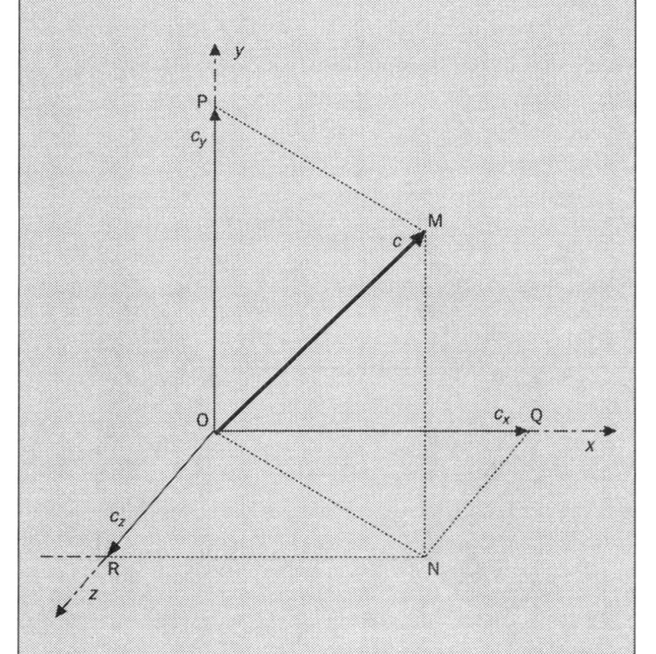

Applying Pythagoras's theorem to triangle OMN:

$$OM^2 = MN^2 + ON^2$$

$$c^2 = c_y^2 + ON^2$$

Applying Pythagoras's theorem to triangle ONQ:

$$ON^2 = OQ^2 + NQ^2$$

$$ON^2 = c_x^2 + c_z^2$$

From which we have that:

$$c^2 = c_x^2 + c_y^2 + c_z^2$$

This relationship also holds true for the **mean square velocities**:

$$\overline{c^2} = \overline{c_x^2} + \overline{c_y^2} + \overline{c_z^2}$$

The speed of gas molecules

The molecules in a gas have a very wide range of different speeds. This is illustrated in the graph shown in Figure 4.38, in which number of molecules is plotted against molecular speed.

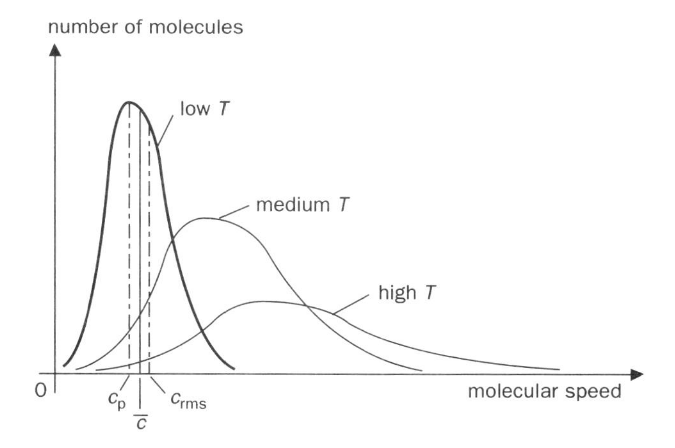

Figure 4.38 Distribution of molecular speeds

You should note that:

- Few molecules move either very fast or very slowly, and none are at rest.
- The variation of gas molecular speed depends on temperature. The higher the temperature, the greater is the **proportion** of high-speed molecules and the lower is the **proportion** of low-speed molecules.
- Three specific molecule speeds are shown on the graph in Figure 4.38. They are:
  - (i) The **most probable speed** (c_p) – This is the speed at which the **greatest number** of molecules are moving.
  - (ii) The **mean speed** ($\bar{c}$) – This is the **average** value of the speeds of **all** the molecules.
  - (iii) The **root-mean-square speed** (c_{rms}) – This is the **square root** of the **mean-square speed** ($\overline{c^2}$) of the molecules.
- The **mean-square speed** ($\overline{c^2}$) of the molecules, which appears in Equation 10, **cannot** be obtained by merely squaring the mean speed of the molecules.

If we consider four molecules having speeds of 5, 10, 15 and 20 units.

Then the mean speed is given by:

$$\bar{c} = \frac{(5 + 10 + 15 + 20)}{4} = 12.5$$

and the **square** of the mean speed is:

$$\bar{c}^2 = \frac{(5 + 10 + 15 + 20)^2}{4} = 156.25$$

whereas the **mean** of the **squares** of the speeds is given by:

$$\overline{c^2} = \frac{5^2 + 10^2 + 15^2 + 20^2}{4} = 187.5$$

Make sure you appreciate the difference between the two terms.

The term **root-mean-square speed** is most often used to quantify the molecular speeds of a gas, but you may also encounter **mean speed**. Although r.m.s. speed (c_{rms}) is of the same order of magnitude as the mean speed of the gas molecules, they have totally different meanings.

For example, if we consider four gas molecules moving with speeds of 650, 700, 750 and 800 m s^{-1}.

Then the mean speed $\bar{c}$ is given by:

$$\bar{c} = \frac{650 + 700 + 750 + 800}{4} = 725 \text{ m s}^{-1}$$

whereas the **root-mean-square** speed (c_{rms}) is given by:

$$c_{rms} = \sqrt{\overline{c^2}} = \sqrt{\frac{650^2 + 700^2 + 750^2 + 800^2}{4}}$$

$$= 727.2 \text{ m s}^{-1}$$

The difference is small, in this case, but important.

- If the gas pressure and density are known, the r.m.s. speed (c_{rms}) of the molecules of a particular gas can be obtained from a rearranged version of Equation 10.

$$c_{rms} = \frac{3p}{\rho}$$

Equation 10(a)

For example, air has a density of about 1.2 kg m^{-3} at room temperature.

Thus the r.m.s. speed (c_{rms}) of air molecules at standard atmospheric pressure (= 1.013 × 10^5 Pa) is given by:

$$c_{rms} = \sqrt{\frac{(3) \times (1.013 \times 10^5 \text{ Pa})}{(1.2 \text{ kg m}^{-3})}} = 502.5 \text{ m s}^{-1}$$

This is of the same order of magnitude as the speed of sound in air (about 340 m s^{-1}) and suggests that the kinetic theory assumptions are reasonable. This is because the speed of sound is the speed at which the disturbance is passed on from one air molecule to the next and it is reasonable to expect this to be of the same magnitude as the r.m.s. speeds of the molecules themselves.

- You should spend some time familiarising yourself with the terms used in these equations. Difficulties are usually caused by confusion between symbols. In this text, n has been used for the number of **moles**, while N represents the number of **molecules** and N_A is the number of molecules in **one mole**. The symbols, and their meanings, may look similar, but they are totally different.

> ## *Guided example (3)*
>
> Calculate the root-mean-square speed of the gas molecules in each of the following cases.
>
> (a) Hydrogen of density 0.009 kg m^{-3} at a pressure of 2.5×10^5 Pa
>
> (b) Helium of density 0.18 kg m^{-3} at a pressure of 3.0×10^5 Pa
>
> (c) Oxygen of density 1.43 kg m^{-3} at a pressure of 1.01×10^5 Pa.
>
> **Guidelines**
>
> Use Equation 10(a) in each case.

Molecular kinetic energy and temperature

The equation of state for (n) moles of an ideal gas of volume (V) at a pressure (p) and an absolute temperature (T) is:

$$pV = nRT \qquad \text{(i)}$$

where (R) is the universal molar gas constant.

From kinetic theory of gases we have that:

$$pV = \frac{1}{3} N m \overline{c^2} \qquad \text{(ii)}$$

where N = number of molecules in volume V of gas
m = mass of each molecule
and $\overline{c^2}$ = the mean-square speed of the gas molecules.

Combining Equations (i) and (ii) we obtain:

$$\frac{1}{3} N m \overline{c^2} = nRT$$

This may be re-written as:

$$\frac{2}{3} N \left(\frac{1}{2} m \overline{c^2}\right) = nRT$$

$$\therefore \quad \frac{1}{2} m \overline{c^2} = \frac{3}{2} \frac{n}{N} RT = \frac{3}{2} \frac{R}{N/n} T$$

But N/n is the number of gas molecules **per mole** (i.e. $N/n = N_A$, the Avogadro constant).

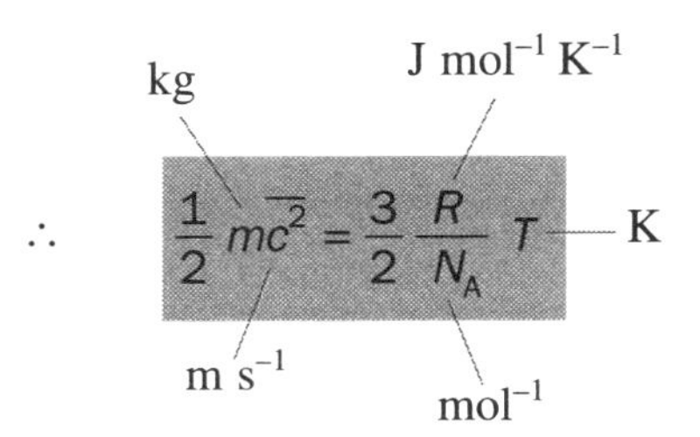

$$\therefore \quad \frac{1}{2} m \overline{c^2} = \frac{3}{2} \frac{R}{N_A} T$$

Equation 11

- R is the gas constant per **mole** of gas molecules (i.e. the **molar** gas constant)
- $R/N_A = k$ is called **Boltzmann's constant** – it is the gas constant per **molecule**.

Thus Equation 11 can be written:

$$\frac{1}{2} m \overline{c^2} = \frac{3}{2} kT$$

Equation 12

which tells us that: **the average translational kinetic energy of a gas molecule is directly proportional to the absolute temperature of the gas.**

The equation of state for n moles of an ideal gas is:

$pV = nRT$.

This can also be expressed in terms of **Boltzmann's constant** (k).

Since $k = \frac{R}{N_A}$ then $R = kN_A$.

Substituting for R in $pV = nRT$ gives: $pV = n(kN_A)T$

But nN_A is the product of the **number of moles of gas** and the **number of molecules per mole**, i.e. it is equal to the total number of molecules in the gas (N).

$$\therefore \quad pV = NkT$$

Equation 13

You should note that from Equation 12, the relationship between **mean-square speed** and **temperature** is:

$$\overline{c^2} \propto T$$

From which the **r.m.s. speed** is related to temperature by:

$$\sqrt{\overline{c^2}} \propto \sqrt{T}$$

i.e. the **root-mean-square speed of gas molecules** is proportional to the **square root of the absolute temperature** of the gas.

Avogadro's law

This law states that:

> Equal volumes of ideal gases at the same temperature and pressure contain equal numbers of molecules.

If we consider two ideal gases x and y we can write:

For gas x : $p_x V_x = N_x k T_x$

For gas y : $p_y V_y = N_y k T_y$

If the gases have equal volumes and are at the same pressure, then:

$$p_x V_x = p_y V_y$$

$$\therefore \quad N_x k T_x = N_y k T_y$$

If two gases are also at the same temperature, then $T_x = T_y$, and so:

$N_x = N_y$

i.e. the number of molecules in gas x = the number of molecules in gas y.

Guided examples (4)

1. The observed speeds of ten particles at a particular instant were as shown in the table below.

number of particles	1	2	4	1	1	1
speed (m s^{-1})	5.0	7.0	9.0	12.0	14.0	15.0

What is (a) the most probable speed

(b) the mean speed

(c) the root-mean-square speed of these particles?

Guidelines

(a) This is the speed at which the **greatest number** of particles are moving.

(b) You must add all the speeds and divide by the total number of particles. Remember that at some speeds there is more than one particle.

(c) The r.m.s. speed is the **square root** of the **mean** of the **squares** of the speeds.

2. Nitrogen gas is kept in a container at a temperature of 27 °C and a pressure of 1.0×10^5 Pa. If the density of nitrogen is 1.25 kg m^{-3}, calculate:

(a) the root-mean-square speed of the nitrogen molecules

Guided examples (4) continued

(b) the temperature at which the r.m.s. speed of the molecules would become twice as great as that calculated in (a).

Guidelines

(a) Use Equation 10.

(b) Make use of the fact that r.m.s. speed is proportional to the square root of the **absolute** temperature.

3. A cylinder of volume 0.25 m^3 contains nitrogen gas at a pressure of 1.5×10^5 Pa and a temperature of 17 °C. Assuming that the gas behaves ideally, calculate:

(a) the number of **moles** of gas in the cylinder

(b) the r.m.s. speed of the gas molecules at 17 °C

(c) the average translational kinetic energy of a nitrogen molecule

(d) the total kinetic energy of the gas contained in the cylinder.

(Molar mass of nitrogen = 0.028 kg mol^{-1}; molar gas constant, R = 8.31 J mol^{-1} K^{-1}; Avogadro constant, $N_A = 6.02 \times 10^{23}$ mol^{-1}.)

Guidelines

(a) Use Equation 4.

(b) Use Equation 10.

(c) Use Equation 12.

(d) Total kinetic energy = kinetic energy per molecule × total number of molecules.

The total number of molecules = number of moles of gas × number of molecules per mole.

Work done by a gas

Consider a mass of an ideal gas at a pressure (p) enclosed in a cylinder by a frictionless piston of area (A). The piston is in equilibrium under the action of an external force (F) and the pressure force (pA) which the gas exerts (see Figure 4.39 overleaf).

If the piston is allowed to move outwards a distance Δx which is small enough for the pressure to be assumed constant, then:

External work done (ΔW) by the expansion is given by:

$\Delta W = pA\Delta x$ (since *work* = *force* × *distance*)

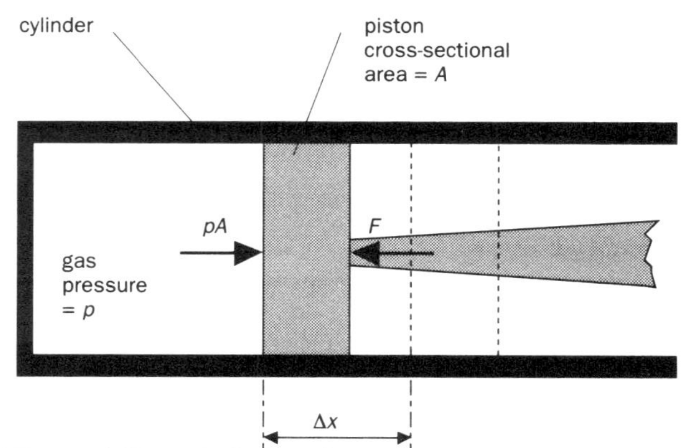

Figure 4.39 Work done by an expanding gas

But $A\Delta x = \Delta V$ (the small increase in volume resulting from the piston's movement)

$$\therefore \quad \Delta W = p\Delta V$$

(ΔW in J; p in N m^{-2} (Pa); ΔV in m^3)

Equation 14

i.e. when a gas expands at constant pressure, it does **work** (in J) on the surroundings, equal to the **pressure** (in Pa) multiplied by the change of **volume** (in m^3).

You should note that:

- The same result can be obtained by considering a compression of the gas in which case work is done **on** the gas rather than **by** it.
- For a finite expansion in which the volume changes from an initial value (V_1) to some final value (V_2), the total work done (W) is given by:

$$W = \int_{V_1}^{V_2} p \, dV$$

Equation 15

This equation is valid for **all** relationships between p and V.

For a process in which p is constant (called an **isobaric** process) we have that:

$$W = p\int_{V_1}^{V_2} dV = [V]_{V_1}^{V_2}$$

$$\therefore \quad W = p(V_2 - V_1)$$

Equation 16

Generally if a pressure–volume graph is plotted for the gas (such a graph is called an **indicator diagram,** as shown in Figure 4.40(a)), then the work done (W) by the gas as the volume changes from (V_1) to (V_2) is equal to the area enclosed by the p–V graph between the two V-values (shown shaded).

For an **isobaric** process the p–V graph is a horizontal line and so

W = area enclosed by p–V graph

$= p(V_2 - V_1)$

as shown in Figure 4.40(b).

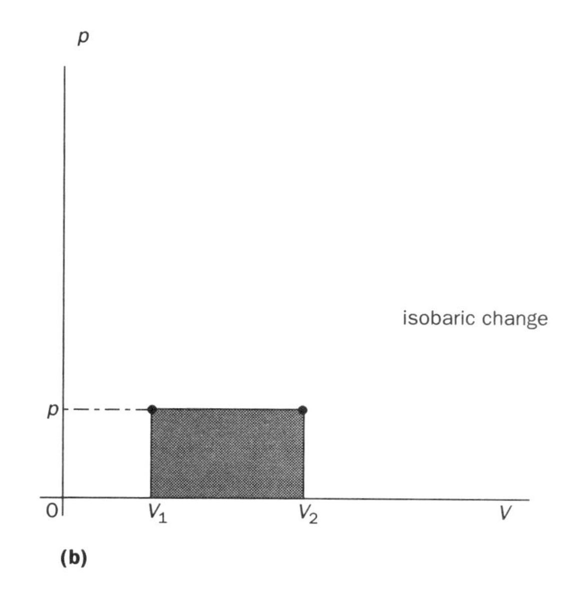

Figure 4.40 Pressure–volume graph for a gas

First law of thermodynamics

Thermodynamics is a branch of physics which deals with processes in which **heat** flows into or out of a system (e.g. the petrol/air mixture in the cylinder of a car engine) while **work** is done on or by the system. The first law of thermodynamics is a specific case of the **principle of conservation of energy** and may be stated as follows:

$$\Delta Q = \Delta U + \Delta W$$

Equation 17

where:

ΔQ = **heat energy** supplied to a system, or flowing out of it

ΔU = change of **internal energy** of the system

ΔW = **work done** by or on the system

The sign convention we adopt here is that:

- ΔQ is **positive** when heat is supplied **to** the system and it is **negative** when heat is transferred **from** the system.
- ΔW is **positive** if the external work is done **by** the system and it is **negative** if the external work is done **on** the system.
- ΔU is **positive** if the internal energy of the system **increases**.

If we consider the case of the gas expansion of Figure 4.40(a), Equation 17 can be written as:

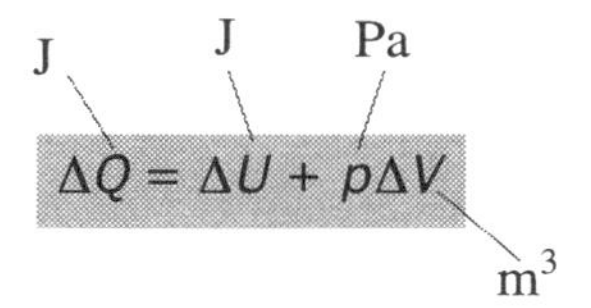

$$\Delta Q = \Delta U + p\Delta V$$

Equation 18

where $p\Delta V$ = the **work done** (in this case **by** the system **on** the surroundings).

Changes of state such as **melting, vaporisation** and **sublimation** can be described in terms of the first law of thermodynamics by stating that:

$$\Delta l = \Delta U + \Delta W$$

Equation 19

where

Δl = the **latent heat energy** supplied to cause the change of state

ΔU = increase in **internal potential energy** resulting from the change of state

ΔW = the external **work done** as a result of the change of state.

This term is most significant in the cases of vaporisation (change from liquid to vapour) and sublimation (change from solid directly to a vapour) because these two processes involve large volume increases and a large proportion of the latent heat is used to do work in pushing back the atmosphere. ΔW is taken as **positive** for an expansion (e.g. in melting and vaporisation) and **negative** for a contraction (e.g. in condensation).

The internal energy of a system can be increased by:

- **supplying heat energy** to the system and/or
- **doing work** on the system.

For example, if we consider the air in a bicycle pump to be the system, its internal energy and hence its temperature can be increased by supplying heat from outside the pump or by pushing the plunger in with the finger held over the nozzle. Figure 4.41 illustrates the two methods.

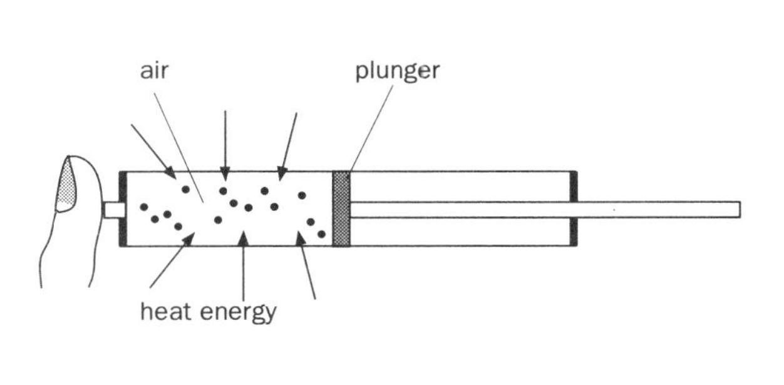

(a) Heat is supplied (the space around the pump is at a higher temperature than the air inside)

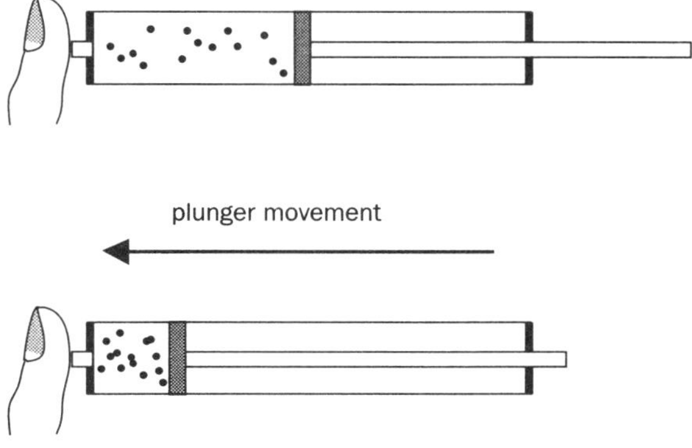

(b) Work is done on the air by movement of the plunger. Potential energy increases

Figure 4.41 Increasing the internal energy of the air in a bicycle pump

This leads us to an alternative way of stating the first law which is preferred by some examination boards.

The **increase** in internal energy of a system = Heat supplied **to** the system + Work done **on** the system

i.e. $\Delta U = \Delta Q + \Delta W$

Equation 20

In the case of an **isolated system** (i.e. one on which there are no external influences), no work is done on or by the system and no heat is supplied to or transferred from the system.

$\therefore \quad \Delta W = 0$ and $\Delta Q = 0$

From which it follows that the change in internal energy $\Delta U = 0$.

Thus **the internal energy of an isolated system is constant**.

Adiabatic processes

An **adiabatic process** is one in which there is **no heat energy** flow into or out of the system. When a system undergoes such a process, $\Delta Q = 0$ and Equation 20 becomes: $\Delta U = \Delta W$.

Thus when a system undergoes an adiabatic process: **the increase in the internal energy of the system is equal to the work done on it**.

The principal heat capacities of a gas

Since only a relatively small expansion is produced when a solid or a liquid is heated, most of the heat energy supplied is used to increase the internal energy. In the case of a gas, however, the expansion can be much greater, enabling the gas to do external work. Thus when a gas is heated the increase in its internal energy and hence the temperature rise produced depends on how much external work the gas is allowed to do. Because of this a gas has an infinite number of heat capacities. The most useful two gas heat capacities, called the **principal heat capacities**, which are used are those which relate to **constant volume** and **constant pressure** conditions. They are defined as follows:

The molar heat capacity at constant volume (C_v) is the heat energy needed to produce a unit temperature rise in one mole of a gas when the **volume** is kept constant.

The molar heat capacity at constant pressure (C_p) is the heat energy needed to produce a unit temperature rise in one mole of a gas when the **pressure** is kept constant.

You should note that:

- The unit of molar heat capacity at constant volume or pressure is J mol^{-1} K^{-1}.
- If we are dealing with unit mass of gas instead of one mole, we use the terms **principal specific heat capacity at constant volume** and **principal specific heat capacity at constant pressure**, and these are denoted by the symbols c_v and c_p.
- $C_p > C_v$ – This can be explained in terms of the first law of thermodynamics. When a gas is heated at **constant pressure** some of the energy is used to raise the internal energy (hence the temperature) of the gas and some is used to do external work. If the gas is heated at **constant volume** we are effectively not allowing the gas to do any work and so all the energy supplied is used to raise the internal energy and therefore the temperature. Thus the amount of heat needed to raise the temperature of a gas at constant pressure is greater than that needed to produce the same temperature rise at constant volume.
- The amount of heat ΔQ needed to raise the temperature of **1 mole** of an ideal gas by an amount ΔT at **constant volume** is given by:

  $\Delta Q = C_v \Delta T$ *Equation 21(a)*

- The amount of heat ΔQ needed to raise the temperature of **1 mole** of an ideal gas by an amount ΔT at **constant pressure** is given by:

  $\Delta Q = C_p \Delta T$ *Equation 21(b)*

- It can be shown that for ideal gases:

  $C_p - C_v = R$ *Equation 22*

- The ratio C_p/C_v is denoted by γ (= 1.67 for a monatomic gas, 1.40 for a diatomic gas and 1.33 for a polyatomic gas).

Methods of producing changes in gases

The **state** of a gas (i.e. the values of p, V and T) can be changed by any of the following processes.

Isothermal process

- Occurs at **constant temperature** (T constant).
- For an **isothermal** expansion or contraction of an ideal gas,
  pV = constant and $p_1V_1 = p_2V_2$ (see Figure 4.42).

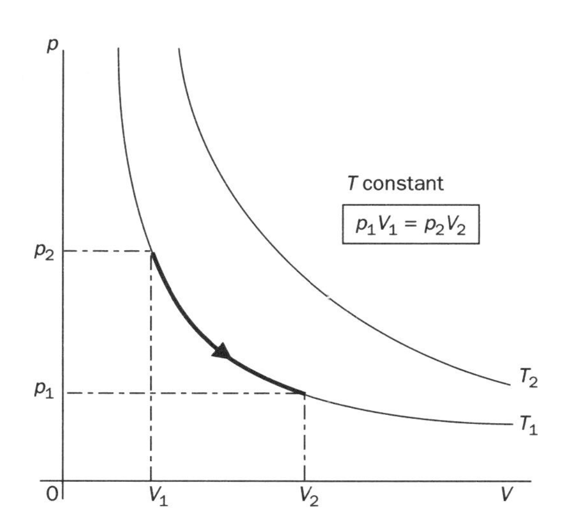

Figure 4.42 Isothermal process

- Since the internal energy of an ideal gas depends only on its temperature, $\Delta U = 0$, and so $\Delta Q = \Delta W$. This means that if gas expands and does work ΔW then an amount of heat ΔQ must be supplied to keep the temperature constant.
- A **reversible** isothermal change is an **ideal** one which would require the gas to be contained in a thin, highly conductive vessel, surrounded by a constant-temperature reservoir. In addition, the change would have to occur very slowly.

Isovolumetric process

- Occurs at **constant volume**.
- For an **isovolumetric** change of an ideal gas,

$$V = \text{constant and } \frac{P_1}{T_1} = \frac{P_2}{T_2}$$

(see Figure 4.43).

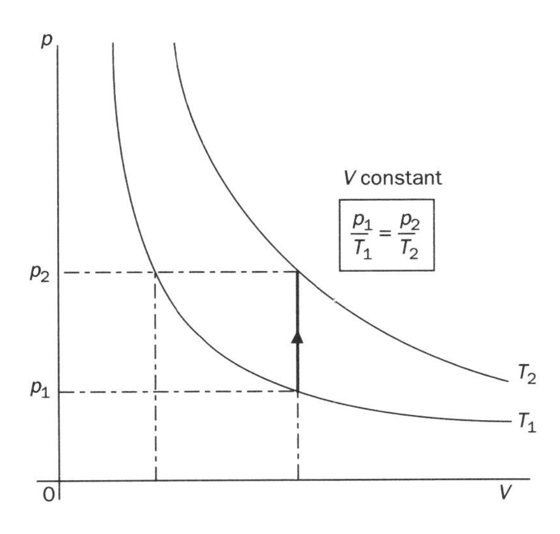

Figure 4.43 Isovolumetric change

- Since $\Delta V = 0$, no work is done and all the heat entering the gas becomes internal energy.

The temperature rises from T_1 to T_2 and the pressure increases from p_1 to p_2.

$$\therefore \quad \Delta Q = \Delta U = C_v(T_2 - T_1).$$

Isobaric process

- Occurs at **constant pressure**.
- For an **isobaric** change of an ideal gas,

$$p = \text{constant and } \frac{V_1}{T_1} = \frac{V_2}{T_2}$$

(see Figure 4.44).

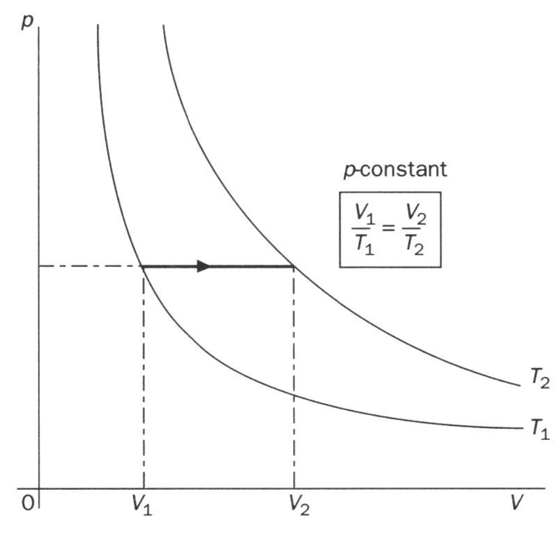

Figure 4.44 Isobaric process

- Some of the heat supplied to the gas increases the **internal energy** (temperature rise T_1 to T_2) and the rest is used to do work.

Adiabatic process

- Occurs with **no heat entering or leaving** the gas.
- For an **adiabatic** change of an ideal gas, $pV^\gamma =$ constant (where $\gamma = C_p/C_v$) and $p_1V_1^\gamma = p_2V_2^\gamma$ (see Figure 4.45).

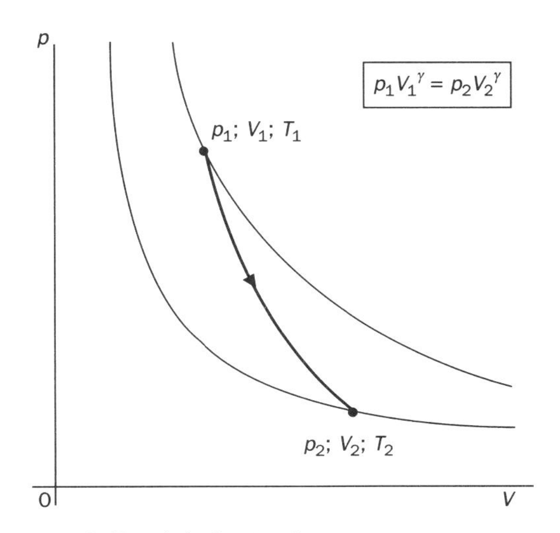

Figure 4.45 Adiabatic change

- Since $\Delta Q = 0$, $\Delta W = -\Delta U$. This means that for an adiabatic expansion all the work is done at the expense of the internal energy of the gas which thus cools (T_2 to T_1). In the case of a compression the gas heats up as the work done increases the internal energy.
- A reversible adiabatic change is an ideal process which requires the gas to be in a thick-walled, poorly conducting vessel and the volume change must be small and rapid.

Work done during a cycle

When a gas undergoes changes which eventually return it to its original state, it is said to have gone through a **cycle** of operations. The p–V (indicator) diagram is then a closed loop whose area represents the net work done by or on the gas. Consider, for example, an ideal gas which undergoes the changes shown in Figure 4.46.

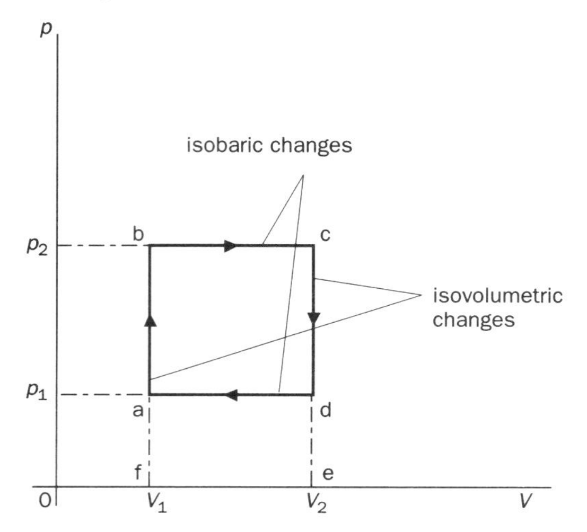

Figure 4.46 Area enclosed by the closed loop resulting from a cycle of changes on a gas = the net work done

Work done by the gas as it expands from V_1 to V_2 = area bcfe = $p_2(V_2 - V_1)$.
(This expansion results from an **isovolumetric** change a to b followed by an **isobaric** change b to c.)

Work done on the gas in compressing it from V_2 to V_1 = area adfe = $p_1(V_2 - V_1)$.
(This compression results from an **isovolumetric** change c to d followed by an **isobaric** change d to a.)

Net work done by the gas = area bcfe – area adfe

$= p_2(V_2 - V_1) - p_1(V_2 - V_1)$

$= (p_2 - p_1)(V_2 - V_1)$

∴ **Net work done by the gas = area abcd**
(the area enclosed by the rectangle)

This can be applied to any **cycle** of operations.

Guided examples (5)

1. A sample of gas is enclosed in a cylinder by a frictionless piston of area 60 cm^2. The cylinder is heated so that 400 J of heat energy is supplied to the gas which then expands against atmosphere pressure and pushes the piston 20 cm along the cylinder. Given that atmospheric pressure is 1.0×10^5 Pa, calculate:

(a) the external work done by the gas and

(b) the change in internal energy of the gas.

Guidelines

(a) Use Equation 16 – Remember SI units throughout.

Alternatively use the fact that the work done by the gas = force exerted by gas on piston × distance moved by piston. (force exerted by gas on piston = force exerted by atmospheric pressure acting on piston).

(b) Use Equation 17.

2. When 0.60 kg of water is converted into steam at 100 °C at a pressure of 1.0×10^5 Pa the amount of heat energy supplied is 2.9×10^6 J.

If during the vaporisation the water increases its volume by 2.2 m^3, calculate:

(a) the work done against the external pressure

(b) the increase in the internal energy of the water.

Guidelines

(a) Use Equation 14.

(b) Use Equation 18.

3. When 22.4 g of nitrogen is heated from 293 K to 393 K in an insulated and freely extensible container, 2.33×10^3 J of heat is required. When the same mass of nitrogen is contained in an insulated, rigid container, 1.66×10^3 J of heat energy is required to give the same temperature rise. Calculate:

(a) the principal molar heat capacities of nitrogen

(b) the value of the universal molar gas constant.

(molar mass of nitrogen = 28.0 g mol^{-1})

Guided examples (5) *continued*

Guidelines

(a) Use Equation 8 to find the number of moles of nitrogen in the sample.

Then use Equation 21(b) to obtain the molar heat capacity at constant pressure, C_p, and Equation 21(a) to obtain the molar heat capacity at constant volume, C_v.

(b) Use Equation 22.

4.

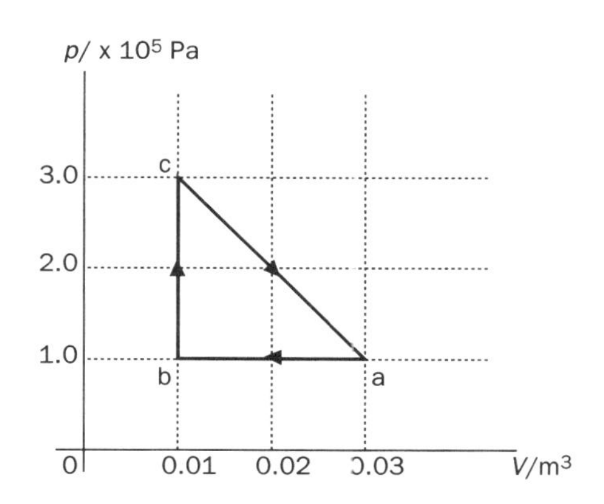

The indicator diagram shows an energy cycle for **1 mole** of an ideal gas. The gas is cooled at constant pressure (a to b), heated at constant volume (b to c) and then returned to its original state (c to a). Calculate:

(a) the gas temperature at a, b and c

(b) the heat removed from the gas during the process a to b

(c) the heat supplied to the gas during the process b to c, and

(d) the net work done in the cycle.

(universal molar gas constant, $R = 8.3$ J mol^{-1} K^{-1}; C_v for the gas $= 5R/2$)

Guidelines

(a) Use Equation 4 for 1 mole of gas – take the values of p and V for each point from the indicator diagram.

(b) Obtain a value for C_p by using Equation 22. Then use Equation 21(b) to find heat removed during the isobaric process (a to b).

(c) Use Equation 21(a) to find heat supplied during the isovolumetric process (b to c).

(d) Use net work done in the cycle = area of triangle abc.

Self-assessment

SECTION A
Qualitative assessment

1. For each of the gas laws (**Boyle's** law, **Charles'** law and the **pressure** law):

(a) give a written statement of the law

(b) express the law as an equation

(c) express the law on a graph relating the quantities involved.

2. (a) Explain what is meant by an **ideal** gas.

(b) Write down the equation of state for N moles of an ideal gas. Define each quantity in the equation.

3. (a) Define the **mole**.

(b) How is the mole related to the **Avogadro constant?**

(c) What is

(i) the **relative molecular mass** and

(ii) the **molar mass** of a substance?

(d) How many moles are there in M kg of a substance of molar mass M_m kg mol^{-1}?

4. (a) What is **Brownian motion?**

(b) How is it explained in terms of molecules?

5. (a) State the main assumptions made in the kinetic theory of gases.

(b) Derive the equation,

$$p = \frac{1}{3}\frac{Nm}{V}\overline{c^2}$$

for an ideal gas of volume (V) at a pressure (p) if the gas contains (N) molecules, each of mass (m) and with mean-square speed ($\overline{c^2}$).

6. Explain what is meant by

(a) **most probable** speed

(b) **mean** speed and

(c) **root-mean-square** speed of gas molecules.

Self-assessment *continued*

7. Derive the expression,

$$\frac{1}{2}m\overline{c^2} = \frac{3}{2}kT$$

relating average molecular kinetic energy and absolute temperature of a gas.

8. State **Avogadro's law** and prove your statement.

9. (a) Consider an ideal gas enclosed in a cylinder by a piston of area (A). What is the **work done** when the piston moves over a small distance Δx as a result of the gas pressure (p)? Assume that the change in volume is so small that the pressure remains constant. What is the work done in terms of the volume change of the gas?

(b) What is the expression for the total work done by the gas for a **finite** expansion in which the volume increases from V_1 to V_2? What is the expression if the expansion occurs at **constant pressure?**

10. (a) State the **first law of thermodynamics**.

(b) Explain what is meant by **internal energy** in the context of the first law.

(c) How can the internal energy of a system be (i) **increased**, (ii) **decreased?**

(d) What is the alternative form of the first law which is based on the change of internal energy of a system?

11. (a) What is an **isolated** system?

(b) Why is the internal energy of such a system constant?

(c) What is an **adiabatic** process?

(d) What do you conclude by applying the first law of thermodynamics to a system which undergoes an adiabatic process?

12. (a) Why can a gas have an infinite number of heat capacities?

(b) Define (i) the **molar heat capacity at constant volume** C_v
and (ii) the **molar heat capacity at constant pressure** C_p, of a gas.

(c) Explain why C_p is greater than C_v.

(d) Give an expression for the amount of heat ΔQ which must be supplied to n moles of an ideal gas so as to produce a temperature rise ΔT if the heating is performed (i) at **constant volume**, and (ii) at **constant pressure**.

13. The state of a gas can be changed by each of the following processes: **isothermal**; **isovolumetric**; **isobaric; adiabatic**. For each of these:

(a) State the condition under which the change occurs.

(b) Show the change on a pressure–volume graph.

SECTION B
Quantitative assessment

(***Answers:*** 1.5×10^{-4}; 2.7×10^{-4}; 2.1×10^{-3}; 8.6×10^{-3}; 0.010; 0.026; 0.80; 1.3; 60; 120; 210; 240; 240; 480; 480; 480; 480; 480; 480; 520; 520; 730; 830; 920; 1.6×10^3; 2.0×10^3; 2.5×10^3; 2.6×10^3; 7.0×10^3; 1.0×10^4; 2.1×10^4; 2.3×10^4; 2.3×10^5; 4.8×10^{23}.)

1. (a) A fixed mass of gas has a volume of 3000 cm^3 at a pressure of 1.0×10^5 Pa. Calculate its volume if the pressure increases to 2.0×10^6 Pa with the temperature remaining constant.

(b) A fixed mass of gas has a volume V when the temperature is 127 °C. To what temperature must the gas be raised so that its volume increases to 2.75 V with the pressure remaining constant?

(c) A fixed mass of gas has a volume of 0.02 m^3 at a temperature of 44 °C and a pressure of 2.02×10^5 Pa. Find the new volume of the gas at standard temperature and pressure (i.e. 0 °C and 1.01×10^5 Pa).

2. (a) At a particular instant five oxygen molecules have speeds of 450, 475, 480, 495 and 510 m s^{-1}.
Calculate their:

(i) mean speed

(ii) mean-square speed and

(iii) root-mean-square speed.

(b) Calculate the r.m.s. speed for the molecules of a gas of density 1.3 kg m^{-3} at a pressure of 1.0×10^5 Pa.

4. A sealed can of volume 2×10^4 cm^3 contains gas at a pressure of 1×10^5 Pa and a temperature of 27 °C. Assuming the gas to be ideal, calculate:

(a) the number of moles of gas in the can (given that the universal molar gas constant, R = 8.31 J mol^{-1} K^{-1})

Self-assessment *continued*

(b) the number of gas molecules in the can (given that the Avogadro constant, $N_A = 6.02 \times 10^{23}\ \text{mol}^{-1}$)

(c) the mass of gas in the can if its relative molecular mass is 32

(d) the gas density

(e) the r.m.s. speed of the gas molecules.

4. The total translational kinetic energy of a certain mass of an ideal gas at a temperature of 57 °C is 550 J. If the relative molecular mass of the gas is 2.0, calculate:

(a) the total kinetic energy of the gas molecules at a temperature of 280 °C

(b) the mass of gas present and

(c) the r.m.s. speed of the molecules at 280 °C

(given that the universal molar gas constant, $R = 8.3\ \text{J mol}^{-1}\ \text{K}^{-1}$).

5.

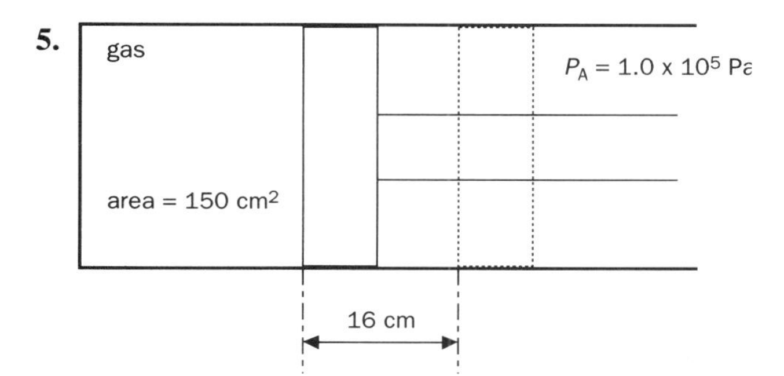

The diagram shows a sample of gas enclosed in a cylinder by a frictionless piston of area 150 cm^2. When 300 J of energy is supplied to the gas, it expands against a constant atmospheric pressure of 1.0×10^5 Pa and pushes the piston out a distance of 16 cm along the cylinder. Calculate:

(a) the work done by the gas and

(b) the increase in internal energy of the gas.

6. When 10 cm^3 of water is boiled at 100 °C and at an atmospheric pressure of 1.0×10^5 Pa, 1.6×10^4 cm^3 of steam is produced. Calculate:

(a) the mass of water boiled

(b) the heat energy needed to produce the vaporisation

(c) the external work done during the vaporisation

(d) the increase in internal energy.

(Density of water = $1.0 \times 10^3\ \text{kg m}^{-3}$; specific latent heat of vaporisation of steam = $2.26 \times 10^6\ \text{J kg}^{-1}$.)

7. An insulated, freely extensible vessel contains 0.5 mol of oxygen at 273 K. If atmospheric pressure is taken as 1.0×10^5 Pa and the principal molar heat capacity at constant pressure of oxygen is 29 J mol^{-1} K^{-1} calculate:

(a) the heat energy needed to raise the temperature of the gas to 323 K

(b) the volume increase of the gas produced by the temperature rise

(c) the external work done by the gas

(d) the increase in the internal energy of the gas

(e) the heat energy needed to cause the same temperature rise at constant volume.

(Universal molar gas constant, $R = 8.3\ \text{J mol}^{-1}\ \text{K}^{-1}$.)

8.

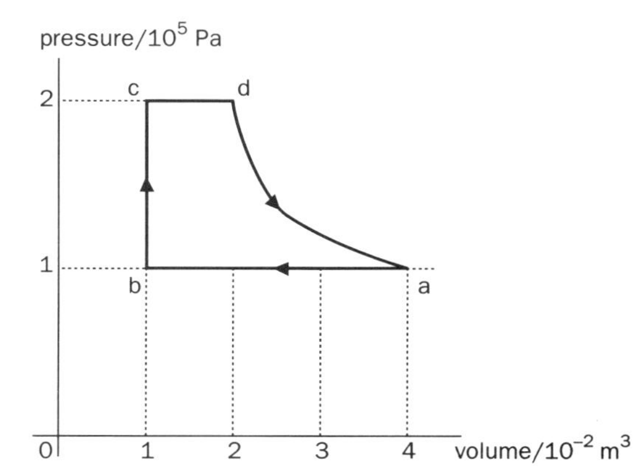

The indicator (pV) diagram shows an energy cycle for **one mole** of an ideal gas which is: cooled at constant pressure (a to b), heated at constant volume (b to c), heated at constant pressure (c to d) and finally returned to its original state by an isothermal expansion (d to a).

Calculate:

(a) the gas temperature at a, b, c and d

(b) the heat energy removed or supplied during the processes a to b, b to c, and c to d

(c) the net work done in 1 cycle (an estimate).

(Universal molar gas constant, $R = 8.3\ \text{J mol}^{-1}\ \text{K}^{-1}$; molar heat capacity at constant pressure of the gas, $C_p = 29\ \text{J mol}^{-1}\ \text{K}^{-1}$.)

4.4 Heat transfer

As we have already seen in Section 4.2, heat energy flows as a result of a temperature difference. We must now consider the three different mechanisms – **conduction, convection and radiation** – by which the heat flow can occur.

Conduction

This is the mechanism in solids (particularly metals), and takes place without any physical transfer of the material. Conduction can also occur in liquids and gases, but not in a vacuum.

Convection

This is the mechanism in fluids, involving physical transfer of fluid from region to region at different temperatures (impossible in solids or a vacuum).

Radiation

Thermal radiation is emitted by all bodies at temperatures above absolute zero. The radiation can pass through matter and through a vacuum.

Perhaps the best way to introduce heat transfer mechanisms is to see them in the context of a familiar situation – such as using an electric cooker fitted with 'halogen' radiant heating elements to warm up some soup in a pan, as shown in Figure 4.47.

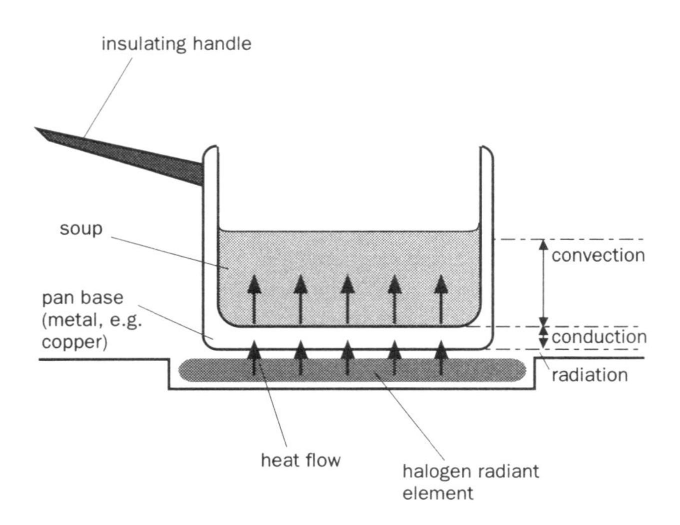

Figure 4.47 Heat transfer mechanisms involved in heating a pan of soup

The heat energy **radiating** from the hot halogen element is absorbed by the outside layer of the base of the pan. This causes a large temperature rise, creating a **temperature gradient** between the outside of the pan and the relatively cold inside, which causes heat energy to be **conducted** through the metal base to the inside. The heat energy passes by **conduction** into the soup which is in immediate contact with the base and is then transferred through the soup, mainly by **convection**. (There will be some conduction, but this will be minimal, since liquids are poor conductors.)

We shall now consider the three heat transfer mechanisms in more detail.

Thermal conduction

We have seen that when two bodies at different temperatures are placed in contact, heat energy flows from the higher to the lower temperature body. In the same way a temperature gradient inside a body will produce a flow of heat from the higher to the lower temperature region which continues until the temperatures equalise.

Thermal conduction is the process by which heat flows from the hotter to the colder regions of a body without any transfer of the material itself.

There are two distinct thermal conduction mechanisms.

Energetic free electron diffusion

Metallic solids contain 'free' electrons which are responsible for electrical conduction. When one end of a metal bar is heated the free electrons at that end gain energy and can then transfer thermal energy by **diffusing** to colder regions. Thus the same property (having conduction electrons) which makes metals good electrical conductors also makes them good thermal conductors.

Molecular or atomic coupling

The vibrational kinetic energy of the molecules or atoms in a solid increases with temperature ($E_k = {}^3/_2kT$). Thus heating one end of a solid increases the vibrational kinetic energy of the atoms at that end. Energy transfer then occurs as these atoms pass on their increased vibration to neighbouring atoms with lower energy.

Heat energy is conducted in metals by both mechanisms and in non-metals only by atomic coupling (since these do not contain free electrons).

Thermal conduction equation

Figure 4.48 shows a sample of material of cross-sectional area (A) through which heat energy flows steadily.

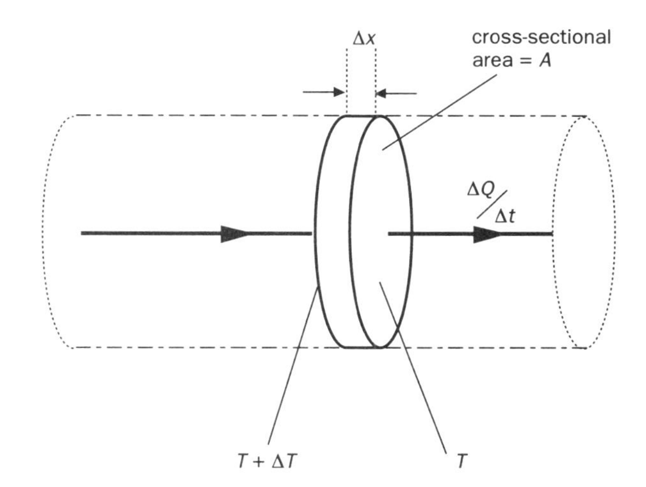

Figure 4.48 Heat flow through a thin sample of a material

A thin disc of the sample of thickness (Δx) has its faces at temperatures ($T + \Delta T$) and T. Assuming no heat is lost from the edge of the disc, then when steady state conditions are obtained, the heat flow rate $\Delta Q / \Delta t$ is directly proportional to:

(i) the cross-sectional area (A) and

(ii) the temperature gradient $\left(\dfrac{\Delta T}{\Delta x}\right)$

i.e. $\dfrac{\Delta Q}{\Delta t} \propto A \dfrac{\Delta T}{\Delta x}$

From which:

W $\quad$ W m^{-1} K^{-1} $\quad$ m^2

$$\frac{\Delta Q}{\Delta t} = -kA \frac{\Delta T}{\Delta x}$$

K m^{-1} *Equation 1*

You should note that:

- The heat flow rate $\dfrac{\Delta Q}{\Delta t}$ is a **rate of transfer of energy** and is, therefore, equivalent to **power**.
- The term **steady state** describes the condition in which the rate of heat flow is **uniform** at all points in the material.
- $\dfrac{\Delta Q}{\Delta t}$ is the heat flow rate **normal** to the faces of the disc (in J s^{-1} or W).
- $\dfrac{\Delta T}{\Delta x}$ is the **temperature gradient** between the faces of the disc (in °C m^{-1} or K m^{-1}).
- k is a constant whose value depends on the material – called the **thermal conductivity**.
- The unit of k is W m^{-1} K^{-1}.
- The negative sign in the equation indicates that when heat flows in the **positive** x-direction, the temperature gradient is **negative**.
- By re-arranging Equation 1, we have that:

$$k = -\frac{\Delta Q/\Delta t}{A \Delta T/\Delta x}$$

- From which:

The **thermal conductivity (*k*)** of a material is defined as the rate of flow of **heat energy** per unit **area**, per unit **temperature gradient** when the heat flow is normal to the faces of a thin, parallel-sided slab of the material under steady state conditions.

Heat flow along a uniform bar

We are now going to look at the flow of heat through a solid bar of uniform cross-sectional area (A) and thermal conductivity (k). Consider first the case in which the bar is perfectly insulated (lagged) so as to prevent heat loss from the surface (see Figure 4.49). Since there is no loss of heat energy from the sides of the bar, the rate of flow of heat energy will be constant along the length of the bar.

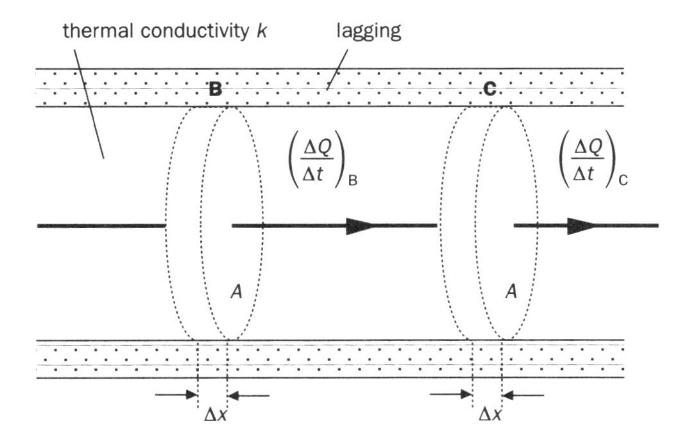

Figure 4.49 Heat flow along a perfectly lagged bar of uniform cross-section

Under steady state conditions the heat flow rate for each of the two thin sections B and C is given by:

$$\left(\frac{\Delta Q}{\Delta t}\right)_B = -kA\left(\frac{\Delta T}{\Delta x}\right)_B \quad \text{for section B}$$

and $\left(\frac{\Delta Q}{\Delta t}\right)_C = -kA\left(\frac{\Delta T}{\Delta x}\right)_C$ for section C

The bar is perfectly lagged, so no heat is lost from the surface:

$$\therefore \quad \left(\frac{\Delta Q}{\Delta t}\right)_B = \left(\frac{\Delta Q}{\Delta t}\right)_C$$

and since both sections have the same thermal conductivity (k) and cross-sectional area (A):

$$\therefore \quad \left(\frac{\Delta T}{\Delta x}\right)_B = \left(\frac{\Delta T}{\Delta x}\right)_C$$

This means that **the temperature gradient is constant along a perfectly lagged, uniform bar**. The graph of **temperature** against **distance** along the bar shown in Figure 4.50 illustrates this deduction.

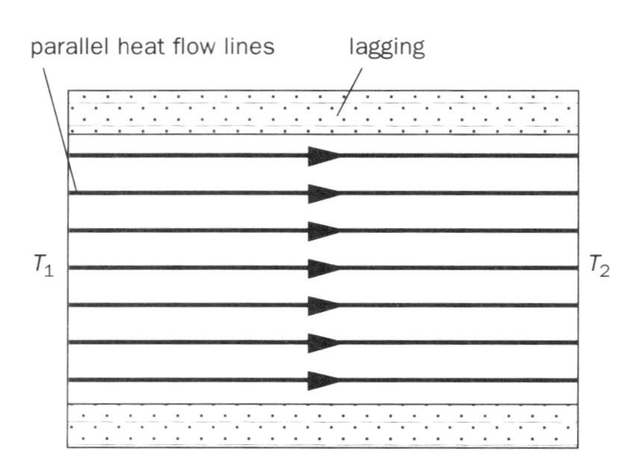

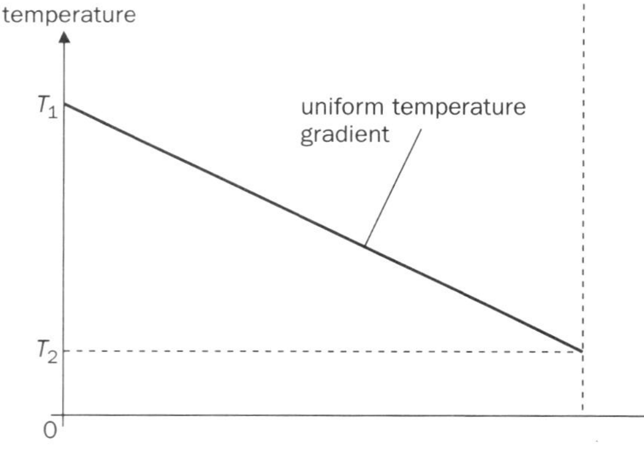

Figure 4.50 Temperature–distance graph for a perfectly lagged uniform bar

In the case of an unlagged, uniform bar some of the heat passing through section B is lost through the surface before it reaches section C.

$$\therefore \quad \left(\frac{\Delta Q}{\Delta t}\right)_B > \left(\frac{\Delta Q}{\Delta t}\right)_C$$

and since both sections have the same thermal conductivity (k) and cross-sectional area (A):

$$\therefore \quad \left(\frac{\Delta T}{\Delta x}\right)_B > \left(\frac{\Delta T}{\Delta x}\right)_C$$

This means that in an **unlagged** uniform bar the **temperature gradient decreases with distance from the hot end of the bar**.

Figure 4.51 illustrates the decreasing temperature gradient obtained when heat flows through an unlagged uniform bar.

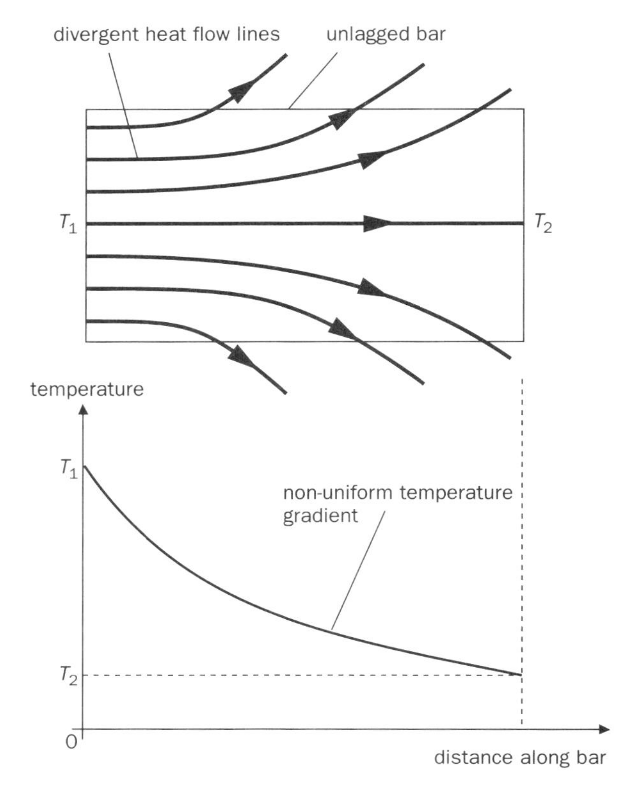

Figure 4.51 Temperature–distance graph for an unlagged uniform bar

The analogy between thermal and electrical conduction

Thermal conduction and electrical conduction are both 'flow' processes and may be linked as follows:

- **Charge** Q is analogous to **heat energy** ΔQ
- **Electric** current $I = \frac{dQ}{dt}$ is analogous to **heat** current $\frac{\Delta Q}{\Delta t}$
- **Electric** charge flows as a result of a **potential** difference
- **Heat** energy flows as a result of a **temperature** difference
- **Potential** difference (V) is analogous to **temperature** difference (ΔT)
- Electric current equation $\quad I = \frac{V}{R} \quad$ (i)

- Heat current equation
$$\frac{\Delta Q}{\Delta t} = -kA\frac{\Delta T}{\Delta x} = -\frac{\Delta T}{\Delta x/kA} \qquad \text{(ii)}$$

Comparing the two phenomena, we see that both are very similar situations. Comparing Equations (i) and (ii), we see in particular that a quantity called **resistance** can be assigned to each. The **thermal resistance** (R_T) of a body is a very useful quantity, given by:

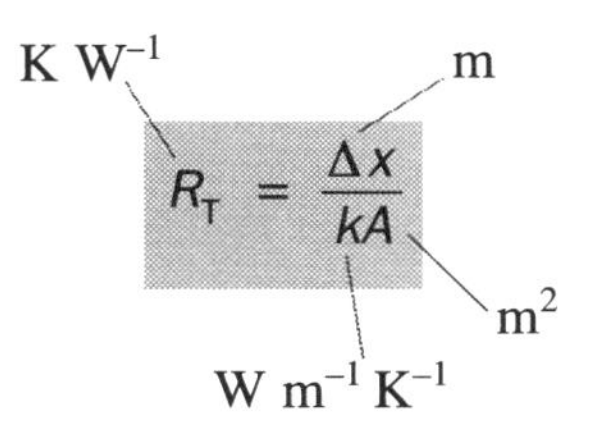

Equation 2

Electrical resistance (R) of a conductor of **length** (l), **cross-sectional area** (A) and **resistivity** (ρ) is given by:

$$R = \frac{\rho l}{A} = \frac{l}{\sigma A}$$ where σ is the electrical **conductivity**

but $R_T = \frac{\Delta x}{kA}$

$\therefore \frac{\Delta x}{kA}$ is analogous to $\frac{l}{\sigma A}$

From which it can be seen that:

- **Thermal conductivity** (k) is analogous to **electrical conductivity** (σ).

The idea of thermal resistance may be used to solve problems in which several conductors are joined in **series** or **parallel**, using techniques borrowed from electrical theory.

Thermal conductors in series

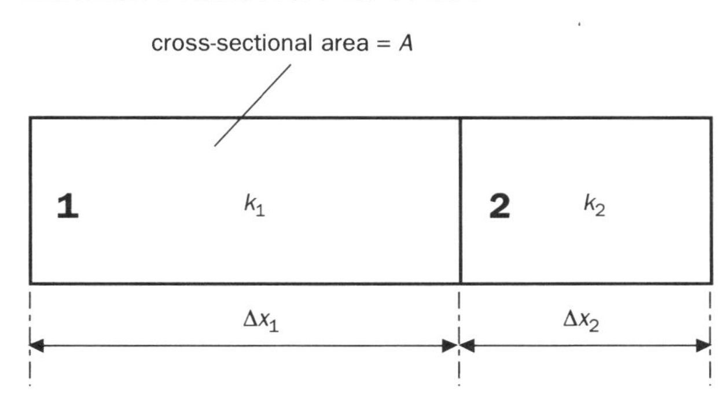

Figure 4.52 Thermal conductors in series

Consider two lagged metal conductors joined end to end as shown in Figure 4.52.

Thermal resistance (R_1) of (1) is given by:

$$R_1 = \frac{\Delta x_1}{k_1 A}$$

Thermal resistance (R_2) of (2) is given by:

$$R_2 = \frac{\Delta x_2}{k_2 A}$$

To find the total thermal resistance we treat thermal conductors in the same way we treat resistors in series – we simply **add** them.

$\therefore$ Total thermal resistance, (R_T) is given by:

$$R_T = R_1 + R_2$$

Summary

	Electrical conduction	**Thermal conduction**
Quantity flowing	Charge Q	Heat energy ΔQ
Flow rate	Electric current $I = \frac{dQ}{dt}$	Heat current $\frac{\Delta Q}{\Delta t}$
Cause of flow	Potential difference V	Temperature difference ΔT
Flow equation	$I = \frac{V}{R}$ (i)	$\frac{\Delta Q}{\Delta t} = -kA\frac{\Delta T}{\Delta x} = -\frac{\Delta T}{\Delta x/kA}$ (ii)
Resistance to flow	Electrical resistance R	Thermal resistance R_T From (i) and (ii) above: $R_T = \frac{\Delta x}{kA}$
Conductivity	Electrical conductivity (σ)	Thermal conductivity (k)

$\therefore$ Heat current $\left(\frac{\Delta Q}{\Delta t}\right)$ is given by:

$$\frac{\Delta Q}{\Delta t} = \frac{\Delta T}{R_T} = \frac{\Delta T}{R_1 + R_2}$$

Guided example (1)

1. A front door contains a double-glazed unit of height 900 mm and width 500 mm. The unit consists of two single panes of glass each of thickness 5.0 mm separated by an air-gap of 20 mm. If the thermal conductivities of the glass and air are 0.10 and 0.024 W m^{-1} K^{-1}, calculate the heat energy per second which is conducted through the unit when the temperature difference across it is 22 K.

Guidelines

Find the thermal resistance for each pane of glass and for the air layer between. Then add to obtain total thermal resistance. (Remember that there are **two** panes of glass.)

Then use

$$\frac{\Delta Q}{\Delta t} = \frac{\Delta T}{R_T} = \frac{\Delta T}{R_1 + R_2}$$

Heat loss in buildings

In order to maintain the inside of a building at a higher temperature than that outside, the rate of heat supply must equal the rate of heat loss. As a result of the **temperature gradient** between the inside and the outside surfaces, heat energy is transferred by conduction through the walls, windows, floor and roof. Draughts due to gaps in window frames, doors, etc. also contribute significantly to the total energy loss by allowing heated air to escape, being replaced by colder air from outside. Figure 4.53 gives an idea of the contribution made by walls, roof, etc. for a typical home in which no heat insulation precautions have been taken.

In today's energy-conscious world we are all encouraged to minimise heat energy losses in our homes. Virtually all modern buildings have cavity wall insulation, roof and loft insulation, carpeting, double glazing and draught exclusion at requisite points.

Calculation of heat energy loss – *U*-values

When designing or modifying buildings to minimise heat losses, heating engineers measure the heat flow rate (or power loss) through specific materials and structures. This is expressed in a quantity called the ***U*-value**.

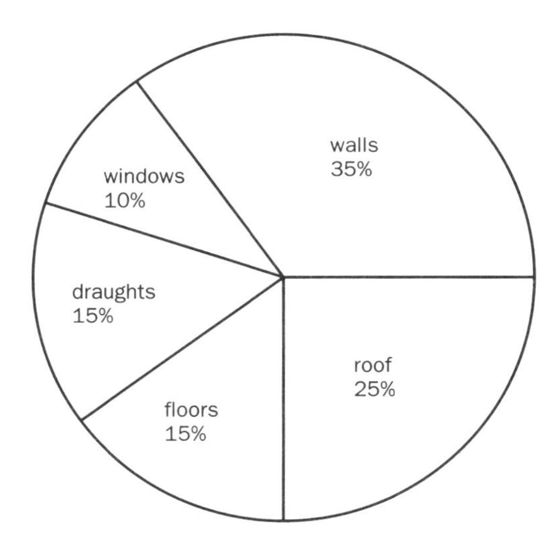

Figure 4.53 Pie chart showing heat energy loss contributions from various parts of a building

The ***U*-value** of a specified thermal conductor (e.g. a window unit, a door, etc.) is defined as the **heat flow rate** through the conductor per **square metre** produced by a 1 K **temperature difference** between its two surfaces (see Figure 4.54).

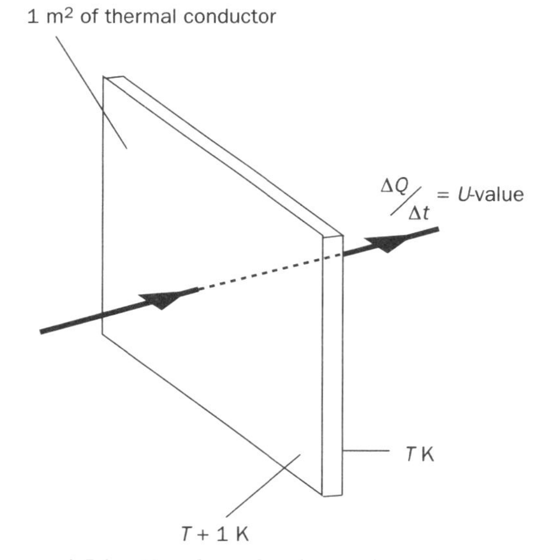

Figure 4.54 U-value of a thermal conductor

Given the area of all the thermal conducting surfaces in a building together with their respective U-values, the total heat flow rate (or power loss) for any known temperature difference may be calculated.

The **power loss** (P) in (J s^{-1} or W) is given by:

$$P = UA\,\Delta T$$

(W; W m^{-2} K^{-1}; m^2; K)

Equation 3

Guided examples (2)

1.

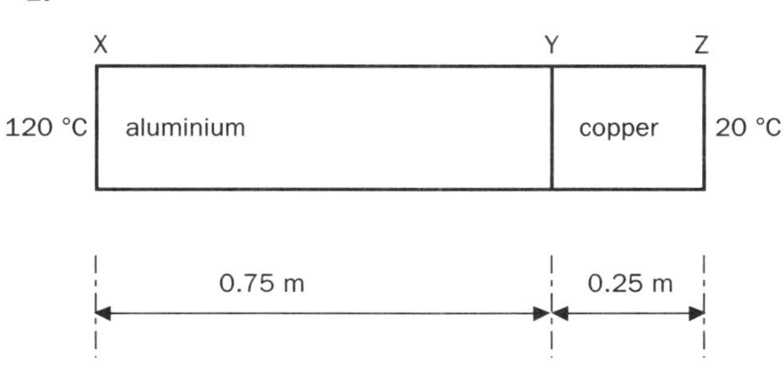

The diagram shows a perfectly lagged composite bar XYZ which consists of 0.75 m of aluminium joined to 0.25 m of copper. Each bar has a radius of 3.0 cm. If the ends X and Z are kept at temperatures of 120 °C and 20 °C respectively, calculate:

(a) the temperature where the two bars are joined

(b) the temperature midway between X and Y and

(c) the heat flow rate through the composite bar.

(Thermal conductivities – aluminium, 210 W m^{-1} K^{-1}; copper, 390 W m^{-1} K^{-1}.)

Guidelines

(a) Since the bar is perfectly lagged, there is no heat loss:

$$\therefore \left(\frac{\Delta Q}{\Delta t}\right) \text{ is constant.}$$

$$\therefore \left(\frac{\Delta Q}{\Delta t}\right)_{Al} = \left(\frac{\Delta Q}{\Delta t}\right)_{Cu}$$

Use Equation 1 and solve for the unknown temperature T.

(b) Same thinking as for (a).

(c) Use Equation 1.

2. Water is boiling steadily in a copper kettle which is being heated continuously on a halogen hob. The heat conducted through the kettle base evaporates 3×10^{-3} kg of water per second. If the base has a uniform thickness of 2.5 mm and an area of 3.5×10^{-2} m^2, calculate the temperature of the surface of the kettle which is closest to the hob. Assume that there are no heat losses.

(Thermal conductivity of copper = 385 W m^{-1} K^{-1}, specific latent heat of vaporisation of water = 2.2×10^6 J kg^{-1})

Guidelines

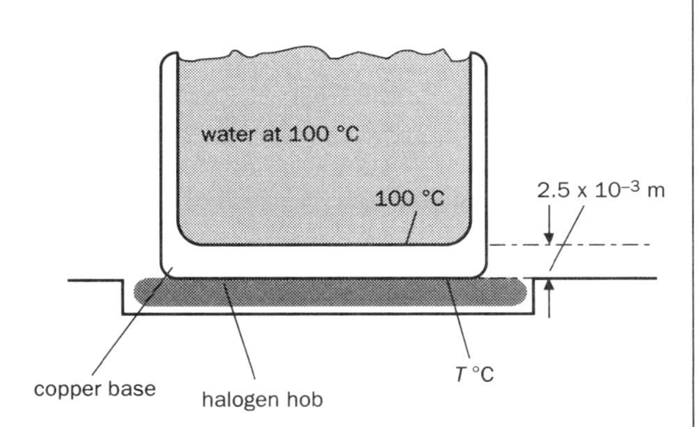

Calculate the heat energy required each second to evaporate the water, i.e. (mass per second) × (specific latent heat)

Then use Equation 1 to calculate the unknown temperature, T.

3. A single-glazed window unit is 7.0 mm thick and has a surface area of 2.6 m^2. It is replaced by a double-glazed unit of the same area which consists of two 7.0 mm thick glass panes with a 15 mm air gap in between. In winter the average temperature difference between the inside and outside of the house is 15 °C. Use the idea of thermal resistance to calculate the heat flow rate for each window unit.

Thermal conductivity of glass = 1.00 W m^{-1} K^{-1}

Thermal conductivity of air = 0.02 W m^{-1} K^{-1}

Guidelines

Use Equation 2 to calculate the **thermal resistance** for the glass panels and for the air gap.

Then for the double glazed unit:

Total $R_T = 2R_{T(glass)} + R_{T(air)}$

Then use

$$\frac{\Delta Q}{\Delta t} = \frac{\Delta T}{R_T}$$

Guided examples (2) *continued*

4.

	Cavity brick walls with insulation	Tiled roof with insulation	Floors (with carpeting)	Windows (double-glazed)	Doors (double-glazed)
U-value (W m^{-2} K^{-1})	0.70	0.55	0.40	2.80	3.20
Total area (m^2)	140	68	45	11	4

The above information is for a modern, open-plan bungalow. Assuming that the average temperature difference between the inside and outside of the bungalow is 16 °C in winter, calculate:

(a) the total heat loss per second and

(b) the number of units (i.e. kW h) of electrical energy which would be used in a 24 h period if the bungalow was electrically heated.

Guidelines

Use Equation 3 to calculate the power loss from each part.

Total power loss = sum of power losses for each part.

No. of kW h = total power loss (in kW) × no. of hours.

Convection

The fact that a fluid (i.e. liquid or gas) which is at a higher temperature than the surrounding fluid will tend to rise is a well-known fact. We can explain this by saying that when the fluid is heated its volume increases while its mass remains the same. The density (= mass/volume) of the heated fluid therefore decreases and the hot fluid floats upwards. Cooler, denser fluid then falls to take the place of the hot fluid. Circulating **convection currents** are thus set up in the fluid and in this way the heat supplied at one point is gradually transmitted throughout the substance. This is the method by which water is heated in a kettle or immersion heater. You should note that, since the hot water rises, in order to heat all of the water, the heating element has to be positioned at the base of the kettle or heater. The fact that heated gases expand much more than liquids coupled with the fact that gases also flow more easily than liquids means that gaseous convection currents are far more vigorous than those in liquids.

The origin of the **force** that causes the fluid to rise, against gravity, needs some further thought. As we saw in Section 1.3, the **pressure** in a liquid **increases** with depth below the surface. The increase of pressure is proportional to the increase of depth. This is also true for gases, although on a large scale, the relationship between pressure and depth is more complex. (Since the atmosphere does not really have a surface, when dealing with convection currents in air, it is more convenient to think in terms of the pressure **decreasing** with height.) If a region of gas or liquid is at the same temperature as its surroundings, it will be in **equilibrium** under the action of the **weight** of the body acting vertically downwards, and the **resultant** of the **pressure** forces, which will act upwards, since the pressure beneath the region of fluid will be slightly greater than that above it. In this condition, it will have **zero** resultant force on it. If it becomes warmer than the surroundings, it will expand slightly. This means that the pressure force will **increase** slightly, while the **weight** stays the same. There would now be a resultant force **upwards,** which will cause the hot fluid to rise.

On a very large scale, convection currents in the atmosphere give rise to all the various wind and weather conditions which we experience. Onshore and offshore breezes, gales, hurricanes, etc. are all due to convection currents produced as a result of **differences** in temperature between regions on the Earth's surface.

Rising convection currents of hot air are used by many birds as well as man. Eagles, vultures, sailplanes, hang gliders and paragliders rely on these currents to gain height. Each must find rising air in order to maintain or gain height. This can happen when the wind blows towards a ridge or when hot air rises in a convection current. Birds and experienced pilots know all the signs to look for in order to find the sustaining thermals which help to keep them afloat for the longest possible time. Rock faces, ploughed fields, towns etc. absorb thermal radiation from the

Sun more effectively than growing crops, forests and water, and so become warmer. Air in contact with these regions becomes hotter, and a large 'bubble' of this warmer air rises through the atmosphere as a 'thermal'.

As the thermal rises, the surrounding pressure decreases, and the thermal expands. This means that it does **work** and so loses energy, causing its temperature to fall. If the temperature of the thermal becomes equal to that of the surroundings, it will once more be in equilibrium, with no resultant force on it. Since the temperature of the atmosphere tends to **decrease** with altitude, the thermal may continue ascending, provided its temperature remains higher than that of the surrounding air. If the air contains water vapour, clouds will form when the temperature falls enough for the water vapour to condense into droplets. This condensation releases **latent heat of condensation**, which gives a further boost to the energy of the thermal. Under the right conditions, the air can go on ascending for thousands of metres, producing tall *cumulonimbus* thunderclouds. At these altitudes, the water droplets freeze into ice, and if these grow sufficiently, they can fall to the ground as hailstones. Friction between rising and falling ice crystals is thought to be largely responsible for the separation of charges which results in lightning.

A hurricane pounds the Florida coastline. Atmospheric convection currents give rise to all the various wind and weather conditions which we experience

In homes having a central heating system, convection currents transfer hot water from a boiler to convective heaters located at various points around the house. These heaters are often called 'radiators' but they should more correctly be called 'convectors' since they transfer heat to a room by creating convection currents in the air around them. As we saw earlier (Figure 4.53) a large percentage of the total heat energy loss in a building is through the roof. This is due to the fact that hot air convects upwards and collects near the ceiling, so the heat passes through ceilings by conduction, into the loft and then out through the roof itself. We try to minimise the amount of heat lost in this way by placing some kind of thermal barrier in the loft. This is called loft insulation and usually takes the form of polystyrene beads or thick glass wool filling the spaces between the joists.

Free and forced convection

The heat energy from a hot object may be carried away by a fluid by means of convection. If the fluid's movement is solely due to its being heated by the object and so rising, the convection is said to be **free** or natural. This is the case for example when we allow a hot cup of tea to cool down by simply letting it stand for a while. If, on the other hand, the convective fluid is caused to move in a stream over the hot object, the convection is said to be **forced**. Thus blowing air over the hot tea is an example of forced convection and, as we know, we do it because it gives a more rapid heat loss than free convection

and so cools the tea a lot faster. Forced convection ensures rapid removal of unwanted heat from a car engine. The car's forward motion gives a steady flow of cool air over the radiator. In a traffic jam when the car is stationary and the engine is running, the heat loss from the radiator is reduced to that which can result from radiation and free convection of the air around the radiator. This heat loss is not sufficient and overheating would inevitably occur if it were not for the forced convection provided by an electric fan located in front of the radiator which operates as soon as the temperature becomes too high. In addition, the cooling water is pumped around the engine, to supplement the natural convection.

Thermal radiation

Thermal radiation consists of electromagnetic waves emitted by objects as a result of their temperature. The wavelength range of the emitted radiation is continuous and extends from the long wavelength end of the infra-red region through the visible and into the ultra-violet region of the electromagnetic spectrum. The way in which the radiated energy is distributed among these wavelengths depends on the nature and temperature of the emitting object. For emitter temperatures below 1000 °C the radiated energy is in the infra-red region (i.e. from around 7.5×10^{-7}m to 4.0×10^{-4} m) but, at temperatures above about 1000 °C, visible light and even ultra-violet wavelengths are involved. The general properties of thermal radiation are the same as those of all other electromagnetic waves in that:

- It travels in straight lines.
- It travels at a constant speed of 3×10^8 m s^{-1} in a vacuum.
- It can be reflected and refracted. This is simply demonstrated by using either a converging lens or mirror to focus the Sun's rays onto our skin. The fact that it feels hot at the focal point shows that the Sun's thermal radiation has been reflected or refracted.
- The intensity from a point source decreases as the square of the distance from the source (i.e. it follows an inverse square law).
- It is unaffected by electric and magnetic fields.

Figure 4.55 shows thermal radiation of intensity (I) incident on a body. Some of the energy is reflected from the surface (R), some is absorbed (A) and some may be transmitted through the body (T) and emerge on the other side.

Since most materials do not transmit thermal radiation we can assume that the radiation incident on a material surface will generally be partly absorbed and partly reflected. How much of the incident radiation is reflected and how much is absorbed depends on the nature of the surface of the material. Experiment shows that:

- Polished, silvered surfaces are the best reflectors
- Matt-black surfaces are the best absorbers.

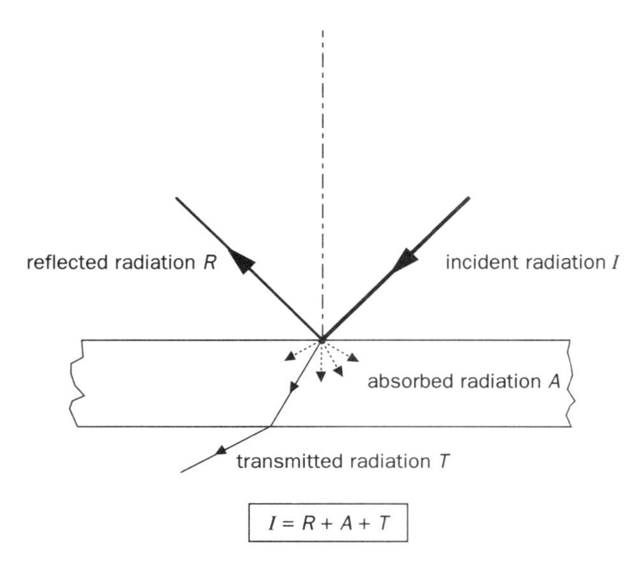

Figure 4.55 Absorption, reflection and transmission of thermal radiation

Surfaces which are good absorbers of thermal radiation are also good emitters of thermal radiation. Thus a polished silver teapot will keep the tea hot much longer than a brown ceramic one. The emission of radiation is much greater from the brown teapot than it is from the silver one and so the tea in the latter retains its heat for longer.

Detectors of thermal radiation

The simplest way of detecting thermal radiation is to hold the back of your hand close to the radiation source. Your skin will absorb the radiation, causing a rise in temperature. You will 'feel' the heat and you will also be able to tell when there is a significant change in intensity, but there are much more sensitive instruments than such a subjective impression.

The bolometer Figure 4.56 shows one type of bolometer which can be used to detect and measure thermal radiation. It contains two identical resistors in the form of long, blackened platinum strips. One of the resistors (the hot resistor) faces the heat radiation being measured and the other is kept at room temperature. As the hot resistor absorbs the thermal radiation, its temperature increases and this causes its resistance to increase. The change in resistance is detected by making the two resistors part of a Wheatstone bridge circuit.

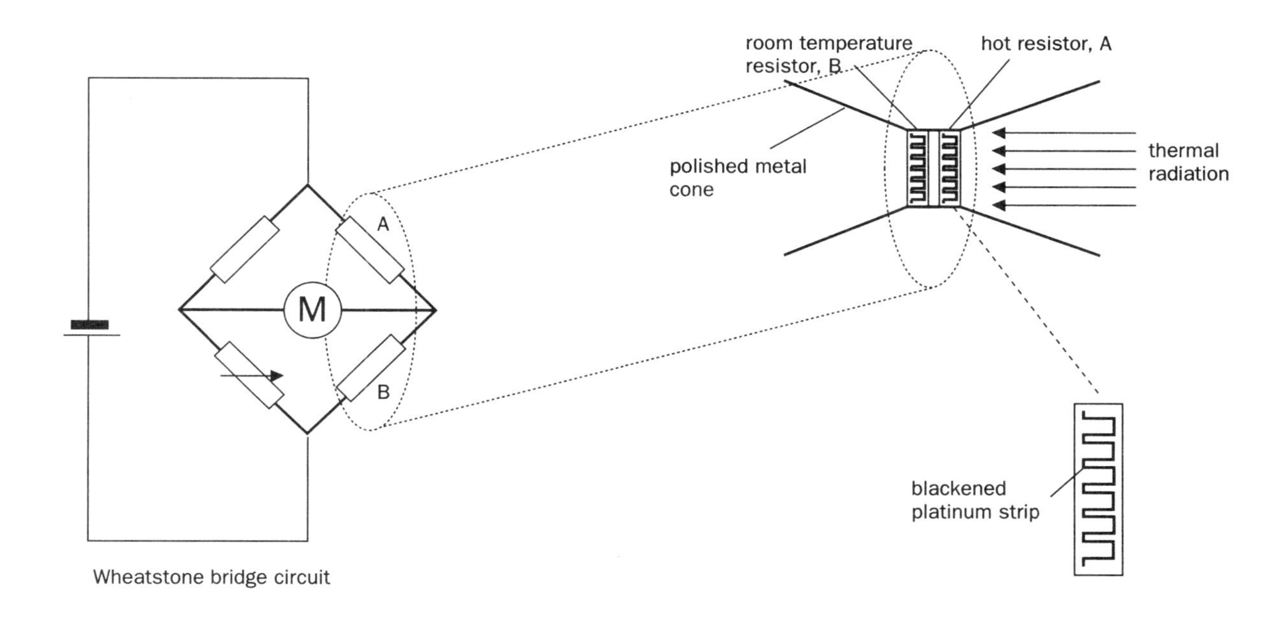

Figure 4.56 Bolometer used to detect thermal radiation

The thermopile This is the more commonly used type of radiation detector. The essential features of a simple thermopile are shown in Figure 4.57.

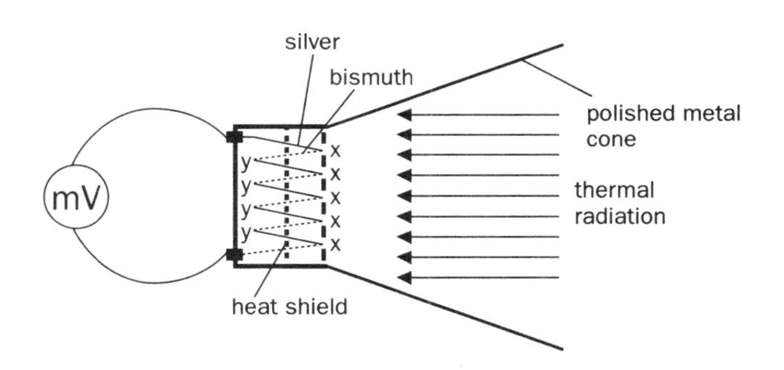

Figure 4.57 Detecting thermal radiation with a thermopile

The thermopile makes use of the thermoelectric e.m.f. produced between the junctions of two different metals when there is a temperature difference between them. It consists of a large number of thermocouples connected in series. The junctions marked x are set facing the heat radiation and the other set of junctions marked y are shielded from the heat. The polished metal cone concentrates the radiation onto the x junctions which are also blackened so as to improve absorption. The millivoltmeter reading is a measure of the intensity of the incident radiation.

Infra-red photography The infra-red radiation emitted by an object can be photographed using a special type of film (infra-red film) whose emulsion contains a dye which absorbs the radiation. If monochrome (i.e. black and white) film is used temperature differences between regions are indicated by varying shades of grey between white and black. With colour infra-red film hot spots appear red and cool regions are blue. In the medical technique of *thermography* the infra-red radiation emitted by a scanned body is detected and analysed to give a thermal image (called a *thermogram*) of the body. This technique can identify hot spots on a patient's body which are generally indicative of some disorder. Thermography is widely used for the location of tumours and in the investigation of hairline bone fractures.

Weather satellites above the Earth's atmosphere monitor surface and cloud temperatures with thermal imaging cameras.

Self-assessment

SECTION A
Qualitative assessment

1. Briefly describe the two thermal conduction mechanisms by which heat is transmitted between two points in a **metallic** solid.

2. Write down the equation for the heat flow rate through a sample of material and state the meaning of each of the terms.

3. (a) Explain what is meant by the **thermal conductivity** of a material.
   (b) What is the unit of thermal conductivity?

4. (a) Sketch graphs to show how the temperature varies with distance along a metal bar when one end is maintained at a high temperature and the other is at room temperature if the bar is:
   (i) uniform and perfectly insulated
   (ii) uniform but not insulated.
   (b) What can you say about the temperature gradient along each of the bars considered in (a) (i) and (ii)?

5. There is a link between thermal and electrical conduction.
   (a) Write down the electrical equivalent for each of the following:
   (i) heat **flow**
   (ii) heat **current**
   (iii) heat current flows as a result of a **temperature difference**.
   (b) By considering the equations for thermal and electric current, show that:
   (i) **thermal resistance** is equivalent to **electrical resistance**
   (ii) **thermal conductivity** is equivalent to **electrical conductivity**.

6. (a) Explain what is meant by the ***U*-value** of a specified thermal conductor.
   (b) Deduce the units of U-value.
   (c) For which will the U-value be larger, a single or a double-glazed window of equal area?
   (d) Apart from the U-value of all the thermal conducting surfaces in a building, what other information would you require in order to calculate the total heat flow rate from the building?

7. (a) Explain how convection currents are formed when a fluid is heated.
   (b) Explain how the transfer of heat by convection contributes to heat loss in a house.
   (c) How is convective heat transfer used in central heating systems?

8. Distinguish between **free** and **forced** convection.

9. (a) What is **thermal radiation**?
   (b) State three properties of thermal radiation.
   (c) Which material surfaces are the best
   (i) reflectors
   (ii) absorbers and
   (iii) emitters of thermal radiation?

10. Give two reasons why you feel cold if you sit near a window in winter.

11. Briefly describe how thermal radiation can be detected with
   (a) a bolometer
   (b) a thermopile.

SECTION B
Quantitative assessment

(***Answers:*** 0.88; 5.1; 6.7; 8.2; 9.3; 11; 23; 61; 90; 150; 1.2×10^3.)

1. A circular ship's porthole is 14 mm thick and has a diameter of 300 mm. It is made of a glass having a thermal conductivity of 1.2 W m^{-1} K^{-1}. If the temperatures of the inner and outer surfaces of the glass are 18 °C and 8 °C respectively, calculate the rate at which heat energy is lost through the porthole.

2. The walls of a picnic cool-box consist of a 1.0 mm thick plastic inner and outer casing separated by polystyrene insulation of thickness 15 mm. If the box has a total surface area of 1.4 m^2 and there is a 15 °C temperature difference between the internal and external surfaces, calculate the rate at which heat energy flows into the box. (The thermal conductivities of plastic and polystyrene are 0.04 W m^{-1} K^{-1} and 0.006 W m^{-1} K^{-1} respectively.)

Self-assessment *continued*

3.

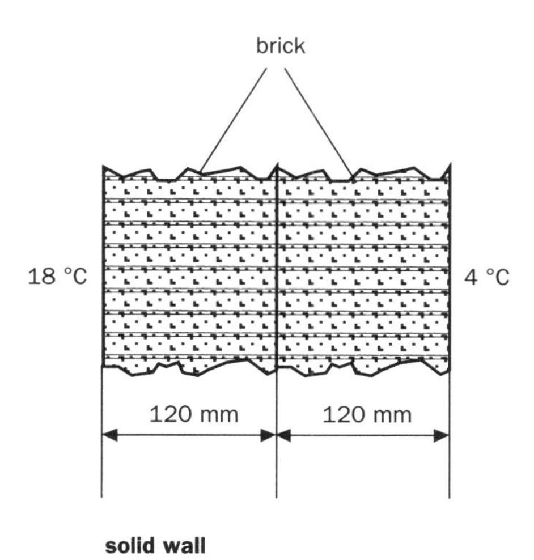

solid wall

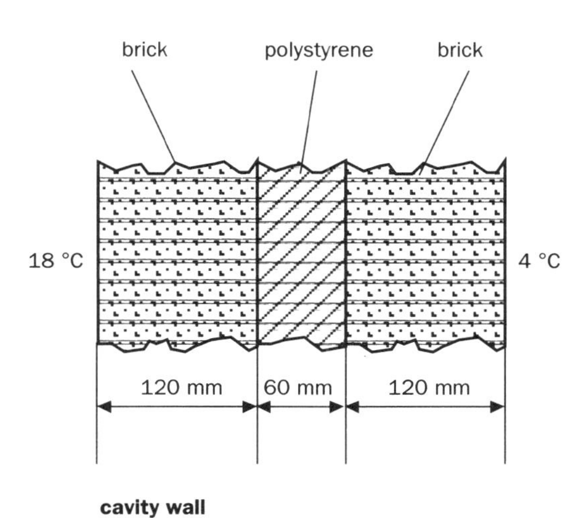

cavity wall

The diagrams show the dimensions of a solid brick wall (two-brick thick) and a cavity wall (with polystyrene insulation). In each case the interior and exterior surface temperatures are 18 °C and 4 °C respectively. Calculate the heat energy conducted per second through 5.0 m^2 of each type of wall.

(The thermal conductivities of brick and polystyrene are 0.50 W m^{-1} K^{-1} and 0.006 W m^{-1} K^{-1} respectively.)

4. A hot-air balloon is made of 1.5 mm thick material of total surface area 600 m^2 and thermal conductivity 0.06 W m^{-1} K^{-1}.

(a) If the material's outer surface is at a temperature of 10 °C and heat energy is lost from the balloon at a rate of 3.0×10^4 J s^{-1}, what is the temperature of the inner surface?

(b) If this temperature is maintained by igniting a 300 kW burner at regular intervals for a time of 10 s, what is the time between each ignition?

5. Calculate the overall *U*-value of a room having dimensions of 3.5 m × 3.0 m × 2.5 m which is kept at a steady temperature of 17 °C by an electric heater of power rating 2.5 kW when the outside temperature is 2 °C. (Assume all the heat loss occurs through the walls.)

6. Calculate the rate of loss of heat energy through a window unit measuring 0.7 m × 0.4 m when the internal and external temperatures are 20 °C and – 6 °C respectively when the unit consists of:

(a) a single 7.0 mm thick pane of glass and

(b) two 2.0 mm thick glass panes with a 20 mm air gap between them.

(The thermal conductivities of glass and air are 1.2 W m^{-1} K^{-1} and 0.·026 W m^{-1} K^{-1} respectively.)

7. A diver's wet-suit has a total surface area of 0.40 m^2. It consists of an outer rubber layer of thickness 10 mm with an inner woollen lining which can be assumed to trap a 3.6 mm thick layer of still air next to the diver's skin. If the diver's skin temperature is 34 °C and he is diving in water at a temperature of 14 °C:

(a) calculate the total thermal resistance of the wet-suit, and

(b) estimate the rate at which the diver loses heat energy.

(The thermal conductivities of the air and rubber are 0.024 and 0.05 W m^{-1} K^{-1} respectively.)

Past-paper questions

1 Silicon dioxide (SiO_2) occurs in several different forms. These include *crystalline* forms, such as quartz, and *amorphous* forms, such as silica glass.

(a) Describe the meanings of the terms *crystalline* and *amorphous* in relation to the molecular structure of the solid.

(b) Describe the use of X-ray diffraction to distinguish and identify the form of a solid.

(c) Explain, in relation to the molecular structure, why the density of a substance in the different solid forms is (i) similar, but (ii) not identical in each form.

(O & C, 1992)

2

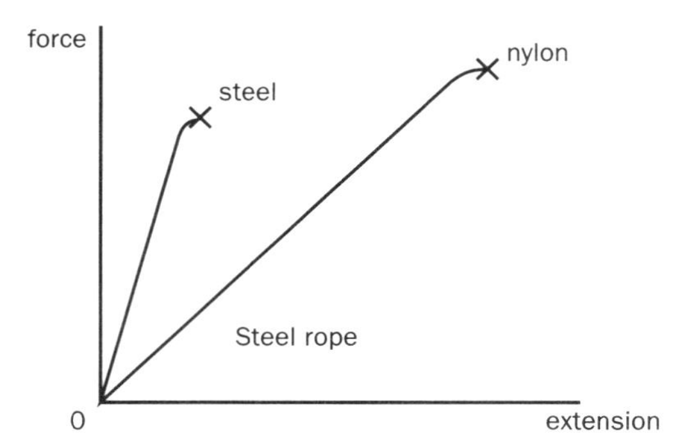

The force–extension graphs for a steel tow rope and a nylon tow rope are shown above. In what regions of the graphs can Hooke's law be applied?

The breaking stress for steel is greater than the breaking stress for nylon. Explain why, in this case, the force needed to break the steel rope is less than that needed to break the nylon rope.

What further information apart from force and extension would be needed in order to calculate the Young modulus of one of the ropes?

Explain how the graph shows that less energy is required to break the steel rope than the nylon rope.

(Lond., January 1993)

3 (a) Explain what is meant by *stress* and *strain* for a material. Why is the *Young modulus* of a material defined in terms of stress and strain rather than load and extension?

(b) A maintenance platform is suspended on the outside of a tall building. The platform and its load of two workmen weigh 2000 N. Two alternative rope systems are available to support the platform, one using steel wire rope and the other nylon rope. Use the properties of steel and nylon rope systems given below to calculate the unstretched length of each suspension system when the platform is in equilibrium 20.0 m below the point of support.

(c) With a load of 2200 N, compare for the two rope systems

(i) the energy stored and

(ii) the breaking load.

(d) The platform is lowered to just above ground and the workmen jump off. State qualitatively what will be observed for each system. Explain what happens to the energy stored.

Data on steel and nylon rope systems.

	Young modulus/Pa	Breaking stress/Pa	Total area of cross-section of ropes /m^2
steel	2.0×10^{11}	3.0×10^{8}	1.0×10^{-4}
nylon	7.0×10^{7}	6.0×10^{7}	5.0×10^{-4}

(O & C, 1992)

4 A cylindrical copper wire and a cylindrical steel wire, each of length 1.000 m and having equal diameters, are joined at one end to form a composite wire 2.000 m long. This composite wire is subjected to a tensile stress until its length becomes 2.0002 m. Calculate the tensile stress applied to the wire.

(The Young modulus for copper = 1.2×10^{11} Pa and for steel 2.0×10^{11} Pa)

(WJEC, 1991)

5 In accurate experiments to measure the Young modulus for a metal in the form of a wire, two identical wires are suspended vertically side by side with loads being added to the first of the wires.

(a) Explain why the wire is chosen to be long and thin.

(b) Why is the second wire suspended alongside the first (test wire)?

(c) Sketch a graph of load against extension which you might obtain when loading a mild steel wire until it breaks. Label any important points or regions.

(WJEC, 1994)

6 The diagram at the top of the next page shows a simplified view of a suspension bridge. The road is supported by cylindrical steel rods attached to the suspension cables. There are two identical cables and sets of rods on each side of the roadway.

Each pair of rods effectively supports 5.0 m of road which has a mass per unit length of 2.0×10^3 kg m^{-1}. The longest rods are 55 m long and have a cross-sectional area of 5.0×10^{-3} m^2.

The Young modulus for steel = 2.0×10^{11} Pa.

Acceleration for free fall, g = 10 m s^{-2}.

(a) Assuming the limit of proportionality is not exceeded, calculate:

(i) the tension in each rod

(ii) the tensile stress in the longest rods

(iii) the tensile strain in the longest rods.

(b) (i) What changes will occur in a rod when a lorry goes over the bridge?

(ii) Explain why the rods used in bridges may fail due to *metal fatigue*.

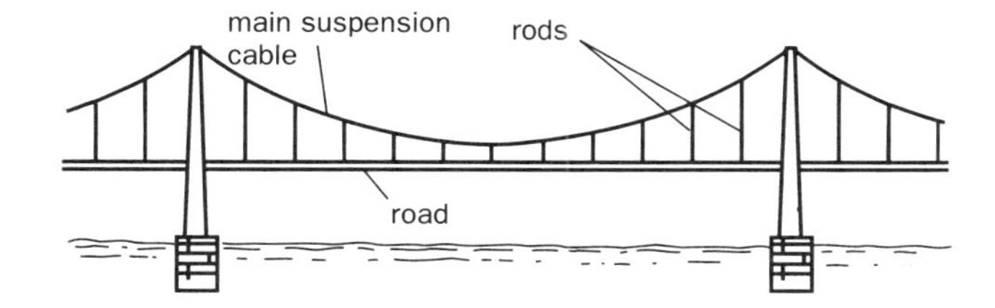

(AEB, 1994)

7 The sketch shows, approximately, how the resultant force between adjacent atoms in a solid depends on r, their distance apart.

(a) Which distance on the graph represents the equilibrium separation of the atoms? Briefly justify your answer.

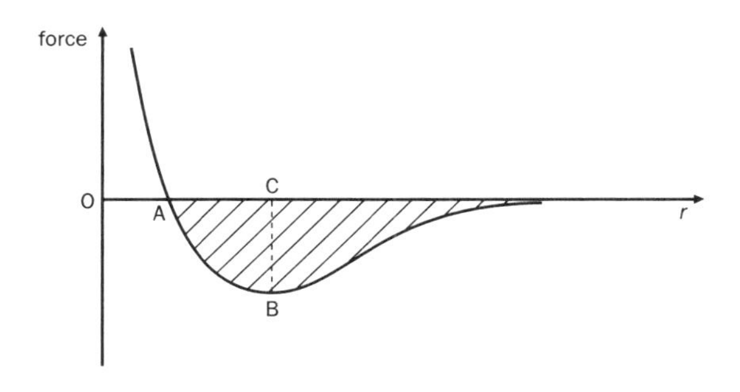

(b) What is the significance of the shaded area?

(c) Use the graph to explain why you would expect the solid to obey Hooke's law for *small* extensions and compressions.

(WJEC, 1992)

8 (a) (i) State **four** assumptions of the kinetic theory of the ideal gas.

(ii) A gas molecule in a cubical box travels with a speed c at right angles to one wall of the box. Show that the average force the molecule exerts on that wall is proportional to c^2.

(b) The table gives measured values of pressure against density for a fixed mass of gas at a constant temperature of 27 °C.

pressure / 10^5 Pa	0.60	0.80	1.00	1.20	1.40
density / kg m^{-3}	0.68	0.91	1.14	1.37	1.60

(i) Plot a graph of pressure against density. Does your graph indicate that the gas behaves as an ideal gas under these conditions? Justify your answer.

(ii) Use your graph to calculate the root mean square speed of the molecules of the gas.

(iii) The temperature of the gas is raised to 57 °C. Calculate the pressure when the density is 1.00 kg m^{-3}, and hence draw the corresponding graph of pressure against density at 57 °C, using the same axes as before.

(c) A container holds a mixture of helium and argon, relative atomic masses 4 and 40 respectively.

(i) For the gases in the container, calculate the ratio

$$\frac{\text{root mean square speed of helium atoms}}{\text{root mean square speed of argon atoms}}$$

(ii) There are approximately equal numbers of helium and argon atoms in a container, when gas starts to leak slowly out of a small hole in the side. After a short time, will the number of argon atoms remaining in the containers be greater or less than the number of helium atoms?

Give a reason for your answer.

(NEAB, 1993)

9 The table shows the distribution of molecular speeds among 15 molecules of an ideal gas.

Number of molecules	2	4	5	3	1
Speed / m s^{-1}	200	300	500	600	700

(a) Calculate the mean square speed $<c^2>$.

(b) Calculate the pressure exerted by the gas if $<c^2>$ for all its molecules is the same as that calculated in (a) and the density of the gas is 1.25 kg m^{-3}.

(WJEC, 1989)

10 (a) State Avogadro's law.

(b) The pressure p of an ideal gas is given by:

$p = {}^1/_3\, nm <c^2>$

where n in the number of molecules per unit volume, m is the mass of one molecule and $<c^2>$ is the mean square speed of the gas molecules.

Use the above equation to deduce Avogadro's law.

(WJEC, 1991)

11 (a) Define the molar heat capacity of a gas at

(i) constant pressure ($C_{\text{v, molar}}$)

(ii) constant volume ($C_{\text{p, molar}}$).

(b) C_p is greater then C_v. Explain carefully why this is so.

(c) Show that for an ideal gas

$C_{\text{p, molar}} - C_{\text{v, molar}} = R$

where R is the molar gas constant.

(WJEC, 1992)

12 A vessel of volume 1.0×10^{-3} m^3 contains helium gas at a pressure of 2.0×10^5 Pa when the temperature is 300 K.

(a) What is the mass of helium in the vessel?

(b) How many helium atoms are there in the vessel?

(c) Calculate the r.m.s. speed of the helium atoms.

Take:

Relative atomic mass of helium = 4.
The Avogadro constant = 6.0×10^{23} mol^{-1}.
The molar gas constant R = 8.3 J mol^{-1} K^{-1}.

(WJEC, 1992)

13 (a) List the assumptions of the kinetic theory of gases.

(b) Prove that, with the usual symbols, the pressure p of a gas is given by

$p = {}^1/_3\, \rho <c^2>$.

(c) (i) By comparing this equation with the equation of state for 1 mole of an ideal gas, find the translational kinetic energy for 1 mole of the monatomic gas.

(ii) Hence show that the average translational kinetic energy per molecule of a monatomic gas is given by ${}^3/_2\, kT$ where

k (called the Boltzmann constant) is $\dfrac{R}{N_A}$.

(d) 5 moles of helium at 27 °C are allowed to mix with 3 moles of neon at 127 °C. Find the temperature of the resulting mixture of gases.

Assume no heat losses during the mixing process.

(WJEC, 1993)

14 2.0 g of helium are contained in a vessel of volume 1.5×10^{-3} m^3 at a pressure of 2.5×10^5 Pa.

(a) What is the temperature of the gas?

(b) How many atoms of helium are present?

(c) What is the r.m.s. speed of the helium atoms?

Relative molecular mass of helium = 4.

(WJEC, 1994)

15 (a) State Boyle's law. Give the assumptions upon which the kinetic theory of gases is based and describe qualitatively how the theory gives an explanation of Boyle's experimental law.

(b) A car tyre of volume 1.6×10^{-2} m^3 contains air at a temperature of 17 °C and a pressure of 2.6×10^5 Pa. A foot pump of volume 3.0×10^{-4} m^3 is used to pump air, at a temperature of 17 °C and an initial pressure of 1.0×10^5 Pa, into the tyre to increase the pressure to 3.1×10^5 Pa. Assume that pumping does not change the temperature or the volume of air in the tyre.

(i) What amount of air in moles does the tyre contain initially?

(ii) Calculate the minimum number of strokes of the pump needed to raise the pressure to 3.1×10^5 Pa.

(iii) Is the assumption of constant pressure justified? Explain your answer.

(iv) Calculate the root-mean-square speed of the air molecules in the tyre. (Assume an average value for the molar mass of air of 0.029 kg mol^{-1}.)

(O & C, 1993)

16 (a) Two equations which relate to electrical conduction,

$$I = \frac{V}{R}$$

and

$$R = \rho \frac{l}{A}$$

can be combined to give a single equation

$$I = \frac{1}{\rho} A \frac{V}{l}$$

This last equation can be written in the form

$$\frac{dQ}{dt} = \sigma A \frac{V}{l}$$

What does the symbol σ represent?

A similar equation describes thermal conduction.

$$\frac{dQ}{dt} = kA \frac{\Delta T}{\Delta x}$$

In the cases of both thermal and electrical conduction, something can be considered as flowing. In each of these cases, identify the quantity which is flowing and give a suitable unit.

Identify, for both types of conduction, the quantity which may be thought of as 'driving the flow'.

State another way in which these equations illustrate similarities between electrical and thermal conduction.

Good electrical conductors are good thermal conductors: this suggests that electrons play a part in both mechanisms. But diamond is a good thermal conductor despite it being an electrical insulator. Suggest a mechanism for thermal conduction in diamond.

(b) An iron bar 0.10 m long and a copper bar 1.2 m long are joined together as shown in the diagram below.

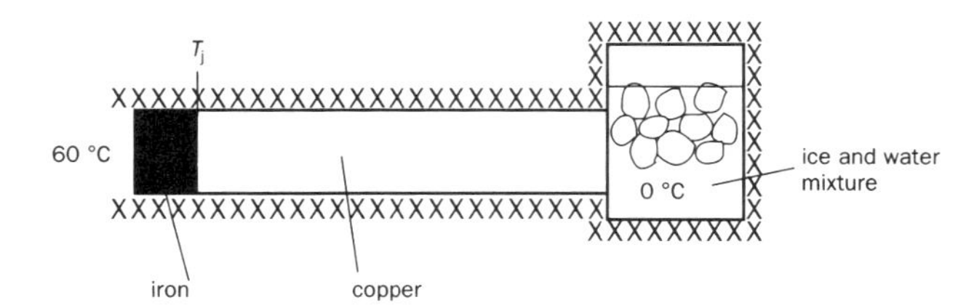

The hot end of the iron bar is kept at 60 °C while the cold end of the copper bar is kept at 0 °C by the mixture of ice and water. The apparatus is lagged. At thermal equilibrium the temperature at the junction of the metals is T_j. Both bars have a diameter of 0.16 m.

Write down expressions for the temperature gradients across each of the bars.

Using these expressions, write down equations for power transfers through each of the bars.

Explain why it is reasonable to assume that the power transfer through each conductor is the same.

Using the above assumption, calculate T_j and the power flow.

Describe how you would check your calculations by experiment.

(Thermal conductivity of copper = 390 W m^{-1} K^{-1}, thermal conductivity of iron = 75 W m^{-1} K^{-1}.)

(Lond., January 1993)

17 (a) Describe and contrast convection and conduction as mechanisms of heat transfer.

(b) In many experiments in physics it is necessary to reduce the rate of transfer of thermal energy between the sample under investigation and any surrounding container as much as possible. Give a brief account of the ways in which this may be achieved.

(c) Define *thermal conductivity* and show that it has an SI unit W m^{-1} K^{-1}.

(d) A small greenhouse consists of 34 m^2 of glass of thickness 3.0 mm and 9.0 m^2 of concrete wall of thickness 0.080 m. On a sunny day, the interior of the greenhouse receives a steady 25 kW of solar radiation. Estimate the difference in temperature between the inside and outside of the greenhouse. The temperatures inside and outside may be assumed uniform and the heat transfer downwards into the ground inside the greenhouse may be neglected.

(Thermal conductivity of glass = 0.85 W m^{-1} K^{-1}; Thermal conductivity of concrete = 1.5 W m^{-1} K^{-1}.)

(O & C, 1992)

18

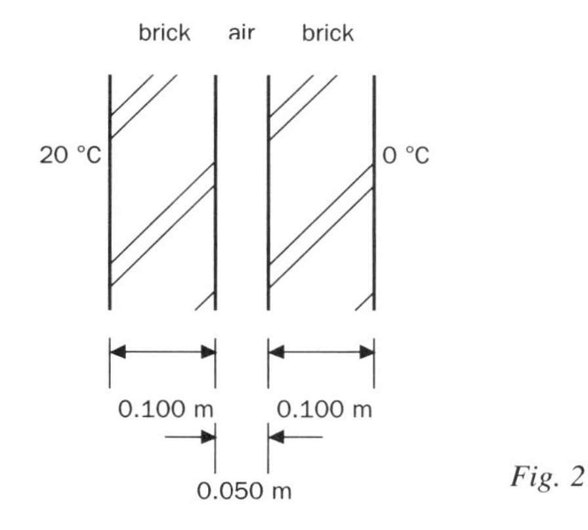

Fig. 1

Fig. 2

Figure 1 and Figure 2 show the dimensions of a solid brick wall and a cavity wall, respectively. In each case the surface temperatures are 20 °C and 0 °C, as shown. Using the values of thermal conductivity given below, calculate the thermal energy conducted per second through 10 m^2 of

(i) the brick wall shown in Figure 1

(ii) the cavity wall shown in Figure 2.

You may assume that the temperature at the centre of the cavity is 10 °C.

(Thermal conductivity of brick = 0.60 W m^{-1} K^{-1}; thermal conductivity of air = 2.4×10^{-2} W m^{-1} K^{-1}.)

(NEAB, 1993)

19 An Arctic explorer has a suit made from thin nylon filled with down (small soft feathers).

In still air the rate of heat loss per square metre of external surface per kelvin temperature difference between the body temperature and the outside temperature is 0.80 W m^{-2} K^{-1}.

The suit has a total exposed area of 2.0 m^2. The suit is tight fitting and the skin temperature is 34 °C. The maximum recommended heat loss from the body is 120 W.

(a) (i) What factors other than temperature difference affect the rate of heat loss per square metre in still air?

(ii) Explain why *down* is a good insulating material.

(b) Determine the lowest outside temperature for which the suit can provide effective insulation. Give your answer in °C.

(c) Explain why the outside surface of the material is not the same temperature as the surroundings.

(AEB, 1993)

20 A block of metal, of mass 103 g, is heated to 100 °C and then transferred to a polystyrene beaker containing 200 g of water at 19.8 °C. When thermal equilibrium has been reached, the water temperature is 21.6 °C.

Calculate the energy gained by the water during the process and the specific heat capacity of the metal.

The experiment as described is not a particularly good one for measuring the specific heat capacity of the metal. Name *two* important sources of error. State a way of reducing *one* of them.

(The specific heat capacity of water is 4.20 kJ kg^{-1} K^{-1}.)

21 (a) A deep-frying vessel holds 20 litres of cooking oil. The oil has to be heated from 15 °C to 200 °C in 30 minutes. Calculate

(i) the mass of cooking oil in the vessel

(ii) its heat capacity and

(iii) the average rate at which energy must be transferred to the oil.

(1 litre = 1000 cm^3, density of cooking oil = 920 kg m^{-3}, specific heat capacity of cooking oil = 1850 J kg^{-1} K^{-1}.)

In practice the rate of energy supply has to be about 30% higher than is needed to heat the oil. Give *two* possible reasons for this.

The rate at which the temperature of oil rises is higher at the beginning of the heating process than at the end. Why should this be so?

(Lond., June 1992)

22 In the hair dryer shown in the diagram air is drawn in through the system by a fan. The air is warmed as it passes the heating elements.

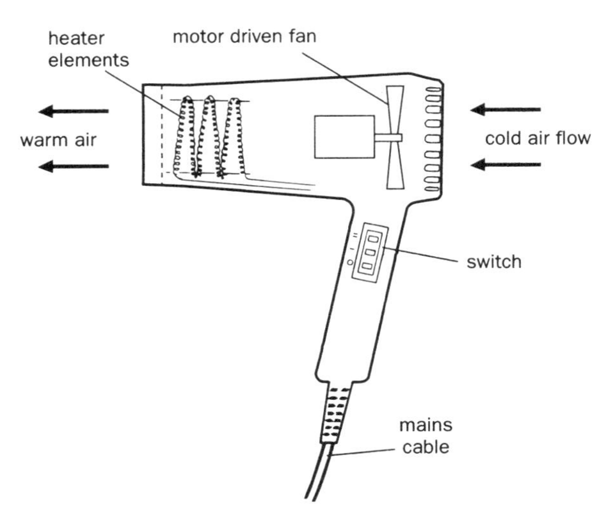

The hair dryer has the option of using one heating element on its own or two heating elements in parallel. Each element provides 600 W when operated at 240 V.

There are also two fan speeds, the volume flow automatically increasing when both heating elements are used.

(The effective specific heat capacity of the air is 990 J kg^{-1} K^{-1}.)
The density of air is 1.25 kg m^{-3} at 60 °C.

(a) In one case the ambient temperature is 20 °C. When one 600 W element is used and the fan is run at its slower speed, the temperature of the air leaving the drier is 60 °C.

Calculate:
(i) the mass flow rate
(ii) the volume flow rate.

(b) When the second element is switched on the temperature of the air from the drier rises to 75 °C.

Calculate the new mass flow rate of the air.

(c) Explain why it would be unwise to design the drier so that both heating elements are used without an increase in the volume flow rate.

(d) Determine the supply current when both heating elements are on, neglecting the current in the fan.

(e) The heating element is cooled by the process of *forced convection.*
(i) How does *forced convection* differ from *natural convection*?
(ii) The equation which relates to cooling by forced convection is

$$\frac{d\theta}{dt} = -k(\theta - \theta_0)$$

State in your own words what this equation expresses in mathematical terms.
(iii) Use the equation to explain what happens to the hot air temperature when the hair dryer is used in colder conditions.

(f) State and explain what would happen if the fan were to stop working.

(AEB, 1994)

23 In an experiment to determine the specific latent heat of vaporisation of ethanol, the following data were collected.

Voltage applied to heater = 9.0 V
Current through heater = 3.2 A
Mass of ethanol evaporated each second = 0.029 g
Temperature at which ethanol boiled = 78 °C

(a) Give two reasons why energy has to be supplied to change the liquid to vapour at its boiling point.

(b) Use the data to determine a value for the specific latent heat of vaporisation of ethanol.

(c) Describe how the data would be obtained. Your description should include
- a diagram of the apparatus including the circuit used for the heater
- the ranges and important specifications of any measuring instruments used
- the procedure you would use to process the data.

(d) The value given for the specific latent heat in data books is 8.57×10^5 J kg^{-1}.
(i) Give a reason why the result for the experiment is too high.
(ii) Explain whether the result using a single set of data would be more reliable or less reliable if the substance being investigated had a similar specific latent heat of vaporisation but a much higher boiling point than ethanol.

(e) The mass of one mole of ethanol is 46 g. Using the data book value given in (d), calculate the energy input required to vaporise one mole of ethanol at its boiling point.

(AEB, 1994, part question)

24 (a) (i) State what is meant by the *internal energy* of a system.
(ii) State two processes which can cause the internal energy of a system to **increase**.
(iii) Give one practical example of each of these processes.

(b) On a Winter's day the air contains no water vapour and is at a temperature of –10 °C. The air is inhaled and exhaled at a rate of 0.45 kg h^{-1}.

When inside the lungs, the air is warmed to 37 °C and gains water vapour as a result of evaporation. The exhaled air contains 0.040 kg of vapour in each kilogramme of air.
(i) Calculate the energy needed each hour to raise the temperature of the air.
(ii) Calculate the energy needed each hour to evaporate the water which is exhaled.

Specific heat capacity of air = 990 J kg^{-1} K^{-1}

Specific latent heat of vaporisation of water at 37 °C = 2.4×10^6 J kg^{-1}

(AEB, 1992)

25

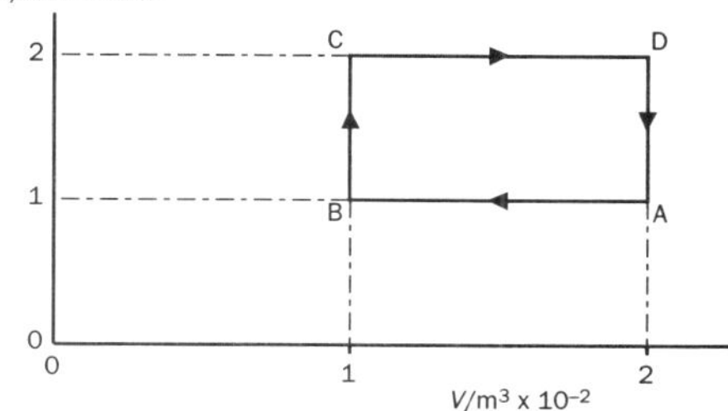

A fixed mass of gas is taken through the closed cycle A → B → C → D → A as shown in the diagram above.

(a) Calculate the work done in the cycle.

(b) How much heat is transferred in the cycle?

(c) Is the heat absorbed or emitted by the gas?

(WJEC, 1989)

26 Air is contained in a cylinder by a frictionless gas-tight piston.

(a) Find the work done by the gas as it expands from a volume of 0.015 m^3 to a volume of 0.027 m^3 at a constant pressure of 2.0×10^5 Pa.

(b) Find the final pressure if, starting from the same initial conditions as in (a), and expanding by the same amount, the change occurs

(i) isothermally

(ii) adiabatically.

(γ for air is 1.40)

(WJEC, 1991)

27 (a) Figure 1 shows, for isothermal conditions, the graph of pressure, p, against the reciprocal of the volume, $1/V$, for a fixed mass of gas.

(i) What is meant by an *isothermal change*?

(ii) Use data from Figure 1 to show that the changes represented by the graph are isothermal.

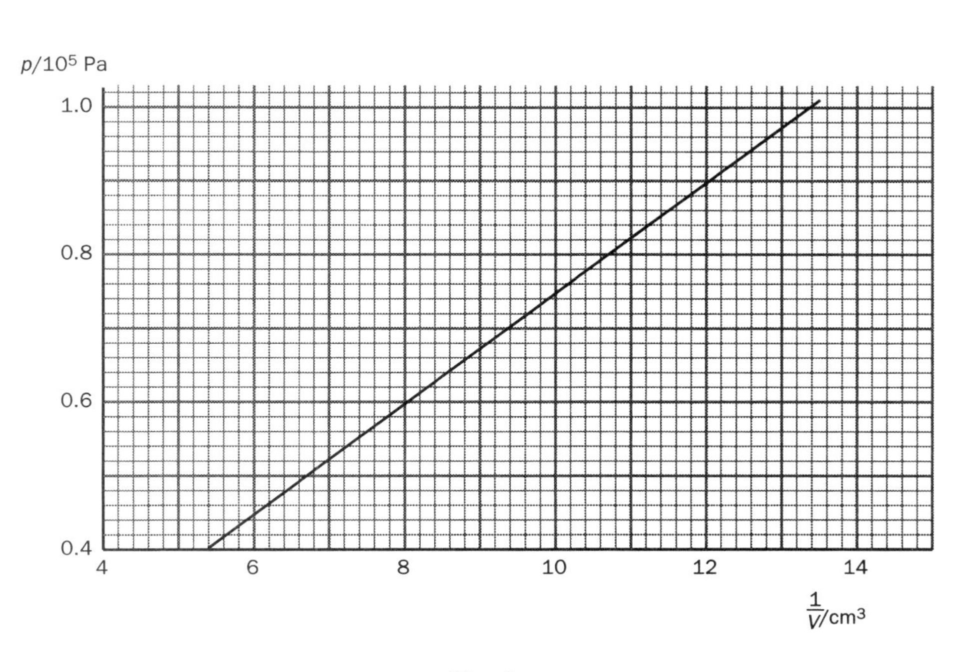

Fig. 1

(b) Figure 2 shows a cycle of processes (**KLMN**) for three moles (3 mol) of an ideal gas.

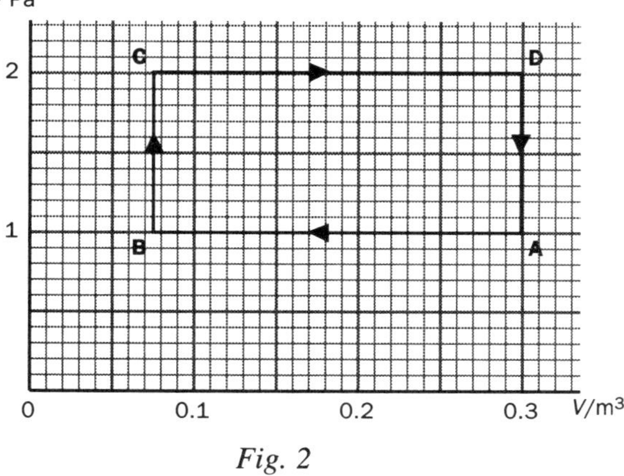

Fig. 2

Determine:

(i) the net (useful) work done by the gas during one cycle of changes

(ii) the maximum temperature of the gas during the cycle.

Molar gas constant, $R = 8.3$ J mol^{-1} K^{-1}.

(AEB, 1993)

5 Atomic and Nuclear Physics

In this section we will examine atoms, the building blocks of the Universe, looking at the fundamental particles from which they are constructed and the forces which control their behaviour and which give rise to all the physical, chemical, electrical, etc. properties of matter. The basis of this idea first surfaced in the 5th century BC, but it was the beginning of the 19th century before supporting evidence was assembled that matter was indeed composed of **atoms**, and that compounds were merely combinations of two or more different atoms from a small number of **elements**. It was another hundred years before Rutherford's famous experiments identified the **nucleus** as a tiny speck of matter occupying only a minute fraction of the volume of the atom, containing all the positive charge and nearly all the mass.

As is the nature of progress, new discoveries came thick and fast – the **proton** was identified in 1914, and the **neutron** in 1932 by James Chadwick. In the same year, the **proton–neutron model** of the nucleus was developed, which agreed with nearly all of the observations of atomic behaviour made up to that date. This model predicted an undiscovered force which only existed in the nucleus – the **strong nuclear force**. For the next 30 years or so, an increasing number of new particles with strange,

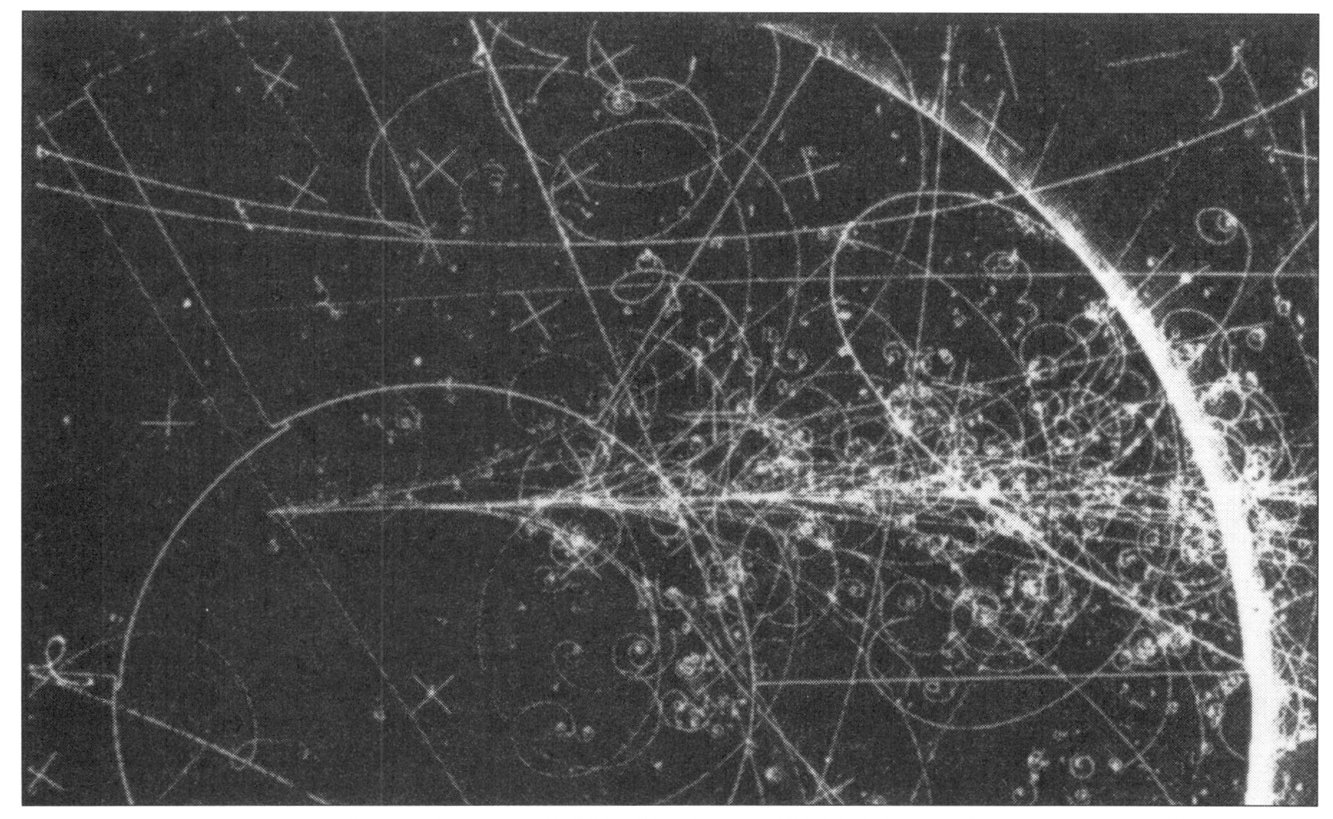

A neutrino reaction seen in the Big European Bubble Chamber at CERN. A beam of neutrinos (invisible in the picture) entered from the left, and interacted with a quark to produce this dramatic spray of particles.

exotic properties was observed as physicists probed ever deeper into the structure of the nucleus. It seemed that there was going to be an infinite number of particles, until Murray Gell-Mann worked out a method for bringing order to this apparent chaos by grouping all the newly-discovered particles into a regular pattern he called the **eightfold way**. He was able to explain the grouping of properties (rather in the manner of Mendleev's Periodic Table of the elements) by proposing that all these particles, including the proton and neutron, were composed from a small number of fundamental particles he named **quarks** (from a line in James Joyce's *Finnegan's Wake*).

5.1 Atomic structure

One of the major foundations of our understanding of the way in which the Universe is made is the concept of atomic structure – that all matter is constructed from tiny, indivisible particles called atoms. All atoms, with the exception of hydrogen (whose nucleus is a single proton), consist of a nucleus composed of protons and neutrons, surrounded by a region of empty space containing electrons. The protons and neutrons are known collectively as **nucleons**. Table 5.1 gives the charge, mass and normal location of these sub-atomic particles.

Nuclear terminology

Different atoms have nuclei which are distinguished by the different numbers of protons and neutrons they contain. The following terms are used to describe these differences.

Proton number (Z)

(Also known as the **atomic number**)
This is simply the number of **protons** in the nucleus of an atom of an element. It has no unit. The number of protons defines the element, and dictates nearly all of its properties. If an atom is electrically neutral (as is usual) then the number of electrons orbiting the atom must be equal to the number of protons in the nucleus, since protons and electrons carry exactly equal and opposite charge.

Nucleon number (A)

(Also known as the **mass number**)
This is the total number of **nucleons** in the nucleus of an atom, i.e. it is the total number of both protons and neutrons. To obtain the number of neutrons in an atom, simply subtract the proton number (Z) from the nucleon number (A). All nuclei, except that of the commonest form of hydrogen, contain neutrons. Protons are positively-charged, and so repel each other. It appears impossible for two or more protons to 'stick' together in a nucleus without another particle – the neutron. Typically there are at least as many neutrons as protons in a nucleus, but there can be up to 50% more neutrons in the case of some of the largest nuclei with over 90 protons.

Table 5.1 Sub-atomic particles

Particle	Location	Rest mass /kg	Charge /C
proton	Inside the nucleus of atoms	1.67×10^{-27} (approx.) (= mass of 1836 electrons)	$+ 1.6 \times 10^{-19}$ (= 1 electronic charge unit)
neutron	Inside the nucleus of atoms	approx. equal to proton	0
electron	Orbiting atomic nuclei, flowing in conductors as an electric current, in cathode rays (in TV tubes), beta-radiation, etc.	9.11×10^{-31} (approx.)	$- 1.6 \times 10^{-19}$

Notation

The proton number (Z) and nucleon number (A) of an element are written as follows:

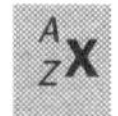

$^{A}_{Z}X$ where X is the chemical symbol of the element.

For example, the commonest form of carbon (C), containing 6 protons and 6 neutrons, would be written: $^{12}_{6}C$

Isotopes

All of the known elements occur in a variety of forms, containing different numbers of neutrons. A **nuclide** is a particular **species** of atom which contains a specified number of protons and a specified number of neutrons.

Isotopes are nuclides with the same **proton number** but different **mass number**; i.e. they contain identical numbers of protons, but different numbers of neutrons. Isotopes are all nuclides of the same element, occupy the same place in the periodic table and have identical chemical properties.

All the isotopes of a particular element share virtually identical physical and chemical properties, since it is the electron structure that determines the way in which one atom interacts with another. Each isotope has the same number of protons, hence their electron structures will be identical. Most elements consist of mixtures of different isotopes. There are usually one or two isotopes of each element which are far more common than the others. Isotopes are not usually given separate names. Exceptions include **deuterium** and **tritium**, which are isotopes of hydrogen with 1 and 2 neutrons respectively. They are treated in this special way because their masses and hence their physical properties are appreciably different from those of hydrogen. (A deuterium atom contains twice as much mass as a hydrogen atom. This difference is much less significant with heavier atoms.)

$^{1}_{1}H$	Hydrogen	1 p		
$^{2}_{1}H$	Deuterium	1 p	1 n	known as 'heavy' hydrogen.
$^{3}_{1}H$	Tritium	1 p	2 n	radioactive gas, used for emergency illumination.

Although there are only around 90 naturally-occurring elements, the number of isotopes known is about 1500. Only about 300 occur in nature, and some of these are radioactive. The rest are man-made and are all radioactive.

Evidence for the existence of the nucleus

Before 1911, the atom was thought of as containing negatively and positively charged matter, evenly distributed. One theory described the electrons as being like the 'plums in a plum-pudding'. The nuclear model of the atom, i.e. the idea that most of the mass of the atom (and all of the positive charge) is contained in a tiny nucleus surrounded by relatively distant electrons, was proposed by Rutherford in 1911.

Ernest Rutherford, the Nobel prize-winning physicist whose incisive ideas on atomic structure gave us the nuclear model of the atom

He based this idea on the observation of α-particles passing through very thin metal foils – some of the α-particles were deflected away from their original direction. This was one of the earliest uses of high-energy particles to probe the nature of matter, and was a forerunner of the huge, multi-national experiments taking place today in the quest to unlock the ultimate secrets of the universe.

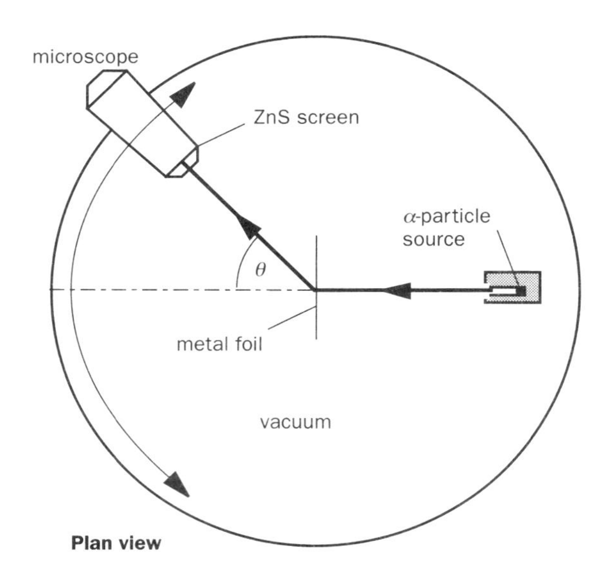

Figure 5.1 Rutherford's α-particle scattering experiment

Geiger and Marsden, two of Rutherford's assistants, made a painstaking study of this effect, using the apparatus shown in Figure 5.1. They aimed a narrow beam of α-particles at a film of gold about 1 μm (1×10^{-6} m) thick. The experiment was performed in a vacuum chamber to avoid collisions between the α-particles and air molecules. The angular deflections of the α-particles were measured by means of the microscope, which was used to observe the scintillations (tiny flashes of light) that occurred whenever an α-particle struck the zinc sulphide screen. The screen and microscope could be rotated around the fixed source and metal foil.

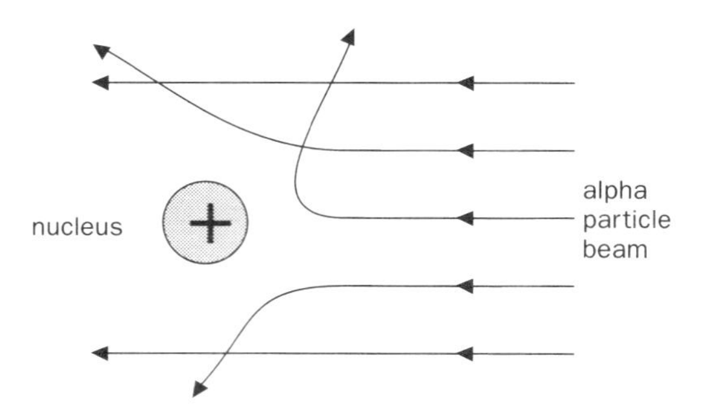

Figure 5.2 Deflection of α-particles by a large, positively charged nucleus

Geiger and Marsden spent many hours in a darkened laboratory observing the scintillations for a wide range of angles. They found that while most α-particles passed straight through, some were scattered considerably, and a small proportion were deflected by more than 90°. A tiny proportion, about 1 in every 8000, were deflected by 180°, i.e. they were reflected back towards the radioactive source (see Figure 5.2).

The α-particles were obviously being deflected by positive charge somewhere in the gold atoms. Since even the strongest electric fields then available could only cause slight deflections of α-particles, clearly something in the gold atoms was capable of exerting extremely powerful fields. Rutherford concluded that this could only be true if the atom contained a dense, positively charged nucleus, and was otherwise mostly empty space! Detailed analysis of Geiger and Marsden's results obtained using a range of α-particle energies and different metal films confirmed the existence of the nucleus, and enabled its mass and diameter to be estimated. This was a major landmark experiment in physics, not only confirming the nuclear model, but also showing the value of bombarding matter with high-speed particles to observe what happens.

You should note that:

- The deflection of the α-particles is explained by the **electrostatic repulsion** between the positively-charged α-particle and the positively-charged nucleus of an atom in the metal foil. The magnitude of the repulsive force is inversely proportional to the square of the distance between the centres of the α-particle and the nucleus.
- The vast majority of the α-particles pass through the foil undeflected. Since the metal foils were thousands of atoms thick, this must mean that the nuclei can only occupy a tiny fraction of the volume of the atom. The undeviated α-particles pass so far from any nuclei that they experience negligible deflecting forces.
- Some α-particles pass close enough to a nucleus to be deflected by a small angle.
- A few α-particles pass close enough to a nucleus to be deflected by angles >90°.
- A very small number of α-particles suffer deflections of between 90° and 180° (i.e. total reflection). These α-particles (approx. 1 in 8000) must have approached a nucleus almost head-on.

Experimental considerations

- The target foil should be as thin as possible so that scattering of the α-particles is due to single interactions, and to prevent too many of the α-particles being absorbed.

- The degree of scattering is related to the charge on the target nuclei, so foil materials of high atomic mass, such as gold, give the most scattering.
- The beam of α-particles must be as narrow and parallel as possible so that the angle of incidence can be determined accurately, enabling precise measurement of the angle of scattering.

Energy considerations

The approaching α-particle has high kinetic energy (gained from the conversion of matter to energy within its parent nucleus). Far from the target nucleus, where the electrostatic repulsion is negligible, the α-particle will travel at constant velocity in the vacuum, so its kinetic energy will be constant. As it approaches the target nucleus it is decelerated by the repulsive electrostatic force and thus it loses kinetic energy. The kinetic energy lost is equal to the gain in electrostatic potential energy (of the system comprising the α-particle and the target nucleus). At all times, the sum of these energies is constant.

In the 'head-on' collision, the α-particle is brought to rest momentarily by the electric field. At this instant its kinetic energy is zero, and the electrostatic potential energy gained is equal to the initial kinetic energy of the α-particle. The distance between the α-particle and the target nucleus at this instant is a minimum value for a head-on collision, and is referred to as the 'closest approach distance'. The electrostatic potential energy is then converted back into kinetic energy as the α-particle is accelerated back along its original path.

Collisions of electrons with atoms

If we are to understand what happens when an electron collides with an atom, we need to know something about the idea of **energy levels** in atoms.

Following the great success of the α–particle scattering experiments, Rutherford's nuclear model of the atom reigned supreme. The atom was pictured as a central nucleus containing the protons and neutrons around which the electrons moved in circular orbits. The centripetal force needed for this circular motion was provided by the electrostatic attraction between the negatively charged electrons and the positively charged nucleus. The model seemed to be a good one, but it was flawed.

Because of their continuous change of direction, the electrons are accelerating. But, according to electromagnetic theory, an accelerating charge emits energy and so the electron must emit electromagnetic radiation as it revolves around the nucleus. This continuous loss of energy would cause the orbit radius to decrease. Thus the electron would spiral into the nucleus and the atom would cease to exist!

It was Niels Bohr who pointed out this difficulty with Rutherford's model, and he solved the problem by proposing that:

> There are certain **'allowed'** orbits in which electrons can exist without emitting energy.

The only 'allowed' orbits are those for which the **angular momentum** of the electrons is an integer multiple of $h/2\pi$ (where h is Planck's constant). The electron angular momentum is thus said to be 'quantised' in that it can only exist with certain discrete values.

The energy associated with a particular orbit is the total energy which an electron has in that orbit. An electron has kinetic energy as a result of its motion and potential energy because of the attraction force of the positive charge in the nucleus. Thus it follows from the first proposal that:

> Electrons can only have certain 'allowed' energy values (called energy levels).

> Electrons can pass from one 'allowed' energy level to another.

If, as a result of being bombarded by other electrons or atoms, the right amount of energy (ΔE) is absorbed by an atom, one of its electrons can 'jump' from a lower energy level (E_1) to one of higher energy (E_2). This is shown diagrammatically in Figure 5.3.

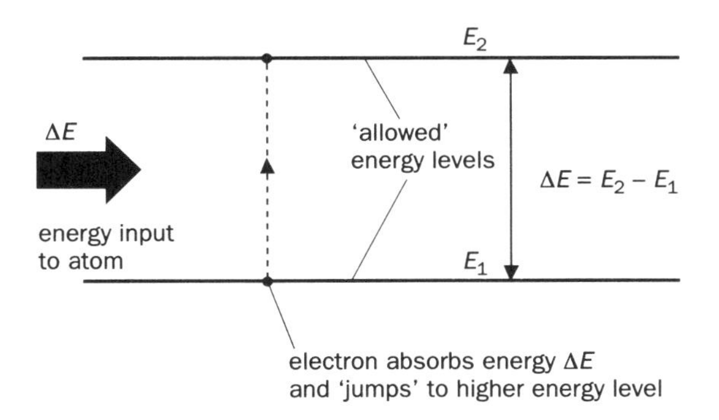

Figure 5.3 Energy absorption by an atom

After absorbing this energy, the atom is said to be in an **excited** state. This is an unstable condition, and the atom will normally return to a lower energy level within a short time (typically less than 1 μs). The electron falls back from E_2 to E_1, and the energy difference ΔE is emitted as a burst of electromagnetic radiation (a **photon**) of a particular frequency (f) or wavelength (λ) (see Figure 5.4).

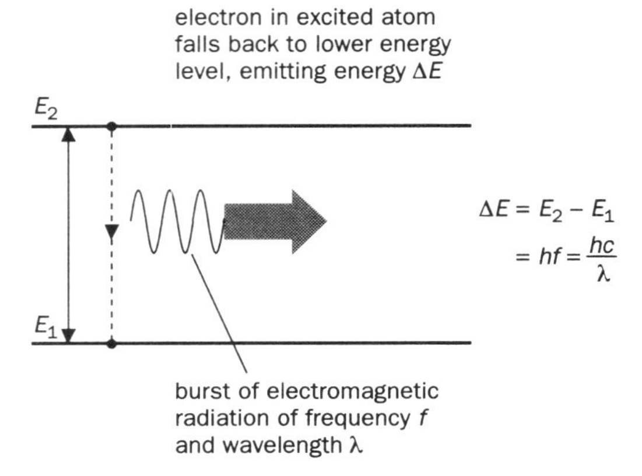

Figure 5.4 Emission of photon of electromagnetic radiation as atom returns to lower energy level

J, J s

$$\Delta E = E_2 - E_1 = hf$$

s^{-1} (or H z)

Equation 1

where h = Planck's constant (6.63×10^{-34} J s)
f = the frequency of the emitted radiation (Hz)

Since $c = f\lambda$, then

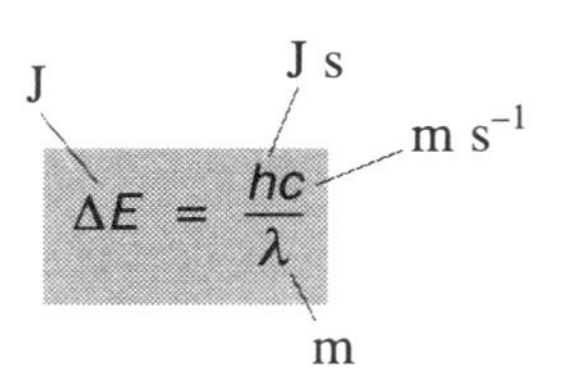

Equation 2

where c = the velocity of electromagnetic radiation in a vacuum (3.00×10^8 m s^{-1})
λ = the wavelength of the emitted radiation (m)

The Bohr model accurately predicts the wavelengths of the light emitted from hydrogen, but fails when it is applied to more complex atoms. Although the current model (based on wave mechanics) considers the electrons to exist in the form of a 'cloud' around the nucleus, the concepts of discrete energy levels and inter-level transitions of electrons are retained. Some of the most direct evidence for the existence of energy levels comes from observations of the emissions of electromagnetic radiation from excited atoms.

Energy levels and spectra

In Section 2.3, we saw how a diffraction grating and a spectrometer could be used to view the **emission spectrum** of light sources. If the light source is, for example, a hot filament in an electric lamp, the spectrum seen is **continuous** (there are no gaps between the colours). If a parallel beam of light from an electric filament lamp passes through a narrow slit and is then focused on a screen, a bright, white image of the slit is seen on the screen, as shown in Figure 5.5. If, however, the light is diffracted by a grating (or refracted by a prism), the individual frequencies which comprise the white light will be deviated by different amounts. Each different frequency arrives at a different point on the screen, producing a different colour. Each colour blends smoothly into the next, giving the appearance of a continuous spread of colour. This is the familiar production of the **visible spectrum**. (A similar spectrum can be produced from sunlight, but if you look at the Sun's spectrum very carefully, you will discover that it is not quite continuous – there are colours 'missing'. We will examine this observation in more detail a little later.)

If the white light source is replaced by a truly monochromatic source, such as a laser, then only a

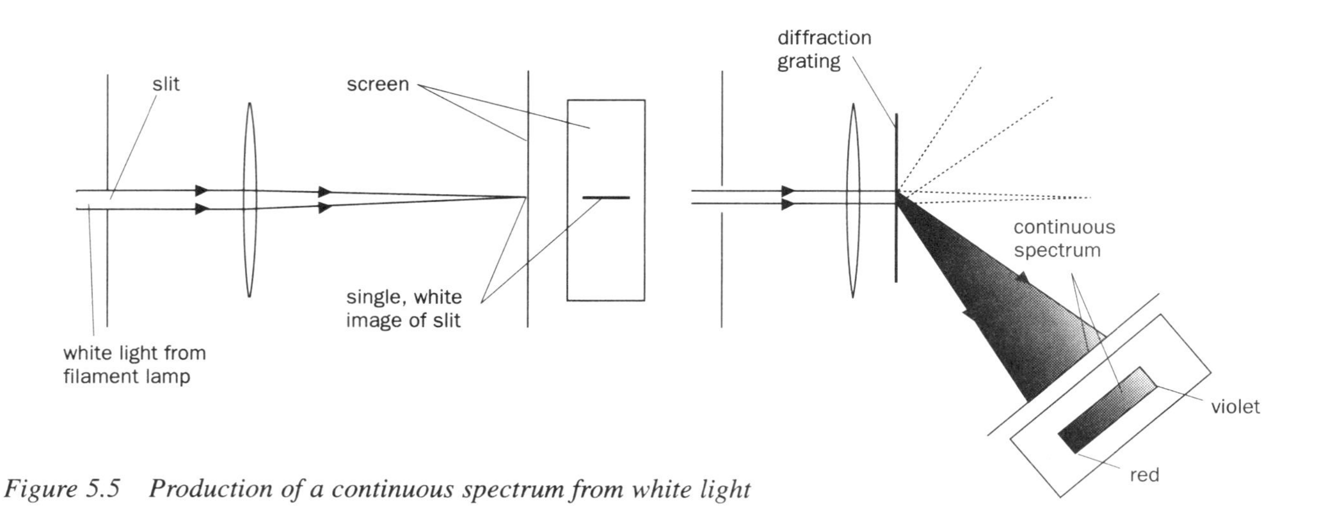

Figure 5.5 Production of a continuous spectrum from white light

single image of the slit is produced, since only one frequency of light is present in the laser beam. The image of the slit is the same colour as the laser light, and its position on the screen corresponds to the position of that colour on the spectrum produced using a white light source (see Figure 5.6 below).

However, if a gas discharge tube is used the observed spectrum consists of a series of separate lines of different colours (and therefore different frequency) against a dark background, as shown in Figure 5.7. Each line is actually an image of the collimator slit; the position of the line in the spectrum corresponds to the specific **frequency** of the light emitted.

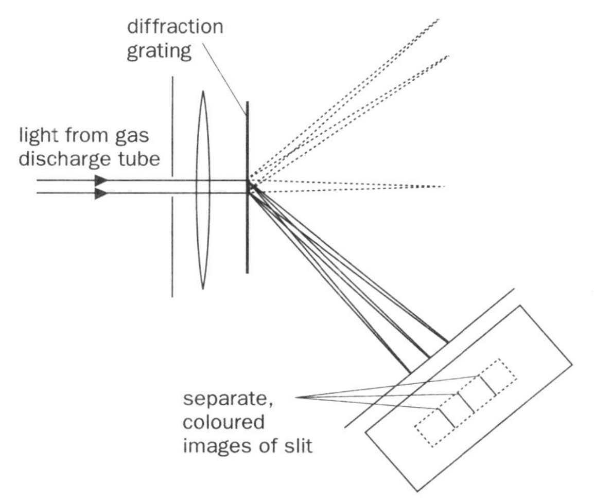

Figure 5.7 Production of a spectrum from a gas discharge lamp

Figure 5.8 shows a more detailed view of the appearance of the spectrum, as seen in the eyepiece of a spectrometer telescope.

Line spectra are evidence for the existence of energy levels in atoms. In a gas discharge tube, the

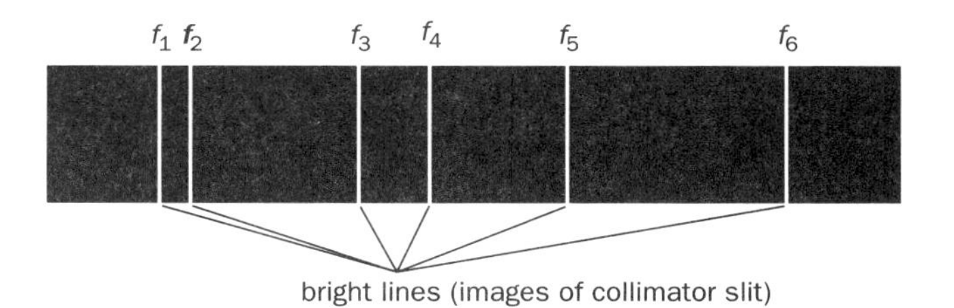

Figure 5.8 Emission line spectrum

atoms of the low-pressure gas in the tube are sufficiently far apart to be acting as individual atoms, free of each other's influence. The atoms are excited by the passage of an electric current through the tube. The gas atoms become excited by collisions with the electrons passing through the tube (carrying the electric current), gaining energy. When the atom returns to a lower energy level, this energy is emitted as a photon of radiation of a specific frequency. The fact that only certain frequencies are present in the emission spectrum shows that only certain energy level transitions are possible. The actual frequencies emitted are a property of the gas atoms in the discharge tube – each element has its own unique line spectrum. Each line in the spectrum is produced by an electronic transition within an atom from one energy level to another.

Line spectra are only observed when the excited atoms behave as **individual atoms**, as is the case in a gas discharge tube. In solids, such as lamp filaments, the presence of neighbouring atoms modifies the energy levels of each atom, so that there is an almost infinite number of possible transitions between energy levels. The result is the **continuous spectrum** seen from such sources.

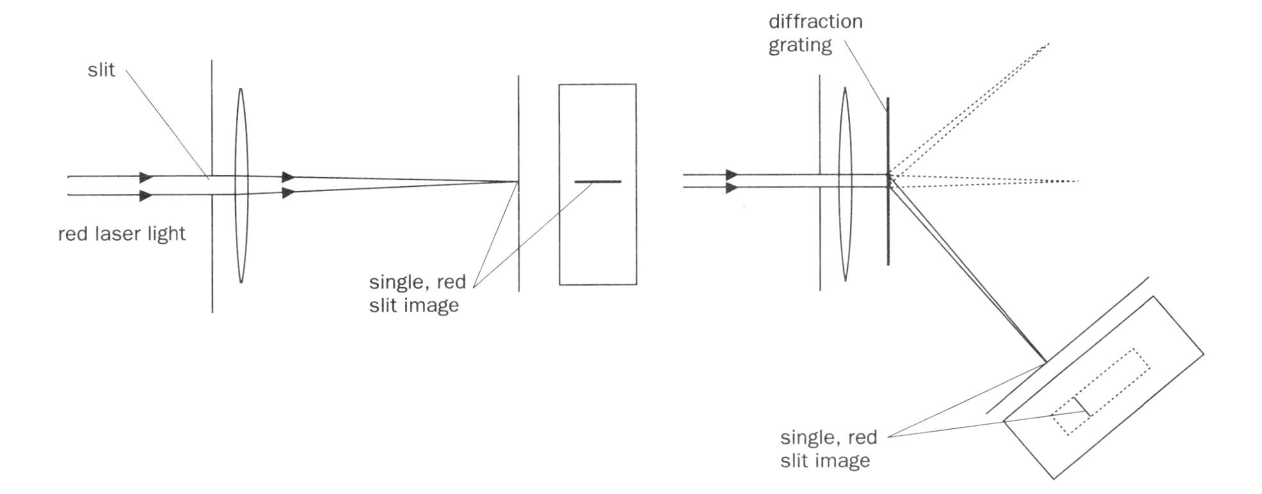

Figure 5.6 Diffraction of a monochromatic light source

The hydrogen spectrum

The simplest atom of all is hydrogen (1 proton, 1 electron), and if we examine its spectrum (Figure 5.9), we can deduce its energy levels.

These lines were first measured by Johann Balmer in 1855. Precise measurement of the spectrum shows that visible light photons are emitted with frequencies and energies shown in Table 5.2.

Table 5.2 Balmer series of hydrogen

Frequency	Photon energy	
/ 10^{14} Hz	/ 10^{-19} J	/ eV
4.6	3.0	1.9
6.2	4.1	2.6
6.9	4.6	2.9
7.3	4.8	3.0

A more complete analysis of the emissions from hydrogen discharge tubes shows that there are many more emissions in non-visible ranges, including the *Lymann* series of ultraviolet radiation, and the *Paschen* series in the infrared region. Each frequency is due to an electron making a transition ('jumping') from one energy level to another (lower) level. Some of these energy levels are shown in Figure 5.10.

If an electron is at an excited level (E_1) and makes a transition to a lower level (E_2), the frequency (f) of the emitted photon will be given by:

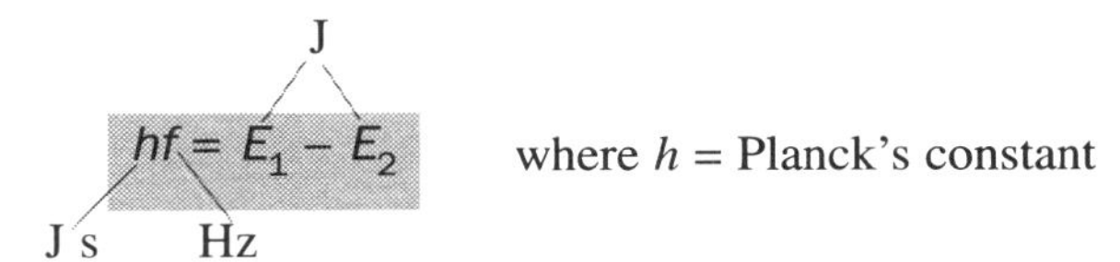

where h = Planck's constant

Equation 3

You should note that:

- The lowest energy level of the atom (–13.6 eV) is also referred to as the **ground state**. This is the normal configuration of the atom. Energy must be supplied to excite the atom into a higher level.
- The highest energy level (0 eV) represents **ionisation**. In this case sufficient energy has been absorbed to remove the electron to infinity.
- The energy levels are not evenly spaced.
- The **series** of lines (named after their discoverer) are grouped according to the final energy level of the transition. The *Lymann* series all end on the ground state, etc.

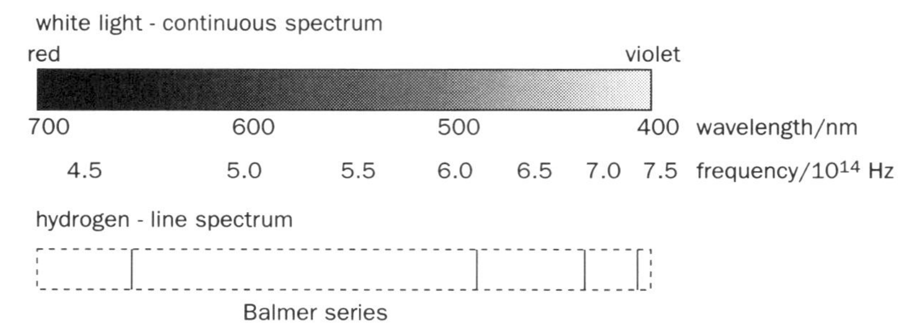

Figure 5.9 Line spectrum of hydrogen

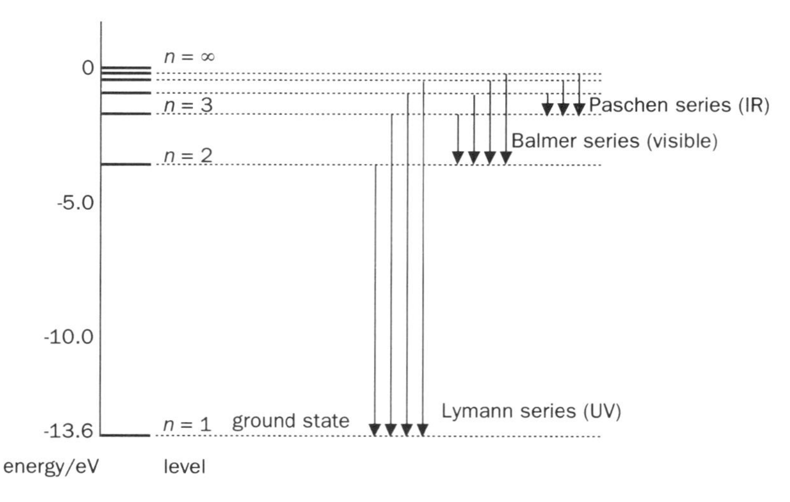

Figure 5.10 Transitions between energy levels in the hydrogen atom

- Energy level diagrams for other atoms or molecules are very much more complicated, with many more levels, due to the interactions between electrons and nuclei.
- Line spectra are produced by individual atoms, acting independently. This is why gas discharge tubes are used – at the low pressures in the tube, the molecules of the gas are sufficiently far apart that they do not interact.

Band spectra

In molecules, the atoms are joined together by bonds formed by the electrons. The interactions between the electrons produce a large number of equally spaced energy levels. Electromagnetic radiation emitted from molecules therefore forms a spectrum of closely spaced lines, which resemble bands.

Absorption spectra

So far we have considered only spectra formed by the emission of radiation. Earlier, the 'missing' radiation from the Sun was mentioned. Close examination of the Sun's spectrum shows numerous dark lines at irregular intervals, as shown in Figure 5.11.

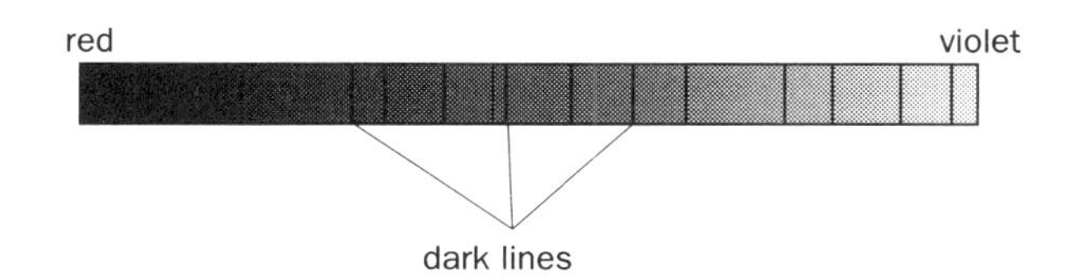

Figure 5.11 Line absorption spectrum of sunlight

The Sun produces a continuous spectrum, containing all the visible frequencies (and a great many more in other regions of the electromagnetic spectrum). The lines are due to the **absorption** of specific frequencies.

The gases around the Sun are relatively cool, and not emitting visible radiation. On its journey to an observer, here on Earth, the light passes through these layers of gas. The atoms of the gas become excited by absorbing photons of the appropriate frequency to cause their electrons to jump to higher energy levels. Since the Sun's light contains all possible frequencies, there will be plenty of photons of the right frequencies to cause excitation. The absorbed photons no longer exist. The gas atoms re-emit photons of the **same frequency** but their directions will be random (see Figure 5.12). Therefore the light reaching the observer will be deficient in photons of the absorbed frequencies, hence the dark lines on the spectrum.

Energy levels

Our discussion so far has yielded the following ideas relating to atomic energy levels:

- The electrons in an atom can only have certain energy values – called the **energy levels** of an atom.
- The energy levels can be drawn as horizontal lines to form an **energy level diagram**.
- All atoms of the same element have the same energy level diagram which is different from that of atoms of another element.

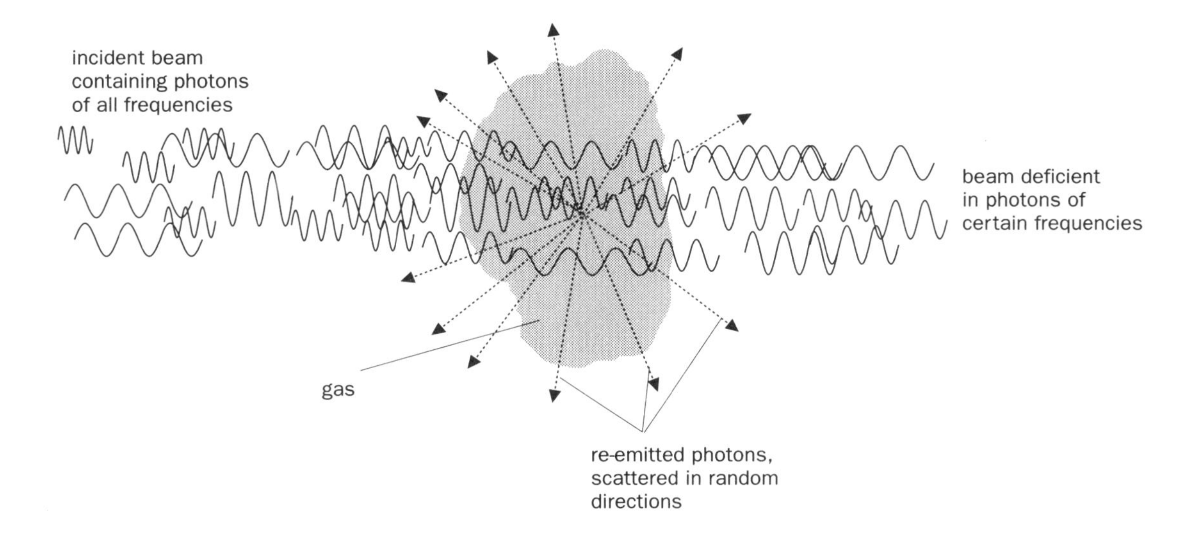

Figure 5.12 Production of a line absorption spectrum

Figure 5.13 shows part of the energy level diagram for the hydrogen atom.

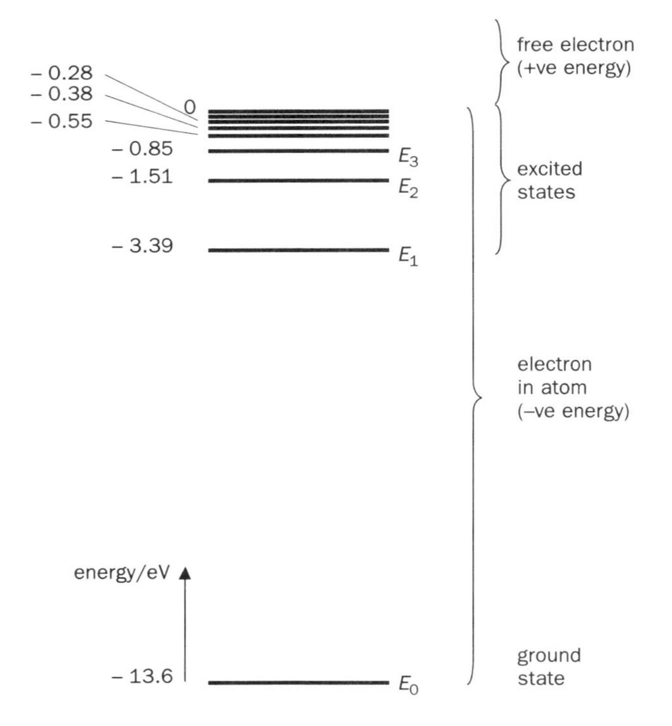

Figure 5.13 Some of the energy levels in the hydrogen atom

You should note that:

- Energy level values are expressed in **electron volts** (eV).
- 1 eV = 1.6×10^{-19} J.
- The levels have negative energy-values. This is because the energy of an electron at rest outside the atom is assumed to be zero, and when an electron falls into the atom energy is emitted in the form of electromagnetic radiation, i.e. the energy levels closest to the nucleus can only be of lower energy than those further away by being more negative.
- Electrons in the higher energy levels are more loosely bound to the nucleus since they require less energy in order to 'escape' from the atom.
- The **ground state** of an atom (energy level E_0) is the most stable and lowest energy state available in that atom. In the case of the hydrogen atom, its single electron normally occupies the lowest energy level and in this condition the atom is said to be in its ground state.
- If energy is absorbed by the atom, the electron may 'jump' into one of the higher energy levels. The atom is now said to be in an excited state.
- If an atom absorbs enough energy so that an electron is raised to the highest energy level (energy = 0), the electron becomes free of the atom, i.e. **ionisation** occurs. The minimum energy which must be absorbed by a hydrogen atom in the ground state if it is to be ionised is 13.6 eV.

Excitation and ionisation potential

Experiments on collisions between atoms and electrons in a gas-discharge tube, performed by Franck and Hertz in 1914, provided direct evidence for the existence of energy levels. What happens when an electron collides with an atom depends very much on the kinetic energy of the electron. In the apparatus shown in Figure 5.14, electrons emitted from a hot cathode are accelerated in a tube containing a low-pressure gas towards a grid. The grid is at a more positive potential than the cathode, and this p.d. (V_C) can be varied.

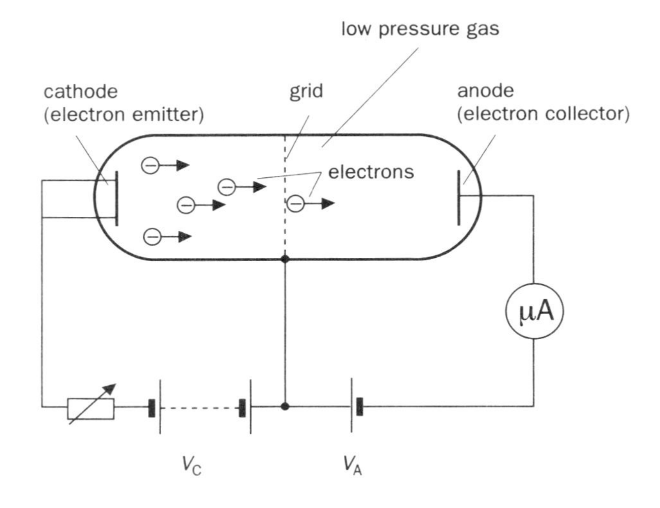

Figure 5.14 Franck–Hertz experiment to measure excitation potentials

Some electrons will pass through the grid, and provided $V_C > V_A$, they will reach the anode, and a tiny current will be registered by the microammeter. Increasing the grid–cathode p.d. and measuring the current flowing produces the graph shown in Figure 5.15.

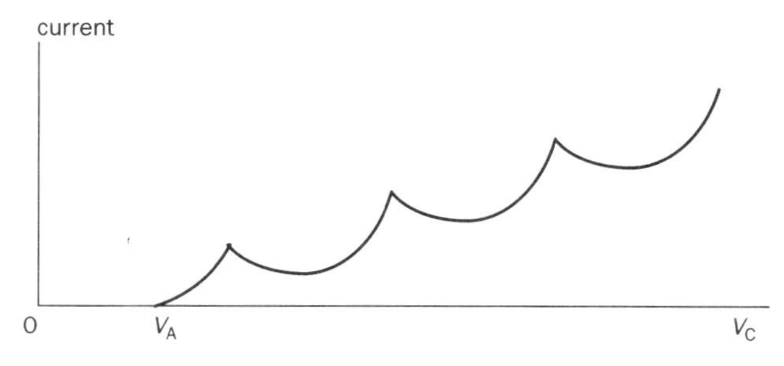

Figure 5.15 Graph of current against grid–cathode voltage in a Franck–Hertz experiment

Once V_C is greater than V_A some electrons pass through the grid and reach the anode, where they register as a current on the microammeter. As V_C is increased, the current increases. At a particular voltage, the current suddenly decreases. Below this voltage, collisions between the electrons and the gas atoms are **elastic**, so no energy is absorbed. At this particular voltage, some of the atoms of the gas have become **excited** by the collisions, since the electrons have just enough energy to raise them to the next energy level. The colliding electrons lose their kinetic energy in an **inelastic** collision, and so fail to pass through the grid. The p.d. at which this first occurs is therefore known as the **first excitation potential**. This is the potential required to give the electrons sufficient kinetic energy to produce the smallest possible electron energy transition within the atom. Increasing V_C gives the electrons even more kinetic energy, so the current again increases. The graph shows peaks which occur at voltages corresponding to excitation potentials of the atoms of the gas in the tube.

The p.d. through which the colliding electron must be accelerated so as to acquire a particular excitation energy is called the **excitation potential**.

You should note that:

- Since different atoms have different energy level diagrams they will also have different excitation energies and potentials.
- The fact that an atom has several distinct excitation potentials provides direct evidence for Bohr's idea of discrete energy levels.

Ionisation is said to occur if the kinetic energy of the colliding electron is sufficient to cause an atomic electron to gain enough energy to escape from the atom. The energy needed to cause ionisation is called the **ionisation energy**. Since ionisation may be caused by lifting electrons from all energy levels including the ground state, an atom has several ionisation energies. The **ionisation potential** is the accelerating p.d. value needed so as to impart the ionisation energy to the bombarding electron.

Guided example

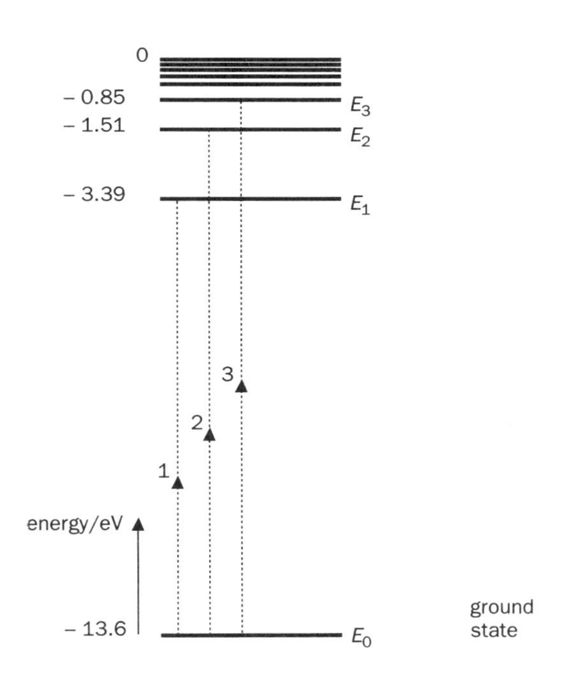

The energy level diagram for a hydrogen atom is shown above.

(a) Calculate the ground state energy in joules.

(b) For each of the electronic transitions (1) (2) and (3) shown, calculate:

(i) the excitation energy in joules

(ii) the excitation potential

(iii) the frequency and wavelength of the radiation which would be emitted by the return of the electron to the ground state.

(c) What is the ionisation energy of the hydrogen atom in the ground state (in joules)?

(Planck's constant, $h = 6.6 \times 10^{-34}$ J s; speed of e.m. radiation, $c = 3 \times 10^{8}$ m s^{-1})

Guidelines

(a) Use the fact that 1 eV = 1.6×10^{-19} J.

(b) (i)

$$\Delta E_1 = E_1 - E_0$$
$$\Delta E_2 = E_2 - E_0$$
$$\Delta E_3 = E_3 - E_0$$

(ii) Remember that excitation potential is the p.d. through which the bombarding electron must be accelerated to produce a given excitation energy.

(iii) Use Equation 1 and Equation 2.

(c) Remember that the ionisation energy is the energy which must be given to an atom to remove its most loosely-bound electron. In the case of the hydrogen atom, there is only the one electron.

Self-assessment

SECTION A
Qualitative assessment

1. State what is meant by the following terms:
   (a) nuclide
   (b) radionuclide
   (c) isotope
   (d) radioisotope.

2. Briefly describe Rutherford's α-particle scattering experiment, and explain how the results of the experiment led to the conclusion that matter is made of atoms with tiny, dense, positively charged nuclei surrounded by empty space containing electrons.

3. What major flaw in Rutherford's nuclear model was pointed out by Niels Bohr, and what modifications did he suggest?

4. (a) Explain what is meant by **energy levels** in an atom.
   (b) Use the idea of **electronic transitions** between energy levels in an atom to explain
      (i) **excitation**
      (ii) the production of **line emission spectra**.
   (c) Explain why the **energy levels** in an energy level diagram are labelled with **negative** energy values.
   (d) Explain what is meant by the **ground state** of an atom.

5. (a) What kind of sources produce **line emission spectra**?
   (b) What are the characteristics of a line emission spectrum?
   (c) Why does each element have a **unique** line emission spectrum?

6. Explain the meaning of each of the following terms:
   (a) excitation energy
   (b) ionisation energy
   (c) ionisation potential.

7. Explain what happens to the energy of an electron which collides
   (a) elastically
   (b) inelastically
   with an atom.

SECTION B
Quantitative assessment

(***Answers:*** 8.6×10^{-20}; 5.1×10^{-19}; 5.4×10^{-19}; 1.6×10^{-18}; 2.2×10^{-18}; 5.8×10^{-8}; -3.4; 3.4; 14; 1.1×10^{6}; 1.1×10^{6}; 2.7×10^{6}; 5.2×10^{15}.)

electronic charge	e	$= 1.6 \times 10^{-19}$ C
electron mass	m_e	$= 9.11 \times 10^{-31}$ kg
speed of e.m. radiation	c	$= 3.0 \times 10^{8}$ m s^{-1}
Planck's constant	h	$= 6.63 \times 10^{-34}$ J s
permittivity	$\frac{1}{4\pi\varepsilon_0}$	$= 9.0 \times 10^{9}$ m F^{-1}

1. In the Rutherford model of the nuclear atom, the electrons are assumed to move in circular orbits around the central nucleus. The necessary centripetal force is provided by the electrostatic attraction between the oppositely-charged electron and nucleus. Assuming an orbital radius of 2.0×10^{-10} m, calculate the velocity of the electron in a hydrogen atom.

2. An electron is initially located in an atomic energy level of value –1.5 eV. Calculate the value in eV for the energy level to which it must fall in order to emit a photon of wavelength 654 nm.

3. One of the ionisation energies of helium is 21.4 eV. Calculate
   (a) the minimum velocity of an electron which causes this ionisation when it collides with a helium atom
   (b) the lowest frequency and corresponding wavelength of the electromagnetic radiation which could cause this ionisation.

4. An electron moving with a velocity of 2.0×10^{6} m s^{-1} collides with a hydrogen atom in its ground state. As a result of this collision, the atom emits a photon of wavelength 152 nm. Calculate
   (a) the kinetic energy and
   (b) the velocity
   of the electron immediately after the collision.

Self-assessment *continued*

5. The diagram shows some of the energy levels in the hydrogen atom. Calculate:
   (a) The energy (in joules) which an atom of hydrogen must absorb for each of the transitions marked A and B.
   (b) The ionisation energy (in joules) for an atom in:
      (i) the ground state
      (ii) the first excited state.
   (c) The ionisation potential in each of the cases in part (b).

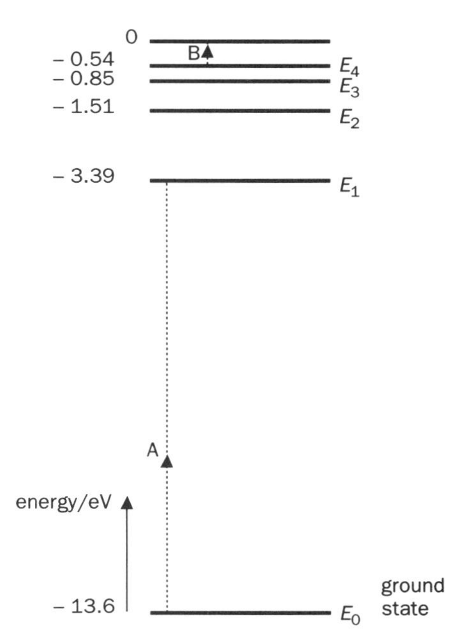

5.2 Radioactivity

In 1896, Antoine-Henri Becquerel noticed that photographic plates stored close to a uranium compound had become fogged, despite being wrapped in light-proof material. Something had penetrated the wrapping of the film, with sufficient energy to cause chemical changes in the light-sensitive emulsion. By careful experimentation, and elimination of other possible causes, he concluded that a highly penetrating, energetic radiation was being emitted by the uranium. This chance observation led to the discovery of what we call **radioactivity**.

Radioactivity is the process in which an unstable **parent** nucleus becomes more stable by **decaying** into a **daughter** nucleus accompanied by the emission of **particles** and/or **energy**.

The process of radioactive decay can be written in the basic form:

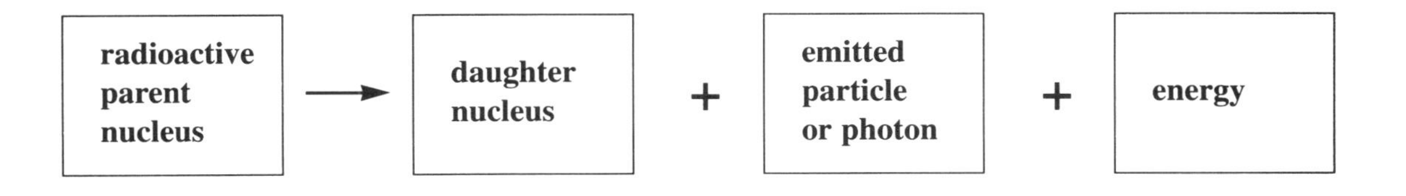

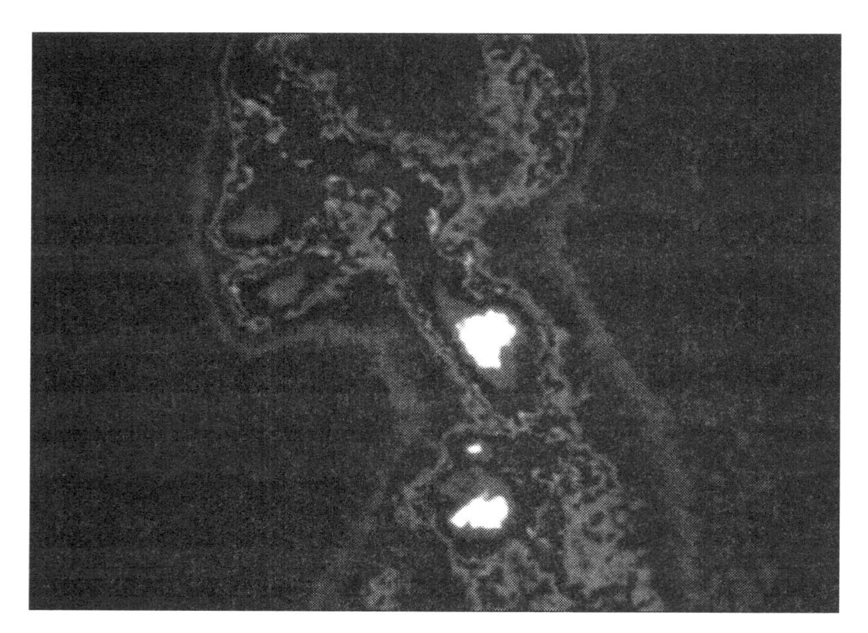

Gamma camera scan of the skull and neck of a person with a secondary cancer in the neck vertebrae. The image shows the distribution and intensity of γ-radiation emitted from a radionuclide injected into the body and which concentrates in the bones. Cancerous bone concentrates the radionuclide more strongly than normal, healthy bone and appears as brighter 'hot spots' on the image

Thus the parent nucleus may be **transmuted** into a different element. If this daughter nuclide is unstable, it may decay further. This process will continue until a **stable** nuclide is created. This is known as a **decay chain**. There are several natural decay chains, and many of the naturally-occurring radioactive elements on earth are long-lasting steps in these chains, as unstable atoms created at the beginning of time decay into more stable forms.

Types of radioactive emission

In this section, we will examine the three basic types of radiation emitted by radioactive substances, which are named **alpha** (α), **beta** (β) and **gamma** (γ) radiation. They can be emitted individually or in any combination by different nuclei, and each type may be identified by observing their different penetrating power, ionising ability and behaviour in a magnetic field. Table 5.3 compares the nature and properties of these three radiations.

Table 5.3 Comparison of radiation properties

Radiation	Penetration	Ionising ability	Effect of magnetic or electric Field	Nature
Alpha (α) radiation	Few cm of air at s.t.p. Thin sheets of paper. Thin layers of mica	Causes intense ionisation, around 10^4 ion-pairs per mm of air	Deflected slightly in same direction as positive charge	Heavy, positively charged particles. Identified as **helium** nuclei ($2p + 2n$)
Beta (β) radiation	Many cm of air at s.t.p. Few mm of light metals such as aluminium	Less intense than α, typically 10^2 ion-pairs per mm of air	Strong deflection in opposite direction to α	Light, negatively charged particle. Identified as **electrons**
Gamma (γ) radiation	Several cm of even dense metals like lead. Several metres of lighter materials such as concrete	Interacts weakly with matter, causing low ionisation, around 1 ion-pair per mm of air	No effect	Electromagnetic radiation of very short wavelength

NB: Penetration and ionisation depend on the initial energy of the particle or radiation. It is possible for a high-energy α-particle to be more penetrating than a low-energy β-particle. The table indicates typical values.

Alpha (α) radiation

Most α-particle emitters are heavy nuclei having proton numbers greater than 82, although some smaller nuclides which are deficient in neutrons can be α-emitters.

Production of the α-particle (α-decay)

It is believed that the α-particle is created some time before its emission – it has a pre-existence in the nucleus. The nuclear equation which describes the decay of a **parent** nuclide into a **daughter** nuclide by α-particle emission is:

$$ {}^{A}_{Z}X \rightarrow {}^{(A-4)}_{(Z-2)}Y + {}^{4}_{2}He + Q $$

This is shown diagramatically in Figure 5.16.

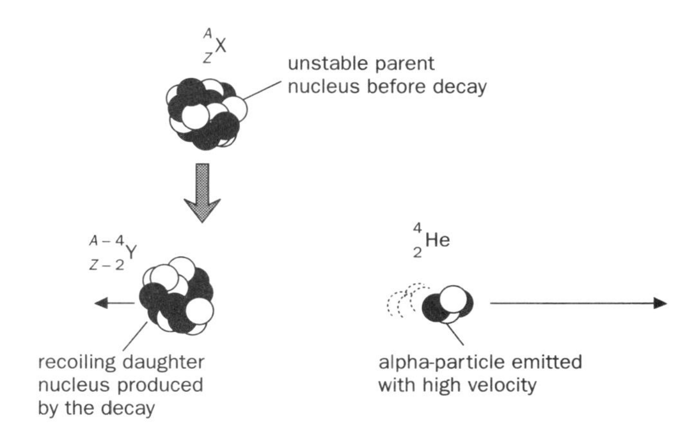

Figure 5.16 α-decay

You should note that:

- ${}^{4}_{2}He$ is the symbol for the α-particle (helium nucleus) and Q is the energy released in the decay.
- When a nucleus undergoes α-decay, it loses a total of **4** nucleons (2 protons and 2 neutrons), thus A decreases by 4 while Z decreases by 2.
- A and Z are **balanced** across the equation – both **charge** and **mass number** are conserved.
- Each decay results in the release of a precise quantity of energy (Q), which is specific to each radioisotope.
- The energy released by the decay (Q) appears as **kinetic energy** of the daughter nucleus and the emitted α-particle. The velocity of the relatively light α-particle is much higher than that of the recoiling daughter nucleus, therefore the majority of the energy is carried away by the α-particle.
- The nuclide **thorium-228** is a typical α-emitter which decays to **radium-224**. The decay equation is:

  $$ {}^{228}_{90}Th \rightarrow {}^{224}_{88}Ra + {}^{4}_{3}He + Q $$

  (Check that A and Z are balanced on each side of the equation)
- The initial velocity of the α-particle depends on the energy released within the parent nucleus, but may be up to 10^7 m s^{-1}, which corresponds to an α-particle energy of around 9 MeV.
- Many nuclides emit α-particles with a single energy value, while others emit α-particles with two or more discrete energies. For example, all the α-particles emitted by thorium-228 have 5.4 MeV of energy, while those emitted by radium-224 have either 4.78 MeV or 4.59 MeV.

Beta$^-$ (β^-) radiation

Neutron-rich nuclei tend to decay by β^--emission. A β^--particle is a high speed **electron** emitted from the nucleus. The minus sign is used to distinguish it from the β^+-particle, which is a **positron** – the antiparticle to the electron (see later).

Production of the β^--particle

A β^--particle is produced when a neutron in the parent nucleus decays. Neutrons are normally stable when contained within the nucleus, but can decay in situations where the nucleus has too much energy, usually due to an excess of neutrons. In this case, the neutron decays into a **proton**, by emitting a β^--particle. The result is transmutation into a nucleus of a different element, with the **same nucleon number**, but with a **proton number increased by one** (since a neutron has effectively turned into a proton). The nuclear equation for the decay of a neutron is:

$$ {}^{1}_{0}n \rightarrow {}^{1}_{1}p + {}^{0}_{-1}e + {}^{0}_{0}\bar{\nu} + Q $$

where n and p are the neutron and proton, e is the β^--particle, $\bar{\nu}$ is an **antineutrino** and Q represents the energy released in the decay.

The overall equation for the decay of a **parent** nuclide (X) into a **daughter** nuclide (Y) by **β^--particle** emission is:

$$ {}^{A}_{Z}X \rightarrow {}^{A}_{(Z+1)}Y + {}^{0}_{-1}e + {}^{0}_{0}\bar{\nu} + Q $$

This is shown diagrammatically in Figure 5.17 overleaf.

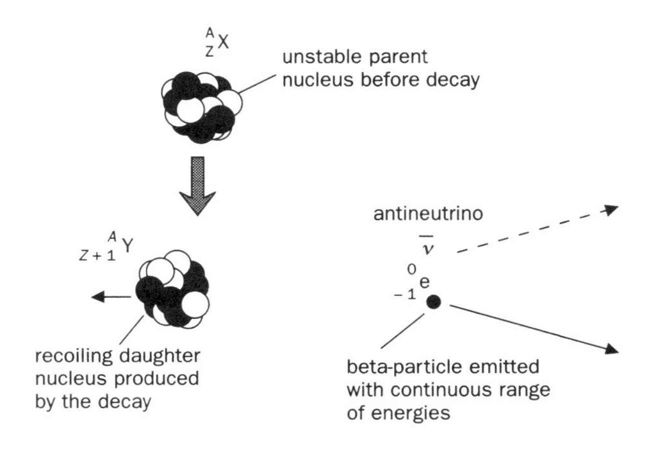

Figure 5.17 β-decay

You should note that:

- The β^--particle is created in the nucleus at the instant of decay, when it is ejected at extremely high velocities, approaching that of light.
- There is no overall change in the **nucleon number** (A), but **proton number** (Z) increases by 1.
- Charge and mass number are conserved on both sides of the equation (A and Z balance).
- β^--particle emission is accompanied by the simultaneous emission of an **antineutrino** – a virtually massless, highly penetrating particle which is very difficult to detect (see later).
- Each decay results in the release of a precise quantity of energy (Q), which is specific to each radioisotope.
- The total energy released in the decay is shared between the daughter nucleus, the β^--particle and the antineutrino. Since there are 2 particles emitted, the proportion of energy taken by each is variable, so the more energy that is taken by the antineutrino, the less there is available for the β^--particle. Consequently, the β^--particles are observed to have a very wide, continuous range of energies from almost zero up to nearly the total Q-value of the decay. The typical distribution of energies of the β^--particles emitted from a single isotope is shown in Figure 5.18.

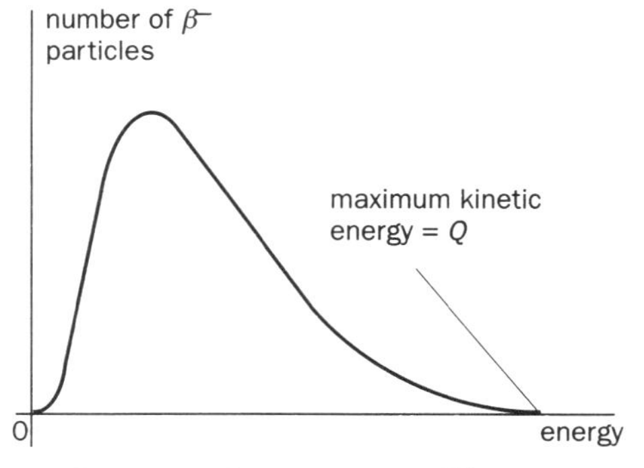

Figure 5.18 Typical spectrum of β^--particle energies

- The nuclide $^{29}_{31}Al$ is a typical β^- particle emitter, which decays to $^{29}_{14}Si$ according to the equation:

  $$^{29}_{13}Al \rightarrow {}^{29}_{14}Si + {}^{0}_{-1}e + {}^{0}_{0}\bar{\nu} + Q$$

  (Check that A and Z balance on each side of the equation)
- If the emission of the β^- particle occurs in a medium in which the speed of light is lower than the speed of the electron, an optical 'shock wave' is produced, a phenomenon similar to the creation of a 'sonic boom' when an aircraft flies faster than the speed of sound. The optical shock wave is seen as a glow of light, known as **Cherenkov radiation**.

Gamma (γ) radiation

Alpha- and beta-decay can often produce a daughter nuclide which is in an unstable, excited state. The daughter nuclide can decay further by emitting a **photon** of **electromagnetic radiation** of very high frequency (from X-ray frequency and beyond). Such a photon is called a **gamma ray**. Since there are no particles of matter involved, the emission of a γ-photon causes no change in the charge, proton number or nucleon number of the parent nuclide.

You should note that:

- The overall effect of γ-ray emission is to reduce the energy of the nucleus. The nucleus remains unaltered physically by the process of γ-emission, apart from having less energy.
- γ-radiation does not consist of particles of matter, but is in fact electromagnetic radiation, part of the electromagnetic spectrum, with very short wavelength ($< 1 \times 10^{-11}$ m) and very high energy.
- The wavelength of the γ-ray photons is characteristic of the nuclide from which they have been emitted.
- γ-radiation is indistinguishable from X-rays or cosmic rays of the same wavelength – it gets a separate name simply because it has a different origin from the other types of radiation. (**Rain** is water falling from sheets of stratus clouds, a **shower** is water falling from lumpy cumulus clouds. Their different names simply reflect their different origins. Both make you just as wet!)
- γ-radiation produces very little ionisation and, consequently, is extremely penetrating.
- In all cases of radioactive decay, the **energy** released in the decay comes from the direct conversion of a quantity of **matter** into energy. Such a large amount of energy is released that there is a measurable change in the mass of the particles involved.

Relative effects of magnetic fields on radiation

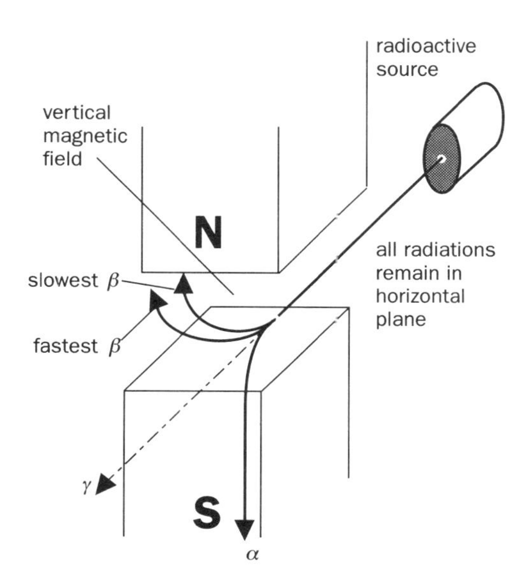

Figure 5.19 Effect of a magnetic field

Figure 5.19 shows the effect of a **vertical** magnetic field on a **horizontal** beam consisting of α- and β-particles and γ-rays. Since the electromagnetic force exerted on a charged particle is perpendicular to the direction of travel, the charged α- and β-particles will follow a circular path in a **horizontal plane**. The direction of bending shows that they have opposite charges. The radius of curvature of the track gives information about the mass and velocity of the particles. The γ-radiation, being uncharged, is unaffected by the magnetic field and continues in a straight line.

This demonstration would only be practicable using separate sources: the magnetic field required to deflect the α-particles by a measurable amount would cause massive deflection to the β^--particles. This is because the α-particles are thousands of times more massive.

Background radiation

Whenever a radiation detector is switched on, it will always indicate the presence of radioactivity, even if there are no radioactive sources nearby. The detector is measuring the **background radiation**, which is always present. This background radiation has many sources, including cosmic radiation, which originates outside the Earth, the presence of natural radioactive materials in the rocks around us and a small amount of emissions from the medical, military and industrial applications of nuclear physics. The background count rate should be constant for a particular place, but will show considerable random fluctuations. It should be determined before the start of any experiment by measuring the count rate with no radioactive sources present. The background count rate will be low, so a long time interval should be used. This background count rate should then be subtracted from all count rates measured during the experiment, to give the **corrected count rate**.

Radiation hazards

The term 'radiation' can be applied to any situation in which energy is distributed from some point. In everyday language, this word is often used to mean only hazardous radiation, although, clearly, most forms of radiation (sound, light, etc.) are purely beneficial. **Nuclear** radiation specifically refers to the emission of particles and photons from the nucleus, which are indeed hazardous. The danger with these radiations is that they cause **ionisation** – electrons are displaced from molecules due to interactions with the radiation. Such ionisation occurring in living tissue can lead to damage to cells, which may eventually cause disruption to the normal function of that tissue, leading to illness and, possibly, death. It must be stressed that α-particles and β-particles are merely pieces of ordinary matter. The danger they pose is entirely due to the energy (due to their velocity) with which they are emitted from their parent nuclei. Once halted by collisions with matter, an α-particle will merely pick up two electrons and become a helium atom – a harmless gas. β-particles are electrons, which are present in every atom. An analogy can be drawn with bullets: lying around on a table, these are harmless; only when fired from a gun do they become dangerous, and even then, only for the short time during which their velocity is high.

When using radioactive materials in the laboratory, the following guidelines must be followed.

Sources must:

- always be handled with forceps, never held in the hand
- always be held so that the open window is pointed away from the body
- never be brought close to the eyes for inspection
- always be returned to their container immediately after use.

Refer also to the safety instructions which should be displayed in your laboratory whenever radioactive sources are being used.

Nuclear radiation is a hazard to the human body either as a result of exposure to external radiation or due to ingestion of radioactive matter. The effect of the radiation depends on:

- the type of radiation
- the site of the body affected and
- the dose of radiation received.

Alpha particles pose little threat from external sources: because of their low penetrating power they cannot penetrate the outer layers of skin. However if an α-emitter is inhaled, swallowed or otherwise ingested, then the α-particles can be very harmful, since they can directly affect living cells, especially in the lungs or intestines. Inside the body, cells do not have the protection of many layers of skin. The threat of ingested radioactive material applies to all radiations, but because of their energies and ionising ability, α-particles are particularly dangerous.

β-particles are more penetrating, but again most of their energy is absorbed in the outer skin layers. Adequate protection is offered by sheets of perspex or aluminium.

γ-radiation is of course highly penetrating and can affect tissues deep within the body. Substantial lead or concrete shielding is necessary for protection from external sources.

The main effects are burns, caused by heating effects, radiation sickness and hair loss, and death, depending on the radiation dose and the organs affected. Damage to the cells is due to ionisation, which can disrupt their delicate chemistry. Long-term illnesses such as cancers, leukaemia and eye cataracts may develop years after exposure. It is also possible that genetic damage may be inherited by future generations. However, to put the situation into context, the body is well-equipped to cope with ionisation. It is estimated that every cell in the body suffers damage of some kind every week, on average. The immune system normally copes with any abnormal cells and it is only when the damage is extensive or the immune system is weak that problems occur.

Every hour, each individual is affected by (on average):

- $^1/_2$ million cosmic rays (neutrons and secondary rays)
- 15 million disintegrations of potassium-40 atoms within the body
- 30 000 disintegrations of various atoms within our lungs
- 200 million γ-rays from rocks and soil.

Detection of radiation

Nuclear radiations are detected by observing their interaction with matter, since energy is transferred from the radiation to the detector, causing:

- **ionisation** of gas: this can give rise to an electric current in a **Geiger–Müller** tube or an **ionisation chamber**, or leave a visible track in a **cloud chamber** or **bubble chamber**. In addition, the creation of ions can encourage the passage of sparks across the gap between electrodes.
- **excitation**: certain substances can absorb energy from the nuclear radiation, re-emitting this energy in the form of a burst of visible light or other electromagnetic radiation, e.g. fluorescence of a phosphor in a scintillation counter (see Geiger and Marsden's α-particle scattering experiment).
- **chemical** changes: this can affect **photographic film**, which can then be developed to show a visible record. Film badges are worn by personnel in radioactive environments. The film is developed at the end of every shift, and the individual's total exposure to harmful radiation is carefully monitored.
- **charge-carrier mobilisation**: in **solid-state detectors**, nuclear radiation causes the release of extra charge-carriers in a semiconductor, which produces a measurable change in the electrical characteristics of the material.

The ionisation chamber

The ionisation chamber is basically a container with a central electrode. A potential difference is maintained between the container and the central electrode. For a current to pass, there must be some ions present in the container. The size of the current will depend on the number of ions available, which in turn will depend on the intensity of the radioactivity present in the chamber. The current may be as small as 1×10^{-12} A, so very sensitive methods need to be used to measure it. The DC (direct coupled) amplifier is one such device. This amplifies the tiny current flowing, so that it can be read on a standard meter (see section on Determining the half-life of radon-220, page 148, for a detailed description of an experiment using the ionisation chamber).

The Geiger–Müller (G–M) tube

Figure 5.20 shows the main features of a G–M tube together with its associated circuitry.

The G–M tube is, in effect, a small ionisation chamber. The major difference is that the **radioactive material** is not placed inside the chamber, as in the previous device, but the **ionising radiations** are able

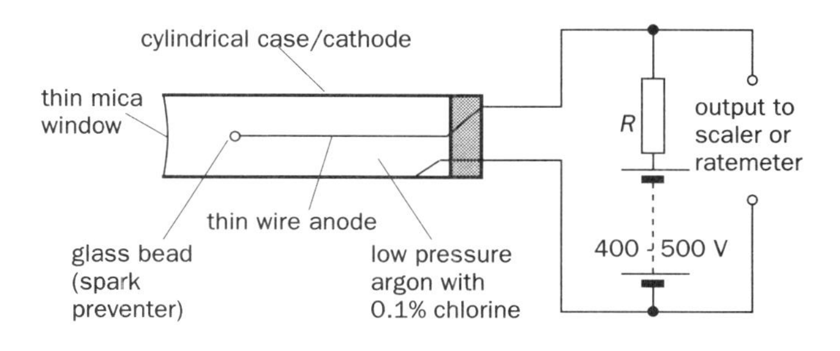

Figure 5.20 Geiger–Müller tube

to enter the tube through a thin mica window at one end. This mica window is gastight, but allows the α, β and γ radiations to enter. γ-rays can also penetrate the metal wall of the tube. This method of detection is sensitive enough to detect single ionising events.

The tube is filled with argon at low pressure (say 100 mm Hg). Inside the tube is a very thin wire central electrode (the anode), which is maintained at a p.d. of 400 to 500 V relative to the outside wall (the cathode), creating an intense electric field due to the small diameter of the wire. An ionising 'particle' entering the tube may produce an ion-pair from an argon atom. The electron of this pair will be rapidly accelerated towards the anode by the strong electric field, thus gaining kinetic energy from the electric field. The accelerated electron may then collide with argon atoms, and so create further ion-pairs. In this way, an 'avalanche' of electrons rushes to the anode. The resistance of the gas is said to have broken down. Meanwhile, the relatively heavy positive argon ions surrounding the anode wire will cancel out the electric field around it, so the avalanche stops, and no further current flows. Thus the effect of a single ionising event in the tube is to create a measurably large burst of electric current which abruptly ceases. (Perhaps 1×10^8 electrons may be liberated in less than 1 ms in this process, which is called **gas amplification**.)

Also in the tube is a small quantity of a '**quenching agent**', typically a halogen such as bromine. When the positive argon ions move away from the anode, they would eventually arrive at the cathode walls, and would have enough energy to liberate electrons, starting unwanted avalanches. The bromine atoms neutralise the argon ions by collisions before they reach the walls of the tube, thus preventing unwanted emissions of charge-carriers.

The G–M tube has a **dead-time** of around 200 μs due to the fact that the argon ions take this long to move far enough from the anode for the electric field to be re-established. During this time, no ionising particles can be detected, since avalanching cannot occur. There is also a **recovery time** (about 100 μs), during which the argon ions are being neutralised by the halogen quenching agent. Pulses are produced during this time, but are too weak to be detected. This limitation places a theoretical limit on the count rate of around 3000 counts per second. In practice, this is reduced to around 1000 counts per second since the emission of radiation is an irregular, random process.

The series resistor R in Figure 5.20 causes a voltage pulse to be generated whenever the anode current flows, and this is applied to a ratemeter (which indicates on a dial the number of counts per second) or a scaler (which simply records the number of counts). For accurate work, the scaler is used in conjunction with a stop-watch.

The characteristic curve of the Geiger–Müller tube
The response of the G–M tube depends on the applied p.d., since this controls the strength of the electric field inside the tube. To investigate the characteristic of the G–M tube, a β-source is placed a fixed distance from the tube, and the count rate measured as the p.d. is varied. Figure 5.21 shows the typical characteristic curve obtained by plotting the count rate against operating p.d. for a fixed source.

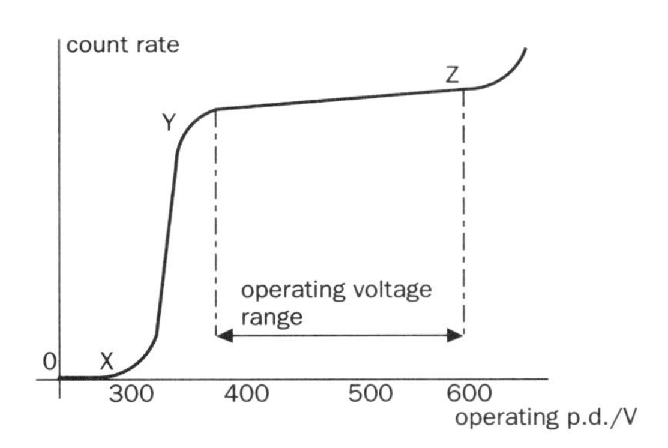

Figure 5.21 Characteristic curve of the G–M tube

When the p.d. is less than the threshold voltage (X) there is not enough gas amplification to produce detectable pulses. The region X–Y is the proportional region, where the strength of the pulse depends on the energy of the initial ionising radiation. The plateau region Y–Z is the normal operating region of the tube, since when the applied p.d. is within this range, every ionising radiation that enters the tube produces a complete avalanche along the length of the anode. This means that all the pulses have the same amplitude. If higher voltages are used, quenching becomes inefficient, and a continuous discharge may occur, due to the electrons liberated from the cathode by the unquenched argon ions.

The G–M tube may be connected to a **scaler**. This is simply a meter which counts the ionising events. We are generally interested in **count rate**, so a scaler must be used in conjunction with some timing device, such as a stopwatch, to enable count rates to be calculated. A **ratemeter** is a meter which indicates count rate, removing the need for a stopwatch.

The diffusion cloud chamber

When ionising radiation passes through a saturated vapour (e.g. alcohol), the ions produced act as centres on which the vapour condenses. Thus the track of a particle is revealed as a thin white trail of tiny liquid droplets, similar to the 'vapour trails' left by high-flying aircraft (these trails are in fact clouds of liquid droplets, not vapour).

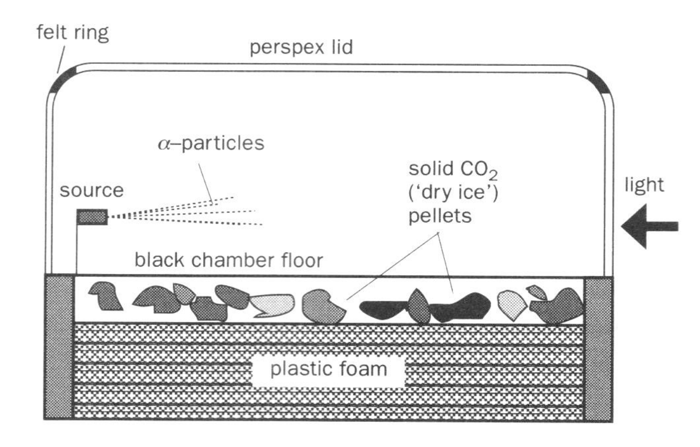

Figure 5.22 Diffusion cloud chamber

Figure 5.22 shows a cross-section of a typical simple diffusion cloud chamber, found in many school laboratories. The felt ring is soaked in alcohol, which vaporises and diffuses downwards due to the temperature difference maintained by the solid CO_2. The radiation from the source is revealed as white tracks just above the contrasting black floor of the chamber. Rubbing the perspex lid creates an electric field which helps to define the tracks clearly.

Nowadays, the cloud chamber has been replaced by the **bubble chamber**. Ionising radiations leave a trail of bubbles in a superheated liquid. Under the influence of a magnetic field, charged particles follow a curved path. Measurements of the curvature of the path yields information about the mass and velocity of the particles (see circular motion theory).

Path of a charged particle in a magnetic field As we have already seen in Section 3.5 on Electromagnetic induction, a moving charged particle experiences a force in a magnetic field. Since the force is always perpendicular to the direction of motion, the path of the particle must be circular, in the absence of any other forces. For a particle of mass (m), charge (q), moving with velocity (v) at a right angle to a magnetic field of strength (B), the force on the particle is given by:

$$F = Bqv$$

Since this force must be equal to the mass multiplied by the centripetal acceleration (mv^2/r):

$$\frac{mv^2}{r} = Bqv$$

So the radius (r) of the path of the particle is given by:

$$r = \frac{mv}{Bq}$$

If the magnetic field strength is uniform, the radius of the path depends on the mass, velocity and charge of the particle. If the trajectory of the particle can be made visible, for example in a cloud chamber or bubble chamber, or plotted by a computer connected to a wire chamber, information about the mass, velocity and charge of the particle can be deduced. This is the principle behind all of the modern discoveries of the particles which make the Universe.

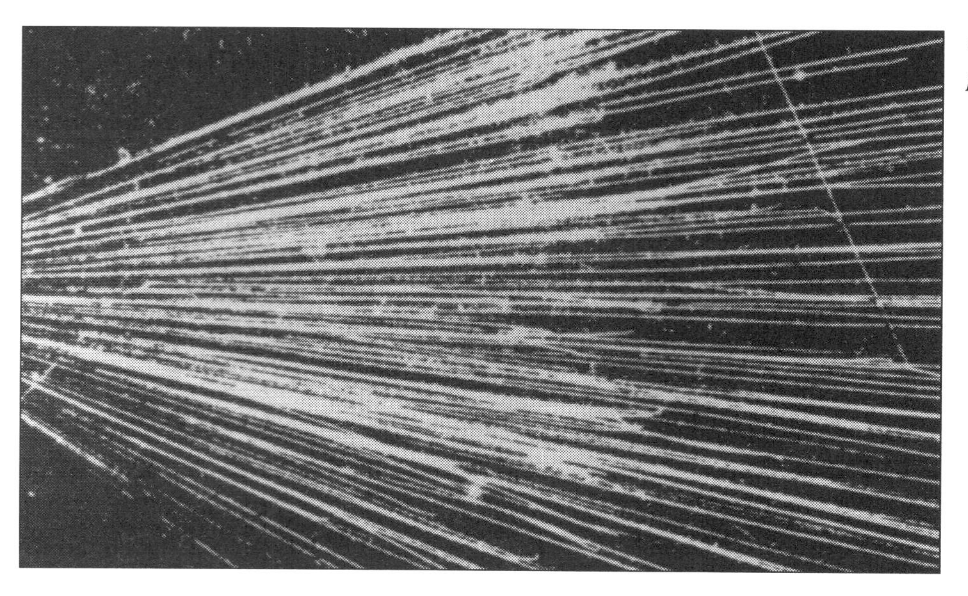

Cloud chamber tracks produced by α-particles

Appearance of cloud-chamber tracks

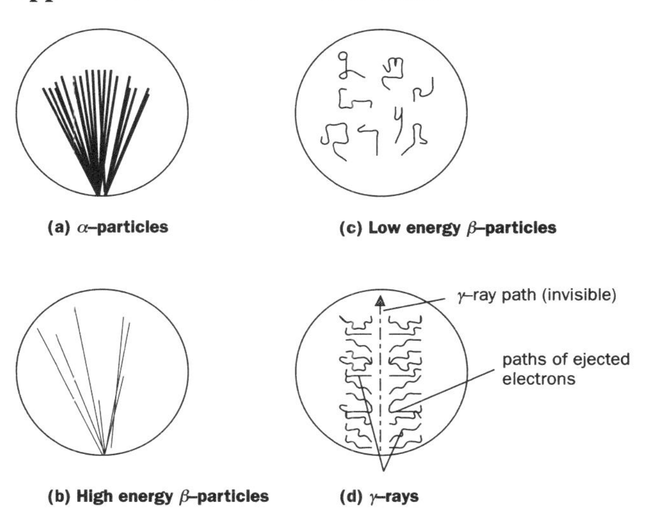

Figure 5.23 Cloud-chamber tracks

Figure 5.23 gives an idea of the types of tracks produced in a cloud chamber by different radiations.

You should note that:

- **α-particle** tracks are short, straight and very clear. This is because the particles produce intense ionisation and therefore dense droplet formation. The tracks from one particular nuclide will all have the same length, since α-particles are all emitted with the same energy from a given nuclide. The length of the track gives information about the initial energy of the particle. Higher energy means a longer track.
- Very fast **β-particles** produce thin, straight tracks as their ionisation is less intense than that of α-particles. The tracks from a given nuclide show great variation in length, and we say that β-emission shows a **continuous spectrum** of energies, unlike α-particles, which are **monoenergetic**. This observation is evidence for the existence of some additional particle, undetected by the cloud chamber. (This is the **neutrino**, which we will study in Section 5.5 on Fundamental forces.) Slower β-particles produce shorter, thicker, more tortuous tracks as they are easily deflected on collision with matter.
- **γ-rays** are uncharged, and only produce ions when they collide with matter (this does not happen very frequently). Electrons ejected by collisions produce ionisation, and the tracks of the ejected electrons may be seen.

The inverse square law for gamma radiation

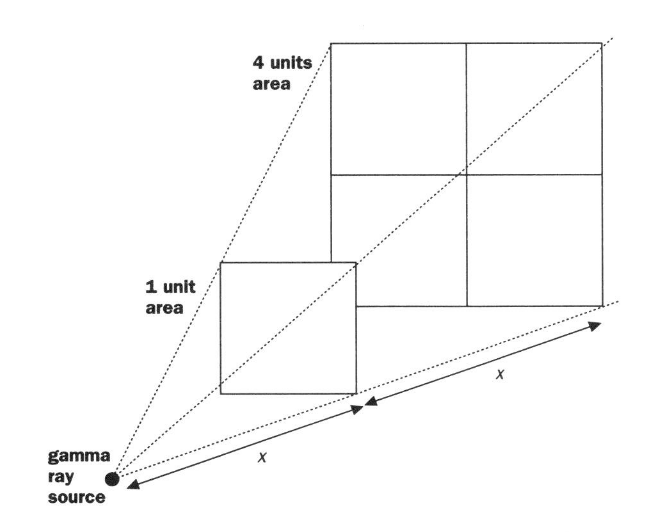

Figure 5.24 Inverse square law for radiation from a point source

Gamma radiation (and all forms of radiated energy) emitted from a point source spreads out as shown in Figure 5.24. Consider the radiation passing through an opening of unit area at a distance x from the source. At a distance $2x$ from the source, this amount of radiation covers an area 4 times as great. Thus its intensity has decreased to a quarter. In general the intensity of the radiation is inversely proportional to the **square** of the distance from the source. The inverse square law can be investigated using the apparatus shown in Figure 5.25.

The background count rate is first determined without the gamma source present. The source is then assembled in the holder, following all radioactive safety procedures. Starting with the source close to the window of the G–M tube, the count rate is measured

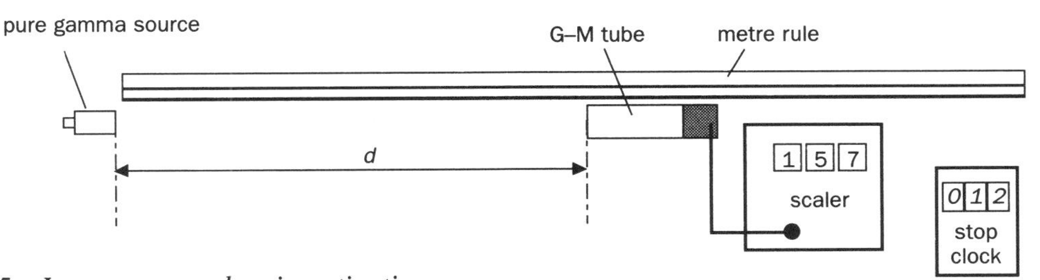

Figure 5.25 Inverse square law investigation

and recorded, using the appropriate techniques for minimising experimental errors. This procedure is repeated, varying d in steps of 4–5 cm until very low count rates are obtained. The sampling time interval should be varied as the count rate falls, to keep the percentage error consistent. Short periods of a minute or two will be fine for very high count rates, but should be increased for low count rates. The **corrected count rate** (R) is calculated by subtracting the background count rate.

If the relationship is of the form: $R \propto \frac{1}{d^2}$

then $R = \frac{k}{d^2}$

Equation 1

where R = corrected count rate

d = distance from source

k = constant of proportionality

then $\sqrt{R} = \frac{\sqrt{k}}{d}$ $\quad \therefore \quad d = \frac{\sqrt{k}}{\sqrt{R}}$

The source is located deep within its lead container (see Figure 5.26), and it would be dangerous to try to locate its position by direct measurement.

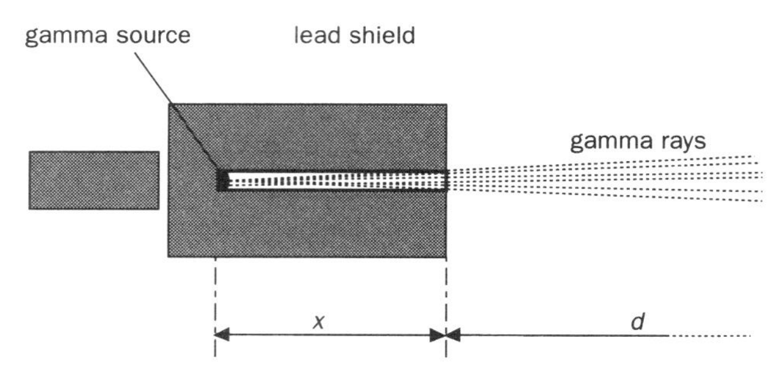

Figure 5.26 Measurement error in inverse square law investigation

Measuring the distance (d) between the end of the source container and the end of the G–M tube gives rise to a significant error, since the true distance required is the distance between the source and the G–M tube ($d + x$). This is a constant or **systematic** error, and can be dealt with mathematically.

Maths window

If there is an error x in measuring d, then d should be replaced by $(d + x)$ in the equation:

$$(d + x) = \frac{\sqrt{k}}{\sqrt{R}}$$

Maths window *continued*

$$\therefore \quad d = k.\frac{\sqrt{1}}{\sqrt{R}} - x$$

which can be compared with:

$$y = mx + c$$

the equation of a straight line.

So if d is plotted on the y-axis against $1/\sqrt{R}$ on the x-axis, a straight line of gradient ($\sqrt{k}$) and intercept ($-x$) on the d axis should be obtained (see Figure 5.27), which shows that the intensity of gamma rays is inversely proportional to the square of the distance from the source, i.e. an 'inverse square relationship'.

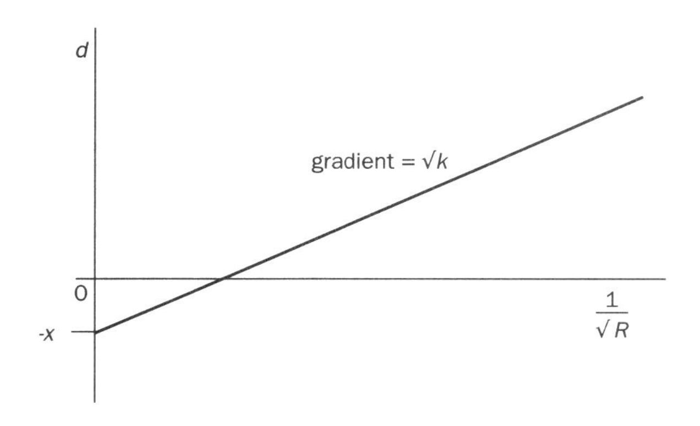

Figure 5.27 Results of inverse square law investigation for γ-radiation

α- and β-radiation would also obey this relationship, if they were emitted from a point source in a vacuum. In the laboratory, absorption of the particles by air means that the intensity of the radiation decreases even more rapidly with distance. Air is virtually 'transparent' to γ-radiation.

Guided example (1)

A sample of a radioisotope emits 1.2×10^{12} γ-photons per second in all directions. Assuming the cross-sectional area of a person is around 1 m^2, estimate the number of photons per second received by someone standing

(a) 2 m

(b) 4 m

(c) 5.5 m

from this sample, if they were accidentally exposed. (The area of a sphere, radius (r), is $4\pi r^2$.)

Guided example (1) *continued*

Guidelines

The radiation is emitted in all directions, so you first need to work out the area of the **spherical surface** at the appropriate distance from the source.

(a) Calculate the area of a sphere of radius 2 m, then calculate what fraction of this area is covered by the person, of area 1 m^2. This is the fraction of the total emitted radiation received.

(b) The distance has doubled. Since the variation of intensity of γ-radiation with distance follows an inverse square law, we can say that the intensity at 4 m will be **one-quarter** of that received at 2 m.

(c) 5.5 m is not a simple multiple of any of the previously-used distances, so a more formal method is needed. Use Equation 1, where d is 2 m and R is the intensity calculated in (a), to calculate the constant of proportionality (k). Then calculate the new value of R at 5.5 m.

Radioactive decay

This is an attempt by an unstable nucleus to become more stable by emitting particles and/or energy. The emission of nuclear radiation is spontaneous, and is unaffected by temperature, pressure, chemical combination or any other external influence. In an emission of nuclear radiation, the energy released by the atom may be millions of times greater than the energy involved in any chemical change.

The disintegration of a nucleus (i.e. the emission of α-, β- or γ-radiation) is random and haphazard – it is impossible to predict when an individual nucleus will decay. However, since there are always huge numbers of nuclei present in any appreciable amount of radioactive material, then the methods of statistics can be applied. This enables us to calculate what fraction of a substance will have decayed, on average, in a given time interval.

The radioactive decay law

The **rate** of disintegration of a given nuclide at any time is directly proportional to the number (N) of nuclei of that nuclide present at that time:

i.e. $$-\frac{dN}{dt} \propto N$$

The minus sign indicates that N decreases as t increases.

$$\therefore \quad \frac{dN}{dt} = -\lambda N$$

Equation 2

where λ is a positive constant, called the **radioactive decay constant**.

If the mass of **1 mole** of a substance is A_r, then a mass (m) of the substance will contain N molecules,

where $$N = \frac{mN_A}{A_r}$$

so Equation 2 can be written:

$$\frac{dN}{dt} = -\lambda m \frac{N_A}{A_r}$$

It can be shown that Equation 2 can be written as:

$$N = N_0 e^{-\lambda t}$$

Equation 3

where: N_0 is the number of nuclei at time $t = 0$
e is the base of natural logarithms (= 2.718)

Alternative forms are:

$$\ln N - \ln N_0 = -\lambda t$$

$$\ln\left(\frac{N}{N_0}\right) = -\lambda t$$

$$\frac{N}{N_0} = e^{-\lambda t}$$

These are all useful forms of Equation 3.

YOU MUST BECOME TOTALLY FAMILIAR WITH THE USE OF THE NATURAL LOGARITHM BUTTONS ON YOUR CALCULATOR!

They may be marked: [ln] or [$\log_e$] for the logarithm button

and [e^x] for the antilog button

You should note that:

- Equation 3 shows that a radioactive substance decays **exponentially** with time.
- The **rate of decay** of a substance (dN/dt), or the number of disintegrations per second of a source, is called its **activity**, and it is expressed in units called **becquerel** (Bq). 1 Bq is equal to 1 decay (or disintegration) per second. (An older unit is the **curie** (Ci). 1 Ci is equal to 3.70×10^{10} disintegrations per second.)

- The fraction N/N_0 is the fraction of the original number of nuclei remaining. It will be identical to
  R/R_0 where R represents count rate; or
  I/I_0 where I represents ionisation current, etc.

Radioactive decay constant (λ) is defined as the **fraction** of the total number of nuclei present which decays per **unit time**, provided the unit time is small.

- For substances with half-lives measured in hours, or longer, λ would be simply the fraction decaying per second, since 1 second is a small unit of time compared to the half-life. Thus the unit of λ would be second^{-1} (s^{-1}). However, many nuclides have half-lives measured in milliseconds or even microseconds, so the value of λ would have to be the fraction decaying per nanosecond (ns^{-1}), or some other even smaller unit of time.
- The exponential nature of radioactive decay can be shown graphically by plotting the **activity** of a sample of material (which is proportional to the number of nuclei remaining) against **time** (see Figure 5.28).

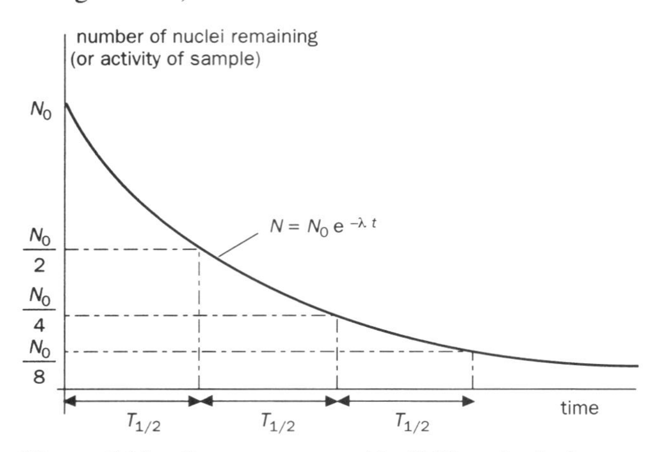

Figure 5.28 Decay curve and half-life calculation

The **half-life** ($T_{1/2}$) of a radioactive sample is defined as either:

the **time** taken for **half** of the **number** of radioactive nuclei present to decay

or:

the **time** taken for the **activity** of a sample to decrease to **half** of some initial value.

- The idea of half-life is incorporated into Figure 5.28 where $T_{1/2}$ is the time taken for the number of active nuclei to reduce from N_0 to $N_0/2$, or from $N_0/2$ to $N_0/4$, etc.
- Half-life values are very wide ranging, from nanoseconds to millions of years and beyond. The half-life is unique to each isotope, and a measurement of this value is useful in identifying specific radioactive isotopes.

Determining the half-life of radon-220

Radon-220 is a radioactive gas, produced as part of the decay chain of the solid material thorium-232, and is a convenient material for laboratory study of half-life. The thorium is contained in a plastic bottle, and the radon-220 produced during its decay collects in the air space in the bottle. Radon-220 emits an α-particle, transmuting into polonium-216. (Polonium-216 is itself an α-emitter, but of such short half-life that it has no effect on the measurement of the half-life of the radon-220. Effectively, the radon behaves as if it emits two α-particles simultaneously.) Typical apparatus for measuring the half-life of radon-220 is shown in Figure 5.29.

The α-particles produced from these decays cause massive ionisation in the air inside the ionisation chamber. The ion-pairs produced migrate to the

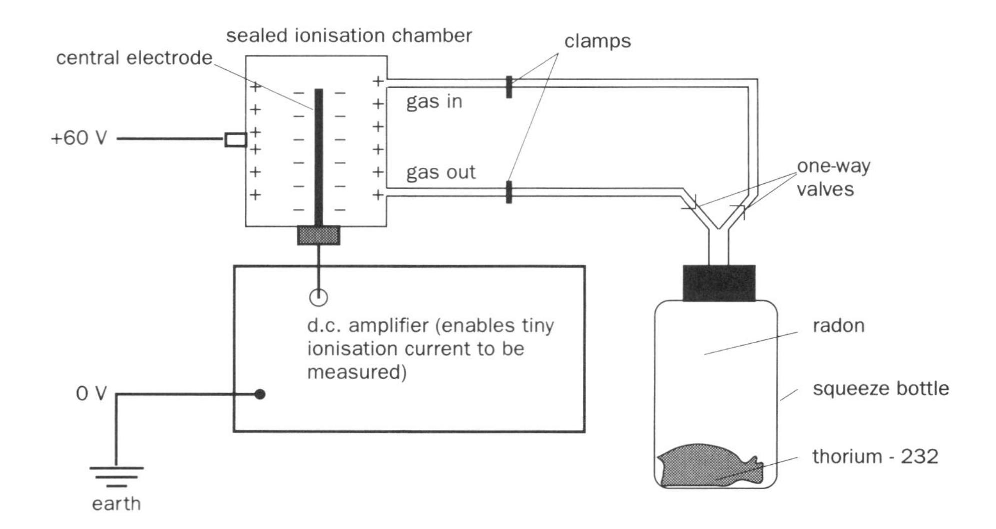

Figure 5.29 Measuring the half-life of radon-220

electrodes because of the applied p.d. This tiny current (as low as 1×10^{-11} A) is measured using the d.c. amplifier. Since the ionisation current will be directly proportional to the activity of the radon, measurements of the ionisation current will also be directly proportional to the number of nuclei of radon-220 present in the chamber at any instant.

The ionisation chamber is supplied with radon gas by squeezing the plastic bottle with the clamps open. Once filled, the clamps are closed to seal the quantity of gas in the chamber. The tiny current flowing between the canister and the central electrode, due to ionisation of the air by the emitted α-particles, is measured using the DC amplifier, and will be seen to decrease with time. The ionisation current is recorded every 10 seconds or so, until a low value is reached, and a graph of ionisation current against time can be plotted. From this graph, several estimates of the half-life can be made in the usual way.

Guided example (2)

The following data were obtained from an ionisation chamber containing radon-220 as described above:

time /s	0	20	40	60	80	100	120	140	160	180
current /nA	1.0	0.79	0.58	0.47	0.36	0.29	0.20	0.16	0.13	0.10

Determine the half-life.

Guidelines

Plot a graph of ionisation current against time. Choose any starting value of current, and determine the time at which this current occurred and the time taken for the current to fall to **half** of the starting value. **Use the curve** to determine the times, not the actual data points. Repeat for two more starting values, and calculate the average value of the half-life.

Relationship between half-life ($T_{1/2}$) and radioactive decay constant (λ)

Consider a sample of material, which contains N_0 undecayed atoms at time $t = 0$.

After 1 half-life ($t = T_{1/2}$), the number of undecayed atoms remaining (N) has been reduced by half:

i.e. $$\frac{N}{N_0} = \frac{1}{2}$$

Since $$\ln \frac{N}{N_0} = -\lambda t$$

then in this case:

$$\ln \frac{1}{2} = -\lambda T_{1/2}$$

$$\therefore \quad -0.693 = -\lambda T_{1/2}$$

$$\therefore \quad T_{1/2} = \frac{0.693}{\lambda}$$

Equation 4

Remember that the units of $T_{1/2}$ and λ must be consistent (e.g. if $T_{1/2}$ is measured in seconds, then λ will be in s^{-1}). In general, it is probably better to work in seconds, converting half-lives where necessary.

Guided examples (3)

1. A radiographer calculates that a patient is to be injected with a tracer containing 1×10^{18} atoms of iodine-131 in order to monitor the activity of his thyroid. If the half-life of iodine-131 is 8 days, calculate

(a) the radioactive decay constant

(b) the initial activity (disintegrations per second)

(c) the number of undecayed atoms of iodine-131 after 24 days

(d) the total activity after 3 days.

Guidelines

(a) Convert the half-life to seconds, then use Equation 4 to calculate the radioactive decay constant (λ).

(b) The initial activity (dN/dt) is calculated using Equation 2. Since λ has been calculated in second^{-1} (as the half-life was converted to seconds), this gives the answer in disintegrations second^{-1}.

Guided examples (3) *continued*

(c) 24 days is exactly **three** half-lives, so the number of atoms of iodine-131 remaining will have halved **three** times. If N_0 is the initial number of atoms, then the number remaining (N) is given by:

$$N = N_0 \times \left(\frac{1}{2}\right)^3 = N_0 \times \frac{1}{8}$$

(d) 3 days is not a convenient multiple of the half-life, so the decay law (Equation 2) must be applied. Convert 3 days to seconds. The new activity (A) is given by:

$$A = A_0 e^{-\lambda t}$$

where A_0 is the initial activity (answer (b)), λ is the radioactive decay constant, t is the time (3 days) converted to seconds.

2. In an experiment to measure the half-life of radon-220, the initial ionisation current is 1.0×10^{-9} A. Assuming that each α-particle is emitted with an energy of 6.3 MeV, and that creating an ion-pair from the molecules of air in the chamber requires 30 eV, calculate

(a) the initial velocity of the emitted α-particles
(b) the number of ion-pairs produced per second
(c) the number of α-particles emitted per second
(d) the number of atoms of radon-220 initially present in the chamber.

Mass of an α-particle = 6.8×10^{-27} kg
Charge on an α-particle = 3.2×10^{-19} C
Charge on an electron = 1.6×10^{-19} C
Half-life of radon-220 = 55 s

Guidelines

(a) The initial **kinetic** energy of the α-particle can be assumed to be 6.3 MeV. Convert this to joules, and calculate the velocity (v) using

$$\text{kinetic energy} = \frac{1}{2} mv^2$$

where m is the mass of the α-particle.

(b) The ionisation current is due to the creation of ions in the chamber. Calculate the charge flowing per second (*current = charge flowing/time taken*) and from this, the number of electrons flowing per second (each electron contributes 1.6×10^{-19} C of charge). This is equal to the number of ion-pairs produced per second.

(c) Each ion-pair requires 30 eV of energy. Assuming all of the α-particle's 6.3 MeV of energy is used in ionisation, calculate the number of ion-pairs produced by each particle, and use this to calculate the number emitted per second.

(d) Answer (c) is the **decay rate** (dN/dt) of the radon-220. Use Equation 2 to calculate N, the number of nuclei present.

Useful applications of radioactivity

Tracers

A tracer is a radioisotope that is added to a non-radioactive material in order to trace or monitor its position or measure its concentration. Medical staff have a constant need to be able to observe what is happening deep within the human body. For example, they may wish to monitor the chemical activity of the brain and other organs, or study the internal blood supply deep inside bones. Surgery is often inappropriate, and what is needed is something which can transmit information from within the organ itself. Radioactive materials emit radiation which can penetrate solid matter. This means that a radioisotope can be used as a 'label', which can enter the tissue being studied. The emitted radiation is then monitored, which can be used to provide information about the function of that part of the body. A common technique is to inject the patient with a measured quantity of an element which is selectively absorbed by the organ in question. A few of the element's atoms will have been replaced by a radioactive isotope of the same element, which has exactly the same chemical properties. Very tiny quantities of the radioisotope are needed. Iodine is absorbed by the thyroid gland, and the radioactive isotope iodine-131 is used. This is a γ-emitter (also a β-emitter) with a half-life of just over 8 days. The radioactivity is measured externally, using a 'gamma camera' (a sophisticated solid-state detector which measures both the energy and direction of the γ-photons), and can indicate over- and under-activity of the thyroid.

To monitor the blood flow and oxygen take-up in the brain, a radioactive 'tracer', such as oxygen-15, is injected. This is a **positron** emitter (see Section 5.5 on Fundamental particles for more details of this mode of emission). The positrons are antimatter, and annihilate completely with any electrons they encounter, emitting γ-photons which can be monitored as above.

Industrial uses of tracers include the detection of leaks in underground pipes. Figure 5.30 below shows how the fluid (containing the radioactive tracer) leaking from the pipe produces a measurable increase in γ-ray detection immediately above the site of the leak. This saves the expense of digging up the entire length of pipe.

To monitor the wear of piston rings in a car engine, the steel rings can include a radioactive isotope of iron. As the rings wear, fragments are carried away by the engine oil. Measuring the radioactivity of the engine oil gives information about the rate of wear of the rings.

Tracers have to be chosen carefully: the radiation emitted must be detectable externally, which means γ-radiation; the half-life must be long enough to enable measurements to be made, which could take several days in the case of tracing leaks, or even weeks when monitoring engine wear, but short enough so that the radiation hazard is reduced to negligible amounts as quickly as possible, especially in medical applications or where there is an environmental pollution hazard.

Radioisotope dating

All organic matter contains carbon, and living plants are constantly exchanging carbon with the atmosphere through photosynthesis and respiration. All other life ingests this carbon from the plants, either directly or indirectly. The radioactive isotope carbon-14 is present in the atmosphere due to the interaction of cosmic radiation with nitrogen. The carbon-14 decays back to nitrogen by β-emission, with a half-life of 5730 years. The rate of production of carbon-14 is in equilibrium with its rate of decay, so the concentration of carbon-14 in the atmosphere, and therefore in living organisms, is constant. Once a plant or animal dies, however, it stops exchanging carbon with the atmosphere, so its carbon-14 content gradually decreases. After 5730 years, the number of carbon-14 atoms will have decreased to half the original amount. Any material manufactured from dead organisms, such as wooden structures, or even the dead organism itself can be dated by measuring the amount of carbon-14 in a sample of material, and comparing this with what would be expected from a fresh sample today. The material is oxidised to produce carbon dioxide, and the number of carbon-14 atoms counted using a mass spectrometer. One gram of carbon from a freshly-prepared source should give a disintegration rate of around 14 disintegrations per minute.

There are several other radioactive materials which can be used for dating. Inorganic materials like rocks can be dated by applying a similar technique to the decay of rubidium-87, which is present naturally in some igneous and metamorphic rocks. Tritium (hydrogen-3) is formed by interaction of hydrogen with cosmic rays. It decays to helium-3, and can be used to date samples of water from deep underground sources.

Guided example (4)

A piece of wood is taken from an archaeological artefact. 10 mg of carbon is extracted from the wood, which yields 66 counts over a period of 12 hours from the carbon-14 present in the sample. If the equilibrium decay rate of carbon-14 is 14 disintegrations per min per gram of freshly-prepared carbon, calculate the age of the wood. (Half-life of carbon-14 = 5730 years.)

Guidelines

The equilibrium decay rate is expressed in disintegrations per **minute**, per **gram** of carbon. You are given the number of counts from a sample of only 10 mg (0.010 g) over a time interval of 12 hours.

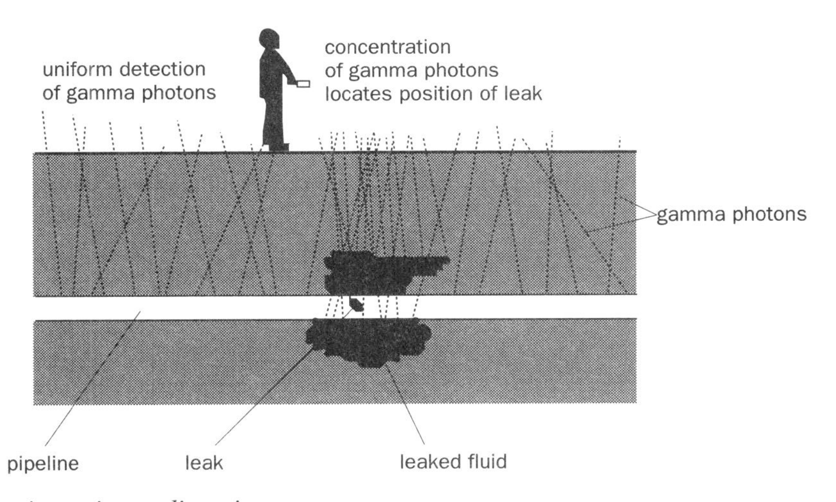

Figure 5.30 Leak location using radioactive tracer

Guided example (4) *continued*

Use these figures to calculate the number of counts per minute which would be expected from 1 g of this sample. Apply Equation 2 in the form

$$\ln\left(\frac{C}{C_0}\right) = -\lambda t \text{ to calculate } t$$

where C_0 = equilibrium count rate
C = actual count rate
λ = radioactive decay constant (calculated from half-life)
t = age of the sample

Cancer treatment (radiotherapy)

Biological tissues can be damaged when ionising radiation breaks the DNA chains in cells, thus disrupting the normal function of the cell. The cell may die completely, or may continue to grow and divide in some abnormal way. In the radiological treatment of cancers, nuclear radiation is directed at the tumour, in order to kill its cells. Rods or pellets containing β-emitting radioisotopes may be inserted directly into the tumour, or beams of gamma radiation from a heavily-shielded source outside the body may be concentrated on the tumour deep within the body. Experiments are continuing into the use of high-energy neutron beams and pions. The results are promising, but the apparatus needed to produce these exotic beams of particles is cumbersome and expensive.

Disposal of radioactive waste

Isotopes with short half-lives are highly active, but their activity decays rapidly, so short-term isolation is all that is required to reduce the radiation hazard to an acceptable level. Those with long half-lives, however, present a different problem: their activity may be low, but in large quantities their emissions will still be a serious threat. Moreover, their activity could take thousands of years to decrease significantly. Consequently the problem is not one of containing a highly-active source, but is that of finding safe storage for thousands of years for a growing quantity of material. The storage facility must of course contain the emitted radiation, but more importantly it must prevent the radioactive material from reaching the environment: e.g. water supplies, the atmosphere or oceans where it could then be circulated. Thus it must be resistant to corrosion, wear and tear etc. and well away from any threat of disruption due to natural phenomena such as earthquakes – a formidable engineering problem! (see also Section 5.4, page 441)

Self-assessment

SECTION A
Qualitative assessment

1. Compare the
   (a) penetrating power
   (b) ionising ability
   (c) behaviour in a magnetic field
   (d) nature
   of α-particles, β-particles and γ-radiation.

2. Write nuclear equations for the following decays:
   (a) emission of a β-particle from carbon-14 $\left({}^{14}_{6}C\right)$
   (b) emission of an α-particle from radon-220 $\left({}^{220}_{86}Rn\right)$.

3. Briefly describe **four** interactions between ionising radiation and matter, which can form the basis of detection instruments.

4. Suggest possible origins of **background radiation** and describe how this is allowed for in radioactivity experiments.

5. (a) Describe the principles behind the Geiger–Müller tube as a detector of ionising radiations.
   (b) Sketch a typical characteristic curve for a G–M tube, identifying the threshold voltage, proportional region and normal operating range.
   (c) State what is meant by **quenching**, **dead time** and **recovery time** of a G–M tube.

6. Describe the principles of operation of a **diffusion cloud chamber**, stating clearly how visible tracks are produced when an ionising radiation passes through the chamber. Sketch the appearance of tracks produced by:
   (a) α-particles
   (b) high energy β-particles

Self-assessment *continued*

(c) low energy β-particles

(d) γ-radiation.

7. A γ-radiation source of a certain activity produces an intensity (I) of radiation at a distance of 1 m from the source. State the intensity produced by a source of **twice** the activity, at a distance of **2 m**.

8. Define the **half-life** of a radioisotope. A sample of a radioisotope has an initial corrected activity of 400 Bq. If the half-life of the isotope is 5 min, sketch a graph of activity against time for this sample, covering a time interval of 20 min.

9. Define the **radioactive decay constant** (λ) of a radionuclide, and state how it is related to its half-life.

10. Explain the meaning of each term in the following equations:

(a) $$\frac{dN}{dt} = -\lambda N$$

(b) $$N = N_0 e^{-\lambda t}$$

11. Describe the effects of ionising radiation on the human body, and describe the relative dangers posed by α-, β- and γ-radiation.

12. Briefly describe the hazards to the environment posed by the disposal of radioactive waste material.

SECTION B
Quantitative assessment

(***Answers:*** 8.5×10^{-9}; 1/8; 1/8; 15/16; 63/64; 2.9; 3.0; 6.0; 10; 12; 13; 41; 1.2×10^3; 4.2×10^3; 6.9×10^5; 3.7×10^{12}; 4.7×10^{12}; 9.2×10^{13}; 1.1×10^{22}.)

1. A Geiger–Müller tube has a window area of 1.8 cm^2. When it is placed 4 cm from a β-emitter, 6200 counts are recorded in one minute (after correction for background count rate). Estimate the activity of the source in counts min^{-1}. Explain why the true activity of the source will be higher than your estimate.

2. 5.0 m from an industrial γ-radiation source, the count rate recorded by a Geiger–Müller tube is 20 s^{-1}. What is the closest allowable distance from the source if the count rate is not to exceed 60 s^{-1}?

3. A radioisotope of silver has a half-life of 20 minutes.

(a) How many half-lives does it have in 1 hour?

(b) What fraction of the original nuclei would remain undecayed after 1 hour?

(c) What fraction of the original nuclei would have decayed after 2 hours?

4. Taking the half-life of radium-226 to be 1600 years:

(a) What fraction of a given sample remains undecayed after 4800 years?

(b) What fraction has decayed after 6400 years?

(c) How many half-lives does it have in 9600 years?

5. In an experiment to determine the half-life of a radioactive gas, the following results were obtained:

Background count rate = 1.8 s^{-1}

Time from start (s)	0	30	60	90	120
Count rate (s^{-1})	32	20	13	8.3	5.8

Determine the half-life of the gas.

6. The half-life of iron-55 is 2.6 years. Find

(a) the radioactive decay constant (λ)

(b) the number of atoms in a sample of 1.0 g of iron-55 (Avogadro constant 6.0×10^{23} mol^{-1})

(c) the initial number of disintegrations per second for 1 g of iron-55.

7. The half-life of bismuth-212 is 61 min. How long does it take for the activity of a sample to decrease to 80% of its initial value?

8. The half-life of copper-64 is 13 hours. Find the percentage loss of activity which occurs in a sample in 2.0 hours.

9. The equilibrium concentration of carbon-14 in living plants gives 16 disintegrations per minute per gram of carbon. If a piece of cloth is thought to have been made in 0 AD, estimate the activity (in disintegrations per minute) of 1 g of carbon prepared from the cloth. The half-life of carbon-14 is 5.6×10^3 years.

10. A tiny fragment of a valuable archaeological artefact yields 0.007 g of carbon. Using the data from Question 9, estimate its age (in years) if this sample of carbon yields 20 counts in 5.0 hours.

Self-assessment *continued*

11. A radioisotope having a half-life of 6 days is used as a medical tracer. Calculate

(a) the number of atoms of the radioisotope present in a patient to give a total disintegration rate of 5×10^6 s^{-1}, and

(b) the number of atoms of the radioisotope which must have been present in a dose prepared 2 days before.

12. A point source of γ-radiation has a half-life of 20 min. A Geiger counter placed at a distance of 4.0 m from the source records a corrected count rate of 460 s^{-1}. One hour later a new corrected count rate of 6.0 s^{-1} is obtained, but the distance between the source and counter had been altered accidentally. Calculate the new distance between the source and counter.

5.3 The nucleus

Measuring energies on the atomic scale

The electronvolt

From the fundamental definition of p.d. we know that it is the *work done* (or energy converted) per unit charge. In TV tubes, X-ray tubes, etc. electrons are accelerated by applying a potential difference. Since the charge on the electron is fixed, the *energy* gained by an electron is proportional to the p.d. applied. Figure 5.31 shows an electron of charge (e) being accelerated by a p.d. (V).

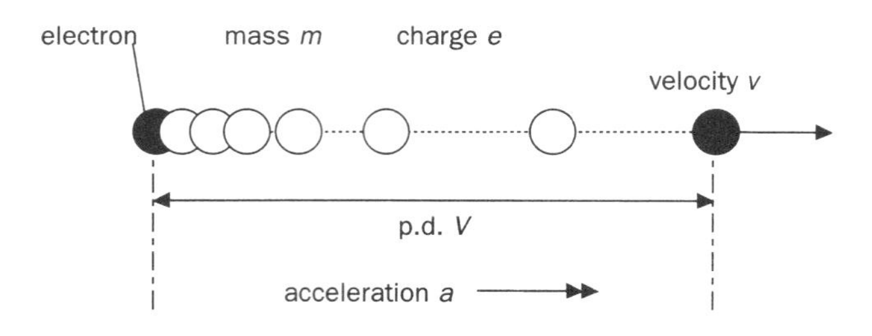

Figure 5.31 An electron being accelerated by an electric field

The work done (W) on the electron (i.e. the energy gained by the electron) is given by:

$W = eV$

W will be in **joule** (J) if e is measured in **coulomb** (C) and V in **volt** (V).

Worked example (1)

An electron is accelerated through a p.d. of 100 kV in an X-ray tube. Calculate the energy gained by the electron.

$W = eV$

$\therefore \quad W = 1.6 \times 10^{-19} \times 100000 = 1.6 \times 10^{-14}$ J

Thus even at very high p.d. the energy gained by an individual electron is very small indeed.

A more convenient unit is obtained by treating the charge on the electron (or proton) as 1 **unit** of charge. This gives us a new unit of energy – the **electronvolt**. In the above example, the energy gained by the electron moving through a p.d. of 100 kV is:

$W = 1\ e \times 100000\ \text{V} = 100000\ \text{eV}$

We can define the **electronvolt** as the quantity of energy gained by a particle carrying a charge equal to the charge on an electron when it is accelerated through a p.d. of 1 V.

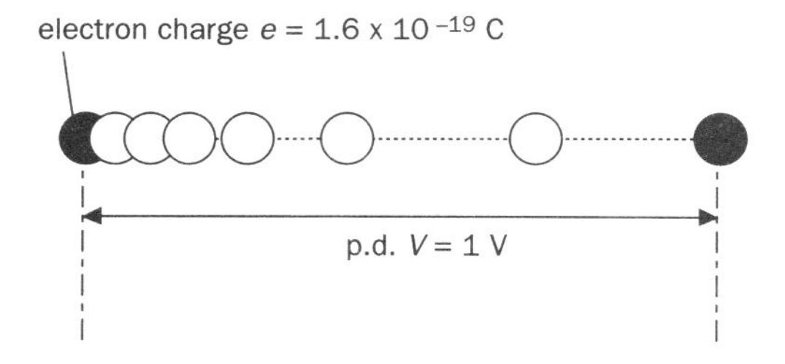

From the diagram, 1 eV is clearly equal to 1.6×10^{-19} J, and is a very convenient unit to use when dealing with atomic and nuclear particles.

How big is the nucleus?

In Section 5.1, we saw how Rutherford's α-particle scattering experiment revealed the structure of the nucleus. If we know how much energy is possessed by an α-particle, we can calculate the **closest distance of approach** of this α-particle to any particular nucleus. The closest possible approach will occur when the α-particle collides head-on with the target nucleus and rebounds along its original path, as shown in Figure 5.32 below. As the α-particle approaches, it will slow down due to the electrostatic repulsion between itself and the nucleus, thus its kinetic energy decreases, while the electric potential energy of the field increases. Ultimately, the α-particle will come to rest momentarily when the kinetic energy has been reduced to zero.

To calculate the closest distance of approach, we need to consider the kinetic energy lost by the α-particle and the electric potential energy gained by the electric field.

At closest approach:

kinetic energy lost by α-particle = potential energy gained by field

The kinetic energy of α-particles is usually expressed in **electronvolts**. If we consider α-particles ejected with an initial energy of 5.0 MeV (5 million electron volts), this is equivalent to:

$$5.0 \times 10^{6} \times 1.6 \times 10^{-19} = 8.0 \times 10^{-13}$$

$$\therefore \quad E_k = 8.0 \times 10^{-13} \text{ J}$$

(this is the kinetic energy lost)

The potential energy gained by the electric field is given by:

$$E_p = \frac{1}{4\pi\varepsilon_0} \frac{Q_\alpha Q_G}{r}$$

where Q_α and Q_G are the charges on the α-particle and gold nucleus respectively and

r is the closest distance of approach of the α-particle to the gold nucleus.

Since $E_p = E_k$

$$\therefore \quad r = \frac{1}{4\pi\varepsilon_0} \frac{Q_\alpha Q_G}{E_k}$$

Guided example (1)

In an α-particle scattering experiment using gold foil, the α-particles have an energy of 5.0 MeV. Given that the mass of the α-particle is 6.7×10^{-27} kg, and that its charge is 3.2×10^{-19} C, calculate:

(a) the velocity of the α-particles;

(b) the closest distance between the centres of a gold nucleus and an α-particle which suffers a head-on collision. (Proton number of gold = 79.)

Guidelines

(a) The kinetic energy (E_k) of the α-particle is given by:

$$E_k = \frac{1}{2} mv^2$$

v is the velocity of the α-particle.
(Remember to calculate the energy in joules)

(b) From above:

$$r = \frac{1}{4\pi\varepsilon_0} \frac{Q_\alpha Q_G}{E_k}$$

(Remember to calculate the charges in coulombs)

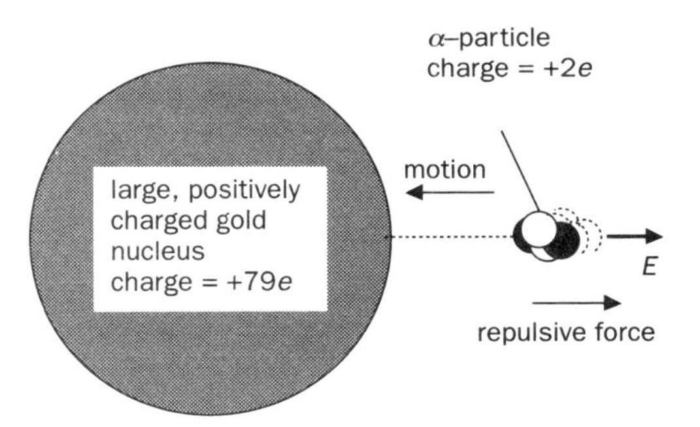

(a) In moving against the field, the charge–field system gains electric potential energy

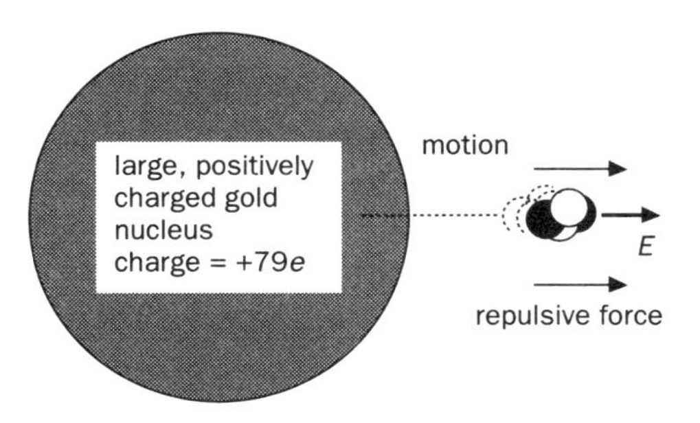

(b) When the charged particle moves in the direction of the electric field, the system loses its electric potential energy and gains kinetic energy

Figure 5.32 Head-on collision between an α-particle and a gold nucleus

You should note that:

- the particles do not physically touch – it is their electric fields which interact
- this method indicates that nuclei have a certain maximum size, but does not tell us the actual size of the nucleus, if such a concept could be applied to the nucleus
- Experiments of this nature indicate that the nucleus must be smaller than around 10^{-14} m.

Electron diffraction

α-particles as probes are limited by the maximum energy they can possess. In order to probe more deeply into the size of the nucleus, an alternative approach is needed. In studies of the sizes of **atoms**, X-ray diffraction is used, down to sizes as small as 10^{-10} m. α-particle scattering indicates that the nucleus is at least 10 000 times smaller than this, so electromagnetic radiation is unlikely to be useful, as wavelengths are too long (γ-radiation is available at such short wavelengths, but it is difficult to collimate such high-energy radiation into the narrow beams necessary for accurate work).

We shall see later (Section 5.6) that beams of particles can be diffracted as if they were beams of waves, and that the wavelength of the particles is inversely proportional to their momentum (i.e. the faster the particle, the shorter the wavelength). In order to achieve a short enough wavelength (less than 10^{-15} m), the electron energy needed is in excess of 1 GeV (10^9 eV). The wavelength (λ) of a high speed electron can be calculated using the equations:

$$\lambda = \frac{h}{p}$$

h = Planck's constant (6.6×10^{-34} J s)

p = the **momentum** of the electron (kg m s^{-1})

$$\lambda = \frac{hc}{E}$$

c = the velocity of light (m s^{-1})

E = the **energy** of the electron (J)

NB: This equation is only valid if $E \gg M_eC^2$ (the mass energy of the electron = 0.511 MeV).

Principles of electron diffraction

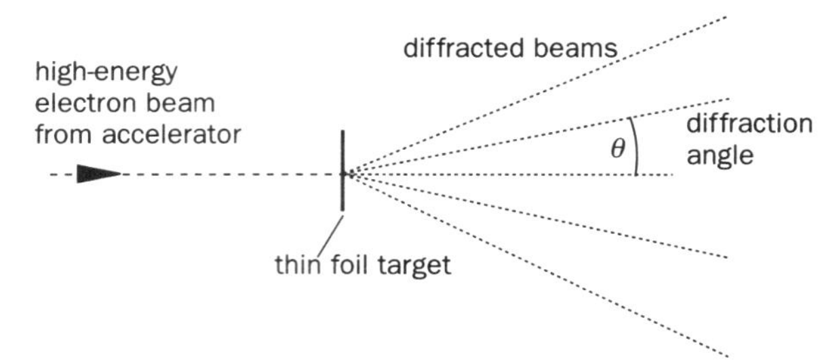

Figure 5.33 Electron diffraction

The intensity of the diffracted electrons is measured at various angles either side of the straight-through position (see Figure 5.33). A graph of intensity against diffraction angle might look like Figure 5.34.

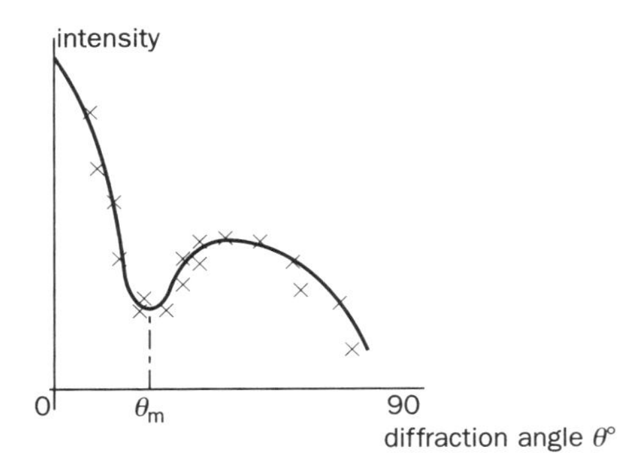

Figure 5.34 Graph of intensity against diffraction angle for electron diffraction

The important measurement is the diffraction angle (θ_m) at which the first **minimum** of intensity occurs. It can be shown that the radius (r) of the spherical nuclei in the target is related to the wavelength (λ) of the electrons by the equation:

$$r = \frac{0.61\lambda}{\sin \theta_m}$$

Equation 1

Guided example (2)

A beam of electrons of energy 450 MeV is produced in an accelerator and directed at a thin metal foil. It is found that the minimum intensity of the resulting diffraction pattern occurs at an angle of 49° to the straight-through direction. Calculate:

(a) the p.d. required to accelerate the electrons from rest in the accelerator

(b) the wavelength of the electrons

(c) the radius of the nuclei in the metal foil.

Velocity of light (c) = 3.0×10^8 m s^{-1}.

Planck's constant (h) = 6.6×10^{-34} J s.

Guidelines

(a) Raising an electron through a p.d. of 1 V increases its energy by 1 eV.

(b) Use $\lambda = hc/E$

(remember to express E in joules)

(c) use Equation 1.

Electron diffraction measurements indicate that nuclear radii are in the order of 2 to 8 **femto**metre (fm) (1 fm = 1×10^{-15} m).

At sufficiently high energies, electron and other particle beams can be used to give information about the internal structure of the nucleus, and even about the internal structure of the neutrons and protons within the nucleus (see Section 5.6 for a fuller discussion of the use of particle beams to probe matter).

Nuclear radius and nucleon number

Clearly, not all nuclei will have equal radii. More particles should mean a bigger nucleus, but the exact relationship between the nuclear radius and the number of particles has important implications for the nature of the forces binding the particles in the nucleus.

Maths window

We can assume that the relationship between nuclear radius (r) and nucleon number (A) may be of the form:

$$r = r_0 A^m$$

where r_0 is a constant of proportionality (presumably related to the size of a nucleon).

To determine the relationship between the two variables, express the equation in logarithmic form:

$$\ln r = m \ln A + \ln r_0$$

which compares with the standard equation of a straight line:

$$y = mx + c$$

If the nuclear radius (r) is measured for a wide range of nuclides, plotting **ln *r*** on the *y*-axis against **ln *A*** on the *x*-axis should give a straight line of gradient ***m*** and intercept **ln r_0** on the *y*-axis.

Analysis of data yields the graphs shown in Figure 5.35 below. Graph (b) shows the logarithmic relationship, confirming the form of the equation.

The gradient (m) of the graph of **ln *r*** against **ln *A*** is found to be:

$$m = 0.33 \quad \text{(i.e. 1/3)}$$

and the intercept on the *y*- axis ($\ln r_0$) is:

$$\ln r_0 = 0.26$$

$$\therefore \quad r_0 = 1.3 \text{ fm}$$

Therefore the relationship between r and A is given by:

$$r = r_0 A^{1/3}$$

where r_0 is effectively the radius of a nucleon.

Equation 2

In words, the radius of a nucleus is proportional to the cube root of the number of nucleons. Since the radius of a sphere is also proportional to the cube root of the volume, we can make the assumption that the nucleus resembles a sphere assembled from particles which are a constant distance apart.

Nuclear density

According to the results obtained from measurements of nuclear radii, we can build up a picture of the nucleus as behaving in a similar way to a **liquid drop**. A drop of liquid grows in volume (and diameter) as more and more molecules are added to it. It grows in an entirely predictable way, and the **density** of the drop is constant, no matter how many molecules are added. Picturing the nucleus in the same way is somewhat surprising, since at the close distances involved, all the protons will exert massive **repulsive** Coulomb forces on each other, as they are all

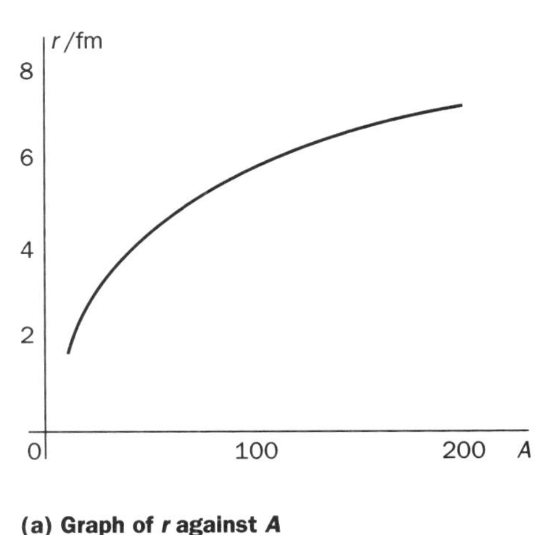

(a) Graph of *r* against *A*

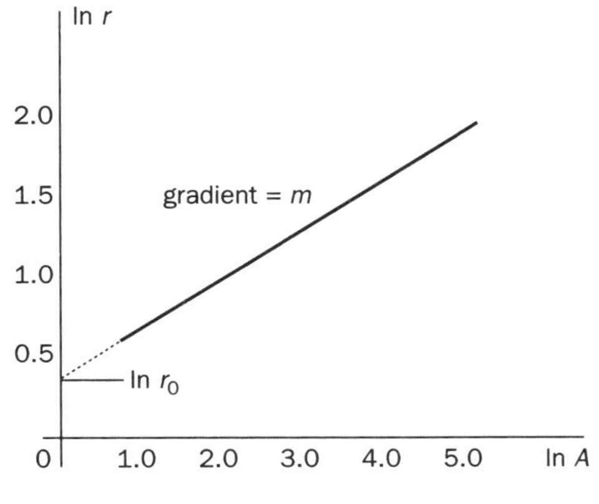

(b) Graph of ln *r* against ln *A*

Figure 5.35 Relationship between nuclear radius and nucleon number

positively charged. We will soon see how we must look for a new force inside the nucleus to account for this behaviour:

The density (ρ) of a nucleus can easily be calculated by assuming it to be a sphere, radius (r), mass (M), composed from a number (A) of nucleons, each of mass (m).

Since *(mass)* = *(volume)* × *(density)*:

$$M = \frac{4}{3}\pi r^3 \times \rho$$

Substituting for r (from Equation 2):

$$M = \frac{4}{3}\pi r_0^3 A \times \rho$$

And since the mass (M) will be (approximately) equal to Am we have:

$$Am = \frac{4}{3}\pi r_0^3 A \times \rho$$

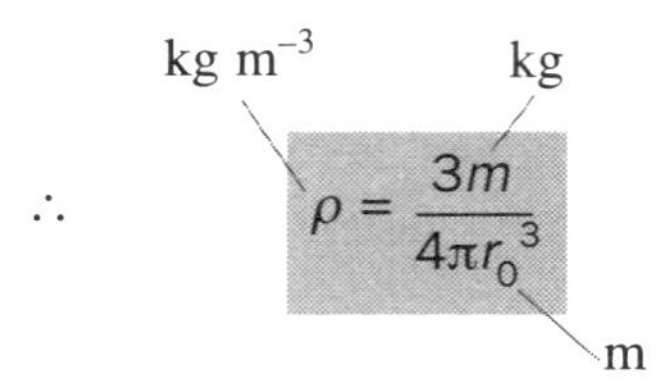

Equation 3

Equation 3 shows that the density of the nucleus does not depend on the number of nucleons, so all nuclei have the same density. Assuming:

nucleon radius (r_0) = 1.3×10^{-15} m and

nucleon mass (m) = 1.7×10^{-27} kg

nuclear density (ρ) is given by:

$$\rho = \frac{3 \times 1.7 \times 10^{-27}}{4 \times \pi \times (1.3 \times 10^{-15})^3} = 1.8 \times 10^{17} \text{ kg m}^{-3}$$

This is around 10 million million times the density of the densest metal. Such a high figure is accounted for by the fact that atoms are almost entirely **empty space**. In a uranium atom, something like 99.8% of the mass is contained in the nucleus, which occupies less than one thousand billionth of the volume of the atom, the rest being empty space, sparsely populated by 92 electrons. If all the matter which makes up the Earth (6×10^{24} kg) were compressed into a cube of nuclear matter, its sides would be only 300 m long. In other words, the matter contained in the entire planet, including oceans, atmosphere and all life, only occupies a volume equivalent to that of a small hill, the rest is empty space.

Proton–neutron model of the nucleus

Coulomb repulsion

Hydrogen-1 is the simplest nucleus with only one proton and no neutrons. All helium nuclei contain two protons and **at least one neutron**. The two protons are positively charged, and therefore repel each other. The magnitude of this repulsive force (F) can be estimated by assuming the protons to be two point charges, each of magnitude $+1.6 \times 10^{-19}$ C, separated by a distance of around one femtometre (1×10^{-15} m):

$$F = \frac{1}{4\pi\varepsilon_0} \frac{(1.6 \times 10^{-19})\,(1.6 \times 10^{-19})}{(1 \times 10^{-15})^2} \approx 200 \text{ N}$$

If you are so inclined, you might like to estimate the initial acceleration this would produce in one of the protons!

If protons are to exist so close together in the nucleus, there must be an attractive force much stronger than the Coulomb repulsion between them. Gravity is always attractive, but the gravitational attraction force F_G between the protons is only tiny:

$$F_G = 6.6 \times 10^{-11} \times \frac{(1.7 \times 10^{-27} \times 1.7 \times 10^{-27})}{(1 \times 10^{-15})^2}$$

$$\approx 2 \times 10^{-34} \text{ N}$$

Clearly, gravity could not hold the protons together; therefore another force must exist within the nucleus.

The strong nuclear force

Studies of the nucleus have shown that the **strong force** (or **strong interaction**) is the force that binds nucleons together, and that it:

- is negligible at distances greater than a few femtometres
- is strongly attractive at distances of 1 to 2 fm
- is strongly repulsive at very small distances (less than 0.5 fm)
- has no effect outside the nucleus
- is unaffected by any charge on the nucleons.

The relationship between the strong force and the distance between nucleons is complex, and does not lead to simple equations. However, the resultant force between nucleons can be represented graphically, as shown in Figure 5.36.

Between uncharged bodies, like two neutrons, the inter-nucleon force is entirely due to the strong force. The effect of the Coulomb repulsion between protons opposes the attractive strong force, reducing the resultant force slightly, as shown by the broken line in the graph.

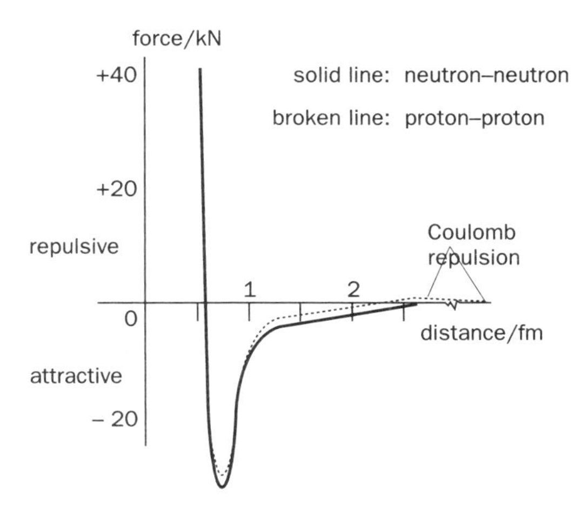

Figure 5.36 Variation of force with distance between nucleons

Energy of the nucleons

If you refer to Section 4.1, on Solids, you will see that the overall shape of the graph of the strong force against the distance between two nucleons is broadly similar to the shape of the intermolecular force against distance curve for two molecules, although the forces are much greater and the distances much smaller. For the two nucleons, there is a distance at which there is zero force. If the nucleons were to approach to a closer distance, the strong force becomes repulsive, tending to push them apart again. If, on the other hand, some external agent tried to pull the nucleons further apart, the strong force would become attractive, tending to pull them closer together again. This means that, at this distance apart, the nucleons are in **equilibrium** – in order to change the distance between them, work would have to be done against the strong force. This explains why nuclear density is constant.

At this separation, the potential energy of the strong force field is a minimum, as there must be an input of work to increase or decrease the distance between the nucleons. It can be shown that the relationship between the force (F) between nucleons and the energy (W) associated with changing the distance (x) between them is given by:

$$F = -\frac{dW}{dx}$$

i.e. force = – *(gradient of the energy–distance graph)*

Putting the two graphs together shows the relationship more clearly (see Figure 5.37).

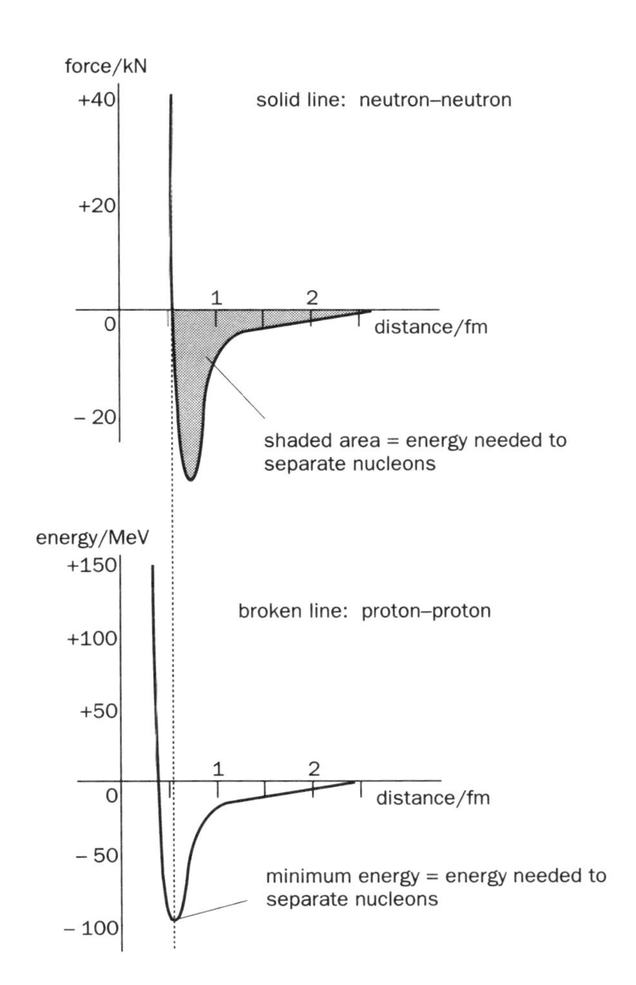

Figure 5.37 Comparison of energy–distance and force–distance graphs for two nucleons

You should note that:

- at the equilibrium separation:

  1. the resultant force between nucleons is zero, and

  2. the nucleons have minimum energy.

- the area between the attractive region of the force–distance curve and the distance axis is equal to the minimum value on the energy–distance graph. This is the energy which would be needed to separate the nucleons to a distance beyond the range of the strong force.

Self-assessment

SECTION A
Qualitative assessment

1. Define the **electronvolt**.
2. Describe in detail the energy changes which occur when an α-particle approaches another nucleus head-on.
3. Explain how electron diffraction can be used to estimate the radius of a nucleus. Explain briefly why this technique is only possible using very high-energy electron beams from accelerators.
4. Sketch a graph of electron intensity against angle of diffraction, and state how this could be used to estimate the radius of the target nucleus.
5. Sketch a graph showing how the radius (r) of a nucleus is related to its nucleon number (A). Explain how the equation relating these quantities can be obtained from the experimental data.
6. Derive an equation for the density of a nucleus in terms of the mass and radius of one of its nucleons.
7. Explain how the experimental observations of the size and constituents of the nucleus indicate the existence of the **strong** force. Describe the main characteristics of this force.
8. Sketch a graph showing how the strong force varies with the distance between a pair of **neutrons**. Indicate approximate value on the axes, and show a separate graph, on the same axes, for a pair of **protons**.
9. Sketch two graphs, one of the strong force against distance between particles, and the other of the energy possessed by the strong force field against distance between particles, for a pair of neutrons. Indicate clearly how the two graphs are related, and explain the significance of the area bounded by the attractive region of the force–distance graph and the distance axis.

SECTION B
Quantitative assessment

(***Answers:*** 6.0×10^{-35}; 2.6×10^{-15}; 3.4×10^{-15}; 5.8×10^{-13}; 1.5×10^{-11}; 1.5×10^{-11}; 5.8×10^{-11}; 1.6; 3.6; 25; 71; 3.6×10^{8}; 2.4×10^{11}; 1.8×10^{17}; 5.7×10^{34}.)

(Unless otherwise stated, use:

mass of a proton $\approx$ mass of a neutron $\approx 1.7 \times 10^{-27}$ kg

charge on proton (and electron) $= 1.6 \times 10^{-19}$ C

mass of an electron $\approx 1 \times 10^{-30}$ kg

Planck's constant (h) $= 6.6 \times 10^{-34}$ J s

permittivity of free space (ε_0) $= 8.8 \times 10^{-12}$ F m^{-1}

gravitational constant $= 6.7 \times 10^{-11}$ N m^2 kg^{-2}.)

1. An electron is accelerated from rest through a p.d. of 360 MV. Calculate the energy gained by the electron:
   (a) in electronvolts
   (b) in joules.
2. Calculate the energy of an α-particle if its closest distance of approach to a nucleus of U-238 is 7.4×10^{-14} m (0.74 fm). (The proton number of uranium is 92.) Express your answer in
   (a) joules
   (b) MeV.
3. Calculate the wavelength of the electrons issuing from the accelerator in Question 1. Hence calculate the angle (θ_m) of the first minimum of intensity due to diffraction by a target in which the nuclei have a radius of 5.0×10^{-15} m.
4. (a) Use a graphical method with the following data to verify the relationship between nuclear radius (r) and nucleon number (A).

A	4	6	12	20	56	100	184	235
r/fm	2.1	2.4	3.0	3.5	5.0	6.0	7.4	8.0

   (b) Estimate the radius of a single nucleon from the graph.
5. Using the data from Question 4(b),
   (a) Calculate the density of nuclear matter.
   (b) Estimate the volume of matter contained in the observable Universe (mass 1×10^{52} kg) if it were condensed to a sphere having the density of nuclear matter.
   (c) Calculate the radius of this sphere, and express it as a fraction of the mean radius of the Earth's orbit around the Sun (1.5×10^{11} m).

Self-assessment *continued*

6. Calculate
   (a) the Coulomb repulsion
   (b) the gravitational attraction

   between two protons separated by a distance of 1.8 fm.

7. Estimate the total shaded area on the graph of force against distance for two neutrons within the nucleus. Hence state the energy required to remove a neutron from the nucleus.

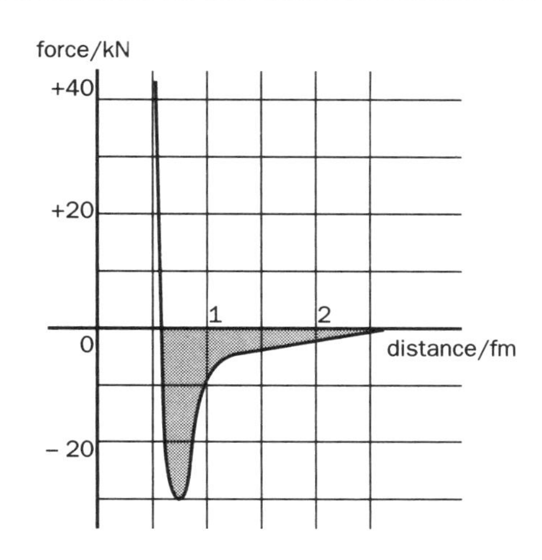

5.4 Nuclear power

Atomic and nuclear masses

Since the earliest days of the study of nuclear reactions, it was realised that accurate measurement of the masses of individual atoms and nuclei was a crucial analytical tool. The **mass spectrometer** was developed to enable the masses of individual atoms to be measured accurately, able to distinguish between isotopes of the same element. Figure 5.38 shows the basic layout of such a device.

The atoms to be measured are vaporised in the ion source where a strong electric field strips electrons from them. A **velocity selector** only allows ions with the same velocity to leave the chamber. The ions are collimated into a narrow beam, which passes through a strong magnetic field. The path of an ion in a magnetic field depends on its mass (as well as its charge, velocity and the strength of the magnetic field) so the point on the detector where it arrives enables precise determination of its mass. Individual atoms can be counted, making this an extremely sensitive device.

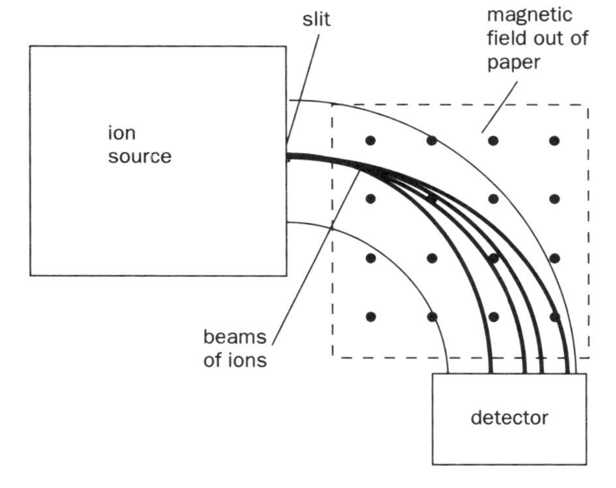

Figure 5.38 Mass spectrometer

The atomic mass unit (u)

The kilogram is an inconvenient unit for measuring atomic masses. The **atomic mass unit** (u) is based on a scale which ranks the atoms according to their masses, using the common isotope of carbon-12 as a reference. On this scale, carbon-12 is assigned a mass of exactly 12 u, and all other atomic masses are expressed relative to this.

The **relative atomic mass** of an atom is defined by the ratio:

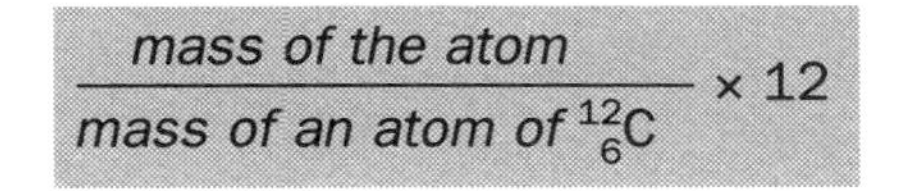

$$\frac{\text{mass of the atom}}{\text{mass of an atom of } {}^{12}_{6}\text{C}} \times 12$$

Equation 1

This is numerically equal to the atomic mass expressed in atomic mass units (u).

Experiment shows that:

$1\ \text{u} \approx 1.661 \times 10^{-27}$ kg

The atomic mass of atoms is usually very close to a whole number of atomic mass units, which is consistent with our idea of a nucleus built up from individual nucleons, each of which has a mass of very nearly 1 u. Table 5.4 shows the atomic masses for some particles and atoms.

Table 5.4 Particle masses

Particle	Mass/u
electron (e)	0.000549
neutron (n)	1.008665
proton (p)	1.007276
hydrogen atom (proton + electron)	1.007825
helium atom (2p + 2n + 2e)	4.002603
α-particle (2p + 2n)	4.001505

You should take care to distinguish between **atomic** and **nuclear** mass. School data books usually quote atomic mass, which is the mass of a neutral atom **complete with electrons**. To calculate the nuclear mass, simply subtract the mass of the appropriate number of electrons.

Numerical work in this topic area requires working to 6 or even 7 significant figures in order to detect the subtle changes in mass involved in nuclear reactions.

'Missing' mass and binding energy

Adding the masses of a proton and an electron gives the mass of a hydrogen atom. However, adding the masses of the constituents of a helium atom (2p + 2n + 2e) produces an atomic mass of 4.03298, which is significantly higher than the measured mass of a helium atom. The helium atom is 'too light' by 0.030377 u. This may not seem very much, but it is equivalent to the mass of 55 electrons!

In general, all **atoms** are lighter than the sum of the masses of their protons, neutrons and electrons. In particular, it is in the **nucleus** that this **mass defect** is particularly apparent. Somehow, protons and neutrons are lighter inside the nucleus than outside it. Protons and neutrons ejected from a nucleus regain their 'missing' mass when they are outside it. The explanation for this discrepancy is the concept of **binding energy**.

The **mass defect** (ΔM) of a nucleus is the difference between the total mass of all its separate nucleons and the mass of the nucleus itself.

The **binding energy** (ΔE) of a nucleus is the energy released when the nucleus is assembled from its constituent nucleons. It is equal to the energy needed to separate the nucleus into individual nucleons.

In the beginning, when nuclei were created from the soup of fundamental particles following the 'Big Bang', the strong nuclear force had to do work in order to bring the particles together. The energy to do this work came from the direct conversion of mass to energy, which is why the assembled nucleus is lighter than its individual nucleons. (Some work is also needed to assemble the electrons around the atom, but this is a negligible amount compared to that required for the nucleons.) In the nucleus, the nucleons are in a lower energy state. In order to separate a nucleus into its nucleons, an amount of energy equal to the work done by the strong force in bringing the nucleons together must be supplied. This energy is converted into matter, restoring the nucleons to their original masses (see Figure 5.39).

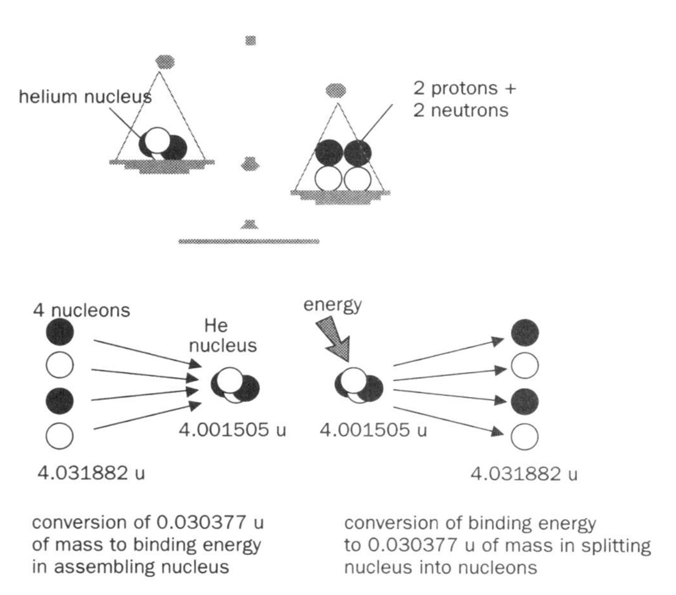

Figure 5.39 Mass and binding energy

It was Einstein who related mass and energy numerically in his famous equation:

$$E = mc^2$$

(J; kg; m s^{-1})

Equation 2

where E = energy (J)
m = mass converted (kg)
c = velocity of light ($m\ s^{-1}$)

If we apply this equation to the 'missing' mass of the helium nucleus, we can calculate its binding energy:

missing mass = 0.030377 u

$\therefore\quad m = 0.030377 \times 1.661 \times 10^{-27}$
(since 1 u $\approx 1.661 \times 10^{-27}$ kg)

$\therefore\quad m = 5.046 \times 10^{-29}$ kg

$\therefore\quad E = 5.046 \times 10^{-29} \times (3.00 \times 10^{8})^{2}$

$\therefore\quad E = 4.541 \times 10^{-12}$ J or 28.38 MeV
(since 1 eV = 1.6×10^{-19} J)

Therefore to split each helium nucleus into its components would require an input of 28.38 MeV of energy, which is a colossal amount of energy on this scale. Mass and energy are considered to be interchangeable, and are often expressed in the same units. The principle of conservation of energy, which states that energy cannot be created or destroyed, is still valid but needs to be modified to take into account the fact that energy may be converted into matter, and vice versa.

A useful conversion factor between mass in atomic mass units and energy in MeV is:

1 u = 931.3 MeV

Guided examples (1)

Atomic Masses:

particle/nuclide	**atomic mass u**
proton (p)	1.007276
neutron (n)	1.008665
electron (e)	0.000549
helium-3 (^{3}He)	3.016030
helium-4 (^{4}He)	4.002603
carbon-12 (^{12}C)	12.000000
krypton-90 (^{90}Kr)	89.9197
barium-143 (^{143}Ba)	142.921
uranium-235 (^{235}U)	235.04394

1 u = 931.3 MeV

1. Calculate the **nuclear mass** of
(a) ^{12}C
(b) ^{3}He.

Guided examples (1) *continued*

Guidelines

As the binding energy of electrons to the atom is negligible on this scale, simply subtract the mass of the correct number of electrons from the **atomic mass** of the atom to obtain the mass of the nucleus.

2. Calculate:
(a) the **mass defect**
(b) the **binding energy**
(c) the **binding energy per nucleon**
of
(i) ^{12}C
(ii) ^{3}He
(iii) ^{4}He.

State which of these you expect to be the most stable and why.

Guidelines

(a) In each case, work out the numbers of protons, neutrons and electrons in the complete atom and add their individual masses together to obtain the total mass of the constituents of the atom. Then subtract the mass of the complete atom (which should be less) to obtain the **mass defect.**

(b) Simply convert the mass defect (u) into the equivalent amount of energy (MeV) using the conversion factor supplied.

(c) Divide the answer to (b) by the nucleon number of the atom.

Maximum stability means the highest binding energy per nucleon.

Nuclear transformations

In the preceding chapters, we have examined radioactivity, and looked briefly at the high energies of the particles emitted from nuclei. We must now consider the source of all this energy. A common laboratory nuclear reaction is the decay of radon-220:

$$^{220}_{86}Rn \rightarrow ^{216}_{84}Po + ^{4}_{2}He + \text{energy}$$

If we measure the masses of all the particles involved, we find that some mass is 'missing'. The following calculation of this missing mass shows that the same answer is obtained in α-particle decay using nuclear mass or atomic mass.

Particle	Nuclear mass/u	Total		Atomic mass/u (includes electrons)	Total
Parent $^{220}_{86}$Rn	219.96417	219.96417		220.01138	220.01138
Daughter $^{216}_{84}$Po	215.95580			216.00191	
α-particle $^{4}_{2}$He	4.001505	219.9573		4.002603 (helium **atom**)	220.00451
Missing mass		0.00687 u			0.00687 u
Energy equivalent		6.39 MeV			6.39 MeV

Therefore when an α-particle is emitted from a radon-220 nucleus, 0.00687 u of mass is converted into 6.39 MeV of energy. If the kinetic energy of the α-particles emitted from radon-220 is measured, each one is found to have an energy of 6.29 MeV, showing that the α-particle carries away over 98.5% of the energy released in the decay. (The other 0.10 MeV is the kinetic energy of the recoiling Po-216 nucleus.)
In all cases of radioactive decay, the energy released in the process is supplied by the direct conversion of matter to energy.

β-decay can be handled in an identical fashion. A common reaction is the decay of carbon-14 by β-emission:

$$^{14}_{6}C \rightarrow ^{14}_{7}N + ^{0}_{-1}e + \overline{\nu}$$

To calculate the energy released:

Particle	Nuclear mass/u	Total	
Parent C-14	13.999948	13.999948	
Daughter N-14	13.999231		
Beta particle	0.000549	13.99978	
Missing mass			0.000168 u
Energy equivalent			0.156 MeV

(Note that you **cannot use atomic masses for β-decay**. This is because the tables give the mass of a complete, neutral atom. The daughter product of a β-emitter is not a neutral atom, since it is created with a missing orbital electron. The safest method to avoid mistakes in your calculation is to use the **nuclear** masses of the parent and daughter.)

The energy released in the decay is 0.156 MeV. The β-particle and the antineutrino absorb almost all of this energy as kinetic energy, since they are so light that there is almost no recoil energy of the daughter nucleus.

Guided examples (2)

particle/nuclide	atomic mass /u
polonium-212 (^{212}Po)	211.988865
lead-208 (^{208}Pb)	207.97666
helium-4 (^{4}He)	4.002603
nitrogen-12 (^{12}N)	12.01864

1. Helium-3 has a very high neutron absorption cross-section (it readily captures neutrons) to become helium-4. The reaction is:

$${}^{3}_{2}He + {}^{1}_{0}n + \rightarrow {}^{4}_{2}He$$

Using the data supplied for the previous guided examples, calculate the energy released in the reaction.

Guidelines

Can you use atomic masses? Remember that atomic mass of an atom includes the mass of its electrons. Count the number of electrons included on each side of the reaction equation. If they are identical, you can use atomic masses.

Add the masses of the helium-3 and the neutron; subtract the mass of the helium-4. A **positive** answer indicates a conversion of matter to energy. Express this mass difference in MeV, using the appropriate formula.

2. Free neutrons decay into protons by emitting β-particles and antineutrinos. Write an equation for this decay and calculate the maximum energy of the emitted β-particles.

Guidelines

Compare the mass of the neutron to the total mass of the proton and electron into which it decays. Calculate the loss in mass and express this in MeV. The β-particles will carry away almost all of this energy as kinetic energy.

3. Polonium-212 (proton number 84) decays into the stable isotope lead-208 by the emission of an α-particle.

(a) Write an equation for this decay.

(b) Calculate the energy released in the reaction.

(c) Calculate the energy carried off by

(i) the α-particle

(ii) the lead-208 nucleus.

Guidelines

(a) Make sure proton number and nucleon number balance.

(b) Can you use atomic masses? Count the numbers of electrons included on each side of the reaction equation, if you treat the α-particle as a helium atom. If they balance, you can use atomic masses.

(c) The energy will be shared between the α-particle and the recoiling lead atom. Ignore the masses of the electrons, and apply the **principle of conservation of momentum**. You should recognise that this is equivalent to an explosion, and that the energy of the explosion is shared among the fragments in **inverse proportion** to their **mass**. Since the lead atom is about 52 times heavier than the α-particle, the lead atom only receives 1/52 of the energy released.

4. Show, by calculation, that carbon-12 cannot decay by β-emission to nitrogen-12.

Guidelines

Write an equation for the decay. Compare the masses before and after the decay, taking great care to account for the electrons included in the **atomic mass** data supplied (if in doubt, the safest method is to work out the **nuclear mass** of all the particles involved). A decay is only possible if there is a **decrease** of mass.

Nuclear power sources

Binding energy

All the nuclei (with the exception of hydrogen-1) possess binding energy and, generally, the more nucleons, the greater the binding energy. The **binding energy per nucleon** is a measure of how tightly a nucleus is bound: more binding energy per nucleon means a more stable, tightly-bound nucleus. If we calculate the amount of binding energy per nucleon for all the stable nuclides, the vast majority have around 8 MeV per nucleon. However, there are variations, revealed by plotting the binding energy per nucleon against nucleon number, as shown in Figure 5.40 overleaf.

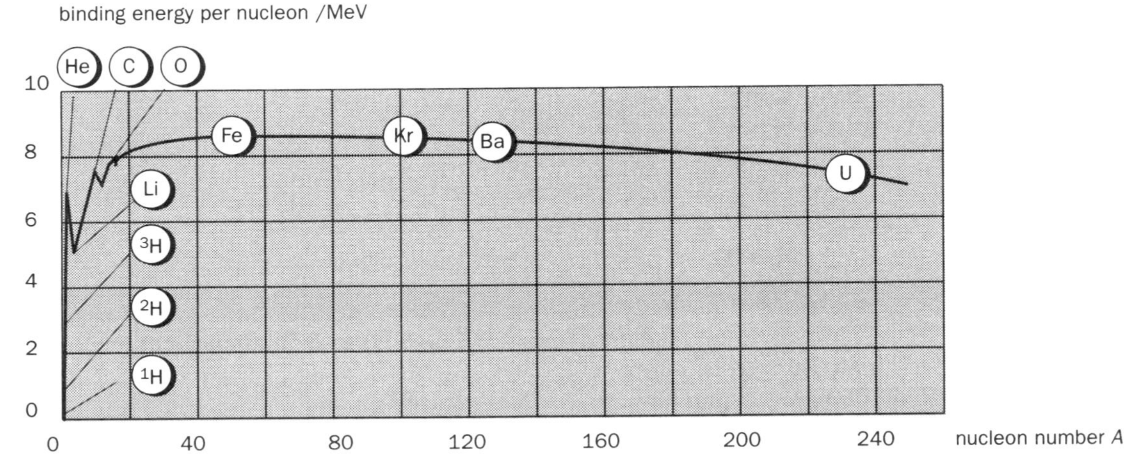

Figure 5.40 Graph of binding energy per nucleon against nucleon number

The lighter nuclei, hydrogen, helium-3, lithium, etc. have low values of binding energy per nucleon. (Helium-4, the stable, common isotope, is a distinct 'blip' on the curve, having a higher binding energy per nucleon than predicted by its nucleon number.) The highest binding energy per nucleon values occur in the middle of the graph, corresponding to nucleon numbers of between 40 and 140. The larger nuclei have slightly lower values.

The differences between values of binding energy per nucleon suggest a means of extracting energy from nuclear reactions. If heavy nuclei can be split, lighter nuclei will be formed, and the difference in binding energy per nucleon released in the process (nuclear fission). Similarly, if two very light nuclei can be fused together, the resulting nucleus will be more stable, and again the difference in binding energy will be released (nuclear fusion). We shall examine these two processes in more detail.

Nuclear fission

In the late 1930s scientists were bombarding uranium ($Z = 92$) with beams of neutrons, hoping to transmute the uranium into plutonium ($Z = 94$). They were amazed to discover that instead of heavier nuclei, traces of lighter nuclei such as barium ($Z = 56$) and krypton ($Z = 36$) were present after bombardment. The conclusion was that the uranium nuclei had split into lighter fragments, a process named **nuclear fission**. A great deal of effort went into the development of this discovery, leading to the world's first self-sustaining nuclear chain reaction in 1942, and the explosion of the first nuclear fission bomb in 1945. Figure 5.41 below shows how fission of the isotope U-235 occurs when struck by a slow neutron.

The U-235 nucleus captures the **slow** neutron, and becomes so unstable that a short time later it splits into two **fission fragments.** These are the nuclei of lighter elements, containing around 30 to 60 protons (the total number of protons in the fragments must be 92). In addition, several **neutrons** (typically three) will be ejected at high speeds. A large amount of matter is converted to energy in the process, which appears as kinetic energy of the fission fragments and neutrons. The high speed neutrons are unlikely to cause further fission, unless they are slowed down to a

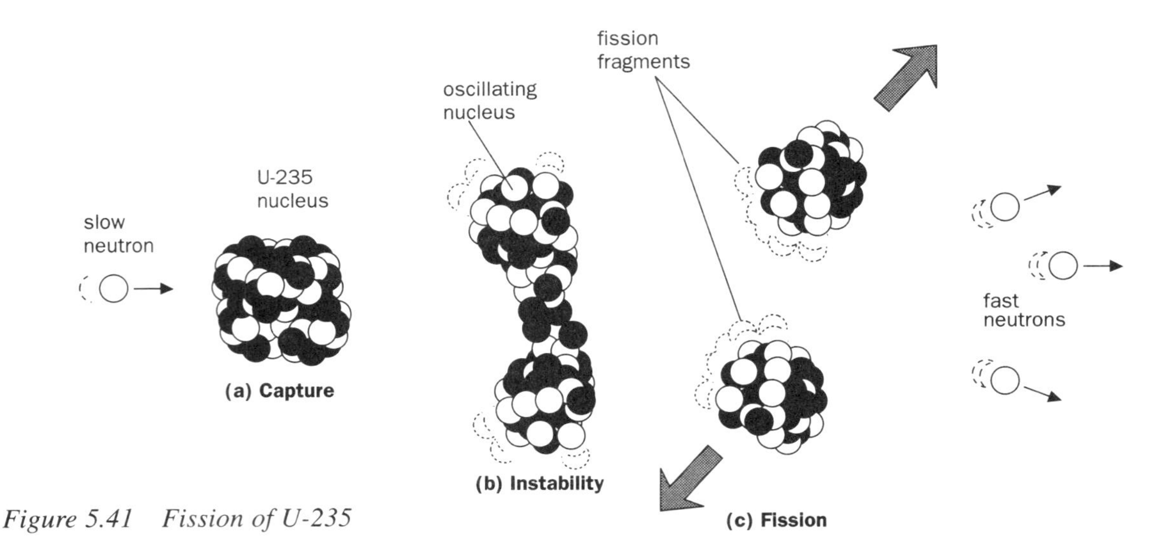

Figure 5.41 Fission of U-235

speed at which U-235 nuclei can capture them. The speeds of neutrons which are readily captured are comparable to the speeds of particles due to thermal vibration, hence such slow neutrons are referred to as **thermal** neutrons. The U-235 nucleus is not 'shattered' by the neutron; it must capture it, which is the trigger for the instability that leads to fission. If the fission-generated neutrons are slowed down to the required speeds, then they can cause further fission, which leads, in turn, to further fission – the so-called **fission chain reaction** (Figure 5.42).

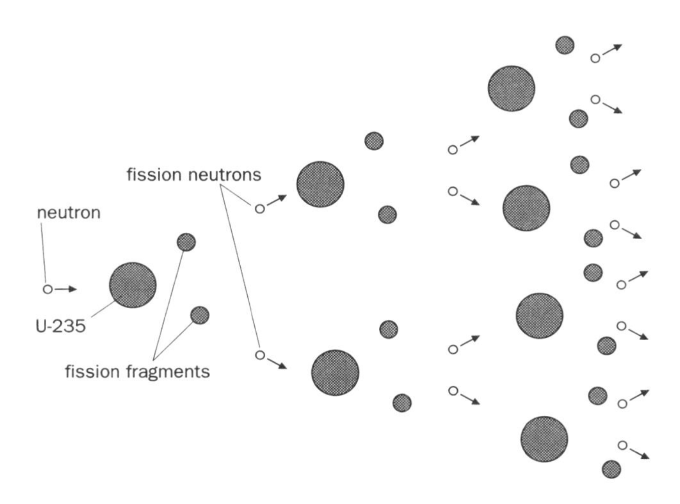

Figure 5.42 Fission chain reaction

Under the right conditions, each fission can trigger further fission, and the number of fissions per second can escalate to the point of explosion. In a nuclear fission reactor, the rate of reaction is allowed to reach the rate at which it is just self-sustaining, and controlled to maintain this rate.

A controlled, self-sustaining chain reaction occurs when only **one** neutron from each fission goes on to produce further fission.

Maintaining and controlling the nuclear fission chain reaction

In a nuclear reactor, energy released from fission in the fuel is extracted, and used to produce steam which drives turbines, either to run generators for electricity, or coupled directly to propellor shafts in nuclear powered ships and submarines. Maintaining and controlling the fission reaction requires the integration of a number of components into a whole.

The right fuel

Uranium is a naturally-occurring metal, mined in several parts of the world. In its natural state, it consists of over 99% of the isotope U-238, and about 0.7% U-235, both isotopes being α-particle emitters with very long half-lives of the order of billions of years, which is why the material is still around on Earth. For a chain reaction to occur in natural uranium, there must be a large amount of the material assembled in a small volume, otherwise the neutron produced in each fission will probably have escaped before encountering another nucleus. Fortunately for the nuclear industry, U-235 is over 200 times more likely to capture a thermal neutron than the common U-238. To improve the ability to produce a chain reaction, natural uranium may be 'enriched' to increase the concentration of U-235. One way of achieving this is to convert the uranium into a gaseous compound, then separate the two isotopes by diffusion, since the slightly lighter U-235 will diffuse at a slightly faster rate. Such a process is technically difficult, slow and expensive. This can increase the concentration of U-235 to between 2 and 3%.

Additionally, there must be sufficient material present for the reaction to proceed. If the mass is too small, too many neutrons will escape. For a given fissile material, there is a **critical mass** of material, which must be present before a chain reaction can proceed. This mass could be as small as 3 kg of weapons-grade plutonium, given the right conditions. The shape of the mass is also important. A sphere has the smallest possible surface area for a given mass. Any other shape has a larger surface area, so a greater amount of the material would be needed to create a non-spherical critical mass.

The U-238 in the uranium fuel can capture neutrons to become U-239, which decays to plutonium-239 by β-emission. Pu-239 is a fissile material, which fissions efficiently with **fast** neutrons of energy 1 MeV or so. Pu-239 is used in **fast** reactors. Used fuel from thermal reactors can be placed inside fast reactors, where the high neutron flux converts the U-238 into Pu-239. In this way, such a reactor **breeds** more Pu-239 from the spent fuel rods than it consumes. Such **breeder** reactors were thought to offer the possibility of extending the lifetime of the world's finite uranium resources by a factor of two or three, but there are still many technical and political problems to be solved.

Moderation

The neutrons released by each fission have around 1 MeV of energy, and are travelling too fast for reliable capture by the U-235 nuclei. They must be slowed by absorbing most of their kinetic energy, so that they have less than 1 eV of energy left. The kinetic energy is absorbed by forcing the neutrons to have multiple collisions with other nuclei. The most reliable way to achieve this is to incorporate into the reactor large amounts of a material whose nuclei have the following properties:

- they must not absorb the neutrons (i.e. they must have a low **absorption cross-section**)
- they must not become radioactive as a result of neutron bombardment
- they should be small (comparable to the size of the neutron) for efficient energy absorption.

Such a material is called a **moderator**, since its function is to reduce the high speeds of fission neutrons to more moderate, thermal speeds.

The need for the nuclei of the moderator to be small is explained by considering collisions. In an elastic collision, energy is transferred most effectively if the two bodies are equal in mass. Carbon nuclei have masses about 12 times the mass of the neutrons, but the other properties of carbon (low neutron absorption cross-section, and its desirable physical properties such as high melting point and ready availability at low cost) make it particularly suitable as a moderator. Water molecules (H_2O) are also small enough to be potentially useful as a moderator. However, the neutron absorption cross-section of the hydrogen nuclei in ordinary water is relatively high. 'Heavy' water (D_2O) is water in which the hydrogen atoms are replaced by **deuterium**, which has a neutron absorption cross-section over 600 times smaller. The physical and chemical properties of heavy water are virtually identical to those of ordinary water, so it could serve a dual function as a coolant fluid. The **pressurised water reactor** (PWR) employs water as a moderator and coolant. Graphite is used in the **Magnox** and **advanced gas cooled** (AGR) reactors.

Control rods

While a reaction cannot proceed unless there is more than the **critical mass** of fissile material present, a slightly greater mass will lead to an escalating reaction, and explosion. To control the reaction, all excess neutrons must be absorbed, so that only one neutron from each fission can go on to produce further fission. **Control rods** containing **boron** or **cadmium** are incorporated into the nuclear reactor. These materials have a very high neutron absorption cross-section (they appear as a large 'target' to the neutrons and absorb them readily). Control rods can be inserted into the reactor to any level, and their position is continuously monitored and controlled to maintain the reaction at a steady rate. In an emergency, they can be pushed right in to absorb most of the neutrons, shutting the reactor down. They are not used to 'control' the reactor, as the reaction is not variable: there is one safe rate of reaction, and no other. If the rods are too far out, the reaction will quickly speed up and run out of control. Too far in and the reaction will stop. In this sense, the reactor is **divergent**, since a small disturbance from the equilibrium position of the control rods will lead either to shut-down or melt-down if not corrected. Figure 5.43 is a simplified view of part of the core of a **thermal** nuclear reactor, showing the layout of the fuel rods and control rods in the graphite blocks which act as the moderator.

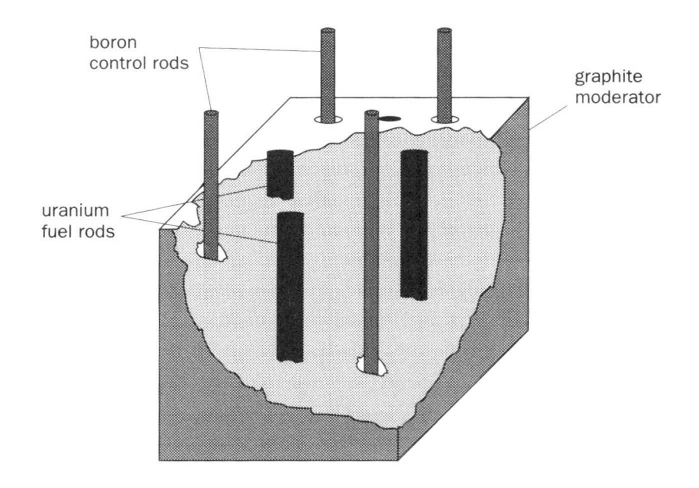

Figure 5.43 Layout of thermal nuclear reactor

Extracting the energy

The energy released in the fission reaction is almost entirely kinetic energy of the fission fragments and neutrons. This kinetic energy of the fission fragments increases the internal energy of all the components of the reactor, resulting in a temperature rise. The neutrons are mostly absorbed, also increasing the internal energy. To extract this heat energy, a **heat exchange fluid** is circulated around the reactor core, as shown in Figure 5.44. The fluid becomes heated by contact with the hot reactor components. Outside the reactor, the hot fluid is used to boil water in a **heat exchanger**. Often called the **coolant**, the primary function of the heat exchange fluid is to extract the heat energy to boil water. Of course it also serves to cool the reactor, but this is secondary to the heat exchange. There must be absolutely no contact between the coolant and the outside environment, since the coolant will be contaminated by passing through the reactor.

The coolant used must have the following properties:

- high specific heat capacity
- non-corrosiveness
- stability at the high temperatures in the reactor.

Water is used as a coolant in the PWR. The reactor vessel is pressurised to raise the boiling point of the water, ensuring that it remains liquid at high temperatures. Another common fluid is carbon dioxide gas, used in the Magnox and AGR reactors, again at high pressure. A few very high temperature reactors employ liquid sodium as the coolant.

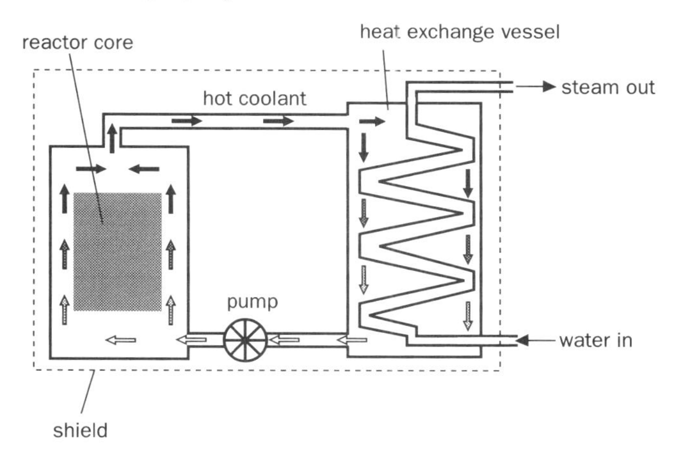

Figure 5.44 Heat exchange in the nuclear reactor

Guided example (3)

Atomic Masses:

particle/nuclide	**atomic mass /u**
uranium-235 (^{235}U)	235.04394

One possible fission reaction is:

$$^{235}_{92}U + ^{1}_{0}n \rightarrow ^{90}_{36}Kr + ^{143}_{56}Ba + ^{1}_{0}n + ^{1}_{0}n + ^{1}_{0}n$$

(a) Calculate the binding energy per nucleon of U-235.

Guided example (3) *continued*

(b) Assuming that the average binding energy per nucleon of the fission products is 8.2 MeV, estimate the energy released per fission.

Guidelines

(a) Calculate the **mass defect** by comparing the mass of the uranium atom with the total mass of its constituents. Express this in MeV, and divide by the number of nucleons.

(b) It is the **difference** between the binding energy per nucleon of the U-235 and that of the fission fragments that is released as energy in the fission reaction. Calculate this difference, and hence estimate the total energy released by each fission.

Nuclear fusion

If you refer to Figure 5.40, you will see that the binding energy per nucleon of light nuclei is much lower than that of the majority of nuclides. This means that if two light nuclei could be forced to **fuse** into a single larger nucleus, the difference in binding energy would be released, and could be used to produce steam to drive turbines in a power station.

A view inside JET (Joint European Torus) showing the ring-shaped chamber where the plasma is confined by powerful magnetic fields (flux density of 3.45 T at plasma centre)

The larger nucleus would have a lower mass than the sum of the masses of the lighter ones, and it is this mass difference that is converted to energy. This is the basis of **nuclear fusion**. A possible fusion reaction involves **deuterium** and **tritium**, isotopes of hydrogen containing one and two neutrons respectively (see Figure 5.45):

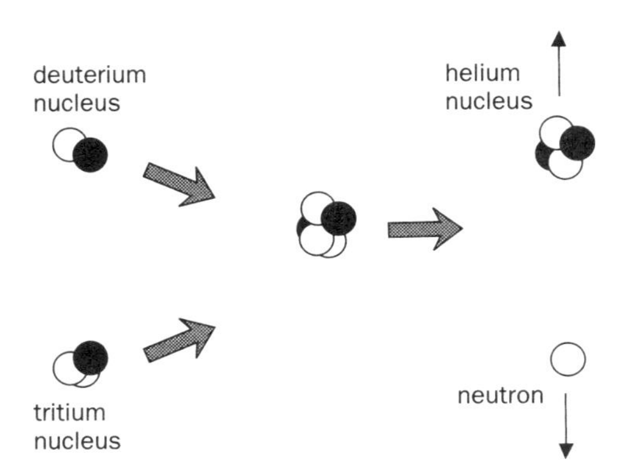

Figure 5.45 Deuterium–tritium fusion reaction

If we look at the particle masses involved:

Particle	Atomic mass /u	Total	
Deuterium	2.014102		
Tritium	3.016050	5.030602	
Helium	4.002603		
Neutron	1.008665	5.011268	
Missing mass			0.019334 u
Energy equivalent			18.01 MeV

This simple reaction releases 18 MeV of energy. (Although only nuclei are involved, you can use atomic masses, since the atomic masses of deuterium and tritium are 'wrong' by one electron each, and the atomic mass of helium is 'wrong' by two electrons, so the two 'errors' cancel out.)

Harnessing this, almost limitless, energy source has been a goal of physicists for decades. Deuterium is a naturally-occurring isotope that can be extracted from water, and tritium is already produced commercially. The major technological problem is to give the nuclei sufficient energy to overcome their electrostatic repulsion. Fusion reactions like this provide the energy of the Sun and all the stars. At the centres of massive bodies like the Sun, the temperature exceeds 20 million degrees, and matter there exists as a **plasma**, in which all nuclei are stripped of their electrons. The density of this plasma is greater than 100 000 kg m^{-3} (due to compression by the enormous gravitational forces produced by such a massive body), and under these conditions, the nuclei collide with enough energy to fuse, releasing more energy to sustain the reaction.

These conditions can be created on Earth, at the centre of a nuclear fission explosion. The high temperature and high pressure combination can be made to create a fusion reaction in light nuclei. The resulting **hydrogen bomb** is a truly awesome weapon, delivering the explosive power of millions of tons of high explosive from a few kilograms of plutonium and a few grams of lithium, deuterium and tritium. Creating these conditions for a controlled fusion reaction, in order to harness the energy for peaceful purposes, is altogether more difficult. The deuterium–tritium plasma must reach a temperature of millions of degrees, but cannot be allowed to come into contact with the walls of a container, since this would instantly cool the plasma and cause the reaction to be halted. Considerable research is being devoted to methods of containment of the reacting plasma. One of the most promising methods is the use of magnetic fields. Since the plasma consists entirely of charged particles, magnetic fields will exert forces on them. Controlling the shape of the field should enable the plasma to be contained as a magnetic 'bottle'.

Fusion promises an almost inexhaustible supply of energy from a reaction that should be inherently safer than the fission reaction, which can so quickly become unstable and escalate out of control. Furthermore by careful selection of materials the radioactive burden can be reduced to very low levels such that storage would be for, at most, 100 years.

Nuclear safety

Containment

Nuclear reactors produce considerable quantities of radiation. Of particular concern is the production of γ-radiation and neutrons, which are highly penetrating. Reactors are built with several levels of **containment**, which include steel casing for the reactor itself and thick concrete walls immediately outside to absorb the radiation. These layers have a secondary, though very important, function of confining any accidental escapes of radioactive gases or contaminated material in the event of an accident.

Nuclear waste

The nuclear industry, and nuclear reactors in particular, produce large quantities of what can be regarded as hazardous material. After some years in the reactor, the proportion of U-235 will have decreased to the point at which a chain reaction can no longer be sustained. During its time in the reactor the nuclei have been transmuting into other elements, which may be highly radioactive. These include the fission fragments and the plutonium-239 created when U-238 absorbs a stray neutron. The fuel is removed and must be re-processed, a hazardous operation in which the plutonium and any other useful isotopes are extracted. The remaining radioactive material must then be stored for later disposal. At the end of its life, the entire nuclear reactor represents a major engineering problem. It cannot be simply torn apart and sold for scrap, due to the dangerous materials present. We are just starting to feel the effects of the costs of de-commissioning the first generation of nuclear reactors.

In addition to the reactor and fuel, anything which has been in contact with radioactive materials, or been exposed to the neutron bombardment of the reactor, may be contaminated or have become radioactive, and must be disposed of safely. The nuclear industry publishes copious information on how they handle the problems of safe handling and disposal of their waste products. You may find it interesting to compare and contrast their literature with that of the environmental lobby groups, such as Greenpeace.

Basically, nuclear waste material is divided into three categories: low-, intermediate- and high-level waste, each being handled differently, as shown in Table 5.5.

Production of artificial radioisotopes

In Section 5.2 (Radioactivity) we examined the important role of radioactive isotopes in medicine and industry. A great many of these isotopes do not exist in nature, and are manufactured artificially. In the nuclear reactor, there is a high level of **neutron flux**. All material within the containment walls of the reactor is subjected to constant neutron bombardment. Samples of elements are placed within the reactor for several weeks, and become transmuted to new, radioactive isotopes. Those used in medicine and as tracers often have short half-lives, so there is a continuous need for production of these valuable substances.

Typical artificial transmutation reactions

Production of ^{65}Zn

$$^{64}_{30}Zn + ^{1}_{0}n \rightarrow ^{65}_{30}Zn + \gamma$$

Production of ^{233}U

$$^{232}_{90}Th + ^{1}_{0}n + \rightarrow ^{233}_{90}Th + \gamma$$

$$^{233}_{90}Th \rightarrow ^{233}_{92}U + 2\beta$$

Table 5.5 Waste disposal methods

Category of waste	Typical waste materials	Disposal methods
Low-level	Used packaging, protective clothing	Burial (land or sea) Liquids and gases may be released into the atmosphere or at sea
Intermediate-level	Irradiated components from the reactor, scrap fuel cladding, fuel re-processing solvents	Storage of materials in concrete warehouses; deep burial
High-level	Non-useful by-products of the fission reaction	Liquefied waste stored in water-cooled tanks; vitrification (sealing in glass blocks)

Self-assessment

SECTION A
Qualitative assessment

1. Explain briefly how a **mass spectrometer** measures the mass of individual atoms.

2. Define the atomic mass unit (u). Distinguish between **atomic mass** and **nuclear mass.**

3. Explain in detail what is meant by **mass defect** and **binding energy**.

4. Complete the following nuclear equations
   (a) α-emission $^{241}_{95}\text{Am} \rightarrow$
   (b) β-emission $^{18}_{7}\text{N} \rightarrow$

5. Sketch a graph of **binding energy per nucleon** against **nucleon number**. Indicate appropriate values on the scales.

6. Explain how this graph shows that energy is available
   (a) by the fusion of light nuclei
   (b) by the fission of heavy nuclei.

7. Describe the fission of U-235 by slow neutrons. Explain why the neutrons must be slow.

8. A particular **thermal** nuclear fission reactor, using **enriched** uranium fuel, uses water as the **coolant**. The **control rods** are made from boron coated steel, and the **moderator** is graphite. Explain the significance of the terms printed in **bold** type. In the case of the **coolant, control rods** and **moderator**, list the desirable properties of materials chosen for these functions and suggest possible alternatives to those listed.

9. Outline the processes involved in a **nuclear fusion** reaction, and explain why there is a release of energy. Describe the conditions necessary for such a reaction to occur and hence explain why it is proving difficult to achieve a controlled, sustained fusion reaction.

10. Describe the hazards posed by waste from nuclear reactors. Give examples of materials likely to be categorised as **low**-level, **intermediate**-level and **high**-level waste, and state the methods used to dispose of them.

11. Explain how radioisotopes may be manufactured artificially in nuclear reactors.

SECTION B
Quantitative assessment

(***Answers:*** 0.034346; 0.09894; 4.87; 5.331; 5.62; 7.679; 7.87; 28.2; 31.972073; 31.99; 92.14; 245.0678; 7.22×10^{12}; 1.98×10^{13}; 5.33×10^{23}; 6.02×10^{23}.)

(1 u = 1.661×10^{-27} kg

1 u = 931.3 MeV

Planck's constant (h) = 6.6×10^{-34} J s

Avogadro constant (N_A) = 6.02×10^{23} mol^{-1}

electron charge (e) = 1.6×10^{-19} C

Atomic masses:

particle /nuclide	**mass/u**
electron (e)	0.000549
neutron (n)	1.008665
proton (p)	1.007276
hydrogen (^{1}H)	1.007825
helium-4 (^{4}He)	4.002603
α-particle	4.001505
lithium-6 (^{6}Li)	6.015124
radium-226 (^{226}Ra)	226.02544
radon-222 (^{222}Rn)	222.01761
phosphorus-32 (^{32}P)	31.973909)

In the answers give energies in MeV, masses in atomic mass units unless otherwise specified in the question.

1. Given that the mass of a carbon-12 atom is exactly 12 u, calculate the number of atoms in 12 g of carbon-12. What is the significance of this number?

2. Calculate the
   (a) mass defect
   (b) binding energy
   (c) binding energy per nucleon of
      (i) lithium-6
      (ii) carbon-12.

Self-assessment *continued*

3. The artificially-created radioisotope fermium-249 has an atomic mass of 249.079 u, and emits an α-particle with the release of 8.0 MeV of energy.
   (a) Identify the daughter product and calculate its atomic mass.
   (b) Calculate the energy carried off by the α-particle.

4. It is proposed to use the energy released by the decay of 200 g of radium-226 as an emergency power source. If its half-life is 1622 years, calculate
   (a) the number of nuclei initially present
   (b) the initial rate of decay
   (c) the energy released by each decay
   (d) the power of the device (in watts).

5. Phosphorus-32 is a commercially-available isotope which emits β-particles with a maximum energy of 1.71 MeV. Write an equation for the decay and calculate the atomic mass of its daughter product.

6. If enriched uranium contains 2.3% U-235, and each fission releases 210 MeV of energy, calculate the total amount of energy released when all the U-235 in 10 kg of natural uranium has fissioned (express your answer in joules).

7. Estimate the energy that could be extracted from the theoretical fusion of two lithium-6 nuclei into a single nucleus.
   (Use the answers to Question 2.)

5.5 Particle physics

Particle physics is science on a grand scale, in that it is an attempt to answer the fundamental question raised when man first started to think – what is everything made of? The idea that matter is made from atoms has been around for centuries, but the question 'what are atoms made of?' has only recently been (partly) answered. So far, you have learned that atoms are simple structures, with almost all their mass contained in a tiny nucleus of protons and neutrons, surrounded by a region of empty space containing electrons. However, there is a much deeper layer of complexity to this structure, and while these particles were being observed and identified, many others were also observed. This section is the story of their discovery and the attempts made by physicists to make sense of the nature of matter.

Observation of β-decay

As early as 1914, physicists knew that radionuclides emitted β-particles with a wide spectrum of energies, up to a specific maximum which was unique to that isotope. Since the energy associated with each decay in the nucleus of that isotope should be the same, all the β-particles should have the same energy. It follows that the varying 'missing' amounts of energy must be carried off in some other way. At first it appeared that β-emission violated the principle of conservation of energy, but by 1931, Pauli had suggested that a neutral, massless particle must be emitted simultaneously with the β-particle. Being uncharged, this particle would be undetectable in cloud and bubble chambers, as it would have very little interaction with matter. This new particle was named the **neutrino** (meaning *little neutral one*) by Fermi in 1932.

Other evidence for the existence of an additional particle comes from observing the decay of a radioactive nucleus by emission of a β-particle in a bubble chamber or wire chamber.

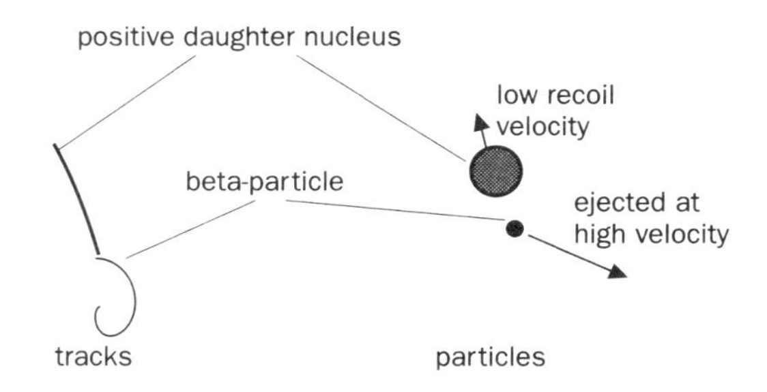

Figure 5.46 Beta-decay

Figure 5.46 shows the tracks made in a bubble chamber when a radioactive nucleus decays by β^--emission. The positive ion which is the daughter nucleus of the decay leaves a dense track which curves away due to the magnetic field applied. The β^--particle leaves a much lighter track of very narrow radius due to its low mass. Its track curves in the opposite direction (due to its negative charge) and its radius decreases into a spiral as the β^--particle loses energy travelling through the bubble chamber liquid, and therefore slows down. The key observation is that the daughter nucleus and the β^--particle do not set off in exactly opposite directions immediately after the decay, thus apparently violating the principle of conservation of momentum. The only possible conclusion from such observations is that there must be a third particle which carries off some of the energy in a different direction.

The neutrino

Symbol	ν
charge	neutral
rest mass	zero (or very small)

The existence of the neutrino was predicted as long ago as 1931 to explain apparent non-conservation of energy in β-decay. It was finally observed in 1956, by which time nuclear fission reactors had been developed. Each fission releases floods of **antineutrinos**. Although they interact only weakly with matter, very occasionally there will be a head-on collision between a neutrino and a proton, producing a neutron and a **positron** (see Matter and antimatter, below). The discovery of antineutrinos confirmed the existence of neutrinos, which were later directly observed in cosmic rays. In 1961, it was shown that there were two distinct neutrinos – the electron neutrino ν_e (associated with β decay) and the muon neutrino ν_μ associated with pion decay (see later).

While the neutrino has resisted all attempts to measure its mass there is nonetheless some speculation that, because there are believed to be huge numbers of neutrinos in the universe, a non-zero mass, even a tiny one, would account for a significant amount of matter. The existence, or otherwise, of this mass has profound implications for the future of the Universe. At present, Doppler shift observations show that the Universe is still expanding. If neutrinos are massless, there is not enough mass in the Universe to halt this expansion, which means that the Universe will continue to grow for ever. If, however, there is a large amount of mass 'hidden' in the neutrinos, then the expansion will be halted at some time in the future by gravitational attraction, and the Universe will then start to contract, finally ending when all matter comes together again with a 'Big Crunch'.

Matter and antimatter

Modern physics has thrown up many astounding discoveries, but antimatter remains one of the most amazing. Each of the nuclear particles (proton, neutron, electron, neutrino, etc.) appears to have an **antiparticle**. Each antiparticle is a charge conjugate of its particle – same mass, spin, etc. but opposite charge. Antiparticles are routinely created in particle accelerators when high energy particles are made to collide with each other. Their lifetimes are short, because when they encounter their equivalent particle, they **annihilate** each other, converting their matter to energy. It is even possible to form **antiatoms** from these antiparticles. These have been produced in high-energy particle accelerators, but only the simplest nuclei have been seen so far, with very short lifetimes of less than 10^{-10} s. Matter appears to dominate over antimatter in the universe. If, at the instant of the Big Bang, there were equal numbers of particles and antiparticles, the slight differences in their decay characteristics would change this equality, resulting in the present excess of matter. It is not impossible that, somewhere in the Universe, there exist substantial quantities of antimatter, even entire antimatter planets and stars, but, so far, none have been detected.

Electron and positron

Symbols:

electron $_{-1}^{0}e$ β^-

positron $_{+1}^{0}e$, $\bar{e}$, β^+

The existence of the positron (a positive electron), was predicted by the English physicist P.A.M. Dirac in 1930, and was first observed in 1932 by C.D. Anderson when a high energy cosmic ray particle was observed in a cloud chamber fitted with a lead partition. Cosmic rays tend to have such high energies that, even with the most powerful magnetic fields, they simply do not curve sufficiently to enable accurate measurements to be made. The lead partition slowed the particles sufficiently for a curved track to be seen. Anderson saw a characteristic electron track, but curving the wrong way. This was the **positron,** a new fundamental particle with the same mass and magnitude of charge as the electron, but **positively**, rather than negatively, charged. Many radioactive nuclei are positron emitters (in the case of radioactive decay, the positron is called a β^+-particle). An example of positron (β^+-particle) emission is the decay of the radioactive isotope sodium-22 into neon-22:

$$_{11}^{22}Na \rightarrow _{10}^{22}Ne + _{+1}^{0}e + _{0}^{0}\nu + Q$$

where Q is the energy released.

Proton and antiproton

Symbols

proton p

antiproton $\overline{p}$

The antiproton was discovered in 1955 by Segre and Chamberlain, an achievement which earned them the 1959 Nobel Physics Prize. High-energy protons from new high-energy particle accelerators just becoming available were directed at a metal target. Some of the energy of the protons was converted into a spray of new particles when they collided with the nuclei in the target. Some of these new particles were observed to have the same mass as protons, but were negatively charged. Antiprotons are stable in isolation, but again annihilate on contact with protons to produce π-**mesons** and other particles, which in turn decay (in microseconds) into positron/electron neutrino/ antineutrino pairs and photons. Antiprotons and protons which pass close to each other (but not close enough to annihilate) can neutralise each other's charges, creating a neutron and an antineutron.

Neutrino and antineutrino

In radioactive decay, the emission of a β^--particle is accompanied by an antineutrino ($\bar{\nu}$), while positron (β^+) emission is accompanied by a neutrino (ν).

Pair production and annihilation

When an antiparticle encounters its equivalent particle, they annihilate each other, and their combined mass is released as photons (energy). At the instant of the Big Bang, it is likely that particles and antiparticles would have been created in equal numbers. Many of these would have annihilated with each other, and it is only due to slight differences in the decay characteristics that there remain the particles of matter which make up our Universe.

Electron–positron pairs can be formed when γ-rays or other very high energy photons of electromagnetic radiation pass through matter. This **pair-production** is a striking example of the interchangeability of matter and energy. When a sufficiently energetic γ-ray photon passes close to a nucleus, it may vanish, creating an electron–positron pair in its place. We can use Einstein's mass–energy relation to calculate the minimum γ-ray photon energy needed for pair production.

Assuming the rest masses of the electron and positron to be 9.1×10^{-31} kg, and the velocity of light to be 3.0×10^{8} m s^{-1}, we have:

minimum γ-ray photon energy

$$= 2 \times (9.1 \times 10^{-31}) \times (3.0 \times 10^{8})^2$$
$$= 8.2 \times 10^{-14} \text{ J}$$
$$= 1.02 \text{ MeV.}$$

If the incident γ-ray photon has greater energy than this, the excess energy appears as kinetic energy shared between the proton and the electron. In the reverse process, called **annihilation**, the electron and positron disappear when they combine, and their mass is transformed into energy in the form of electromagnetic radiation.

Exotic particles

In the 1930s, theories were advanced to explain the forces which bound the nucleus together (see Fundamental forces, on page 449), which required the existence of a set of particles, known as **bosons**, which were exchanged between particles whenever they exerted forces on each other. During the search for these bosons, an increasing number of other particles with a wide range of masses and unusual properties was discovered. Just when the world of physics had started to believe that it had discovered a simple framework for the structure of matter, things were once again getting out of hand. It was not until the 1960s that Murray Gell-Mann and Yuval Ne'emen independently worked out the simplicity of the underlying structure of these exotic particles.

Quarks and antiquarks

When high-energy electron beams collide with protons and neutrons, the resulting collisions reveal that the electrons are being scattered by particles within the protons and neutrons themselves. These particles are called **quarks** and **gluons**. When matter is probed with radiation, its constituents are likely to be ejected. Atoms become ionised when electrons are knocked out of them, and at higher energies, neutrons and protons can be knocked out of nuclei. However, this is not the case with neutrons and protons – quarks cannot be knocked out of them. While there is now no doubt that quarks exist, it appears that they must be confined in pairs or triplets, and cannot exist separately. They are currently regarded as fundamental particles, as we do not have the ability to probe deeper into any underlying structure they may have. Of the quarks which have been identified, three are sufficient to account for the structure of matter at our level of interest. These are named the **up, down**, and **strange** quarks. The names have no significance, they are just labels to distinguish them from each other. Table 5.6 shows the charge and strangeness of these quarks and their antiquarks, together with the abbreviations and symbols used in this book.

In addition, there are three other quarks, but knowledge of their properties is only required for much deeper levels of understanding. These are:

top t (also called 'truth')
bottom b (also called 'beauty')
charm c

These also have corresponding antiquarks.

Particle classification

Physicists use the different properties of subatomic particles to group them into various types and classes. The most basic classification is made according to particle **mass**.

- The lightest particles are called **leptons**.
- The heaviest particles are called **hadrons**.
- The hadron group is further sub-divided into **baryons** (which are the heaviest particles) and **mesons**, which are somewhat less massive.

In addition to their masses, these groups of particles also differ in the type of fundamental force through

Table 5.6 Quarks and antiquarks

Quark	Abbreviation		Charge/e	Strangeness s
up	u	(↑)	+ 2/3	0
down	d	(↓)	– 1/3	0
strange	s	(?)	– 1/3	– 1
Antiquark			**Charge/e**	**Strangeness s**
antiup	$\bar{u}$	(↑) filled	– 2/3	0
antidown	$\bar{d}$	(↓) filled	+ 1/3	0
antistrange	$\bar{s}$	(?) filled	+ 1/3	+ 1

which they interact. The **baryons** and **mesons** mainly interact through the **strong** nuclear force whereas the **leptons** only take part in interactions involving the **weak** nuclear force.

Leptons

The name lepton means 'light one'. The electron, the muon (μ) and the tau (τ), together with the neutrino associated with each, belong to this 'family' of light particles. Like the quark family there are six members (and corresponding antiparticles). These light particles are regarded as **fundamental** particles, and are associated with the **weak** force (see Fundamental forces, on page 449). Table 5.7 shows the members of the lepton family, together with their respective particle and antiparticle symbols.

Table 5.7 Some of the particles in the lepton family

Name	Particle symbol	Antiparticle symbol
electron	e^-	e^+
electron neutrino	ν_e	$\overline{\nu}_e$
muon	μ^-	μ^+
muon neutrino	ν_μ	$\overline{\nu}_\mu$

In addition, the **tau** (τ) and the **tau neutrino** (ν_τ) complete the lepton family.

Hadrons

These include the neutron and proton and other, generally heavy, particles which respond to the **strong** force. They are not regarded as fundamental particles in themselves, but as being made up of **quarks** (which are regarded as fundamental). When quarks combine together in triplets, the resulting particles (including neutrons and protons) are called **baryons**. Pairs of quarks produce particles called **mesons** (for example the **pi-meson** or pion). The hadrons, or strongly interacting particles, are the only ones to be influenced by all four of the forces of nature.

Baryons

Baryons are particles formed by triplets of quarks, and their charge, mass and strangeness are governed by the properties of the quarks they contain. Table 5.8 shows how quark triplets are arranged to form some of the more commonly known baryons.

The neutron and proton are the two best-known baryons, but there are many more, very short-lived baryons which have been observed in detectors as a result of cosmic ray interactions or particle collision experiments.

At this point, it is useful to introduce the concept of **baryon number** (B). In classifying elementary particles, the proton and neutron are both assigned a baryon number of +1, and their antiparticles a baryon number of –1. All the baryons have a baryon number of +1 or –1, while the mesons all have baryon number 0. In the same way that **charge** and **mass number** are conserved during nuclear reactions, **baryon number** is also conserved during interactions between elementary particles. For example, in the interaction between a proton and an antiproton, **π-mesons** (baryon number = 0) may be formed. One possible reaction is:

$$\overline{p} + p \rightarrow \pi^+ + \pi^-$$

Table 5.8 Arrangement of quarks in particles

	Neutron	Proton	Lambda
configuration of quarks	↑ ↓↓ udd	↓ ↑↑ uud	↓ ↑ ? uds
charge/*e*	$+\frac{2}{3}-\frac{1}{3}-\frac{1}{3}=0$	$+\frac{2}{3}+\frac{2}{3}-\frac{1}{3}=+1$	$+\frac{2}{3}-\frac{1}{3}-\frac{1}{3}=0$
strangeness	0	0	– 1
mass/u	1	1	1.1

On the left-hand side, the total baryon number is zero (+1 and –1). On the right-hand side the total baryon number must also be zero, which is satisfied as the **π-mesons** have baryon number = 0.

Table 5.9 below shows examples of the lowest mass members of the baryon group, all with baryon number = 1.

The property of **strangeness** needs some mention here. It describes a very fundamental property of the quarks, but our understanding of it parallels scientists' concept of charge in the late 19th century. They knew that matter possessed this property that caused attractive and repulsive forces, they knew that there were two varieties of charge, named positive and negative, yet they had no clue what this property was. (Indeed, even now, there are few physicists who would attempt to say what charge actually is… .)
In a similar way, we know that there are fundamental attributes possessed by particles which govern their behaviour, but we do not know what these properties are. The fact that they have such whimsical names as 'charm', 'beauty', etc. does not detract from the powerful insight and predictive power they give to our knowledge of matter. Strangeness relates to the behaviour of particles whose decay time is slower than is expected, given the forces involved. All the leptons, and some hadrons, including the nucleons (proton and neutron), have zero strangeness. The lambda baryon contains a strange quark, which is slightly more massive than the up or down. The lambda is highly unstable, with a lifetime in the order of 2×10^{-10} s, decaying to a proton and a pion, or a neutron and a pion. The strange quark is transmuted to an up or down quark, and the difference in mass is carried away by the pion which is created in the decay. This is analogous to β^- decay, in which a down quark in the neutron transmutes to an up quark, releasing an electron and an antineutrino.

Mesons

Mesons are formed from quark–antiquark pairs. For mesons which carry charge, the quark and antiquark are of different types: the positive pi-meson (pion) is made from an up quark and a down antiquark. Neutral mesons are made from quark–antiquark pairs of the same type. The oppositely charged particles orbit around each other, but are inevitably short-lived compared to neutral mesons. All particles belonging to the meson group have baryon number = 0. About 50 mesons have been identified, and the lowest mass members of three of the meson families are shown in Table 5.10.

It should be appreciated that these groupings of elementary particles may not stand the test of time. Research in this field forms one of the main cutting edges of scientific endeavour, and while there is growing confidence in these classifications, new discoveries are being made all the time, which need to be incorporated into these ideas.

Table 5.9 Examples of baryons

Family	Symbol	Charge	Strangeness
nucleon	N	$p = 1$; $n = 0$	0
lambda	Λ	0	– 1
delta	Δ	2, 1, 0, – 1	0
sigma	Σ	1, 0, – 1	– 1
xi	Ξ	0, – 1	– 2
omega	Ω	– 1	– 3

Table 5.10 Examples of mesons

Family	Symbol	Charge	Strangeness
pion	π	1, 0, – 1	0
eta	η	0	0
kaon	κ	1, 0	1

Fundamental forces

The interactions between bodies occur due to the forces they exert on each other. Physicists have discovered four types of force, which may, ultimately, prove to be manifestations of one, fundamental force. On an everyday basis, we need only consider two of these forces – **gravity** and the **electromagnetic force**. Gravity is responsible for the assembly of the planets and the motion of the Universe, while electromagnetic forces control the assembly of atoms and all the chemical, mechanical and electrical properties of matter, as well as being the means by which bodies in contact exert forces on each other. At a deeper level, what goes on inside the nucleus can only be explained by considering the forces at work in there.

Two such forces have been identified – named the **strong** force, and the **weak** force. Each of these forces can be described in terms of a field which is a region in space defining the strength and direction of the force at any point. A key element in the development of these field theories is that there must exist particles which 'carry' the force and transmit it from one particle to the other. These messenger particles are called **bosons**, and they are quite distinct from the fundamental matter particles. Table 5.11 summarises the properties of the four forces.

Gravity

This force is an attractive force only, is exerted between all particles with mass, has a range which extends to infinity and is the weakest of the four fundamental forces, in terms of its magnitude at the range of a proton diameter. However, it is responsible for the forces that caused the planets and stars to be assembled from the matter created in the Big Bang, the forces which dictate the motions of the bodies in the Universe and, ultimately, the fate of the Universe itself. This topic is covered in some detail in Sections 1.10 and 1.11 on Gravitation and Gravitational fields. Surprisingly little is known about the origins of this force, and the search for the **graviton**, the boson responsible for the interaction between masses, has so far proved fruitless.

Electromagnetic force

This force is exerted between particles which possess the property which we call electric **charge**, can be

Table 5.11 The fundamental forces

Force	Particles affected	Range	Relative strength at a range of 1 proton diameter	Bosons (particles exchanged)	Role in universe
Strong	Quarks (and particles made from quarks)	10^{-15} m	1	Gluons between quarks Mesons between hadrons	Binds quarks within protons, neutrons (and other hadrons)
Electro-magnetic	All charged particles, including quarks	Infinity	10^{-2}	Photons	Binds atoms and molecules, determines mechanical, chemical and electrical properties of matter
Weak	Quarks and leptons	10^{-17} m	10^{-5}	*W*, *Z*, particles	Controls nuclear decay, fuels Sun and stars
Gravitational	All particles with mass	Infinity	10^{-40}	Gravitons (as yet undetected)	Assembles matter into planets, stars and galaxies, and controls their motion

attractive or repulsive, and has a range which extends to infinity. The exchange particle in this case is the **photon**. The photon is emitted by one particle and absorbed by the other. The electromagnetic force is responsible for binding electrons around atoms and controls the chemical, mechanical and electrical properties of matter. It is a force which has been extensively studied and the full theory is described in the quantum electrodynamics (QED) theory, which is somewhat beyond the scope of this book.

Strong nuclear force

In order to understand the behaviour of the nucleus, it is necessary to consider the forces acting within it. Positively-charged protons repel each other due to their repulsive electrostatic forces, and gravity is too weak to bind them together. Therefore there must exist a force which is attractive at extremely short range, and strong enough to overcome the repulsive electrostatic force between the charged particles. Furthermore, the force must not be affected by charge, and it must become repulsive at extremely short range (otherwise the nucleons would be pulled together into a vanishingly small volume). This is the strong nuclear force between nucleons, and it is transmitted through the exchange of mesons, for example **pi-mesons** (or **pions**). The strong force which binds the quarks within the hadrons is carried by exchange bosons called **gluons**. The property that creates the force is something akin to charge, but there are three types, not just two. These are called 'colours' and even named red, green and blue. Quarks can only associate in groups that give zero colour – red, green and blue, as in neutrons and protons, or quark–antiquark pairs formed as mesons. Since gluons act on each other, the strong force actually gets stronger at greater distances within the subatomic particles, which constrains the quarks within.

Weak nuclear force

This has been incorporated into a larger theory which explains the electromagnetic force and the weak force simultaneously – the **electroweak** theory. The weak force acts within the nucleus by exchanging *W* and *Z* bosons. The *Z* is neutral, while *W* can be positive or negative. Quarks and leptons exert forces on each other by exchanging these bosons, which are much heavier than protons (90 – 100 × heavier). In doing this the principle of energy conservation is violated. This is allowed at the quantum level, but only for a very short time, which is why this force is so weak, and has such short range (100 000 times weaker than the electromagnetic force).

Unifying the forces of nature

The great dream of the modern theoretical physicist is to create a unified theory which describes all four fundamental forces. The first steps towards achieving this were taken in the 1960s when Abdus Salam and Steven Weinberg unified the electromagnetic and weak forces within a single theory, which predicted the existence of the *W* and *Z* bosons. In 1983, *W* and *Z* bosons were actually observed in the proton–antiproton collider at CERN, confirming the validity of the unified **electroweak** force. The next step is the so-called **grand unification** in which the electroweak and strong forces will be combined. This link requires that quarks and leptons have the ability to turn into each other by the exchange of an extremely massive boson – the *X*-particle. Searching for these massive particles requires accelerators capable of much higher energies than anything so far constructed. The final step in which the gravitational force would be unified with the other three fundamental forces would give what physicists call a 'theory of everything', and would indicate that all four fundamental forces can be explained by a common framework of understanding.

Exchange bosons and Feynman diagrams

Each of the fundamental forces of nature causes interactions through the exchange of messenger particles called **exchange bosons**. The bosons can be thought of as occurring in clouds around each particle, able to act as messengers carrying the force to other particles within range. Graphical representations of these interactions are known as **Feynman diagrams** after the physicist Richard Feynman who devised them as a simple way of visualising the complex mathematics of the interactions. A Feynman diagram is a space–time representation in which the paths of particles in space and time are represented as lines. An interaction between particles is represented as a point where lines meet. By convention, time is represented as proceeding from the bottom of the diagram to the top, so incoming particles arrive from the bottom of the diagram. If the ideas behind the Feynman diagrams are applied in full, they can give a complete picture of the range of possible outcomes (and their relative probabilities) from any given starting combination of particles. We are only concerned here with the simple use of the Feynman diagram to represent specific interactions. Some examples should help to clarify their use.

The electromagnetic force

In this case, the exchange particle is the **photon.**

Figure 5.47(a) shows the **repulsive** force between two **similarly** charged particles (electrons) caused by the exchange of a **photon**. After emission of the photon by one particle, it is re-absorbed by the other.

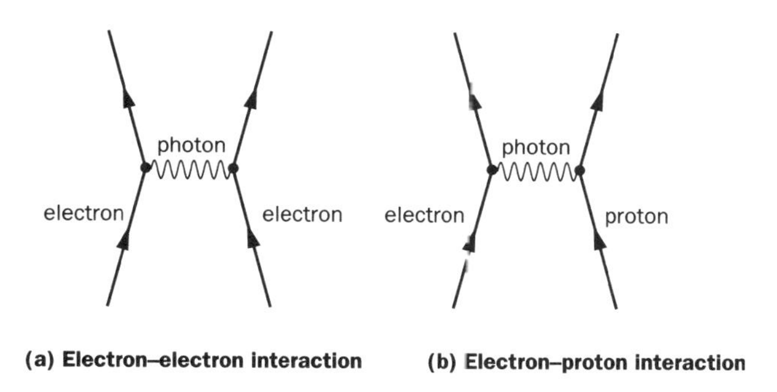

Figure 5.47 Feynman diagrams for electromagnetic force

Figure 5.47(b) shows the **attractive** force between particles carrying **opposite** charges. In the modern quantum field theory, the appearance of the attractive or repulsive force is regarded as a secondary consequence of the primary process of exchange of the photon.

The strong nuclear force

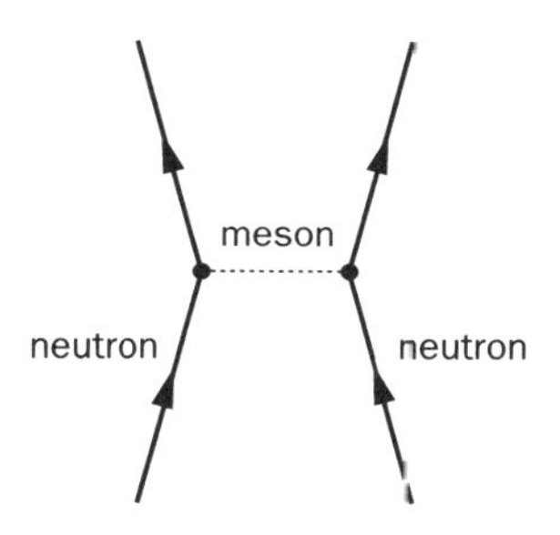

Figure 5.48 Feynman diagram showing the strong nuclear force between two neutrons

Figure 5.48 shows the **strong** nuclear force between two neutrons resulting from the exchange of a **meson**. Other possible exchanges exist, depending on the distance between the nucleons, and may involve single pions, two π-mesons etc. and possibly include an excited intermediate state for one of the nucleons.

The weak nuclear force

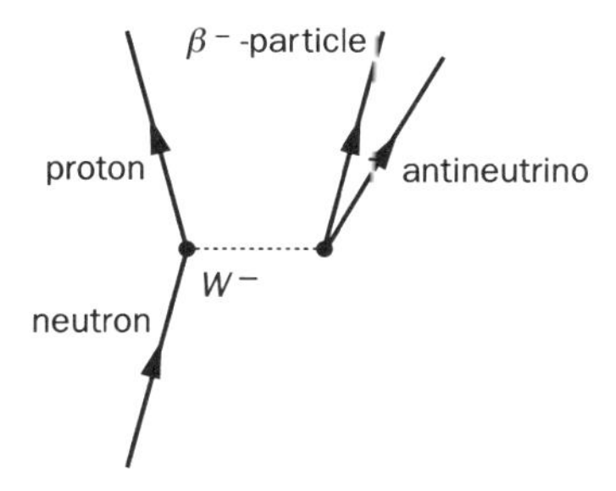

Figure 5.49 Feynman diagram of β^- decay

Figure 5.49 shows a neutron decaying into a proton by emitting a β^--particle and an antineutrino by the exchange of a W^- boson.

Gravitational force

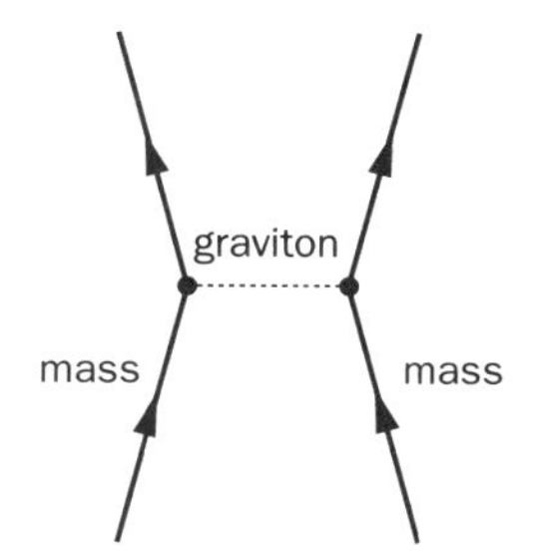

Figure 5.50 Feynman diagram for gravitational force

Figure 5.50 shows how any two particles possessing **mass** exert a gravitational force on each other by exchange of a **graviton**. Unlike the other gauge bosons, there is at present no experimental evidence for the existence of the graviton.

Exploring matter

To observe something, three things are needed: illumination (some form of radiation), the object (to reflect the radiation, or change it in some way) and a detector. You are reading this page by seeing with your eye (the detector) the effect of visible light (the radiation) being reflected differently by the ink and the paper (the object). Microscopes allow objects that are close together to be distinguished, but visible light has a limitation: you cannot distinguish objects which are closer together than the wavelength of the light. Light wavelengths are much too long to distinguish atoms. Shorter wavelengths can be used, but the eye can no longer detect them. Atoms are typically separated from each other by distances of the order of 10^{-10} m in solids, so X-rays of similar wavelengths are diffracted by the crystal structure of solids, and this is the basis of X-ray crystallography. Beams of electrons can be accelerated and behave as very short-wave radiation. This is the principle of the electron microscope. More acceleration means greater power, and a bigger microscope. These allow molecular structure to be distinguished, but not individual atoms. To 'see' the structure of the atom needs an accelerator several metres long, but to resolve the nucleus, several kilometres of accelerator are needed. Only the most powerful of accelerators can peer into the structure of the neutrons and protons in the nucleus, and the quarks inside these are the limit of our ability to resolve detail at present. The resolving power of particle beams is therefore related to the energy of the beam. To 'see' levels of detail, the following energies are needed:

100 eV (1×10^2 eV)

the electron cloud around the nucleus

100 MeV $(1 \times 10^8$ eV)
the nucleus itself is revealed
10 GeV $(1 \times 10^{10}$ eV)
quarks within the nucleus are resolved

Bombardment by high-energy particles

The structure of matter can be analysed by bombarding it with energy, and observing the result. Even the simple chemical flame test works in this way – a sample of the substance is excited by heating it, and the colour of the light emitted is used to infer the chemical nature of the elements comprising the substance. Spectroscopy is merely a more precise development of this technique.

To probe more deeply into the finer structure of atoms themselves requires much shorter wavelengths, or high energy beams of particles. When Rutherford probed the nature of the atom by bombarding thin metal foils with α-particles, he was employing a similar technique: by observing the effect of the atoms on the α-particles, he could draw precise conclusions about the nature of the matter. Using α-particles, he was able to 'see' the structure of the atom as a tiny nucleus surrounded by a comparatively huge region of empty space. However, he could only perceive the nucleus as a point of matter. Studying the structure of the nucleus itself requires particles of extremely high energy.

Particle accelerators

Charged particles (electrons, protons, etc.) may be accelerated by a potential difference. In particle accelerators, beams of particles in a vacuum are accelerated across gaps between electrodes maintained at a high potential difference. Accelerators differ in the manner in which they produce the p.d. and in the arrangement of the electrodes.

Van de Graaff accelerator

The basic principle of the Van der Graaff generator is shown in Figure 5.51. A rubber belt moves continuously past a brush, connected to one terminal of a high-voltage supply (the other terminal is earthed). Charge is 'sprayed' onto the belt by the brush, and is carried to the dome of the generator where it is removed by the second brush. Charge (positive or negative) builds up on the dome, which is heavily insulated, and can reach very high potential. Charged particles can be accelerated to energies up to 20 MeV by this voltage.

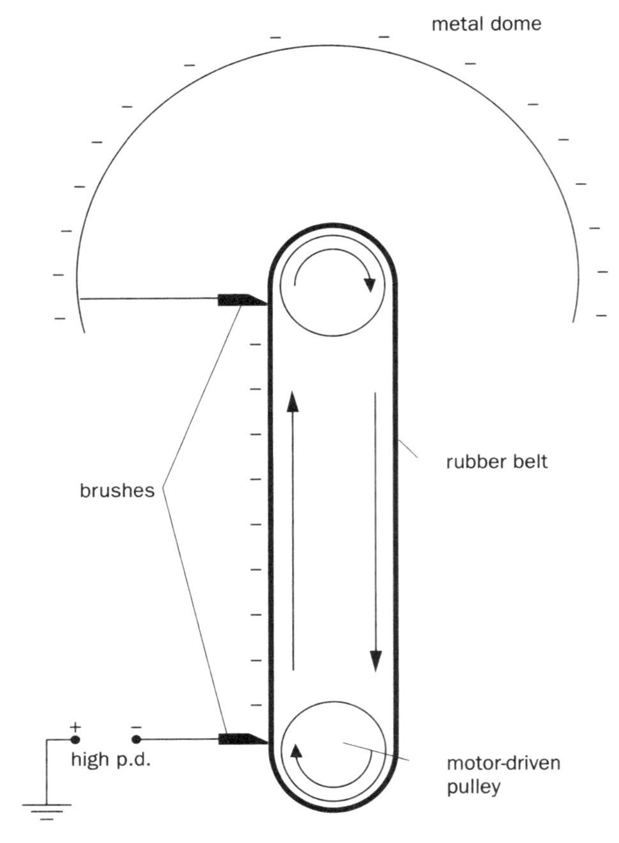

Figure 5.51 The Van der Graaff generator

Linear accelerator

Charged particles are accelerated in a straight, evacuated tube by a series of electrodes which are supplied with a high frequency alternating p.d., as shown in Figure 5.52. An ion would be attracted towards an oppositely charged electrode, and repelled from it on the next phase of its cycle when it became similarly charged.

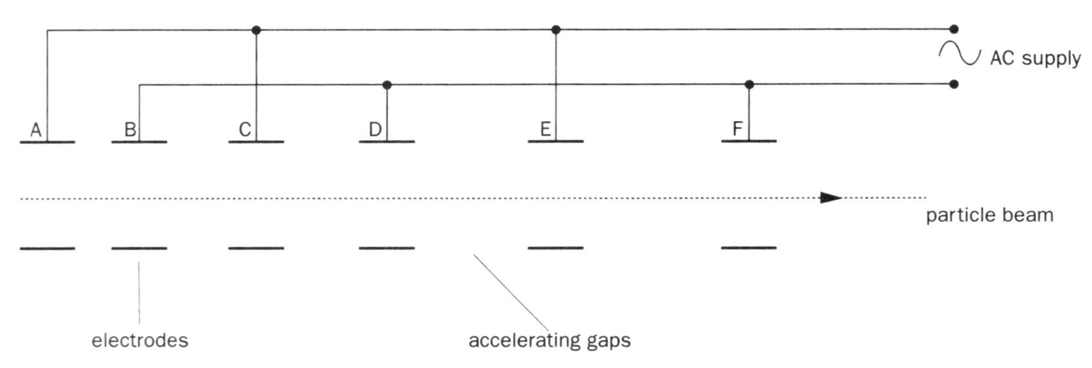

Figure 5.52 Linear accelerator

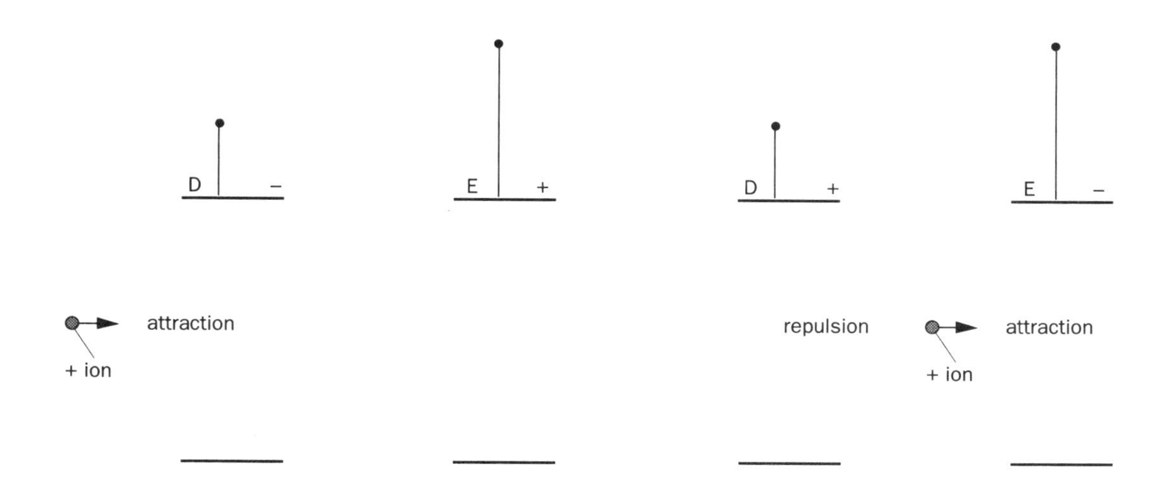

Figure 5.53 Action of electrodes

The timing of the alternating p.d. is such that the ion is attracted towards an opposite charge, which reverses as the ion passes through it. Figure 5.53 shows a positive ion being attracted towards electrode D, which is negatively charged. As the ion passes through D, the polarity reverses and D becomes positive. The ion is then repelled by this electrode, and attracted towards the next, oppositely charged electrode in the chain. The sequence is repeated at each electrode, so the ion is accelerated in a series of steps. The gaps between electrodes must increase in length as the acceleration of the ions means that at each gap they will travel further during the time taken for the polarity to reverse. The energy achieved is limited by the maximum length of accelerator which can be built. For example, the 3 km long Stanford linear accelerator, built in 1960, achieved energies of 30 GeV by accelerating electrons (emitted from a conventional electron gun at around 40 MeV) through 960 electrodes.

Cyclotron

The length limitation of the linear accelerator can be overcome by making the ions follow a circular or spiral path. This is the principle of the cyclotron, in which a magnetic field is directed at right angles to the plane of the ions, causing them to follow a curved path. Only two electrodes are needed, and these are in the form of two, hollow D-shapes, with a gap between them. Inside the Ds is an evacuated chamber in which the charged particles are accelerated.

Ions are inserted near the centre of the evacuated cylindrical chamber, as shown in Figure 5.54. Positive ions will be attracted towards whichever of the Ds is negative. The perpendicular magnetic field will cause the ions to move in a circle, and emerge into the gap. If the polarity of the Ds is reversed at the instant an ion reaches the gap, the ion will accelerate across it. The ion will continue to move in a circle, but this time with a wider radius, due to its increased velocity. If

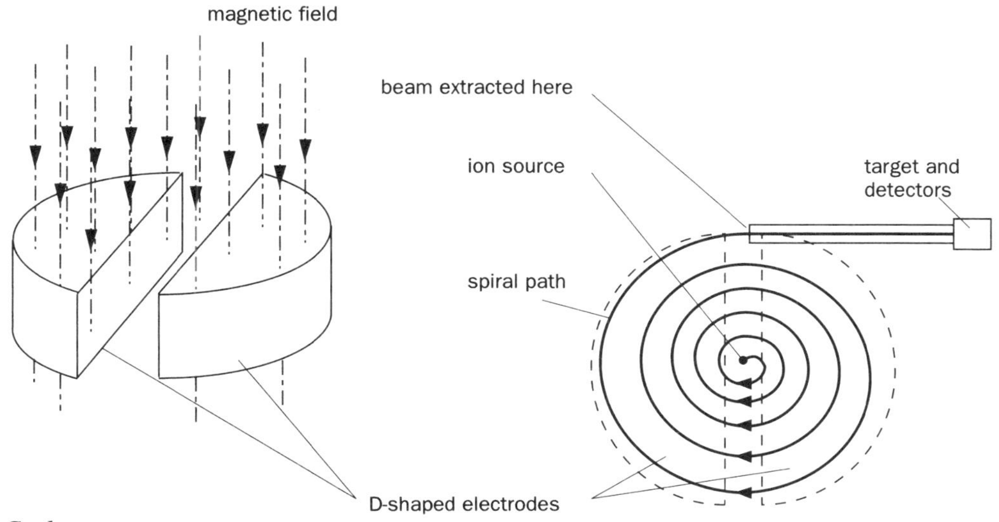

Figure 5.54 Cyclotron

the polarity of the Ds alternates every time the ion reaches the gap, it will increase its velocity every time, and so move on a path which spirals outwards towards the edge of the Ds. At the appropriate instant, the ions will be attracted towards the exit chamber by an appropriate potential. Once free of the magnetic field, they will move in a straight line.

Energies up to 1 GeV are achievable in this way. The energy is limited by the effects of relativity – as the particles approach the speed of light, their mass increase becomes significant enough to inhibit their acceleration, so they reach the gaps out of step with the frequency of the alternating supply. Also, the magnetic field is not uniform, but is barrel shaped to keep the particles contained.

Synchrotron

In the synchrotron, ions are made to follow the same circular path millions of times. On each circuit, they pass through a copper cavity which is given an accelerating field of a few thousand volts by radio frequency waves. The ions are guided through a pipe only a few centimetres across, evacuated to an extremely high vacuum. The enormous energies reached by the particles mean that the pipe must follow a very large radius, since even superconducting magnets can only achieve a gentle curvature at these energies.

Figure 5.55 shows the basic layout of a synchrotron. The strength of the magnets has to be increased as the ions gain energy, and this increase must be **synchronised** with the increase in energy on each circuit. Additional focusing magnets must be provided to maintain the beam cross-section and prevent particles colliding with the walls of the pipe. Ions are injected at high energies from the linear accelerator. Each time they pass the accelerating cavity, their energy is increased. This increase in energy manifests itself as an increase in mass as the particles approach the speed of light. The **super proton synchrotron** at CERN, near Geneva, produces particle energies of up to 500 GeV from a ring 2 km in diameter, using conventional magnets. Ten thousand billion protons can be injected every six seconds into the ring. In the four seconds it takes them to reach 400 GeV, they will travel over one million kilometres! The Fermilab **Tevatron** at Chicago achieves energies of 1 TeV (1×10^{12} eV) by using superconducting magnets. It is possible to make beams of particles and their antiparticles travel in opposite directions, a short distance apart, inside the pipe. A slight adjustment at the appropriate time brings the matter and antimatter together in a head-on collision. This effectively doubles the energy of the collision. In addition, no energy is wasted simply moving particles in a stationary target, so this method greatly enhances the

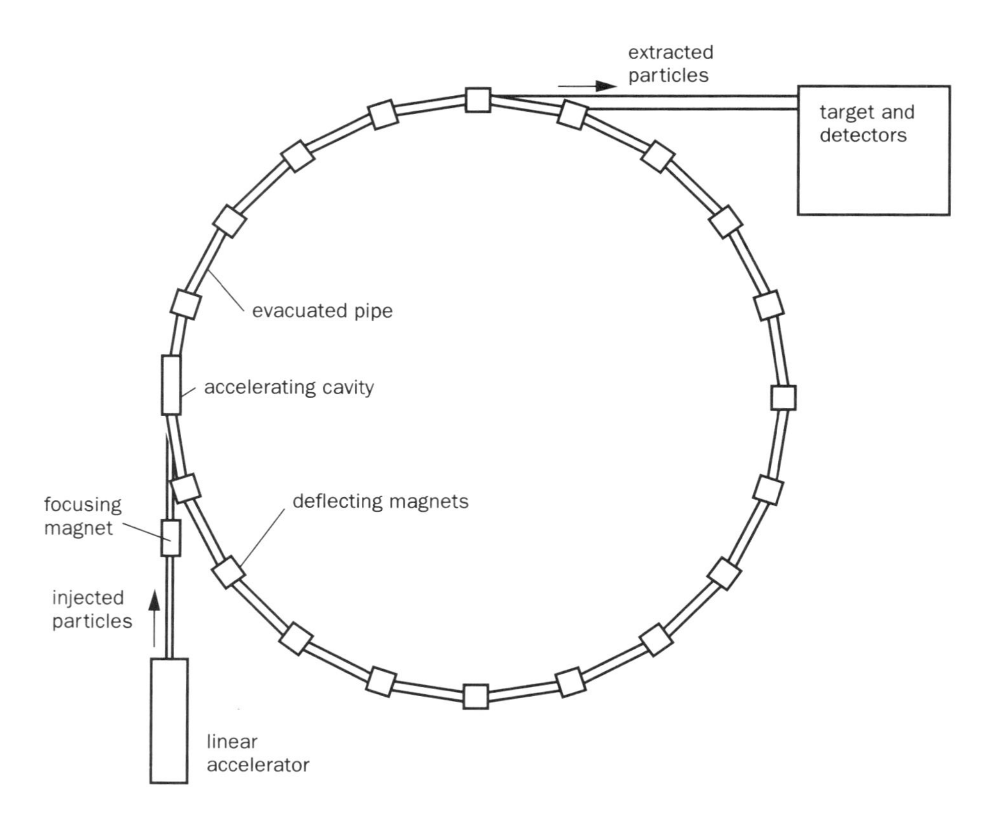

Figure 5.55 Synchrotron

energy and therefore resolving power of the synchrotron as a tool for probing the secrets of matter.

Synchrotrons and cyclotrons suffer from energy losses – so-called **synchrotron radiation**. Charged particles which are moving in circular paths are accelerating, and therefore must be emitting radiation. Much of the energy supplied to the particles is dissipated in this way. Synchrotron radiation sources, e.g. at Daresbury, Cheshire, or the European facility at Grenoble, are used as intense sources of short wave radiation for work in solid state, bio and chemical physics.

(Left) *Aerial view of CERN and the surrounding region. The smallest ring shows the underground position of the proton synchroton, the middle ring is the super proton synchroton with a circumference of 7 km and the largest ring (27 km) that of the European laboratory for particle physics (LEP)*

(Below) *Inside the LEP tunnel*

Self-assessment

Qualitative assessment

1. (a) Which are the **two** fundamental principles which **appear** to be violated in β^--particle emission?

(b) Explain how this observation led to the discovery of a new elementary particle, and describe its properties.

2. (a) What is an **antiparticle**?

(b) Why do antiparticles generally have very short lifetimes?

3. With reference to interactions between elementary particles, explain what is meant by

(a) **pair production**

(b) **annihilation**.

4. (a) Name the **six** types of quark.

(b) Describe the combination of quarks which produces **baryons**, and name three baryons.

(c) Explain, with an example, the principle of **conservation of baryon number**.

(d) Describe the combination of quarks which produces mesons, and name three mesons.

5. (a) Name two particles belonging to the **lepton** family.

(b) With which fundamental force are leptons associated?

6. (a) Name the **four** fundamental forces of nature, name the **exchange boson** associated with each and the particles most strongly affected.

(b) Describe the relative **strength** and **range** of each force.

7. A moving charged particle experiences a force in a magnetic field which causes it to move in a circular path. Explain how the radius of the path depends on

(a) the strength of the field

Self-assessment *continued*

(b) the charge/mass ratio of the particle
(c) the velocity of the particle.

8. Explain the principles of the
(a) linear accelerator
(b) cyclotron
(c) synchrotron
suggesting the sort of particle energies which can be achieved and what limits the maximum energy.

9. (a) Explain how Feynman diagrams are used to show the interactions between particles.
(b) Sketch Feynman diagrams for the interactions between:
(i) two electrons at rest
(ii) two neutrons.
(c) Sketch a Feynman diagram for β^+-particle (positron) decay.

5.6 Quantum effects

In the modern world, we take for granted the interaction between electromagnetic radiation and matter. The giant *Hubble* space telescope runs on electricity generated by sunlight falling on its solar panels; automatic cameras calculate their exposure settings by measuring the current produced when light shines on *photo-sensitive* material; the huge world photocopying industry is based on the ability of light to make a material conduct electricity. In this section we will explore the phenomenon of the *photoelectric effect*, and examine the implications this has for our understanding of the nature of electromagnetic radiation. We will conclude this section with a brief study of Louis de Broglie's proposal concerning the wave nature of matter.

Photoelectric emission

In the late 19th century, before Planck's quantum theory became accepted, certain observations of the electrical effects of light were puzzling scientists. Hertz had discovered that sparks jumped gaps much more easily if ultra-violet radiation were directed onto the electrodes. In 1902, it was discovered that electrons could be ejected from metals by electromagnetic radiation (the fact that identical electrons were ejected from different metals supported the idea that electrons were a fundamental part of all matter).

Any metal will emit electrons if it is given energy. Heated coils in electron guns give off copious quantities of electrons. This is the process known as **thermionic emission**.

Photoelectric emission is the emission of electrons from the surface of a metal when it is exposed to electromagnetic radiation of sufficiently high frequency.

Metals like zinc emit photoelectrons when illuminated by certain radiations. Table 5.12 illustrates the response of some metals when exposed to different frequencies of electromagnetic radiation.

Table 5.12 Comparison of effects of radiation on metals

Metal	X-rays	Ultra-violet	Blue light	Red light
zinc	yes	yes	no	no
sodium	yes	yes	yes	no
caesium	yes	yes	yes	yes

Simple demonstration of the effect

Ultra-violet radiation (e.g. from a mercury vapour lamp) is directed onto a freshly cleaned zinc plate connected to an electroscope as shown in Figure 5.56. If the initial charge on the electroscope is **positive**, then **no effect** is observed. If, however, the initial charge is **negative**, the gold leaf rapidly falls, showing that the electroscope is **discharging**. Inserting a sheet of glass between the lamp and the zinc plate stops the discharge. Glass does not transmit ultra-violet radiation, thus showing that it is the ultra-violet radiation that is causing the discharge of the electroscope. We can explain these observations in terms of the electron emission which occurs from the zinc plate as a result of the incident ultra-violet radiation.

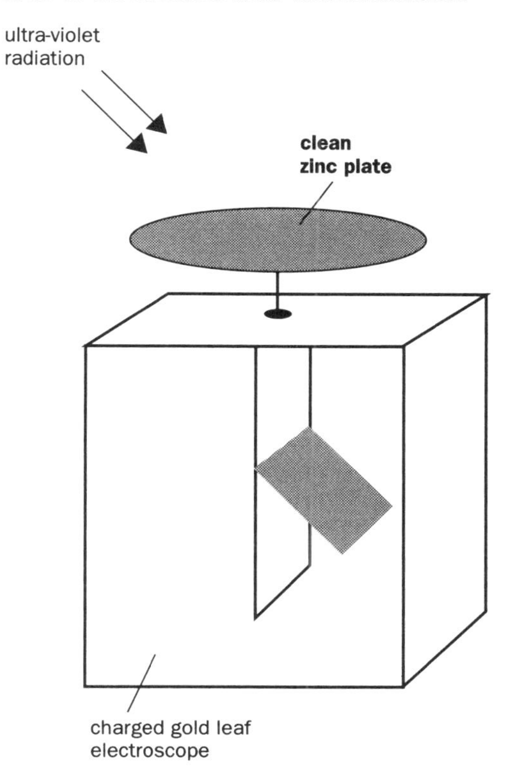

Figure 5.56 Demonstration of photoelectric emission

In Figure 5.57(a), when the initial charge is **positive**, the emitted electrons are immediately **attracted** back to the zinc plate, so although photoemission may occur, the electroscope is not discharged since the electrons are re-captured.

When the initial charge is **negative**, as shown in Figure 5.57(b), any emitted electrons will escape as they are immediately **repelled** by the electric field surrounding the charged plate. It is this loss of electrons that causes the electroscope to discharge. It is important to remember that the emission is due to the ultra-violet radiation; the charged electroscope is used simply to observe the effect.

Observations based on measurements made of the charge, mass and energy of the emitted electrons lead to the following **laws of photoelectric emission**:

- The **number** of electrons emitted per second from any metal is directly proportional to the **intensity** of the radiation falling on it. (Intensity is a measure of the energy per unit surface area. In the case of light, it is equivalent to the **brightness** of the light.)
- The photoelectrons are emitted from a given metal with a range of **kinetic energies**, from zero up to a maximum. The **maximum** energy increases with the **frequency** of the radiation and is independent of the **intensity** of the radiation. (Shining a brighter light of the same colour produces more photoelectrons per second, but does not increase their kinetic energies.)
- For each metal, there is a minimum frequency required to produce emission. This is called the **threshold frequency** (f_0). (Similarly, there is a corresponding **threshold wavelength**, above which no emission can occur.) Radiation below this frequency cannot produce emission, no matter how **intense** the radiation. So even a bright industrial laser (visible light) cannot cause photoemission from zinc, whereas a weak ultra-violet source can.

The **threshold frequency** for a metal is the **minimum** frequency of electromagnetic radiation which can cause photoemission.

The 'photoelectrons' emitted from a surface by electromagnetic radiation are identical to any other electrons such as those found in cathode rays or ejected from nuclei as β^--particles. They have the same mass, charge, etc.

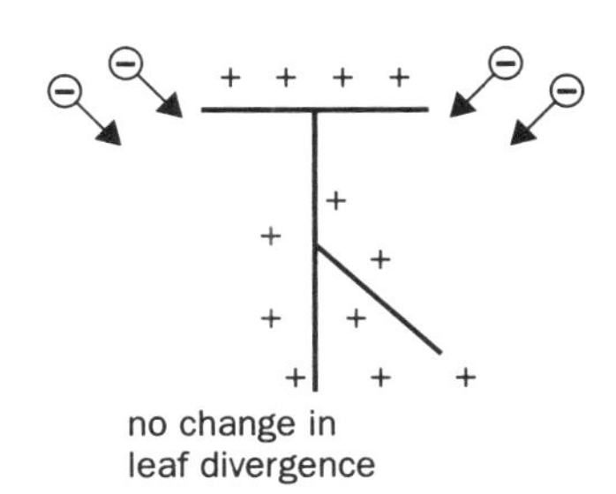

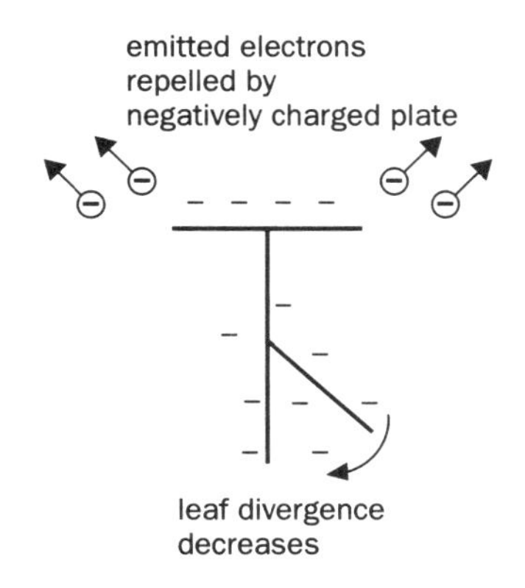

(a) Positively-charged electroscope **(b) Negatively-charged electroscope**

Figure 5.57 Explanation of photoelectric discharge of gold leaf electroscope

Implications of the photoelectric effect

The discovery of the photoelectric effect at the end of the 19th century left scientists in a dilemma – they knew that it occurred, and could quantify its effects, but they could not explain it! The wave theory of electromagnetic radiation predicts that emission of photoelectrons should happen at all frequencies. Electrons in the metal would absorb energy continuously from radiation of any frequency, and be emitted when they had absorbed enough. This process would take longer at lower frequencies, but should still happen. Also there should be no maximum kinetic energy of the emitted electrons.

The apparent failure of the accepted models to describe this phenomenon led the world of science to explore other ways of explaining electromagnetic radiation, namely the **quantum theory**. In 1905, at the age of 26, Albert Einstein published a paper explaining the photoelectric effect in terms of the quantum theory of electromagnetic radiation. Photoelectricity is completely explained by the quantum theory, which led to the complete acceptance of what had started as a rather vague mathematical concept. Einstein extended Planck's quantum theory by supposing that radiation was not only **emitted** in quanta, but also **absorbed** as quanta. Thus was born the idea that electromagnetic radiation had a dual nature: some of its properties could be adequately explained by its wave-like nature (reflection, refraction, interference, diffraction, polarisation) but other effects, particularly the photoelectric effect, could only be explained by the particle-like behaviour. Other quantum effects include **line** emission and absorption spectra.

The photon model of electromagnetic radiation

In Section 3.5, we examined the nature of electromagnetic radiation, discovering that there was a continuous range from the very low frequencies of long-wave radio to the very high frequencies of γ-radiation and cosmic rays – the **electromagnetic spectrum**. The way in which the energy of the radiation was related to its frequency was explained by Max Planck in 1900. Before this, electromagnetic energy was assumed to be continuous, not 'grainy' or 'lumpy'. Planck produced a theory that energy was emitted in very small, but separate, 'packets', which he called quanta.

Thus the essential points of the quantum theory are:

- Light, and all forms of electromagnetic radiation, is emitted in brief bursts or 'packets' of energy, i.e. it is **quantised**.
- The packets of energy, which are called **photons**, travel in one direction only in a straight line.
- When an atom emits a photon, its energy changes by an amount equal to the photon energy.
- The amount of energy (E) contained in each **quantum** is directly proportional to the **frequency** (f) of the radiation, so the relationship between energy and frequency is given by:

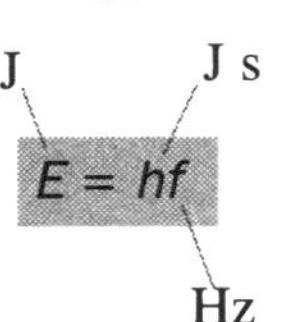

Equation 1

You should note that:

- h is **Planck's constant** = 6.6×10^{-34} J s.
- Equation 1 can be expressed as

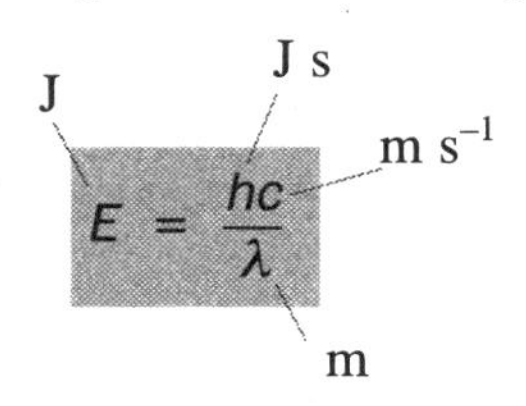

where c = velocity of electromagnetic radiation in a vacuum (m s^{-1})
λ = wavelength of the photon (m).

- E is expressed in joules, but it is often more convenient (easier numbers!) to convert these energies to electronvolts.

A good way to think about the meaning of Planck's constant is to realise that its unit is really the **joule per hertz** (J Hz^{-1}). In other words, it is the amount of energy that a quantum of radiation of frequency 1 Hz would have. Double the frequency, double the energy of each quantum, and so on. This is a very tiny quantity of energy. Even at a frequency of around 2.4×10^{14} Hz, the energy of each quantum is still only equal to one electronvolt (1.6×10^{-19} J), which is still a tiny amount of energy. In everyday life, we are simply unaware of this 'graininess' of energy, just as we do not observe that matter is made of discrete particles. For most purposes, there is no need to consider this 'particle-like' behaviour of electromagnetic radiation, but if we are to explore the fundamental nature of matter and energy, we need to incorporate this deeper understanding into our thinking. Planck's **quantum theory**, as it became known, was one of the giant leaps in the development of our knowledge of the interactions of the Universe. It won him a Nobel prize, and so radically changed thinking that physics before 1900 is referred to as **classical** physics, while subsequent discoveries, incorporating quantum theory, are called **modern** physics.

Guided example (1)

Calculate the photon energy (a) in joules and (b) in electronvolts for each of the following:

(i) red light of wavelength 640 nm

(ii) X-rays of frequency 2.0×10^{19} Hz

(iii) ultra-violet radiation of wavelength 2.6×10^{-7} m.

Guidelines

Simply apply Equation 1 or its equivalent in each case. Use SI units to calculate energies in joules. To express your answers in electronvolts, remember that 1 eV = 1.6×10^{-19} J.

It is important to realise that the wave theory is not *wrong* in the conventional sense, simply that it is not adequate to explain certain phenomena such as the photoelectric effect. The photon model becomes useful when the event being studied happens on the scale of individual particles; it is only then that we observe the *quantised* nature of the universe. The wave theory and the photon theory sit quite happily side-by-side; each has a useful role in helping us to understand the interactions between matter and energy.

Work function

Common symbols W, Φ
SI unit joule (J)

If we could observe the sequence of events leading up to photoemission, we might see it as follows:

Photons are pictured arriving at the surface of a metal, where some of them are **completely absorbed** by electrons near the surface. The electron's kinetic energy, and therefore its velocity, is greatly increased by the interaction. If it is moving towards the surface of the metal, it may escape from the attractive electric field of the metal ions. Figure 5.58 shows some of the possible results.

Electron 1 at the surface requires the least possible energy to liberate it, so it escapes with the maximum possible kinetic energy.

Electron 2 deep in the metal has lost too much kinetic energy by the time it reaches the surface and is attracted back.

Electron 3 escapes, but has less kinetic energy than electron 1.

Electron 4 gained enough kinetic energy to escape, but was moving in the wrong direction and is absorbed in the metal.

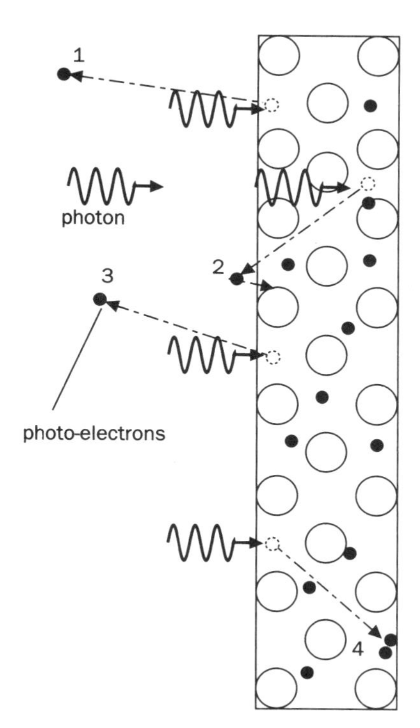

Figure 5.58 Picturing photoelectric emission from a metal surface

It is important to realise that although the above explanation is a gross over-simplification, it is an adequate explanation at this level, and should enhance your understanding of this phenomenon, in particular the existence of a range of kinetic energies and a maximum kinetic energy of the ejected electrons.

Each electron has to lose some of its kinetic energy in order to overcome the attractive electric fields of the positive ions. Electrons from deep layers lose more energy than those from the surface.

The **work function** (W or Φ) of a material is defined as the **work** necessary to remove an electron from the **surface** of the material.

Einstein's photoelectric equation

Einstein suggested that when a photon causes photoelectric emission from a metal surface, some of the photon energy is used to overcome the work function (i.e. it provides the energy needed to liberate a surface electron) while the remainder appears as kinetic energy of the ejected electron. This is very succinctly expressed in **Einstein's photoelectric equation**:

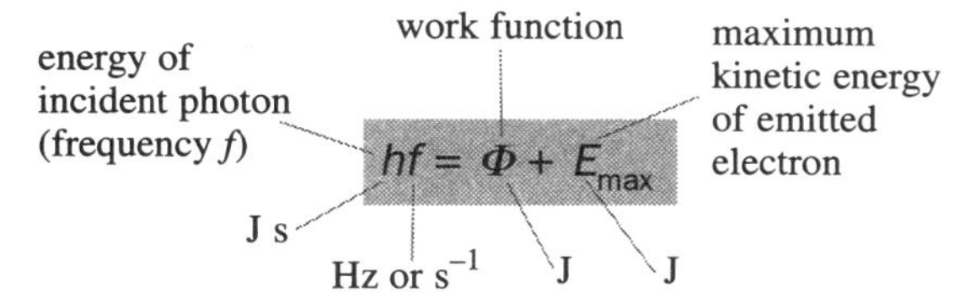

Equation 2

You should note that:

- the maximum kinetic energy of the emitted photoelectron (E_{max}) is given by:

  $$E_{max} = \frac{1}{2} m v_{max}^2$$

  where v_{max} is the maximum velocity of the electron.
- Most photoelectrons will have energies less than the maximum value. This is because electrons emitted from atomic layers beneath the metal surface lose energy in collisions with atoms on their way out of the metal.
- Equation 2 may also be expressed in terms of the **wavelength** (λ) of the incident photons:

  $$\frac{hc}{\lambda} = \Phi + E_{max}$$

  (h: J s; c: m s^{-1}; λ: m; Φ: J; E_{max}: J)

  Equation 3
- Equation 2 shows that the maximum kinetic energy of a photoelectron depends only on the **frequency** of the photon. A more **intense** beam (i.e. **brighter** light) simply contains more photons per second, and will produce more photoelectrons, but will not affect their maximum kinetic energy.
- Photoelectrons can only escape if the maximum kinetic energy is greater than zero:

  i.e. $hf - \Phi > 0$

  $\therefore \quad f > \frac{\Phi}{h}$

  So the **minimum** frequency (f_0) (also called the **threshold frequency**) required to cause photoemission from a metal of work function (Φ) is given by:

  $$f_0 = \frac{\Phi}{h}$$

  (f_0: Hz or s^{-1}; Φ: J; h: J s)

  Equation 4

Stopping potential

The above equations apply to a metal surface at **zero** potential. If, however, the surface is held at a **positive** potential, the photoelectrons will have to lose more of their kinetic energy in order to overcome the increased attraction. Increasing the positive potential will reduce the number of electrons escaping per second as a smaller proportion will have the necessary energy to overcome the increased attraction. Also, the maximum kinetic energy will be reduced, and Equation 2 can be written:

$$E_{max} = hf - \Phi - eV$$

where e is the electronic charge
V is the applied potential

If the potential is increased further, a value will be reached at which photoemission is just prevented. This value is known as the **stopping potential** (V_s).

Since the maximum kinetic energy has been reduced to zero:

$$eV_s = hf - \Phi$$

from which:

$$V_s = \left(\frac{h}{e}\right) f - \frac{\Phi}{e}$$

(V_s: V; h: J s; Φ: J; e: C; f: Hz or s^{-1})

Equation 5

If we compare this to the standard equation for a straight line:

$$y = mx + c$$

then plotting measured values of V_s against f for photons of different frequencies should give a straight line of gradient h/e and intercept Φ/e on the V_s axis.

Alternatively, since $\Phi = hf_0$, Equation 5 can be written:

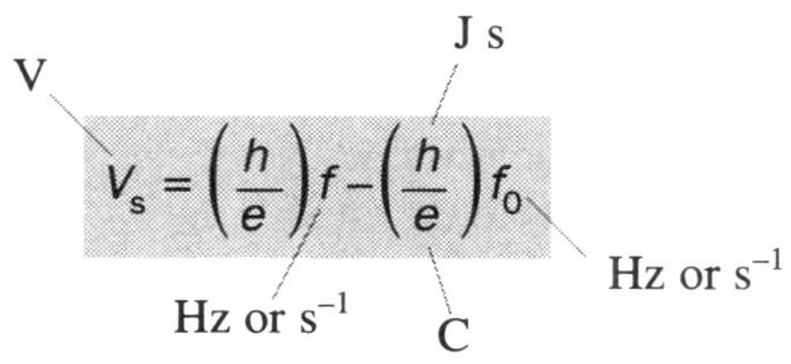

$$V_s = \left(\frac{h}{e}\right) f - \left(\frac{h}{e}\right) f_0$$

Equation 6

Measuring the stopping potential

The need here is for a source of photons of variable frequency, and a means of measuring the potential required to prevent photoemission. The experiment must be performed in a vacuum.

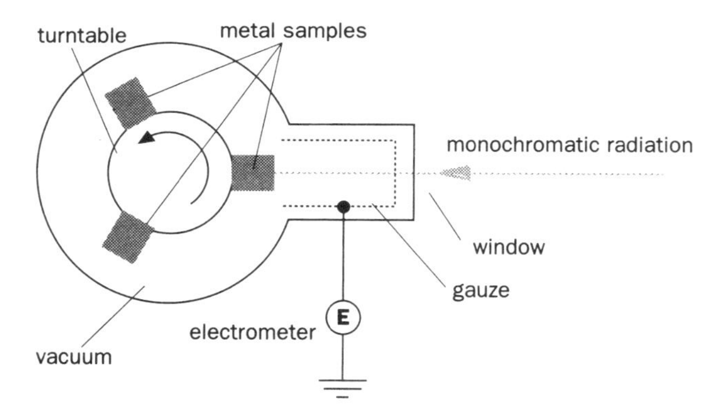

Figure 5.59 Measuring the stopping potential

Figure 5.59 shows a simplified version of the apparatus Millikan used to measure stopping potential, and so determine the value of Planck's constant. Three metal samples were arranged on a turntable in an evacuated glass chamber. A motor-driven 'knife' in the chamber could clean each metal surface. Monochromatic radiation, selected by an external spectrometer, was allowed to fall on one of the metal surfaces. The radiation entered through a quartz window in order that radiation in the ultra-violet region of the electromagnetic spectrum could penetrate the apparatus. Photoelectrons were emitted from the metal surface, and collected on the gauze nearby, where they flowed to Earth. This tiny current was registered by the sensitive electrometer. The metal sample could be given a positive potential, and this was increased until no current could be detected by the electrometer. This potential was the **stopping potential** (V_s). Millikan measured the stopping potential for a wide range of frequencies of incident radiation. He also repeated the experiment with different metal samples. Typical results are shown in Figure 5.60, the straight line confirming Einstein's predictions.

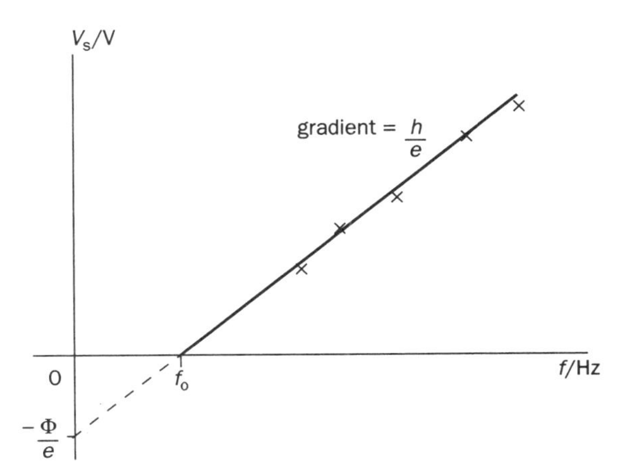

Figue 5.60 Graph of stopping potential against radiation frequency for a given metal

You should note that:

- the gradient is always equal to h/e, regardless of the metal being tested
- the intercept on the frequency axis is equal to the **threshold frequency** (f_0).

The intercept on the V_s axis is equal to Φ/e, from which the **work function** (Φ) can be obtained.

Guided example (2)

A metal surface having a work function of 3.0 eV is illuminated with radiation of wavelength 350 nm.

Calculate:

(a) The minimum frequency and maximum wavelength of incident radiation which would cause electron emission.

(b) The maximum kinetic energy of the emitted photoelectrons.

(c) The minimum retarding or 'stopping' potential for these photoelectrons.

Guided example (2) *continued*

(1 eV = 1.6×10^{-19} J;
Planck's constant, $h = 6.63 \times 10^{-34}$ J s;
electron charge, $e = 1.6 \times 10^{-19}$ C;
speed of electromagnetic radiation in a vacuum, $c = 3.0 \times 10^8$ m s^{-1})

Guidelines

(a) Work function in **joules** = Φ (in eV) $\times 1.6 \times 10^{-19}$
Threshold frequency (f_0) given by Equation 4.
Use $c = f\lambda$ to calculate threshold wavelength.

(b) Maximum kinetic energy (E_{max}) of photoelectrons is obtained from Equation 2.

(c) Stopping potential (V_s) is obtained from Equation 5.

Wave–particle duality

The phenomena of reflection, refraction, interference and diffraction can all be explained using the idea of light as a wave motion. Furthermore, the fact that light can be polarised indicates that the waves are transverse. But when it comes to explaining the photoelectric effect we need to think of the electromagnetic radiation as having particle properties. So which of our two models is a correct description of light? Is it a wave motion or a particle motion? The answer of course is that both ideas are merely different models which help us to explain how electromagnetic radiation behaves in different circumstances; neither is a perfect or full description.

Based on the idea that light and all other electromagnetic radiation is both particle and wave, Louis de Broglie suggested that the same kind of dual nature must also be applicable to matter. He proposed that any particle of matter of momentum (p) has an associated wavelength (λ) given by:

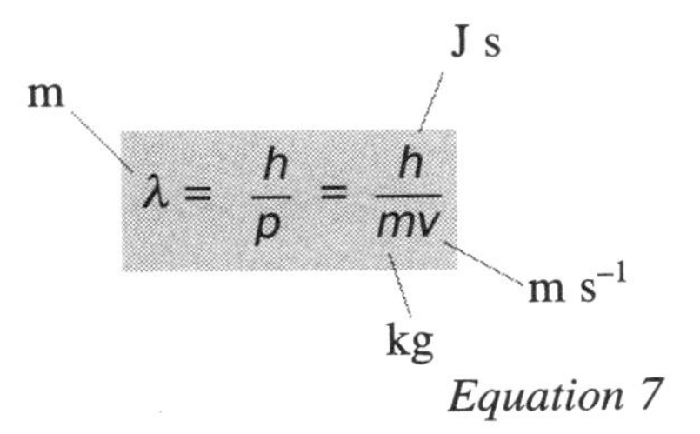

Equation 7

You should note that:

- λ is known as the **de Broglie** wavelength.
- v is the particle **velocity** and m is the particle **mass**. (Strictly speaking, this should be the particle's

relativistic mass. Since the concept of matter having a wavelength only becomes significant at very high velocities, allowance should be made for the increase in mass which occurs with increased velocity as the speed of light is approached. However, such a treatment is beyond the scope of this book.)

The de Broglie wavelength of an electron which has been accelerated by an electric field can be obtained as follows:

Consider an electron of mass (m) and charge (e) accelerated from rest to a final velocity (v) by a p.d. (V).

kinetic energy gained by electron = work done by accelerating p.d.

i.e. $\frac{1}{2}mv^2 = eV$

$\therefore \quad v = \sqrt{\frac{2eV}{m}}$

The **momentum** (p) of the electron is given by $p = mv$.

$\therefore \quad p = m\sqrt{\frac{2eV}{m}} = \sqrt{2eVm}$

The de Broglie wavelength (λ) is given by:

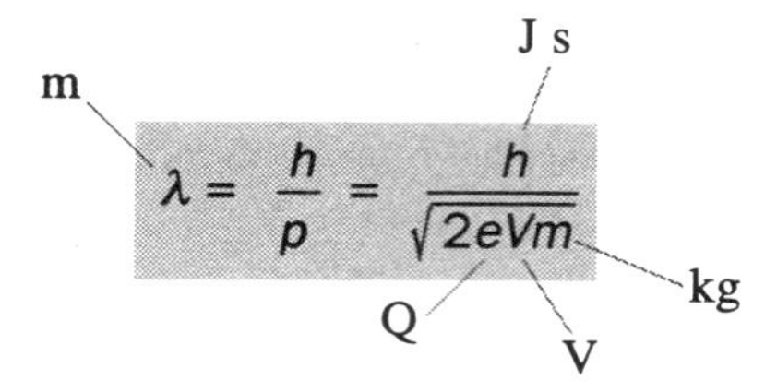

Equation 8

Prince Louis de Broglie who proposed a wave–particle duality for electrons

For accelerating potentials of about 100 V, the associated wavelength of an electron is about 10^{-10} m. This is the same order of magnitude as X-ray wavelengths. Since diffraction of X-rays (and other electromagnetic radiation) is explained by considering their **wave** properties, it follows that diffraction of beams of electrons would confirm that they too had wave properties. This was confirmed in 1926 by Davison and Germer, who demonstrated diffraction of a beam of electrons by a single crystal of nickel. The distance between the planes of atoms in the crystal is of the same order of magnitude as the de Broglie wavelength of the electron beam, thus producing significant diffraction.

Diffraction effects have since been observed with beams of protons, neutrons and even α-particles. However, as Equation 8 indicates, the associated wavelength decreases as mass increases, making diffraction effects more difficult to observe with more massive particles. Diffraction of a beam of electrons directed at a crystal or a thin metal foil is shown in Figure 5.61.

The diffraction pattern produced is a set of concentric circles. The angular deflection of the diffracted electrons can be calculated, to provide information about the spacing between the atoms in the metal foil, in much the same way as X-ray diffraction. This wave-like behaviour of electrons is used in electron microscopes, in which beams of electrons are used instead of light to provide magnified images. At high accelerating voltages, the wavelengths of the electrons are much shorter than light wavelengths, so much finer detail can be resolved. The field ion microscope uses beams of helium ions, with an even smaller wavelength, leading to the ability to resolve individual molecules.

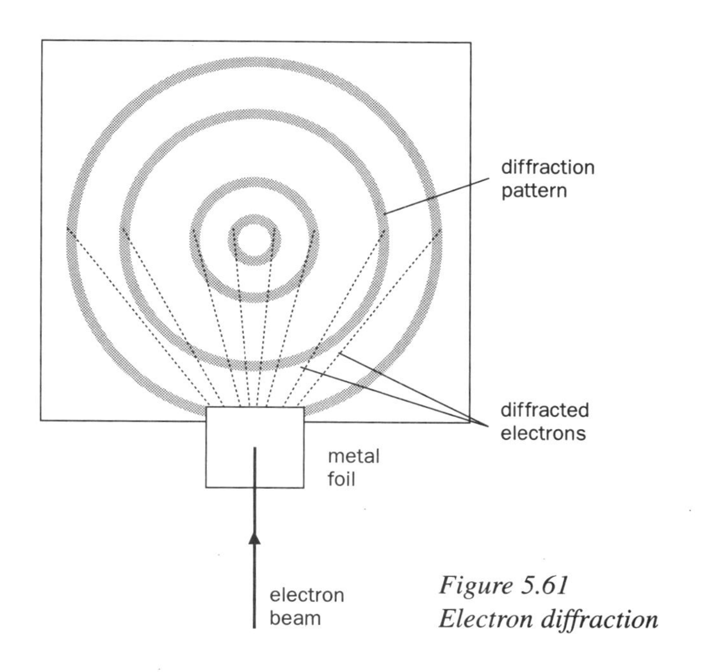

Figure 5.61
Electron diffraction

It should be appreciated that all physical entities can be described as waves or particles. The two ideas or models are linked by the following relations:

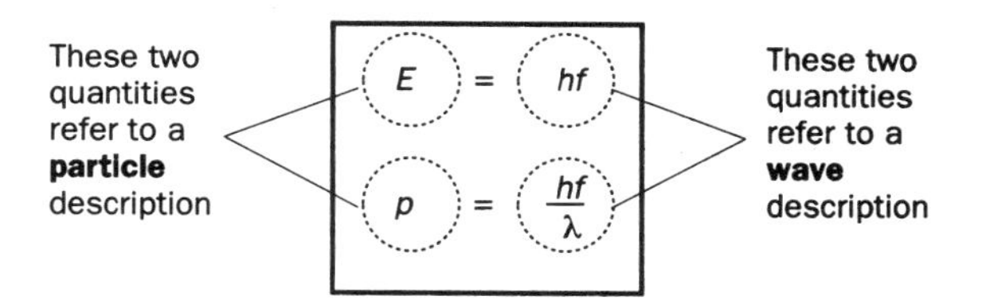

Guided examples (3)

1. Calculate the de Broglie wavelength of a bullet of mass 30 g if it is moving at a velocity of 250 m s^{-1}.

Guidelines

Simply apply Equation 7.

2. Calculate the associated (de Broglie) wavelength of the electrons in an electron beam which has been accelerated through a p.d. of 4000 V. (Electron charge, $e = -1.6 \times 10^{-19}$ C; electron mass = 9.1×10^{-31} kg; Planck's constant, $h = 6.63 \times 10^{-34}$ J s)

Guidelines

Calculate the energy acquired by electrons from the electric field.

From this, electron velocity and momentum are obtained.

Then use Equation 7.

3. An α-particle emitted from a radon-220 nucleus is found to have a de Broglie wavelength of 5.7×10^{-15} m. Use the following data to calculate the α-particle energy in MeV.

charge on an electron	$= 1.6 \times 10^{-15}$ C
α-particle mass	$= 6.7 \times 10^{-27}$ kg
Planck's constant	$= 6.63 \times 10^{-34}$ J s

Guidelines

Use Equation 7 to calculate the α-particle velocity and use this to calculate its kinetic energy in joules, then convert this to MeV.

Self-assessment

SECTION A
Qualitative assessment

1. (a) What is a **photon**?
   (b) Write down an expression for the energy of a photon in terms of its
      (i) frequency
      (ii) wavelength.

   Explain the meaning of any symbols used and state the units of the physical quantities involved.

2. Explain what is meant by **photoelectric emission**?

3. A freshly cleaned zinc plate is connected to a charged gold-leaf electroscope. Describe and explain what happens to the leaves of the electroscope when the plate is exposed to ultra-violet radiation if the initial charge on the electroscope is
   (a) positive
   (b) negative.

Self-assessment *continued*

4. Electrons are emitted from a zinc surface when it is exposed to a faint source of **ultraviolet** radiation. Explain why a very intense source of **visible** radiation does not cause photoemission from the zinc surface.

5. A metallic surface is illuminated with monochromatic radiation which causes photoemission. Explain how to increase
   (a) the number of photoelectrons emitted per second
   (b) the maximum kinetic energy of the photoelectrons.

6. Explain what is meant by the terms **work function** and **threshold wavelength** as used to describe the photoelectric effect.

7. State Einstein's photoelectric equation, and explain the meaning of each term.

8. Describe an experiment to verify Einstein's photoelectric equation, and explain how the work function and Planck's constant can be determined from the results.

9. An electron is accelerated from rest by an electric field. Derive an expression for the de Broglie wavelength of the electron in terms of its charge, mass and accelerating potential.

SECTION B
Quantitative assessment

(***Answers:*** 6.4×10^{-34}; 1.2×10^{-19}; 1.7×10^{-19}; 2.1×10^{-19}; 2.6×10^{-19}; 5.9×10^{-19}; 1.3×10^{-15}; 2.1×10^{-14}; 1.6×10^{-12}; 4.9×10^{-11}; 3.0×10^{-7}; 7.7×10^{-7}; 1.6; 2.4; 6.7×10^{5}; 1.5×10^{7}; 3.9×10^{14}; 1.2×10^{15}; 1.8×10^{17}; 1.4×10^{22}.)

(Planck's constant $h = 6.63 \times 10^{-34}$ J s

Speed of electromagnetic radiation in a vacuum
$c = 3.0 \times 10^{8}$ m s^{-1}

electronic charge $e = 1.6 \times 10^{-19}$ C

electron mass $m_e = 9.1 \times 10^{-31}$ kg

proton mass (and neutron mass)
$m_p \approx m_n = 1.7 \times 10^{-27}$ kg)

1. (a) Calculate the energy of a photon of
   (i) infra-red radiation of wavelength 1.2 μm
   (ii) X-radiation of frequency 2.0×10^{18} Hz.
   (b) Calculate the frequency and wavelength of a γ-ray photon of energy 9.5×10^{12} J.

2. The beam from an argon laser has a power of 7.5×10^{-2} W and a wavelength of 490 nm. Calculate the number of photons emitted per second by the laser.

3. (a) Calculate the work function (in eV) for a magnesium surface if the minimum frequency of electromagnetic radiation which causes photoemission from the metal is 8.9×10^{14} Hz.
   (b) If the same surface were illuminated with radiation of wavelength 250 nm, calculate
   (i) the maximum kinetic energy and
   (ii) the maximum velocity
   of the emitted photoelectrons.

4. A brightly polished metal surface has a work function of 4.2 eV. Calculate
   (a) the threshold wavelength for this metal
   (b) the frequency of radiation which produces photoelectrons from the metal with a maximum kinetic energy of 1.3×10^{-19} J.

5. Calculate the maximum kinetic energy of photoelectrons emitted from a surface if a stopping potential of 1.6 V just prevents emission.

6. When electromagnetic radiation of wavelength 400 nm is incident on a metal surface, a stopping potential of 0.75 V is just sufficient to prevent photoemission. Calculate
   (a) the maximum kinetic energy of the emitted photoelectrons when the stopping potential is zero
   (b) the work function of the metal in eV.

Self-assessment *continued*

7. Frequency/10^{14} Hz

5.2 6.0 7.2 8.0 8.8 10.8

Max. kinetic energy/10^{-19} J

0.8 1.3 2.1 2.6 3.1 4.4

The above results were obtained in an experiment to investigate photoelectricity in which the maximum kinetic energy of photoelectrons was measured at different frequencies of incident radiation. Plot a graph of maximum kinetic energy against frequency and use your graph to obtain values for

(a) Planck's constant

(b) the minimum frequency of radiation that would cause photoemission from this surface

(c) the work function of the surface (in eV)

(d) the threshold wavelength.

8. (a) Calculate the de Broglie wavelength of

(i) a β^--particle travelling with a velocity of 1.5×10^7 m s^{-1}

(ii) a proton which has been accelerated from rest to a velocity of 2.5×10^5 m s^{-1}.

(b) Calculate the velocity of an α-particle for which the de Broglie wavelength is 6.4×10^{-15} m.

Past-paper questions

1 (a) (i) Briefly describe the principles involved in estimating the size of a nucleus from experiments in which alpha particles are scattered by target materials such as gold foil.

(ii) Calculate the initial kinetic energy, in MeV, required by an alpha particle if it is to reach the closest distance of approach when travelling head-on towards the nucleus of a gold atom ($^{197}_{79}$Au). The radius of the gold nucleus may be taken to be 7.6 fm, and that of the alpha particle to be 2.0 fm.

(iii) What are the values of the energy, in MeV, of naturally occurring alpha particles from radioactive sources, such as those used in the original scattering experiments? Suggest a method by which alpha particles could be made to be more penetrating. Also explain why electrons are generally preferred to alpha particles when probing the nucleus.

(NEAB, 1991, part question)

2 (a) For a nucleus of proton number Z and nucleon number A, explain what is meant by (i) *mass difference*, (ii) *binding energy*.

(b) Use values from the data booklet to calculate for the nuclide $^{60}_{28}$Ni

(i) the mass of the nucleus in u

(ii) the mass difference in u

(iii) the binding energy per nucleon in MeV.

(c) Sketch a graph to show how the binding energy per nucleon depends on the nucleon number for all known nuclides. Indicate appropriate values on the axes of your graph.

(NEAB, 1993, part question)

3 (a) The graph shows how the potential energy of a pair of neutrons in a nucleus may be considered to vary with their separation.

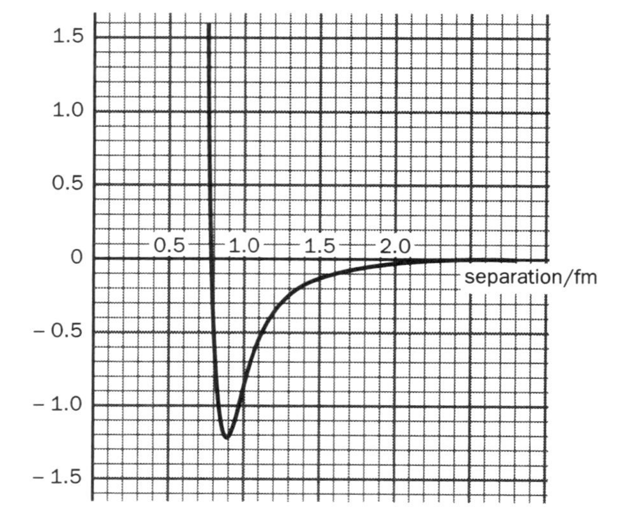

(i) If the neutrons were stationary in the nucleus, at what separation ought they to be in equilibrium according to the graph? Explain your answer.

(ii) Estimate the energy, in MeV, which would have to be supplied to a pair of stationary neutrons in order to separate them completely, starting from their equilibrium position.

(iii) What value of binding energy per neutron is suggested by your answer to (ii)? How does this compare with the actual value of binding energy per nucleon for most of the nuclei where each nucleon is surrounded by several near neighbours? Suggest an explanation for the discrepancy between these values.

(iv) What physical quantity is represented by the gradient of the graph? Explain the significance of the different sign and magnitude of the gradients on either side of the equilibrium position.

(v) Use the graph to estimate the magnitude of the nuclear force when the separation is 1.25 fm.

(b) (i) State **two** reasons why high energy electrons are suitable for probing the nucleus by diffraction techniques.

(ii) When electron waves of wavelength λ are incident on nuclei of radius R, the first minimum of the scattered intensity occurs at an angle θ from the direct beam, where

$$\sin\theta = \frac{0.61\,\lambda}{R}$$

Assuming the electrons obey the relationship energy = (momentum) × (speed of light in a vacuum) show that

$$\sin\theta = \frac{0.61\,hc}{E\,r_0\,A^{1/3}}$$

where E is the energy of the incident electrons, A is the nucleon number of the target nuclei and r_0 has its usual meaning.

(iii) Use the results from (ii) to decide the maximum and minimum energy values, in MeV, which would be suitable for investigating the first minimum of the $^{12}_{6}$C nucleus by this technique. The value of the constant r_0 in the equation $R = r_0\,A^{1/3}$ may be taken to be 1.3 fm.

(NEAB, 1994)

4 (a) The radius, R, of an atomic nucleus can be determined by bombardment with high energy electrons. Name an apparatus suitable for producing electrons of the required energy, and explain why this energy has to be high.

State the equation which relates R to nucleon number, A. If the radius of the nucleus of $^{93}_{37}$Rb is 5.89 × 10^{-15} m, calculate the radius of the nucleus of $^{143}_{55}$Cs.

(b) When ^{235}U is bombarded with neutrons, $^{93}_{37}$Rb and $^{143}_{55}$Cs are possible fission products. Assuming that these two nuclei are initially in contact, calculate the total kinetic energy they would acquire due to Coulomb repulsion.

charge of a proton = 1.6×10^{-19} C
permittivity of free space, $\varepsilon_0 = 8.9 \times 10^{-12}$ F m^{-1}

(NEAB, 1988, part question)

5 Radioisotopes contribute to *background radioactivity*. These isotopes can have *natural radioactivity* or *artificially induced radioactivity*.

Explain the terms *natural radioactivity* and *artificially induced radioactivity*. Give another important contributor to background radioactivity other than radioisotopes.

How would the precautions needed when handling an alpha emitter differ from those needed for a gamma emitter? Explain the reasons for these differences.

(Lond., January 1993)

6 The first example of an induced nuclear transformation detected in the laboratory resulted from the bombardment of nitrogen by alpha particles:

$^{14}_{7}\text{N} + \alpha \rightarrow {}^{17}_{8}\text{O} + \text{p}$

Use the following data to calculate the increase in rest mass which occurs.

Mass of nitrogen-14 atom = 23.2530×10^{-27} kg
Mass of oxygen-17 atom = 28.2282×10^{-27} kg
Mass of alpha particle = 6.6442×10^{-27} kg
Mass of proton = 1.6725×10^{-27} kg

State the source of this increased rest mass.

Calculate the minimum kinetic energy of the incident alpha particle for the transformation to be possible. You may neglect the recoil energy of the oxygen-17 atom and the proton.

(Speed of light in vacuum, $c = 3.0 \times 10^8$ m s^{-1})

(Lond., June 1994)

7 (a) The various *isotopes* of an element X are distinguished by using the notation $^{A}_{Z}$X. Explain the meaning of *A*, *Z* and of the term *isotope*.

(b) Radioactive sources which might be used in schools are ^{226}Ra which emits α, β and γ rays, and ^{90}Sr which emits β rays only.

(i) List **three** safety precautions which need to be taken into account when using such sources.

(ii) The half-life of ^{90}Sr is 28 years. When its activity falls to 25% of its original value it should be replaced. After how many years should it be replaced?

(c) (i) Explain briefly how β-particles are emitted by the nucleus.

(ii) When $^{226}_{88}Ra$ emits an α-particle it decays to radon (*Rn*). Write down a balanced equation for this change.

(d) Radioactive isotopes have many applications merely by virtue of being isotopes. Describe and explain **one** such application.

(e) (i) Describe Rutherford's α-particle scattering experiment and summarise the evidence that it provided for the nuclear model of the atom.

(ii) How would the results be different if aluminium foil were used instead of the gold foil in such an experiment?

(WJEC, 1990)

8 (a) Explain the meaning of the symbol $^{238}_{92}$U. What is meant by an isotope?

(b) Certain types of nucleus may spontaneously lose a small amount of mass by the process known as radioactivity.

(i) Describe the nature of the radiations which may be emitted during this process.

(ii) Define *radioactive decay* and *briefly* explain how this leads to the equation

$$\frac{dN}{dt} = -\lambda N.$$

Why is there a negative sign?

(iii) Sketch the variation of *N* with time and explain what is meant by half-life.

(c) (i) Draw a sketch of the variation of binding energy per nucleon with mass number. Use the sketch to explain why nuclear fusion occurs in some circumstances and nuclear fission in others.

(ii) Explain why very high temperatures are required for nuclear fusion.

(iii) A future fusion reactor might use the reaction

$^{2}_{1}\text{H} + {}^{2}_{1}\text{H} \rightarrow {}^{4}_{2}\text{He} + \text{energy}$

to produce useful energy. From the following data calculate the number of reactions required to produce 1 J of energy.

(iv) Calculate the mass of $^{2}_{1}$H required to provide 1 J of energy

Mass of $^{2}_{1}$H = 2.0136 amu
Mass of $^{4}_{2}$He = 4.0015 amu
1 amu = 1.661×10^{-27} kg
Velocity of light $c = 3{\cdot}00 \times 10^8$ m s^{-1}

(WJEC, 1992)

9 (a) Explain the significance of the characters H, 1 and 2 in the expression

$^{2}_{1}$H

and hence state what is meant by an isotope.

(b) Copy and complete the following equations:

(i) $^{238}_{92}\text{U} \rightarrow {}^{234}_{90}\text{Th} + \ldots.$

(ii) $^{234}_{90}\text{Th} \rightarrow {}^{234}_{91}\text{Pa} + \ldots.$

(c) Sketch a graph of the variation of binding energy per nucleon with atomic mass number. Use your sketch to explain why (i) fusion and (ii) fission occur.

(WJEC, 1991)

10 (a) Carefully explain what is meant by the terms *mass defect* and *binding energy*.

(b) Can the nuclear reaction represented by the equation shown below occur naturally?

$^{206}_{82}\text{Pb} \rightarrow ^{202}_{80}\text{Hg} + \alpha\text{-particle.}$

(c) Justify your answer to (b).

(Take as nuclear masses: $^{206}_{82}$Pb = 205.969 u, $^{202}_{80}$Hg = 201.971 u, α-particle = 4.004 u.)

(WJEC, 1993)

11 The radioactive isotope of iodine ^{131}I has a half-life of 8.0 days and is used as a tracer in medicine.
Calculate

(a) the number of atoms of ^{131}I which must be present in the patient when she is tested to give a disintegration rate of 6.0×10^5 s^{-1}

(b) the number of atoms of ^{131}I which must have been present in a dose prepared 24 hours before.

(NEAB, 1991)

12 A tube containing an isotope of radon, $^{222}_{86}$Rn, is to be implanted in a patient. The radon has an initial activity of 1.6×10^4 Bq, a half-life of 4 days and it decays by alpha emission. To provide the correct dose, the tube, containing a freshly prepared sample of the isotope, is to be implanted for 8 days.

(a) (i) What are the proton (atomic) number and the nucleon (mass) number of the daughter nucleus produced by the decay of the radon?

(ii) State one reason why an alpha emitter is preferred to a beta or gamma emitter for such purposes.

(b) Determine

(i) the decay constant for radon in s^{-1}

(ii) the initial number of radioactive radon atoms in the tube.

(c) The operation to implant the tube has to be delayed.

Ignoring the effects of any daughter products of the decay, determine the maximum delay possible if the patient is to receive the prescribed dose using the source.

(AEB, 1992)

13 The proportion of unstable carbon-14 to stable carbon-12 in a growing plant is approximately 1:10^{12}. Calculate the number of carbon-14 atoms in a 1 g sample of carbon from a growing plant.

Carbon-14 has a half-life of 5730 years. Calculate the initial number of disintegrations per second in this sample of carbon.

In carbon-14 dating, the proportion of carbon-14 to carbon-12 in a sample of material is measured. Explain why carbon-14 dating might be unsuitable for estimating the age of samples that are more than 20 000 years old.

The Avogadro constant = 6.0×10^{23} mol^{-1}.

(Lond., January 1994)

14 Excavations near Welshpool, Powys, have recently revealed the remains of a large wooden monument two-thirds the size of Stonehenge. Archaeological finds suggest that the site is about 4000 years old. If carbon dating were to be used to check this estimate, what activity would you expect to find, if the activity of 1 g of comparable living wood is 19 counts min^{-1}?

(Half-life of ^{14}C = 5700 years)

(WJEC, 1992)

15 A small maintenance-free energy source is needed to provide 5.0 W of power to a remote isolated weather station. The radioactive isotope ^{90}Sr , which is a β-emitter with a half-life of 28 years, is chosen. The β-particles are stopped by the surrounding absorbing material, heating it. The mean energy of the β-particles is 0.40 MeV. The efficiency of absorption and conversion of the internal energy of the absorber to electricity by thermocouples is 20%.

(a) Show that the initial activity from the source required to produce 5.0 W of electrical power is 3.9×10^{14} Bq.

(b) Hence find the mass of strontium in the source to provide 5.0 W.

(c) Sketch the power against time curve for the strontium source for the first 28 years of its life. Use the curve to make a rough estimate of the total energy of the radiation from the source over this time. Explain your method.

(d) The minimum power on which the weather station will function is 2.0 W. Find the maximum time for which the station can operate before the power source must be replaced.

(e) Another isotope, suggested for use as the energy source, is plutonium-239, an α-emitter of half-life 2.4×10^4 years and energy 5.1 MeV. Suggest one possible advantage or disadvantage that this isotope may have compared to strontium. Justify your answer.

(Take 1 year to be 3.1×10^7 seconds)

(O & C, 1993)

16 (a) (i) What are alpha, beta and gamma rays?

(ii) Describe briefly one method whereby they may be distinguished from one another experimentally.

(b) Explain what is meant by

(i) radioactive decay

(ii) radioactive decay constant

(iii) half-life

(iv) the becquerel.

(c) (i) A newspaper article stated that the NASA Galileo space probe to Jupiter ‘contained 49 lb. of plutonium to provide 285 watts of electricity through its radioactive thermonuclear generator (RTG)’.

(Note: An RTG is a device for converting

thermal energy produced by fission into electrical energy.)

Assuming that the plutonium is ^{239}Pu which is built into a small nuclear reactor and that the efficiency of the RTG is 10%, what is the maximum time for which the RTG will supply the required energy output?

(Take the energy emitted for each nuclear disintegration of the ^{239}Pu to be 32 pJ.)

($N_A = 6.0 \times 10^{23}$ mol^{-1}, 1 lb. = 0.45 kg)

(ii) What factors will tend to

(I) increase

(II) decrease your estimate of the time?

(WJEC, 1991)

17 Explain why a negatively charged zinc plate will be discharged if it is illuminated with ultraviolet radiation.

It has been suggested that ultraviolet radiation causes the air around the plate to conduct so that the negative charge can escape. Describe an experimental test you could perform to show that this is incorrect.

Explain why visible light causes photoelectric emission to occur in only a few metals.

(Lond., January 1994)

18 (a) In the photoelectric effect, electrons may be emitted from a metal when electromagnetic radiation falls on the surface. Monochromatic radiation from source A is shone in turn on metal P and then on metal Q. This procedure is repeated with source B and then source C. You may assume that the apparatus is arranged so that emission of electrons can be detected. The results are summarised in the table below.

	wavelength/nm	metal P	metal Q
Source A	300	electrons emitted	electrons emitted
Source B	600	electrons emitted	no electrons emitted
Source C	800	no electrons emitted	no electrons emitted

(i) Visible light has a range of wavelengths from 400 to 700 nm. State in which region of the electromagnetic spectrum lies each of the radiations from source A and from source C.

(ii) By considering the Einstein photoelectric equation, explain the pattern of the results obtained.

(iii) How would you expect the results to be different, if at all, if the light from source B is doubled in intensity (i.e. is made twice as bright)? Explain your answer.

(b) Electrons are known to show wave properties, as illustrated by electron diffraction.

(i) When electrons are accelerated from rest by a potential difference, V, to a speed, v, in a vacuum, these quantities are related by the equation

$$\frac{1}{2}mv^2 = eV,$$

where e is the charge of an electron and m is its mass.

Explain how the equation is an application of the principle of conservation of energy.

(ii) Using data from a standard data book, calculate the speed of electrons accelerated from rest through a p.d. of 2000 V.

(c) In a laboratory demonstration of electron diffraction, electrons are accelerated in a vacuum tube and pass through a thin disc of graphite onto a fluorescent screen at the end of the tube, where a pattern of concentric rings is seen.

(i) Use the momentum–wavelength equation to show that the wavelength associated with the electrons referred to in (b) (ii) above is 2.7×10^{-11} m.

(ii) By reference to your answer in (c) (i), explain why electrons can be diffracted through quite large angles by passing them through a thin sheet of graphite.

(iii) Describe one simple test you could carry out using the above apparatus to support the idea that these rings are produced by beams of **negatively charged** particles. Explain with the aid of a diagram, how you would reach this conclusion from the observations made.

(NEAB, 1994)

19 A photoelectric cell is illuminated with monochromatic light. The graph below shows how the current through the cell varies with the potential difference across the cell.

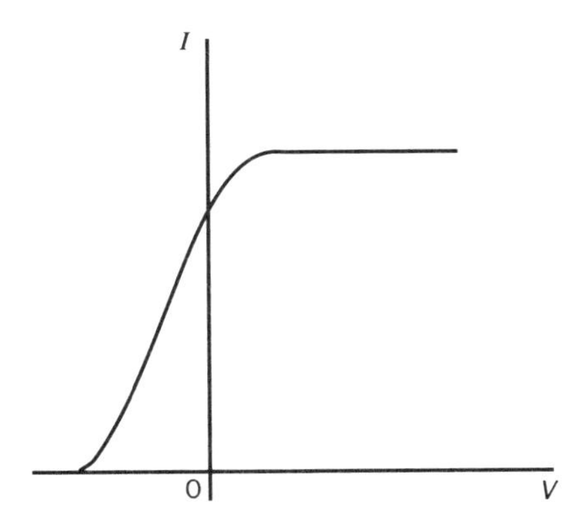

Draw a diagram to show the circuit you would use to obtain the values needed to plot the graph.

How would you find the stopping potential from the graph?

Explain one feature of the graph which you would expect to change if there were an increase in

(a) the wavelength of the light illuminating the cell and

(b) the intensity of the light illuminating the cell.

(Lond., January 1993)

20 The work function for sodium metal is 3.7×10^{-19} J.

Explain the term *work function*.

A fresh sodium surface is irradiated with UV light of

wavelength 319 nm. Calculate the maximum kinetic energy of the emitted photoelectrons.

What determines

(i) the number of photoelectrons released per second

(ii) the maximum photoelectron energy?

(The Planck constant, $h = 6.63 \times 10^{-34}$ J s; speed of electromagnetic radiation *in vacuo*, $c = 3.0 \times 10^{8}$ m s^{-1}.)

(Lond., June 1994)

21 Electrons with a maximum kinetic energy of 4.0 eV are ejected from a metal surface by UV radiation of 150 nm wavelength. Calculate

(a) the work function of the metal in electronvolts

(b) the threshold wavelength of the metal.

(WJEC, 1994)

22 The graph shows how the maximum kinetic energy (E_k) of an electron emitted from the surface of potassium by the photoelectric effect varies with the frequency f of the incident radiation.

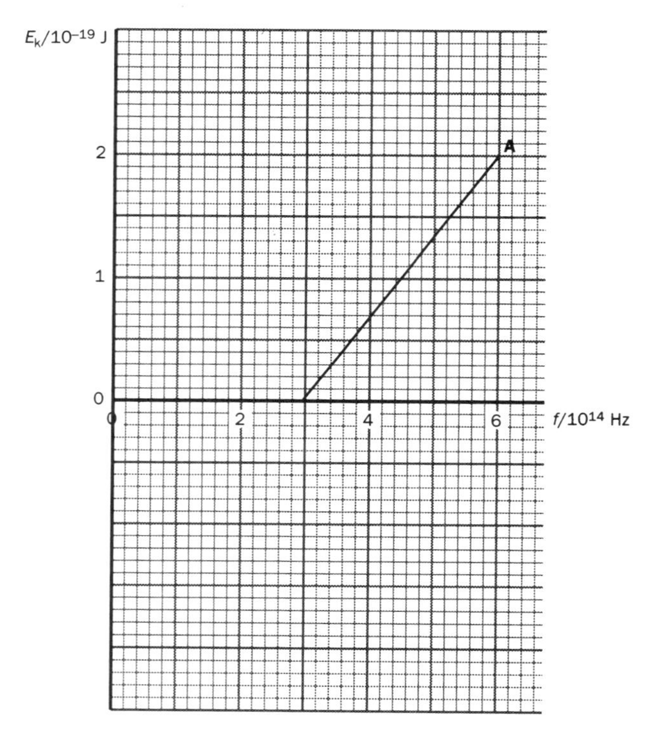

(a) The maximum kinetic energy of the photoelectron is given by $(E_k)_{max} = hf - \Phi$.

(i) State the significance of hf.

(ii) Use the graph to determine a value for the Planck constant.

(b) Determine the minimum photon energy required to emit an electron. Give your answer in eV.

Charge on an electron $= -1.6 \times 10^{-19}$ C.

(c) Sketch a copy of the graph labelling the line **A**. On the same axes draw a second graph showing the result of using a material with a higher work function.
Label this line **B**.

(AEB, 1994)

23 The ionisation energy for a hydrogen atom is 13.6 eV if the atom is in its ground state. It is 3.4 eV if the atom is in the first excited state.

Explain the terms *ionisation energy* and *excited state.*

Calculate the wavelength of the photon emitted when a hydrogen atom returns to the ground state from the first excited state. Name the part of the electromagnetic spectrum to which this wavelength belongs.
(Electronic charge, $e = -1.6 \times 10^{-19}$ C,
the Planck constant, $h = 6.63 \times 10^{-34}$ J s,
speed of light, $c = 3.0 \times 10^{8}$ m s^{-1}.)

(Lond., June 1992)

24 Some of the energy levels of the hydrogen atom are shown (not to scale) in the diagram.

1 eV $= -1.6 \times 10^{-19}$ J

energy/eV
0.00
− 0.54
− 0.85
− 1.51
− 3.39
− 13.58 ground state

(a) Why are the energy levels labelled with negative energies?

(b) Mark on a copy of the diagram a transition which will result in the emission of radiation of wavelength 487 nm. Justify your answer by a suitable calculation.

(c) What is likely to happen to a beam of photons of energy

(i) 12.07 eV

(ii) 5.25 eV

when passed through a vapour of atomic hydrogen?

(O & C, 1992)

25 (a) Explain the presence of dark lines in the otherwise continuous spectrum of the Sun in terms of the atomic processes involved.

(b) In a complete solar eclipse the light from the main body of the Sun is cut off and only light from the corona of hot gases forming the outer layer reaches the Earth.

State what kind of spectrum you would expect to see for this light. How would it compare with the spectrum referred to in (a)?

(NEAB, 1991)

26 (a) (i) Explain what is meant by *photoelectric emission.*

(ii) Briefly describe a simple experiment to

demonstrate this effect qualitatively.

(b) Write down Einstein's photoelectric equation and explain the meaning of each term in it.

(c) In an experiment in photoelectricity, the maximum kinetic energy of the photoelectrons was determined for different wavelengths of the incident radiation. The following results were obtained:

wavelength /nm	300	375	500
maximum kinetic energy/eV	2.03	1.20	0.36

Use the results to determine

(i) the work function for the metal

(ii) a value for Planck's constant.

(WJEC, 1990)

27 The decay scheme for $^{27}_{12}$Mg is shown in the figure.

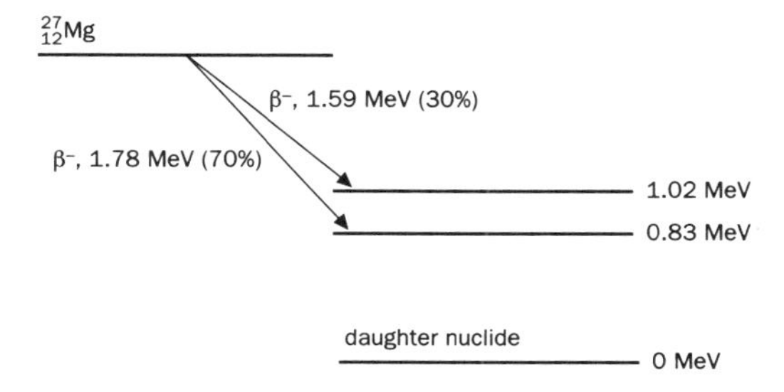

(a) (i) Identify the daughter nuclide.

(ii) Briefly describe the decay sequence represented by the figure.

(b) (i) Write down an equation to represent the initial decay of a $^{27}_{12}$Mg nucleus.

(ii) Name another form of radiation to be expected in this decay scheme in addition to beta radiation and give **three** possible values for its energy.

(iii) Calculate the wavelength of the photon with the highest energy that might be expected to occur in this decay.

(c) (i) Explain why it became necessary, in the development of theories about the nucleus, to propose the existence of the neutrinos and the antineutrinos.

(ii) State **one** difference between the physical properties of the neutrino and the antineutrino.

(NEAB, 1991)

28 (a) (i) Show that a particle of mass m and charge Q moving with speed v at right angles to a magnetic field of flux density B travels in a circular arc of radius r given by

$$r = \frac{mv}{QB}$$

(ii) Calculate this radius for an α-particle of energy 6.4×10^{-13} J travelling at right angles to a magnetic field of flux density 0.50 T.

Mass of α-particle = 6.7×10^{-27} kg

Magnitude of charge of electron = 1.6×10^{-19} C

(b) The deflection of β-particles in a magnetic field can be readily demonstrated in the laboratory using a permanent magnet and a Geiger–Müller tube and counter.

Discuss the difficulties involved in demonstrating the deflection of α-particles in a magnetic field.

(NEAB, 1993)

29 This question is about the acceleration of protons to high energies in a proton synchrotron.

(a) Before injection into the synchrotron, the protons are accelerated from rest in an electrostatic accelerator through a potential difference of 7.0 MV. Calculate their speed on injection into the synchrotron.

(b) In the synchrotron the protons move in a circular orbit of radius r, in the horizontal plane. The orbit is in a region of uniform vertical magnetic field B. Show that the orbital period of the protons is independent of the speed of the protons.

(c) According to the theory of relativity, at very high proton energies an increase in energy results in an increase in the mass of the protons. The speed of the protons is almost constant at nearly the speed of light. Suggest how a group of high energy protons may be kept in an orbit of constant radius and periodic time as their energy is increased.

(O & C, 1993)

Index